# FIELD THEORY OF GUIDED WAVES

INTERNATIONAL SERIES IN PURE AND APPLIED PHYSICS

LEONARD I. SCHIFF, *Consulting Editor*

---

*Allis and Herlin* Thermodynamics and Statistical Mechanics
*Becker* Introduction to Theoretical Mechanics
*Clark* Applied X-rays
*Collin* Field Theory of Guided Waves
*Evans* The Atomic Nucleus
*Finkelnburg* Atomic Physics
*Ginzton* Microwave Measurements
*Green* Nuclear Physics
*Gurney* Introduction to Statistical Mechanics
*Hall* Introduction to Electron Microscopy
*Hardy and Perrin* The Principles of Optics
*Harnwell* Electricity and Electromagnetism
*Harnwell and Livingood* Experimental Atomic Physics
*Harnwell and Stephens* Atomic Physics
*Houston* Principles of Mathematical Physics
*Hund* High-frequency Measurements
*Kennard* Kinetic Theory of Gases
*Leighton* Principles of Modern Physics
*Middleton* An Introduction to Statistical Communication Theory
*Morse* Vibration and Sound
*Morse and Feshbach* Methods of Theoretical Physics
*Muskat* Physical Principles of Oil Production
*Present* Kinetic Theory of Gases
*Read* Dislocations in Crystals
*Richtmyer, Kennard, and Lauritsen* Introduction to Modern Physics
*Schiff* Quantum Mechanics
*Seitz* The Modern Theory of Solids
*Slater* Introduction to Chemical Physics
*Slater* Quantum Theory of Atomic Structure, Volume I
*Slater* Quantum Theory of Atomic Structure, Volume II
*Slater* Quantum Theory of Matter
*Slater and Frank* Electromagnetism
*Slater and Frank* Introduction to Theoretical Physics
*Slater and Frank* Mechanics
*Smythe* Static and Dynamic Electricity
*Stratton* Electromagnetic Theory
*Thorndike* Mesons: A Summary of Experimental Facts
*Townes and Schawlow* Microwave Spectroscopy
*White* Introduction to Atomic Spectra

---

The late F. K. Richtmyer was Consulting Editor of the series from its inception in 1929 to his death in 1939. Lee A. DuBridge was Consulting Editor from 1939 to 1946, and G. P. Harnwell from 1947 to 1954.

# FIELD THEORY OF GUIDED WAVES

**ROBERT E. COLLIN**

*Associate Professor of Electrical Engineering*
*Case Institute of Technology*

McGRAW-HILL BOOK COMPANY, INC.

New York Toronto London

1960

FIELD THEORY OF GUIDED WAVES

 *Library of Congress Catalog Card Number* 60-11953

THE MAPLE PRESS COMPANY, YORK, PA.

11802

TO KATHLEEN

# PREFACE

Our knowledge of electromagnetic wave propagation in cylindrical waveguides and the guiding and radiation of electromagnetic fields in general has now reached a state of maturity where it is possible to give a reasonably coherent and logical account of the theory. Twenty years ago wave propagation in a hollow conducting waveguide was the subject matter of research papers, but today it is standard textbook material and presented to senior students in most engineering schools. Surprisingly enough, however, a comprehensive account of guided-wave theory at the graduate level has not appeared in textbook form. The mathematical techniques and solutions of the more elaborate waveguiding structures, waveguide discontinuities, waveguide antennas and coupling apertures, and so forth, are for the most part still only to be found in the periodical literature or covered in a fragmentary way in a number of books. This presents the graduate student with a distinct handicap in that it is difficult for him to obtain a clear understanding of the basic theory and mathematical techniques involved by reading periodical literature. This book is an attempt to bridge the gap between the usual senior–first-year-graduate treatment of guided electromagnetic waves and the more advanced topics which are of prime importance to the graduate student and the research worker in the field. In addition, the book has been planned so as to provide a reference work for the engineer or physicist who has not had the opportunity to follow the development of the field over the past two decades. The treatment is essentially self-contained and presupposes a prior knowledge, at the introductory level, of field theory and propagation in waveguides. A familiarity with the standard separation of variables method of solving partial differential equations and some of the elementary concepts of Fourier series and eigenfunction expansions, and an acquaintance with Fourier and Laplace transforms are also assumed. With this as a background, a student should find little difficulty in following the presentation.

The emphasis in this book is on the field solutions in contrast to a description of microwave structures in terms of equivalent-circuit parameters. On the other hand, the final results are often better interpreted in terms of equivalent-circuit elements. For this reason, a sufficient amount of microwave circuit theory is included in order to make a clear and logical interpretation of field solutions in terms of equivalent circuits possible.

Since the treatment places the electromagnetic field in the foreground, the first chapter begins with a survey of that portion of classical electromagnetic theory that is pertinent to the remainder of the book. In addition to the usual topics, a discussion of boundary conditions at the edge of a current sheet and the edge of a conducting wedge, equivalent magnetic current sheets, and some field equivalence principles, including Babinet's principle, are covered. Chapter 2 provides an introduction to Green's functions for Poisson's equation, dyadic Green's functions for the vector Helmholtz equation, and the solution of boundary-value problems by means of Green's functions. A particular example on alternative representations for a scalar Green's function and a formal demonstration of the equivalence of the two representations are also included. The material of this chapter is expanded in several other places throughout the remaining chapters.

Chapter 3 begins with a discussion of TEM waves in rectangular coordinates, and is followed by a treatment of TEM waves in general orthogonal curvilinear coordinates, in order to determine in what coordinate systems solutions for TEM waves are possible. Next comes an introductory treatment of wave-transmission, scattering, impedance, and admittance matrices. This matrix theory is applied to the problem of transmission through dielectric sheets and provides a compact solution in terms of equivalent transmission-line circuits. The latter part of the chapter is devoted to a study of TEM wave propagation in anisotropic dielectric material and in ferrite media which exhibit Faraday rotation and nonreciprocity.

In Chap. 4 the solutions for the parameters of TEM wave transmission lines are developed, and the relation between the field and circuit quantities is established. The determination of the characteristic impedance of transmission lines by conformal-mapping techniques is covered and applied to a typical strip-line problem. Also included is a detailed account of variational methods for obtaining upper and lower bounds for the characteristic impedance.

The basic theory of propagation in hollow cylindrical guides is given in the first part of Chap. 5. The remainder of the chapter deals with mode completeness, construction of Green's functions for waveguides, analogy with transmission lines, and the tangent method for the experimental determination of the parameters of a waveguide discontinuity. Chapter 6 treats the propagation of waves in rectangular guides inhomogeneously filled with dielectric and ferrite slabs. The Rayleigh-Ritz method for determining upper bounds on the propagation constants and for constructing approximate eigenfunctions is presented and applied to the problem of determining an equivalent circuit for a junction of an empty guide and an inhomogeneously filled guide.

Radiation from probe and loop antennas in a rectangular guide is analyzed in Chap. 7. In addition, the very useful theory of coupling by small apertures is developed. The polarizabilities of circular and elliptical apertures are worked out in detail. In the last part of the chapter, a short introduction to the transient behavior of a rectangular guide is given. Chapter 8 is devoted to an account of variational methods for the solution of waveguide discontinuity problems. A number of possible formulations are outlined and compared. Examples on capacitive and inductive diaphragms are included. Chapter 9 considers propagation in periodic-loaded guides and introduces Floquet's theorem and other fundamental properties common to all periodic structures. Both the field and the circuit approach are used. Higher-order mode interaction in a capacitively loaded guide is examined in the latter portion of the chapter.

Chapter 10 gives an introduction to integral transform and function-theoretic techniques for the solution of two-part boundary-value problems. Three closely related but distinct methods are presented. These are: (1) solution of an infinite set of linear algebraic equations, (2) direct construction of a suitable transform function, and (3) Wiener-Hopf technique. Applications to parallel-plate media, bifurcations, and a semidiaphragm in a rectangular guide are included.

The theory of surface waves and surface-wave guiding structures is presented in Chap. 11. Excitation of surface waves by a line source above a grounded dielectric slab is examined and gives an introduction to some of the properties of the continuous eigenvalue spectrum associated with open-boundary waveguides. The radiation field is evaluated by means of the method of steepest descents. Special care is taken with the specification of the branch cuts for evaluation of the inversion integral in order to give a clear understanding of the factors which must be considered.

The final chapter discusses the basic theory of artificial dielectric media, introduces the Lorentz theory, the static field theory, and the equivalent transmission-line approaches. Detailed solutions for a disk medium and a two-dimensional strip medium are given.

In addition a Mathematical Appendix summarizes a number of useful mathematical topics with which the student should have some familiarity. Vector analysis, properties of vector fields, elementary concepts of dyadic analysis, basic principles from the calculus of variations, and other miscellaneous topics are included. An account of methods for summing Fourier series in closed form is also given, together with a table of summed series.

The above range of topics is by no means exhaustive. It would be easy to fill another volume of equal size with additional material. However, it is felt that the essential theory and techniques which are basic to the whole field are adequately covered in most respects. Notable in their absence are a chapter dealing with electronic waveguides and one covering the theory of electromagnetic resonators. The former is perhaps better suited for a book dealing with microwave tubes, and, since several excellent books in this area have appeared in recent years, no attempt is made to deal with this topic here. It was initially planned to include a chapter on resonators, but unfortunately ambition dwindled and died before it could be written.

A number of problems are given at the end of each chapter. These have been designed to give the student a chance to apply the mathematical techniques developed in the text. In some cases the problems supplement the theory given in the preceding chapter. There is only one satisfactory way to gain a good understanding of a particular mathematical technique, and that is to apply it to a problem and work through the details. This procedure almost invariably brings to light a number of obscure points that, somehow or other, never seem to receive adequate treatment in a general theoretical discussion. The working of problems cannot be overemphasized if more than a casual understanding of the methods of analysis is to be obtained.

The original concept for this book arose out of a set of lecture notes prepared for a graduate course in guided-wave theory and given at Laval University, Quebec, in 1955. After two years of additional work carried out at Case Institute of Technology, the present volume was arrived at. Needless to say, the analysis leans heavily on the work of a number of people. To give full recognition to the various contributors is a rather dismaying task because of the large volume of literature in existence. As a general rule, reference is made only to those articles with which the author has the greatest familiarity and which are more readily obtainable. Several people have helped, in different ways, to bring the manuscript to its final form. Acknowledgment is made to Mr. R. Vaillancourt, Canadian Armament Research and Development Establishment, for correcting many of the errors in the original Laval University lecture notes. I also wish to acknowledge the assistance of Mr. Hwei-Piao Hsu for proofreading many of the chapters of the first draft of the completed manuscript. My colleagues Professors R. Plonsey and F. Brammer have also indirectly contributed a great deal to the book through numerous illuminating discussions. Finally, I wish to express my thanks to Professor Samuel Seely, Chairman, Electrical Engineering Department, Case Institute of Technology, for generous provision of facilities and encouragement, and to Misses Ruth Hudak and Toni Alvarez for careful and cheerful typing of both the first draft and final copy of the manuscript.

*Robert E. Collin*

# CONTENTS

CHAPTER 1

# BASIC ELECTROMAGNETIC THEORY

The early history of guided waves and waveguides dates back to the period around the end of the nineteenth century. In 1897 Lord Rayleigh published an analysis of electromagnetic-wave propagation in dielectric-filled rectangular and circular conducting tubes, or waveguides as they are now called. This was followed by an analysis of wave propagation along dielectric cylinders by Hondros and Debye in 1910. Several other workers contributed in both theory and experiment to our knowledge of waveguides during the same period. However, the true birth of the waveguide as a useful device for transmission of electromagnetic waves did not come about before 1936. At this time, Carson, Mead, Schelkunoff, and Southworth from the Bell Telephone Laboratories published an extensive account of analytical and experimental work on waveguides which had been under way since 1933. The result of other work was published the same year by Barrow.

The period 1936 to the early 1940s saw a steady growth in both the theory and experiment of waveguides as practical communication elements. Even though this early growth was very important, our present state of knowledge of guided electromagnetic waves owes much to the tremendous efforts of a multitude of physicists, mathematicians, and engineers who worked in the field during World War II. The desire for radars operating at microwave frequencies put high priority on work related to waveguides, waveguide devices, antennas, etc. A variety of devices were developed to replace conventional low-frequency lumped-circuit elements. Along with these devices, methods of analysis were also being developed. The formulation and solution of intricate boundary-value problems became of prime importance.

The postwar years has seen a continued development and refinement in both theory and techniques. Our efforts in this book are devoted to a presentation of the major theoretical results and mathematical techniques, obtained to date, that underlie the theory of guided waves. The treatment places the electromagnetic field in the foreground, and for this reason we begin with a survey of that portion of classical electromagnetic field theory which is pertinent to those devices and problems under examination. It is important to note from the beginning that only a few of the boundary-value problems that we shall study can be solved rigorously. In the majority of cases we must be content with approximate solutions. The development of approximate solutions, in turn, depends strongly on a knowledge of the physical principles involved, and this, of course, leads us back to the physics of the electromagnetic field.

It is assumed that the student has a background in field theory at the level of the book "Fields and Waves in Modern Radio" by Ramo and Whinnery.† A certain maturity in mathematical ability is also assumed. For convenience an appendix summarizes the major mathematical topics the student should have a knowledge of in order to pursue the present book without too much hindrance.

## 1.1. Maxwell's Equations

Provided we restrict ourselves to the macroscopic domain, the electromagnetic field is governed by the classical field equations of Maxwell. In derivative form these are

$$\nabla \times \mathcal{E} = -\frac{\partial \mathcal{B}}{\partial t} \tag{1a}$$

$$\nabla \times \mathcal{H} = \frac{\partial \mathcal{D}}{\partial t} + \mathcal{J} \tag{1b}$$

$$\nabla \cdot \mathcal{D} = \rho \tag{1c}$$

$$\nabla \cdot \mathcal{B} = 0 \tag{1d}$$

$$\nabla \cdot \mathcal{J} = -\frac{\partial \rho}{\partial t} \tag{1e}$$

The above equations are not all independent since, if the divergence of Eq. (1*a*) is taken, we get the result that

$$\nabla \cdot \frac{\partial \mathcal{B}}{\partial t} = 0$$

because the divergence of the curl of any vector is identically zero. By interchanging the order of the time and space derivatives and integrating with respect to time, (1*d*) is obtained, provided the constant of integration is taken as zero. In a similar fashion (1*c*) follows from (1*b*) with the assistance of (1*e*). Equation (1*a*) is just the differential form of Faraday's law of induction, since the curl of a vector is the line integral of this vector around an infinitesimal contour divided by the area of the surface bounded by the contour. Equation (1*b*) is a generalization of Ampère's circuital law (also referred to as the law of Biot-Savart) by the addition of the term $\partial \mathcal{D}/\partial t$, which is called the displacement current

† S. Ramo and J. R. Whinnery, "Fields and Waves in Modern Radio," 2d ed., John Wiley & Sons, Inc., New York, 1953.

density. Equation (1*c*) is the differential form of Gauss' law, while (1*d*) expresses the fact that the magnetic lines of flux form a system of closed loops and nowhere terminate on "magnetic" charge. Finally, (1*e*) is just the equation of conservation of charge; the amount of charge diverging away per second from an infinitesimal volume element must equal the time rate of decrease of the charge contained within.

At this time it is convenient to restrict ourselves to fields which vary with time according to the complex exponential function $e^{j\omega t}$, where $\omega$ is the radian frequency. There is little loss in generality in using such a time function since any physically realizable time variation can be decomposed into a spectrum of such functions by means of a Fourier integral, i.e.,

$$f(t) = \int_{-\infty}^{\infty} g(\omega)e^{j\omega t}\,d\omega$$

and

$$g(\omega) = \frac{1}{2\pi}\int_{-\infty}^{\infty} f(t)e^{-j\omega t}\,dt$$

With the above assumed time variation all time derivatives may be replaced by $j\omega$. We will not include the factor $e^{j\omega t}$ explicitly as this factor always occurs as a common factor in all terms. The field vectors are now represented by boldface roman type and are complex functions of the space coordinates only.

The basic field equations may be expressed in equivalent integral form also. An application of Stokes' theorem to (1*a*) and (1*b*) gives

$$\iint_S \nabla \times \mathbf{E} \cdot d\mathbf{a} = \oint_C \mathbf{E} \cdot d\mathbf{l} = -j\omega \iint_S \mathbf{B} \cdot d\mathbf{a} \tag{2a}$$

$$\oint_C \mathbf{H} \cdot d\mathbf{l} = \iint_S (j\omega\mathbf{D} + \mathbf{J}) \cdot d\mathbf{a} \tag{2b}$$

Using the divergence theorem, the remaining three equations become

$$\iint_S \mathbf{D} \cdot d\mathbf{a} = \iiint_V \rho\,dV \tag{2c}$$

$$\iint_S \mathbf{B} \cdot d\mathbf{a} = 0 \tag{2d}$$

$$\iint_S \mathbf{J} \cdot d\mathbf{a} = -j\omega \iiint_V \rho\,dV \tag{2e}$$

In (2*a*) and (2*b*) the surface $S$ is an open surface bounded by a closed contour $C$, while in (2*c*) through (2*e*) the surface $S$ is a closed surface with an interior volume $V$. The vector element of area $d\mathbf{a}$ is directed outward. This latter description of the field equations is of particular usefulness in deriving the boundary conditions to be applied to the field vectors at a boundary between two mediums having different electrical properties.

Equations (1) contain considerably more information than we might anticipate from a cursory glance at them. A good deal of insight into the meaning of these equations can be obtained by decomposing each field vector into the sum of an irrotational or lamellar part plus a rotational (solenoidal) part. The lamellar part has an identically vanishing curl, while the rotational part always has a zero divergence. A vector field is completely specified only when both the lamellar and solenoidal parts are given.† Let the subscript $l$ denote the lamellar part and the subscript $r$ denote the rotational or solenoidal part. For an arbitrary vector field $\mathbf{C}$ we have

$$\mathbf{C} = \mathbf{C}_l + \mathbf{C}_r \tag{3a}$$

where

$$\nabla \times \mathbf{C}_l \equiv 0 \tag{3b}$$

and

$$\nabla \cdot \mathbf{C}_r \equiv 0 \tag{3c}$$

When the field vectors are decomposed into their basic parts and the results of (3b) and (3c) are used, the field equations (1) become

$$\nabla \times \mathbf{E}_r = -j\omega\mathbf{B}_r \tag{4a}$$
$$\nabla \times \mathbf{H}_r = j\omega(\mathbf{D}_l + \mathbf{D}_r) + \mathbf{J}_l + \mathbf{J}_r \tag{4b}$$
$$\nabla \cdot \mathbf{D}_l = \rho \tag{4c}$$
$$\nabla \cdot \mathbf{B}_r = 0 \tag{4d}$$
$$\nabla \cdot \mathbf{J}_l = -j\omega\rho \tag{4e}$$

From (4c) and (4e) we find that the divergence of the lamellar part of the displacement current is related to the divergence of the lamellar part of the conduction current. The rotational part of the conduction current has zero divergence and hence exists in the form of closed current loops. It does not terminate on a distribution of charge. On the other hand, the lamellar part $\mathbf{J}_l$ is like a fragment of $\mathbf{J}_r$ with the ends terminating in a collection of charge. The lamellar part of the displacement current provides the continuation of $\mathbf{J}_l$, so that the total lamellar current $\mathbf{J}_l + j\omega\mathbf{D}_l$ forms the equivalent of a rotational current. Equations (4c) and (4e) show at once that

$$\nabla \cdot \mathbf{J}_l + j\omega\nabla \cdot \mathbf{D}_l = 0 \tag{5}$$

In view of this latter result, we are tempted to integrate (5) to obtain $\mathbf{J}_l + j\omega\mathbf{D}_l = 0$ and to conclude that (4b) is equivalent to

$$\nabla \times \mathbf{H}_r = j\omega\mathbf{D}_r + \mathbf{J}_r$$

or, in words, that the rotational part of $\mathbf{H}$ is determined only by the rotational part of the total current (displacement plus conduction).

† See Sec. A.1e.

This line of reasoning is incorrect since the integration of (5) gives, in general,

$$\mathbf{J}_l + j\omega\mathbf{D}_l = \mathbf{C}_r$$

where $\mathbf{C}_r$ is a rotational vector. Comparing with (4*b*), we see that $\mathbf{C}_r$ is the difference between $\nabla \times \mathbf{H}_r$ and $j\omega\mathbf{D}_r + \mathbf{J}_r$. It is the total current consisting of both the lamellar part and the rotational part that determines $\nabla \times \mathbf{H}_r$ and not vice versa. Furthermore, the total lamellar part of the current, that is, $\mathbf{J}_l + j\omega\mathbf{D}_l$, is a rotational current in its behavior, and so there is no reason to exclude this in preference for the separate rotational parts $\mathbf{J}_r$ and $j\omega\mathbf{D}_r$.

Equations (4*a*) and (4*d*) are more or less self-explanatory since **B** does not have a lamellar part.

## 1.2. Relation between Field Intensity Vectors and Flux Density Vectors

The field equations cannot be solved before the relationship between **B** and **H** and that between **E** and **D** are known. In vacuum or free space the relationship is a simple proportion, i.e.,

$$\mathbf{B} = \mu_0\mathbf{H} \tag{6a}$$

$$\mathbf{D} = \epsilon_0\mathbf{E} \tag{6b}$$

where $\mu_0 = 4\pi \times 10^{-7}$ henry per meter and $\epsilon_0 = (36\pi)^{-1} \times 10^{-9}$ farad per meter. The constant $\mu_0$ is called the magnetic permeability of vacuum, and $\epsilon_0$ is the electric permittivity of vacuum. For material bodies the relationship is generally much more complicated with $\mu_0$ and $\epsilon_0$ replaced by tensors of rank 2, or dyadics. Letting a bar above the quantity signify a dyadic, we have

$$\mathbf{B} = \bar{\mu} \cdot \mathbf{H} \tag{7a}$$

$$\mathbf{D} = \bar{\epsilon} \cdot \mathbf{E} \tag{7b}$$

where $$\bar{\mu} = \mu_{xx}\mathbf{a}_x\mathbf{a}_x + \mu_{xy}\mathbf{a}_x\mathbf{a}_y + \mu_{xz}\mathbf{a}_x\mathbf{a}_z + \mu_{yx}\mathbf{a}_y\mathbf{a}_x + \mu_{yy}\mathbf{a}_y\mathbf{a}_y + \mu_{yz}\mathbf{a}_y\mathbf{a}_z + \mu_{zx}\mathbf{a}_z\mathbf{a}_x + \mu_{zy}\mathbf{a}_z\mathbf{a}_y + \mu_{zz}\mathbf{a}_z\mathbf{a}_z$$

with a similar definition for $\bar{\epsilon}$. The unit vectors along the coordinate axis are represented by $\mathbf{a}_x$, etc. The dot product of a dyadic and a vector or of two dyadics is defined as the resulting vector or dyadic that is obtained by taking the dot product of the adjacent unit vectors which appear on either side of the dot. Dyadic algebra has a one-to-one correspondence with matrix algebra with the advantage of a somewhat more condensed notation. In general the dot product is not commutative, i.e.,

$$\bar{\mathbf{A}} \cdot \bar{\mathbf{C}} \neq \bar{\mathbf{C}} \cdot \bar{\mathbf{A}}$$

unless $\bar{\mathbf{A}}$ and $\bar{\mathbf{C}}$ are both symmetric, so that the component $A_{ij} = A_{ji}$, etc. With the definition of the dot product established, we find that the $x$ component of (7$a$) is $B_x = \mu_{xx}H_x + \mu_{xy}H_y + \mu_{xz}H_z$, with similar expressions existing for the $y$ and $z$ components. Thus, each component of **B** is related to all three components of **H**. The coefficients $\mu_{ij}$ may also be functions of **H** so that the relationship is not a linear one. With a suitable rotation of axes to the principal axes $u$, $v$, $w$, the dyadic $\bar{\boldsymbol{\mu}}$ becomes a diagonal dyadic of the form

$$\bar{\boldsymbol{\mu}} = \mu_{uu}\mathbf{a}_u\mathbf{a}_u + \mu_{vv}\mathbf{a}_v\mathbf{a}_v + \mu_{ww}\mathbf{a}_w\mathbf{a}_w$$

Although we introduced a somewhat involved relationship above, in practice we find that for many materials $\bar{\boldsymbol{\mu}}$ and $\bar{\boldsymbol{\epsilon}}$ reduce to simple scalar quantities and, furthermore, are essentially independent of the field strength. When $\mu$ and $\epsilon$ are constant, the divergence equations for **D** and **B** give

$$\nabla \cdot \mathbf{D} = \rho \qquad \text{or} \qquad \nabla \cdot \mathbf{E} = \frac{\rho}{\epsilon} \tag{8a}$$

$$\nabla \cdot \mathbf{B} = 0 \qquad \text{or} \qquad \nabla \cdot \mathbf{H} = 0 \tag{8b}$$

and **H** is seen to be a solenoidal field also. When $\mu$ and $\epsilon$ vary with position, we have

$$\nabla \cdot \mathbf{D} = \nabla \cdot \epsilon\mathbf{E} = \epsilon\nabla \cdot \mathbf{E} + \mathbf{E} \cdot \nabla\epsilon = \rho$$

or

$$\nabla \cdot \mathbf{E} = \frac{\rho}{\epsilon} - \mathbf{E} \cdot \nabla \ln \epsilon \tag{9a}$$

and similarly

$$\nabla \cdot \mathbf{H} = -\mathbf{H} \cdot \nabla \ln \mu \tag{9b}$$

since

$$\nabla \ln \epsilon = \frac{1}{\epsilon} \nabla\epsilon \qquad \text{etc.}$$

When the material under consideration has finite losses, $\mu$ and $\epsilon$ are complex with negative imaginary parts.

The physical meaning of Eqs. (9) will require a consideration of the properties of matter which gives rise to relationships of the form expressed by (7) with $\mu$ and $\epsilon$ scalar functions of position. Equations (7) may be regarded as defining the quantities $\bar{\boldsymbol{\mu}}$ and $\bar{\boldsymbol{\epsilon}}$ and as such require the definitions of **H** and **D** to be known. Of course, when $\bar{\boldsymbol{\mu}}$ and $\bar{\boldsymbol{\epsilon}}$ for a particular medium are known, then we might just as well regard Eqs. (7) as defining **H** and **D**.

When an electric field is applied to a dielectric material, the electron orbits of the various atoms and molecules involved become perturbed, resulting in a dipole polarization **P** per unit volume. Some materials have a permanent dipole polarization also, but, since these elementary dipoles are randomly oriented, there is no net resultant field in the absence of an applied external field. The application of an external field tends to align the dipoles with the field, resulting in a decrease in the electric field intensity in the material. The displacement flux vector

**D** is defined by

$$\mathbf{D} = \epsilon_0 \mathbf{E} + \mathbf{P} \tag{10}$$

in the interior of a dielectric material. In (10), **E** is the net field intensity in the dielectric, i.e., the vector sum of the applied field and the field arising from the dipole polarization. For many materials **P** is collinear with the applied field and also proportional to the applied field and hence proportional to **E** also. Thus we have

$$\mathbf{P} = \chi_e \epsilon_0 \mathbf{E} \tag{11}$$

where $\chi_e$ is called the electric susceptibility of the material and is a dimensionless quantity. For the scalar case we have

$$\mathbf{D} = \epsilon_0(\chi_e + 1)\mathbf{E} = \epsilon \mathbf{E}$$

and so

$$\epsilon = (1 + \chi_e)\epsilon_0 \tag{12}$$

The relative dielectric constant is $\kappa = \epsilon/\epsilon_0$ and is equal to $1 + \chi_e$. Equation (11) is not universally applicable, since **P** will be in the direction of the applied field only in materials having a high degree of symmetry in their crystal structure. In general, $\bar{\chi}_e$ is a dyadic quantity, and hence $\epsilon$ becomes the dyadic quantity $(\bar{\mathbf{I}} + \bar{\chi}_e)\epsilon_0$, where $\bar{\mathbf{I}}$ is the unit dyadic.

The situation for materials having a permeability different from $\mu_0$ is somewhat analogous, with **H** being defined by the following relation:

$$\mathbf{H} = \frac{\mathbf{B}}{\mu_0} - \mathbf{M} \tag{13}$$

where **M** is the magnetic dipole polarization per unit volume. Logically we should now take **M** proportional to **B** for those materials for which such a proportion exists, but convention has adopted the following:

$$\mathbf{M} = \chi_m \mathbf{H} \tag{14}$$

where $\chi_m$ is the magnetic susceptibility of the medium. Equation (14) leads at once to the following result for $\mu$:

$$\mu = (1 + \chi_m)\mu_0 \tag{15}$$

As in the dielectric case, **M** is not always in the direction of **B**, and hence $\chi_m$ and $\mu$ are dyadic quantities in general.

Materials for which $\mu$ and $\epsilon$ are scalar constants are referred to as homogeneous isotropic materials. When $\mu$ and $\epsilon$ vary with position, the material is no longer homogeneous but is still isotropic. When $\mu$ and $\epsilon$ are dyadics, the material is said to be anisotropic. In the latter case the material may again be homogeneous or nonhomogeneous, depending on whether $\bar{\mu}$ and $\bar{\epsilon}$ vary with position or not.

In material bodies with $\mu$ and $\epsilon$ scalar functions of position, Maxwell's curl equations may be written as

$$\nabla \times \mathbf{E} = -j\omega\mu_0\mathbf{H} - j\omega\mu_0\chi_m\mathbf{H} \tag{16a}$$
$$\nabla \times \mathbf{H} = j\omega\epsilon_0\mathbf{E} + j\omega\epsilon_0\chi_e\mathbf{E} + \mathbf{J} \tag{16b}$$

In Eq. (16*b*) the term $j\omega\epsilon_0\chi_e\mathbf{E}$ may be regarded as an equivalent polarization current $\mathbf{J}_e$. Similarly in (16*a*) it is profitable at times to consider the term $j\omega\mu_0\chi_m\mathbf{H}$ as the equivalent of a magnetic polarization current $\mathbf{J}_m$.

Introducing the polarization vectors **P** and **M** into the divergence equations gives

$$\nabla \cdot \mathbf{D} = \epsilon_0\nabla \cdot \mathbf{E} + \nabla \cdot \mathbf{P} = \rho \tag{17a}$$
$$\nabla \cdot \mathbf{B} = \mu_0\nabla \cdot \mathbf{H} + \mu_0\nabla \cdot \mathbf{M} = 0 \tag{17b}$$

The additional terms $\nabla \cdot \mathbf{P}$ and $\nabla \cdot \mathbf{M}$ may be interpreted as defining an electric-polarization charge density $-\rho_e$ and a magnetic-polarization charge density $-\rho_m$, respectively. The latter definition is a mathematical one and should not be interpreted as showing the existence of a physical magnetic charge. Introducing the parameters $\chi_e$ and $\chi_m$, we obtain the following equations giving the polarization charge densities:

$$\begin{aligned}\rho_e &= -\nabla \cdot \mathbf{P} = -\epsilon_0\,\nabla \cdot \chi_e\mathbf{E} \\ &= -\epsilon_0\chi_e\nabla \cdot \mathbf{E} - \epsilon_0\mathbf{E} \cdot \nabla\chi_e \end{aligned} \tag{18a}$$
$$\begin{aligned}\rho_m &= -\nabla \cdot \mathbf{M} = -\nabla \cdot \chi_m\mathbf{H} \\ &= -\chi_m\nabla \cdot \mathbf{H} - \mathbf{H} \cdot \nabla\chi_m \end{aligned} \tag{18b}$$

When the susceptibilities are dyadic functions of position, we may obviously make a similar interpretation of the additional terms which arise in Maxwell's equations. In the analysis of scattering by material bodies such an interpretation is often convenient since it leads directly to an integral equation for the scattered field in terms of the polarization currents.

At high frequencies, the interior of the atomic system causes the polarization vectors **P** and **M** to lag behind the applied fields. As a result, $\epsilon$ and $\mu$ must be represented by complex quantities in order to account for this difference in time phase. The real parts will be designated by $\epsilon'$, $\mu'$ and the imaginary parts by $\epsilon''$, $\mu''$; thus

$$\epsilon = \epsilon' - j\epsilon''$$
$$\mu = \mu' - j\mu''$$

For many materials $\epsilon''$ and $\mu''$ are negligible, i.e., for materials with small losses.

For many material bodies the conduction current, which flows as a result of the existence of a field **E**, is directly proportional to **E**, so that we may put

$$\mathbf{J} = \sigma \mathbf{E} \tag{19}$$

where $\sigma$ is a parameter of the medium known as the **conductivity**. Equation (19) is not true in general, of course, since $\sigma$ may depend on both the magnitude and direction of $\mathbf{E}$, and hence is a nonlinear dyadic quantity, as for example in semiconductors at a rectifying boundary. When Eq. (19) holds, the curl equation for $\mathbf{H}$ becomes

$$\begin{aligned} \nabla \times \mathbf{H} &= (j\omega\epsilon + \sigma)\mathbf{E} = j\omega\epsilon'\left(1 - j\frac{\epsilon''}{\epsilon'} - j\frac{\sigma}{\omega\epsilon'}\right)\mathbf{E} \\ &= j\omega\epsilon'(1 - j\tan\delta)\mathbf{E} \end{aligned} \tag{20}$$

The effective dielectric permittivity is now complex, even if $\epsilon''$ is zero, and is given by

$$\epsilon_e = \epsilon'(1 - j\tan\delta) \tag{21}$$

where $\tan\delta$ is the loss tangent for the material.

If a material body has an appreciable conductivity, the density $\rho$ of free charge in the interior may be taken as zero. Any initial free-charge density $\rho_0$ decays exponentially to zero in an extremely short time. The existence of a charge distribution $\rho$ implies the existence of a resultant electric field $\mathbf{E}_0$ and hence a current $\mathbf{J}_0 = \sigma\mathbf{E}_0$. Now $\nabla \cdot \mathbf{J}_0 = -\partial\rho/\partial t$ and $\nabla \cdot \epsilon\mathbf{E}_0 = \rho$. When $\epsilon$ and $\sigma$ are constants, we find, upon eliminating $\mathbf{E}_0$ and $\mathbf{J}_0$, that

$$\frac{\partial\rho}{\partial t} = -\frac{\sigma}{\epsilon}\rho$$

and hence

$$\rho = \rho_0 e^{-(\sigma/\epsilon)t} \tag{22}$$

For metals the decay time $\tau = \epsilon/\sigma$ is of the order of $10^{-18}$ second, so that for frequencies well beyond the microwave range the decay time is extremely short compared with the period of the impressed fields.

The above result means that in conducting bodies we may take $\nabla \cdot \mathbf{D}$ equal to zero. If $\epsilon$ and $\sigma$ are constants, the divergence of $\mathbf{E}$ and $\mathbf{J}$ will also be zero. In practice, we often simplify our boundary conditions at a conducting surface by assuming that $\sigma$ is infinite. It is then found that the conduction current flows on the surface of the conducting boundary. Both lamellar and solenoidal surface currents may exist. The lamellar current terminates in a surface charge distribution. The normal component of the displacement vector $\mathbf{D}$ also terminates in this charge distribution and provides the continuation of $\mathbf{J}$ into the region external to the conducting surface. The perfectly conducting surface is a mathematical idealization, and we find at such a surface a discontinuity in the field vectors which does not correspond to physical reality. The nature of the fields at a conducting surface as $\sigma$ approaches infinity will be considered in more detail in a later section.

### 1.3. Electromagnetic Energy and Power Flow

The time-average amount of energy stored in the electric field which exists in a volume $V$ is given by

$$W_e = \tfrac{1}{4}\,\mathrm{Re} \iiint_V \mathbf{D} \cdot \mathbf{E}^* \, dV \tag{23a}$$

The time-average energy stored in the magnetic field is

$$W_m = \tfrac{1}{4}\,\mathrm{Re} \iiint_V \mathbf{B} \cdot \mathbf{H}^* \, dV \tag{23b}$$

The asterisk (*) signifies the complex conjugate value, and the additional factor $\frac{1}{2}$, which makes the numerical coefficient $\frac{1}{4}$, arises because of the averaging over one period in time. We are using the convention that **D**, etc., represents the peak value of the complex field vector. The imaginary parts of (23) correspond to power loss in the material.

The time-average complex power flow across a surface $S$ is given by the integral of the complex Poynting vector over that surface,

$$P = \frac{1}{2} \iint_S \mathbf{E} \times \mathbf{H}^* \cdot d\mathbf{a} \tag{24}$$

The real power is given by the real part of $P$ while the imaginary part represents energy stored in the electric and magnetic fields. The integral of the complex Poynting vector over a closed surface yields a result which may be regarded as a fundamental theorem in the theory of equivalent circuits for waveguide discontinuities. It is also of fundamental importance in the problem of evaluating the input impedance of an antenna. The integral of $\mathbf{E} \times \mathbf{H}^*$ over a closed surface $S$ gives

$$\frac{1}{2} \oint\!\!\oint_S \mathbf{E} \times \mathbf{H}^* \cdot d\mathbf{a} = -\frac{1}{2} \iiint_V \nabla \cdot \mathbf{E} \times \mathbf{H}^* \, dV$$

if the vector element of area is taken as directed into the volume $V$. Now $\nabla \cdot \mathbf{E} \times \mathbf{H}^* = \mathbf{H}^* \cdot \nabla \times \mathbf{E} - \mathbf{E} \cdot \nabla \times \mathbf{H}^*$. The curl of **E** and $\mathbf{H}^*$ may be replaced by $-j\omega\mu\mathbf{H}$ and $-j\omega\epsilon^*\mathbf{E}^* + \sigma\mathbf{E}^*$ from Maxwell's equations. Thus we get

$$\begin{aligned} \frac{1}{2} \oint\!\!\oint_S \mathbf{E} \times \mathbf{H}^* \cdot d\mathbf{a} &= \frac{1}{2} \iiint_V [j\omega(\mathbf{B} \cdot \mathbf{H}^* - \mathbf{D}^* \cdot \mathbf{E}) + \sigma \mathbf{E} \cdot \mathbf{E}^*] \, dV \\ &= 2j\omega(W_m - W_e) + P_L \end{aligned} \tag{25}$$

It is in the interpretation of (25) that the expressions (23) for the energy stored in the field are obtained. Since ordinary materials are passive, it follows from (25) that $\epsilon$ and $\mu$ have negative imaginary parts in order to represent a loss of energy. The power loss $P_L$ in $V$ is the sum of the conduction loss and $j\omega$ times the imaginary parts of (23). The imaginary part of the integral of the complex Poynting vector over the closed surface $S$ gives a quantity proportional to the difference between the time-average energy stored in the magnetic and electric fields. The right-hand side of (25), when the fields are properly normalized, can be related to or essentially used to define a suitable input impedance if the complex power flows into the volume $V$ through a waveguide or along a transmission line. We shall have several occasions to make use of this result later.

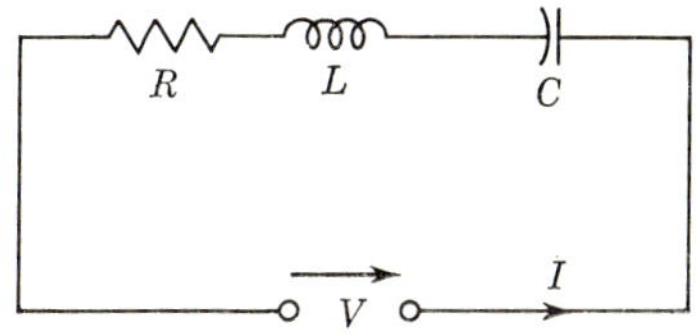

FIG. 1.1. An $RLC$ circuit.

Consider a simple $RLC$ circuit as illustrated in Fig. 1.1. The input impedance to this circuit is $Z = R + j\omega L + 1/j\omega C$. The complex time-average power flow into the circuit is given by

$$P = \tfrac{1}{2}VI^* = \tfrac{1}{2}ZII^* = \tfrac{1}{2}j\omega\left(LII^* - \frac{II^*}{\omega^2 C}\right) + \tfrac{1}{2}II^*R$$

The time-average reactive energy stored in the magnetic field around the inductor is $\frac{1}{4}LII^*$, while the time-average electric energy stored in the capacitor is $\frac{C}{4}\frac{I}{\omega C}\frac{I^*}{\omega C}$, since the peak voltage across $C$ is $\frac{I}{\omega C}$. We therefore obtain the result that

$$P = \tfrac{1}{2}ZII^* = 2j\omega(W_m - W_e) + P_L \tag{26}$$

which at least suggests the usefulness of (25) in impedance calculations. In (26) the average energy dissipated in $R$ per second is $P_L = \frac{1}{2}RII^*$.

A significant property of the energy functions $W_e$ and $W_m$ is that they are positive functions. For the electrostatic field an important theorem known as Thomson's theorem states that the charges which reside on conducting bodies and which give rise to the electric field $\mathbf{E}$ will distribute themselves in such a way that the energy function $W_e$ is minimized. Since we shall use this result to obtain a variational expression for the characteristic impedance of a transmission line, a proof of this property of the electrostatic field will be given below.

Consider a system of conductors $S_1, S_2, \ldots, S_n$ as illustrated in Fig. 1.2, which are held at potentials $\phi_1, \phi_2, \ldots, \phi_n$. The electric field between the various $S_i$ may be obtained from the negative gradient of a suitable scalar potential function $\phi$, where $\phi$ reduces to $\phi_i$ on $S_i$, with $i = 1, 2, \ldots, n$. The energy stored in the electric field is given by

$$W_e = \frac{\epsilon}{2} \iiint_V \nabla\phi \cdot \nabla\phi \, dV \qquad (27)$$

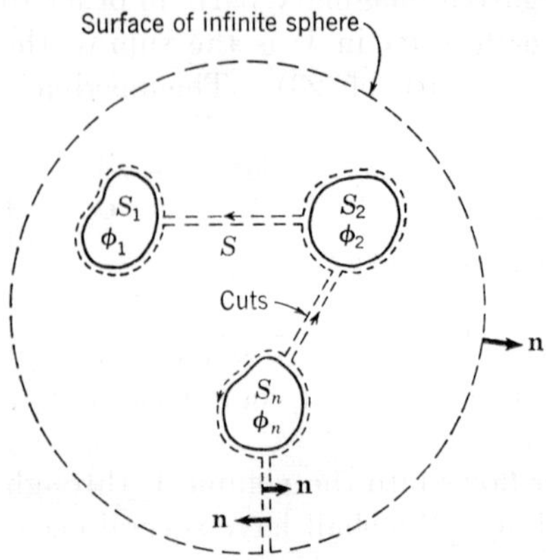


FIG. 1.2. Illustration of cuts to make surface $S$ simply connected.

Let the charges be displaced from their equilibrium positions by a small amount but still keep the potentials of the conductors constant. The potential distribution around the conductors will change to $\phi + \delta\phi$, where $\delta\phi$ is a small change in $\phi$. The change in $\phi$, that is, $\delta\phi$, is, of course, a function of the space coordinates but will vanish on the conductors $S_i$ since $\phi_i$ is maintained constant. The change in the energy function $W_e$ is

$$\delta W_e = \frac{\epsilon}{2} \iiint_V \nabla(\phi + \delta\phi) \cdot \nabla(\phi + \delta\phi) \, dV - \frac{\epsilon}{2} \iiint_V \nabla\phi \cdot \nabla\phi \, dV$$

$$= \frac{\epsilon}{2} \iiint_V 2\nabla\phi \cdot \nabla\delta\phi \, dV + \frac{\epsilon}{2} \iiint_V \nabla\delta\phi \cdot \nabla\delta\phi \, dV$$

The first term may be shown to vanish as follows. Let $u \nabla v$ be a vector function, and, using the divergence theorem, we get

$$\iiint_V \nabla \cdot u \nabla v \, dV = \iiint_V (\nabla u \cdot \nabla v + u \nabla^2 v) \, dV$$

$$= \oiint_S u \nabla v \cdot d\mathbf{a} = \oiint_S u \frac{\partial v}{\partial n} \, da$$

where $\mathbf{n}$ is the outward normal to the surface $S$ which surrounds the volume $V$. The above result is known as Green's first identity. If we let $\phi = v$ and $\delta\phi = u$, and recall that $\phi$ is a solution of Laplace's equation $\nabla^2\phi = 0$, in the region surrounding the conductors we obtain the result

$$\iiint_V \nabla\phi \cdot \nabla\delta\phi \, dV = \oiint_S \delta\phi \frac{\partial \phi}{\partial n} \, da$$

The surface $S$ is chosen as the surface of the conductors, plus the surface of an infinite sphere, plus the surface along suitable cuts joining the various conductors and the infinite sphere. The integral along the cuts does not contribute anything since the integral is taken along the cuts twice but with the normal **n** oppositely directed. Also the potential $\phi$ vanishes at infinity at least as rapidly as $r^{-1}$, and so the integral over the surface of the infinite sphere also vanishes. On the conductor surfaces $\delta\phi$ vanishes so that the complete surface integral vanishes. The change in the energy function $W_e$ is, therefore, of second order, i.e.,

$$\delta W_e = \frac{1}{2} \iiint_V \nabla \delta\phi \cdot \nabla \delta\phi \, dV \tag{28}$$

The first-order change is zero, and hence $W_e$ is a stationary function for the equilibrium condition. Since $\delta W_e$ is a positive quantity, the true value of $W_e$ is a minimum since any change from equilibrium increases the energy function $W_e$.

## 1.4. Boundary Conditions

In the solution of Maxwell's field equations in a region of space which is divided into two or more parts which form the boundaries across which

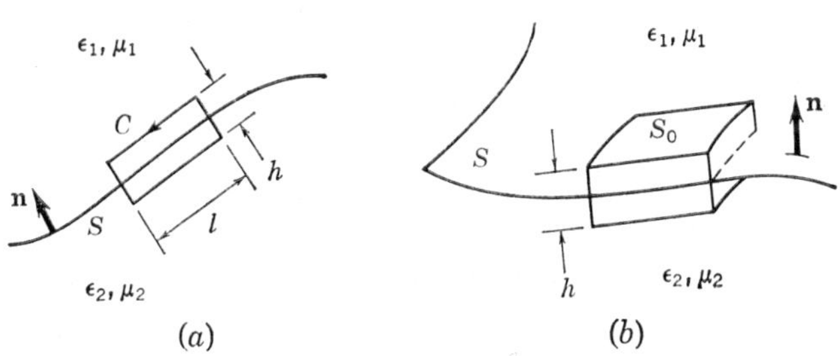


FIG. 1.3. Contour $C$ and surface $S_0$ for deriving boundary conditions.

the electrical properties of the medium, that is, $\mu$ and $\epsilon$, change discontinuously, it is necessary to know the relationship between the field components on adjacent sides of the boundaries. This is required in order to properly continue the solution from one region into the next, such that the whole final solution will indeed be a solution of Maxwell's equations. These boundary conditions are readily derived using the integral form of the field equations.

Consider a boundary surface $S$ separating a region with parameters $\epsilon_1$, $\mu_1$ on one side and $\epsilon_2$, $\mu_2$ on the adjacent side. The surface normal **n** is assumed directed from medium 2 into medium 1. We construct a small contour $C$, which runs parallel to the surface $S$ and an infinitesimal distance on either side as in Fig. 1.3. Also we construct a small closed surface $S_0$, which consists of a "pillbox" with faces of equal area, parallel to the surface $S$ and an infinitesimal distance on either side. The line integral of the electric field around the contour $C$ is equal to the negative time rate of change of the total magnetic flux through $C$. In the limit as $h$ tends to zero the total flux through $C$ vanishes since $\mathbf{B}$ is a bounded function. Thus

$$\lim_{h\to 0} \oint_C \mathbf{E} \cdot d\mathbf{l} = 0 = (E_{1t} - E_{2t})l$$

or

$$E_{1t} = E_{2t} \tag{29}$$

where the subscript $t$ signifies the component of $\mathbf{E}$ tangential to the surface $S$, and $l$ is the length of the contour. The negative sign preceding $E_{2t}$ arises because the line integral is in the opposite sense in medium 2 as compared with its direction in medium 1. Furthermore, the contour $C$ is chosen so small that $\mathbf{E}$ may be considered constant along the contour.

The line integral of $\mathbf{H}$ around the same contour $C$ yields a similar result, provided the displacement flux $\mathbf{D}$ and current density $\mathbf{J}$ are bounded functions. Thus we find that

$$H_{1t} = H_{2t} \tag{30}$$

Equations (29) and (30) state that the tangential components of $\mathbf{E}$ and $\mathbf{H}$ are continuous across a boundary for which $\mu$ and $\epsilon$ change discontinuously. A discontinuous change in $\mu$ and $\epsilon$ is again a mathematical idealization of what really consists of a rapid but finite rate of change of parameters.

The integral of the outward flux through the closed surface $S_0$ reduces in the limit as $h$ approaches zero to an integral over the end faces only. Provided there is no surface charge on $S$, the integral of $\mathbf{D}$ through the surface $S_0$ gives

$$\lim_{h\to 0} \oiint_{S} \mathbf{D} \cdot d\mathbf{a} = 0 = (D_{1n} - D_{2n})S_0$$

or

$$D_{1n} = D_{2n} \tag{31}$$

Similarly we find that

$$B_{1n} = B_{2n} \tag{32}$$

where the subscript $n$ signifies the normal component. It is seen that the normal components of the flux vectors are continuous across the discontinuity boundary $S$.

The above relations are not completely general since they are based on the assumption that the current density $\mathbf{J}$ is bounded and that a surface charge density does not exist on the boundary $S$. If a surface density of charge $\rho_s$ does exist, then the integral of the normal outward component of $\mathbf{D}$ over $S_0$ is equal to the total charge contained within $S_0$.

Under these conditions the boundary condition on **D** becomes

$$\mathbf{n} \cdot (\mathbf{D}_1 - \mathbf{D}_2) = \rho_s \tag{33}$$

The field vectors remain finite on $S$, but a situation, for which the current density **J** becomes infinite in such a manner that

$$\lim_{h \to 0} h\mathbf{J} = \mathbf{J}_s$$

a surface distribution of current, arises at the boundary of a perfect conductor for which the conductivity $\sigma$ is infinite. In this case the boundary condition on the tangential magnetic field is modified. Let $\boldsymbol{\tau}$ be the unit tangent to the contour $C$ in medium 1. The positive normal $\mathbf{n}_1$ to the plane surface bounded by $C$ is then given by $\mathbf{n}_1 = \mathbf{n} \times \boldsymbol{\tau}$. The normal current density through $C$ is $\mathbf{n}_1 \cdot \mathbf{J}_s$. The line integral of **H** around $C$ now gives

$$(\mathbf{H}_1 - \mathbf{H}_2) \cdot \boldsymbol{\tau} = \mathbf{n}_1 \cdot \mathbf{J}_s = \mathbf{n} \times \boldsymbol{\tau} \cdot \mathbf{J}_s$$

Rearranging gives

$$(\mathbf{H}_1 - \mathbf{H}_2) \cdot \boldsymbol{\tau} = -\boldsymbol{\tau} \cdot \mathbf{n} \times \mathbf{J}_s$$

and, since the orientation of $C$ and hence $\boldsymbol{\tau}$ is arbitrary, we must have

$$\mathbf{H}_1 - \mathbf{H}_2 = -\mathbf{n} \times \mathbf{J}_s$$

which becomes, when both sides are vector-multiplied by **n**,

$$\mathbf{n} \times (\mathbf{H}_1 - \mathbf{H}_2) = \mathbf{J}_s \tag{34}$$

If medium 2 is the perfect conductor, then the fields in medium 2 are zero, so that

$$\mathbf{n} \times \mathbf{H}_1 = \mathbf{J}_s \tag{35}$$

At a perfectly conducting boundary the boundary conditions on the field vectors may, therefore, be specified as follows:

$$\mathbf{n} \times \mathbf{H} = \mathbf{J}_s \tag{36a}$$
$$\mathbf{n} \times \mathbf{E} = 0 \tag{36b}$$
$$\mathbf{n} \cdot \mathbf{B} = 0 \tag{36c}$$
$$\mathbf{n} \cdot \mathbf{D} = \rho_s \tag{36d}$$

where the normal **n** points outward from the conductor surface.

At a discontinuity surface it is sufficient to match the tangential field components only, since, if the fields satisfy Maxwell's equations, this automatically makes the normal components of the flux vectors satisfy the correct boundary conditions. The vector differential operator $\nabla$ may be written as the sum of two operators as follows:

$$\nabla = \nabla_t + \nabla_n$$

where $\nabla_t$ is that portion of $\nabla$ which represents differentiation with respect to the coordinates on the discontinuity surface $S$, and $\nabla_n$ is the operator which differentiates with respect to the coordinate normal to $S$. The curl equation for the electric field now becomes

$$\nabla_t \times (\mathbf{E}_t + \mathbf{E}_n) + \nabla_n \times (\mathbf{E}_t + \mathbf{E}_n) = -j\omega\mu(\mathbf{H}_t + \mathbf{H}_n)$$

The normal component of the left-hand side is $\nabla_t \times \mathbf{E}_t$ since $\nabla_n \times \mathbf{E}_n$ is zero, and the remaining two terms lie in the surface $S$. If $\mathbf{E}_t$ is continuous across $S$, then the derivative of $\mathbf{E}_t$ with respect to a coordinate on the surface $S$ is also continuous across $S$. Hence the normal component of $\mathbf{B}$ is continuous across $S$. A similar result applies for $\nabla_t \times \mathbf{H}_t$ and the normal component of $\mathbf{D}$. If $\mathbf{H}_t$ is discontinuous across $S$ because of a surface current density $\mathbf{J}_s$, then $\mathbf{n} \cdot \mathbf{D}$ is discontinuous by an amount equal to the surface density of charge $\rho_s$. Furthermore, the surface current density and charge density satisfy the surface continuity equation.

In the solution of certain diffraction problems it is convenient to introduce fictitious current sheets of density $\mathbf{J}_s$. Since the tangential magnetic field is not necessarily zero on either side, the correct boundary condition on $\mathbf{H}$ at such a current sheet is that given by (34). The tangential electric field is continuous across such a sheet, but the normal component of $\mathbf{D}$ is discontinuous. The surface density of charge is given by the continuity equation as

$$\rho_s = \frac{j}{\omega} \nabla \cdot \mathbf{J}_s$$

and $\mathbf{n} \cdot \mathbf{D}$ is discontinuous by this amount.

For the same reasons it is convenient at times to introduce fictitious magnetic current sheets across which the tangential electric field and normal magnetic flux vectors are discontinuous [1.1, 1.15].† If a fictitious magnetic current $\mathbf{J}_m$ is introduced, the curl equation for $\mathbf{E}$ is written as

$$\nabla \times \mathbf{E} = -j\omega\mu\mathbf{H} - \mathbf{J}_m \tag{37}$$

The divergence of the left-hand side is identically zero, and since $\nabla \cdot \mathbf{J}_m$ is not necessarily taken as zero, we must introduce a fictitious magnetic charge density as well. From (37) we get

$$\nabla \cdot \mathbf{J}_m = -j\omega\nabla \cdot \mathbf{B} = -j\omega\rho_m \tag{38}$$

An analysis similar to that carried out for the discontinuity in the tangential magnetic field yields the following boundary conditions on the electric field at a magnetic current sheet:

$$\mathbf{n} \times (\mathbf{E}_1 - \mathbf{E}_2) = -\mathbf{J}_m \tag{39}$$

† Bracketed numbers are keyed to References given at the end of the chapter.

The negative sign arises by virtue of the way $\mathbf{J}_m$ was introduced into the curl equation for $\mathbf{E}$. The normal component of $\mathbf{B}$ is discontinuous according to the following relation:

$$\mathbf{n} \cdot (\mathbf{B}_1 - \mathbf{B}_2) = \rho_m \tag{40}$$

where $\rho_m$ is determined by (38). At a magnetic current sheet, the tangential components of $\mathbf{H}$ and the normal component of $\mathbf{D}$ are continuous across the sheet. To avoid introducing new symbols, $\mathbf{J}_m$ and $\rho_m$ in (39) and (40) are used to represent a surface density of magnetic current and charge, respectively. It should be emphasized again that such current sheets are introduced only to facilitate the analysis and do not represent a physical reality. Also the magnetic current and charge introduced here do not correspond to those associated with the magnetic polarization of material bodies which was introduced in Sec. 1.2.

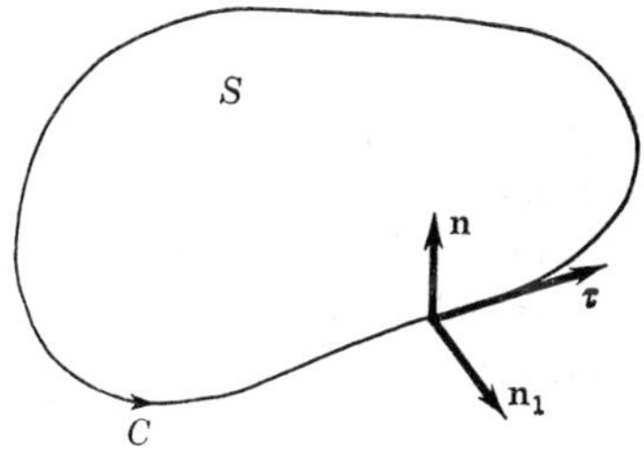

FIG. 1.4. Coordinate system at boundary of open surface $S$.

If the fictitious current sheets which we introduced do not form closed surfaces and the current densities normal to the edge are not zero, we must include a distribution of charge around the edge of the current sheet to take account of the discontinuity of the normal current at the edge [1.15]. Let $C$ be the boundary of the open surface $S$ on which a current sheet of density $\mathbf{J}$ exists. Let $\boldsymbol{\tau}$ be the unit tangent to $C$, and $\mathbf{n}$ the normal to the surface $S$, as in Fig. 1.4. The normal $\mathbf{n}$ to the surface $S$ is defined as having the direction of the normal to $S$ as the boundary $C$ is approached from the interior. The unit normal to the boundary curve $C$ which is perpendicular to $\mathbf{n}$ and directed out from $S$ is given by

$$\mathbf{n}_1 = \boldsymbol{\tau} \times \mathbf{n}$$

The charge flowing toward the boundary $C$ per second is $\mathbf{n}_1 \cdot \mathbf{J}_s$ per unit length of $C$. Replacing $\mathbf{J}_s$ from (34) by the discontinuity in $\mathbf{H}$ gives

$$\begin{aligned} j\omega\sigma_e &= (\boldsymbol{\tau} \times \mathbf{n}) \cdot \mathbf{n} \times (\mathbf{H}_1 - \mathbf{H}_2) \\ &= -\mathbf{n} \times (\boldsymbol{\tau} \times \mathbf{n}) \cdot (\mathbf{H}_1 - \mathbf{H}_2) \\ &= -\boldsymbol{\tau} \cdot (\mathbf{H}_1 - \mathbf{H}_2) \end{aligned} \tag{41}$$

where $j\omega\sigma_e$ is the time rate of increase of electric charge density per unit length along the boundary $C$. A similar analysis gives the following expression for the time rate of increase of magnetic charge along the boundary $C$ of an open surface containing a magnetic surface current:

$$j\omega\sigma_m = \boldsymbol{\tau} \cdot (\mathbf{E}_1 - \mathbf{E}_2) \tag{42}$$

These line charges are necessary in order that the continuity equations relating charge and current shall hold at the edge of the sheet.

We now turn to a consideration of the behavior of the field vectors at the edge of a conducting wedge of internal angle $\phi$, as illustrated in Fig. 1.5. The solutions to diffraction problems are not always unique unless the behavior of the fields at the edges of infinite conducting bodies is specified [1.7, 1.11]. In particular, it is found that some of the field components become infinite as the edge is approached. However, the order of singularity allowed is such that the energy stored in the vicinity of the diffracting edge is finite. The situation illustrated in Fig. 1.5 is not the most general by any means, but will serve our purpose as far as the remainder of this volume is concerned. A more complete discussion may be found in the references listed at the end of this chapter.

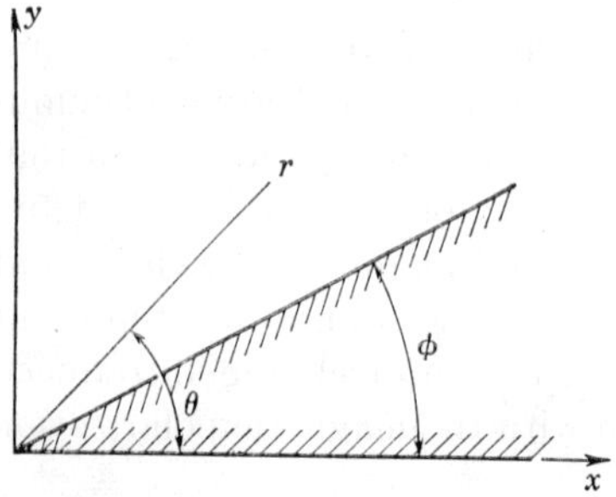

FIG. 1.5. Conducting wedge.

The field equations may be written in cylindrical coordinates $r$, $\theta$, $z$ as follows:

$$
\begin{aligned}
\frac{1}{r}\frac{\partial}{\partial r}(rE_\theta) - \frac{1}{r}\frac{\partial E_r}{\partial \theta} &= -j\omega\mu H_z \\
\frac{\partial E_r}{\partial z} - \frac{\partial E_z}{\partial r} &= -j\omega\mu H_\theta \\
\frac{1}{r}\frac{\partial E_z}{\partial \theta} - \frac{\partial E_\theta}{\partial z} &= -j\omega\mu H_r
\end{aligned} \tag{43a}
$$

$$
\begin{aligned}
\frac{1}{r}\frac{\partial}{\partial r}(rH_\theta) - \frac{1}{r}\frac{\partial H_r}{\partial \theta} &= +j\omega\epsilon E_z \\
\frac{\partial H_r}{\partial z} - \frac{\partial H_z}{\partial r} &= +j\omega\epsilon E_\theta \\
\frac{1}{r}\frac{\partial H_z}{\partial \theta} - \frac{\partial H_\theta}{\partial z} &= +j\omega\epsilon E_r
\end{aligned} \tag{43b}
$$

In the vicinity of $r = 0$ we may expand the field components in a power series in $r$ with coefficients which are functions of $\theta$ and $z$. Thus let

$$
\begin{aligned}
H_z &= r^\alpha(a_0 + a_1 r + a_2 r^2 + \cdots) \\
H_\theta &= r^\alpha(b_0 + b_1 r + b_2 r^2 + \cdots) \\
H_r &= r^\alpha(c_0 + c_1 r + c_2 r^2 + \cdots)
\end{aligned} \tag{44a}
$$

$$
\begin{aligned}
E_z &= r^\alpha(A_0 + A_1 r + A_2 r^2 + \cdots) \\
E_\theta &= r^\alpha(B_0 + B_1 r + B_2 r^2 + \cdots) \\
E_r &= r^\alpha(C_0 + C_1 r + C_2 r^2 + \cdots)
\end{aligned} \tag{44b}
$$

A system of equations involving the unknown expansion coefficients is obtained when the power series (44*a*) and (44*b*) are substituted into (43*a*) and (43*b*) and the coefficients of like powers of $r$ equated. To avoid dealing with an unduly large number of coefficients, we will determine the minimum allowed value of $\alpha$ first, such that the electromagnetic energy in the vicinity of the edge is finite in any given finite domain. The sum of the electric and magnetic energy in a small region surrounding the edge is proportional to

$$\iiint_0^r (\epsilon \mathbf{E} \cdot \mathbf{E}^* + \mu \mathbf{H} \cdot \mathbf{H}^*) r \, d\theta \, dr \, dz \tag{45}$$

Since each term is positive, the integral of each term must remain finite. Thus each term must not increase any faster than $r^{-1}$ as $r$ approaches zero. Hence each field component must increase no faster than $r^{-1/2}$ as $r$ approaches zero. The minimum allowed value of $\alpha$ is, therefore, $-\frac{1}{2}$. For $r$ sufficiently small, the fields may thus be represented by the first three terms of their series expansions. The resulting relations between the expansion coefficients are found to be

$$\begin{aligned}
(\alpha+1)B_0 &= \frac{\partial C_0}{\partial \theta} & -j\omega\mu a_0 &= (\alpha+2)B_1 - \frac{\partial C_1}{\partial \theta} \\
-j\omega\mu b_0 &= \frac{\partial C_0}{\partial z} - (\alpha+1)A_1 & -j\omega\mu b_1 &= \frac{\partial C_1}{\partial z} \qquad A_0 = 0 \\
-j\omega\mu c_0 &= \frac{\partial A_1}{\partial \theta} - \frac{\partial B_0}{\partial z} & j\omega\mu c_1 &= \frac{\partial B_1}{\partial z} \qquad \frac{\partial A_0}{\partial \theta} = 0
\end{aligned} \tag{46a}$$

as obtained from (43*a*). By symmetry the set of equations (43*b*) give

$$\begin{aligned}
(\alpha+1)b_0 &= \frac{\partial c_0}{\partial \theta} & j\omega\epsilon A_0 &= (\alpha+2)b_1 - \frac{\partial c_1}{\partial \theta} \\
j\omega\epsilon B_0 &= \frac{\partial c_0}{\partial z} - (\alpha+1)a_1 & j\omega\epsilon B_1 &= \frac{\partial c_1}{\partial z} \qquad a_0 = 0 \\
j\omega\epsilon C_0 &= \frac{\partial a_1}{\partial \theta} - \frac{\partial b_0}{\partial z} & -j\omega\epsilon C_1 &= \frac{\partial b_1}{\partial z} \qquad \frac{\partial a_0}{\partial \theta} = 0
\end{aligned} \tag{46b}$$

There are additional equations involving higher-order coefficients which, however, we will not require for the following analysis. We find at once that the tangential components $E_z$ and $H_z$ are finite at the edge since $A_0$ and $a_0$ are zero.

To determine the influence of the wedge angle $\phi$ on the parameter $\alpha$, the coefficients must be expressed as functions of the angle $\theta$ explicitly. By successive differentiation with respect to $\theta$ and $z$ and elimination of unwanted terms by substitution, the following set of equations are obtained:

$$\frac{\partial^2 \psi}{\partial \theta^2} + (\alpha + 1)^2 \psi = 0 \qquad \psi = b_0, c_0, B_0, C_0 \tag{47}$$

Once solutions for $b_0$, $c_0$, $B_0$, and $C_0$ are found, the solutions for $a_1$ and $A_1$ as functions of $\theta$ are readily written down from the third equations in (46*a*) and (46*b*). The remaining coefficients $b_1$, $c_1$, $B_1$, and $C_1$ are not related to the others to the order of approximation we are considering here and do not concern us at the moment. The solution to (47) is of the form

$$\psi = h \sin (\alpha + 1)\theta + g \cos (\alpha + 1)\theta \tag{48}$$

where $h$ and $g$ are functions of $z$. Now $E_r$ must vanish at $\theta = \phi$, $2\pi$, independent of $r$, and hence $C_0(z,\phi) = C_0(z,2\pi) = 0$. Since $C_0$ is of the form (48), we obtain the result

$$\tan (\alpha + 1)\phi = \tan (\alpha + 1)2\pi \tag{49}$$

and hence $(\alpha + 1)(2\pi - \phi) = n\pi$, where $n$ is any integer. Solving for $\alpha$ gives

$$\alpha = \frac{n\pi}{2\pi - \phi} - 1 \tag{50}$$

When $\phi = 0$, the minimum allowed value of $\alpha$ is $-\frac{1}{2}$, since $\alpha$ cannot be less than this in order not to violate the energy condition. When $\phi = \pi/2$, the minimum allowed value of $\alpha$ is $-\frac{1}{3}$, while for $\phi = \pi$, $\alpha = 0$, and none of the field components become infinite at the edge, which in actual fact does not exist any more.

## 1.5. The Wave Equation

The electric and magnetic field vectors **E** and **H** are solutions of the inhomogeneous vector wave equations. These vector wave equations will be derived here, but first we must make a distinction between the impressed currents and charges which are the sources of the field and the currents and charges which arise because of the presence of the field in a medium having finite conductivity. The latter current is proportional to the electric field and is given by $\sigma\mathbf{E}$. This conduction current may readily be accounted for by replacing the permittivity $\epsilon$ by the complex permittivity $\epsilon(1 - j \tan \delta)$ as given by (21) in Sec. 1.2. The density of free charge, apart from that associated with the impressed currents, may be assumed to be zero. Thus the charge density $\rho$ and current density **J** appearing in the field equations will be taken as the impressed charges and currents. Any other currents will be accounted for by the complex electric permittivity, although we shall still write just simply $\epsilon$ for this permittivity.

The wave equation for **E** is readily obtained by taking the curl of the curl of **E** and substituting for the curl of **H** on the right-hand side. Referring back to the field equations (1), it is readily seen that

$$\nabla \times \nabla \times \mathbf{E} = -j\omega\mu\nabla \times \mathbf{H} = \omega^2\mu\epsilon\mathbf{E} - j\omega\mu\mathbf{J}$$

where it is assumed that $\mu$ and $\epsilon$ are scalar constants. Using the vector identity $\nabla \times \nabla \times \mathbf{E} = \nabla\nabla \cdot \mathbf{E} - \nabla^2\mathbf{E}$ and replacing $\nabla \cdot \mathbf{E}$ by $\rho/\epsilon$ gives the desired result

$$\nabla^2\mathbf{E} + k^2\mathbf{E} = j\omega\mu\mathbf{J} + \frac{\nabla\rho}{\epsilon} \tag{51a}$$

where $k = \omega(\mu\epsilon)^{1/2}$ and will be called the wave number. The wave number is equal to $\omega$ divided by the velocity of propagation of electromagnetic waves in a medium with parameters $\mu$ and $\epsilon$ when $\epsilon$ is real. It is also equal to $2\pi/\lambda$, where $\lambda$ is the wavelength of plane waves in the same medium. The impressed charge density is related to the impressed current density by the equation of continuity so that (51*a*) may be rewritten as

$$\nabla^2\mathbf{E} + k^2\mathbf{E} = j\omega\mu\mathbf{J} - \frac{\nabla\nabla \cdot \mathbf{J}}{j\omega\epsilon} \tag{51b}$$

In a similar fashion we find that **H** is a solution of the following equation:

$$\nabla^2\mathbf{H} + k^2\mathbf{H} = -\nabla \times \mathbf{J} \tag{51c}$$

The magnetic field is determined by the rotational part of the current density only, while the electric field is determined by both the rotational and lamellar parts of **J**. This is as it should be, since **E** has both rotational and lamellar parts while **H** is a pure solenoidal field when $\mu$ is constant. The operator $\nabla^2$ may, in rectangular coordinates, be interpreted as the Laplacian operator operating on the individual rectangular components of each field vector. In a general curvilinear coordinate system this operator is replaced by the grad div − curl curl operator since

$$\nabla^2 \equiv \nabla\nabla \cdot - \nabla \times \nabla \times$$

The wave equations given above are not valid when $\mu$ and $\epsilon$ are scalar functions of position or dyadics. However, we will postpone the consideration of the wave equation in such cases until we take up the problem of propagation in inhomogeneous and anisotropic media. In a source-free region both the electric and magnetic fields satisfy the homogeneous equation

$$\nabla^2\mathbf{E} + k^2\mathbf{E} = 0 \tag{52}$$

and similarly for **H**. Equation (52) is also known as the vector Helmholtz equation.

Since the current density vector enters into the inhomogeneous wave equations in a rather complicated way, the integration of these equations is usually performed by the introduction of auxiliary potential functions which serve to simplify the mathematical analysis. These auxiliary potential functions may or may not represent clearly definable physical entities (especially is this so in the absence of sources), and so we prefer to adopt the viewpoint that these potentials are just useful mathematical functions from which the electromagnetic field may be derived. We shall discuss some of these potential functions in the following section.

### 1.6. Auxiliary Potential Functions

The first auxiliary potential functions we will consider are the vector and scalar potential functions **A** and $\phi$, which are just extensions of the magnetostatic vector potential and the electrostatic scalar potential. These functions are here functions of time and vary with time according to the function $e^{j\omega t}$, which, however, for convenience will be deleted. Again we will assume that $\mu$ and $\epsilon$ are constants.

The flux vector **B** is always solenoidal and may, therefore, be derived from the curl of a suitable vector potential function **A** as follows:

$$\mathbf{B} = \nabla \times \mathbf{A} \tag{53}$$

since $\nabla \cdot \nabla \times \mathbf{A} = 0$, and this makes $\nabla \cdot \mathbf{B} = 0$ also. Thus one of Maxwell's equations is satisfied identically. The vector potential **A** may have both a solenoidal and a lamellar part. At this stage of the analysis the lamellar part is entirely arbitrary since $\nabla \times \mathbf{A}_l = 0$. Substituting (53) into the curl equation for **E** gives

$$\nabla \times (\mathbf{E} + j\omega\mathbf{A}) = 0$$

Since $\nabla \times \nabla\phi = 0$, the above result may be integrated at once to give

$$\mathbf{E} = -j\omega\mathbf{A} - \nabla\phi \tag{54}$$

where $\phi$ is a scalar function of position and is called the scalar potential. So far we have two of Maxwell's field equations satisfied, and it now remains to find the relation between $\phi$ and **A** and the condition on $\phi$ and **A** so that the two remaining equations $\nabla \cdot \mathbf{D} = \rho$ and $\nabla \times \mathbf{H} = j\omega\epsilon\mathbf{E} + \mathbf{J}$ are satisfied. The curl equation for **H** gives

$$\begin{aligned} \nabla \times \mu\mathbf{H} &= \nabla \times \nabla \times \mathbf{A} = \nabla\nabla \cdot \mathbf{A} - \nabla^2\mathbf{A} = j\omega\epsilon\mu\mathbf{E} + \mu\mathbf{J} \\ &= k^2\mathbf{A} - j\omega\epsilon\mu\nabla\phi + \mu\mathbf{J} \end{aligned} \tag{55}$$

Since $\phi$ and the lamellar part of **A** are as yet arbitrary, we are at liberty to choose a relationship between them. With a view toward simplifying the equation satisfied by the vector potential **A**, i.e., (55), we choose

$$\nabla \cdot \mathbf{A} = -j\omega\epsilon\mu\phi \tag{56}$$

This particular choice is known as the Lorentz condition and is by no means the only one possible. Using (56), we find that (55) reduces to

$$\nabla^2\mathbf{A} + k^2\mathbf{A} = -\mu\mathbf{J} \tag{57}$$

In order that the divergence equation for **D** shall be satisfied it is necessary that

$$\nabla \cdot \mathbf{E} = \frac{\rho}{\epsilon} = -j\omega\nabla \cdot \mathbf{A} - \nabla^2\phi \qquad \text{from (54)}$$

Using the Lorentz condition to eliminate $\nabla \cdot \mathbf{A}$ gives the following equation to be satisfied by the scalar potential $\phi$:

$$\nabla^2\phi + k^2\phi = \frac{-\rho}{\epsilon} \tag{58}$$

The vector potential must be a solution of the inhomogeneous vector wave equation (57) while the scalar potential must be a solution of the scalar inhomogeneous wave equation (58). Clearly some advantage has been gained inasmuch as the wave equations for **A** and $\phi$ are simpler than those for **E** and **H**. Because of the continuity equation relating current and charge, the potentials are related to each other by the Lorentz condition (56). Alternatively, (56) may be regarded as the condition on **A** and $\phi$ in order that the current and charge shall satisfy the continuity equation. These potentials are not unique since we may add to them any solution of the homogeneous wave equations provided the new potentials still obey the Lorentz condition. Consider a new vector potential $\mathbf{A}' = \mathbf{A} - \nabla\psi$, where $\psi$ is an arbitrary scalar function satisfying the scalar equation

$$\nabla^2\psi + k^2\psi = 0$$

This transformation leaves the magnetic field **B** unchanged since $\nabla \times \nabla\psi \equiv 0$. In order that the electric field, as given by (54), shall remain invariant under this transformation, it is necessary that the scalar potential $\phi$ be transformed to a new scalar potential $\phi'$ according to the relation $\phi' = \phi + j\omega\psi$. Thus we have

$$\mathbf{E} = -j\omega\mathbf{A}' - \nabla\phi' = -j\omega\mathbf{A} + j\omega\,\nabla\psi - \nabla\phi - j\omega\,\nabla\psi = -j\omega\mathbf{A} - \nabla\phi$$

The transformation condition on $\phi'$ is just the condition that the new potentials shall satisfy the Lorentz condition.† The above transforma-

† In the particular gauge where $\nabla \cdot \mathbf{A}$ is zero, the Lorentz condition can no longer be applied since, in general, both $\nabla \cdot \mathbf{A}$ and $\phi$ cannot be equal to zero.

tion to the new potentials $\mathbf{A}'$ and $\phi'$ is known as a gauge transformation, and the property of the invariance of the field vectors under such a transformation is known as gauge invariance.

Using the Lorentz condition, the fields may be written in terms of the vector potential alone as follows:

$$\mathbf{B} = \nabla \times \mathbf{A} \tag{59a}$$

$$\mathbf{E} = -j\omega\mathbf{A} + \frac{\nabla\nabla \cdot \mathbf{A}}{j\omega\epsilon\mu} \tag{59b}$$

where $\mathbf{A}$ is a solution of (57), and $\phi$ is a solution of (58) as well as being related to $\mathbf{A}$ by (56). We may include in $\mathbf{A}$ any arbitrary solution to the homogeneous scalar Helmholtz equation, that is, $-\nabla\psi$, and, provided the new scalar potential is taken as $\phi + j\omega\psi$, we will always derive the same electromagnetic field by means of the operations given in (53) and (54). Consequently, $\psi$ may be chosen in such a manner that, if possible, a simplification in the solution of Maxwell's equations is obtained. For example, $\psi$ may be chosen so that $\nabla \cdot \mathbf{A}' = 0$. Then we find that

$$\mathbf{B} = \nabla \times \mathbf{A}' \tag{60a}$$

$$\mathbf{E} = -j\omega\mathbf{A}' - \nabla\phi' \tag{60b}$$

and the fields are explicitly separated into their solenoidal and lamellar parts. The vector potential is now a solution of

$$\nabla^2\mathbf{A}' + k^2\mathbf{A}' = -\mu\mathbf{J} + j\omega\epsilon\mu\,\nabla\phi' \tag{61a}$$

and the scalar potential is a solution of

$$\nabla^2\phi' = -\frac{\rho}{\epsilon} \tag{61b}$$

Thus the lamellar part of the electric field is determined by a scalar potential which is a solution of Poisson's equation with the exception that both $\phi'$ and $\rho$ vary with time. Since current and charge are related by the continuity equation, that is, $\nabla \cdot \mathbf{J} = -j\omega\rho$, we find that the divergence of the right-hand side of (61$a$) vanishes, in view of the relation between $\phi'$ and $\rho$ expressed by (61$b$). The term $j\omega\epsilon\mu\,\nabla\phi'$ may be regarded as canceling the lamellar part of $\mathbf{J}$ in (61$a$), and hence permits the solution for a vector potential having zero divergence, i.e., no lamellar part. In a region where $\rho$ is zero, we may take $\phi'$ zero also, and then both the electric and magnetic fields are determined by the vector potential $\mathbf{A}'$ alone.

The arbitrariness in the vector potential is only in the lamellar part. In a source-free region it is usually more convenient to find a vector potential function with a nonzero divergence, and the fields are then determined by (59). Since $\mathbf{E}$ has zero divergence in a source-free region, it is quite apparent that **E** could equally well be placed equal to the curl of a vector potential function. The details for this case will now be considered under the subject of Hertzian potentials.

In a homogeneous isotropic source-free region $\nabla \cdot \mathbf{E} = 0$, and so **E** may be derived from the curl of an auxiliary vector potential function which we will call a magnetic type of Hertzian vector potential. The reason for the terminology "magnetic" will be explained later. Thus we take

$$\mathbf{E} = -j\omega\mu \nabla \times \mathbf{\Pi}_h \tag{62}$$

where $\mathbf{\Pi}_h$ is the magnetic Hertzian potential. The curl equation for **H** gives

$$\nabla \times \mathbf{H} = k^2 \nabla \times \mathbf{\Pi}_h$$

so that

$$\mathbf{H} = k^2\mathbf{\Pi}_h + \nabla\phi \tag{63}$$

where $\phi$ is as yet an arbitrary scalar function. Substituting in the curl equation for **E** gives

$$\nabla \times \nabla \times \mathbf{\Pi}_h = \nabla\nabla \cdot \mathbf{\Pi}_h - \nabla^2\mathbf{\Pi}_h = k^2\mathbf{\Pi}_h + \nabla\phi \tag{64$a$}$$

Since both $\phi$ and $\nabla \cdot \mathbf{\Pi}_h$ are still arbitrary, we make them satisfy a Lorentz type of condition, i.e., put $\nabla \cdot \mathbf{\Pi}_h = \phi$, so that the equation for $\mathbf{\Pi}_h$ becomes

$$\nabla^2\mathbf{\Pi}_h + k^2\mathbf{\Pi}_h = 0 \tag{64$b$}$$

Since $\nabla \cdot \mathbf{B} = 0$, we find, by taking the divergence of (63) and remembering that $\mu$ is a constant, that $\phi$ is a solution of

$$\nabla^2\phi + k^2\phi = 0 \tag{65}$$

The fields are now given by the following relations:

$$\mathbf{E} = -j\omega\mu \nabla \times \mathbf{\Pi}_h \tag{66$a$}$$

$$\begin{aligned}\mathbf{H} &= k^2\mathbf{\Pi}_h + \nabla\nabla \cdot \mathbf{\Pi}_h \\ &= \nabla \times \nabla \times \mathbf{\Pi}_h\end{aligned} \tag{66$b$}$$

with $\mathbf{\Pi}_h$ a solution of the vector Helmholtz equations (64). In practice, we do not need to determine the scalar function $\phi$ explicitly since it is equal to $\nabla \cdot \mathbf{\Pi}_h$ and has been eliminated from the equation determining the magnetic field **H**.

In a source-free region where $\mu \neq \mu_0$ and $\epsilon = \epsilon_0$, the Hertzian potential $\mathbf{\Pi}_h$ will be shown to be the solution of the inhomogeneous vector Helmholtz equation with the source function being the magnetic polarization **M** per unit volume. It is for this reason that $\mathbf{\Pi}_h$ is called a magnetic Hertzian potential. When the magnetic polarization is taken into account explicitly, we rewrite (62) as

$$\mathbf{E} = -j\omega\mu_0 \nabla \times \mathbf{\Pi}_h \tag{67}$$

Equation (63) now becomes

$$\mathbf{H} = k_0^2\mathbf{\Pi}_h + \nabla\phi \tag{68}$$

where $k_0^2 = \omega^2\mu_0\epsilon_0$. In place of (64), we get

$$\nabla\nabla \cdot \mathbf{\Pi}_h - \nabla^2\mathbf{\Pi}_h = k_0^2\mathbf{\Pi}_h + \nabla\phi + \mathbf{M}$$

since $\mathbf{B} = \mu_0(\mathbf{H} + \mathbf{M})$. Again, equating $\phi$ to $\nabla \cdot \mathbf{\Pi}_h$, we get the final result

$$\nabla^2\mathbf{\Pi}_h + k_0^2\mathbf{\Pi}_h = -\mathbf{M} \tag{69}$$

Usually it is more convenient to absorb $\mathbf{M}$ into the parameter $\mu$. An exception arises when we wish to determine the field scattered from a magnetic body, in which case the use of (69) can, at times, be fruitfully applied to set up an integral equation for the scattered field. This approach depends on our ability to solve the resultant integral equation, by either rigorous means or a perturbation technique, if it is to be of any value to us.

The electric Hertzian potential $\mathbf{\Pi}_e$ is introduced in somewhat the same manner as the vector potential $\mathbf{A}$. Thus let

$$\mathbf{H} = j\omega\epsilon\,\nabla \times \mathbf{\Pi}_e \tag{70}$$

Carrying through an analysis along lines which by now are no doubt quite familiar, we find that $\mathbf{\Pi}_e$ is a solution of

$$\nabla^2\mathbf{\Pi}_e + k^2\mathbf{\Pi}_e = 0 \tag{71}$$

and that the electric field is given by

$$\mathbf{E} = k^2\mathbf{\Pi}_e + \nabla\nabla \cdot \mathbf{\Pi}_e = \nabla \times \nabla \times \mathbf{\Pi}_e \tag{72}$$

In a source-free medium where $\mu = \mu_0$ and $\epsilon \neq \epsilon_0$, the source for $\mathbf{\Pi}_e$ is readily shown to be the dielectric polarization $\mathbf{P}$ per unit volume. If $\epsilon$ is replaced by $\epsilon_0$ in (70) and the analysis carried out again, we find that $\mathbf{\Pi}_e$ is a solution of

$$\nabla^2\mathbf{\Pi}_e + k_0^2\mathbf{\Pi}_e = -\frac{\mathbf{P}}{\epsilon_0} \tag{73}$$

and $\mathbf{E}$ is now given by

$$\begin{aligned}\mathbf{E} &= k_0^2\mathbf{\Pi}. + \nabla\nabla \cdot \mathbf{\Pi}_e \\ &= \nabla \times \nabla \times \mathbf{\Pi}_e - \frac{\mathbf{P}}{\epsilon_0}\end{aligned} \tag{74}$$

In the study of wave propagation in cylindrical waveguides we will find that the transverse electric modes may be derived from a Hertzian potential $\mathbf{\Pi}_h$ having a single component along the axis of propagation, while the transverse magnetic modes may be derived from the electric Hertzian potential $\mathbf{\Pi}_e$ with a single component directed along the axis of propagation. In other words, the modes with a longitudinal electric field component ($E$ modes) will be derived from $\mathbf{\Pi}_e$, while the modes with a longitudinal magnetic field component ($H$ modes) will be derived from $\mathbf{\Pi}_h$. In a similar manner, the analysis of the modes of propagation in an inhomogeneously filled rectangular waveguide is facilitated by the use of Hertzian vector potential functions.

When $\epsilon$ and $\mu$ vary with position, the auxiliary potential functions discussed above satisfy more complicated partial differential equations. It is difficult to give a general formulation, so that we are forced to say that, in practice, it is best to treat each case separately in order to take advantage of any simplification which may be possible.

## 1.7. Some Field Equivalence Principles

The term "field equivalence principle" will be interpreted in a fairly broad sense to include any principle which permits the solution of a given field boundary-value problem by replacing it by another equivalent problem for which we either know the solution or can find the solution with less difficulty. The definition also includes the artifice of replacing the actual sources by a system of virtual sources without affecting the field in a certain region exterior to that containing the sources. The classical method of images used for the solution of potential problems comes under our definition of a field equivalence principle. The following four topics will be considered in turn:

1. Love's field equivalence theorem
2. Babinet's principle
3. The Schwartz reflection principle
4. Method of images

*Love's Field Equivalence Theorem* [1.12 to 1.14]

Let a closed surface $S$ separate a homogeneous isotropic medium into two regions $V_1$ and $V_2$ as in Fig. 1.6. All the sources are assumed to be located in the region $V_1$ so that both $V_2$ and $S$ are source-free. The theorem states that the field in $V_2$ produced by the given sources in $V_1$ is the same as that produced by a system of virtual sources on the surface $S$. Furthermore, if the field produced by the original sources is $\mathbf{E}$, $\mathbf{H}$, then the virtual sources on $S$ consist of an electric current sheet of density $\mathbf{n} \times \mathbf{H}$ and a magnetic current sheet of density $\mathbf{E} \times \mathbf{n}$, where the normal $\mathbf{n}$ points from $V_1$ into

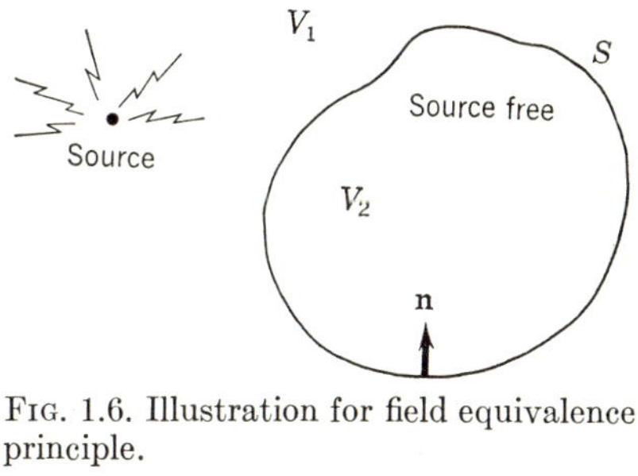


FIG. 1.6. Illustration for field equivalence principle.

$V_2$. In the region external to $S$, that is, in $V_1$, these virtual sources produce a null field. The theorem is readily proved by an application of the uniqueness theorem for electromagnetic fields. We will prove the uniqueness theorem first.

Let $\mathbf{E}_1$, $\mathbf{H}_1$ be an electromagnetic field in a region $V$ and arising from a system of source external to $V$. Let $S$ be a closed surface enclosing $V$ and excluding sources. Let $\mathbf{E}_2$, $\mathbf{H}_2$ be another possible field in the region $V$. Since both sets of fields satisfy Maxwell's equations, the difference field $\mathbf{E}_2 - \mathbf{E}_1$, $\mathbf{H}_2 - \mathbf{H}_1$ does also. If this difference field is used in the integral of the complex Poynting vector over $S$, as given by (25) in Sec. 1.3, we get

$$\oiint_S (\mathbf{E}_2 - \mathbf{E}_1) \times (\mathbf{H}_2 - \mathbf{H}_1)^* \cdot d\mathbf{a} = j\omega \left( \iiint_V \mu|\mathbf{H}_2 - \mathbf{H}_1|^2 \, dV - \iiint_V \epsilon|\mathbf{E}_2 - \mathbf{E}_1|^2 \, dV \right) + \iiint_V \sigma|\mathbf{E}_2 - \mathbf{E}_1|^2 \, dV$$

If either the tangential components of $\mathbf{E}_2$ and $\mathbf{E}_1$ or the tangential components of $\mathbf{H}_2$ and $\mathbf{H}_1$ are equal over the surface $S$, than either $(\mathbf{E}_2 - \mathbf{E}_1) \times \mathbf{n}$ or $(\mathbf{H}_2 - \mathbf{H}_1) \times \mathbf{n}$ vanishes and the surface integral vanishes. Thus the real and imaginary parts of the right-hand side above must vanish also. This is possible only if $\mathbf{E}_2 - \mathbf{E}_1 = 0$ so that the real part vanishes, and if $\mathbf{H}_2 - \mathbf{H}_1 = 0$ so that the remaining part of the imaginary term vanishes. Thus, provided either the tangential component of $\mathbf{E}$ or the tangential component of $\mathbf{H}$ satisfies the required values on the boundary $S$, the field interior to the surface $S$ is unique.

To prove the equivalence theorem, let $\mathbf{E}_1$, $\mathbf{H}_1$ be the field $\mathbf{E}$, $\mathbf{H}$ in $V_2$ and a null field in $V_1$. The field $\mathbf{E}_1$, $\mathbf{H}_1$ as defined is discontinuous across the surface $S$ by the amount $\mathbf{n} \times \mathbf{H}_1 = \mathbf{n} \times \mathbf{H}$ and $\mathbf{E}_1 \times \mathbf{n} = \mathbf{E} \times \mathbf{n}$. Since this field $\mathbf{E}_1$, $\mathbf{H}_1$ satisfies the correct boundary conditions at the electric and magnetic current sheets on $S$, it follows from the uniqueness theorem that the field $\mathbf{E}_1$, $\mathbf{H}_1$ as defined is the unique field generated by the virtual sources on $S$.

The field in $V_2$ is readily found from the vector potential functions, which, in turn, are determined by the virtual sources on $S$. In addition to the usual vector potential function $\mathbf{A}_e$ arising from an electric current distribution, we must also introduce a vector potential function $\mathbf{A}_m$ which has as its source the magnetic current distribution. The fields obtained from $\mathbf{A}_e$ are given by

$$\mathbf{B} = \nabla \times \mathbf{A}_e \tag{75a}$$

$$\mathbf{E} = -j\omega \mathbf{A}_e + \frac{\nabla\nabla \cdot \mathbf{A}_e}{j\omega\epsilon\mu} \tag{75b}$$

A similar analysis to that used for obtaining (75) shows that the fields obtained from a vector potential function $\mathbf{A}_m$ are given by

$$\epsilon\mathbf{E} = -\nabla \times \mathbf{A}_m \tag{76a}$$

$$\mathbf{H} = -j\omega\mathbf{A}_m + \frac{\nabla\nabla \cdot \mathbf{A}_m}{j\omega\epsilon\mu} \tag{76b}$$

The potential $\mathbf{A}_m$ is closely related to the Hertzian potential $\mathbf{\Pi}_h$, as may be seen from a comparison of (76*a*) and (62). The total field is the sum of that given by (75) and (76). The vector potentials are given in terms of the virtual sources as follows:

$$\mathbf{A}_e = \oiint_S \frac{\mu\mathbf{n} \times \mathbf{H}}{4\pi r} e^{-jkr}\, da \tag{77a}$$

$$\mathbf{A}_m = \oiint_S \frac{\epsilon\mathbf{E} \times \mathbf{n}}{4\pi r} e^{-jkr}\, da \tag{77b}$$

Equations (77) are derived in Sec. 1.8. If part of the surface $S$ consists of a perfectly conducting boundary $S_e$, the tangential electric field vanishes on this part of $S$, and hence the only virtual source (in this case actually a real source) on $S_e$ is an electric current sheet. The counterpart of a perfectly conducting surface or electric wall is called a magnetic wall and is defined as a surface over which the tangential magnetic field vanishes.

*Babinet's Principle* [1.15 to 1.17]

Maxwell's equations in a source-free region are

$$\nabla \times \mathbf{E} = -j\omega\mu\mathbf{H} \qquad \nabla \times \mathbf{H} = j\omega\epsilon\mathbf{E} \qquad \nabla \cdot \mathbf{D} = 0 \qquad \nabla \cdot \mathbf{B} = 0$$

and exhibit a high degree of symmetry in the expressions for **E** and **H**. If a field $\mathbf{E}_1$, $\mathbf{H}_1$ satisfies the above equations, then a second field $\mathbf{E}_2$, $\mathbf{H}_2$, obtained from $\mathbf{E}_1$, $\mathbf{H}_1$ by the following transformations:

$$\mathbf{E}_2 = \mp \left(\frac{\mu}{\epsilon}\right)^{1/2} \mathbf{H}_1 \tag{78a}$$

$$\mathbf{H}_2 = \pm \left(\frac{\epsilon}{\mu}\right)^{1/2} \mathbf{E}_1 \tag{78b}$$

is also a solution to the source-free field equations. The above equations are a statement of the duality property of the electromagnetic field and constitute the basis of Babinet's principle. We must, of course, consider also the effect of the transformation (78) on the boundary conditions to be satisfied by the new field. Clearly, since the roles of **E** and **H** are interchanged, we must interchange all electric walls for magnetic walls and magnetic walls for electric walls. At a boundary where the conditions on $\mathbf{E}_1$, $\mathbf{H}_1$ were that the tangential components are continuous across a surface with $\epsilon$ and $\mu$ equal to $\epsilon_1$, $\mu_1$ on one side and $\epsilon_2$, $\mu_2$ on the other side, the new boundary conditions on $\mathbf{E}_2$ and $\mathbf{H}_2$ are still the continuity of the tangential components, and this changes the relative amplitudes of the fields on the two sides of the discontinuity surface. This is permissible since $\mathbf{E}_2$ and $\mathbf{H}_2$ in (78) can be multiplied by an arbitrary constant and still remain solutions of the field equations. When

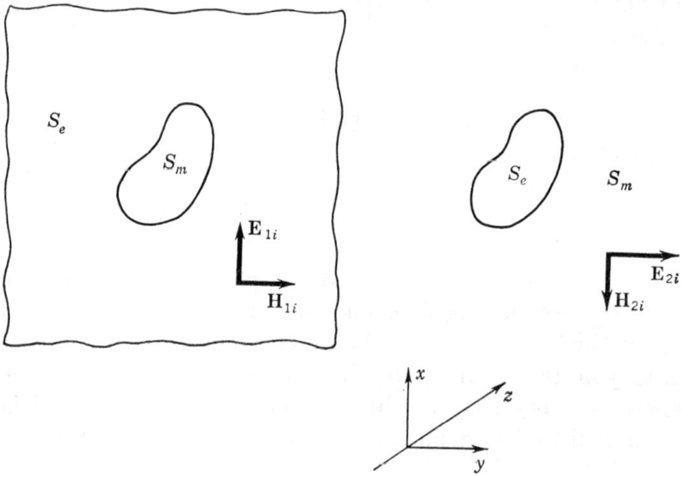


FIG. 1.7. Illustration of Babinet's principle.

a parallel-polarized wave is incident on a plane-dielectric interface, we obtain a reflected and a transmitted wave. The solution for the case of a perpendicular-polarized incident wave is readily deduced from the solution of the parallel-polarized case by transforming the incident, reflected, and transmitted waves according to (78) and matching these new field components at the interface. That the field $\mathbf{E}_2$, $\mathbf{H}_2$ is indeed a solution to the field equations is readily demonstrated. For example,

$$\nabla \times \mathbf{E}_2 = \left(\frac{\mu}{\epsilon}\right)^{1/2} \nabla \times \mathbf{H}_1 = j\omega\epsilon \left(\frac{\mu}{\epsilon}\right)^{1/2} \mathbf{E}_1 = -j\omega\mu \mathbf{H}_2$$

and similarly for the other field equations.

As an example, we will apply the duality principle to show the equivalence between the problem of diffraction by an aperture in a perfectly conducting infinite-plane screen and the complementary problem of diffraction by an infinitely thin perfectly conducting disk having the same shape as the aperture opening in the first problem. Let the screen be located in the $xy$ plane, and denote the aperture surface by $S_m$ and the conducting screen surface by $S_e$ as in Fig. 1.7. The complementary problem is also illustrated in the same figure.

In order to use the transformations (78), we must reduce the problem to one where the aperture surface $S_m$ corresponds to a magnetic wall. Consider first the case of unsymmetrical excitation with the incident fields chosen as follows:

$$\begin{aligned} \mathbf{E}_i &= \mathbf{a}_x A e^{-jkz} \\ \mathbf{H}_i &= \mathbf{a}_y A Y e^{-jkz} \end{aligned} \qquad z < 0 \tag{79a}$$

$$\begin{aligned} \mathbf{E}_i &= -\mathbf{a}_x A e^{jkz} \\ \mathbf{H}_i &= \mathbf{a}_y A Y e^{jkz} \end{aligned} \qquad z > 0 \tag{79b}$$

where $Y = (\epsilon/\mu)^{1/2}$. From the unsymmetrical nature of the incident electric field and the symmetry of the screen about the $z = 0$ plane, we conclude that the transverse components of the scattered electric field in the region $z < 0$ are the negative of those in the region $z > 0$. The transverse part of the scattered electric field is therefore an odd function of $z$. Since the divergence of **E** is zero, we conclude that the $z$ component of the scattered electric field is an even function of $z$ for

$$\nabla \cdot \mathbf{E}_s = \nabla_t \cdot \mathbf{E}_{st} + \frac{\partial E_{sz}}{\partial z} = 0$$

and the relation can hold on both sides of the screen only if $\partial E_{sz}/\partial z$ is an odd function of $z$, and hence $E_{sz}$ is an even function of $z$. The subscript $t$ signifies the transverse, that is, $x$ and $y$, components while the subscript $s$ denotes the scattered field. From Maxwell's equation for the curl of **E** it is found that $-j\omega\mu\mathbf{H}_t = \nabla_t \times \mathbf{E}_z + \mathbf{a}_z \times \partial\mathbf{E}_t/\partial z$, and hence, if the transverse part of the scattered electric field is an odd function of $z$, the transverse part of the scattered magnetic field must be an even function of $z$. The transverse part of the scattered field may be represented as follows in view of the above considerations:

$$\begin{aligned} \mathbf{E}_{st} &= \mathbf{E}_0(x,y,z) \\ \mathbf{H}_{st} &= \mathbf{H}_e(x,y,z) \end{aligned} \qquad z < 0 \tag{80a}$$

$$\begin{aligned} \mathbf{E}_{st} &= -\mathbf{E}_0(x,y,-z) \\ \mathbf{H}_{st} &= \mathbf{H}_e(x,y,-z) \end{aligned} \qquad z > 0 \tag{80b}$$

where the negative sign in front of $z$ for $z > 0$ is required since $z$ is of opposite sign on the two sides of the screen. In the aperture surface the transverse part of the electric field must be continuous, and so from (79) and (80) we get

$$\mathbf{a}_x A + \mathbf{E}_0(x,y) = -\mathbf{a}_x A - \mathbf{E}_0(x,y)$$

which is a possible result only if both sides are equal to zero. For this case we then get the result that the electric field tangential to the aperture vanishes in the aperture, and hence the aperture surface can be replaced by an electric wall. The problem is now simply that of reflection from an infinite perfectly conducting plane, and hence the scattered field is given by

$$\begin{aligned} \mathbf{E}_{st} &= -\mathbf{a}_x A e^{jkz} \\ \mathbf{H}_{st} &= \mathbf{a}_y A Y e^{jkz} \end{aligned} \qquad z < 0 \tag{81a}$$

$$\begin{aligned} \mathbf{E}_{st} &= \mathbf{a}_x A e^{-jkz} \\ \mathbf{H}_{st} &= \mathbf{a}_y Y A e^{-jkz} \end{aligned} \qquad z > 0 \tag{81b}$$

Consider next the case of symmetrical excitation with incident fields given by

$$\begin{aligned} \mathbf{E}_i &= \mathbf{a}_x B e^{-jkz} \\ \mathbf{H}_i &= \mathbf{a}_y B Y e^{-jkz} \end{aligned} \qquad z < 0 \tag{82a}$$

$$\begin{aligned} \mathbf{E}_i &= \mathbf{a}_x B e^{jkz} \\ \mathbf{H}_i &= -\mathbf{a}_y B Y e^{jkz} \end{aligned} \qquad z > 0 \tag{82b}$$

A consideration of symmetry conditions this time leads us to conclude that the transverse part of the scattered electric field is now an even function of $z$ with the transverse part of the scattered magnetic field an odd function of $z$. Thus the scattered fields can be represented as follows:

$$\begin{aligned} \mathbf{E}_{ts} &= B\mathbf{E}_e(x,y,z) \\ \mathbf{H}_{ts} &= B\mathbf{H}_0(x,y,z) \end{aligned} \qquad z < 0 \tag{83a}$$

$$\begin{aligned} \mathbf{E}_{ts} &= B\mathbf{E}_e(x,y,-z) \\ \mathbf{H}_{ts} &= -B\mathbf{H}_0(x,y,-z) \end{aligned} \qquad z > 0 \tag{83b}$$

where the amplitude $B$ is inserted only for convenience. Since the transverse magnetic field is continuous across the aperture surface, we must have

$$\mathbf{a}_y Y + \mathbf{H}_0(x,y) = -\mathbf{a}_y Y - \mathbf{H}_0(x,y)$$

which is possible only if both sides vanish. In this case we may replace the aperture surface by a magnetic wall since the tangential magnetic field vanishes in the aperture. The scattered field, however, cannot be determined by elementary means since we have reflection from a nonhomogeneous screen in this case. The solution to the diffraction problem with a wave incident from $z < 0$ is readily obtained from a superposition of the above two solutions with $B = A$ in order to cancel the incident wave in $z > 0$. This solution is

$$\begin{aligned} \mathbf{E}_t &= \mathbf{a}_x(2Ae^{-jkz} - Ae^{jkz}) + A\mathbf{E}_e(x,y,z) \\ \mathbf{H}_t &= \mathbf{a}_y Y(2Ae^{-jkz} + Ae^{jkz}) + A\mathbf{H}_0(x,y,z) \end{aligned} \qquad z < 0 \tag{84a}$$

$$\begin{aligned} \mathbf{E}_t &= \mathbf{a}_x A e^{-jkz} + A\mathbf{E}_e(x,y,-z) \\ \mathbf{H}_t &= \mathbf{a}_y Y A e^{-jkz} - A\mathbf{H}_0(x,y,-z) \end{aligned} \qquad z > 0 \tag{84b}$$

We may handle the complementary problem in the same way. With unsymmetrical excitation we get again simply the solution for reflection from a perfectly conducting infinite plane. For symmetrical excitation the solution is readily obtained using the transformation (78), since the new boundary conditions are simply an interchange of magnetic and electric walls. The solution for symmetrical excitation is thus

$$\left.\begin{aligned} \mathbf{E}_i &= \mathbf{a}_y \left(\frac{\mu}{\epsilon}\right)^{1/2} YBe^{-jkz} \\ \mathbf{H}_i &= -\mathbf{a}_x \left(\frac{\epsilon}{\mu}\right)^{1/2} Be^{-jkz} \end{aligned}\right\} \quad z < 0 \tag{85a}$$

$$\left.\begin{aligned} \mathbf{E}_i &= \mathbf{a}_y \left(\frac{\mu}{\epsilon}\right)^{1/2} YBe^{+jkz} \\ \mathbf{H}_i &= \mathbf{a}_x \left(\frac{\epsilon}{\mu}\right)^{1/2} Be^{jkz} \end{aligned}\right\} \quad z > 0 \tag{85b}$$

$$\left.\begin{aligned} \mathbf{E}_{ts} &= \left(\frac{\mu}{\epsilon}\right)^{1/2} B\mathbf{H}_0(x,y,z) \\ \mathbf{H}_{ts} &= -\left(\frac{\epsilon}{\mu}\right)^{1/2} B\mathbf{E}_e(x,y,z) \end{aligned}\right\} \quad z < 0 \tag{85c}$$

$$\left.\begin{aligned} \mathbf{E}_{ts} &= \left(\frac{\mu}{\epsilon}\right)^{1/2} B\mathbf{H}_0(x,y,-z) \\ \mathbf{H}_{ts} &= \left(\frac{\epsilon}{\mu}\right)^{1/2} B\mathbf{E}_e(x,y,-z) \end{aligned}\right\} \quad z > 0 \tag{85d}$$

For $z < 0$ the lower signs in the transformation (78) are used, while for $z > 0$ the upper signs are used to obtain (85) from (82) and (83). This is necessary in order to obtain a solution which is continuous across the open portion of the screen. We cannot have a discontinuous solution since the magnetic wall does not have physical existence. It is only by choosing special forms of excitation for the two complementary problems that they can be reduced to the duals of each other in the half space $z < 0$. The solution must be continued analytically into the region $z > 0$ bearing in mind the nonexistence of physical magnetic walls.

Superimposing a simple solution of the problem with unsymmetrical excitation with an amplitude sufficient to cancel the incident wave for $z > 0$ yields the following solution to the problem of diffraction by a disk:

$$\left.\begin{aligned} \mathbf{E}_t &= \mathbf{a}_y(2Be^{-jks} - Be^{jkz}) + ZB\mathbf{H}_0(x,y,z) \\ \mathbf{H}_t &= -\mathbf{a}_xY(2Be^{-jkz} + Be^{jkz}) - YB\mathbf{E}_e(x,y,z) \end{aligned}\right\} \quad z < 0 \tag{86a}$$

$$\left.\begin{aligned} \mathbf{E}_t &= \mathbf{a}_yBe^{-jkz} + ZB\mathbf{H}_0(x,y,-z) \\ \mathbf{H}_t &= -\mathbf{a}_xYBe^{-jkz} + YB\mathbf{E}_e(x,y,-z) \end{aligned}\right\} \quad z > 0 \tag{86b}$$

where $Z = (\mu/\epsilon)^{1/2} = Y^{-1}$. The duality principle thus allows us to readily obtain the solution to diffraction by a disk from the solution to the complementary problem of diffraction by an aperture or vice versa.

For the aperture problem let $\mathbf{E}_{1i}$, $\mathbf{H}_{1i}$ be the incident field for $z < 0$, and let $\mathbf{E}_{1s}$, $\mathbf{H}_{1s}$ be the scattered field in the region $z > 0$. For the complementary problem, let the complementary incident field for $z < 0$ be $\mathbf{E}_{2i} = Z\mathbf{H}_{1i}$, $\mathbf{H}_{2i} = -Y\mathbf{E}_{1i}$, and let the scattered field for $z > 0$ be $\mathbf{E}_{2s}$, $\mathbf{H}_{2s}$. These fields are given in detail by (84) and (86), and, using these expressions, we find that, for $z > 0$,

$$\mathbf{E}_{1s} - Z\mathbf{H}_{2s} = \mathbf{E}_{1i} \tag{87a}$$

$$\mathbf{H}_{1s} + Y\mathbf{E}_{2s} = \mathbf{H}_{1i} \tag{87b}$$

Equations (87) constitute the mathematical statement of Babinet's principle for complementary screens, which is a result of the dual property of the electromagnetic field. It should be noted that the duality principle is often referred to as Babinet's principle.

*The Schwartz Reflection Principle* [1.18]

The Schwartz reflection principle states that the analytic continuation of a function $f(x,y)$ across a surface $S$, say the plane $y = 0$, on which $f(x,0)$ vanishes is given by an odd reflection of $f$ in the plane $y = 0$. The analytic continuation of a function $f(x,y)$ for which $\partial f/\partial y$ vanishes on the plane $y = 0$ is given by an even reflection of $f$ in the plane $y = 0$. It should be noted at this time that we made use of this principle several times in the discussion of Babinet's principle. This principle has been used to demonstrate the equivalence between the problem of a semi-infinite bifurcation of a waveguide and the problem of a semidiaphragm in a waveguide. Since we shall consider these problems in a later section, we will postpone further comments on the reflection principle until then.

*Method of Images*

The method of images is closely akin to the reflection principle except that here we are concerned with replacing a source within a given boundary by a system of sources, which are images of the original source in the boundary. The underlying ideas are best explained by considering a particular application. We choose the particular problem of a linear current element in a rectangular waveguide as illustrated in Fig. 1.8. The field radiated by the linear current element is such that the tangential electric field vanishes on the boundary $C$. The method of images shows that the field radiated by the source $I$ in the presence of the boundary $C$ is the same as that radiated by the source and the image sources $I$ located at $x = \pm 2n's$, $z = d$; $x = \pm(2n + 1)s$, $z = -d$; and image sources $-I$ at $x = \pm(2n + 1)s$, $z = d$; $x = \pm 2ns$, $z = -d$; where $n$ is any integer from zero to infinity, and the prime on $n$ in the first term means exclusion of the value $n = 0$, since this corresponds to the original

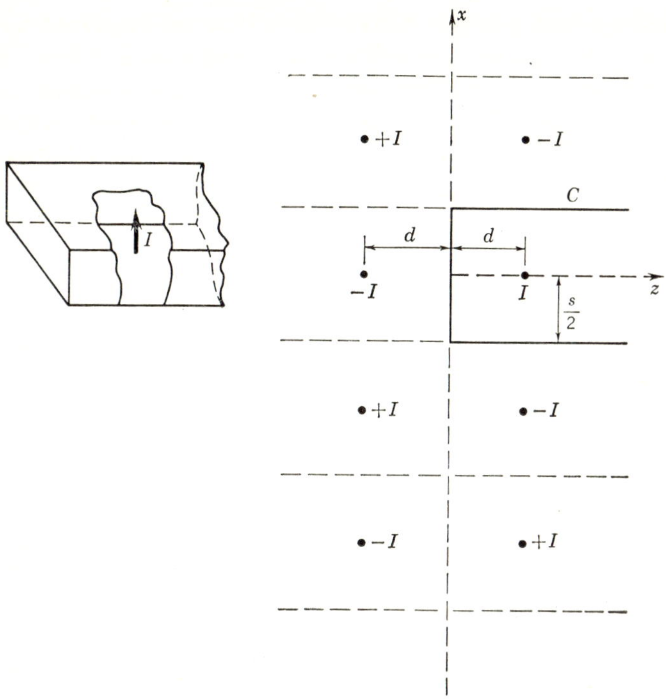


FIG. 1.8. Illustration for method of images.

source. The proof follows at once by symmetry considerations. The field radiated by the real source and all the virtual sources is such that it vanishes on the boundary $C$.

## 1.8. Integration of Inhomogeneous Helmholtz Equation

In this section we consider the integration of the inhomogeneous Helmholtz equation having the form

$$\nabla^2\mathbf{A} + k^2\mathbf{A} = -\mu\mathbf{J} \tag{88}$$

As a preliminary step we prove the following theorem first:

$$I = \iiint\limits_V \mathbf{P}(x_0,y_0,z_0)\nabla^2 \frac{1}{r}\, dV_0 = \begin{cases} 0 & (x,y,z) \text{ not in } V \\ -4\pi\mathbf{P}(x,y,z) & (x,y,z) \text{ in } V \end{cases} \tag{89}$$

where $\mathbf{P}$ is an arbitrary vector function which is continuous at the point $(x,y,z)$, and $r$ is the distance from the point $(x,y,z)$ to the point $(x_0,y_0,z_0)$; that is, $r^2 = (x - x_0)^2 + (y - y_0)^2 + (z - z_0)^2$. Direct differentiation shows that $\nabla^2(1/r) = 0$ at all points except $r = 0$, where it has an integrable singularity. The first part of the theorem follows at once since

$r \neq 0$ if $(x,y,z)$ is not in $V$. For the second part of the theorem we need be concerned only about the value of the integral in the vicinity of the point $r = 0$. We surround the point $(x,y,z)$ by a small sphere of volume $V_0$ and surface $S_0$. We choose $V_0$ so small that $\mathbf{P}$ may be considered constant throughout, and in particular equal to its value $\mathbf{P}(x,y,z)$ at the center. This is always possible in view of the postulated continuity of $\mathbf{P}$ at the point $(x,y,z)$. We may now take $\mathbf{P}$ outside the integral sign and, using the result that $\nabla^2(1/r) = \nabla_0^2(1/r)$, where $\nabla_0$ indicates differentiation with respect to the $x_0$, $y_0$, $z_0$ coordinates, we get

$$I = \mathbf{P}(x,y,z) \iiint\limits_{V_0} \nabla_0 \cdot \nabla_0 \frac{1}{r}\, dV_0 = \mathbf{P}(x,y,z) \oint\limits_{S_0} \nabla_0 \frac{1}{r} \cdot d\mathbf{a}_0$$

upon using the divergence theorem. Now $\nabla_0(1/r) = \mathbf{r}/r^3$, and

$$d\mathbf{a}_0 = -r\mathbf{r}\, d\Omega$$

where $d\Omega$ is an element of solid angle, and $\mathbf{r}$ points from the $x_0$, $y_0$, $z_0$ coordinates to the $x$, $y$, $z$ coordinates. Using this result shows at once that $I = -4\pi\mathbf{P}(x,y,z)$ and proves the second part of the theorem.

Consider the following result:

$$\nabla^2 \frac{e^{-jkr}}{r} = e^{-jkr}\, \nabla^2 \frac{1}{r} - \frac{k^2}{r} e^{-jkr}$$

which may be readily derived by direct differentiation. Rearranging terms gives the useful result

$$\nabla^2 \frac{e^{-jkr}}{r} + k^2 \frac{e^{-jkr}}{r} = e^{-jkr}\, \nabla^2 \frac{1}{r} \tag{90}$$

Now $\nabla^2(1/r) = 0$ except at $r = 0$, and so (90) shows at once that $(e^{-jkr}/r)$ is a solution to the scalar Helmholtz equation with a source function $e^{-jkr}\nabla^2(1/r)$ or $\nabla^2(1/r)$, since the factor $e^{-jkr}$ may be dropped because it equals unity at $r = 0$, the only point where $\nabla^2(1/r)$ differs from zero. The source function $-\nabla^2(1/4\pi r)$ has all the properties of a unit source or a delta function by virtue of its properties as expressed by (89). For an arbitrary source distribution the solution follows by superposition; for example, for a source function $-\rho/\epsilon$ the solution is

$$\iiint\limits_{V} \frac{\rho}{\epsilon} \frac{e^{-jkr}}{4\pi r}\, dV_0$$

The vector inhomogeneous Helmholtz equation is readily integrated in a similar fashion to yield

$$\mathbf{A} = \frac{\mu}{4\pi} \iiint\limits_{V} \frac{\mathbf{J}e^{-jkr}}{r}\, dV_0 \tag{91}$$

That this is the solution of (88) is readily demonstrated as follows:

$$\nabla^2\mathbf{A} + k^2\mathbf{A} = \frac{\mu}{4\pi}\left[\iiint\limits_V \mathbf{J}(x_0, \ldots)\, \nabla^2 \frac{e^{-jkr}}{r}\, dV_0 + k^2 \iiint\limits_V \mathbf{J}\frac{e^{-jkr}}{r}\, dV_0\right]$$

where the $\nabla^2$ operator could be brought inside the integral sign since it affects only the $x$, $y$, $z$ coordinates, while the integration is over the $x_0$, $y_0$, $z_0$ coordinates. Using (90), we find that the right-hand side reduces to

$$\frac{\mu}{4\pi}\iiint\limits_V \mathbf{J}(x_0, \ldots)\, \nabla^2 \frac{1}{r}\, dV_0$$

which from (89) equals $-\mu\mathbf{J}(x,y,z)$. If the current source is a surface or line distribution, then the solution is, of course, given by a surface or a line integral.

We now take up the problem of determining the vector and scalar potentials from an electric current sheet distributed over an open surface $S$ and possibly having a line distribution of charge on the boundary $C$. If the normal current density on the boundary $C$ is zero, there are no line charges on the boundary, and then the vector and scalar potentials are solutions of (57) and (58) given in Sec. 1.6. These solutions are simply

$$\mathbf{A} = \frac{\mu}{4\pi}\iint\limits_S \frac{\mathbf{J}_s(x_0, \ldots)}{r}\, e^{-jkr}\, da_0 \tag{92a}$$

$$\phi = \frac{1}{4\pi\epsilon}\iint\limits_S \frac{\rho_s(x_0, \ldots)}{r}\, e^{-jkr}\, da_0 \tag{92b}$$

where $\rho_s$ is given by $-\nabla \cdot \mathbf{J}_s/j\omega$. The electric and magnetic fields are given by

$$\mathbf{B} = \nabla \times \mathbf{A} \tag{93a}$$

$$\mathbf{E} = -j\omega\mathbf{A} - \nabla\phi = -j\omega\mathbf{A} + \frac{\nabla\nabla \cdot \mathbf{A}}{j\omega\epsilon\mu} \tag{93b}$$

with $\mathbf{A}$ and $\phi$ obeying the Lorentz condition since $\mathbf{J}_s$ and $\rho_s$ satisfy the continuity equation. When a distribution of line charge $\sigma_e$ is located on the boundary $C$, the new scalar potential is given in Ref. 1.14,

$$\phi_1 = \frac{1}{4\pi}\iint\limits_S \frac{\rho_s(x_0, \ldots)}{r}\, e^{-jkr}\, da_0 + \frac{1}{4\pi}\oint_C \sigma_e \frac{e^{-jkr}}{r}\, dl_0 \tag{94}$$

where $j\omega\sigma_e = J_n$, and $J_n$ is the normal component of the current at the boundary $C$. The equation for the electric field must now be changed to the following:

$$\mathbf{E} = -j\omega\mathbf{A} - \nabla\phi_1 = -j\omega\mathbf{A} + \frac{\nabla\nabla \cdot \mathbf{A}}{j\omega\epsilon\mu} \tag{95}$$

The validity of (94) and (95) depends on $\mathbf{A}$ and $\phi_1$ obeying the Lorentz condition $\nabla \cdot \mathbf{A} + j\omega\epsilon\mu\phi = 0$. If the line charge $\sigma_e$ were not placed on the boundary $C$, then $\mathbf{A}$ and $\phi_1$ would not satisfy the Lorentz condition, and the integral for $\mathbf{A}$ would not be given by (92$a$). It is not surprising to find that inclusion of a line charge makes $\mathbf{A}$ and $\phi_1$ satisfy the Lorentz condition since the Lorentz condition is equivalent to the condition that current and charge should obey a continuity equation. We now prove that $\nabla \cdot \mathbf{A} = -j\omega\epsilon\mu\phi_1$ with $\phi_1$ given by (94). We have

$$\nabla \cdot \mathbf{A} = \frac{\mu}{4\pi} \iint_S \nabla \cdot \mathbf{J}_s(x_0, \ldots) \frac{e^{-jkr}}{r} \, da_0$$

Now

$$\nabla \cdot \frac{\mathbf{J}(x_0, \ldots)}{r} e^{-jkr} = \mathbf{J} \cdot \nabla \frac{e^{-jkr}}{r} = -\mathbf{J} \cdot \nabla_0 \frac{e^{-jkr}}{r}$$
$$= -\nabla_0 \cdot \frac{\mathbf{J}(x_0, \ldots)}{r} e^{-jkr} + \frac{e^{-jkr}}{r} \nabla_0 \cdot \mathbf{J}$$

where $\nabla_0$ indicates differentiation with respect to the $x_0$, $y_0$, $z_0$ coordinates. Inserting this result into the integral, we get

$$\nabla \cdot \mathbf{A} = \frac{\mu}{4\pi} \left( \iint_S \frac{e^{-jkr}}{r} \nabla_0 \cdot \mathbf{J} \, da_0 - \iint_S \nabla_0 \cdot \frac{\mathbf{J}}{r} e^{-jkr} \, da_0 \right)$$

Now $\nabla_0 \cdot \mathbf{J} = -j\omega\rho_s$, and the second integral may be converted to a line integral around $C$ by using the two-dimensional form of the divergence theorem. Thus we finally get

$$\nabla \cdot \mathbf{A} = -j\omega\mu\epsilon \left( \iint_S \frac{\rho_s e^{-jkr}}{4\pi\epsilon r} \, da_0 + \frac{1}{j\omega\epsilon} \oint_C J_n \frac{e^{-jkr}}{4\pi r} \, dl_0 \right)$$
$$= -j\omega\epsilon\mu\phi_1$$

upon comparison with (94) and recognizing that the line-charge density $\sigma_e$ is given by $J_n/j\omega$, with $J_n$ being the normal outward-directed current on the boundary $C$. The significance of this result is that since $\mathbf{A}$ and $\phi_1$ satisfy the Lorentz condition, the field may bed etermined from the vector potential alone. In general this requires a line charge $\sigma_e$ on the boundary $C$, but this in no way enters into the evaluation of $\mathbf{A}$ as given by (92$a$).

The evaluation of the potentials from a magnetic current sheet proceeds in a similar manner. The normal component of magnetic current density at the boundary must be terminated in a magnetic line-charge distribution. This line-charge distribution must be taken account of in evaluating the scalar potential $\phi_m$.

## 1.9. Lorentz Reciprocity Theorem

One of the most useful theorems in electromagnetic field theory is the Lorentz reciprocity theorem. We will derive this theorem for the particular case where $\bar{\bar{\mu}}$ and $\bar{\bar{\epsilon}}$ are symmetrical constant tensors of rank 2, or dyadics.

Let $\mathbf{E}_1$, $\mathbf{H}_1$ be the field generated, in the volume $V$ bounded by a closed surface $S$, by a volume distribution of electric current $\mathbf{J}_{e1}$ and magnetic current $\mathbf{J}_{m1}$. Magnetic currents do not exist physically, but we will include them for generality because of their use in various field equivalence principles. Let $\mathbf{E}_2$, $\mathbf{H}_2$ be the field generated by a second system of sources $\mathbf{J}_{e2}$ and $\mathbf{J}_{m2}$. The fields are solutions of Maxwell's equations:

$$\begin{aligned} \nabla \times \mathbf{E}_1 &= -j\omega\bar{\bar{\mu}} \cdot \mathbf{H}_1 - \mathbf{J}_{m1} \\ \nabla \times \mathbf{H}_1 &= j\omega\bar{\bar{\epsilon}} \cdot \mathbf{E}_1 + \mathbf{J}_{e1} \\ \nabla \times \mathbf{E}_2 &= -j\omega\bar{\bar{\mu}} \cdot \mathbf{H}_2 - \mathbf{J}_{m2} \\ \nabla \times \mathbf{H}_2 &= j\omega\bar{\bar{\epsilon}} \cdot \mathbf{E}_2 + \mathbf{J}_{e2} \end{aligned}$$

Expanding the quantity $\nabla \cdot (\mathbf{E}_1 \times \mathbf{H}_2 - \mathbf{E}_2 \times \mathbf{H}_1)$ gives

$$\begin{aligned} \nabla \cdot (\mathbf{E}_1 \times \mathbf{H}_2 - \mathbf{E}_2 \times \mathbf{H}_1) = \mathbf{H}_2 \cdot \nabla \times \mathbf{E}_1 - \mathbf{E}_1 \cdot \nabla \times \mathbf{H}_2 \\ - \mathbf{H}_1 \cdot \nabla \times \mathbf{E}_2 + \mathbf{E}_2 \cdot \nabla \times \mathbf{H}_1 \end{aligned}$$

Substituting for the curl of the field vectors from Maxwell's equations and noting that $\bar{\bar{\mu}} \cdot \mathbf{H} = \mathbf{H} \cdot \bar{\bar{\mu}}$, $\bar{\bar{\epsilon}} \cdot \mathbf{E} = \mathbf{E} \cdot \bar{\bar{\epsilon}}$ because of the symmetry of the dyadics, we get

$$\nabla \cdot (\mathbf{E}_1 \times \mathbf{H}_2 - \mathbf{E}_2 \times \mathbf{H}_1) = -\mathbf{H}_2 \cdot \mathbf{J}_{m1} - \mathbf{E}_1 \cdot \mathbf{J}_{e2} + \mathbf{H}_1 \cdot \mathbf{J}_{m2} + \mathbf{E}_2 \cdot \mathbf{J}_{e1} \tag{96}$$

All the other terms in the right-hand side of the expansion of the divergence cancel. Integrating (96) over the volume $V$ and converting the volume integral of the divergence to a surface integral give

$$\begin{aligned} \oiint_S (\mathbf{E}_1 \times \mathbf{H}_2 - \mathbf{E}_2 \times \mathbf{H}_1) \cdot d\mathbf{a} = \iiint_V (\mathbf{H}_1 \cdot \mathbf{J}_{m2} - \mathbf{E}_1 \cdot \mathbf{J}_{e2} \\ - \mathbf{H}_2 \cdot \mathbf{J}_{m1} + \mathbf{E}_2 \cdot \mathbf{J}_{e1})\, dV \end{aligned} \tag{97}$$

Equation (97) is the general form of the Lorentz reciprocity theorem for media characterized by symmetrical tensor permeability and permittivity. Various special forms of the theorem may be obtained by considering a source-free volume, a volume bounded by a perfectly conducting surface, no magnetic current sources, etc.

As an example, consider a volume $V$ which is bounded by a perfectly conducting surface $S$ and in which two linear current elements $\mathbf{J}_{e1}$, $\mathbf{J}_{e2}$ exist. From (97) we get

$$\mathbf{E}_1 \cdot \mathbf{J}_{e2} = \mathbf{E}_2 \cdot \mathbf{J}_{e1}$$

since $\mathbf{n} \times \mathbf{E}_1 = \mathbf{n} \times \mathbf{E}_2 = 0$ on $S$, and the surface integral vanishes. The above equation states that a current element $\mathbf{J}_{e1}$ generates a field $\mathbf{E}_1$ with a component along $\mathbf{J}_{e2}$ which is equal to the component of $\mathbf{E}_2$ along $\mathbf{J}_{e1}$, where $\mathbf{E}_2$ is the field generated by $\mathbf{J}_{e2}$.

The Lorentz reciprocity theorem is useful for deriving reciprocal properties of wave-guiding devices and antennas, for deriving orthogonality properties of waveguide modes, for constructing the field radiated from currents in waveguides, and so forth. The theorem will be used on several occasions throughout the remainder of this book.

## REFERENCES

*Electromagnetic theory*

1.1. J. A. Stratton: "Electromagnetic Theory," McGraw-Hill Book Company, Inc., New York, 1941.
1.2. W. K. H. Panofsky and M. Phillips: "Classical Electricity and Magnetism," Addison-Wesley Publishing Company, Reading, Mass., 1955.
1.3. J. D. Kraus: "Electromagnetics," McGraw-Hill Book Company, Inc., New York, 1953.
1.4. S. A. Schelkunoff: "Electromagnetic Waves," D. Van Nostrand Company, Inc., Princeton, N.J., 1943.
1.5. W. R. Smythe: "Static and Dynamic Electricity," 2d ed., McGraw-Hill Book Company, Inc., New York, 1950.
1.6. S. Ramo and J. R. Whinnery: "Fields and Waves in Modern Radio," 2d ed., John Wiley & Sons, Inc., New York, 1953.

*Boundary conditions at an edge*

1.7. A. E. Heins and S. Silver: The Edge Conditions and Field Representation Theorems in the Theory of Electromagnetic Diffraction, *Proc. Cambridge Phil. Soc.*, vol. 51, pp. 149–161, 1955.
1.8. C. J. Bouwkamp: A Note on Singularities at Sharp Edges in Electromagnetic Theory, *Physica*, vol. 12, pp. 467–474, October, 1946.
1.9. J. Meixner: The Edge Condition in the Theory of Electromagnetic Waves at Perfectly Conducting Plane Screens, *Ann. Physik*, vol. 6, pp. 2–9, September, 1949.
1.10. J. Meixner: The Behavior of Electromagnetic Fields at Edges, *N.Y. Univ. Inst. Math. Sci. Research Rept. EM-72*, December, 1954.
1.11. D. S. Jones: Note on Diffraction by an Edge, *Quart. J. Mech. and Appl. Math.*, vol. 3, pp. 420–434, December, 1950.

*Field equivalence principles*

1.12. A. E. H. Love: The Integration of the Equations of Propagation of Electric Waves, *Phil. Trans. Roy. Soc. London*, ser. A, vol. 197, pp. 1–45, 1901.

1.13. S. A. Schelkunoff: Some Equivalence Theorems of Electromagnetics and Their Application to Radiation Problems, *Bell System Tech. J.*, vol. 15, pp. 92–112, 1936.

1.14. J. A. Stratton and L. J. Chu: Diffraction Theory of Electromagnetic Waves, *Phys. Rev.*, vol. 56, pp. 99–107, 1939.

1.15. B. B. Baker and E. T. Copson: "The Mathematical Theory of Huygens' Principle," 2d ed., Oxford University Press, New York, 1950.

1.16. H. G. Booker: Slot Aerials and Their Relationships to Complementary Wire Aerials, *Jour. IEE (London)*, vol. 93, part IIIA, pp. 620–626, 1946.

1.17. E. T. Copson: An Integral Equation Method for Solving Plane Diffraction Problems, *Proc. Roy. Soc. (London)*, ser. A, vol. 186, pp. 100–118, 1946.

1.18. S. N. Karp and W. E. Williams: Equivalence Relations in Diffraction Theory, *Proc. Cambridge Phil. Soc.*, vol. 53, pp. 683–690, 1957.

1.19. N. Marcuvitz and J. Schwinger: On the Representation of the Electric and Magnetic Fields Produced by Currents and Discontinuities in Waveguides, *J. Appl. Phys.*, vol. 22, pp. 806–819, June, 1951.

*Auxiliary potential functions*

1.20. P. M. Morse and H. Feshbach: "Methods of Theoretical Physics," part I, McGraw-Hill Book Company, Inc., New York, 1953. (See also Refs. 1.1 to 1.6.)

## PROBLEMS

**1.1.** The $E_{nm}$ modes in a rectangular waveguide of cross-sectional dimensions $a$ by $b$ can be derived from the following vector potential function, $\mathbf{A} = \mathbf{a}_z \sin (n\pi x/a) \sin (m\pi y/b)e^{\pm\Gamma_{nm}z}$, where $\Gamma_{nm}^2 = n^2\pi^2/a^2 + m^2\pi^2/b^2 - k^2$. Show that $\mathbf{A}$ is a solution of the vector Helmholtz equation. Use the Lorentz relation to find the scalar potential $\phi$ associated with $\mathbf{A}$. Find a gauge transformation which will give a new set of potentials $\mathbf{A}'$ and $\phi'$ such that $\nabla \cdot \mathbf{A}' = 0$. Find also the solenoidal and lamellar parts of the electric field.

**1.2.** A magnetic current sheet of density $\mathbf{J}_m = \mathbf{a}_x \sin x \sin y$ is located on the surface defined by $z = 0$ and the four lines $x = 0$, $x = 1$, and $y = 0$, $y = 1$. Find the discontinuity in the tangential electric field and normal magnetic flux across the magnetic current sheet. Also determine the density of magnetic line charge which must be placed on the boundary in order that the magnetic current and charge shall obey the continuity equation at the boundary. Set up the integral expressions for the vector and scalar potentials $\mathbf{A}_m$ and $\phi_m$.

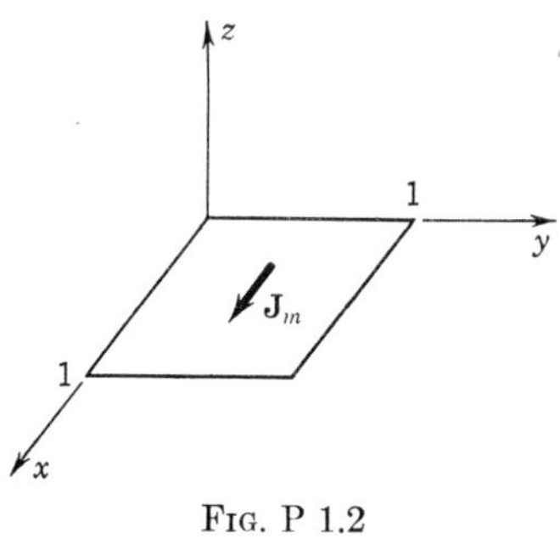


FIG. P 1.2

**1.3.** Rework in detail the example on the application of Babinet's principle given in Sec. 1.7, and include suitable functions for the $z$ components of the scattered electric and magnetic fields. Thus show in detail that Eqs. (87) hold.

**1.4.** Consider a uniform line current $\mathbf{a}_x J_0$ along the $x$ axis and extending from $-1 \leq x \leq 1$. Set up the expressions for the vector and scalar potentials in integral form. The scalar potential arises from the two point charges $+J_0/j\omega$ at $x = 1$ and $-J_0/j\omega$ at $x = -1$. Show that $\nabla \cdot \mathbf{A} = -j\omega\mu\epsilon\phi$. The inclusion of the two point charges is the one-dimensional form of a line charge along a surface boundary. The electric field is thus seen to be derivable from the vector potential $\mathbf{A}$ alone.

**1.5.** Consider a system of sources radiating a field $\mathbf{E}_0$, $\mathbf{H}_0$ into a region in which a volume $V$ consisting of inhomogeneous material exists. Let all sources lie external to the surface $S$ which bounds $V$. In the absence of the inhomogeneous material in $V$, let the sources radiate a field $\mathbf{E}_i$, $\mathbf{H}_i$. In the presence of the material in $V$, the total field radiated or transmitted into $V$ is $\mathbf{E}_t$, $\mathbf{H}_t$, while external to $V$ the field is the sum of the incident field and a reflected field, that is, $\mathbf{E}_i + \mathbf{E}_r$, $\mathbf{H}_i + \mathbf{H}_r$. Show that the scattered field $\mathbf{E}_t$, $\mathbf{H}_t$ and $\mathbf{E}_r$, $\mathbf{H}_r$ can be derived from a system of virtual sources on the boundary $S$, in particular, from an electric current sheet $\mathbf{n} \times \mathbf{H}_i$ and a magnetic current sheet $\mathbf{E}_i \times \mathbf{n}$, where $\mathbf{n}$ is the inward normal to $S$. This is Schelkunoff's induction theorem [1.13]. When the inhomogeneous material in $V$ is removed, the reflected field vanishes, and the induction theorem becomes Love's equivalence theorem.

**1.6.** Use the Lorentz reciprocity theorem to show that a tangential electric current or a normal magnetic current adjacent to a perfectly conducting surface does not radiate an electromagnetic field. Will a normal electric current or a tangential magnetic current adjacent to a perfectly conducting surface radiate a field?

HINT: Consider the field components, at the conductor surface, radiated by arbitrary currents remote from the surface.

**1.7.** Derive the inhomogeneous Helmholtz equations satisfied by the vector potential functions $\mathbf{A}_e$ and $\mathbf{A}_m$ when the electric and magnetic polarizations $\mathbf{P}$ and $\mathbf{M}$ of a material body are considered as the source functions. The results will be similar to that expressed by Eqs. (69) and (73).

**1.8.** Consider an infinite-plane perfectly conducting screen containing an aperture. For the case of odd excitation show that the normal electric field and tangential magnetic field in the aperture are equal to twice that associated with one of the incident waves. For the case of even excitation show that the normal electric field and tangential magnetic field in the aperture are zero. Next superimpose the two cases and show that, for a field incident on only one side of the screen, the normal electric field and tangential magnetic field in the aperture are equal to the corresponding field components of the incident wave.

CHAPTER 2

# GREEN'S FUNCTIONS

Basically a Green's function is the solution to a given differential equation, together with specified boundary conditions, with the source function being a source of unit strength and localized at a point in space. A more satisfactory definition is difficult to give at the present time. The following sections will be devoted to an examination of the solution of boundary-value problems by employing Green's functions. The method of solution makes extensive use of Green's identities† as developed by the mathematician George Green in 1828. Having examined the techniques involved, we will have a far better feeling for what a Green's function is than any definition of such a function could give.

## 2.1. Green's Functions for Poisson's Equation

The scalar potential $\phi$ in electrostatics is a solution of Poisson's equation

$$\nabla^2\phi = -\frac{\rho}{\epsilon} \tag{1}$$

where $\rho$ is the density of the charge distribution which gives rise to the scalar potential $\phi$, and $\epsilon$ is assumed constant. In the absence of any boundaries the solution is given by the following integral:

$$\phi(x,y,z) = \iiint\limits_V \frac{\rho(x_0,y_0,z_0)}{4\pi\epsilon r}\, dV_0 \tag{2}$$

where $r = [(x - x_0)^2 + (y - y_0)^2 + (z - z_0)^2]^{1/2}$ and is the distance from the source point to the point in space where the potential is evaluated. A source of unit strength located at the point $(x_0,y_0,z_0)$ can be conveniently represented by a three-dimensional delta function $\delta(x - x_0)$ $\delta(y - y_0)\delta(z - z_0)$, where the delta function is zero at all points except $x_0 = x$, $y_0 = y$, $z_0 = z$. At these latter points it becomes infinite in such a manner that

$$\iiint\limits_V \delta(x - x_0)\delta(y - y_0)\delta(z - z_0)\, dV_0 = \begin{cases} 0 & (x,y,z) \text{ not in } V \\ 1 & (x,y,z) \text{ in } V \end{cases} \tag{3a}$$

† Section A.1c.

If $\mathbf{P}(x_0,y_0,z_0)$ is an arbitrary vector function which is continuous at $(x,y,z)$, then

$$\iiint_V \mathbf{P}(x_0, \ldots)\delta(x - x_0)\delta(y - y_0)\delta(z - z_0)\, dV_0 = \begin{cases} 0 & (x,y,z) \text{ not in } V \\ \mathbf{P}(x,y,z) & (x,y,z) \text{ in } V \end{cases} \tag{3b}$$

A similar result holds for a scalar function also. Our discussion of the solution of the inhomogeneous Helmholtz equation in Sec. 1.8 showed that $-\nabla^2(4\pi r)^{-1}$ has the properties given by (3) above. The three-dimensional solution of the following inhomogeneous differential equation defines the free-space three-dimensional Green's function for Poisson's equation:

$$\nabla^2 G(x,y,z|x_0,y_0,z_0) = -\delta(x - x_0)\delta(y - y_0)\delta(z - z_0) \tag{4}$$

The Green's function $G$ is a function of both the point of observation or field point and the source point. From the known properties of the function $(4\pi r)^{-1}$ we see at once that this is the solution of (4), and hence

$$G(x,y,z|x_0,y_0,z_0) = \frac{1}{4\pi r} \tag{5}$$

The Green's function is symmetrical in the variables $x_0$, $y_0$, $z_0$ and $x$, $y$, $z$. This property will be found to be true in most cases which we will encounter, and is equivalent to the field obeying a reciprocity principle. Once the solution for a unit source is known, the solution for any given arbitrary source distribution follows at once by superposition, i.e.,

$$\phi(x, \ldots) = \frac{1}{\epsilon} \iiint_V G(x,y,z|x_0,y_0,z_0)\rho(x_0,y_0,z_0)\, dV_0 \tag{6}$$

which, with the aid of (5), is certainly recognizable as the standard solution of Poisson's equation.

In addition to the notation used above, it will be convenient, in what follows, to use the following abbreviated notation as well. The source point $(x_0,y_0,z_0)$ will be designated by the vector $\mathbf{r}_0$, the field point $(x,y,z)$ by the vector $\mathbf{r}$, and the distance between the field point and the source point by $|\mathbf{r} - \mathbf{r}_0|$. In addition, the three-dimensional delta function will be designated by $\delta(\mathbf{r} - \mathbf{r}_0)$. In this notation the Green's function is written as $G(\mathbf{r}|\mathbf{r}_0)$, and (6) becomes

$$\phi(\mathbf{r}) = \frac{1}{\epsilon} \iiint_V G(\mathbf{r}|\mathbf{r}_0)\rho(\mathbf{r}_0)\, dV_0$$

When $G$ is symmetrical, $G(\mathbf{r}|\mathbf{r}_0) = G(\mathbf{r}_0|\mathbf{r})$.

Let $\phi$ and $\psi$ be two scalar functions of position which are continuous with continuous partial derivatives. Let the region of space $V$ contain conducting bodies with surfaces $S_1, S_2, \ldots, S_n$, as illustrated in Fig. 2.1. By introducing suitable cuts, the region of interest can be made simply connected. Let $S$ be a closed surface consisting of the surfaces $S_1, \ldots, S_n$, the surfaces along the cuts, and the surface of an infinite sphere which surrounds all the conducting bodies.

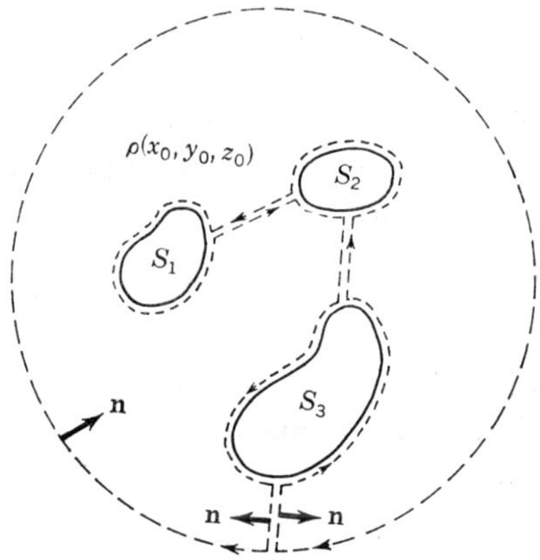


FIG. 2.1. Illustration for Green's theorem.

Applying Green's second identity† to the functions $\phi$ and $\psi$ gives

$$\iiint_V (\phi \nabla_0^2\psi - \psi \nabla_0^2\phi)\, dV_0 = \oiint_S \left(\psi \frac{\partial \phi}{\partial n} - \phi \frac{\partial \psi}{\partial n}\right) da_0$$

where $\mathbf{n}$ is the normal to $S$ and directed into the volume $V$. We now let $\phi$ be the required solution of Poisson's equation, and let $\psi$ be the Green's function as given by (5). Replacing $\nabla_0^2\phi$ by $-\rho/\epsilon$ and $\nabla_0^2 G$ by the negative delta function, we find that

$$-\iiint_V \phi(\mathbf{r}_0)\delta(\mathbf{r} - \mathbf{r}_0)\, dV_0 + \frac{1}{\epsilon}\iiint_V \rho(\mathbf{r}_0)G(\mathbf{r}|\mathbf{r}_0)\, dV_0$$
$$= \oiint_S \frac{\partial \phi(\mathbf{r}_0)}{\partial n} G(\mathbf{r}|\mathbf{r}_0)\, da_0 - \oiint_S \phi(\mathbf{r}_0) \frac{\partial G(\mathbf{r}|\mathbf{r}_0)}{\partial n}\, da_0 \quad (7a)$$

where $$\frac{\partial G}{\partial n} = (\nabla_0 G) \cdot \mathbf{n} \qquad \frac{\partial \phi}{\partial n} = (\nabla_0 \phi) \cdot \mathbf{n}$$

and $\nabla_0$ designates differentiation with respect to $x_0, y_0, z_0$. It is important to note that, if $x_0, y_0, z_0$ are used as the variables of integration in Green's identity, then all derivatives of functions in the integrands must

† Section A.1c.

also be with respect to $x_0$, $y_0$, $z_0$. Using the property of the delta function as given by (3) shows that the required solution for $\phi$ is

$$\phi(\mathbf{r}) = \frac{1}{\epsilon} \iiint\limits_V \rho(\mathbf{r}_0) G(\mathbf{r}|\mathbf{r}_0)\, dV_0 + \oiint\limits_S \left( \phi \frac{\partial G}{\partial n} - G \frac{\partial \phi}{\partial n} \right) da_0 \tag{7b}$$

The surface integrals along the cuts do not contribute since they are traversed twice in the opposite sense and with the normal $\mathbf{n}$ oppositely directed. The three terms contributing to $\phi$ are the volume distribution of charge $\rho$, a surface distribution of charge $-\epsilon(\partial\phi/\partial n)$, and a surface distribution of electric dipoles of dipole moment $\epsilon\phi$ per unit area.†

In a two-dimensional space the Green's function is defined by the following differential equation:

$$\frac{\partial^2 G}{\partial x^2} + \frac{\partial^2 G}{\partial y^2} = -\delta(x - x_0)\delta(y - y_0) \tag{8a}$$

The source function in this case is an infinite line source of unit density. The solution for $G$ is readily obtained by rewriting (8a) in cylindrical coordinates $r$, $\theta$ as follows:

$$\frac{1}{r}\frac{\partial}{\partial r} r \frac{\partial G}{\partial r} + \frac{1}{r^2}\frac{\partial^2 G}{\partial \theta^2} = -\delta(r) \tag{8b}$$

where the origin for the $r$, $\theta$ coordinate system has been chosen coincident with the line charge, i.e., at $x_0$, $y_0$. From symmetry considerations we see that $G$ is not a function of $\theta$, and hence depends only on the radial

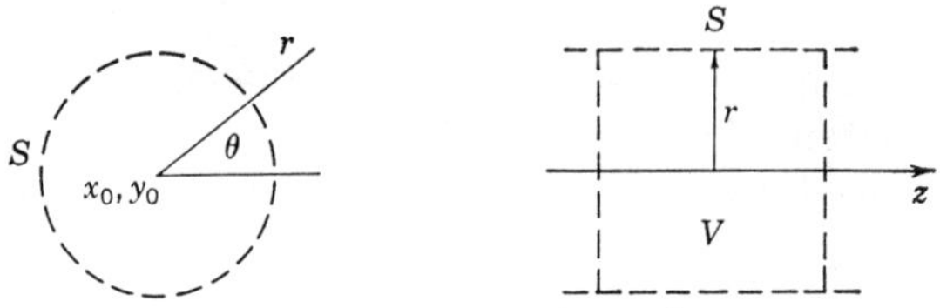


FIG. 2.2. Surface $S$ surrounding a line source.

distance $r$ from the line source. If we surround the line source by a cylindrical surface $S$ as illustrated in Fig. 2.2 and integrate (8b) throughout the volume enclosed by a unit length of $S$, we get

$$\iiint\limits_V \nabla \cdot \nabla G\, dV = \oiint\limits_S \nabla G \cdot d\mathbf{a} = -\iiint\limits_V \delta(r)\, dV = -1$$

by virtue of the properties of the delta function. Since $G$ is a function of $r$ only, we get $(\partial G/\partial r)\, 2\pi r = -1$, and a further integration gives the

† J. A. Stratton, "Electromagnetic Theory," chap. 3, McGraw-Hill Book Company, Inc., New York, 1941.

desired result, which is

$$G = -\frac{\ln r}{2\pi} \tag{9a}$$

Transforming back to our $x$, $y$ coordinate system, we have

$$G(x,y|x_0,y_0) = -\frac{\ln [(x - x_0)^2 + (y - y_0)^2]^{1/2}}{2\pi} \tag{9b}$$

## 2.2. Modified Green's Functions

The solution for $\phi$ given by (7b) is not too useful in practice since usually we know $\phi$ on part of the boundary and $\partial\phi/\partial n$ on the remainder but rarely do we know both on the complete surface $S$. However, by a suitable modification of our Green's function it is possible, in principle at least, to obtain a solution for $\phi$ from a knowledge of either $\phi$ or $\partial\phi/\partial n$ on the boundary. If the required boundary conditions specify the value of $\phi$ on the complete boundary, the potential $\phi$ is said to satisfy Dirichlet boundary conditions. If the normal derivative $\partial\phi/\partial n$ is specified on the whole boundary, the corresponding boundary conditions are referred to as Neumann boundary conditions. Finally, if $\phi$ is specified on part of the boundary and $\partial\phi/\partial n$ on the remainder, we have mixed boundary conditions.

If $\phi$ is known on the complete boundary, we see from an examination of the solution (7b) that the offending term involving $\partial\phi/\partial n$ can be eliminated by making the Green's function vanish on the complete boundary, i.e., on the surface $S$. Thus we define the Green's function of the first kind as a solution to the following equation:

$$\nabla^2 G_1(\mathbf{r}|\mathbf{r}_0) = -\delta(\mathbf{r} - \mathbf{r}_0) \tag{10a}$$

and obeying the boundary condition

$$G_1 = 0 \qquad \text{on } S \tag{10b}$$

The solution for the potential $\phi$ is thus given by

$$\phi(\mathbf{r}) = \iiint\limits_V \frac{\rho}{\epsilon} G_1 \, dV_0 + \oint\!\!\!\oint\limits_S \phi \frac{\partial G_1}{\partial n} \, da_0 \tag{11}$$

When $\partial\phi/\partial n$ is given everywhere on $S$, we modify our Green's function to make $\partial G/\partial n$ vanish on the boundary. The Green's function of the second kind is defined as a solution of

$$\nabla^2 G_2(\mathbf{r}|\mathbf{r}_0) = -\delta(\mathbf{r} - \mathbf{r}_0) \tag{12a}$$

subject to the boundary condition

$$\frac{\partial G_2}{\partial n} = 0 \qquad \text{on } S \tag{12b}$$

The solution for $\phi$ in this case is given by

$$\phi(\mathbf{r}) = \iiint\limits_V \frac{\rho}{\epsilon} G_2 \, dV_0 - \oint\!\!\!\oint\limits_S \frac{\partial \phi}{\partial n} G_2 \, da_0 \tag{13}$$

When we have mixed boundary conditions, it is desirable, if possible, to find a Green's function which vanishes on that part of $S$ for which we know $\phi$, and which has a normal derivative equal to zero over that part of $S$ for which we know $\partial\phi/\partial n$. From (11) and (13) we see that the Green's functions are useful for solving boundary-value problems, when the volume distribution of charge $\rho$ is zero, as well. Needless to say, quite often in practice it is about as difficult to find a solution for the Green's functions as it is to solve the original boundary-value problem. The use of Green's functions converts a differential equation together with the boundary conditions to an integral equation, which in many cases is more readily attacked by approximate techniques.

We will now show that the Green's function is symmetric in the variables $x$, $y$, $z$ and $x_0$, $y_0$, $z_0$, that is,

$$G(\mathbf{r}|\mathbf{r}_0) = G(\mathbf{r}_0|\mathbf{r}) \tag{14}$$

This property might have been anticipated since the delta function is symmetrical, i.e., $\delta(\mathbf{r} - \mathbf{r}_0) = \delta(\mathbf{r}_0 - \mathbf{r})$. This reciprocity principle states that the potential at $(x,y,z)$ from a unit charge at $(x_0,y_0,z_0)$ is the same as the potential at $(x_0,y_0,z_0)$ due to a unit charge at $(x,y,z)$. Let $G_1(\mathbf{r}|\mathbf{r}_{01})$ be the Green's function of the first kind for a point charge at $(x_{01},y_{01},z_{01})$, and let $G_1(\mathbf{r}|\mathbf{r}_{02})$ be the Green's function for a point charge at $(x_{02},y_{02},z_{02})$. If these two functions are used in Green's second identity, we get

$$\iiint\limits_V [G_1(\mathbf{r}|\mathbf{r}_{01}) \nabla^2 G_1(\mathbf{r}|\mathbf{r}_{02}) - G_1(\mathbf{r}|\mathbf{r}_{02}) \nabla^2 G_1(\mathbf{r}|\mathbf{r}_{01})] \, dV$$

$$= - \oint\!\!\!\oint\limits_S \left[ G_1(\mathbf{r}|\mathbf{r}_{01}) \frac{\partial G_1(\mathbf{r}|\mathbf{r}_{02})}{\partial n} - G_1(\mathbf{r}|\mathbf{r}_{02}) \frac{\partial G_1(\mathbf{r}|\mathbf{r}_{01})}{\partial n} \right] da$$

Since $G_1$ vanishes on $S$ and

$$\nabla^2 G_1(\mathbf{r}|\mathbf{r}_{01}) = -\delta(\mathbf{r} - \mathbf{r}_{01})$$
$$\nabla^2 G_1(\mathbf{r}|\mathbf{r}_{02}) = -\delta(\mathbf{r} - \mathbf{r}_{02})$$

we obtain the following results from the volume integral:

$$G_1(\mathbf{r}_{02}|\mathbf{r}_{01}) = G_1(\mathbf{r}_{01}|\mathbf{r}_{02})$$

Since $(x_{01},y_{01},z_{01})$ and $(x_{02},y_{02},z_{02})$ are arbitrary points, the stated reciprocity principle is proved. A similar proof holds for the Green's function of the second kind since $\partial G_2/\partial n$ equals zero in this case and the surface integral again vanishes.

## 2.3. Solution of an Inhomogeneous Differential Equation

Before considering the detailed derivation of the Green's function for some practical problems, we will investigate a general method for finding the solution to an inhomogeneous differential equation.

Consider the general second-order differential equation

$$\frac{d^2\psi}{dx^2} + P\frac{d\psi}{dx} + Q\psi = f(x) \tag{15}$$

where $P$ and $Q$ are known functions of $x$ or constants. We will show that the solution of this equation can be obtained at once from a knowledge of two independent solutions to the homogeneous equation

$$\frac{d^2\psi}{dx^2} + P\frac{d\psi}{dx} + Q\psi = 0 \tag{16}$$

Let $\psi_1$ and $\psi_2$ be two linearly independent solutions of (16). The condition that $\psi_1$ and $\psi_2$ should be independent is given by the statement that the Wronskian should not vanish. The Wronskian is, by definition, the following determinant:

$$W(\psi_1,\psi_2) = \begin{vmatrix} \psi_1 & \psi_2 \\ \dfrac{d\psi_1}{dx} & \dfrac{d\psi_2}{dx} \end{vmatrix} = \psi_1\psi_2' - \psi_2\psi_1' \tag{17}$$

The proof is as follows:

If $\psi_1$ and $\psi_2$ are linearly *dependent*, constants $a$ and $b$ exist such that

$$a\psi_1 + b\psi_2 = 0 \tag{18a}$$

Differentiating gives

$$a\psi_1' + b\psi_2' = 0 \tag{18b}$$

Equations (18*a*) and (18*b*) are two homogeneous equations which have a solution for $a$ and $b$ only if the determinant vanishes, i.e., if

$$\psi_1\psi_2' - \psi_2\psi_1' = 0$$

If the determinant *does not* vanish, then a solution for the constants $a$ and $b$ does not exist, and the two solutions $\psi_1$ and $\psi_2$ are then linearly independent.

Before proceeding to solve (15), we will need to derive two results involving the Wronskian.

First consider

$$\frac{d}{dx}\left(\frac{\psi_2}{\psi_1}\right) = \frac{\psi_1\psi_2' - \psi_2\psi_1'}{\psi_1{}^2} = \frac{W}{\psi_1{}^2} \tag{19}$$

and this is one of the results we will need. The equation satisfied by $\psi_1$ is

$$\psi_1'' + P\psi_1' + Q\psi_1 = 0 \tag{20}$$

where a prime means differentiation with respect to $x$. Multiply by $\psi_2$ to get

$$\psi_2\psi_1'' + \psi_2 P\psi_1' + Q\psi_2\psi_1 = 0 \tag{21}$$

Beginning with the equation for $\psi_2$ and multiplying by $\psi_1$, we obtain a similar result, i.e.,

$$\psi_1\psi_2'' + \psi_1 P\psi_2' + Q\psi_2\psi_1 = 0 \tag{22}$$

Subtracting these latter two gives

$$\psi_1\psi_2'' - \psi_2\psi_1'' + P(\psi_1\psi_2' - \psi_2\psi_1') = 0 \tag{23}$$

Now $$\psi_1\psi_2'' - \psi_2\psi_1'' = \frac{d}{dx}(\psi_1\psi_2' - \psi_2\psi_1') = \frac{dW}{dx} = W'$$

Thus $dW/dx = -PW$, and hence $dW/W = -P\,dx$. Integrating gives $\ln W = -\int^x P\,dx$, or

$$W = \exp\left(-\int^x P\,dx\right)$$

This is the other result we require. In the indefinite integral $\int^x P\,dx$ we may take the constant of integration equal to zero without any loss in generality for the application of $W$ to follow. If $P$ is zero, the Wronskian is equal to a constant.

Let us now try to find a solution to the inhomogeneous equation of the form

$$\psi = U\psi_1 \tag{24}$$

where $U$ is a function of $x$ to be determined. Equation (24) gives, of course, only the particular solution, and the general solution is obtained by adding to the particular solution a linear combination of the solutions to the homogeneous equation. Substituting (24) into (15), we get

$$U(\psi_1'' + P\psi_1' + Q\psi_1) + U''\psi_1 + U'(P\psi_1 + 2\psi_1') = f \tag{25}$$

since $$\psi' = \psi_1 U' + U\psi_1'$$

and $$\psi'' = U''\psi_1 + U\psi_1'' + 2U'\psi_1'$$

The first term in (25) vanishes since $\psi_1$ is a solution of the homogeneous equation. The remainder of (25) is of the first order in $U'$ and may be integrated by multiplying both sides by the integrating factor

$$\psi_1 \exp \int^x P\,dx$$

that is,

$$U''\psi_1^2 \exp \int^x P\,dx + U'\psi_1 \exp \int^x P\,dx\,(P\psi_1 + 2\psi_1')$$
$$= \frac{d}{dx}\left(U'\psi_1^2 \exp \int^x P\,dx\right) = \psi_1 f \exp \int^x P\,dx$$

Now $\exp\left(-\int^x P\,dx\right) = W$, and so we get

$$\frac{d}{dx}\left(U'\frac{\psi_1^2}{W}\right) = f\frac{\psi_1}{W} \tag{26a}$$

and

$$U' = \frac{W}{\psi_1^2}\int^x f\frac{\psi_1}{W}\,dx \tag{26b}$$

Also $\dfrac{W}{\psi_1^2} = \dfrac{d}{dx}\left(\dfrac{\psi_2}{\psi_1}\right)$ from (19), and using the result

$$\frac{d}{dx}\left(\frac{\psi_2}{\psi_1}\int^x \frac{\psi_1 f}{W}\,dx\right) = \left(\int^x \frac{\psi_1 f}{W}\,dx\right)\frac{d}{dx}\left(\frac{\psi_2}{\psi_1}\right) + \frac{\psi_2}{\psi_1}\frac{f\psi_1}{W}$$

we may replace

$$\frac{W}{\psi_1^2}\int^x \frac{\psi_1 f}{W}\,dx \qquad \text{by} \qquad \frac{d}{dx}\left(\frac{\psi_2}{\psi_1}\int^x \frac{\psi_1 f}{W}\,dx\right) - \frac{\psi_2 f}{W}$$

and (26b) becomes

$$U' = \frac{dU}{dx} = \frac{d}{dx}\left(\frac{\psi_2}{\psi_1}\int^x \frac{\psi_1 f}{W}\,dx\right) - \frac{\psi_2 f}{W} \tag{26c}$$

Finally, upon integrating once, we get

$$U = \frac{\psi_2}{\psi_1}\int^x \frac{\psi_1 f}{W}\,dx - \int^x \frac{\psi_2 f}{W}\,dx \tag{27}$$

Our particular solution is, therefore,

$$\psi = \psi_1 U = \psi_2 \int^x \frac{\psi_1 f}{W}\,dx - \psi_1 \int^x \frac{\psi_2 f}{W}\,dx \tag{28a}$$

or, replacing the variable of integration by $x'$, we have

$$\psi(x) = \int^x \left[\frac{\psi_2(x)\psi_1(x') - \psi_1(x)\psi_2(x')}{\psi_1(x')\psi_2'(x') - \psi_2(x')\psi_1'(x')}\right] f(x')\,dx' \tag{28b}$$

The function in brackets may be regarded as the Green's function for the differential equation (15). The constant of integration in the above solution may be taken as zero, since it is taken care of by the addition of a linear combination of the solutions to the homogeneous equation to obtain the general solution. Sometimes it is convenient to choose definite limits on the integration so as to make the solution satisfy some particular boundary condition. The details will become clear from the example to be discussed in the next section.

## 2.4. Green's Function for a Rectangular Trough Region

Consider the following inhomogeneous equation defining a Green's function $G(x,y|x_0,y_0)$:

$$\frac{\partial^2 G}{\partial x^2} + \frac{\partial^2 G}{\partial y^2} = -\delta(x - x_0)\delta(y - y_0) \tag{29}$$

subject to the boundary conditions

$$G = 0 \qquad \begin{cases} x = 0 & 0 \le y \le b \\ y = 0,\ b & x \ge 0 \\ x \to \infty \end{cases} \tag{30}$$

Apart from a factor $1/\epsilon$, this Green's function gives the potential from a unit line source at $(x_0,y_0,z_0)$ inside the rectangular trough region of Fig. 2.3. In order to apply our previously developed theory, we must suppress one of the variables in the partial differential equation. We know that a solution of the form $\psi(x)U(y)$ exists for Laplace's equation. Appropriate solutions for $U(y)$ which satisfy the required boundary conditions are

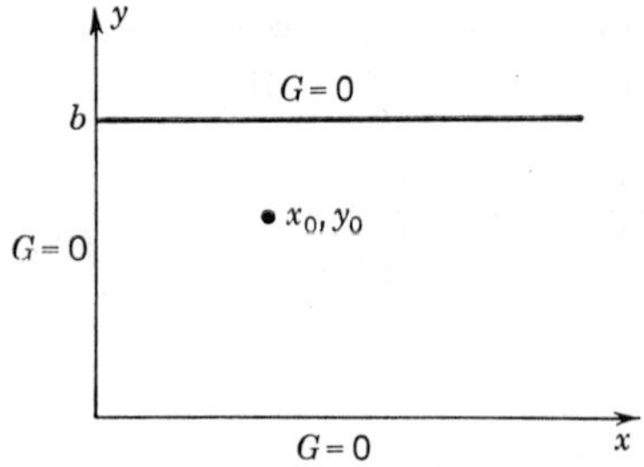


FIG. 2.3. Line source in a rectangular trough region.

$$\sin\frac{n\pi y}{b} \qquad n = 1, 2, 3, \ldots$$

For each value of $n$, let the solution for $\psi(x)$ be $\psi_n(x)$. Thus

$$G = \sum_{n=1}^{\infty} \psi_n(x) \sin\frac{n\pi y}{b} \tag{31}$$

Substituting into the differential equation gives

$$\sum_{n=1}^{\infty} \left(\frac{\partial^2}{\partial x^2} - \frac{n^2\pi^2}{b^2}\right) \psi_n(x) \sin\frac{n\pi y}{b} = -\delta(x - x_0)\delta(y - y_0) \tag{32}$$

The differential equation for $\psi_n(x)$ is obtained by multiplying both sides by $\sin(n\pi y/b)$ and integrating over $0 \le y \le b$ and using the orthogonal properties of the sine functions to get

$$\left(\frac{d^2}{dx^2} - \frac{n^2\pi^2}{b^2}\right)\psi_n(x) = -\frac{2}{b}\sin\frac{n\pi y_0}{b}\,\delta(x - x_0) \tag{33}$$

For the two solutions to the homogeneous equation

$$\frac{d^2\psi_n}{dx^2} - \frac{n^2\pi^2}{b^2}\psi_n = 0 \tag{34}$$

choose for $\psi_{1n}$ a solution which satisfies the boundary condition at $x = 0$, that is,

$$\psi_{1n} = \sinh\frac{n\pi x}{b} \tag{35a}$$

and for $\psi_{2n}$ a solution which vanishes as $x \to \infty$, that is,

$$\psi_{2n} = e^{-n\pi x/b} \tag{35b}$$

For these solutions the Wronskian is

$$W = \psi_{1n}\psi'_{2n} - \psi_{2n}\psi'_{1n} = -\frac{n\pi}{b}\sinh\frac{n\pi x}{b}\,e^{-n\pi x/b} - \frac{n\pi}{b}\cosh\frac{n\pi x}{b}\,e^{-n\pi x/b} = -\frac{n\pi}{b} \neq 0 \tag{36}$$

The particular solution for $\psi_n(x)$ is thus [from Eq. (28$a$)]

$$\psi_n(x) = e^{-n\pi x/b}\int_{x_1}^{x}\frac{\sinh(n\pi x'/b)}{n\pi/b}\,\frac{2}{b}\left[\sin\frac{n\pi y_0}{b}\,\delta(x' - x_0)\right]dx' - \sinh\frac{n\pi x}{b}\int_{x_2}^{x}\frac{e^{-n\pi x'/b}}{n\pi/b}\,\frac{2}{b}\left[\sin\frac{n\pi y_0}{b}\,\delta(x' - x_0)\right]dx' \tag{37}$$

By a judicious choice of the lower limits $x_1$ and $x_2$, we can make our solution fit the boundary conditions at $x = 0$ and $x \to \infty$ without having to add solutions of the homogeneous equation. We note that both integrals are zero unless $x_0$ lies in the intervals $x_1 < x_0 < x$, $x_2 < x_0 < x$. If we choose $x_1 < x_0$ and $x_2 > x_0$, then the first integral is zero unless $x > x_0$, while the second integral is zero unless $x < x_0$; that is, $x$ must lie on the opposite side of $x_0$ compared with the lower limit in order that $x_0$ should lie in between and the integral should not vanish. Using these limits and the properties of the delta functions, we get

$$\psi_n(x) = \begin{cases} \dfrac{2}{n\pi}\sin\dfrac{n\pi y_0}{b}\sinh\dfrac{n\pi x_0}{b}\,e^{-n\pi x/b} & x \geq x_0 \\[2ex] \dfrac{2}{n\pi}\sin\dfrac{n\pi y_0}{b}\,e^{-n\pi x_0/b}\sinh\dfrac{n\pi x}{b} & x \leq x_0 \end{cases} \tag{38}$$

and these solutions are seen to satisfy the proper boundary conditions at $x = 0$ and $x \to \infty$ and are also solutions of the inhomogeneous equation.

The Green's function for this problem is, therefore,

$$G(x,y|x_0,y_0) = \frac{2}{\pi}\sum_1^\infty \frac{1}{n}\sin\frac{n\pi y}{b}\sin\frac{n\pi y_0}{b}\begin{cases} e^{-n\pi x_0/b}\sinh\dfrac{n\pi x}{b} & x \le x_0 \\ e^{-n\pi x/b}\sinh\dfrac{n\pi x_0}{b} & x \ge x_0 \end{cases} \tag{39}$$

We can obtain an alternative solution by suppressing the variable $x$ in our partial differential equation. A solution for $\psi(x)$ which vanishes at $x = 0$ is $A(k)\sin kx$, where $A$ is a constant depending on the particular value of the separation constant $k$. The corresponding solution for $U(y)$ is a linear combination of $\sinh ky$ and $\cosh ky$. The integral $\int_0^\infty A(k)\sin kx\,dk$ is also a solution for $\psi(x)$. This integral may be considered as the Fourier sine transform of $A(k)$; thus, since

$$\psi(x) = \int_0^\infty A(k)\sin kx\,dk \tag{40a}$$

the Fourier inversion of this gives

$$A(k) = \frac{2}{\pi}\int_0^\infty \psi(x)\sin kx\,dx \tag{40b}$$

Now $$\frac{d\psi}{dx} = \int_0^\infty kA(k)\cos kx\,dk$$

and $$\frac{d^2\psi}{dx^2} = -\int_0^\infty k^2A(k)\sin kx\,dk$$

Thus, substituting in the partial differential equation, we get

$$\int_0^\infty A(k_0)\sin k_0x(-k_0^2U + U'')\,dk_0 = -\delta(x - x_0)\delta(y - y_0) \tag{41}$$

where $U'' = d^2U(y)/dy^2$, and $k_0$ is a dummy variable of integration. Multiply both sides by $(2/\pi)\sin kx$, and integrate over $x$ from $0 \le x \le \infty$ to get

$$\begin{aligned}\frac{2}{\pi}\int_0^\infty\left[\int_0^\infty A(k_0)\sin k_0x(U'' - k_0^2U)\,dk_0\right]\sin kx\,dx \\ = \int_0^\infty A(k_0)\left(\frac{2}{\pi}\int_0^\infty \sin k_0x\sin kx\,dx\right)(U'' - k_0^2U)\,dk_0 \\ = A(k)(U'' - k^2U) = -\frac{2}{\pi}\sin kx_0\,\delta(y - y_0)\end{aligned} \tag{42}$$

since $$\frac{2}{\pi}\int_0^\infty \sin k_0x\sin kx\,dx = \delta(k - k_0)$$

The differential equation for $U(y)$ is thus

$$U'' - k^2U = -\frac{2}{\pi}\frac{\sin kx_0}{A(k)}\,\delta(y - y_0) \tag{43}$$

For the two independent solutions to the homogeneous equation choose

$$U_1 = \sinh ky \tag{44a}$$
$$U_2 = \sinh k(b - y) \tag{44b}$$

These solutions satisfy the boundary conditions

$$U_1(0) = U_2(b) = 0 \tag{44c}$$

The Wronskian is

$$-k \sinh ky \cosh k(b - y) - k \cosh ky \sinh k(b - y) = -k \sinh kb \neq 0 \tag{45}$$

Using Eq. (28*a*), the particular solution is found to be

$$U(y) = \frac{2}{\pi}\frac{\sin kx_0}{kA(k)}\left[\frac{\sinh k(b - y)}{\sinh kb}\int_{y_1}^{y} \sinh ky'\,\delta(y' - y_0)\,dy' + \frac{\sinh ky}{\sinh kb}\int_{y}^{y_2} \sinh k(b - y')\,\delta(y' - y_0)\,dy'\right] \tag{46}$$

To satisfy the boundary conditions at $y = 0$ and $b$, we now choose $y_1 < y_0$ and $y_2 > y_0$, so that

$$U(y) = \frac{2 \sin kx_0}{\pi kA(k) \sinh kb}\begin{cases}\sinh k(b - y_0) \sinh ky & y \leq y_0\\ \sinh ky_0 \sinh k(b - y) & y \geq y_0\end{cases} \tag{47}$$

The partial solution for the Green's function corresponding to a particular value of $k$ is $U(y)A(k) \sin kx$, and the complete solution is the integral over $0 \leq k \leq \infty$ and is

$$G(x,y|x_0,y_0) = \frac{2}{\pi}\int_0^\infty \frac{\sin kx_0 \sin kx}{k \sinh kb}\begin{Bmatrix}\sinh k(b - y_0) \sinh ky & y \leq y_0\\ \sinh ky_0 \sinh k(b - y) & y \geq y_0\end{Bmatrix} dk \tag{48}$$

This solution is equivalent to the previous one although in an entirely different form. The equivalence can be demonstrated by converting the integral to a contour integral and evaluating it in terms of its residues. We note that the integrand is an even function of $k$, and thus

$$\int_0^\infty (\cdot\;\cdot\;\cdot)\,dk = \frac{1}{2}\int_{-\infty}^\infty (\cdot\;\cdot\;\cdot)\,dk$$

The integrand has poles whenever $\sinh kb = 0$, or $kb = \pm jn\pi$, $n = 1, 2, \ldots$. There is no pole for $k = 0$ since the numerator vanishes as $k^4$ as $k \to 0$.

In the complex $k = u + jv$ plane we will choose a square contour $C$ closed in the upper half plane as illustrated in Fig. 2.4. If we can show that the integrand vanishes along $C_1$, $C_2$, and $C_3$, then the value of the integral along the real axis is given by $2\pi j\Sigma$ residues of integrand inside $C$.

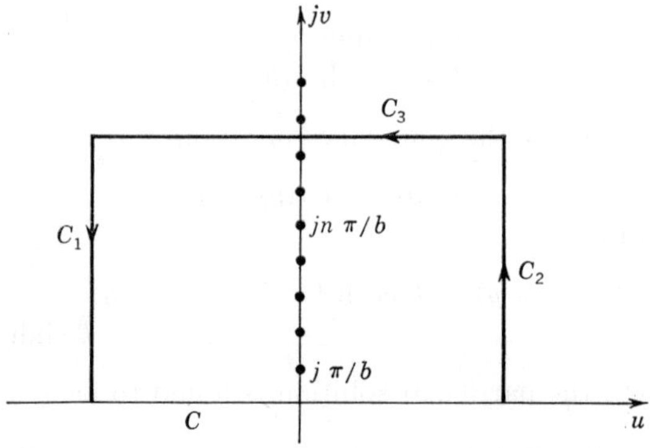


FIG. 2.4. Contour for showing equivalence between Green's functions.

On $C_1$ and $C_2$ when $u \to +\infty$ and $u \to -\infty$, respectively, the following limits exist:

$$|\sinh (b - y_0)k| \to \tfrac{1}{2}e^{(b-y_0)|u|}$$
$$|\sinh ky| \to \tfrac{1}{2}e^{|u|y}$$
$$|\sinh kb| \to \tfrac{1}{2}e^{|u|b}$$

and hence
$$\left|\frac{\sinh k(b - y_0) \sinh ky}{k \sinh kb}\right| \to \frac{1}{2}\frac{e^{|u|(y-y_0)}}{|u|} \to 0$$

since $y \leq y_0$. Similarly

$$\left|\frac{\sinh k(b - y) \sinh ky_0}{k \sinh kb}\right| \to \frac{1}{2}\frac{e^{|u|(y-y_0)}}{|u|} \to 0$$

since for this term $y \geq y_0$.

The term $\sin kx \sin kx_0$, however, causes trouble as $jv \to \infty$ since

$$|\sin kx \sin kx_0| \to \tfrac{1}{4}e^{v(x+x_0)} \to \infty$$

Therefore, we must rewrite the integral in a different form. We note that

$$\int_{-\infty}^{\infty} \sin kx \sin kx_0 (\cdot\ \cdot\ \cdot)\, dk = \int_{-\infty}^{\infty} \frac{e^{jkx_0} - e^{-jkx_0}}{2j} \sin kx (\cdot\ \cdot\ \cdot)\, dk$$
$$= \frac{1}{2j}\int_{-\infty}^{\infty} e^{jkx_0} \sin kx (\cdot\ \cdot\ \cdot)\, dk - \frac{1}{2j}\int_{-\infty}^{\infty} e^{-jkx_0} \sin kx (\cdot\ \cdot\ \cdot)\, dk \quad (49)$$

The term in parentheses which has not been written out is an even function of $k$. In the second integral make the substitution $l = -k$, and it becomes

$$-\frac{1}{2j}\int_{\infty}^{-\infty} e^{jlx_0}\sin lx\,(\cdot\cdot\cdot)\,dl = \frac{1}{2j}\int_{-\infty}^{\infty} e^{jlx_0}\sin lx\,(\cdot\cdot\cdot)\,dl$$

which is equal to the first integral. Thus we may write

$$G = \frac{2}{\pi}\int_{-\infty}^{\infty}\frac{e^{jkx_0}}{2j}\sin kx\,(\cdot\cdot\cdot)\,dk \tag{50}$$

where the term in parentheses which has not been written out is the same as that in braces in (48), excluding the $\sin kx_0 \sin kx$ term. On the upper part of contour $C$ we now have for $jv$ large

$$|e^{jkx_0}\sin kx| \to \frac{e^{-v(x_0-x)}}{2} \to 0 \qquad \text{for } x < x_0$$

For $x > x_0$ we write the integral as

$$-\frac{2}{\pi}\int_{-\infty}^{\infty}\frac{e^{jkx}}{2j}\sin kx_0\,(\cdot\cdot\cdot)\,dk$$

since for $jv$ large

$$|e^{jkx}\sin kx_0| \to \frac{e^{-v(x-x_0)}}{2} \to 0 \qquad \text{for } x > x_0$$

We now have written the integral in such a form that, as the contour $C$ recedes to infinity, the integrand vanishes faster then $|k|^{-1}$, and hence we can evaluate the integral in terms of its residues at $k = jn\pi/b$; that is, $v = n\pi/b$, $n = 1, 2, \ldots, \infty$. All factors are regular at $v = n\pi/b$ except $\sinh kb$. We have

$$\begin{aligned}\sinh kb = \sinh\left[\left(k - j\frac{n\pi}{b}\right)b + jn\pi\right] &= \cos n\pi \sinh\left(k - j\frac{n\pi}{b}\right)b \\ &\to (-1)^n b\left(k - j\frac{n\pi}{b}\right) \qquad \text{as } k \to j\frac{n\pi}{b}\end{aligned}$$

and the residue is $(-1)^n 1/b$ for this factor. Hence

$$\begin{aligned}G(x,y|x_0,y_0) &= \frac{1}{\pi j}\int_{-\infty}^{\infty}\frac{e^{jkx_0}\sin kx \sinh k(b-y_0)\sinh ky}{k\sinh kb}\,dk \\ &= 2\pi j\sum_{n=1}^{\infty}\frac{1}{\pi j}\frac{e^{-n\pi x_0/b}j\sinh(n\pi x/b)}{(-1)^n bj(n\pi/b)}\,j\sin\frac{n\pi}{b}(b-y_0)j\sin\frac{n\pi y}{b} \\ &= \frac{2}{\pi}\sum_{n=1}^{\infty}\frac{e^{-n\pi x_0/b}}{n}\sinh\frac{n\pi x}{b}\sin\frac{n\pi y}{b}\sin\frac{n\pi y_0}{b}\end{aligned} \tag{51}$$

for $x \leq x_0$, $y \leq y_0$. A similar result holds for $x \leq x_0$ and $y \geq y_0$, while for $x \geq x_0$ we get

$$G = \frac{2}{\pi} \sum_{n=1}^{\infty} \frac{e^{-n\pi x/b}}{n} \sinh \frac{n\pi x_0}{b} \sin \frac{n\pi y}{b} \sin \frac{n\pi y_0}{b} \tag{52}$$

This completes the demonstration of the equivalence of the two forms for the Green's function. In the integral form for the Green's function the poles of the integrand determine the eigenvalues or natural modes for the problem. This is analogous to the behavior of network transfer functions where the poles determine the natural resonant frequencies of the circuit.

## 2.5. Free-space Green's Dyadic Function

The vector potential **A** is a solution of the inhomogeneous vector Helmholtz equation

$$\nabla^2\mathbf{A} + k^2\mathbf{A} = -\mu\mathbf{J} \tag{53}$$

In Sec. 1.8 it was shown that a particular solution is

$$\mathbf{A}(\mathbf{r}) = \frac{\mu}{4\pi} \iiint\limits_V \mathbf{J}(\mathbf{r}_0) \frac{e^{-jk|\mathbf{r}-\mathbf{r}_0|}}{|\mathbf{r} - \mathbf{r}_0|} \, dV_0 \tag{54}$$

If $\mu\mathbf{J}$ is replaced by a unit vector source $(\mathbf{a}_x + \mathbf{a}_y + \mathbf{a}_z)\delta(\mathbf{r} - \mathbf{r}_0)$, we find at once by substituting into Eq. (54) that the solution is

$$(\mathbf{a}_x + \mathbf{a}_y + \mathbf{a}_z) \frac{e^{-jk|\mathbf{r}-\mathbf{r}_0|}}{4\pi|\mathbf{r} - \mathbf{r}_0|} \tag{55}$$

and this may be considered as a vector Green's function. For a general current distribution **J**, the solution would be obtained by a superposition integral provided we make the rule that the $x$ component of the current $J_x$ is to be associated with the unit vector $\mathbf{a}_x$ in the Green's function, and similarly for $J_y$ and $J_z$. This rule is readily taken account of by introducing the dyadic Green's function, which is defined as a solution of the following equation:

$$\nabla^2\bar{\mathbf{G}}(\mathbf{r}|\mathbf{r}_0) + k^2\bar{\mathbf{G}} = -\bar{\mathbf{I}}\delta(\mathbf{r} - \mathbf{r}_0) \tag{56}$$

where $\bar{\mathbf{I}}$ is the unit dyadic or idemfactor $\mathbf{a}_x\mathbf{a}_x + \mathbf{a}_y\mathbf{a}_y + \mathbf{a}_z\mathbf{a}_z$. Since multiplying the source by $\bar{\mathbf{I}}$ simply multiplies the solution by $\bar{\mathbf{I}}$ also, we have the result

$$\bar{\mathbf{G}}(\mathbf{r}|\mathbf{r}_0) = \bar{\mathbf{I}} \frac{e^{-jk|\mathbf{r}-\mathbf{r}_0|}}{4\pi|\mathbf{r} - \mathbf{r}_0|} \tag{57}$$

Each component of the current gives rise to a vector field, and hence the general linear relation between the current and the field is a dyadic relation. Therefore, for vector fields the Green's function is a dyadic quantity. Equation (57) is the free-space dyadic Green's function for the vector Helmholtz equation. For an arbitrary current distribution the solution for the vector potential is

$$\mathbf{A}(\mathbf{r}) = \iiint_V \bar{\mathbf{G}}(\mathbf{r}|\mathbf{r}_0) \cdot \mathbf{J}(\mathbf{r}_0)\, dV_0 \tag{58}$$

since the scalar product of $\mathbf{J}$ with $\bar{\mathbf{G}}$ associates $J_x$ with $\mathbf{a}_x$, etc. The above dyadic Green's function is a symmetrical dyadic so that $\mathbf{J} \cdot \bar{\mathbf{G}} = \bar{\mathbf{G}} \cdot \mathbf{J}$. It is also symmetrical in the variables $\mathbf{r}$ and $\mathbf{r}_0$. The use of dyadic Green's functions for the vector Helmholtz equation results in a considerable advantage in notation.

## 2.6. Modified Dyadic Green's Functions

In the presence of boundaries on which the electromagnetic field must satisfy prescribed boundary conditions, the use of the free-space dyadic Green's function suffers from the same shortcomings that the free-space Green's function for Poisson's equation does in the solution of electrostatic-potential boundary-value problems. The dyadic Green's functions are, therefore, modified so as to facilitate the solution of the given boundary-value problem. The sort of modification we should make is best determined by an examination of the type of solution we get by using the free-space dyadic Green's function.

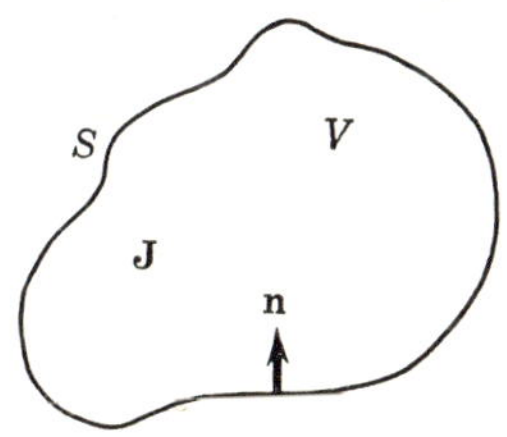

FIG. 2.5. Region $V$ with source function $\mathbf{J}$.

Consider a region of space $V$ bounded by a closed surface $S$, within which a current distribution $\mathbf{J}$ exists. Let $\mathbf{n}$ be the unit inward normal to $S$ as illustrated in Fig. 2.5. Consider the following dyadic quantities:

$$\nabla_0 \cdot [\mathbf{A} \times (\nabla_0 \times \bar{\mathbf{B}})] \tag{59a}$$
$$\nabla_0 \cdot [(\nabla_0 \times \mathbf{A}) \times \bar{\mathbf{B}}] \tag{59b}$$
$$\nabla_0 \cdot [(\nabla_0 \cdot \mathbf{A})\bar{\mathbf{B}}] \tag{59c}$$
$$\nabla_0 \cdot [\mathbf{A}(\nabla_0 \cdot \bar{\mathbf{B}})] \tag{59d}$$

where $\bar{\mathbf{B}}$ is a symmetric dyadic. By expanding the above expressions, we shall be able to obtain a combined vector and dyadic equivalent of Green's second identity. The dyadic $\bar{\mathbf{B}}$ will be replaced by the dyadic Green's function later on. Although $\bar{\mathbf{B}}$ is a symmetric dyadic, $\nabla_0 \times \bar{\mathbf{B}}$ is not symmetric, and special care must be taken in expanding the above expressions, and also in the order in which the terms are written. Perhaps the easiest way to proceed is to write the dyadic $\bar{\mathbf{B}}$ as the sum of three vectors associated with the unit vectors $\mathbf{a}_x$, $\mathbf{a}_y$, $\mathbf{a}_z$, that is,†

$$\bar{\mathbf{B}} = \mathbf{B}_1\mathbf{a}_x + \mathbf{B}_2\mathbf{a}_y + \mathbf{B}_3\mathbf{a}_z$$

Thus, for (59*a*) we have

$$\nabla_0 \cdot [\mathbf{A} \times (\nabla_0 \times \bar{\mathbf{B}})] = \nabla_0 \cdot (\mathbf{A} \times \nabla_0 \times \mathbf{B}_1)\mathbf{a}_x + \nabla_0 \cdot (\mathbf{A} \times \nabla_0 \times \mathbf{B}_2)\mathbf{a}_y + \nabla_0 \cdot (\mathbf{A} \times \nabla_0 \times \mathbf{B}_3)\mathbf{a}_z$$

Each term may now be expanded according to the rules of vector calculus to obtain

$$\begin{aligned}\nabla_0 \cdot (\mathbf{A} \times \nabla_0 \times \mathbf{B}_1) &= (\nabla_0 \times \mathbf{A}) \cdot (\nabla_0 \times \mathbf{B}_1) - (\nabla_0 \times \nabla_0 \times \mathbf{B}_1) \cdot \mathbf{A} \\ &= (\nabla_0 \times \mathbf{A}) \cdot (\nabla_0 \times \mathbf{B}_1) - \mathbf{A} \cdot \nabla_0\nabla_0 \cdot \mathbf{B}_1 + \mathbf{A} \cdot \nabla_0^2\mathbf{B}_1 \qquad (60)\end{aligned}$$

and similarly for the other two terms. By arranging the terms in the expansion so that the vector $\mathbf{B}_1$ always appears on the right-hand side of each term, we may again associate $\mathbf{B}_1$ with $\mathbf{a}_x$, and hence we can combine the end results for all three associated vectors into a dyadic again. Thus the expansion of (59*a*) gives

$$\nabla_0 \cdot [\mathbf{A} \times (\nabla_0 \times \bar{\mathbf{B}})] = (\nabla_0 \times \mathbf{A}) \cdot (\nabla_0 \times \bar{\mathbf{B}}) - \mathbf{A} \cdot (\nabla_0\nabla_0 \cdot \bar{\mathbf{B}}) + \mathbf{A} \cdot (\nabla_0^2\bar{\mathbf{B}}) \qquad (61a)$$

Expanding (59*b*) to (59*d*) in a similar fashion, we obtain

$$\nabla_0 \cdot [(\nabla_0 \times \mathbf{A}) \times \bar{\mathbf{B}}] = (\nabla_0\nabla_0 \cdot \mathbf{A}) \cdot \bar{\mathbf{B}} - (\nabla_0^2\mathbf{A}) \cdot \bar{\mathbf{B}} - (\nabla_0 \times \mathbf{A}) \cdot (\nabla_0 \times \bar{\mathbf{B}}) \qquad (61b)$$

$$\nabla_0 \cdot [(\nabla_0 \cdot \mathbf{A})\bar{\mathbf{B}}] = (\nabla_0 \cdot \mathbf{A})(\nabla_0 \cdot \bar{\mathbf{B}}) + (\nabla_0\nabla_0 \cdot \mathbf{A}) \cdot \bar{\mathbf{B}} \qquad (61c)$$

$$\nabla_0 \cdot [\mathbf{A}(\nabla_0 \cdot \bar{\mathbf{B}})] = (\nabla_0 \cdot \mathbf{A})(\nabla_0 \cdot \bar{\mathbf{B}}) + \mathbf{A} \cdot (\nabla_0\nabla_0 \cdot \bar{\mathbf{B}}) \qquad (61d)$$

Adding (61*a*), (61*b*), and (61*d*) and subtracting (61*c*) from the sum gives

$$\nabla_0 \cdot [\mathbf{A} \times (\nabla_0 \times \bar{\mathbf{B}}) + (\nabla_0 \times \mathbf{A}) \times \bar{\mathbf{B}} + \mathbf{A}(\nabla_0 \cdot \bar{\mathbf{B}}) - (\nabla_0 \cdot \mathbf{A})\bar{\mathbf{B}}] = \mathbf{A} \cdot (\nabla_0^2\bar{\mathbf{B}}) - (\nabla_0^2\mathbf{A}) \cdot \bar{\mathbf{B}} \qquad (62)$$

which may now be used in the divergence theorem to give us the equivalent of Green's second identity. Integrating (62) over the volume $V$ and converting the volume integral of the divergence to a surface integral gives

$$\iiint_V [\mathbf{A} \cdot \nabla_0^2\bar{\mathbf{B}} - (\nabla_0^2\mathbf{A}) \cdot \bar{\mathbf{B}}]\, dV_0 = -\oint\!\!\oint_S \mathbf{n} \cdot [\mathbf{A} \times (\nabla_0 \times \bar{\mathbf{B}}) + (\nabla_0 \times \mathbf{A}) \times \bar{\mathbf{B}} + \mathbf{A}(\nabla_0 \cdot \bar{\mathbf{B}}) - (\nabla_0 \cdot \mathbf{A})\bar{\mathbf{B}}]\, da_0 \qquad (63a)$$

where $\mathbf{n}$ is the unit inward normal to $S$. A term such as $\nabla_0^2\mathbf{A} \cdot \bar{\mathbf{B}}$ may be written as $\bar{\mathbf{B}} \cdot \nabla_0^2\mathbf{A}$ since $\bar{\mathbf{B}}$ is a symmetric dyadic. The first term in the

† Section A.2.

surface integral in (63*a*) may be written as

$$(\mathbf{n} \times \mathbf{A}) \cdot (\nabla_0 \times \bar{\mathbf{B}})$$

the second term as

$$[\mathbf{n} \times (\nabla_0 \times \mathbf{A})] \cdot \bar{\mathbf{B}}$$

and the third and fourth terms as

$$\mathbf{n} \cdot \mathbf{A}\nabla_0 \cdot \bar{\mathbf{B}} \qquad \mathbf{n} \cdot \bar{\mathbf{B}}\nabla_0 \cdot \mathbf{A}$$

Hence, we obtain the following final result, which is the basic starting point for the application of dyadic Green's functions to the solution of the vector Helmholtz equation:

$$\iiint_V [(\nabla_0^2\mathbf{A}) \cdot \bar{\mathbf{B}} - \mathbf{A} \cdot \nabla_0^2\bar{\mathbf{B}}]\, dV_0 = \oiint_S [(\mathbf{n} \times \mathbf{A}) \cdot (\nabla_0 \times \bar{\mathbf{B}}) + (\mathbf{n} \times \nabla_0 \times \mathbf{A}) \cdot \bar{\mathbf{B}} + \mathbf{n} \cdot \mathbf{A}\nabla_0 \cdot \bar{\mathbf{B}} - \mathbf{n} \cdot \bar{\mathbf{B}}\nabla_0 \cdot \mathbf{A}]\, da_0 \qquad (63b)$$

This result is also valid when $\bar{\mathbf{B}}$ is not a symmetric dyadic.

If the dyadic $\bar{\mathbf{B}}$ is now replaced by a dyadic Green's function which is a solution of (56) and $\mathbf{A}$ is the vector potential satisfying (53), then $\nabla_0^2\bar{\mathbf{B}}$ and $\nabla_0^2\mathbf{A}$ may be replaced by $-k^2\bar{\mathbf{B}} - \bar{\mathbf{I}}\delta(\mathbf{r} - \mathbf{r}_0)$ and $-k^2\mathbf{A} - \mu\mathbf{J}$, respectively, in (63*b*). The terms in $k^2$ cancel, and, by using the property of the three-dimensional delta function, we obtain the following solution for the vector potential $\mathbf{A}$:

$$\mathbf{A}(\mathbf{r}) = \iiint_V \mu\mathbf{J}(\mathbf{r}_0) \cdot \bar{\mathbf{G}}(\mathbf{r}_0|\mathbf{r})\, dV_0 + \oiint_S [(\mathbf{n} \times \mathbf{A}) \cdot (\nabla_0 \times \bar{\mathbf{G}}) + (\mathbf{n} \times \nabla_0 \times \mathbf{A}) \cdot \bar{\mathbf{G}} + \mathbf{n} \cdot \mathbf{A}\nabla_0 \cdot \bar{\mathbf{G}} - \mathbf{n} \cdot \bar{\mathbf{G}}\nabla_0 \cdot \mathbf{A}]\, da_0 \qquad (64)$$

The electric field is given by

$$\mathbf{E} = -j\omega\mathbf{A} + \frac{\nabla\nabla \cdot \mathbf{A}}{j\omega\epsilon\mu} = -j\omega\mathbf{A} - \nabla\phi \qquad (65a)$$

while the magnetic field is given by

$$\mu\mathbf{H} = \nabla \times \mathbf{A} \qquad (65b)$$

The scalar potential $\phi$ is related to the vector potential $\mathbf{A}$ by the Lorentz condition

$$-j\omega\epsilon\mu\phi = \nabla \cdot \mathbf{A} \qquad (66)$$

If the boundary conditions specify that the tangential component of $\mathbf{E}$ vanishes on $S$, then $\mathbf{n} \times \mathbf{E} = 0$ on $S$, and hence from (65*a*) we see that we must have $\mathbf{n} \times \mathbf{A} = 0$ on $S$, $\mathbf{n} \times \nabla\phi = 0$ on $S$. The condition $\mathbf{n} \times \nabla\phi = 0$ on $S$ means that the derivatives of $\phi$ on the surface $S$ are zero, and hence $\phi$ is constant, which may be taken as zero, on the surface $S$. If $\phi$ vanishes on $S$, then $\nabla_0 \cdot \mathbf{A} = 0$ on $S$ also by virtue of (66).

An examination of (64) shows that in this case the first and last terms in the surface integral vanish. If we are able to modify our Green's function so that $\mathbf{n} \times \bar{\mathbf{G}}$ and $\nabla_0 \cdot \bar{\mathbf{G}}$ vanish on the surface $S$, then the whole surface integral vanishes, and the solution for $\mathbf{A}$ is given by the volume integral of $\mu \mathbf{J} \cdot \bar{\mathbf{G}}$ only.

Sometimes it is more convenient to deal directly with the equation

$$\nabla \times \nabla \times \mathbf{E} - k^2 \mathbf{E} = -j\omega\mu \mathbf{J} \tag{67}$$

which the electric field $\mathbf{E}$ satisfies. In this case the Green's function is taken as a solution of the corresponding equation

$$\nabla \times \nabla \times \bar{\mathbf{G}} - k^2 \bar{\mathbf{G}} = -\bar{\mathbf{I}}\delta(\mathbf{r} - \mathbf{r}_0) \tag{68}$$

The appropriate Green's identity to use in this case is obtained by adding (61*a*) and (61*b*) and converting the volume integral of the divergence to a surface integral again to get

$$\iiint\limits_V [(\nabla_0 \times \nabla_0 \times \mathbf{A}) \cdot \bar{\mathbf{B}} - \mathbf{A} \cdot (\nabla_0 \times \nabla_0 \times \bar{\mathbf{B}})]\, dV_0$$
$$= -\oiint\limits_S \{[\mathbf{n} \times (\nabla_0 \times \mathbf{A})] \cdot \bar{\mathbf{B}} + (\mathbf{n} \times \mathbf{A}) \cdot (\nabla_0 \times \bar{\mathbf{B}})\}\, da_0 \tag{69}$$

Replacing $\mathbf{A}$ by $\mathbf{E}$ and $\bar{\mathbf{B}}$ by $\bar{\mathbf{G}}$ and using (67) and (68), the following solution for $\mathbf{E}$ is obtained:

$$\mathbf{E}(\mathbf{r}) = j\omega\mu \iiint\limits_V \mathbf{J}(\mathbf{r}_0) \cdot \bar{\mathbf{G}}(\mathbf{r}_0|\mathbf{r})\, dV_0 - \oiint\limits_S [(\mathbf{n} \times \nabla_0 \times \mathbf{E}) \cdot \bar{\mathbf{G}}$$
$$+ (\mathbf{n} \times \mathbf{E}) \cdot \nabla_0 \times \bar{\mathbf{G}}]\, da_0 \tag{70}$$

Since $\nabla \times \mathbf{E} = -j\omega\mu\mathbf{H}$, the first term in the surface integral may be interpreted as a contribution from a surface current of density $\mathbf{n} \times \mathbf{H}$ on $S$. The second term may be considered as a contribution from a magnetic current sheet of density $\mathbf{E} \times \mathbf{n}$. If the tangential components of $\mathbf{E}$, that is, $\mathbf{n} \times \mathbf{E}$, are known or specified on $S$, then, if possible, we should modify our Green's function so that $\mathbf{n} \times \bar{\mathbf{G}}$ vanishes on $S$. Thus only the second term, the one which we know, contributes to the surface integral. In practice, the particular boundary values we are given dictate the choice of Green's function to use.

### *Reciprocity Relation for Dyadic Green's Functions*

In free space, the dyadic Green's function for the vector Helmholtz equation is a symmetrical dyadic. In the presence of boundaries on which $\bar{\mathbf{G}}$ satisfies prescribed boundary conditions, the dyadic Green's function is not a symmetrical dyadic in general. The nonsymmetry is brought about by the boundary conditions that $\bar{\mathbf{G}}$ must satisfy. Any number of examples may be constructed to demonstrate the validity of the above statements. With reference to Fig. 2.6, let $\bar{\mathbf{G}}$ be the Green's function for the half space $z > 0$ and subject to the boundary conditions $\mathbf{n} \times \bar{\mathbf{G}} = 0$ on the perfectly conducting plane at $z = 0$. The Green's dyadic is a function of the coordinates of the field point, the coordinates $x_0$, $y_0$, $z_0$ of the source point, and the coordinates $x_0$, $y_0$, $-z_0$ of the image source. The Green's dyadic may be constructed from the free-space Green's dyadic by using image theory. However, we will not require the analytical form of $\bar{\mathbf{G}}$ to demonstrate the nonsymmetry of the dyadic.

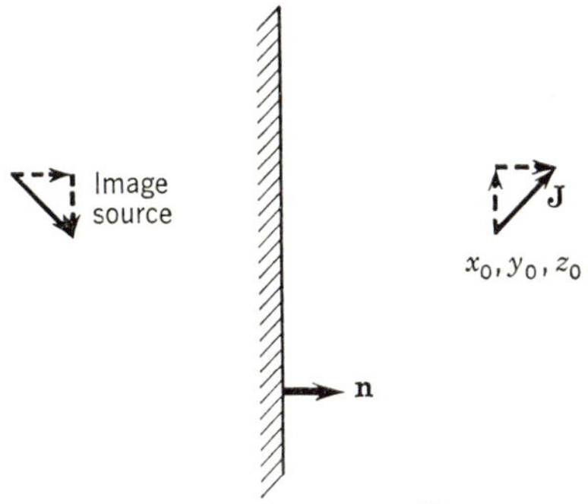


FIG. 2.6. Boundary-value problem for which $\bar{\mathbf{G}}$ is not a symmetric dyadic.

A current element $\mathbf{a}_x J_x$ produces an electric field with a $z$ component given by $J_x G_{xz}$, while a current element $\mathbf{a}_z J_z$ produces a field with an $x$ component given by $J_z G_{zx}$. If $G_{xz} = G_{zx}$, that is, if $\bar{\mathbf{G}}$ is a symmetric dyadic, these components of the field would be equal when $J_x = J_z$. For the problem under consideration, the current $\mathbf{a}_x J_x$ produces a field with a nonzero $z$ component while $\mathbf{a}_z J_z$ produces a field with an $x$ component which vanishes on the conducting plane. Therefore, $G_{xz} \neq G_{zx}$, and the Green's dyadic is not a symmetrical dyadic in general.

Even though $\bar{\mathbf{G}}$ is not a symmetric dyadic in general, a reciprocity principle does hold for $\bar{\mathbf{G}}$. Let $\bar{\mathbf{G}}(\mathbf{r}_1|\mathbf{r})$ be the Green's dyadic such that a current $\mathbf{J}_1(\mathbf{r}_1)$ at $\mathbf{r}_1$ produces an electric field $\mathbf{E}_1(\mathbf{r}_2) = \mathbf{J}_1(\mathbf{r}_1) \cdot \bar{\mathbf{G}}(\mathbf{r}_1|\mathbf{r}_2)$ at the point $\mathbf{r}_2$. Similarly a current element $\mathbf{J}_2(\mathbf{r}_2)$ produces a field $\mathbf{E}_2(\mathbf{r}_1) = \mathbf{J}_2(\mathbf{r}_2) \cdot \bar{\mathbf{G}}(\mathbf{r}_2|\mathbf{r}_1)$ at the point $\mathbf{r}_1$. According to the Lorentz reciprocity theorem (Sec. 1.9 and assuming that the surface integral is zero),

$$\mathbf{E}_1(\mathbf{r}_2) \cdot \mathbf{J}_2(\mathbf{r}_2) = \mathbf{E}_2(\mathbf{r}_1) \cdot \mathbf{J}_1(\mathbf{r}_1)$$

and hence $$\mathbf{J}_1(\mathbf{r}_1) \cdot \bar{\mathbf{G}}(\mathbf{r}_1|\mathbf{r}_2) \cdot \mathbf{J}_2(\mathbf{r}_2) = \mathbf{J}_2(\mathbf{r}_2) \cdot \bar{\mathbf{G}}(\mathbf{r}_2|\mathbf{r}_1) \cdot \mathbf{J}_1(\mathbf{r}_1) \tag{71}$$

Since $\mathbf{J}_1$ and $\mathbf{J}_2$ are arbitrary, we must have for an arbitrary current element

$$\mathbf{J}(\mathbf{r}_0) \cdot \bar{\mathbf{G}}(\mathbf{r}_0|\mathbf{r}) = \bar{\mathbf{G}}(\mathbf{r}|\mathbf{r}_0) \cdot \mathbf{J}(\mathbf{r}_0) = \mathbf{J}(\mathbf{r}_0) \cdot \bar{\mathbf{G}}_t(\mathbf{r}|\mathbf{r}_0) \tag{72}$$

where $\bar{\mathbf{G}}_t$ is the transposed dyadic. Thus the Green's dyadic with the variables $\mathbf{r}$ and $\mathbf{r}_0$ interchanged is equal to the transpose of the original dyadic. If $\mathbf{J}_1 = \mathbf{a}_x$ and $\mathbf{J}_2 = \mathbf{a}_y$, we obtain from (71) the result $G_{xy}(\mathbf{r}_1|\mathbf{r}_2) = G_{yx}(\mathbf{r}_2|\mathbf{r}_1)$, which is a special case of (72). A formal proof of (72) may be obtained by using the vector form of Green's identity instead of the Lorentz reciprocity theorem. The details are left as a problem at the end of the chapter.

## REFERENCES

2.1. P. M. Morse and H. Feshbach: "Methods of Theoretical Physics," parts I and II, McGraw-Hill Book Company, Inc., New York, 1953.
2.2. A. G. Webster: "Partial Differential Equations of Mathematical Physics," Hafner Publishing Company, New York, 1950.
2.3. R. Courant and D. Hilbert: "Methods of Mathematical Physics," vol. 1, English ed., Interscience Publishers, Inc., New York, 1953.
2.4. A. Sommerfeld: "Partial Differential Equations in Physics," Academic Press, Inc., New York, 1949.
2.5. B. Friedman: "Principles and Techniques of Applied Mathematics," John Wiley & Sons, Inc., New York, 1956.

## PROBLEMS

**2.1.** Find the Green's function for the following differential equation and boundary conditions:

$$\frac{d^2G}{dx^2} + \lambda^2 G = -\delta(x - x') \qquad G = 0 \text{ at } x = 0, a$$

where $\lambda^2$ is a separation constant.

HINT: Assume

$$G = \sum_{n=1}^{\infty} a_n \sin \frac{n\pi x}{a}$$

and find the unknown coefficients $a_n$.

*Answer*

$$G(x|x') = \frac{2}{a} \sum_{n=1}^{\infty} \frac{\sin \lambda_n x \sin \lambda_n x'}{\lambda_n^2 - \lambda^2} \qquad \text{where } \lambda_n = \frac{n\pi}{a}$$

**2.2.** By using the method of Sec. 2.3, show that another solution for the Green's function of Prob. 2.1 is

$$G(x|x') = \frac{+1}{\lambda \sin \lambda a} \begin{cases} \sin \lambda x \sin \lambda(a - x') & x < x' \\ \sin \lambda(a - x) \sin \lambda x' & x > x' \end{cases}$$

By summing the series in Prob. 2.1 by the methods given in the Mathematical Appendix, demonstrate the equivalence between the two forms. Convert the sum into two sums, and use the summation formula for

$$\sum_{n=1}^{\infty} \frac{\cos n\alpha}{n^2 - \beta^2}$$

The two cases $x < x'$ and $x > x'$ must be treated separately so as to keep $\alpha$ within the range $0 \leq \alpha \leq 2\pi$.

**2.3.** In Prob. 2.2 show that the Green's function is continuous across the source point at $x'$, but that its derivative is discontinuous by an amount equal to the source strength, i.e., by a unit amount.

**2.4.** By using Laplace transforms, find the Green's function for the following differential equation:

$$L\left(\frac{d^2G}{dt^2} + \frac{R}{L}\frac{dG}{dt} + \frac{1}{LC}G\right) = \delta(t - t')$$

$$G = \frac{dG}{dt} = 0 \qquad \text{at } t = 0$$

This Green's function represents the charge flowing in a series $RLC$ circuit with a unit voltage impulse input at time $t = t'$.

*Answer*

$$G(t|t') = \frac{1}{\omega L} e^{-\alpha(t-t')} \sin \omega(t - t') \qquad t > t'$$

where

$$\alpha = \frac{R}{2L}$$

$$\omega = (LC)^{-1/2}\left(1 - \frac{R^2C}{4L}\right)^{1/2}$$

**2.5.** For Prob. 2.4 show that

$$\int_{-\infty}^{t} e^{j\omega_0 t'} G(t|t')\, dt'$$

gives the usual steady-state solution for the charge in the circuit due to an impressed voltage source $e^{j\omega_0 t}$.

**2.6.** Use the method of images to construct an alternative representation for the Green's function of Sec. 2.4.

**2.7.** Let $\bar{\mathbf{G}}(\mathbf{r}|\mathbf{r}_0)$ be a solution of $\nabla \times \nabla \times \bar{\mathbf{G}} - k^2\bar{\mathbf{G}} = \delta(\mathbf{r} - \mathbf{r}_0)$ in a volume $V$. On the surface $S$ surrounding $V$ let $\bar{\mathbf{G}}$ satisfy one of the following boundary conditions, $\mathbf{n} \times \bar{\mathbf{G}}$ or $\mathbf{n} \times \nabla \times \bar{\mathbf{G}} = 0$. Use the vector form of Eq. (69) to show that

$$\mathbf{J}_1(\mathbf{r}_1) \cdot \bar{\mathbf{G}}(\mathbf{r}_1|\mathbf{r}_2) \cdot \mathbf{J}_2(\mathbf{r}_2) = \mathbf{J}_2(\mathbf{r}_2) \cdot \bar{\mathbf{G}}(\mathbf{r}_2|\mathbf{r}_1) \cdot \mathbf{J}_1(\mathbf{r}_1)$$

HINT: Replace the dyadic $\bar{\mathbf{B}}$ in (69) by a vector function $\mathbf{B}$. Next choose for $\mathbf{A}$ and $\mathbf{B}$ the following:

$$\mathbf{A} = \bar{\mathbf{G}}(\mathbf{r}|\mathbf{r}_1) \cdot \mathbf{J}_1(\mathbf{r}_1)$$
$$\mathbf{B} = \bar{\mathbf{G}}(\mathbf{r}|\mathbf{r}_2) \cdot \mathbf{J}_2(\mathbf{r}_2)$$

**2.8.** Show how the Green's dyadic function may be constructed by constructing the components of $\bar{\mathbf{G}}$ by means of the Lorentz reciprocity theorem.

HINT: To find the $x$ component of $\mathbf{E}$ at $\mathbf{r}$ due to a current element $\mathbf{J}$ at $\mathbf{r}_0$, place a unit current element $\mathbf{a}_x$ at $\mathbf{r}$, and find the components of the field $\mathbf{E}_0$ that it produces at $\mathbf{r}_0$. By the Lorentz reciprocity theorem,

$$\mathbf{E} \cdot \mathbf{a}_x = \mathbf{E}_0 \cdot \mathbf{J}$$

and must also be equal to

$$\mathbf{J}(\mathbf{r}_0) \cdot \bar{\mathbf{G}}(\mathbf{r}_0|\mathbf{r}) \cdot \mathbf{a}_x$$

Placing $\mathbf{J}$ equal to $\mathbf{a}_x$, $\mathbf{a}_y$, $\mathbf{a}_z$ in turn, we get $G_{xx}$, $G_{yx}$, $G_{zx}$.

**2.9.** Consider a rectangular region bounded by the lines $y = 0$, $y = b$, and $x = 0$, $x = l$. Show that

$$G_1 = \frac{4}{bl} \sum_{n=1}^{\infty} \sum_{m=1}^{\infty} \frac{\sin (n\pi y/b) \sin (n\pi y_0/b) \sin (m\pi x/l) \sin (m\pi x_0/l)}{(n\pi/b)^2 + (m\pi/l)^2}$$

is the Green's function of the first kind for the given region. Sum the series over the variable $m$ to obtain an alternative representation as in Prob. 2.1. Next let $l$ tend to infinity, and replace the sum over $n$ by an integral. Compare with the Green's function of Sec. 2.4.

**2.10.** An infinitely long unit current line source is located midway between two infinite parallel perfectly conducting planes. The current has a time variation $e^{j\omega t}$ but does not have a spatial variation along its length. Obtain the solution for the vector potential $\mathbf{A}$ by the following three methods: (*a*) expansion of $\mathbf{A}$ in terms of a suitable set of eigenfunctions, (*b*) method of images, (*c*) Fourier transform.

**2.11.** By a suitable contour integration, demonstrate the equivalence between the Fourier-transform solution and the eigenfunction expansion of Prob. 2.10.

CHAPTER 3

# TRANSVERSE ELECTROMAGNETIC WAVES

A transverse electromagnetic or TEM wave is a wave for which the electric and magnetic field vectors lie in a plane which is transverse or perpendicular to the axis of propagation. This is the principal type of wave which exists on an ideal transmission line. There may be other types such as spherical and cylindrical transverse electromagnetic waves as well as plane TEM waves. Far away from radiating structures the field is essentially a TEM wave. In the discussion to follow a time factor $e^{j\omega t}$ will be assumed.

## 3.1. Plane TEM Waves

We will refer our wave to an $xyz$ rectangular coordinate frame and take the $z$ axis as the direction along which the wave propagates. Maxwell's curl equations are

$$\nabla \times \mathbf{E}_t = -j\omega\mu\mathbf{H}_t \tag{1a}$$

$$\nabla \times \mathbf{H}_t = j\omega\epsilon\mathbf{E}_t \tag{1b}$$

where we have attached a subscript $t$ to the field vectors to indicate that they have *transverse*, that is, $x$ and $y$ components, only. The medium is assumed to be lossless, isotropic, and homogeneous with electrical parameters $\mu$, $\epsilon$. The vector operator $\nabla$ may be split into two parts, a transverse part

$$\nabla_t = \mathbf{a}_x \frac{\partial}{\partial x} + \mathbf{a}_y \frac{\partial}{\partial y}$$

and a longitudinal part

$$\nabla_z = \mathbf{a}_z \frac{\partial}{\partial z}$$

Now $\nabla_t \times \mathbf{E}_t$ is a vector perpendicular to the transverse plane, and similarly with $\nabla_t \times \mathbf{H}_t$. Since the fields are purely transverse, Eqs. (1) reduce to the following:

$$\mathbf{a}_z \times \frac{\partial \mathbf{E}_t}{\partial z} = -j\omega\mu\mathbf{H}_t \tag{2a}$$

$$\mathbf{a}_z \times \frac{\partial \mathbf{H}_t}{\partial z} = j\omega\epsilon\mathbf{E}_t \tag{2b}$$

$$\nabla_t \times \mathbf{E}_t = 0 \tag{2c}$$

$$\nabla_t \times \mathbf{H}_t = 0 \tag{2d}$$

The general theory of vector fields† shows that, when the curl of a vector is identically zero, it may be derived from the gradient of a suitable scalar function since the curl of the gradient is identically zero. Thus Eqs. (2c) and (2d) will be automatically satisfied if we take

$$\mathbf{E}_t = g_1(z)\ \nabla_t\phi \tag{3a}$$
$$\mathbf{H}_t = g_2(z)\ \nabla_t\psi \tag{3b}$$

where $g_1$, $g_2$, $\phi$, and $\psi$ are scalar functions to be determined. In (3), $\phi$ and $\psi$ are functions of the transverse coordinates $x$ and $y$ only, since it is only the transverse curl which vanishes. The form of solution given in (3) is permissible because the vector Helmholtz equation is separable in rectangular coordinates. Since we are considering a source-free medium, the divergence of the field vectors is zero. From (3) it is then seen that both $\phi$ and $\psi$ are solutions of Laplace's equation in the transverse plane; that is, $\nabla_t^2\phi = 0$, and similarly for $\psi$.

The scalar functions $\phi$ and $\psi$ are not independent since $\mathbf{E}_t$ and $\mathbf{H}_t$ are related by (2a) and (2b). Differentiating (2a) with respect to $z$ and substituting from (2b) after cross-multiplying both equations by $\mathbf{a}_z$ gives

$$\mathbf{a}_z \times \left(\mathbf{a}_z \times \frac{\partial^2\mathbf{E}_t}{\partial z^2}\right) - \omega^2\mu\epsilon\mathbf{E}_t = 0 \tag{4}$$

Expanding this result gives

$$\frac{\partial^2\mathbf{E}_t}{\partial z^2} + k^2\mathbf{E}_t = 0 \tag{5}$$

where $k^2 = \omega^2\mu\epsilon$. Equation (5) follows from (4) since $\mathbf{a}_z \cdot (\partial^2\mathbf{E}_t/\partial z^2) = 0$ and $\mathbf{a}_z \cdot \mathbf{a}_z = 1$. In a similar fashion it may be shown that $\mathbf{H}_t$ satisfies the same type of equation. The solution to (5) gives the function $g_1(z)$ as $(A_\pm)e^{\mp jkz}$, where $A_+$ and $A_-$ are suitable amplitude coefficients. Since a time function $e^{j\omega t}$ was assumed, $A_+e^{-jkz}$ represents propagation along the positive $z$ direction, and $A_-e^{jkz}$ represents propagation along the negative $z$ direction. The propagation constant $k$ is equal to

$$\omega(\mu\epsilon)^{1/2} = \frac{\omega}{v_c} = \frac{2\pi f}{v_c} = \frac{2\pi}{\lambda},$$

where $v_c$ is the velocity of light in a medium with parameters $\mu$, $\epsilon$, and $\lambda$ is the wavelength corresponding to the frequency $f$ in the same medium. The solution for the electric field may be written as follows:

$$\mathbf{E}_t = (A_\pm)\ \nabla_t\phi(x,y)e^{\mp jkz} \tag{6}$$

From (2a) the solution for $\mathbf{H}_t$ is found to be

$$\mathbf{H}_t = \pm\left(\frac{\epsilon}{\mu}\right)^{1/2} \mathbf{a}_z \times \nabla_t\phi(A_\pm)e^{\mp jkz} \tag{7}$$

† Section A.1e.

From (7) the appropriate solution for $g_2(z)$ and $\psi$ in terms of $g_1(z)$ and $\phi$ may be found. However, we will not require explicit expressions for these functions since both $\mathbf{E}_t$ and $\mathbf{H}_t$ are uniquely defined by $g_1(z)$ and $\phi$ alone. Equation (7) may be rewritten as

$$\mathbf{H}_t = \pm \left(\frac{\epsilon}{\mu}\right)^{1/2} \mathbf{a}_z \times \mathbf{E}_t \tag{8}$$

or

$$H_x = \mp \left(\frac{\epsilon}{\mu}\right)^{1/2} E_y$$

$$H_y = \pm \left(\frac{\epsilon}{\mu}\right)^{1/2} E_x$$

The factor $(\mu/\epsilon)^{1/2}$ has the dimensions of ohms and is called the characteristic or intrinsic impedance of the medium. It is numerically equal to the ratio of the transverse electric field to the mutually perpendicular transverse magnetic field in a plane TEM wave. The reciprocal quantity $(\epsilon/\mu)^{1/2}$ is called the characteristic admittance of the medium. In free space $Z = (\mu_0/\epsilon_0)^{1/2} = 120\pi = 377$ ohms. The ratio of the transverse electric field to the mutually perpendicular transverse magnetic field is called the wave impedance of the wave. For plane TEM waves the wave impedance is equal to the intrinsic impedance of the medium in which the wave propagates.

The scalar function $\psi$ is related to $\phi$ as follows:

$$\nabla_t\psi = \pm Y\mathbf{a}_z \times \nabla_t\phi$$

where $Y = (\epsilon/\mu)^{1/2}$, and hence $\psi$ is a function whose gradient is everywhere perpendicular to the gradient of $\phi$. The family of curves $\phi =$ constant are mutually orthogonal to the family $\psi =$ constant. Thus the electric and magnetic field vectors are also mutually perpendicular. The transverse magnetic field is consequently related to the transverse electric field by a vector equation such as (7). For this reason it is convenient to introduce a dyadic admittance and impedance which we shall take as follows:

$$\bar{\mathbf{Y}} = Y(\mathbf{a}_y\mathbf{a}_x - \mathbf{a}_x\mathbf{a}_y) \tag{9a}$$

$$\bar{\mathbf{Z}} = Z(\mathbf{a}_x\mathbf{a}_y - \mathbf{a}_y\mathbf{a}_x) \tag{9b}$$

where the bar on top signifies a dyadic quantity. We may now rewrite (8) as

$$\mathbf{H}_t = \pm\bar{\mathbf{Y}} \cdot \mathbf{E}_t \tag{10a}$$

and also

$$\mathbf{E}_t = \pm\bar{\mathbf{Z}} \cdot \mathbf{H}_t \tag{10b}$$

By definition, the dot product of a dyadic with a vector is the vector formed by dotting the unit vectors immediately adjacent to and on either side of the dot.† The product is not, in general, commutative,

† Section A.2.

that is, $\bar{\mathbf{Y}} \cdot \mathbf{E}_t \neq \mathbf{E}_t \cdot \bar{\mathbf{Y}}$. In the above case $\bar{\mathbf{Y}} \cdot \mathbf{E}_t = -\mathbf{E}_t \cdot \bar{\mathbf{Y}}$, but this is not true in general. We have chosen the signs in (10) so that the positive sign applies for propagation in the positive $z$ direction. The dyadic may be thought of as an operator that changes a vector into another vector. In the application above, the dyadic $\mathbf{a}_x\mathbf{a}_y - \mathbf{a}_y\mathbf{a}_x$ is a rotation operator which rotates a vector through 90° in a clockwise direction. This operation is required in order to relate the magnetic field components to the mutually orthogonal electric field components.

The simplest solution to $\nabla_t^2\phi = 0$ is $\nabla_t\phi =$ a constant vector, which may be chosen as a vector along the $x$ axis. This leads to the solution for a plane, uniform, linearly polarized TEM wave. This solution is as follows:

$$\mathbf{E}_t = (A_\pm)\mathbf{a}_x e^{\mp jkz} \tag{11a}$$
$$\mathbf{H}_t = \pm(A_\pm)\mathbf{a}_y Y e^{\mp jkz} \tag{11b}$$

The time-average power flow along the $z$ axis per unit area in the $xy$ plane is given by the dot product of the complex Poynting vector with the unit vector $\mathbf{a}_z$:

$$P = \tfrac{1}{2}(\mathbf{E} \times \mathbf{H}^* \cdot \mathbf{a}_z) = \tfrac{1}{2}(Y|A_\pm|^2) \tag{12}$$

The above wave is called a linearly polarized wave. The plane of polarization is defined as the plane which contains the **E** vector and the axis of propagation, in the above case the $xz$ plane. A more general type of TEM wave is one composed of two mutually perpendicular linearly polarized waves differing in time phase. The resultant wave is called an elliptically polarized wave for reasons to be discussed below.

Consider two mutually perpendicular linearly polarized waves with a time phase of $\theta$:

$$E_x = \text{Re } E_1 e^{-jkz+j\omega t} \tag{13a}$$
$$E_y = \text{Re } E_2 e^{-jkz+j\omega t+j\theta} \tag{13b}$$

where Re denotes the real part. The imaginary part is another possible solution. In a transverse plane, say $z = 0$, we have

$$E_x = E_1 \cos \omega t \tag{14a}$$
$$E_y = E_2 \cos (\omega t + \theta) = E_2(\cos \omega t \cos \theta - \sin \omega t \sin \theta) \tag{14b}$$

In the $xy$ plane the total instantaneous electric field is

$$\mathbf{E} = \mathbf{a}_x E_1 \cos \omega t + \mathbf{a}_y E_2(\cos \omega t \cos \theta - \sin \omega t \sin \theta) \tag{15}$$

Since $E_x$ and $E_y$ are functions of time, we may express $E_x$ as a function of $E_y$ by eliminating the time $t$ between them. From (14a) we get

$$\frac{E_x}{E_1} = \cos \omega t \qquad \left(1 - \frac{E_x^2}{E_1^2}\right)^{1/2} = \sin \omega t$$

Substituting in Eq. (14b), we get

$$\frac{E_y}{E_2} - \cos\theta \frac{E_x}{E_1} = -\sin\theta \left(1 - \frac{E_x^2}{E_1^2}\right)^{1/2}$$

Squaring both sides gives

$$\left(\frac{E_x}{E_1}\right)^2 + \left(\frac{E_y}{E_2}\right)^2 - 2\frac{E_xE_y}{E_1E_2}\cos\theta = \sin^2\theta \tag{16}$$

This is the equation of an ellipse, and thus, in general, the combination of two linearly polarized waves differing in time phase results in an elliptically polarized wave, i.e., a wave whose electric vector describes the locus of an ellipse in a transverse plane. The electric vector rotates around the axis of propagation at an average angular rate $\omega$, and varies in absolute magnitude at the same time so as to trace out an ellipse. If, when the wave is viewed along the direction of propagation, the electric vector rotates clockwise, the wave is called right elliptically polarized. If the sense of rotation is anticlockwise, it is called left elliptically polarized.

When the time phase $\theta$ is 0, $\pi$, $2\pi$, etc., Eqs. (14) show that the resultant wave is linearly polarized, i.e.,

$$\mathbf{E} = (\mathbf{a}_xE_1 \pm \mathbf{a}_yE_2)\cos\omega t$$

or

$$|\mathbf{E}| = (E_1^2 + E_2^2)^{1/2} \qquad \angle\mathbf{E} = \alpha = \tan^{-1}\frac{\pm E_2}{E_1} \tag{17}$$

where $\alpha$ is the angle that the $\mathbf{E}$ vector makes with the $x$ axis. When $E_1 = E_2$ and $\theta = \pi/2 + n\pi$, Eq. (16) reduces to

$$E_x^2 + E_y^2 = E_1^2 \tag{18}$$

which is the equation of a circle. Under these conditions the wave is circular-polarized. For $\theta = \pi/2$, $\pi/2 + 2\pi$, etc., it is left-circular-polarized. For $\theta = 3\pi/2$, $3\pi/2 + 2\pi$, etc., it is right-circular-polarized. The direction of rotation is easily obtained as follows. The angle $\alpha$ which the electric vector makes with the $x$ axis is $\tan^{-1}(E_y/E_x) = \tan^{-1}[\cos(\omega t + \theta)/\cos\omega t]$. When $\theta = \pi/2$, $\alpha = -\tan^{-1}\tan\omega t = -\omega t$, and the wave is left-circular-polarized. When $\theta = 3\pi/2$, $\alpha = \omega t$, the electric vector rotates clockwise, and the wave is called right-circular-polarized. For circular polarization the rate of rotation is uniform and equal to $\omega$. For an elliptical-polarized wave, $\alpha = \tan^{-1}[E_2\cos(\omega t + \theta)/E_1\cos\omega t]$, and the angular rate of rotation of the $\mathbf{E}$ vector is nonuniform. When $E_1 \neq E_2$ but $\theta = \pi/2 + n\pi$, we again obtain an elliptical-polarized wave since (16) becomes

$$\left(\frac{E_x}{E_1}\right)^2 + \left(\frac{E_y}{E_2}\right)^2 = 1 \tag{19}$$

This ellipse has major and minor axes defined by $E_1$ and $E_2$, and these axes are parallel to the $x$ and $y$ axes. By referring $\mathbf{E}$ to a new coordinate frame $uvz$ obtained by rotating the $xyz$ frame about the $z$ axis through an angle $\beta$ in a counterclockwise direction, (16) can be reduced to the form of (19). The angle of rotation $\beta$ is found to be

$$\beta = \tfrac{1}{2} \tan^{-1} \left( 2E_1E_2 \frac{\cos \theta}{E_1^2 - E_2^2} \right) \tag{20a}$$

and (16) reduces to

$$E_u^2 \left( \frac{\cos^2 \beta}{E_1^2} + \frac{\sin^2 \beta}{E_2^2} + \frac{\sin 2\beta \cos \theta}{E_1E_2} \right) + E_v^2 \left( \frac{\cos^2 \beta}{E_2^2} + \frac{\sin^2 \beta}{E_1^2} - \frac{\sin 2\beta \cos \theta}{E_1E_2} \right) = \sin^2 \theta \tag{20b}$$

## 3.2. TEM Waves in Orthogonal Curvilinear Coordinate Systems

In this section we will investigate the solution for transverse electromagnetic waves in orthogonal curvilinear coordinate systems. Let $u_1$, $u_2$, $u_3$ form a right-handed curvilinear coordinate frame with unit vectors $\mathbf{a}_1$, $\mathbf{a}_2$, $\mathbf{a}_3$ along the coordinate curves and with scale factors $h_1$, $h_2$, and $h_3$. We will attempt to find solutions for $\mathbf{E}$ and $\mathbf{H}$ which have no components along the $u_3$ coordinate axis. This latter axis is thus the axis along which the wave propagates. In order to proceed without an undue amount of mathematical difficulties, we shall impose certain restrictions on the type of coordinate system we will deal with. These restrictions will be explained during the development of the theory. Our basic approach will be based on a generalization of the results presented in the previous section.

Let us attempt to find a solution for $\mathbf{E}_t$ of the following form:

$$\mathbf{E}_t = \nabla_t \phi(u_1,u_2)g(u_3) \tag{21}$$

where the subscript $t$ indicates the transverse components. The operator $\nabla_t$ is given by $\dfrac{\mathbf{a}_1}{h_1}\dfrac{\partial}{\partial u_1} + \dfrac{\mathbf{a}_2}{h_2}\dfrac{\partial}{\partial u_2}$ in orthogonal curvilinear coordinates.†

The electric field $\mathbf{E}$ is a solution of the vector Helmholtz equation

$$\nabla \times \nabla \times \mathbf{E} - k^2\mathbf{E} = 0 \tag{22}$$

A solution of the form (21) is only possible in those coordinate systems for which the vector Helmholtz equation is separable in the form of (21). Since $\nabla \cdot \mathbf{E}_t = 0$, we must have

$$\nabla \cdot \nabla_t \phi g = g \, \nabla_t^2 \phi = \frac{g}{h_1h_2h_3} \left( \frac{\partial}{\partial u_1} \frac{h_2h_3}{h_1} \frac{\partial \phi}{\partial u_1} + \frac{\partial}{\partial u_2} \frac{h_1h_3}{h_2} \frac{\partial \phi}{\partial u_2} \right) = 0 \tag{23}$$

† Section A.1*d*.

so $\phi$ is a solution of Laplace's equation in the $u_1u_2$ curvilinear plane. Using the expression for the curl in curvilinear coordinates, we obtain†

$$\nabla \times \mathbf{E}_t = \frac{-\mathbf{a}_1}{h_2h_3}\frac{\partial\phi}{\partial u_2}\frac{\partial g}{\partial u_3} + \frac{\mathbf{a}_2}{h_1h_3}\frac{\partial\phi}{\partial u_1}\frac{\partial g}{\partial u_3} + \frac{\mathbf{a}_3}{h_1h_2}\left(\frac{\partial^2\phi}{\partial u_1\,\partial u_2} - \frac{\partial^2\phi}{\partial u_2\,\partial u_1}\right)g \qquad (24)$$

An examination of (24) shows that the component along the $u_3$ axis vanishes, and hence (24) determines $-j\omega\mu\mathbf{H}_t$ as a vector function with no component along the $u_3$ axis. Taking the curl of (24) and substituting in (22) gives

$$\frac{\mathbf{a}_1}{h_2h_3}\frac{\partial\phi}{\partial u_1}\frac{\partial}{\partial u_3}\left(\frac{h_2}{h_1h_3}\frac{\partial g}{\partial u_3}\right) + \frac{\mathbf{a}_2}{h_1h_3}\frac{\partial\phi}{\partial u_2}\frac{\partial}{\partial u_3}\left(\frac{h_1}{h_2h_3}\frac{\partial g}{\partial u_3}\right)$$
$$- \frac{\mathbf{a}_3}{h_1h_2}\frac{\partial g}{\partial u_3}\left(\frac{\partial}{\partial u_1}\frac{h_2h_3}{h_1h_3^2}\frac{\partial\phi}{\partial u_1} + \frac{\partial}{\partial u_2}\frac{h_1h_3}{h_2h_3^2}\frac{\partial\phi}{\partial u_2}\right)$$
$$+ k^2g\left(\frac{\mathbf{a}_1}{h_1}\frac{\partial\phi}{\partial u_1} + \frac{\mathbf{a}_2}{h_2}\frac{\partial\phi}{\partial u_2}\right) = 0 \qquad (25)$$

Since the term multiplied by $k^2$, that is, $\mathbf{E}_t$, has no $\mathbf{a}_3$ component, the component along $\mathbf{a}_3$ in (25) must vanish. An examination of this term shows that, apart from the $h_3^2$ factor, it is proportional to $\nabla_t^2\phi$. Hence, provided $h_3$ is not a function of the $u_1$ and $u_2$ coordinates, we may bring this factor outside the differentiation operators, and the whole term vanishes, provided $\nabla_t^2\phi = 0$. Equating separately the components along $\mathbf{a}_1$ and $\mathbf{a}_2$ to zero gives

$$\frac{\partial\phi}{\partial u_1}\left(\frac{\partial}{\partial u_3}\frac{h_2}{h_1h_3}\frac{\partial g}{\partial u_3} + \frac{h_2h_3}{h_1}k^2g\right) = 0 \qquad (26a)$$

$$\frac{\partial\phi}{\partial u_2}\left(\frac{\partial}{\partial u_3}\frac{h_1}{h_2h_3}\frac{\partial g}{\partial u_3} + \frac{h_1h_3}{h_2}k^2g\right) = 0 \qquad (26b)$$

Since $g$ cannot satisfy two different differential equations simultaneously, (26*a*) and (26*b*) must reduce to the same form. This will be the case provided $h_1/h_2$ is not a function of $u_3$. In this case Eqs. (26) become

$$\frac{1}{h_3}\frac{\partial}{\partial u_3}\frac{1}{h_3}\frac{\partial g}{\partial u_3} + k^2g = 0 \qquad (27)$$

This is not the only condition for which a solution for $g$ may be found. $\phi$ may be a function of only one of the transverse coordinates $u_1$ or $u_2$. If $\phi$ is not a function of $u_2$, $\partial\phi/\partial u_2$ equals zero, and $g$ is then determined by (26*a*) only. In this case there is no restriction on $h_1/h_2$. Similarly, when $\phi$ is a function of $u_2$ only, the propagation function $g$ is determined by (26*b*) only. If $\phi$ is a function of one coordinate only, then, since $\nabla_t^2\phi = 0$, the only permissible solution for $\nabla_t\phi$ is a constant vector. This

† Section A.1*e*.

solution, therefore, always gives a homogeneous TEM wave, i.e., a wave whose properties do not vary with the transverse coordinates. A solution for a nonhomogeneous TEM wave exists only in those coordinate systems for which $h_1/h_2$ is not a function of $u_3$. The orthogonal curvilinear coordinate systems for which this latter condition is true are:

1. Rectangular coordinates
2. All cylindrical coordinates with propagation along the $z$ axis
3. Spherical coordinates with propagation along the radial axis
4. Conical coordinates with propagation along the radial axis

The conical coordinates have spheres and cones of elliptical cross section as constant coordinate surfaces. The coordinate systems for which an axis, say along $u_3$, can be found, and whose scale factor $h_3$ is a function of $u_3$ only, are those listed above and, in addition, the circular cylindrical coordinates with propagation along the radial axis.

In a cylindrical coordinate system $\theta$, $z$, $r$, the scale factors are $h_1 = r$, $h_2 = 1$, $h_3 = 1$. If we take $\phi$ as a function of $\theta$ only, then $\mathbf{E}_t$ has a $\theta$ component and $\mathbf{H}_t$ a $z$ component only. For one solution the propagation function $g$ must satisfy the following equation:

$$\frac{d}{dr}\frac{1}{r}\frac{dg}{dr} + \frac{1}{r}k^2g = 0 \tag{28a}$$

as obtained from (26a). When $\phi$ is chosen as a function of $z$ only, then $\mathbf{E}_t$ has a $z$ component and $\mathbf{H}_t$ a $\theta$ component only. The function $g$ is now a solution of

$$\frac{d}{dr}r\frac{dg}{dr} + rk^2g = 0 \tag{28b}$$

We will consider this latter solution first. If the substitution $kr = x$ is made in (28b), this equation becomes

$$\frac{d^2g}{dx^2} + \frac{1}{x}\frac{dg}{dx} + g = 0$$

which is the standard form for Bessel's equation of order 0.† The solutions to this equation are linear combinations of the Bessel and Neumann functions of order 0 as follows:

$$g = AJ_0(kr) + BY_0(kr) \tag{29}$$

where $A$ and $B$ are suitable amplitude coefficients, and $x$ has been replaced

† N. W. McLachlan, "Bessel Functions for Engineers," 2d ed., chaps. 1 and 2, Oxford University Press, New York, 1946.

F. Bowman, "Introduction to Bessel Functions," Dover Publications, New York, 1958.

by $kr$. When $|kr|$ is very large, the following asymptotic forms are valid:

$$J_0(kr) \sim \left(\frac{2}{\pi kr}\right)^{1/2} \cos\left(kr - \frac{\pi}{4}\right) \tag{30a}$$

$$Y_0(kr) \sim \left(\frac{2}{\pi kr}\right)^{1/2} \sin\left(kr - \frac{\pi}{4}\right) \tag{30b}$$

If we want to make $g$ represent an outward-propagating wave, the coefficient $B$ must be chosen as $-jA$, and the asymptotic form of (29) becomes

$$g \sim A\left(\frac{2}{\pi kr}\right)^{1/2} e^{-j(kr-\pi/4)} \qquad |kr| \gg 1 \tag{31}$$

The particular combinations of the functions $J_0$ and $Y_0$ such as $J_0 \pm jY_0$ are called Hankel functions of the first and second kind, respectively, i.e.,

$$H_0{}^1 = J_0 + jY_0 \tag{32a}$$

$$H_0{}^2 = J_0 - jY_0 \tag{32b}$$

The Hankel function of the first kind represents an inward-propagating wave, while the Hankel function of the second kind represents an outward-propagating wave.

The solution for our TEM wave is

$$E_z = AH_0{}^2(kr) \tag{33a}$$

$$H_\theta = \frac{1}{j\omega\mu}\frac{\partial E_z}{\partial r} = -\frac{kA}{j\omega\mu} H_1{}^2(kr) \tag{33b}$$

where $H_1{}^2$ is the Hankel function of the second kind and order 1. The asymptotic form of this function is

$$H_1{}^2 \sim \left(\frac{2}{\pi kr}\right)^{1/2} e^{-j(kr-3\pi/4)}$$

and hence, for $|kr|$ large, we get

$$\frac{-E_z}{H_\theta} \approx \left(\frac{\mu}{\epsilon}\right)^{1/2}$$

Far away from the origin the wave impedance of a cylindrical TEM wave is the same as for a plane TEM wave.

We return now to Eq. (26a) and make the substitutions $g = rG$, $x = kr$. The equation for $G$ becomes

$$\frac{d^2G}{dx^2} + \frac{1}{x}\frac{dG}{dx} + \left(1 - \frac{1}{x^2}\right)G = 0$$

which is Bessel's equation of order 1. Appropriate solutions for $g$ in this case are the Hankel functions of order 1 multiplied by $r$, that is,

$$g = \begin{cases} rA_1H_1^1(kr) \\ rA_2H_1^2(kr) \end{cases} \tag{34}$$

This leads to a similar type of TEM wave but with the roles of $\mathbf{E}_t$ and $\mathbf{H}_t$ interchanged. The solution for $\mathbf{E}_t$ is of the form

$$E_\theta \propto H_1^i(kr) \qquad i = 1, 2, \ldots$$

since $\nabla_t\phi(\theta) = \dfrac{\mathbf{a}_\theta}{r}\dfrac{\partial\phi}{\partial\theta}$, and hence a factor $1/r$ occurs which cancels the factor $r$ in (34). The rest of the details for this solution are left for the reader as one of the problems at the end of this chapter.

## 3.3. Reflection and Transmission at a Discontinuity Interface

Let the half space $z > 0$ be filled with a lossless homogeneous dielectric with a relative dielectric constant $\kappa = \epsilon/\epsilon_0$, and let the half space $z < 0$ be free space as in Fig. 3.1. From the region $z < 0$, let a TEM wave be incident on the interface. When this wave strikes the discontinuity interface, a part is transmitted into the region $z > 0$. Let the reflected and transmitted waves be

$$\mathbf{E}_R = \mathbf{e}_R e^{jk_0z} \tag{35a}$$

$$\mathbf{E}_T = \mathbf{e}_T e^{-jkz} \tag{35b}$$

while the incident wave is

$$\mathbf{E}_I = \mathbf{e}_I e^{-jk_0z} \tag{35c}$$

where $k_0 = \omega(\mu_0\epsilon_0)^{1/2} = 2\pi/\lambda_0$, $k = \omega(\mu_0\epsilon)^{1/2} = \kappa^{1/2}k_0 = 2\pi/\lambda$. The transverse magnetic field components are

$$\mathbf{h}_I = \bar{\mathbf{Y}}_0 \cdot \mathbf{e}_I \tag{36a}$$

$$\mathbf{h}_R = -\bar{\mathbf{Y}}_0 \cdot \mathbf{e}_R \tag{36b}$$

$$\mathbf{h}_T = \bar{\mathbf{Y}} \cdot \mathbf{e}_T \tag{36c}$$

where

$$\bar{\mathbf{Y}}_0 = \left(\frac{\epsilon_0}{\mu_0}\right)^{1/2}(\mathbf{a}_y\mathbf{a}_x - \mathbf{a}_x\mathbf{a}_y)$$

$$\bar{\mathbf{Y}} = \kappa^{1/2}\bar{\mathbf{Y}}_0$$

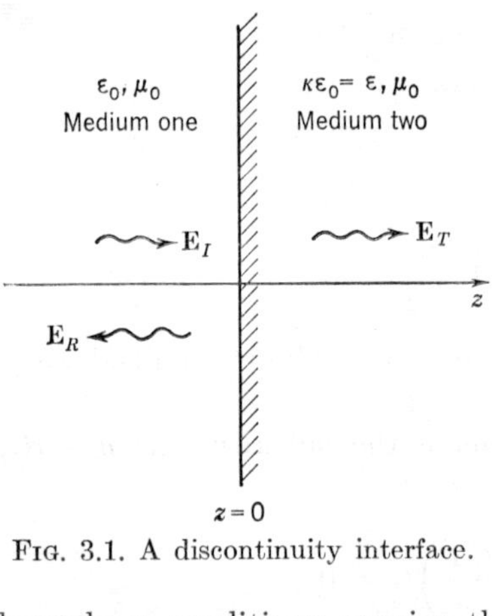


FIG. 3.1. A discontinuity interface.

The quantities $Y_0$, $Y$ and $Z_0$, $Z$ are the wave admittances and impedances, respectively. The vectors $\mathbf{e}$ and $\mathbf{h}$ are used to represent the transverse fields without the propagation factor $e^{\pm jkz}$. At the discontinuity interface, the boundary conditions require that the tangential field components be continuous. The discontinuity interface is uniform in the $xy$ plane, and, consequently, there will be no waves excited other than those given by Eqs. (35*a*) and (35*b*). The condition for the continuity of the tangential electric field is

$$\mathbf{e}_I + \mathbf{e}_R = \mathbf{e}_T \tag{37a}$$

while

$$\mathbf{h}_I + \mathbf{h}_R = \mathbf{h}_T \tag{37b}$$

is the required condition for continuity of the transverse magnetic field components. Using (36), we may write (37*b*) as follows:

$$\bar{\mathbf{Y}}_0 \cdot (\mathbf{e}_I - \mathbf{e}_R) = \bar{\mathbf{Y}} \cdot \mathbf{e}_T$$

or

$$\bar{\mathbf{Y}}_0 \cdot (\mathbf{e}_I - \mathbf{e}_R - \kappa^{1/2}\mathbf{e}_T) = 0 \tag{37c}$$

In order for Eqs. (37) to hold for all values of $x$ and $y$, it is necessary for all the fields to have the same functional form in the $xy$ plane. The reflection coefficient $R_i$ is defined as the ratio of reflected to incident electric field wave amplitudes, and the transmission coefficient $T_{ij}$ as the ratio of transmitted to incident wave amplitudes; thus

$$\mathbf{e}_R = R_1\mathbf{e}_I \qquad \text{at } z = 0 \tag{38a}$$

$$T_{12}\mathbf{e}_I = \mathbf{e}_T \qquad \text{at } z = 0 \tag{38b}$$

With these definitions in mind we may write (37*a*) and (37*c*) as follows:

$$1 + R_1 = T_{12} \tag{39a}$$

$$1 - R_1 = \kappa^{1/2}T_{12} \tag{39b}$$

or

$$Y_0(1 - R) = YT_{12}$$

since $\bar{\mathbf{Y}}_0 = Y_0(\mathbf{a}_y\mathbf{a}_x - \mathbf{a}_x\mathbf{a}_y)$ and $\bar{\mathbf{Y}} = Y(\mathbf{a}_y\mathbf{a}_x - \mathbf{a}_x\mathbf{a}_y)$. Solving Eqs. (39) for $R_1$ and $T_{12}$ gives

$$R_1 = \frac{Y_0 - Y}{Y_0 + Y} = \frac{Z - Z_0}{Z + Z_0} = \frac{1 - \kappa^{1/2}}{1 + \kappa^{1/2}} \tag{40a}$$

$$T_{12} = \frac{2Y_0}{Y_0 + Y} = \frac{2Z}{Z + Z_0} = \frac{2}{1 + \kappa^{1/2}} \tag{40b}$$

We use the subscript 1 to indicate that $R_1$ is the reflection coefficient for a wave incident in region 1. Similarly $T_{12}$ indicates transmission from medium 1 into medium 2. For a wave incident in region 2 we find that

$$R_2 = \frac{Y - Y_0}{Y + Y_0} = -R_1 \tag{41a}$$

$$T_{21} = 1 + R_2 = \frac{2Y}{Y_0 + Y} \tag{41b}$$

The input admittance and impedance (normalized with respect to the characteristic admittance and impedance of medium 1) looking into region 2 at the plane $z = 0$ are $Y/Y_0$ and $Z/Z_0$, respectively, and are given by the following relations, derivable from (40*a*):

$$\frac{Y}{Y_0} = \frac{1 - R_1}{1 + R_1} = \frac{Z_0}{Z} \tag{42}$$

At a plane a distance $l$ from the interface, the ratio of the reflected and incident waves is

$$R_1' = \left.\frac{|\mathbf{e}_R|e^{jk_0z}}{|\mathbf{e}_I|e^{-jk_0z}}\right|_{z=-l} = R_1e^{-2jk_0l} \tag{43}$$

since $z = -l$. We define the normalized input impedance at $z = -l$ by Eq. (42) but with $R_1$ replaced by $R_1'$; thus

$$\begin{aligned} Z_{\text{in}} &= \frac{1 + R_1'}{1 - R_1'} = \frac{1 + R_1e^{-2jk_0l}}{1 - R_1e^{-2jk_0l}} = \frac{e^{jk_0l} + R_1e^{-jk_0l}}{e^{jk_0l} - R_1e^{-jk_0l}} \\ &= \frac{(1 + R_1)/(1 - R_1) + j \tan k_0l}{1 + [j(1 + R_1)/(1 - R_1)] \tan k_0l} = \frac{Z/Z_0 + j \tan k_0l}{1 + j(Z/Z_0) \tan k_0l} \end{aligned} \tag{44}$$

The input impedance transforms with distance $l$ according to the usual transmission-line formula. The quantity $(1 + |R_1|)/(1 - |R_1|)$ is called the standing-wave ratio (SWR) and is equal to the ratio of the maximum to minimum total electric field amplitudes.

The definition of wave impedance may be applied to many types of wave motion. For electromagnetic waves (both TEM waves and waves with longitudinal components), the wave impedance is defined as the ratio of the transverse electric field component to the mutually perpendicular transverse magnetic field component. In the above statement, the adjective transverse is used to mean the components of the field vectors in the plane which is perpendicular to the axis along which the wave is considered to propagate. This latter axis may not necessarily coincide with the wave normal. The concept of wave impedance is a very useful one and greatly facilitates the analysis of many wave-propagation problems, and also provides the basis for a formal analogy between the theory of wave-guiding structures and conventional low-frequency transmission-line theory. In Sec. 1.4 it was demonstrated that it was sufficient to match only the tangential field components across a discontinuity interface in order to ensure proper continuity of the whole solution across the interface. It is for this reason that the concept of wave impedance is so useful in practice, since this concept, together with transmission-line theory, provides the proper solution of Maxwell's equations to ensure continuity of the transverse field components. For electromagnetic waves with the $z$ axis considered as the axis of propagation, the wave impedance is defined by

$$Z = \pm\frac{E_x}{H_y} = \mp\frac{E_y}{H_x} \tag{45}$$

with the sign chosen so that the real part is positive. An imaginary part arises in the case of a medium with finite losses and for waves that are evanescent in the direction of propagation. It should be noted that, in the application of the definition (45), a single wave propagating in either the positive $z$ direction or the negative $z$ direction is considered.

### 3.4. Wave Matrices

With reference to Fig. 3.1 in Sec. 3.3, let a TEM wave of amplitude $c_1$ be incident from the left and a wave of amplitude $b_2$ be incident from the right on the plane-discontinuity interface. In the region $z < 0$ there will be a wave, made up partly of a reflected wave and partly of a wave transmitted past the discontinuity from the right, propagating in the negative $z$ direction. In the region $z > 0$ a similar wave propagating in the positive $z$ direction will exist. Let the amplitudes of the latter two waves be $b_1$ and $c_2$, respectively, as in Fig. 3.2. With reference to Fig. 3.2 we may write

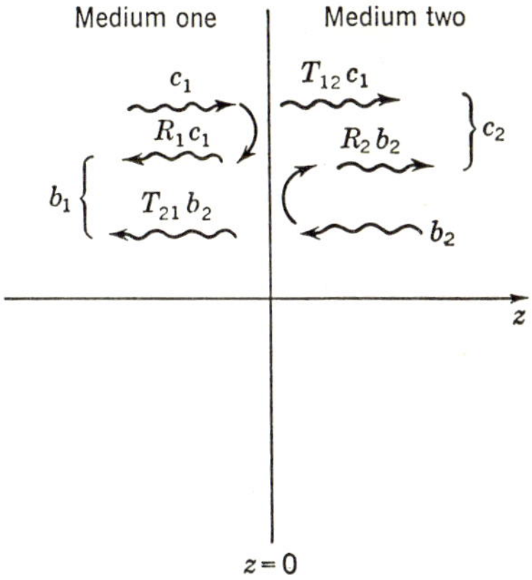


FIG. 3.2. Reflected and transmitted waves at a discontinuity interface.

$$b_1 = R_1c_1 + T_{21}b_2 \tag{46a}$$
$$c_2 = R_2b_2 + T_{12}c_1 \tag{46b}$$

or

$$b_1 = \left(T_{21} - \frac{R_1R_2}{T_{12}}\right)b_2 + \frac{R_1}{T_{12}}c_2 \tag{47a}$$
$$c_1 = \frac{c_2}{T_{12}} - \frac{R_2b_2}{T_{12}} \tag{47b}$$

where $R_1$ and $R_2$ are the interface reflection coefficients, and $T_{12}$ and $T_{21}$ are the transmission coefficients. These equations may be written in matrix form as follows:

$$\begin{bmatrix} c_1 \\ b_1 \end{bmatrix} = \frac{1}{T_{12}}\begin{bmatrix} 1 & -R_2 \\ R_1 & T_{12}T_{21} - R_1R_2 \end{bmatrix}\begin{bmatrix} c_2 \\ b_2 \end{bmatrix} = \begin{bmatrix} A_{11} & A_{12} \\ A_{21} & A_{22} \end{bmatrix}\begin{bmatrix} c_2 \\ b_2 \end{bmatrix} \tag{48}$$

We will call the matrix $[A]$ a wave-transmission chain matrix since it relates the amplitudes of the forward- and backward-propagating waves on the output side to those on the input side. The element $A_{11}$ in the first row and column is equal to $\left.\frac{c_1}{c_2}\right|_{b_2=0} = \frac{1}{T_{12}}$, which is the reciprocal of the transmission coefficient from medium 1 to medium 2. Although Eqs. (46) to (48) were derived with particular reference to TEM waves and a plane discontinuity, they obviously hold for any wave-transmission system for which only a forward- and a backward-traveling wave exist on either side of the discontinuity and equivalent reflection and transmission coefficients can be defined. The equivalent reflection and transmission coefficients are given by the following equations:

$$R_1 = \frac{b_1}{c_1}\bigg|_{b_2=0} \tag{49a}$$

$$R_2 = \frac{c_2}{b_2}\bigg|_{c_1=0} \tag{49b}$$

$$T_{12} = \frac{c_2}{c_1}\bigg|_{b_2=0} \tag{49c}$$

$$T_{21} = \frac{b_1}{b_2}\bigg|_{c_1=0} \tag{49d}$$

These equations together with (48) define the elements $A_{ij}$ in the general case.

When there are evanescent modes or a localized diffraction field near the discontinuity, it is convenient to refer the reflection and transmission coefficients to suitably chosen terminal planes as in Fig. 3.3. The transmission coefficients then refer to transmission from one terminal plane to another. The reflection coefficients transform with distance from the terminal planes according to Eq. (43), that is, $R' = Re^{-2jkl}$, and are seen to be invariant to a shift of a multiple of $\lambda/2$ in the terminal-plane positions. The incident wave varies as $e^{-jk_0z}$, and the transmitted wave as $e^{-jkz}$. The transmission coefficients will, therefore, vary as $e^{\pm j(kl_2-k_0l_1)}$ with shifts of $l_1$ and $l_2$ in the terminal-plane positions in the input and output regions, respectively. As long as $k_0l_1 \pm kl_2$ is equal to a multiple of $2\pi$, the transmission coefficients remain invariant.

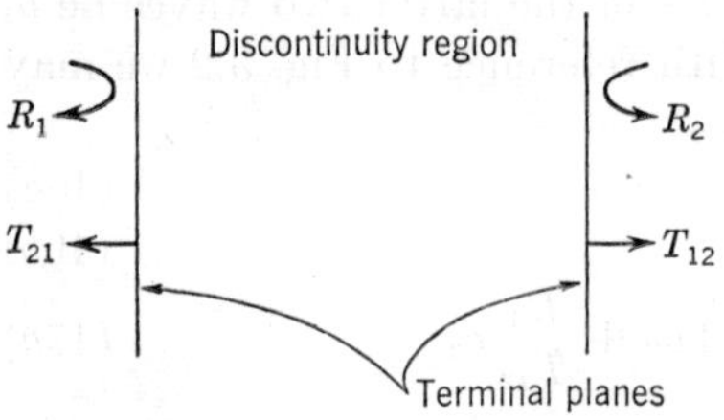


FIG. 3.3. Terminal planes for a discontinuity.

Whenever the regions of space under consideration do not contain nonreciprocal anisotropic media, we may show that the transmission coefficients $T_{12}$ and $T_{21}$ are equal if we normalize the amplitudes $c_1$, $b_1$, $c_2$, $b_2$ so that the power carried by these waves is proportional to $c_1c_1^*$, $b_1b_1^*$, etc. The equality of $T_{12}$ and $T_{21}$ is just the condition for the reciprocity theorem to hold. We prove this result as follows. Let $\mathbf{E}_1$, $\mathbf{H}_1$ and $\mathbf{E}_2$, $\mathbf{H}_2$ be the fields which are solutions to Maxwell's equations for the given problem under two different terminal conditions. These fields are, therefore, not linearly related. According to the Lorentz reciprocity principle,

$$\oiint_S \mathbf{E}_1 \times \mathbf{H}_2 \cdot d\mathbf{a} = \oiint_S \mathbf{E}_2 \times \mathbf{H}_1 \cdot d\mathbf{a} \tag{50}$$

when there are now sources inside $S$. We now let $\mathbf{E}_1$, $\mathbf{H}_1$ be the solution for a wave incident from medium 1 only, and $\mathbf{E}_2$, $\mathbf{H}_2$ be the field for a wave incident from medium 2 only. Specializing to our original problem involving TEM waves, we choose for our surface $S$ a cylindrical surface parallel to the $z$ axis and two plane end surfaces in the $xy$ plane at $k_0z = 2\pi$

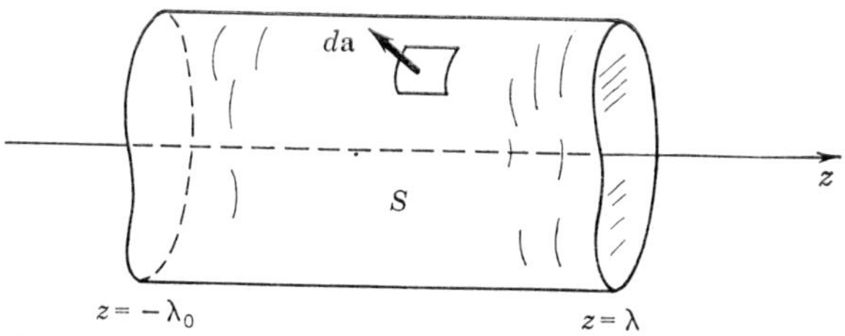


FIG. 3.4. Closed surface $S$ for demonstration of reciprocity.

and $kz = 2\pi$, that is, at $z = -\lambda_0$ and $z = \lambda$ as in Fig. 3.4. The contribution to the surface integral over the cylindrical surface is zero since both $\mathbf{E}$ and $\mathbf{H}$ are transverse, and consequently $\mathbf{E} \times \mathbf{H}$ is perpendicular to $d\mathbf{a}$. We are left with surface integrals over the end faces only. For an incident wave,

$$\mathbf{E}_1 = c_1\mathbf{e}e^{-jk_0z} \tag{51a}$$
$$\mathbf{H}_1 = c_1\bar{\mathbf{Y}}_0 \cdot \mathbf{e}e^{-jk_0z} \tag{51b}$$

the power flow is proportional to $\mathbf{E}_1 \times \mathbf{H}_1^* \cdot \mathbf{a}_z = Y_0c_1c_1^*\mathbf{e} \cdot \mathbf{e}^*$. If we normalize the coefficient $c_1$ by choosing a new amplitude $c_1'$ equal to $c_1(Y_0)^{1/2}$, the power flow is proportional to $c_1'c_1'^*\mathbf{e} \cdot \mathbf{e}^*$. The coefficient $c_2$ may be normalized in a similar manner. Using normalized amplitudes, we may write the following expressions for the fields:

$$\left.\begin{aligned} \mathbf{E}_1 &= Y_0^{-1/2}c_1'(1 + R_1)\mathbf{e} \\ \mathbf{H}_1 &= Y_0^{1/2}c_1'(1 - R_1)(\mathbf{a}_y\mathbf{a}_x - \mathbf{a}_x\mathbf{a}_y) \cdot \mathbf{e} \end{aligned}\right\} \quad z = -\lambda_0 \tag{52a}$$

$$\left.\begin{aligned} \mathbf{E}_1 &= Y_0^{-1/2}c_1'\left(\frac{Y_0}{Y}\right)^{1/2} T_{12}'\mathbf{e} \\ \mathbf{H}_1 &= Y^{1/2}c_1'T_{12}'(\mathbf{a}_y\mathbf{a}_x - \mathbf{a}_x\mathbf{a}_y) \cdot \mathbf{e} \end{aligned}\right\} \quad z = \lambda \tag{52b}$$

$$\left.\begin{aligned} \mathbf{E}_2 &= Y^{-1/2}c_2'\left(\frac{Y}{Y_0}\right)^{1/2} T_{21}'\mathbf{e} \\ \mathbf{H}_2 &= -Y_0^{1/2}c_2'T_{21}'(\mathbf{a}_y\mathbf{a}_x - \mathbf{a}_x\mathbf{a}_y) \cdot \mathbf{e} \end{aligned}\right\} \quad z = -\lambda_0 \tag{52c}$$

$$\left.\begin{aligned} \mathbf{E}_2 &= Y^{-1/2}c_2'(1 + R_2)\mathbf{e} \\ \mathbf{H}_2 &= -Y^{1/2}c_2'(1 - R_2)(\mathbf{a}_y\mathbf{a}_x - \mathbf{a}_x\mathbf{a}_y) \cdot \mathbf{e} \end{aligned}\right\} \quad z = \lambda \tag{52d}$$

where the normalized transmission coefficients $T'_{12}$, $T'_{21}$ are given by

$$T'_{12} = \left(\frac{Y}{Y_0}\right)^{1/2} T_{12} \qquad T'_{21} = \left(\frac{Y_0}{Y}\right)^{1/2} T_{21}$$

Substituting into the Lorentz reciprocity theorem (50), we get

$$\begin{aligned} -\iint\limits_{z=-\lambda_0} c'_1(1+R_1)c'_2T'_{21}[\mathbf{e} \times (\mathbf{a}_y\mathbf{a}_x - \mathbf{a}_x\mathbf{a}_y) \cdot \mathbf{e}] \cdot d\mathbf{a} \\ -\iint\limits_{z=\lambda} c'_1c'_2T'_{12}(1-R_2)[\mathbf{e} \times (\mathbf{a}_y\mathbf{a}_x - \mathbf{a}_x\mathbf{a}_y) \cdot \mathbf{e}] \cdot d\mathbf{a} \\ = \iint\limits_{z=-\lambda_0} c'_1c'_2T'_{21}(1-R_1)[\mathbf{e} \times (\mathbf{a}_y\mathbf{a}_x - \mathbf{a}_x\mathbf{a}_y) \cdot \mathbf{e}] \cdot d\mathbf{a} \\ + \iint\limits_{z=\lambda} c'_1c'_2(1+R_2)T'_{12}[\mathbf{e} \times (\mathbf{a}_y\mathbf{a}_x - \mathbf{a}_x\mathbf{a}_y) \cdot \mathbf{e}] \cdot d\mathbf{a} \end{aligned}$$

which may be written as follows:

$$\begin{aligned} T'_{21}(1 - R_1 + 1 + R_1) \iint\limits_{z=-\lambda_0} [\mathbf{e} \times (\mathbf{a}_y\mathbf{a}_x - \mathbf{a}_x\mathbf{a}_y) \cdot \mathbf{e}] \cdot d\mathbf{a} \\ = -T'_{12}(1 + R_2 + 1 - R_2) \iint\limits_{z=\lambda} [\mathbf{e} \times (\mathbf{a}_y\mathbf{a}_x - \mathbf{a}_x\mathbf{a}_y) \cdot \mathbf{e}] \cdot d\mathbf{a} \end{aligned} \tag{53}$$

The integral at $z = \lambda$ is equal to the negative of that at $z = -\lambda_0$ because of the difference in the direction of $d\mathbf{a}$. Thus (53) gives $T'_{21} = T'_{12}$, and the stated reciprocity principle is proved. The determinant of the $[A]$ matrix is

$$A_{11}A_{22} - A_{12}A_{21} = \frac{T'_{21}}{T'_{12}} = 1 \tag{54}$$

when the amplitudes are normalized. We shall not as a general rule use normalized amplitudes since this does not give any simplification in the mathematics for the type of problems we shall treat.

For the plane-discontinuity problem the matrix elements $A_{ij}$ reduce to

$$A_{11} = \frac{1}{T_{12}} = \frac{Z + Z_0}{2Z} \tag{55a}$$

$$A_{12} = -\frac{R_2}{T_{12}} = \frac{Z - Z_0}{2Z} = \frac{R_1}{T_{12}} \tag{55b}$$

$$A_{21} = \frac{R_1}{T_{12}} \tag{55c}$$

$$A_{22} = \frac{T_{12}T_{21} - R_1R_2}{T_{12}} = \frac{1}{T_{12}} \tag{55d}$$

since $R_2 = -R_1$, $1 + R_1 = T_{12}$, $1 + R_2 = T_{21}$. Equation (48) may, therefore, be written as follows:

$$\begin{bmatrix} c_1 \\ b_1 \end{bmatrix} = \frac{1}{T_{12}} \begin{bmatrix} 1 & R_1 \\ R_1 & 1 \end{bmatrix} \begin{bmatrix} c_2 \\ b_2 \end{bmatrix} \tag{56}$$

Before considering the cascade connection of $n$ sections, we need to develop the wave-transmission matrix for a "length of unbounded space." Consider a forward-propagating wave $c_1e^{-jkz}$ and a backward-propagating wave $b_1e^{jkz}$. At $z = 0$ the amplitudes of the forward- and backward-traveling waves are $c_1$ and $b_1$, respectively. At $z = z_1$ the complex wave amplitudes are $c_1e^{-jkz_1}$ and $b_1e^{jkz_1}$. Let these latter two amplitudes be $c_2$ and $b_2$, respectively. Then we may write

$$c_1 = c_2e^{jkz_1} \tag{57a}$$
$$b_1 = b_2e^{-jkz_1} \tag{57b}$$

The electrical length of the distance from $z = 0$ to $z = z_1$ is $kz_1 = \theta$. (The electrical length of a transmission line or of any medium is the angle by which the phase of a wave changes in propagating through the medium.)

Introducing the electrical length $\theta$, (57) may be written in matrix form as follows:

$$\begin{bmatrix} c_1 \\ b_1 \end{bmatrix} = \begin{bmatrix} e^{j\theta} & 0 \\ 0 & e^{-j\theta} \end{bmatrix} \begin{bmatrix} c_2 \\ b_2 \end{bmatrix} \tag{58}$$

The diagonal matrix $\begin{bmatrix} e^{j\theta} & 0 \\ 0 & e^{-j\theta} \end{bmatrix}$ is the matrix which relates the amplitudes of the forward- and backward-traveling waves at one plane to the complex amplitudes of these same waves at another terminal plane an electrical distance $\theta$ away.

Consider $n$ sections in cascade with electrical constants $\epsilon_i$, $\mu_i$ for the $i$th section. The propagation constant for the $i$th section is $k_i$. If the length of the $i$th section is $l_i$, its electrical length is $\theta_i = k_il_i$. All the sections may differ in electrical properties and also in electrical lengths, or some may be the same without invalidating the results which we shall derive. With reference to Fig. 3.5 let $R_i$ and $T_i$ be the reflection and transmission coefficients looking into the $i$th section. That is, $R_i$ is the reflection coefficient for a wave incident from section $i - 1$, and $T_i$ is the amplitude transmission coefficient from section $i - 1$ to section $i$. The amplitudes $c_2$, $b_2$ at the input to section 2 are related to $c_1$ and $b_1$ as follows:

$$\begin{aligned} \begin{bmatrix} c_1 \\ b_1 \end{bmatrix} &= \frac{1}{T_1} \begin{bmatrix} 1 & R_1 \\ R_1 & 1 \end{bmatrix} \begin{bmatrix} e^{j\theta_1} & 0 \\ 0 & e^{-j\theta_1} \end{bmatrix} \begin{bmatrix} c_2 \\ b_2 \end{bmatrix} \\ &= \frac{1}{T_1} \begin{bmatrix} e^{j\theta_1} & R_1e^{-j\theta_1} \\ R_1e^{j\theta_1} & e^{-j\theta_1} \end{bmatrix} \begin{bmatrix} c_2 \\ b_2 \end{bmatrix} \end{aligned} \tag{59}$$

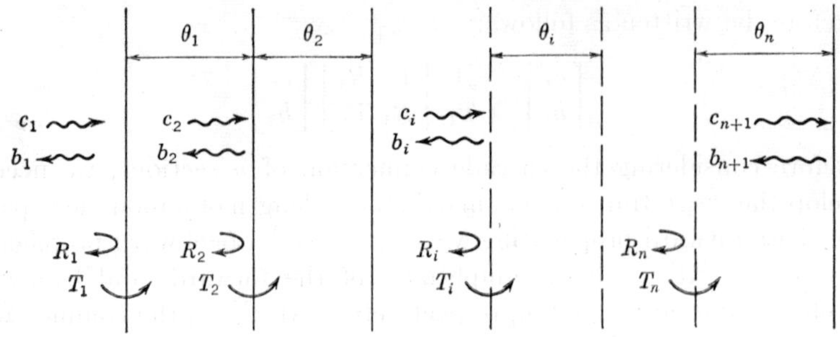


FIG. 3.5. Cascade connection of $n$ sections.

If we express the amplitudes $c_2$ and $b_2$ in terms of $c_3$, $b_3$; $c_3$, $b_3$ in terms of $c_4$, $b_4$; and so on, we find that the output amplitudes $c_{n+1}$, $b_{n+1}$ are related to $c_1$ and $b_1$ by the matrix product of the $n$ matrices which characterize each section, i.e.,

$$\begin{bmatrix} c_1 \\ b_1 \end{bmatrix} = \prod_{i=1}^{n} \frac{1}{T_i} \begin{bmatrix} e^{j\theta_i} & R_i e^{-j\theta_i} \\ R_i e^{j\theta_i} & e^{-j\theta_i} \end{bmatrix} \begin{bmatrix} c_{n+1} \\ b_{n+1} \end{bmatrix} = \begin{bmatrix} A_{11} & A_{12} \\ A_{21} & A_{22} \end{bmatrix} \begin{bmatrix} c_{n+1} \\ b_{n+1} \end{bmatrix} \tag{60}$$

where $\prod_{i=1}^{n}$ means the product of the $n$ matrices, and $A_{ij}$ are coefficients of the resultant $2 \times 2$ matrix. In carrying out this matrix product, it is important to keep the order of the matrices the same as the order of the sections since matrix multiplication is not commutative. The coefficient $A_{11}$ is the reciprocal of the over-all transmission coefficient $T_{1,n+1}$ from the input side of section 1 to the output side of section $n$. When the wave incident from the right on section $n$ is zero, that is, $b_{n+1} = 0$, then

$$\begin{aligned} c_1 &= A_{11} c_{n+1} \\ b_1 &= A_{21} c_{n+1} \end{aligned}$$

and hence the reflection coefficient at the input terminal is

$$\frac{b_1}{c_1} = \frac{A_{21}}{A_{11}} \tag{61}$$

The above wave-transmission matrices may be applied to any type of cascade connection of transmission lines, waveguides, etc., provided there are only a forward- and a backward-traveling wave in each section, and each section is long enough so that any evanescent fields present at each discontinuity do not react with the evanescent fields of the adjacent discontinuities. The theory may also be generalized to include the case of several modes of propagation as well as the case of interacting evanescent modes.

Having considered the wave-transmission matrix in some detail, we now turn to a brief discussion of the following matrices and some of their properties:

1. Scattering matrix
2. Impedance and admittance matrices
3. Voltage- and current-transmission matrix, i.e., conventional *ABCD* matrix

It will be assumed that the wave-amplitude coefficients have been normalized so that the power carried by a given wave is proportional to the square of the absolute value of the amplitude coefficient. In order to have a definite example to consider, we choose the discontinuity interface discussed earlier. However, for convenience the normalized wave amplitudes will be denoted by $c_1$, $b_1$, $c_2$, and $b_2$ without the primes. Also $b_2$ will be chosen to represent a reflected wave and $c_2$ an incident wave.

The scattering matrix $[S]$ relates the amplitudes of the reflected waves $b_1$, $b_2$ to the amplitudes of the incident waves $c_1$, $c_2$ as follows:

$$\begin{bmatrix} b_1 \\ b_2 \end{bmatrix} = \begin{bmatrix} S_{11} & S_{21} \\ S_{12} & S_{22} \end{bmatrix} \begin{bmatrix} c_1 \\ c_2 \end{bmatrix} \tag{62}$$

An examination of the above equation for the two conditions $c_1 = 0$ and $c_2 = 0$ separately shows that $S_{11}$ is the reflection coefficient in medium 1, $S_{22}$ is the reflection coefficient in medium 2, and $S_{12}$, $S_{21}$ are transmission coefficients from medium 1 into medium 2 and vice versa, respectively. When the reciprocity principle holds, $S_{12} = S_{21}$ provided normalized amplitude coefficients are used. For general types of discontinuities the scattering-matrix elements $S_{ij}$ are complex. If no energy loss occurs in the vicinity of the discontinuity, the energy conservation principle may be applied to derive some basic relationships among the scattering-matrix elements. The difference between the incident and reflected power must equal the transmitted power, and hence

$$1 - S_{11}S_{11}^* = S_{12}S_{12}^* \tag{63a}$$
$$1 + S_{22}S_{22}^* = S_{21}S_{21}^* = S_{12}S_{12}^* \tag{63b}$$

From (63) we find that $|S_{11}|$ is equal to $|S_{22}|$. The above equations expressing the conservation of energy for lossless discontinuities may be written as follows:

$$\sum_{n=1}^{2} (c_n c_n^* - b_n b_n^*) = [c_1 \quad c_2] \begin{bmatrix} c_1 \\ c_2 \end{bmatrix}^* - [c_1 \quad c_2][S]_t [S]^* \begin{bmatrix} c_1 \\ c_2 \end{bmatrix}^* = 0 \tag{64}$$

where $[S]_t$ is the transposed matrix. The scattering matrix thus satisfies the equation

$$[U] = [S]_t[S]^*$$

or

$$[S]_t^{-1} = [S]^* \tag{65}$$

where $[U]$ is the unit matrix. Equation (65) is the condition that $[S]$ should be a unitary matrix.† When the discontinuity absorbs energy, the sum of (64) is not zero, and $[S]$ is no longer a unitary matrix.

A discontinuity may be described by an impedance or admittance matrix by defining suitable equivalent terminal voltages and currents. The equivalent voltages are chosen proportional to the total transverse electric field (both incident and reflected waves), while the equivalent currents are taken proportional to the total transverse magnetic field as follows:

$$V_1 = c_1 + b_1 \tag{66a}$$
$$I_1 = c_1 - b_1 \tag{66b}$$
$$V_2 = c_2 + b_2 \tag{66c}$$
$$I_2 = c_2 - b_2 \tag{66d}$$

The voltages $V_i$ and currents $I_i$ are related by the impedance and admittance matrices $[Z]$ and $[Y]$ as follows:

$$\begin{bmatrix} V_1 \\ V_2 \end{bmatrix} = \begin{bmatrix} Z_{11} & Z_{12} \\ Z_{21} & Z_{22} \end{bmatrix} \begin{bmatrix} I_1 \\ I_2 \end{bmatrix}$$

or

$$[V] = [Z][I] \tag{67a}$$

$$\begin{bmatrix} I_1 \\ I_2 \end{bmatrix} = \begin{bmatrix} Y_{11} & Y_{12} \\ Y_{21} & Y_{22} \end{bmatrix} \begin{bmatrix} V_1 \\ V_2 \end{bmatrix}$$

or

$$[I] = [Y][V] \tag{67b}$$

When reciprocity holds, $Z_{12} = Z_{21}$ and $Y_{12} = Y_{21}$. From (66) we get $[V] + [I] = 2[c]$ and $[V] - [I] = 2[b]$, where

$$[c] = \begin{bmatrix} c_1 \\ c_2 \end{bmatrix} \quad \text{and} \quad [b] = \begin{bmatrix} b_1 \\ b_2 \end{bmatrix}$$

Replacing $[V]$ by $[Z][I]$ in the above relations gives

$$([Z] + [U])[I] = 2[c]$$
$$([Z] - [U])[I] = 2[b] = 2[S][c]$$

Finally, by substituting for $[c]$ from the first equation into the second, we get

$$([Z] - [U])[I] = [S]([Z] + [U])[I]$$

and hence

$$[S] = ([Z] - [U])([Z] + [U])^{-1} \tag{68a}$$

In a similar fashion the following relation between the scattering and

† Section A.3.

admittance matrices may be derived:

$$[S] = ([U] - [Y])([U] + [Y])^{-1} \tag{68b}$$

When analyzing a cascade connection of discontinuities on a voltage and current basis, it is convenient to relate the voltage and current on the output side to those on the input side as follows:

$$\begin{bmatrix} V_1 \\ I_1 \end{bmatrix} = \begin{bmatrix} A & B \\ C & D \end{bmatrix} \begin{bmatrix} V_2 \\ I_2 \end{bmatrix} \tag{69}$$

The matrix used here is analogous to the *ABCD* matrix employed in conventional low-frequency circuit analysis. The *ABCD* matrix is given in terms of the impedance-matrix elements as follows:

$$\begin{bmatrix} A & B \\ C & D \end{bmatrix} = \frac{1}{Z_{21}} \begin{bmatrix} Z_{11} & Z_{12}Z_{21} - Z_{11}Z_{22} \\ 1 & -Z_{22} \end{bmatrix} \tag{70}$$

### 3.5. Transmission through Dielectric Sheets

The transmission of plane electromagnetic waves through dielectric sheets at oblique angles of incidence has been treated by several methods. However, most treatments of the subject are unnecessarily complicated. By means of wave matrices, we shall be able to give a very simple approach to the subject. We shall refer to transmission-line theory and replace the dielectric sheet by an equivalent transmission-line circuit. We have not discussed transmission lines as such yet, but the concepts involved in this treatment are given in Sec. 3.3, in particular, Eqs. (40) to (45). The case of lossless sheets is treated first.

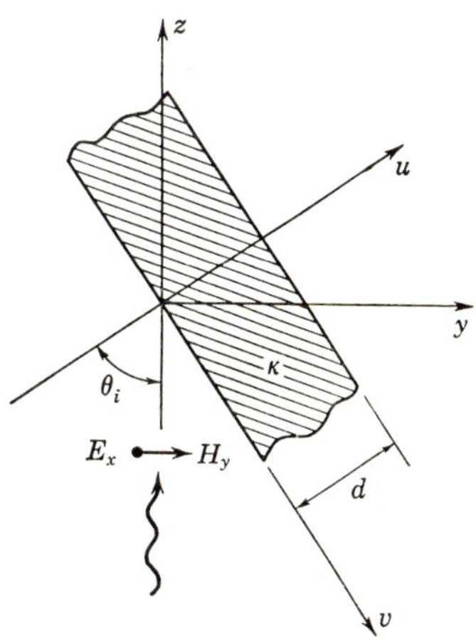

FIG. 3.6. Dielectric-sheet discontinuity.

Let a plane TEM wave be incident on an infinite-plane lossless dielectric sheet at an angle $\theta_i$ measured from the interface normal as in Fig. 3.6. The dielectric constant of the sheet is $\kappa$, and the index of refraction is $\eta = \kappa^{1/2}$.

In the *xy* plane this wave may be considered as a uniform wave propagating in the *z* direction, while in the *uv* coordinate frame it is a nonuniform wave propagating in the direction of the interface normal. It is necessary to distinguish between the two cases where the electric vector lies in the plane containing the interface normal and the direction of propagation (parallel polarization), and where the electric vector is perpendicular to this plane (perpendicular polarization). The case of perpendicular-polarized radiation will be treated first. The incident wave will be taken as

$$E_x = e^{-jk_0z} \qquad H_y = \frac{k_0}{\omega\mu_0} e^{-jk_0z}$$

where $k_0 = \omega(\mu_0\epsilon_0)^{1/2} = 2\pi/\lambda_0$. This wave will be referred to a new coordinate system $xvu$ obtained by rotating the old coordinate frame about the $x$ axis through an angle $\theta_i$. Simple geometry shows that the coordinate transformation is given by

$$\begin{aligned} x &= x \\ y &= v \cos \theta_i + u \sin \theta_i \\ z &= -v \sin \theta_i + u \cos \theta_i \end{aligned}$$

In this new coordinate system,

$$E_x = e^{-jk_0(-v \sin \theta_i + u \cos \theta_i)} \tag{71}$$

The components of $\mathbf{H}$ are obtained from $\nabla \times \mathbf{E} = -j\omega\mu_0\mathbf{H}$, where the curl is taken in the new coordinate frame; thus

$$j\omega\mu_0 H_u = jk_0 \sin \theta_i E_x \tag{72a}$$
$$j\omega\mu_0 H_v = jk_0 \cos \theta_i E_x \tag{72b}$$

The wave impedance in the direction of increasing $u$ (that is, normal to the surface) is

$$\frac{E_x}{H_v} = \left(\frac{\mu_0}{\epsilon_0}\right)^{1/2} \sec \theta_i = Z_1 \tag{73}$$

This is the characteristic impedance of free space measured in a direction at an angle $\theta_i$ to the direction of propagation of the plane uniform perpendicular-polarized wave.

When this wave passes through the air-dielectric surface it is bent toward the normal and propagates at an angle $\theta_r$ to the interface normal. Referred to the $xvu$ coordinate frame, the form of the wave will be the same with $\theta_i$ replaced by $\theta_r$ and $k_0$ replaced by $\eta k_0$. In order that the tangential field components may be matched for all values of $v$ along the air-dielectric interface, the variation with $v$ must be the same for the wave that exists in the dielectric sheet and the wave that exists in the free-space region. This requires that

$$k_0 \sin \theta_i = \eta k_0 \sin \theta_r \qquad \text{or} \qquad \eta = \frac{\sin \theta_i}{\sin \theta_r} \tag{74}$$

which is Snell's law of refraction. The wave impedance in the $u$ direction in the dielectric for perpendicular polarization may be written down at once by analogy with the expression for the wave impedance in free space. It is given by

$$Z_2 = \left(\frac{\mu_0}{\kappa\epsilon_0}\right)^{1/2} \sec\theta_r \tag{75}$$

Usually it is convenient to normalize this impedance by dividing by the impedance for free space. Hence, we may write for the normalized characteristic impedance of the dielectric, with a perpendicularly polarized plane wave incident at an angle $\theta_i$,

$$Z = \frac{\cos\theta_i}{(\kappa - \sin^2\theta_i)^{1/2}} \tag{76}$$

where $\cos\theta_r$ has been replaced by $(\kappa - \sin^2\theta_i)^{1/2}/\eta$ as obtained from Snell's law. The equivalent circuit of a dielectric sheet of thickness $d$ is illustrated in Fig. 3.7.

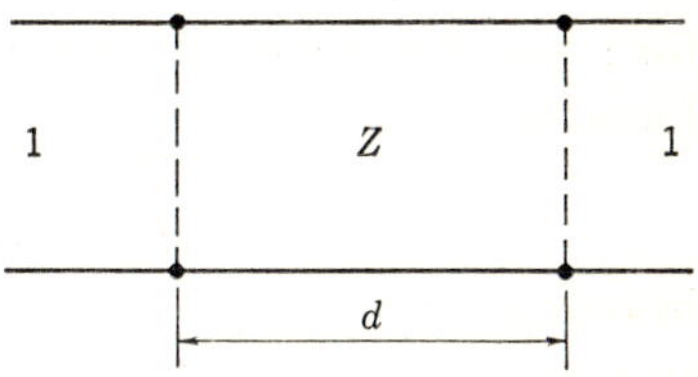

Fig. 3.7. Equivalent circuit for a dielectric sheet.

The transmission properties of such a sheet may be obtained by ordinary transmission-line theory, using the above expression for $Z$. It should be noted that the propagation factor in the dielectric sheet is $jk_0\eta\cos\theta_r = j(2\pi/\lambda_0)(\kappa - \sin^2\theta_i)^{1/2}$, where $\lambda_0$ is the free-space wavelength.

The reflection coefficient for the air-dielectric interface is given by

$$R_1 = \frac{Z-1}{Z+1} \tag{77}$$

The input impedance at the air-dielectric interface is given by

$$Z_{in} = \frac{Z(1+jZt)}{Z+jt} \tag{78}$$

where

$$t = \tan\left[\frac{2\pi}{\lambda_0}(\kappa - \sin^2\theta_i)^{1/2} d\right]$$

For a reflectionless sheet this normalized input impedance must reduce to unity, as it is, provided $t = 0$. (Under these conditions the sheet looks like a continuation of free space for the incident wave, and no reflection takes place.) For $t = 0$, we must have

$$d = \frac{m\lambda_0}{2}(\kappa - \sin^2\theta_i)^{-1/2} \qquad m = 0, 1, \ldots \tag{79}$$

For $m = 1$, the thickness is equal to a half wavelength measured in the dielectric sheet for that particular angle of oblique incidence $\theta_i$.

The analysis for a parallel-polarized incident wave proceeds in a similar fashion and results in an equation like (78) for the input impedance, but with $Z$ replaced by the normalized wave impedance $Z'$ of the dielectric for a parallel-polarized wave. The wave impedance $Z'$ may be easily determined by an application of the following transformations:

$$\mathbf{E}' = \left(\frac{\mu}{\epsilon}\right)^{1/2} \mathbf{H}$$

$$\mathbf{H}' = -\left(\frac{\epsilon}{\mu}\right)^{1/2} \mathbf{E}$$

as discussed in Sec. 1.7 in connection with Babinet's principle. The unnormalized wave impedance of free space and the dielectric medium for a parallel-polarized wave incident from free space at an angle $\theta_i$ are found as follows:

$$-\frac{E'_v}{H'_x} = \frac{(\mu_0/\epsilon_0)^{1/2} H_v}{(\epsilon_0/\mu_0)^{1/2} E_x} = \left(\frac{\mu_0}{\epsilon_0}\right)^{1/2} \cos \theta_i$$

in free space and

$$-\frac{E'_v}{H'_x} = \frac{(\mu_0/\kappa\epsilon_0)^{1/2} H_v}{(\kappa\epsilon/\mu_0)^{1/2} E_x} = \left(\frac{\mu_0}{\epsilon_0}\right)^{1/2} \frac{(\kappa - \sin^2 \theta_i)^{1/2}}{\kappa}$$

in the dielectric medium. The normalized wave impedance $Z'$ is thus given by

$$Z' = \frac{(\kappa - \sin^2 \theta_i)^{1/2}}{\kappa \cos \theta_i} \tag{80}$$

The propagation factor is the same as for perpendicular polarization, and, hence, the dielectric sheet will be reflectionless when $d$ has the value given by (79). The reflection coefficient at the air-dielectric interface is

$$R'_1 = \frac{Z' - 1}{Z' + 1} \tag{81}$$

When the appropriate values for the normalized impedances $Z$ and $Z'$ are substituted into Eqs. (77) and (81), respectively, we obtain the well-known Fresnel reflection coefficients for lossless dielectric media. An examination of (80) and (81) shows that, for some angle $\theta_i$, the reflection coefficient vanishes since $Z'$ becomes equal to unity. Let this value of $\theta_i$ be $\theta_b$; thus from Eq. (80)

$$\sin \theta_b = \left(\frac{\kappa}{\kappa + 1}\right)^{1/2} \tag{82}$$

which defines the Brewster angle. A similar phenomenon does not occur for perpendicular polarization unless the two media have different permeabilities.

We can now readily evaluate the transmission coefficient for a cascade section of two or more parallel but different dielectric sheets by applying the wave-matrix theory developed in Sec. 3.4. We must, of course, use the appropriate expressions for the reflection coefficients as given here. The transmission coefficients are given by

$$T_{12} = 1 + R_1 \tag{83a}$$
$$T'_{12} = 1 + R'_1 \tag{83b}$$

for the free-space–dielectric interface.

As a typical example, consider a dielectric sheet of thickness $d$ in front of an infinite dielectric medium as in Fig. 3.8. A parallel-polarized wave

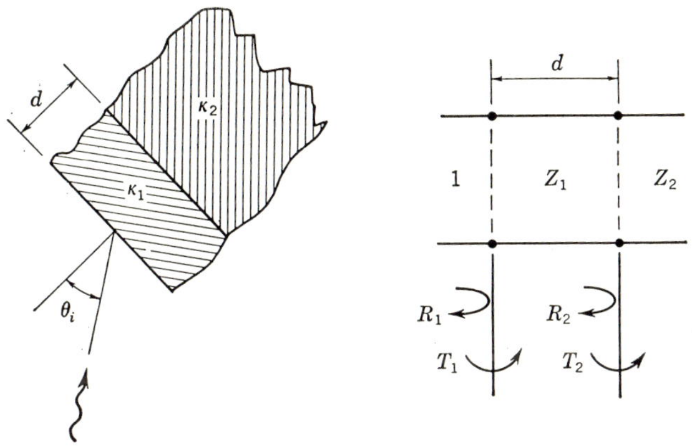


FIG. 3.8. A dielectric-sheet discontinuity and its equivalent circuit.

is assumed incident from free space at an angle $\theta_i$. The dielectric constants are $\kappa_1$ and $\kappa_2$. The electrical length of the sheet of thickness $d$ is $\phi = k_0 d(\kappa_1 - \sin^2 \theta_i)^{1/2}$. The reflection and transmission coefficients at the two interfaces are

$$R_1 = \frac{Z_1 - 1}{Z_1 + 1} \qquad T_1 = 1 + R_1$$
$$R_2 = \frac{Z_2 - Z_1}{Z_2 + Z_1} \qquad T_2 = 1 + R_2$$

where

$$Z_1 = \frac{(\kappa_1 - \sin^2 \theta_i)^{1/2}}{\kappa_1 \cos \theta_i}$$
$$Z_2 = \frac{(\kappa_2 - \sin^2 \theta_i)^{1/2}}{\kappa_2 \cos \theta_i}$$

Let the incident wave have an amplitude $c_1$, the reflected wave an amplitude $b_1$, and the wave transmitted into the medium with dielectric constant $\kappa_2$ an amplitude $c_3$. Using the wave-chain-matrix theory gives

$$\begin{bmatrix} c_1 \\ b_1 \end{bmatrix} = \frac{1}{T_1 T_2} \begin{bmatrix} e^{j\phi} & R_1 e^{-j\phi} \\ R_1 e^{j\phi} & e^{-j\phi} \end{bmatrix} \begin{bmatrix} 1 & R_2 \\ R_2 & 1 \end{bmatrix} \begin{bmatrix} c_3 \\ 0 \end{bmatrix} \tag{84}$$

Multiplying the two matrices together gives the following result for the element $A_{11}$ in the first row and column of the resultant $2 \times 2$ matrix,

$$A_{11} = \frac{e^{j\phi} + R_1R_2e^{-j\phi}}{T_1T_2} \tag{85}$$

This element is equal to the reciprocal of the over-all transmission coefficient $T$. The power-transmission coefficient is given by $TT^*/Z_2$ since $T$ is the nonnormalized amplitude-transmission coefficient. From (85) we get

$$\frac{TT^*}{Z_2} = (A_{11}A_{11}^*Z_2)^{-1} = \frac{T_1^2T_2^2}{Z_2(1 + R_1^2R_2^2 + 2R_1R_2 \cos 2\phi)} \tag{86}$$

For a sheet which is effectively a quarter wavelength thick, $\phi = \pi/2$ and (86) reduces to

$$\frac{TT^*}{Z_2} = \frac{4Z_1^2Z_2}{(Z_1^2 + Z_2)^2}$$

All the incident power is transmitted into the final medium when

$$TT^* = Z_2$$

or

$$Z_1 = Z_2^{1/2} \tag{87}$$

This is the condition for matching one dielectric sheet to free space by means of another sheet of electrical length $\pi/2$ and a normalized wave impedance equal to the square root of the normalized wave impedance of the sheet to be matched. Obviously a perfect match is obtained only at one frequency and one angle of incidence. This is the electromagnetic basis of the technique used in optics to match the surfaces of lenses to free space and is known as lens blooming. Such a matching sheet will not match both polarizations simultaneously except for normal incidence when the distinction between the two polarizations vanishes. The required value of the dielectric constant can be determined from the expressions for the impedances $Z_1$ and $Z_2$. The results for the case of perpendicular polarization may be obtained by using the appropriate expressions for the impedances.

Many dielectrics used at microwave frequencies have such small losses that the losses may be neglected. However, there are many applications where moderately lossy dielectrics must be used, e.g., in radome construction. Dielectric loss may be due to a finite conductivity of the material or to the fact that the applied frequency is too near a resonant frequency of the molecules which compose the dielectric material, so that the damping forces are such that the molecular polarization lags behind the applied field. When the material has a finite conductivity $\sigma$, the curl equation for the magnetic field vector **H** becomes

$$\begin{aligned}\nabla \times \mathbf{H} &= j\omega\epsilon_0\kappa\mathbf{E} + \sigma\mathbf{E} \\ &= j\omega\epsilon_0\kappa\left(1 - j\frac{\sigma}{\omega\kappa\epsilon_0}\right)\mathbf{E} \\ &= j\omega\epsilon_0\kappa(1 - j\tan\delta)\mathbf{E}\end{aligned} \tag{88}$$

where $\tan\delta = \sigma/\omega\epsilon_0\kappa$ and is called the loss tangent of the material. We shall consider only the case where $\sigma \ll \omega\epsilon_0\kappa$, so that $\tan\delta \approx \delta$ and $\delta \ll 1$. The other forms of dielectric loss may be accounted for by introducing a suitably defined loss tangent also. If we replace $\kappa$ in the previously derived solutions to Maxwell's equations by the complex dielectric constant

$$\kappa(1 - j\delta)$$

we obtain at once the effect of small losses in the dielectric medium.

Using the approximation $(1 - j\delta)^{1/2} \approx 1 - j\delta/2$, the propagation constant becomes $j(2\pi/\lambda_0)\cos\theta_r\,(1 - j\delta/2)$. From Snell's law,

$$\begin{aligned}\cos\theta_r &= \frac{[(1 - j\delta)\kappa - \sin^2\theta_i]^{1/2}}{[1 - j(\delta/2)]\eta} = \frac{\{(\kappa - \sin^2\theta_i)[1 - j\kappa\delta/(\kappa - \sin^2\theta_i)]\}^{1/2}}{[1 - j(\delta/2)]\eta} \\ &\approx \frac{S(1 - j\kappa\delta/2S^2)}{\eta(1 - j\delta/2)}\end{aligned} \tag{89}$$

where $S = (\kappa - \sin^2\theta_i)^{1/2}$. Hence the propagation constant becomes

$$j\frac{2\pi S}{\lambda_0} + \frac{2\pi}{\lambda_0}\frac{\kappa\delta}{2S} = j\beta + \alpha \tag{90}$$

where $j\beta$ is the phase factor, and $\alpha$ is the attenuation factor. For perpendicular polarization, the normalized wave impedance becomes

$$Z = \frac{\cos\theta_i}{\eta[1 - j(\delta/2)]\cos\theta_r} \approx \frac{\cos\theta_i}{S}\left(1 + j\frac{\kappa\delta}{2S^2}\right) \tag{91}$$

For parallel polarization, the normalized wave impedance becomes

$$Z' = \frac{\cos\theta_r}{\eta[1 - j(\delta/2)]\cos\theta_i} \approx \frac{S[1 - j(\kappa/2S^2 - 1)\delta]}{\kappa\cos\theta_i} \tag{92}$$

The wave impedances are complex whenever the medium has finite losses. We may treat the propagation of electromagnetic waves through lossy dielectric sheets formally the same as through nonlossy sheets. The only difference is that we must use the wave impedances and the expression for the propagation constant which are appropriate for lossy dielectric sheets. We may again define an electrical length, but this angle will now be complex in order to account for the attenuation that the wave undergoes.

### 3.6. Reflection from a Finite Conducting Plane

The theory developed for reflection from dielectric sheets, and, in particular, lossy dielectric sheets, may be applied to the problem of reflection from finite conducting metal planes also. For metals commonly used at microwave frequencies, the conduction current is normally much greater than the displacement current; that is, $\sigma \gg \omega\epsilon$. For copper, at 10,000 megacycles, $\sigma/\omega\epsilon \approx 10^8$. Consequently Maxwell's curl equation for $\mathbf{H}$ may be taken as $\nabla \times \mathbf{H} = j\omega\epsilon\mathbf{E} + \sigma\mathbf{E} \approx \sigma\mathbf{E}$ in metals. In a metal we may, therefore, obtain the solutions to Maxwell's equations by replacing $j\omega\epsilon_0$ in the solutions for lossless dielectric sheets by $\sigma$. The propagation constant $k$ becomes $(-j\omega\mu_0\sigma)^{1/2} = (\omega\mu_0\sigma/2)^{1/2}(1 - j)$, where it is assumed that the permeability of the metal may be taken as that of free space.

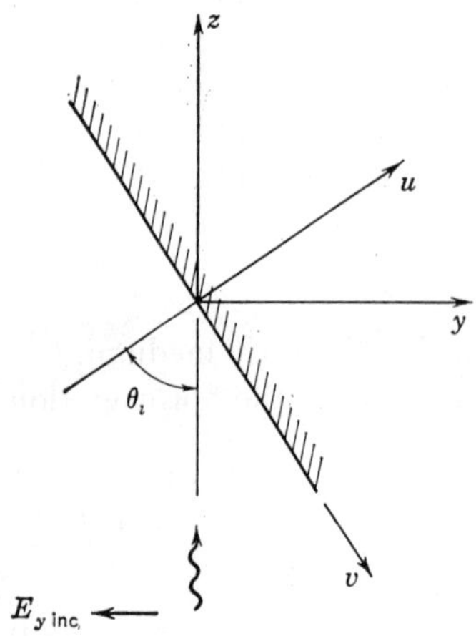

FIG. 3.9. An infinite conducting plane.

Let the parallel-polarized wave be incident on the metallic sheet at an angle $\theta_i$, as in Fig. 3.9. The electrical constants of the metal are assumed to be $\mu_0$, $\epsilon_0$, and $\sigma$. The incident magnetic field may be taken as

$$H_{xI} = H_x e^{-jk_0 z} = H_x e^{jk_0(v \sin \theta_i - u \cos \theta_i)} \tag{93}$$

Using the curl equation for $\mathbf{H}$, we find that the incident electric field in the $vux$ coordinate frame is

$$j\omega\epsilon_0 E_{uI} = -jk_0 \sin \theta_i H_{xI} \tag{94a}$$
$$j\omega\epsilon_0 E_{vI} = -jk_0 \cos \theta_i H_{xI} \tag{94b}$$

The wave impedance measured in a direction normal to the surface, i.e., along $u$, is

$$-\frac{E_{vI}}{H_{xI}} = \left(\frac{\mu_0}{\epsilon_0}\right)^{1/2} \cos \theta_i = Z_1 = Y_1^{-1} \tag{95}$$

By analogy, the wave transmitted into the metal is, by Eqs. (36) to (39) in Sec. 3.3,

$$E_{vT} = -TZ_1 H_x e^{jk(v \sin \theta_r - u \cos \theta_r)} \tag{96a}$$
$$H_{xT} = -Y_m E_{vT} \tag{96b}$$
$$\sigma E_{uT} = -jk \sin \theta_r H_{xT} = jkY_m \sin \theta_r E_{vT} \tag{96c}$$

where $k = (\omega\mu_0\sigma/2)^{1/2}(1 - j)$, and $T$ is the amplitude-transmission coefficient. The wave impedance of the metal measured in the $u$ direction is

$$Z_m = Y_m^{-1} = -\frac{E_{vT}}{H_{xT}} = \left(\frac{\omega\mu_0}{2\sigma}\right)^{1/2}(1 + j)\cos\theta_r \tag{97}$$

When $\sigma$ approaches infinity, $Z_m$ approaches zero, and the metal becomes an ideal short circuit. The attenuation factor along the $u$ direction in the metal is $(\omega\mu_0\sigma/2)^{1/2}\cos\theta_r$ and becomes infinite with $\sigma$. Consequently, for a perfect conductor, the field cannot penetrate into the metal.

The reflected wave is given by the following expressions:

$$E_{vR} = -RZ_1H_xe^{jk_0(v\sin\theta_i + u\cos\theta_i)} \tag{98a}$$
$$H_{xR} = Y_1E_{vR} \tag{98b}$$
$$j\omega\epsilon_0E_{uR} = -jk_0\sin\theta_iH_{xR} = -jk_0\sin\theta_iY_1E_{vR} \tag{98c}$$

where $R$ is the reflection coefficient. The reflection and transmission coefficients are given in Sec. 3.3, Eqs. (40), as follows:

$$T = 1 + R = \frac{2Z_m}{Z_m + Z_1} \tag{99a}$$
$$R = \frac{Z_m - Z_1}{Z_m + Z_1} \tag{99b}$$

By Snell's law, $k_0 \sin\theta_i = k\sin\theta_r$, and hence

$$k\cos\theta_r = (k^2 - k_0^2\sin^2\theta_i)^{1/2} \approx k$$

since $\sigma \gg \omega\epsilon_0$. This is a result of fundamental importance for it shows that, to a very high degree of approximation, the propagation constant normal to the interface may be taken as $k$ in the conductor independent of the angle of incidence $\theta_i$. From (97) we find that

$$Z_m = \left(\frac{\omega\mu_0}{2\sigma}\right)^{1/2}(1 + j)$$

again independent of $\theta_i$. Since $k = (\omega\mu_0\theta/2)^{1/2}(1 - j)$, the attenuation coefficient in the conductor is $(\omega\mu_0\sigma/2)^{1/2}$. In a distance $\delta_s = (2/\omega\mu_0\sigma)^{1/2}$ called the skin depth, the field has decayed to $e^{-1}$ of its value at the surface. Using the definition of $\delta_s$ shows that the wave impedance of the metal may be expressed as follows:

$$Z_m = \frac{1 + j}{\delta_s\sigma} \tag{100}$$

The resistive part is seen to be equal to the low-frequency resistance of a conducting sheet of thickness $\delta_s$ and conductivity $\sigma$. Since these latter results are independent of $\theta_i$ we may treat the boundary of an arbitrary-shaped conducting obstacle as a discontinuity surface with a surface impedance $Z_m$, even though the field incident on the obstacle is not a plane wave. An arbitrary field may be constructed from a superposition of plane waves. The approximation, however, does break down in the vicinity of surfaces which have a radius of curvature not much greater than the skin depth.

Returning to the problem of the conducting plane, we find that at the boundary surface the following relations must hold between the tangential components of the incident, reflected, and transmitted fields:

$$E_{vI} + E_{vR} = -(1 + R)Z_1H_x = -TZ_1H_x = E_{vT} \tag{100a}$$

$$H_{xI} + H_{xR} = (1 - R)H_x = \frac{Z_1}{Z_m} TH_x = H_{xT} \tag{100b}$$

$$j\omega\epsilon_0(E_{uI} + E_{uR}) = \sigma E_{uT} \tag{100c}$$

Equation (100*c*) shows that the normal displacement flux is continuous across the surface. The total current flowing per unit width in the metal is

$$I = \int_0^\infty \sigma E_{vT}\, du = -\sigma TZ_1H_x e^{jkv \sin\theta_r} \int_0^\infty e^{-[(1+j)/\delta_s]u\cos\theta_r}\, du$$
$$= -\frac{2Z_1H_x}{Z_1 + Z_m} e^{jk_0v\sin\theta_i} \tag{101}$$

The total current flowing is independent of $\sigma$ when $\sigma$ approaches infinity, since under these conditions $Z_m$ tends to zero. Thus the current is forced to flow on the surface of the metal when $\sigma \to \infty$. The total current flowing is $2H_xe^{jk_0v\sin\theta_i}$ and is equal to the tangential magnetic field at the surface. Since the field vanishes inside the metal, the normal component of **D** must now terminate in a surface charge distribution. These two boundary conditions are easily expressed as follows:

$$\mathbf{J}_s = \mathbf{n} \times \mathbf{H} \tag{102a}$$

$$\frac{\rho_s}{\epsilon_0} = \mathbf{n} \cdot \mathbf{E} \tag{102b}$$

where **n** is the unit normal pointing away from the surface. On the surface the current density $\mathbf{J}_s$ is related to the charge density $\rho_s$ by the equation of continuity. For the above problem,

$$\nabla_s \cdot \mathbf{J}_s = \frac{\partial}{\partial v}(-2H_xe^{jk_0v\sin\theta_i}) = -2jk_0H_x \sin\theta_i e^{jk_0v\sin\theta_i}$$

and $$-j\omega\rho_s = j\omega\epsilon_0(E_{uI} + E_{uR}) = -2jk_0H_x\sin\theta_i e^{jk_0v\sin\theta_i}$$

so that $\nabla_s \cdot \mathbf{J}_s = -\partial\rho_s/\partial t$, where $\nabla_s$ is the del operator on the discontinuity surface.

The analysis for a perpendicular-polarized incident wave does not give any additional new results, and so the details will not be given.

### 3.7. TEM Waves in Anisotropic Dielectric Media

Anisotropic media are those characterized by tensor permittivities, tensor permeabilities, or perhaps both. Initially we will consider some of the properties of TEM waves in anisotropic dielectric materials. This will be followed by a brief treatment of propagation in ferrite material which exhibits a tensor permeability. By a suitable choice of axes, the tensor permeability or permittivity can be reduced to a diagonal form. The normal coordinates for the two tensors will, in general, not coincide. Also, the direction cosines of the normal coordinates relative to the original coordinates are not necessarily real, so that diagonalization of the constitutive tensors cannot always be obtained by a simple rotation of coordinates. A familiar example is an infinite ferrite medium for which the normal coordinates are clockwise- and counterclockwise-rotating coordinates (see Prob. 3.7).

Let us consider a lossless dielectric material with principal dielectric constants $\kappa_x$, $\kappa_y$, $\kappa_z$ along the principal axes, which are assumed to coincide with the $x$, $y$, $z$ axes. The electric field for a plane uniform TEM wave propagating in a direction defined by the unit normal

$$\mathbf{n} = \mathbf{a}_x n_x + \mathbf{a}_y n_y + \mathbf{a}_z n_z$$

may be represented as follows:

$$\mathbf{E} = \mathbf{E}_0 e^{-j\beta\mathbf{n}\cdot\mathbf{r}} \tag{103}$$

where $\mathbf{E}_0$ is a constant vector, $\beta$ the propagation constant, as yet undetermined, and $\mathbf{r}$ is the position vector $\mathbf{a}_x x + \mathbf{a}_y y + \mathbf{a}_z z$. Using Maxwell's equations, we find that

$$-j\omega\mu_0\mathbf{H} = \nabla \times \mathbf{E} = \nabla \times \mathbf{E}_0 e^{-j\beta\mathbf{n}\cdot\mathbf{r}} = -\mathbf{E}_0 \times \nabla e^{-j\beta\mathbf{n}\cdot\mathbf{r}} = -j\beta\mathbf{n} \times \mathbf{E} \tag{104$a$}$$

$$j\omega\mathbf{D} = \nabla \times \mathbf{H} = \frac{j\beta^2}{\omega\mu_0}[\mathbf{E} - \mathbf{n}(\mathbf{n}\cdot\mathbf{E})] \tag{104$b$}$$

Equation (104$b$) shows that $\mathbf{D}$ and $\mathbf{E}$ are not collinear unless $\mathbf{n}\cdot\mathbf{E} = 0$. In general, $\mathbf{n}\cdot\mathbf{E} \neq 0$, and hence replacing the components of $\mathbf{E}$ in the first term in (104$b$) by

$$\frac{D_x}{\kappa_x\epsilon_0} \qquad \frac{D_y}{\kappa_y\epsilon_0} \qquad \frac{D_z}{\kappa_z\epsilon_0}$$

and solving for the components of $\mathbf{D}$ gives

$$\mathbf{D} = \frac{\beta^2}{k_0^2}\epsilon_0\mathbf{n}\cdot\mathbf{E}\left(\frac{\mathbf{a}_x n_x k_x^2}{\beta^2 - k_x^2} + \frac{\mathbf{a}_y n_y k_y^2}{\beta^2 - k_y^2} + \frac{\mathbf{a}_z n_z k_z^2}{\beta^2 - k_z^2}\right) \tag{105}$$

where

$$k_0^2 = \omega^2\mu\epsilon_0$$
$$k_x^2 = \kappa_x k_0^2$$

and similarly for $k_y$ and $k_z$. From (104*b*) it is seen that $\mathbf{n} \cdot \mathbf{D}$ equals zero, so that $\mathbf{D}$ is always normal to $\mathbf{n}$. The dot product of (105) with $\mathbf{n}$ gives

$$\frac{n_x^2k_x^2}{\beta^2 - k_x^2} + \frac{n_y^2k_y^2}{\beta^2 - k_y^2} + \frac{n_z^2k_z^2}{\beta^2 - k_z^2} = 0 \tag{106}$$

which is the eigenvalue equation for determining the propagation factor $\beta^2$. This is a quadratic equation for $\beta^2$ and determines two distinct values for $\beta^2$ in general. At this time it is convenient to assume that the axes have been chosen so that $k_x > k_y > k_z$. For (106) to be zero, at least one term must be negative; at the same time not all three can be negative, and so $k_z^2 < \beta^2 < k_x^2$. If the principal velocities

$$v_x^2 = \frac{v_c^2k_0^2}{k_x^2} \qquad v_y^2 = \frac{v_c^2k_0^2}{k_y^2} \qquad v_z^2 = \frac{v_c^2k_0^2}{k_z^2}$$

and the wave velocity $v^2 = v_c^2k_0^2/\beta^2$ are introduced, (106) becomes

$$\frac{n_x^2}{v_x^2 - v^2} + \frac{n_y^2}{v_y^2 - v^2} + \frac{n_z^2}{v_z^2 - v^2} = 0 \tag{107}$$

where $v_c$ is the velocity of light in free space. For each direction, in the $xyz$ coordinate frame, defined by the direction cosines $n_x$, $n_y$, $n_z$, two values of $v^2$ are determined by (107). Thus $v^2$ traces out a double-sheeted surface, which is called the Fresnel wave-velocity surface. The trace of the intersection of these surfaces in the coordinate planes consists of a circle and an oval. In the coordinate plane containing the greatest and smallest principal velocity these curves intersect at four points. According to the way in which we choose our axes, this plane, in our case, is the $xz$ plane. Thus, putting $n_y = 0$ in (107), we get

$$(v^2 - v_y^2)[v^2 - n_x^2v_z^2 - (1 - n_x^2)v_x^2] = 0$$

The two curves traced out are the circle $v = v_y$ and the oval

$$v^2 = n_x^2v_z^2 + (1 - n_x^2)v_x^2$$

with semiaxes $v_x$ and $v_z$. Since $v_x < v_y < v_z$, these curves intersect as illustrated in Fig. 3.10. The points of intersection define two directions, known as the optic axes, along which only a single value for $v^2$ and $\beta^2$ exists. If the positive directions of these axes are chosen as in Fig. 3.10, then the unit normals $\mathbf{N}_1$ and $\mathbf{N}_2$ along these axes are

$$\mathbf{N}_1 = \mathbf{a}_xN_x + \mathbf{a}_zN_z \tag{108a}$$

$$\mathbf{N}_2 = \mathbf{a}_xN_x - \mathbf{a}_zN_z \tag{108b}$$

where

$$N_x^2 = \frac{v_y^2 - v_x^2}{v_z^2 - v_x^2} \tag{108c}$$

$$N_z^2 = \frac{v_z^2 - v_y^2}{v_z^2 - v_x^2} \tag{108d}$$

These are the only directions for which a single value for $\beta^2$ exists. The oval and circle do not intersect in the $xy$ or $yz$ plane since the oval's semiaxes are either both greater or both smaller than the radius of the circle, which is either $v_x$ or $v_z$ in these planes.

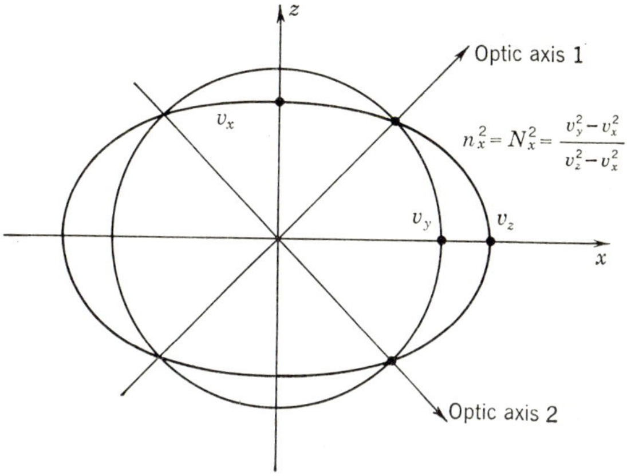


FIG. 3.10. Wave-velocity surface traces in $xz$ plane.

The direction cosines between the wave normal $\mathbf{n}$ and the optic axes are

$$\cos \theta_1 = \mathbf{n} \cdot \mathbf{N}_1 \tag{109a}$$
$$\cos \theta_2 = \mathbf{n} \cdot \mathbf{N}_2 \tag{109b}$$

In terms of these, we find, after an hour or so of algebraic juggling, that the two solutions to (107) are $v_1^2$ and $v_2^2$, where

$$2v_1^2 = v_z^2 + v_x^2 + (v_z^2 - v_x^2) \cos (\theta_1 - \theta_2) \tag{110a}$$
$$2v_2^2 = v_z^2 + v_x^2 + (v_z^2 - v_x^2) \cos (\theta_1 + \theta_2) \tag{110b}$$

The interested reader may readily verify that $(v^2 - v_1^2)(v^2 - v_2^2) = 0$ is identical with (107). The two values of $\beta^2$ are given by

$$\beta_i^2 = k_0^2 \frac{v_c^2}{v_i^2} \qquad i = 1, 2$$

Having determined the two values of the wave velocity, we return to (105) and find that the direction of $\mathbf{D}$ is fixed, i.e.,

$$D_{ix}:D_{iy}:D_{iz} = \frac{n_x}{v_i^2 - v_x^2}:\frac{n_y}{v_i^2 - v_y^2}:\frac{n_z}{v_i^2 - v_z^2} \tag{111}$$

where $\mathbf{D}_i$ is the solution for $\mathbf{D}$ corresponding to the root $v_i^2$. The directions of $\mathbf{D}$, in turn, determine the allowed directions of $\mathbf{E}_0$ and $\mathbf{H}$. The two solutions for **D** are mutually orthogonal since

$$\mathbf{D}_1 \cdot \mathbf{D}_2 \propto \frac{n_x^2}{(v_1^2 - v_x^2)(v_2^2 - v_x^2)} + \frac{n_y^2}{(v_1^2 - v_y^2)(v_2^2 - v_y^2)} + \frac{n_z^2}{(v_1^2 - v_z^2)(v_2^2 - v_z^2)}$$

which, when the identity

$$\frac{n_x^2}{(v_1^2 - v_x^2)(v_2^2 - v_x^2)} = \frac{n_x^2}{v_2^2 - v_1^2}\left(\frac{1}{v_1^2 - v_x^2} - \frac{1}{v_2^2 - v_x^2}\right)$$

(and similarly for the other terms) is used, is seen to vanish by virtue of (107), of which $v_1^2$ and $v_2^2$ are roots.

The direction of energy flow is that of the complex Poynting vector $\mathbf{E} \times \mathbf{H}^*$ and is not in the direction of the wave normal $\mathbf{n}$ in general. In contrast to this, the cross product $\mathbf{D} \times \mathbf{H}^*$ is in the direction of $\mathbf{n}$ since both **D** and **H** are perpendicular to **n**. By forming the products $\mathbf{D} \cdot \mathbf{E}/|\mathbf{D}|\,|\mathbf{E}|$ and $\mathbf{n} \cdot \mathbf{E} \times \mathbf{H}^*/|\mathbf{E} \times \mathbf{H}^*|$ we find that the direction cosines (which are given by the above terms) between **D** and **E** and the wave normal **n** and the Poynting vector $\mathbf{E} \times \mathbf{H}^*$ are equal and, in particular, given by $\left(1 - \left|\frac{\mathbf{n} \cdot \mathbf{E}}{\mathbf{E}}\right|^2\right)^{1/2}$. The situation is illustrated in Fig. 3.11. The wave propagates along the normal **n** according to $e^{-j\beta\mathbf{n}\cdot\mathbf{r}}$. If we let **s** be a unit vector along the direction of the Poynting vector, then

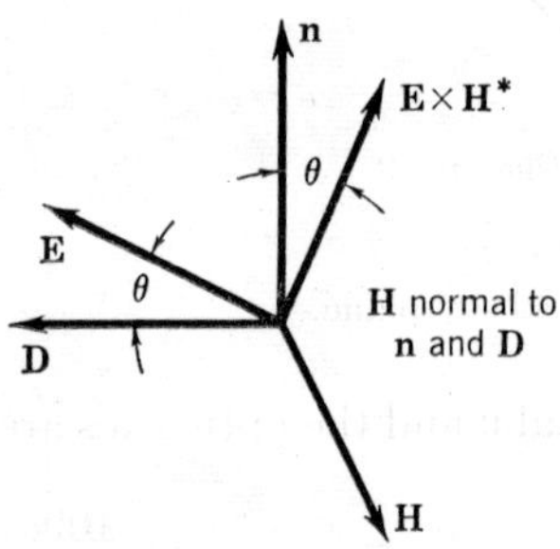


FIG. 3.11

$$\mathbf{n} = (\mathbf{n} \cdot \mathbf{s})\mathbf{s} + [\mathbf{n} - (\mathbf{n} \cdot \mathbf{s})\mathbf{s}] = (\mathbf{n} \cdot \mathbf{s})\mathbf{s} + \boldsymbol{\tau}$$

where $\boldsymbol{\tau}$ is defined by this equation and is a vector perpendicular to **s**. Introducing these vectors, the propagation factor becomes $e^{-j\beta[(\mathbf{n}\cdot\mathbf{s})\mathbf{s}\cdot\mathbf{r}+\boldsymbol{\tau}\cdot\mathbf{r}]}$. In the direction of **s**, the velocity of propagation is

$$u = \frac{v_c k_0}{\beta \mathbf{n} \cdot \mathbf{s}} = \frac{v}{\mathbf{n} \cdot \mathbf{s}} \tag{112}$$

and is the velocity of energy propagation for the wave.

The analysis of plane-wave propagation in the interior of an anisotropic dielectric is relatively straightforward. However, the problem of determining the fields transmitted into an anisotropic medium when a plane wave is incident on an interface separating an isotropic and an anisotropic medium is much more complicated, since in this case neither **n**, $\beta$, or the direction of **D** is known. A medium with two optical axes is known as a biaxial medium. When two of the principal dielectric constants are equal, say $\kappa_y = \kappa_z$, then only a single optical axis exists, and the medium is said to be uniaxial. Referring to (108*c*) and (108*d*) with $v_z^2 = v_y^2$ shows that this axis coincides with the $x$ axis in the present case. Reflection from and transmission through a uniaxial anisotropic interface is considerably easier to analyze, and is the next topic which we will take up.

*An Anisotropic Dielectric Interface*

Consider a uniaxial medium with principal dielectric constants $\kappa_1$, $\kappa_2$, and $\kappa_2$ along the $x$, $y$, and $z$ axes, respectively. With reference to Eq. (104*b*), we find that, if $\mathbf{n} \cdot \mathbf{E} = 0$, **D** and **E** are collinear, and

$$\beta^2 = k_2^2 = \kappa_2 k_0^2$$

In this case **E** must lie in the $xy$ plane for which no anisotropic effect is present. This solution is called the ordinary wave since its behavior is the same as that of a plane wave in an isotropic medium. When

$$\mathbf{n} \cdot \mathbf{E} \neq 0$$

we find from (106) that

$$\beta^2 = \frac{k_1^2 k_2^2}{n_x^2 k_1^2 + (1 - n_x^2) k_2^2} \tag{113}$$

where $k_i^2 = \kappa_i k_0^2$, $i = 1, 2$. For this case **D** has components along all three axes in general. With reference to Fig. 3.12, let $\mathbf{D}_1$, $\mathbf{H}_1$ be the solution for the ordinary wave, and $\mathbf{D}_2$, $\mathbf{H}_2$ be the solution corresponding to the value of $\beta^2$ given by (113). This latter wave is called the extraordinary wave.

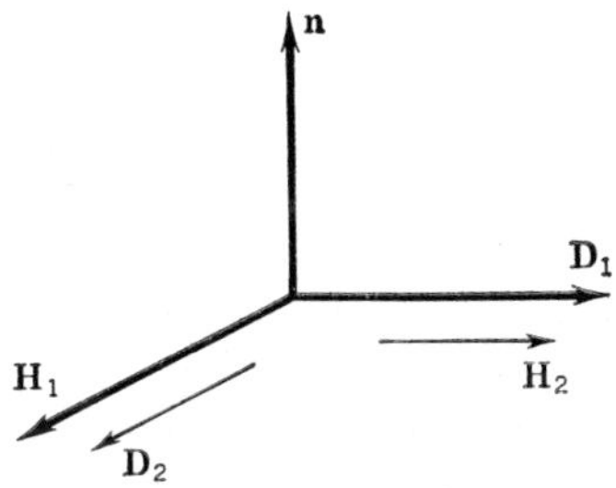


FIG. 3.12

Since $\mathbf{D}_1$ and $\mathbf{H}_1$ are at right angles to each other and are both perpendicular to **n**, and $\mathbf{D}_2$ is perpendicular to $\mathbf{D}_1$ and **n**, then $\mathbf{D}_2$ must be along the direction of $\mathbf{H}_1$. Consequently, $\mathbf{H}_2$, which is perpendicular to $\mathbf{D}_2$ and **n**, is collinear with $\mathbf{D}_1$, and hence lies in the $yz$ plane, i.e., has no $x$ component. These properties of the two solutions, namely, that $\mathbf{D}_1$ and $\mathbf{H}_2$ have no $x$ components, permit us to derive the fields for the two modes of propagation from a magnetic and an electric Hertzian potential function having only an $x$ component. The ordinary wave may be found from the magnetic Hertzian potential $\mathbf{\Pi}_h = \Pi_h \mathbf{a}_x$ as follows:

$$\mathbf{E} = -j\omega\mu_0 \nabla \times \mathbf{\Pi}_h \tag{114a}$$

$$\mathbf{H} = \nabla \times \nabla \times \mathbf{\Pi}_h \tag{114b}$$

and $\mathbf{\Pi}_h$ is a solution of

$$\nabla^2 \mathbf{\Pi}_h + k_2^2 \mathbf{\Pi}_h = 0 \tag{114c}$$

The extraordinary wave is found from the electric Hertzian potential $\mathbf{\Pi}_e = \mathbf{a}_x \Pi_e$ as follows:

$$\mathbf{E} = k_0^2 \mathbf{\Pi}_e + \kappa_2^{-1} \nabla\nabla \cdot \mathbf{\Pi}_e \tag{115a}$$

$$\mathbf{H} = j\omega\epsilon_0 \nabla \times \mathbf{\Pi}_e \tag{115b}$$

and $\mathbf{\Pi}_e$ is a solution of

$$\nabla^2 \mathbf{\Pi}_e + k_1^2 \mathbf{\Pi}_e + \frac{\kappa_1 - \kappa_2}{\kappa_2} \frac{\partial^2 \mathbf{\Pi}_e}{\partial x^2} = 0 \tag{115c}$$

Equation (115$c$) may be derived as follows. Since $\nabla \times \mathbf{H} = j\omega \mathbf{D}$, we get

$$j\omega \mathbf{D} = j\omega\epsilon_0 \nabla \times \nabla \times \mathbf{\Pi}_e = j\omega\epsilon_0 (\nabla\nabla \cdot \mathbf{\Pi}_e - \nabla^2 \mathbf{\Pi}_e)$$

Also, since $\nabla \times \mathbf{E} = -j\omega\mu_0 \mathbf{H} = k_0^2 \nabla \times \mathbf{\Pi}_e$, we have $\mathbf{E} = k_0^2 \mathbf{\Pi}_e + \nabla\phi$, where $\phi$ is as yet arbitrary. Substituting for $\mathbf{D}$ in the above expression gives

$$\begin{aligned} j\omega \mathbf{D} &= j\omega[\kappa_2\epsilon_0 \mathbf{E} + (\kappa_1 - \kappa_2)\epsilon_0 \mathbf{a}_x E_x] \\ &= j\omega\kappa_2\epsilon_0 \left[ k_0^2 \mathbf{\Pi}_e + \nabla\phi + \frac{\kappa_1 - \kappa_2}{\kappa_2} \left( k_0^2 \Pi_e + \frac{\partial \phi}{\partial x} \right) \mathbf{a}_x \right] \\ &= j\omega\epsilon_0 (\nabla\nabla \cdot \mathbf{\Pi}_e - \nabla^2 \mathbf{\Pi}_e) \end{aligned}$$

If we now let $\nabla \cdot \mathbf{\Pi}_e = \kappa_2 \phi = \partial \Pi_e / \partial x$, we obtain (115$c$) and (115$a$). This latter step is permissible since $\nabla \cdot \mathbf{\Pi}_e$ and $\phi$ have not been specified up to this point. Equation (114$a$) gives a solution for $\mathbf{E}$ with no $x$ component, while (115$b$) gives a solution for $\mathbf{H}$ with no $x$ component. Furthermore, (115$c$) yields the eigenvalue equation (113) when a form such as $e^{-j\beta \mathbf{n} \cdot \mathbf{r}}$ is assumed for $\Pi_e$. When these two modes of propagation are excited by a plane wave incident from free space onto the anisotropic medium, the two modes do not, in general, have the same wave normal.

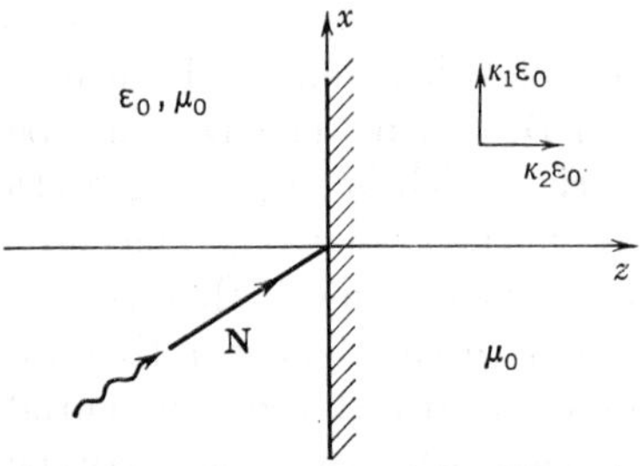


FIG. 3.13. An anisotropic dielectric interface.

Let a uniaxial anisotropic medium occupy the whole half space $z > 0$, and let the optic axis be along the $x$ axis, which in turn lies in the plane infinite interface as in Fig. 3.13. A perpendicular- or parallel-polarized plane wave incident on this interface along a direction defined by the unit vector or wave normal $\mathbf{N}$ can be represented by a suitable superposition of the two modes derivable from $\mathbf{\Pi}_h$ and $\mathbf{\Pi}_e$. Thus we are concerned with two distinct modes of propagation and will find that the interface can be represented by a four-terminal-pair network. An isotropic dielectric interface was found to be representable by a two-terminal-pair network and was consequently much simpler to analyze.

Let the Hertzian potentials for the incident and reflected plane waves in the region $z < 0$ be

$$\mathbf{\Pi}_h = \mathbf{a}_x(A_0 e^{-jk_0\mathbf{N}\cdot\mathbf{r}} + A_1 e^{-jk_0\mathbf{N}'\cdot\mathbf{r}}) \tag{116a}$$
$$\mathbf{\Pi}_e = \mathbf{a}_x(B_0 e^{-jk_0\mathbf{N}\cdot\mathbf{r}} + B_1 e^{-jk_0\mathbf{N}'\cdot\mathbf{r}}) \tag{116b}$$

where $k_0^2 = \omega^2\mu_0\epsilon_0$, and the unit wave normals are given by

$$\mathbf{N} = N_x\mathbf{a}_x + N_y\mathbf{a}_y + N_z\mathbf{a}_z$$
$$\mathbf{N}' = N_x\mathbf{a}_x + N_y\mathbf{a}_y - N_z\mathbf{a}_z$$

In the region $z > 0$, suitable forms for the potentials are

$$\mathbf{\Pi}_h = \mathbf{a}_x(C_1 e^{-jk_2\mathbf{n}_1\cdot\mathbf{r}} + C_0 e^{-jk_2\mathbf{n}'_1\cdot\mathbf{r}}) \tag{117a}$$
$$\mathbf{\Pi}_e = \mathbf{a}_x(D_1 e^{-j\beta\mathbf{n}_2\cdot\mathbf{r}} + D_0 e^{-j\beta\mathbf{n}'_2\cdot\mathbf{r}}) \tag{117b}$$

where the unit wave normals $\mathbf{n}_1'$ and $\mathbf{n}_2'$ have the same relation to $\mathbf{n}_1$ and $\mathbf{n}_2$ as $\mathbf{N}'$ does to $\mathbf{N}$, that is, a change in sign for the $z$ component.

At the interface $z = 0$, the tangential electric and magnetic fields must be continuous for all values of $x$ and $y$. This is possible only if the potentials have the same variation with $x$ and $y$ on both sides of the interface. Hence, the following relations must hold:

$$k_0N_x = k_2n_{1x} = \beta n_{2x} \tag{118a}$$
$$k_0N_y = k_2n_{1y} = \beta n_{2y} \tag{118b}$$

In addition, $\beta$ is a solution of (113). From (113) and (118$a$) we get

$$\beta = k_0\left[\kappa_1 + N_x^2\left(1 - \frac{\kappa_1}{\kappa_2}\right)\right]^{1/2} \tag{118c}$$

From (118$b$) and (118$c$) the component $n_{2y}$ may be found. The components of $\mathbf{n}_1$ may be found at once from (118$a$) and (118$b$), together with the condition $n_{1z}^2 = 1 - n_{1x}^2 - n_{1y}^2$. We see that a knowledge of the wave normal $\mathbf{N}$ for the incident waves permits us to determine both the normals $\mathbf{n}_1$, $\mathbf{n}_2$ and $\beta$.

The fields are obtained from the Hertzian potentials by means of (114) and (115). When the tangential components are equated in the interface plane $z = 0$, the following four equations are obtained:

$$H_x = k_0^2(1 - N_x^2)(A_0 + A_1) = k_2^2(1 - n_{1x}^2)(C_1 + C_0) \tag{119a}$$
$$E_x = k_0^2(1 - N_x^2)(B_0 + B_1) = k_0^2\left(1 - n_{2x}^2\frac{\beta^2}{k_2^2}\right)(D_1 + D_0) \tag{119b}$$
$$\begin{aligned} H_y &= -k_0^2N_xN_y(A_0 + A_1) + k_0^2(1 - N_x^2)Y_{1e}(B_0 - B_1) \\ &= -k_2^2n_{1x}n_{1y}(C_1 + C_0) + k_0^2\left(1 - \frac{\beta^2}{k_2^2}n_{2x}^2\right)Y_{2e}(D_1 - D_0) \end{aligned} \tag{119c}$$
$$\begin{aligned} E_y &= k_0^2(1 - N_x^2)Z_{1m}(A_1 - A_0) - k_0^2N_xN_y(B_0 + B_1) \\ &= k_2^2(1 - n_{1x}^2)Z_{2m}(C_0 - C_1) - k_0^2\frac{\beta^2}{k_2^2}n_{2x}n_{2y}(D_1 + D_0) \end{aligned} \tag{119d}$$

where

$$Y_{1m} = Z_{1m}^{-1} = \frac{1 - N_x^2}{N_z} Y_0 \qquad = \text{wave admittance for } \Pi_h \text{ mode for } z < 0$$

$$Y_{1e} = Z_{1e}^{-1} = \frac{N_z}{1 - N_x^2} Y_0 \qquad = \text{wave admittance for } \Pi_e \text{ mode for } z < 0$$

$$Y_{2m} = Z_{2m}^{-1} = \frac{k_2}{k_0} \frac{1 - n_{1x}^2}{n_{1z}} Y_0 \qquad = \text{wave admittance for } \Pi_h \text{ mode for } z > 0$$

$$Y_{2e} = Z_{2e}^{-1} = \frac{\beta n_{2z} Y_0}{k_0[1 - (\beta^2/k_2^2)n_{2x}^2]}$$

$$= \text{wave admittance for } \Pi_e \text{ mode for } z > 0$$

$$Y_0 = Z_0^{-1} = \left(\frac{\epsilon_0}{\mu_0}\right)^{1/2}$$

The wave admittances are defined as the ratio of the transverse magnetic field to the mutually orthogonal transverse electric field. These equations take on a particularly significant form when suitable equivalent transmission-line voltages and currents are introduced. The equivalent voltages are chosen proportional to the total transverse electric field, while the equivalent currents are chosen proportional to the total mutually perpendicular transverse magnetic field. The characteristic impedance of each line may be arbitrarily chosen and so we will choose the characteristic impedances equal to the corresponding mode wave impedances. The equivalent voltages may be arbitrarily chosen subject to the restriction that the power carried by a given mode be equal to $K(V_iV_i^*/Z_i)$, where the constant of proportionality $K$ is the same for all modes, $V_i$ is the voltage on the $i$th line, and $Z_i$ is the characteristic impedance of the $i$th line. The following choice of equivalent voltages and currents leads to a very simple equivalent circuit for the interface:

| Mode | Voltage | Current |
|---|---|---|
| | Input Side, $z < 0$ | |
| $\Pi_h$ | $V_1 = (1 - N_x^2)Z_{1m}(A_1 - A_0)$ | $I_1 = (1 - N_x^2)(-A_1 - A_0)$ |
| $\Pi_e$ | $V_2 = (1 - N_x^2)(B_0 + B_1)$ | $I_2 = (1 - N_x^2)(B_0 - B_1)Y_{1e}$ |
| | Output Side, $z > 0$ | |
| $\Pi_h$ | $V_3 = \kappa_2 Z_{2m}(1 - n_{1x}^2)(C_0 - C_1)$ | $I_3 = \kappa_2(1 - n_{1x}^2)(C_1 + C_0)$ |
| $\Pi_e$ | $V_4 = (1 - n_{1x}^2)(D_1 + D_0)$ | $I_4 = (1 - n_{1x}^2)(D_0 - D_1)Y_{2e}$ |

Replacing $\beta n_{2x}$ by $k_2 n_{1x}$ and $\beta n_{2y}$ by $k_2 n_{1y}$ from (118), and also introducing the above voltages and currents, the set of equations (119) becomes

$$I_1 = -I_3 \tag{120a}$$
$$V_2 = V_4 \tag{120b}$$
$$\frac{N_x N_y}{1 - N_x^2} I_1 + I_2 = -\frac{n_{1x} n_{1y}}{1 - n_{1x}^2} I_3 - I_4 \tag{120c}$$
$$V_1 - \frac{N_x N_y}{1 - N_x^2} V_2 = V_3 - \frac{n_{1x} n_{1y}}{1 - n_{1x}^2} V_4 \tag{120d}$$

Equation (120*a*) shows that lines 1 and 3 are series-coupled, while (120*b*) shows that lines 2 and 4 are parallel-coupled. A suitable equivalent

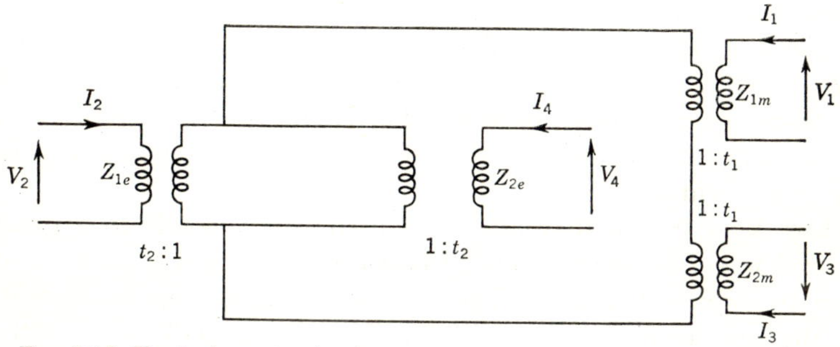


FIG. 3.14. Equivalent circuit of a uniaxial anisotropic dielectric interface.

circuit which gives rise to a set of equations of the above form is illustrated in Fig. 3.14. The transformers illustrated are considered as ideal ones with turns ratios $t_1$:1 and $t_2$:1. Only one of the turns ratios is independent, for example, $t_2$:1 may be chosen as 1:1.† The above equivalent circuit obeys the following equations:

$$I_1 = -I_3 \tag{121a}$$
$$V_2 = V_4 \tag{121b}$$
$$I_2 + I_4 = \frac{t_1}{t_2} I_3 = -\frac{t_1}{t_2} I_1 \tag{121c}$$
$$V_1 - V_3 = \frac{t_1}{t_2} V_2 = \frac{t_1}{t_2} V_4 \tag{121d}$$

Equations (120*c*) and (120*d*) may be put into the form of (121*c*) and

† An ideal transformer is one with infinite primary and secondary inductance and unity coupling. The voltages and currents on side 1 with $n_1$ turns are always related to those on side 2 with $n_2$ turns as follows: $V_1/V_2 = I_2/I_1 = n_1/n_2$. Thus, if the secondary is open-circuited, the primary is effectively an open circuit also because of its infinite self-impedance.

(121*d*) by using (120*a*) and (120*b*). Thus, for (120*c*), we get

$$I_2 + I_4 = \left( \frac{N_x N_y}{1 - N_x^2} - \frac{n_{1x} n_{1y}}{1 - n_{1x}^2} \right) I_3$$

and, hence, comparing with (121*c*) shows that

$$\frac{t_1}{t_2} = \frac{N_x N_y}{1 - N_x^2} - \frac{n_{1x} n_{1y}}{1 - n_{1x}^2} \tag{122}$$

Equation (120*d*) gives the same result.

If we are interested in the solution for a plane perpendicular-polarized wave incident from the region $z < 0$, the ratio of $A_0$ and $B_0$ must be chosen as follows: $A_0/B_0 = N_x N_z / N_y$, in order to make $E_z = 0$. Similarly for an incident parallel-polarized wave $H_z = 0$, and we must choose $A_0/B_0 = N_y / N_x N_z$. When either $N_x$ or $N_y$ equals zero, Eqs. (119) show that there is no coupling between the $\Pi_e$ and the $\Pi_h$ modes. In any case, the amount of coupling is small when $\kappa_1$ and $\kappa_2$ do not differ by more than 50 per cent or so.

When $N_y = 0$, the $\Pi_e$ mode, which now corresponds to a parallel-polarized wave, is directly coupled between lines 2 and 4, with no interaction from lines 1 and 3. The normalized input impedance looking into the anisotropic medium is $Z_{2e}/Z_{1e}$. Using (113) and (118) to eliminate $\beta n_{2z} = \beta(1 - n_{2x}^2)^{1/2}$ gives

$$Z_{\text{in}} = \frac{Z_{2e}}{Z_{1e}} = \frac{(\kappa_2 - \sin^2 \theta_i)^{1/2}}{(\kappa_1 \kappa_2)^{1/2} \cos \theta_i} \tag{123}$$

where $\theta_i$ is the angle of incidence in the $xz$ plane measured from the interface normal; that is, $\sin \theta_i = N_x$. A Brewster angle exists such that $Z_{\text{in}} = 1$, and no reflection takes place. From (123) this angle is given by

$$\sin \theta_b = \left[ \frac{\kappa_2(\kappa_1 - 1)}{\kappa_1 \kappa_2 - 1} \right]^{1/2} \tag{124}$$

and differs from that derived in Sec. 3.5 for an isotropic dielectric medium. A value of $\theta_b$ less than 45° may exist. For isotropic dielectrics $\theta_b$ is always greater than 45° in the less dense medium.

## 3.8. TEM Waves in a Ferrite Medium

It is not our purpose here to give a theory for the magnetic properties of ferrites. Nevertheless, considerable insight into the electromagnetic-wave-propagation properties of ferrites is obtained by considering a simple semiclassical model of the magnetization mechanism which takes place in ferromagnetic materials. Ordinary ferromagnetic materials such as iron and nickel are of little use at microwave frequencies because of their high conductivities. A class of materials known collectively as ferrites has, however, been developed which is quite suitable for use at microwave frequencies. Ferrites are a ceramiclike material with specific resistivities over a million times greater than those of metals, with relative permeabilities ranging up to several thousand and relative dielectric constants in the range 5 to 25. The magnetic properties of these materials arise principally from the magnetic moment associated with the

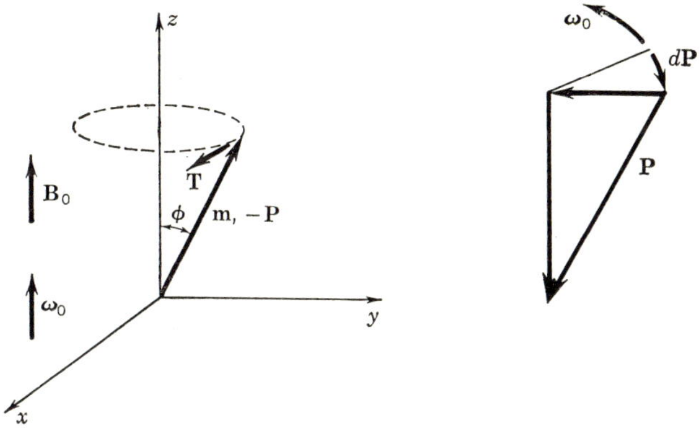


FIG. 3.15. Free precession of spinning electron.

spinning electron. By treating the spinning electron as a gyroscopic top, a classical picture of the magnetization process and, in particular, of the anisotropic magnetic properties may be obtained.†

The electron has an angular momentum $\mathbf{P}$ equal to $\frac{1}{2}\hbar$ or $0.527 \times 10^{-34}$ joule-meters, where $\hbar$ is Planck's constant divided by $2\pi$. The magnetic moment associated with the spinning electron is one Bohr magneton or $m = e\hbar/2w = 9.27 \times 10^{-24}$ ampere-meter², where $m$ is the magnetic dipole moment and $w$ is the mass of the electron. The angular-momentum vector $\mathbf{P}$ and magnetic moment $\mathbf{m}$ are antiparallel for the electron. The ratio of the magnetic moment to the angular momentum is called the gyromagnetic ratio $\gamma$, that is, $\gamma = m/P$. If a spinning electron is located in a constant magnetic field $\mathbf{B}_0$, and $\mathbf{m}$ and $\mathbf{B}_0$ are not collinear, a torque $\mathbf{T} = \mathbf{m} \times \mathbf{B}_0$ is exerted on the system as in Fig. 3.15. The rate of change of angular momentum is equal to the applied torque, and

† The material presented here is based essentially on the following references:

G. P. Harnwell, "Principles of Electricity and Electromagnetism," 2d ed., chaps. 10 and 11, McGraw-Hill Book Company, Inc., New York, 1949.

C. Kittel, On the Theory of Ferromagnetic Resonance Absorption, *Phys. Rev.*, vol. 73, pp. 155–161, January, 1948.

D. Polder, On the Theory of Ferromagnetic Resonance, *Phil. Mag.*, vol. 40, pp. 99–114, January, 1949.

C. L. Hogan, The Microwave Gyrator, *Bell System Tech. J.*, vol. 31, pp. 1–31, January, 1952.

hence†

$$\mathbf{T} = \frac{d\mathbf{P}}{dt} = \mathbf{m} \times \mathbf{B}_0 = \boldsymbol{\omega}_0 \times \mathbf{P} = -\gamma^{-1}\boldsymbol{\omega}_0 \times \mathbf{m} \tag{125}$$

or
$$mB_0 \sin \phi = \omega_0 P \sin \phi = \gamma^{-1}\omega_0 m \sin \phi$$

where $\boldsymbol{\omega}_0$ is the vector precession angular velocity directed along $\mathbf{B}_0$, and $P \sin \phi$ is the component of $\mathbf{P}$ which is changing in direction. At first

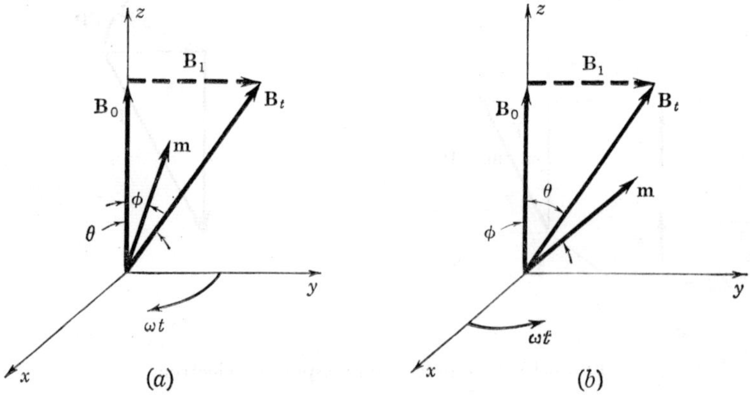


FIG. 3.16. Forced precession of spinning electron.

sight, Fig. 3.15 appears to give rotation in a direction opposite to that which should be produced by the acting torque, but this is only because $\mathbf{m}$ and $\mathbf{P}$ are antiparallel. From (125) the following expression for the free-precession angular velocity (Larmor frequency) is obtained:

$$\omega_0 = \gamma B_0 \tag{126}$$

For free precession, the precession angle $\phi$ is arbitrary.

To obtain a feeling for what happens when a spinning electron is placed in a field $\mathbf{B}_t$ which is a static field plus a small sinusoidally varying component, consider the following situation. With reference to Fig. 3.16, let $\mathbf{B}_0$ be a static field along the $z$ axis, and let $\mathbf{B}_1$ be a constant counterclockwise-rotating vector in the $xy$ plane. The rate of rotation is $\omega$, and, hence, $\mathbf{B}_1$ is the same type of field which would exist in a plane negative circularly polarized wave propagating in the direction of $\mathbf{B}_0$. The total field vector $\mathbf{B}_t$ rotates about the $z$ axis at a rate $\omega$ and with a constant angle $\theta$. When equilibrium has been established between the field and the spinning electron, the magnetic dipole axis must precess

† In the literature some authors include the negative sign in the definition of $\gamma$, whereas others omit the negative sign completely. This leads to an inconsistency which amounts to changing the sign in the off-diagonal terms in the susceptibility tensor or reversing the direction of the magnetic field $\mathbf{B}_0$.

around $\mathbf{B}_0$ in synchronism with $\mathbf{B}_t$. Thus the precession angle $\phi$ must be less than $\theta$ in order to obtain a torque acting in the proper direction to produce counterclockwise precession. The equation of motion (125) gives $mB_t \sin(\theta - \phi) = \omega P \sin \phi$ or $\gamma B_t(\sin \theta \cos \phi - \cos \theta \sin \phi) = \omega \sin \phi$. Replacing $B_t \sin \theta$ by $B_1$ and $B_t \cos \theta$ by $B_0$ and solving for $\sin \phi$ gives

$$\sin \phi = \frac{\gamma B_1}{[(\gamma B_0 + \omega)^2 + (\gamma B_1)^2]^{1/2}}$$

The component of $\mathbf{m}$ in the $xy$ plane which rotates in synchronism with $\mathbf{B}_1$ is $m_- = m \sin \phi = m \cos \phi \tan \phi = m_0 \tan \phi$, where $m_0$ is the component of $\mathbf{m}$ along $\mathbf{B}_0$. Substituting for $\tan \phi$ gives

$$m_- = \frac{\gamma \mu_0 H_1 m_0}{\gamma B_0 + \omega} \tag{127}$$

If we have a positive circularly polarized field $\mathbf{B}_1$, the precession angle $\phi$ must be greater than $\theta$, as in Fig. 3.16*b*, in order to produce a torque which will make $\mathbf{m}$ precess about $\mathbf{B}_0$ in step with the total field $\mathbf{B}_t$. A significant feature is at once apparent: that, since the precession angle $\phi$ is different for the two cases, the effective permeability for negative and positive circularly polarized waves propagating in the direction along $\mathbf{B}_0$ will be different. For the latter case the equation of motion gives

$$\sin \phi = \frac{\gamma B_1}{[(\gamma B_0 - \omega)^2 + (\gamma B_1)^2]^{1/2}}$$

and hence the magnetization $m_+$ associated with a positive circularly polarized field $\mathbf{H}_1$ is

$$m_+ = \frac{\gamma \mu_0 m_0 H_1}{\gamma B_0 - \omega} \tag{128}$$

A ferrite material will consist of a collection of spinning electrons. If the effective number of spinning electrons per unit volume is $N$, the magnetic dipole polarization per unit volume is $\mathbf{M} = N\mathbf{m}$. If a constant uniform magnetic field $\mu_0 \mathbf{H}_0$ is applied externally to an arbitrary-shaped sample of ferrite material, the interior field in the sample will be nonuniform and will differ from $\mathbf{H}_0$ because of the demagnetizing effects set up by the free magnetic poles on the surface of the sample. For ellipsoidal-shaped samples (and, of course, the degenerate forms such as spheroids, spheres, disks, and long rods) the interior field is uniform when a uniform exterior field is applied. The interior field may be evaluated quite readily and is conveniently expressed in the following form:†

$$\mathbf{H}_i = \mathbf{H}_o - \bar{\mathbf{L}} \cdot \mathbf{M}$$

† J. A. Stratton, "Electromagnetic Theory," chap. 4, McGraw-Hill Book Company, Inc., New York, 1941.

where $\bar{\mathbf{L}}$ is a suitable demagnetization dyadic. When the coordinate axis coincides with the principal axis of the ellipsoid, $\bar{\mathbf{L}}$ is a diagonal dyadic with elements $L_{xx}$, $L_{yy}$, and $L_{zz}$. The sum of these elements is equal to

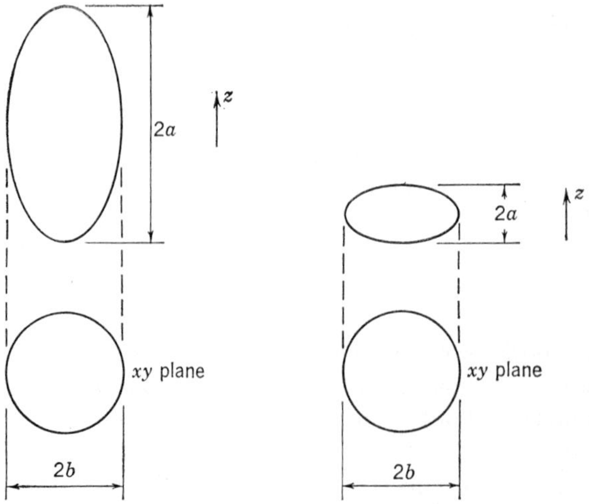


(a) Prolate spheroid (b) Oblate spheroid

FIG. 3.17. Illustration of spheroid-shaped samples.

unity. For prolate and oblate spheroids as illustrated in Fig. 3.17 we have, respectively,

$$L_{zz} = \frac{b^2}{2a^2}\frac{1}{e^3}\left(\ln\frac{1+e}{1-e} - 2e\right) \tag{129a}$$

$$L_{xx} = L_{yy} = \tfrac{1}{2}(1 - L_{zz}) \tag{129b}$$

where $$e = \left(1 - \frac{b^2}{a^2}\right)^{1/2}$$

and $$L_{zz} = \frac{ab^2}{(b^2 - a^2)^{3/2}}\left[\frac{(b^2 - a^2)^{1/2}}{a} - \tan^{-1}\frac{(b^2 - a^2)^{1/2}}{a}\right] \tag{130}$$

with $L_{xx}$ and $L_{yy}$ given by (129b) again. For a sphere,

$$L_{xx} = L_{yy} = L_{zz} = \tfrac{1}{3}$$

For an ellipsoidal sample with a constant external field $H_o$ applied along the $z$ axis, the effective internal $H$ field is then $H_i = H_o - L_{zz}M$. The total magnetic induction field in the sample when both a static internal field $\mathbf{H}_i$ and an internal time-varying field $\mathbf{H}$ exist in the sample is $\mathbf{B} = \mu_0(\mathbf{H}_i + \mathbf{H} + \mathbf{M})$. This is not the local induction field which each spinning electron sees, however. Each spinning electron sees an induction field arising from the applied field, the demagnetization field which depends on sample size and shape, and the field due to all the neighboring spins. A reasonable approximation to make, in practice, is to assume that the partial field contributed by all the neighboring electron spins is collinear and proportional to the total magnetization **M**. This partial field will not produce a torque on the magnetic moment of the electron, provided the static field is large enough to saturate the sample so that all the electron spins are aligned. It is only the fields which are perpendicular to **M** which will produce a torque. In view of these considerations, we may treat the ferrite sample as a single magnetic dipole of strength $\mathbf{M}_s$ per unit volume where the subscript $s$ signifies the saturation dipole moment. The field producing a torque on the system is the partial field $\mathbf{B} = \mu_0(\mathbf{H}_i + \mathbf{H})$. The problem is formally the same as that of a single spinning electron with **m** replaced by $\mathbf{M}_s$ and with $\mathbf{H}_i$ and **H** as the internal applied fields. If $\mathbf{H}_i$ is directed along the $z$ axis, and $|\mathbf{H}| \ll |\mathbf{H}_i|$, then the component of magnetization $M_0$ along the $z$ axis is essentially equal to the saturation dipole moment $M_s$.

For a negative circularly polarized internal field $\mu_0\mathbf{H}_1$ in the $xy$ plane, the negative circularly polarized component of magnetization will be

$$M_- = \frac{\gamma\mu_0 H_1 M_0}{\omega_0 + \omega} \tag{131a}$$

where $\omega_0 = \gamma\mu_0 H_i$. Hence, the intrinsic susceptibility for a negative circularly polarized field is

$$\chi_- = \frac{\mu_0 M_0}{\omega_0 + \omega} \tag{131b}$$

while the permeability is given by

$$\mu_- = \mu_0 \frac{\gamma\mu_0 M_0 + \omega_0 + \omega}{\omega_0 + \omega} \tag{131c}$$

For a positive circularly polarized field, the corresponding quantities are

$$\chi_+ = \frac{\gamma\mu_0 M_0}{\omega_0 - \omega} \tag{132a}$$

$$\mu_+ = \mu_0 \frac{\gamma\mu_0 M_0 + \omega_0 - \omega}{\omega_0 - \omega} \tag{132b}$$

As the applied angular frequency $\omega$ approaches the free-precession angular frequency $\omega_0$, the precession angle $\phi$ approaches $\pi/2$, and the magnetization $M_+$ approaches† $Nm = M_s$. For circularly polarized fields the effective permeabilities are scalar constants.

† Damping forces are always present, and so this never happens in practice. The resonant frequency is $\omega_0$. In terms of the external fields, the resonant frequency is given by a much more involved relation, which depends on sample shape through the demagnetization factors. For arbitrary-shaped samples the internal field is not uniform, in general, and ferromagnetic resonance does not occur simultaneously throughout the whole sample.

The solution for the magnetization when a time-varying field $(\mathbf{a}_x H_x + \mathbf{a}_y H_y)e^{j\omega t}$ in addition to the static field $\mathbf{a}_z B_0$ is applied is readily found from the equation of motion

$$\frac{d\mathbf{M}_s}{dt} = -\gamma \mathbf{M}_s \times \mathbf{B} \tag{133}$$

where $\mathbf{M}_s = N\mathbf{m}$ and is the dipole moment per unit volume. If a solution for $\mathbf{M}_s$ of the form $\mathbf{a}_z M_0 + (\mathbf{a}_x M_x + \mathbf{a}_y M_y + \mathbf{a}_z M_z)e^{j\omega t}$ is assumed, the following three equations are obtained from (133):

$$j\omega M_x = \gamma(\mu_0 M_0 H_y - M_y \mu_0 H_i + \mu_0 M_z H_y e^{j\omega t}) \tag{134a}$$
$$j\omega M_y = \gamma(M_x \mu_0 H_i - \mu_0 M_0 H_x - \mu_0 M_z H_x e^{j\omega t}) \tag{134b}$$
$$j\omega M_z = \mu_0 \gamma(M_y H_x - M_x H_y) e^{j\omega t} \tag{134c}$$

If $H_x$ and $H_y$ are very small compared with $H_i$, then $M_z$ is small in comparison with $M_x$ and $M_y$. Also from (134*c*) it is seen that $M_z$ varies at a rate $\omega$ relative to the applied field, and, hence, we may neglect $M_z$ for a first approximation. For the same reason the terms multiplied by $e^{j\omega t}$ in (134*a*) and (134*b*) may be neglected. With this simplification the solutions for the time-varying magnetization components obtained from (134) are

$$M_x = \frac{\mu_0 \gamma \omega_0 M_0 H_x + j\omega\gamma\mu_0 M_0 H_y}{\omega_0^2 - \omega^2} \tag{135a}$$

$$M_y = \frac{\mu_0 \gamma \omega_0 M_0 H_y - j\omega\gamma\mu_0 M_0 H_x}{\omega_0^2 - \omega^2} \tag{135b}$$

$$M_z = 0 \tag{135c}$$

Thus the effective susceptibility tensor or dyadic is

$$\bar{\bar{\chi}}_m = \mathbf{a}_x\mathbf{a}_x \frac{\gamma\mu_0\omega_0 M_0}{\omega_0^2 - \omega^2} + \mathbf{a}_x\mathbf{a}_y \frac{j\omega\gamma\mu_0 M_0}{\omega_0^2 - \omega^2} - \mathbf{a}_y\mathbf{a}_x \frac{j\omega\gamma\mu_0 M_0}{\omega_0^2 - \omega^2} + \mathbf{a}_y\mathbf{a}_y \frac{\gamma\mu_0\omega_0 M_0}{\omega_0^2 - \omega^2} \tag{136}$$

and is antisymmetric. This latter property accounts for the nonreciprocal properties of wave propagation in ferrites. The permeability dyadic is given by

$$\bar{\bar{\mu}} = \mu_0(\bar{\bar{\mathbf{I}}} + \bar{\bar{\chi}}_m) \tag{137}$$

where $\bar{\bar{\mathbf{I}}}$ is the unit dyadic or idemfactor. When $H_y = 0$ and

$$H_x = \text{Re}\,(H_1 e^{j\omega t}) = H_1 \cos \omega t$$

the magnetization components are given by

$$M_x = \chi_{xx} \cos \omega t \qquad \text{and} \qquad M_y = \frac{\omega}{\omega_0} \chi_{xx} \sin \omega t$$

where $\chi_{xx} = \gamma\mu_0\omega_0 M_0/(\omega_0^2 - \omega^2)$. The two components $M_x$ and $M_y$ constitute a positive elliptically polarized magnetization vector, as in Fig. 3.18, and this is the projection of the total precessing magnetization vector onto the $xy$ plane. The projection of **M** along the $z$ axis is readily seen to result in a small $z$ component of magnetization varying at a rate $2\omega$ in accordance with (134$c$). Under the influence of large time-varying fields, harmonics are generated which have appreciable amplitudes, and ferrites may be employed as harmonic generators.†

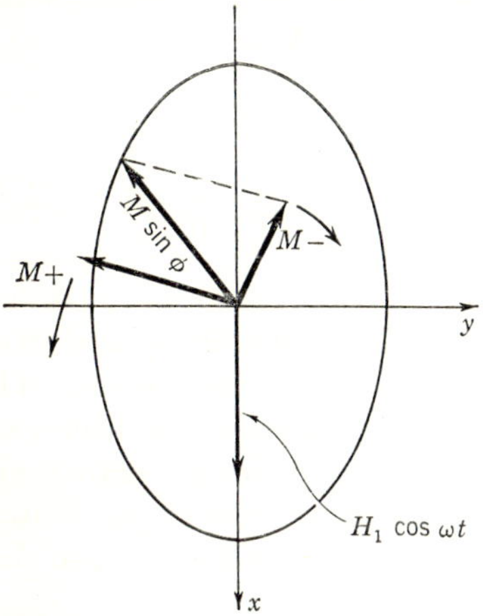


FIG. 3.18. Projection of **M** on $xy$ plane.

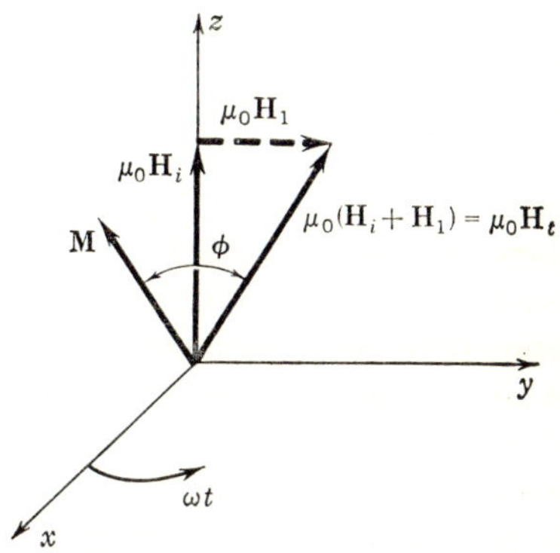


FIG. 3.19. Space relation between **M** and $\mu_0\mathbf{H}_t$ for $\omega > \omega_0$.

The magnetization vector may be decomposed into a positive circularly polarized component $M_+ = (\chi_{xx}/2)(1 + \omega/\omega_0)$ and a negative circularly polarized component $M_- = (\chi_{xx}/2)(1 - \omega/\omega_0)$ in the $xy$ plane. Similarly $\mathbf{H}_x$ may be decomposed into oppositely rotating vectors each of amplitude $H_1/2$. The effective susceptibilities for the two circularly polarized components of $\mathbf{H}_x$ are $\chi_+ = \chi_{xx}(1 + \omega/\omega_0) = \gamma\mu_0 M_0/(\omega_0 - \omega)$ and $\chi_- = \chi_{xx}(1 - \omega/\omega_0) = \gamma\mu_0 M_0/(\omega_0 + \omega)$, and have the same form as given earlier by (132$a$) and (131$b$). Below the resonant frequency, that is, $\omega < \omega_0$, a magnetic field $\mathbf{H}_x$ results in a magnetization along the $y$ axis which lags the applied field by one quarter of a period.

Above the resonant frequency, i.e., for $\omega > \omega_0$, $\chi_{xx}$ is negative and, hence, $\chi_+$ is negative but $\chi_-$ is still positive. This means that **m** and the positive circularly polarized component of the field are 180° out of phase as indicated in Fig. 3.19 rather than as shown in Fig. 3.16$b$. This

† J. L. Melchor et al., Microwave Frequency Doubling from 9 to 18 kmc in Ferrites, *Proc. IRE*, vol. 45, pp. 643–646, May, 1957.

See also Frequency Doubling in Ferrites, *J. Appl. Phys.*, vol. 27, p. 188, 1956, by the same authors.

orientation is permissible since it gives a torque to cause **M** to precess in the same direction as $\mathbf{H}_t$. For the negative circularly polarized field, **M** and $\mathbf{H}_t$ are always in phase. This change in polarity of $\chi_+$ for $\omega > \omega_0$ causes the major axis of the magnetization ellipse to be along the $y$ axis when a linearly polarized sinusoidally varying field is applied along the $x$ axis. Below the resonant frequency $\mu_+$ is greater than $\mu_-$, while the opposite is true for $\omega > \omega_0$.

For practical ferrite materials, damping forces which tend to reduce the precession angle $\phi$ are always present. The general form of the permeability matrix must now be represented as follows:

$$[\mu] = \begin{bmatrix} \mu' - j\mu'' & j(\kappa' - j\kappa'') & 0 \\ -j(\kappa' - j\kappa'') & \mu' - j\mu'' & 0 \\ 0 & 0 & \mu_0 \end{bmatrix} = \begin{vmatrix} \mu & j\kappa & 0 \\ -j\kappa & \mu & 0 \\ 0 & 0 & \mu_0 \end{vmatrix} \tag{138}$$

where $\mu = \mu' - j\mu''$, etc. The terms which are double-primed arise because of the damping forces which give rise to finite losses. The damping forces are proportional to the precession angle $\phi$ and, hence, are greater for a positive circularly polarized field than for a negative circularly polarized field. This leads to selective absorption of the positive circularly polarized component, particularly so if $\omega$ is close to $\omega_0$ so that the precession angle $\phi$ is large.

Our earlier discussion on the magnetization resulting from circularly polarized fields leads us to expect that the only solutions to Maxwell's equations for plane waves propagating along the direction of the applied static field $\mathbf{H}_i$ will be circularly polarized waves, since for such waves the effective permeability is a scalar quantity. A linearly polarized wave may be decomposed into positive and negative circularly polarized waves, and, since the effective permeability for the two polarizations are different, the propagation constants will also be different. Thus, as the wave propagates along the direction of $\mathbf{H}_i$, a linearly varying phase difference between the two polarizations results. This causes the plane of polarization of the resultant wave to rotate, a phenomenon known as Faraday rotation. In ferrites Faraday rotation is a nonreciprocal effect, since the plane of polarization always rotates in the same direction looking in the direction of $\mathbf{H}_i$ whether the wave is propagating parallel or antiparallel to $\mathbf{H}_i$.

Consider an unbounded ferrite medium with an internal static magnetic field $\mathbf{H}_i$ applied along the $z$ axis. For plane TEM waves propagating in the $z$ direction, Maxwell's equations reduce to $\Gamma\mathbf{E}_t e^{-\Gamma z} = -j\omega\mathbf{a}_z \times \mathbf{B}_t e^{-\Gamma z}$ and $\mathbf{H}_t e^{-\Gamma z} = j\omega\epsilon\mathbf{a}_z \times \mathbf{E}_t e^{-\Gamma z}$, where a propagation constant $\Gamma$ has been assumed and $\mathbf{E}_t$, etc., are vectors in the $xy$ plane. Introducing the relation $\bar{\mathbf{\mu}} \cdot \mathbf{H}_t = \mathbf{B}_t$, we get

$$\Gamma^2\bar{\mathbf{\mu}} \cdot \mathbf{H}_t = \Gamma^2\mathbf{B}_t = j\omega\epsilon\Gamma\bar{\mathbf{\mu}} \cdot \mathbf{a}_z \times \mathbf{E}_t = j\omega\epsilon\bar{\mathbf{\mu}} \cdot \mathbf{a}_z \times (-j\omega\mathbf{a}_z \times \mathbf{B}_t) = -\omega^2\epsilon\bar{\mathbf{\mu}} \cdot \mathbf{B}_t \quad (139)$$

Equating the $x$ and $y$ components separately gives

$$(\Gamma^2 + k^2)B_x + j\omega^2\epsilon\kappa B_y = 0 \quad (140a)$$
$$-j\omega^2\epsilon\kappa B_x + (\Gamma^2 + k^2)B_y = 0 \quad (140b)$$

where $k^2 = \omega^2\epsilon\mu$. For a solution, the determinant must vanish; hence, $(\Gamma^2 + k^2)^2 - (\omega^2\epsilon\kappa)^2 = 0$, and thus the solutions for $\Gamma$ are

$$\Gamma_{\pm} = j\omega\epsilon^{1/2}(\mu \pm \kappa)^{1/2} \quad (141)$$

For the root $\Gamma_+$, (140$b$) gives $B_y = -jB_x$, which represents a positive circularly polarized wave with an effective permeability $\mu + \kappa$. The root $\Gamma_-$ gives $B_y = jB_x$, which is a negative circularly polarized wave with an effective permeability $\mu - \kappa$. The effective permeabilities are of the same form as those derived earlier for circularly polarized fields, provided the losses are negligible.

At the plane $z = 0$, let the time-varying magnetic field be

$$\mathbf{B} = \mathbf{a}_x 2B_1 \operatorname{Re} e^{j\omega t}$$

where Re signifies the real part. The field $\mathbf{B}$ may be decomposed into a positive circularly polarized field

$$\mathbf{B}_+ = B_1 \operatorname{Re} (\mathbf{a}_x e^{j(\omega t - \beta_+ z)} - j\mathbf{a}_y e^{j(\omega t - \beta_+ z)}) \quad (142a)$$

and a negative circularly polarized field

$$\mathbf{B}_- = B_1 \operatorname{Re} (\mathbf{a}_x e^{j(\omega t - \beta_- z)} + j\mathbf{a}_y e^{j(\omega t - \beta_- z)}) \quad (142b)$$

where $j\beta_{\pm} = \Gamma_{\pm}$. At a plane $z = l$, the resultant field is

$$\begin{aligned}\mathbf{B} &= B_1 \operatorname{Re} [\mathbf{a}_x(e^{-j\beta_+ l} + e^{-j\beta_- l}) - j\mathbf{a}_y(e^{-j\beta_+ l} - e^{-j\beta_- l})]e^{j\omega t} \\ &= 2B_1 \operatorname{Re} e^{j\omega t - j(\beta_+ + \beta_-)l/2} \left[\mathbf{a}_x \cos (\beta_+ - \beta_-) \frac{l}{2} - \mathbf{a}_y \sin (\beta_+ - \beta_-) \frac{l}{2}\right]\end{aligned}$$

If the losses are negligible, $\beta_{\pm}$ is real, and the resultant field makes an angle $\theta$ given by

$$\theta = (\beta_- - \beta_+) \frac{l}{2} \quad (143)$$

with the $x$ axis, and, hence, the plane of polarization has been rotated through an angle $\theta$. The Faraday rotation per unit length is thus $(\frac{1}{2})(\beta_- - \beta_+)$. For $\omega > \omega_0$, the effective permeability for the positive circularly polarized wave is very small and $\beta_- > \beta_+$; hence, the Faraday rotation is in a clockwise sense about $\mathbf{H}_i$. Below the resonant frequency,

$\beta_+ > \beta_-$ and the sense of rotation is reversed. For a wave propagating in the negative $z$ direction, the sign in front of $\beta_\pm$ and $\mathbf{a}_y$ in (142*a*) and (142*b*) must be changed. A similar analysis then shows that the plane of polarization again undergoes rotation through an angle given by (143). This nonreciprocal Faraday-rotation property has been used to construct nonreciprocal microwave components called gyrators, isolators, or circulators.† It should be noted that, for $\omega > \omega_0$ and a sufficiently strong applied static field, $\mu_+$ may be negative; hence, $\beta_+$ is imaginary, and the positive circularly polarized wave does not propagate. In practice, finite losses are always present, and so $\Gamma_+$ will always have an imaginary part, although this imaginary part may be very small under certain conditions. The region $\omega \gg \omega_0$ and weak applied static fields are favored for gyrators because of the large Faraday rotation together with small losses which may be obtained under these conditions.

## REFERENCES

*TEM waves in general*

3.1. J. A. Stratton: "Electromagnetic Theory," McGraw-Hill Book Company, Inc., New York, 1941.

3.2. S. Ramo and J. R. Whinnery: "Fields and Waves in Modern Radio," 2d ed., John Wiley & Sons, Inc., New York, 1953.

*Wave matrices and wave-impedance concepts*

3.3. C. G. Montgomery, R. H. Dicke, and E. M. Purcell: "Principles of Microwave Circuits," vol. 8 of MIT Rad. Lab. Series, McGraw-Hill Book Company, Inc., New York, 1948.

3.4. E. F. Bolinder: Note on the Matrix Representation of Linear Two Port Networks, *IRE Trans.*, vol. CT-4, p. 337, December, 1957.

3.5. S. A. Schelkunoff: The Impedance Concept and Its Application to Problems of Reflection, Refraction, Shielding and Power Absorption, *Bell System Tech. J.*, vol. 17, pp. 17–48, January, 1938.

3.6. H. G. Booker: The Elements of Wave Propagation Using the Impedance Concept, *Jour. IEE (London)*, vol. 94, part IIIA, pp. 171–198, May, 1947.

*TEM waves in anisotropic dielectric media*

3.7. G. Joos: "Theoretical Physics," 2d ed., Blackie & Sons, Ltd., Glasgow, 1951.

3.8. J. Valasek: "Introduction to Theoretical and Experimental Optics," John Wiley & Sons, Inc., New York, 1949.

3.9. R. E. Collin: A Simple Artificial Anisotropic Dielectric Medium, *IRE Trans.*, vol. MTT-6, pp. 206–209, April, 1958.

The literature on ferrites is extensive. Two particularly good sources of review and original papers on the subject are:

3.10. Ferrites Issue, *Proc. IRE*, vol. 44, October, 1956.

3.11. *IRE Trans.*, vol. MTT-6, January, 1958.

† Hogan, *loc. cit.*

## PROBLEMS

**3.1.** Find the solutions for a uniform cylindrical TEM wave with components $E_\theta$, $H_z$. Obtain the same solution by using the transformation $\mathbf{E}' = Z_0\mathbf{H}$, $\mathbf{H}' = -Y_0\mathbf{E}$, where $\mathbf{E}$, $\mathbf{H}$ is the solution for a uniform cylindrical TEM wave with components $H_\theta$, $E_z$.

*Answer*

$$E_\theta = AH_1{}^2(k_0r)$$
$$H_z = \frac{jk}{\omega\mu_0} AH_0{}^2(k_0r)$$

**3.2.** Consider an infinite biconical transmission line as illustrated. For TEM waves a solution for $\mathbf{E}_t$ of the form $\mathbf{E}_t = -\nabla_t\psi(\theta,\phi)g(r)$ may be found. In the spherical coordinate system $r\theta\phi$, the function $\psi$ is a solution of

$$\sin\theta \frac{\partial}{\partial\theta}\left(\sin\theta \frac{\partial\psi}{\partial\phi}\right) + \frac{\partial^2\psi}{\partial\phi^2} = 0$$

Make the transformation $u = \ln\tan(\theta/2)$, and show that the function $\psi$ is now a solution of $\partial^2\psi/\partial u^2 + \partial^2\psi/\partial\phi^2 = 0$. The above transformation maps the cross sections $S_1$ and $S_2$ into the rectangular coordinate plane $u\phi$ as illustrated. A solution to Laplace's equation for the field between $S_1$ and $S_2$ in the $u\phi$ plane thus provides the solution for $\psi$. Show that, when the cross sections $S_1$ and $S_2$ are the same, then the mapping is symmetrical about $u = 0$. When $S_1$ and $S_2$ are circles defined by $\theta = \theta_0$ and $\theta = \pi - \theta_0$, find a solution for the capacitance between $S_1$ and $S_2$ in the $u\phi$ plane. (There is no fringing effect since the mapping is periodic in $\phi$.) Using the relation $Z_c = (v_cC)^{-1}$, where $v_c = (\mu_0\epsilon_0)^{-1/2}$ is the velocity of light, and $C$ is the capacitance per unit length, obtain the characteristic impedance $Z_c$ of the biconical line.

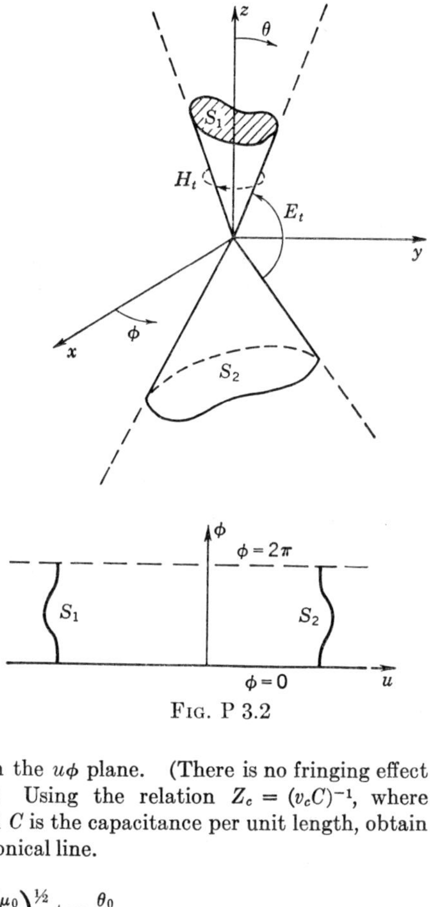

FIG. P 3.2

*Answer*

$$Z_c = \frac{1}{\pi}\left(\frac{\mu_0}{\epsilon_0}\right)^{1/2} \tan\frac{\theta_0}{2}$$

**3.3.** Find the solutions for the field of a TEM wave with components $E_\theta$, $H_\phi$ for a biconical line with circular cross section as in Prob. 3.2. Integrate $H_\phi$ around one conductor to find the total radial current $I$ on the conductor. Also integrate $E_\theta$ along a path $r$ = constant from one conductor to the next to obtain the potential difference $V$. Show that $V/I = Z_c$ is the same as in Prob. 3.2. Show that the line integral of $E_\theta$ along any closed curve on the surface of a sphere $r = r_0$ is zero.

**3.4.** Show that the impedance matrix is always symmetrical independently of whether normalized or unnormalized amplitudes are used, provided the scattering matrix is symmetrical when normalized amplitudes are used.

**3.5.** Consider an infinite anisotropic dielectric medium located in the half space $z > 0$. Let the principal dielectric constants be $\kappa_1$ along the $x$ axis and $\kappa_2$ along the $y$ and $z$ axes. For a parallel-polarized wave incident from $z < 0$ along a direction defined by $N_x = N_y = \frac{1}{2}$, find the power transmitted as an ordinary wave and that transmitted as an extraordinary wave in the dielectric. Find also the reflected power in the reflected parallel- and perpendicular-polarized waves. The interface is in the $xy$ plane at $z = 0$.

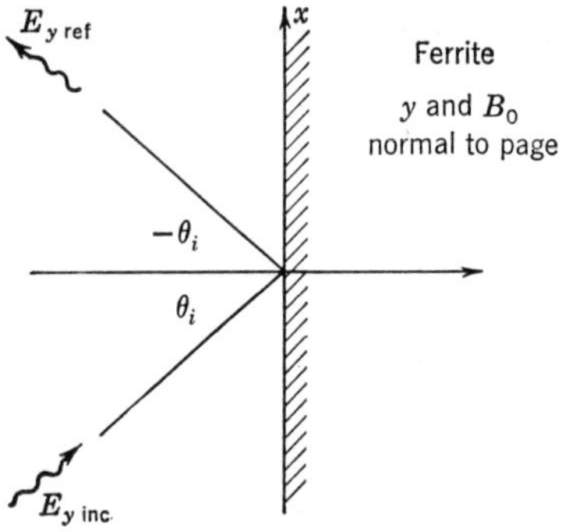


FIG. P 3.6

**3.6.** A ferrite medium occupies the half space $z > 0$ and is magnetized along the $y$ axis by a static magnetic field. Let a TEM wave with the electric vector along the $y$ axis be incident from $z < 0$ at an angle $\theta_i$. Find the amplitude of the wave reflected in the direction $-\theta_i$ and also the field transmitted into the ferrite medium. Show that, for a wave incident from the direction $-\theta_i$, the reflected field in the direction $\theta_i$ is different from that obtained for the first case. This is an example of nonreciprocal behavior. Assume a permeability tensor of the form

$$[\mu] = \begin{bmatrix} \mu & 0 & -j\kappa \\ 0 & \mu_0 & 0 \\ +j\kappa & 0 & \mu \end{bmatrix} \qquad \text{with } \mu \text{ and } \kappa \text{ real}$$

*Answer.* The propagation factor in ferrite is $e^{-jk(x \sin \theta_r + z \cos \theta_r)}$, where $k^2 = \omega^2\epsilon\mu(1 - \kappa^2/\mu^2)$. For a wave incident from $\theta_i$, the reflection coefficient in the direction $-\theta_i$ is $R(\theta_i, -\theta_i) = (Z_{\text{in}} - 1)/(Z_{\text{in}} + 1)$, where

$$Z_{\text{in}} = \frac{(\mu^2 - \kappa^2)k_0 \cos \theta_i}{\mu_0[\mu(k^2 - k_0^2 \sin^2 \theta_i)^{1/2} - jk_0 \sin \theta_i]}$$

and $k_0^2 = \omega^2\mu_0\epsilon_0$. The reflection coefficient $R(-\theta_i, \theta_i)$ is equal to the complex conjugate of $R(\theta_i, -\theta_i)$.

**3.7.** Find the principal axis (normal coordinates) for the permeability tensor given in Prob. 3.6. Show that these may be interpreted as rectangular coordinates rotating clockwise and counterclockwise around the $y$ axis. Transform Maxwell's equations into this new coordinate system, and obtain the solutions for TEM waves propagating along the $y$ axis in an infinite ferrite medium.

**3.8.** Obtain the reflection and transmission coefficients for right and left circularly polarized TEM waves incident normally on an infinite ferrite-air interface. Assume the static magnetic field to be applied in the direction of the interface normal.

**3.9.** A medium is characterized by a permeability $\mu_1$ along the $x$ axis and $\mu_2$ along the $y$ and $z$ axes. Determine suitable potential functions from which the ordinary and extraordinary plane-wave solutions may be found. Obtain the eigenvalue equations for the propagation phase constants. The dielectric constant is $\kappa$ along all three axes.

## CHAPTER 4

# TRANSMISSION LINES

One of the most familiar wave-guiding structures is the conventional transmission line such as the two-wire line and the coaxial line. The fundamental mode of propagation on a transmission line is essentially a transverse electromagnetic wave. In the ideal case, when the conductors can be considered to have infinite conductivity, the basic mode is a TEM mode. Practical lines have finite conductivity, and this results in a perturbation or change from a TEM mode to a mode which has a small axial component of electric field. Many practical lines have such small losses that we may analyze their behavior by considering them as ideal lines, and then make a simple perturbation calculation to obtain the effects of the finite conductivity. Transmission-line theory has two aspects. In one case we are given the characteristic parameters of the line, and our problem is to study the propagation of electromagnetic waves along the line. In the other case we know only the conductor configuration, and our problem is to determine the line parameters such as characteristic impedance, attenuation constant, propagation constant, and shunt conductance. The object of this latter aspect is ultimately to determine the suitability of the transmission line for a given application. It is this second aspect of transmission-line theory that we shall be most concerned with in this chapter.

Since the mode of transmission on an ideal transmission line is a TEM wave, the relations for input impedance, reflection coefficient, wave matrices, impedance matrices, and so forth, given in Chap. 3 are applicable to transmission lines. For this reason we will not consider transmission-line circuit theory here.

## 4.1. General Transmission-line Theory

### *The Ideal Two-conductor Line*

Consider two separate perfectly conducting conductors with uniform cross sections, infinitely long and oriented parallel to the $z$ axis as in Fig. 4.1. The cross sections of the two conductors are denoted by the two open or closed curves $S_1$ and $S_2$ in the $xy$ plane. The medium surrounding the conductors is assumed homogeneous and isotropic with electrical parameters $\mu$ and $\epsilon$. A further assumption will be made that all other material bodies are sufficiently remote so as not to perturb the field around the conductors.

For an ideal transmission line as illustrated, a TEM mode of propagation is possible. In the preceding chapter it was found that for TEM waves Maxwell's equations reduced to the following:

$$\mathbf{E}_t = -\nabla_t\phi(x,y)e^{\pm jkz} \tag{1a}$$

$$\mathbf{H}_t = \mp\bar{\mathbf{Y}} \cdot \mathbf{E}_t \tag{1b}$$

where $\mathbf{E}_t$ and $\mathbf{H}_t$ are transverse vectors, $\nabla_t$ is the transverse operator

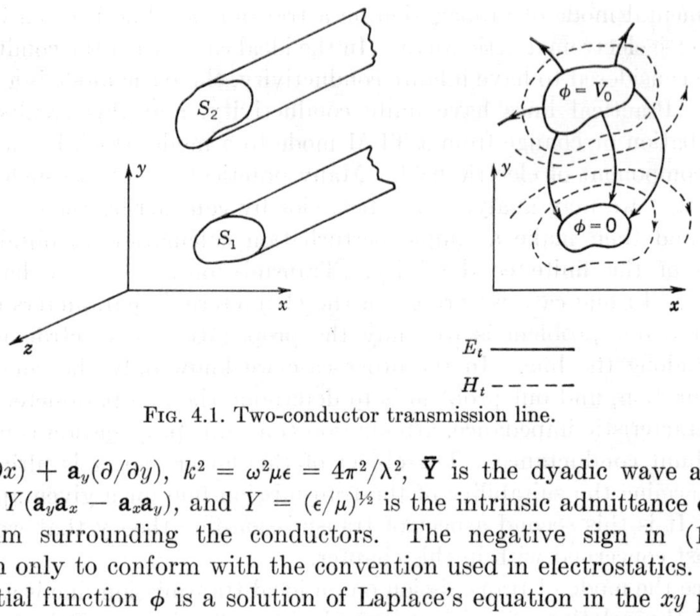


FIG. 4.1. Two-conductor transmission line.

$\mathbf{a}_x(\partial/\partial x) + \mathbf{a}_y(\partial/\partial y)$, $k^2 = \omega^2\mu\epsilon = 4\pi^2/\lambda^2$, $\bar{\mathbf{Y}}$ is the dyadic wave admittance $Y(\mathbf{a}_y\mathbf{a}_x - \mathbf{a}_x\mathbf{a}_y)$, and $Y = (\epsilon/\mu)^{1/2}$ is the intrinsic admittance of the medium surrounding the conductors. The negative sign in (1a) is chosen only to conform with the convention used in electrostatics. The potential function $\phi$ is a solution of Laplace's equation in the $xy$ plane. A solution for $\phi$ exists only if one conductor is at a potential which is different from that of the other one. If the system as a whole is electrically neutral, we may take one conductor at potential $V_0/2$ and the other at potential $-V_0/2$. Since the field $\mathbf{E}_t$ is invariant to a change in $\phi$ corresponding to the addition of a constant, we may, for convenience, take the conductor $S_1$ at zero potential and $S_2$ at a potential $V_0$ instead. In the transverse plane, $\nabla_t \times \mathbf{E}_t$ is identically zero, and, hence, the line integral of $\mathbf{E}_t$ from $S_1$ to $S_2$ is unique and does not depend on the path of integration. It is this property which permits us to introduce a unique potential between the two conductors. Physically this property arises because of the absence of axial magnetic flux, and, hence, there is no induced potential around any closed curve in the $xy$ plane. If we let $\mathbf{e}$

be equal to $\mathbf{E}_t$ without the factor $e^{\pm jkz}$, we have

$$V_0 = -\int_{S_1}^{S_2} \mathbf{e} \cdot d\mathbf{l} = \int_{S_1}^{S_2} \nabla_t \phi \cdot d\mathbf{l} = \int_{S_1}^{S_2} d\phi \tag{2}$$

Associated with the transverse electric field wave $\mathbf{E}_t = -\nabla_t \phi e^{-jkz}$ is a unique voltage wave $V = V_0 e^{-jkz}$.

Since the conductors $S_1$ and $S_2$ are assumed to be perfectly conducting, the normal component of $\mathbf{e}$ is discontinuous at these surfaces, and a surface distribution of charge given by $\rho_s = \epsilon \mathbf{n} \cdot \mathbf{e}$ exists on the conductors, where $\mathbf{n}$ is the unit outward normal from the conductors. For the same reason the tangential magnetic field is discontinuous at the conductor surfaces, and a surface current density given by $\mathbf{n} \times \mathbf{H}_t = \mathbf{J}_s$ exists on the conductors. The direction of $\mathbf{J}_s$ is along the $z$ axis since both $\mathbf{n}$ and $\mathbf{H}_t$ are vectors in the $xy$ plane. At the conductor surfaces, the normal component of $\mathbf{H}_t$ is zero, and the tangential component of $\mathbf{E}_t$ is zero, and, hence, $|\mathbf{J}_s| = |\mathbf{H}_t| = Y|\mathbf{n} \cdot \mathbf{E}_t| = Y\rho_s/\epsilon = \rho_s/(\mu\epsilon)^{1/2}$. The total current flowing on the conductor $S_2$ is given by the line integral of the current density around $S_2$, and, hence,

$$I_0 = \oint_{S_2} |\mathbf{J}_s| \, dl = \oint_{S_2} \frac{\rho_s dl}{(\mu\epsilon)^{1/2}} = Qv_c \tag{3}$$

where $I_0$ is the magnitude of the total axial current on $S_2$, $Q$ is the charge on $S_2$ per unit length, and $v_c$ is the velocity of light in a medium with parameters $\mu$, $\epsilon$. Since the unit normal is oppositely directed on the two conductors, the current $I_0$ flows down the line on one conductor and back on the other. The total current on each conductor is the same since the total charge (positive on one conductor and negative on the other) is the same on both. Thus, associated with the magnetic field wave $\mathbf{H}_t$ is a unique current wave $I = I_0 e^{-jkz}$. The ratio of the potential difference between the conductors to the total current flowing on the conductors for a wave propagating in either the positive or negative $z$ direction is defined as the characteristic impedance $Z_c$ of the transmission line, i.e.,

$$Z_c = \frac{V_0}{I_0} = \frac{V_0}{Q} (\mu\epsilon)^{1/2} = \frac{1}{Cv_c} = \frac{\epsilon}{C} Z \tag{4}$$

where $Z = (\mu/\epsilon)^{1/2}$ is the intrinsic impedance of the medium surrounding the conductors, and $C$ is the electrostatic capacitance between the two conductors per unit length. The characteristic impedance differs from the intrinsic or wave impedance by a factor which is a function of the geometry of the line only. Even though the frequency of the wave may be thousands of megacycles, the capacitance function in the expression for $Z_c$ is that for a static electric field existing between $S_1$ and $S_2$. This latter property follows again only because $\nabla_t \times \mathbf{E}_t$ is identically zero.

The ideal transmission line is unique among wave-guiding structures in that its parameters are those associated with static field distributions in the transverse plane.

The time-average electric energy per unit length of line in a single propagating TEM wave is given by

$$W_e = \tfrac{1}{4}\epsilon \iint \mathbf{E}_t \cdot \mathbf{E}_t^* \, dx\, dy = \tfrac{1}{4}\epsilon \iint \nabla_t\phi \cdot \nabla_t\phi \, dx\, dy \qquad (5)$$

where the integral is over the whole $xy$ plane, and $\epsilon$ is assumed real. The two-dimensional form of Green's first identity is†

$$\iint_S (\nabla_t\phi \cdot \nabla_t\phi + \phi\, \nabla_t^2\phi)\, da = \oint_C \phi \frac{\partial \phi}{\partial n}\, dl$$

where $C$ is a closed contour bounding the surface $S$, and $\mathbf{n}$ is the unit outward normal to the contour $C$. By introducing suitable cuts as in Fig. 4.2 and using Green's first identity, the expression for $W_e$ becomes

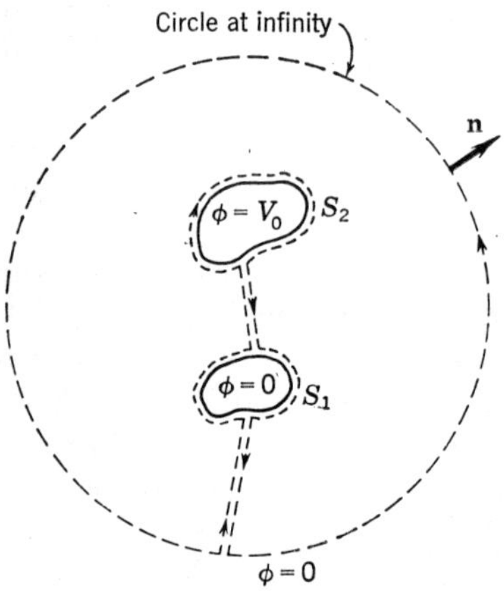


FIG. 4.2. Contour for evaluation of $W_e$.

$$W_e = \tfrac{1}{4}\epsilon \oint_{S_2} V_0 \frac{\partial \phi}{\partial n}\, dl \qquad (6)$$

since $\nabla_t^2\phi = 0$, $\phi = 0$ on $S_1$, $\phi = V_0$ on $S_2$, and, as $r$ becomes large, $\phi(\partial\phi/\partial n)$ vanishes at least as fast as $r^{-2}$, when the net charge on the two conductors is zero. On the conductor $S_2$ we have $\partial\phi/\partial n = \rho_s/\epsilon$, since $\mathbf{n}$ points toward the conductor and, hence, $\partial\phi/\partial n = |\mathbf{e}|$, where $\mathbf{e}$ is the outward-directed normal electric field at the surface. Using this result in (6) shows that

$$W_e = \tfrac{1}{4}\epsilon \oint_{S_2} \frac{V_0}{\epsilon} \rho_s\, dl = \tfrac{1}{4} V_0 Q = \tfrac{1}{4} V_0^2 C \qquad (7)$$

where $C$ is the electrostatic capacitance per unit length. The additional factor $\tfrac{1}{2}$ arises because of averaging over one period in time.

The time-average magnetic energy in the wave per unit length is

$$W_m = \tfrac{1}{4}\mu \iint \mathbf{H}_t \cdot \mathbf{H}_t^*\, da = \tfrac{1}{4}\mu Y^2 \iint \mathbf{E}_t \cdot \mathbf{E}_t^*\, da = W_e \qquad (8)$$

since $\mathbf{H}_t = \bar{\mathbf{Y}} \cdot \mathbf{E}_t$, and $\mu Y^2 = \epsilon$. The inductance $L$ per unit length of

† Section A.1c.

line is most suitably defined by the following relation:

$$\tfrac{1}{4}LI_0^2 = W_m \tag{9}$$

Since $I_0Z_c = V_0$, we get $LI_0^2 = LV_0^2/Z_c^2 = 4W_m = 4W_e = V_0^2C$, and, hence,

$$Z_c = \left(\frac{L}{C}\right)^{1/2} \tag{10a}$$

and, using the relation $Z_c = (Cv_c)^{-1}$, we also get

$$(LC)^{1/2} = (\mu\epsilon)^{1/2} = v_c^{-1} \tag{10b}$$

The definition of inductance given in (9) may be easily shown to give the same results as the low-frequency definition in terms of flux linkages per unit current.

The power flowing along the line is given by

$$\begin{aligned} P &= \tfrac{1}{2}\,\mathrm{Re} \iint \mathbf{E}_t \times \mathbf{H}_t^* \cdot \mathbf{a}_z \, dx\, dy \\ &= \tfrac{1}{2}\,\mathrm{Re} \iint Y\mathbf{E}_t \cdot \mathbf{E}_t^* \, dx\, dy \\ &= \frac{2YW_e}{\epsilon} \\ &= v_c(W_e + W_m) \end{aligned} \tag{11}$$

and is seen to be equal to the sum of the electric and magnetic energy per unit length multiplied by the velocity of propagation. This result is to be expected since the wave is propagating with a velocity $v_c$ and power is a rate of flow of energy. Equation (11) may be put into the following form as well:

$$P = \tfrac{1}{2}Z_cI_0^2 = \tfrac{1}{2}V_0I_0 \tag{12}$$

The result obtained so far has demonstrated that the ideal transmission line behaves in many respects as a low-frequency circuit element. This accounts for the great success of the conventional transmission-line theory based on the concepts of distributed parameters and using a circuit-analysis approach. The voltage and current waves $V = V_0e^{-jkz}$, $I = I_0e^{-jkz}$ are readily shown to be solutions to the differential equations given below:

$$\frac{d^2V}{dz^2} + \omega^2LCV = 0 \tag{13a}$$

$$\frac{d^2I}{dz^2} + \omega^2LCI = 0 \tag{13b}$$

These are the conventional telegraphists' or transmission-line equations for an ideal line.

If we have a transmission line with $N$ separate conductors as in Fig. 4.3, there will be $N - 1$ basic TEM modes of propagation. A suitable set

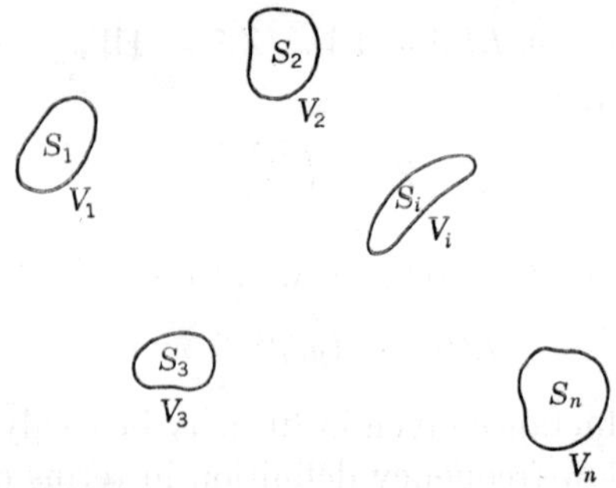


FIG. 4.3. Cross section of an $N$-conductor transmission line.

of modes is obtained by solving Laplace's equation in the $xy$ plane for the following boundary conditions:

$$
\begin{array}{lcll}
\phi = V_1 \text{ on } S_1 & \text{and} & \phi = 0 \text{ on } S_i & i = 2, 3, \ldots, N \\
\phi = V_2 \text{ on } S_2 & \text{and} & \phi = 0 \text{ on } S_i & i = 1, 3, \ldots, N \\
\cdots & \cdots & \cdots & \cdots \\
\phi = V_N \text{ on } S_N & \text{and} & \phi = 0 \text{ on } S_i & i = 1, 2, \ldots, N - 1
\end{array}
$$

Since the potential is arbitrary to within an additive constant, any one of the above solutions can be expressed in terms of the remainder, and so there are only $N - 1$ independent solutions. A superposition of

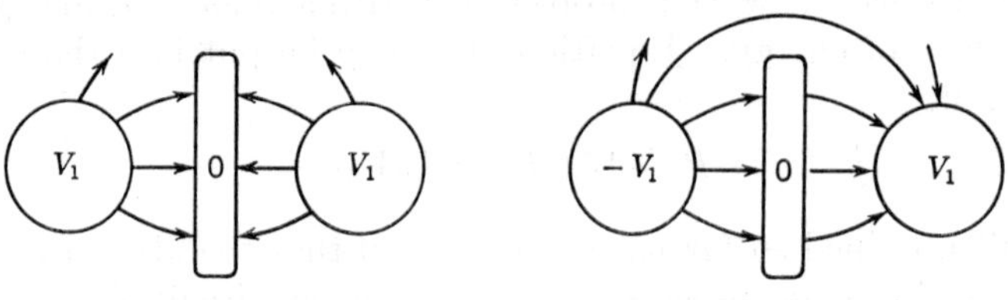


FIG. 4.4. Basic modes of propagation for a symmetrical three-wire transmission line.

these basic modes will permit the potential of all the conductors to be arbitrarily specified. For a three-wire line there are just two independent TEM modes of propagation, as shown in Fig. 4.4.

*The Two-conductor Line with Small Losses*

If the medium in which the conductors are located has finite losses, then $\epsilon$ must be replaced by $\epsilon(1 - j \tan \delta_l)$, where $\tan \delta_l$ is the loss tangent of the material. The effective conductivity of the medium is $\sigma$, where

$$\sigma = \omega\epsilon \tan \delta_l \tag{14}$$

If the conductors are still assumed to be perfect conductors, the fundamental mode of propagation is still a pure TEM mode, since making $\epsilon$ complex does not alter the solution to Maxwell's equations, apart from the change in $\epsilon$ throughout. The total shunt current per unit length of line flowing from conductor $S_2$ across to the conductor $S_1$ is

$$\begin{aligned} I_s &= \oint_{S_2} \sigma \mathbf{n} \cdot \mathbf{e}\, dl \\ &= \oint_{S_2} \frac{\sigma}{\epsilon} \rho_s \, dl = \frac{\sigma Q}{\epsilon} \end{aligned} \tag{15}$$

The shunt conductance $G$ is given by

$$G = \frac{I_s}{V_0} = \frac{\sigma C}{\epsilon} = \omega C \tan \delta_l \tag{16}$$

Although $G$ is given by an expression which is similar to that for a static field problem, it is a function of frequency because of the factor $\omega$ arising in the expression for $\sigma$, and also because $\tan \delta_l$ is a function of frequency. The propagation constant $k$ now becomes

$$\begin{aligned} jk &= j\omega[\mu\epsilon(1 - j \tan \delta_l)]^{1/2} = j\omega \left[ \mu\epsilon \left(1 - j \frac{G}{\omega C}\right)\right]^{1/2} \\ &\approx j\omega(\mu\epsilon)^{1/2}(1 - j\tfrac{1}{2} \tan \delta_l) = j\beta + \alpha \end{aligned} \tag{17a}$$

since $\tan \delta_l \ll 1$ in any practical case and where $\beta$ is the phase constant and $\alpha$ the attenuation constant. Using (10*b*) and (16) shows that

$$j\beta + \alpha = j\omega(LC)^{1/2} + \tfrac{1}{2}\left(\frac{L}{C}\right)^{1/2} G \tag{17b}$$

The characteristic impedance now becomes

$$Z_c = \left[\frac{L}{C(1 - jG/\omega C)}\right]^{1/2} = \left(\frac{j\omega L}{j\omega C + G}\right)^{1/2}$$

When the conductivity of the conductors is finite, power is absorbed in the conductors, and, hence, there must be a component of the Poynting vector directed into the conductors. This, in turn, implies at least a longitudinal component of electric field. In general, longitudinal components of both electric and magnetic fields will exist. A longitudinal component of magnetic field will have associated with it transverse currents on the conductors. Since these transverse currents arise only because of the perturbation of the TEM mode into a mode with longitudinal field components, they are small in comparison with the longitudinal currents. Thus the losses associated with the transverse currents are also small in comparison with the power loss arising from the longitudinal current. For a first approximation, we may assume that the transverse currents are negligible. This qualitative argument is supported to some extent by the rigorous analysis of some simple conductor configurations such as the single-wire line.† The single-wire line has circular symmetry and does allow a solution for a mode with no axial magnetic field component, but such symmetry is not present in all transmission lines used in practice, and we may expect small transverse currents to exist, in general.

If we assume that the current on the conductors flows only in the $z$ direction, and also that the medium surrounding the conductors is homogeneous and isotropic with constant parameters $\mu$ and $\epsilon$, the field may be derived from a vector potential function **A**, where **A** has only a $z$ component. From Sec. 1.6, the relevant equations are found to be

$$\mathbf{B} = \nabla \times \mathbf{A} \tag{18a}$$

$$\mathbf{E} = -j\omega\mathbf{A} + \frac{\nabla\nabla \cdot \mathbf{A}}{j\omega\epsilon\mu} \tag{18b}$$

$$\nabla^2\mathbf{A} + k^2\mathbf{A} = 0 \tag{18c}$$

Equation (18$c$) holds only in the region external to the conductors. The potential **A** has only a $z$ component, because the current is only in the $z$ direction, and no other components of **A** are required to satisfy the boundary conditions. It is seen from (18$a$) that **B** and, hence, **H** have transverse components only.

In Sec. 3.6 on the discussion of reflection of plane waves from a conducting plane, it was demonstrated that the effect of finite conductivity in good conductors may be accounted for by introducing the surface impedance $Z_m$ of the metal, where

$$Z_m = \frac{1+j}{\sigma\delta_s} = R_m + jX_m$$

If the metal is assumed to have the same permeability as free space, the skin depth $\delta_s$ is given by $(2/\omega\mu_0\sigma)^{1/2}$. At the conductor surface, the current density $\mathbf{J}_s$ and longitudinal electric field are related as follows:

$$Z_m J_s = E_z \tag{19}$$

The current $\mathbf{J}_s$ is given by

$$\begin{aligned}\mathbf{J}_s &= \mathbf{n} \times \mathbf{H}_t = \mu^{-1}\mathbf{n} \times (\nabla \times \mathbf{A}) \\ &= \mu^{-1}\,\nabla\mathbf{n} \cdot \mathbf{A} - (\mathbf{n} \cdot \nabla)\mathbf{A} = -\mu^{-1}\frac{\partial \mathbf{A}}{\partial n}\end{aligned} \tag{20}$$

where **n** is the unit outward normal from the conductors, and $\mathbf{n} \cdot \mathbf{A} = 0$,

† J. A. Stratton, "Electromagnetic Theory," chap. 9, McGraw-Hill Book Company, Inc., New York, 1941.

since **n** and **A** are perpendicular. From (18*b*) it is seen that

$$\mathbf{a}_z E_z = \frac{k^2 + \gamma^2}{j\omega\epsilon\mu} \mathbf{A} \tag{21}$$

if it is assumed that **A** propagates along the line according to the factor $e^{-\gamma z}$. Combining (19), (20), and (21) shows that **A** must satisfy the nonhomogeneous boundary condition

$$\frac{j\omega\epsilon Z_m}{k_c^2} \frac{\partial \mathbf{A}}{\partial n} = -\mathbf{A} \tag{22}$$

where $k_c^2 = k^2 + \gamma^2$. A solution for **A** which satisfies Eq. (18*c*) and the boundary condition (22) is extremely difficult if not impossible to obtain for a general two-conductor transmission line. We are thus forced to consider techniques which will give us approximate solutions for the quantities in which we are interested. A variational principle for obtaining an approximate solution for the propagation constant $\gamma$ will be developed, but first we will give the usual derivation of the parameters of a line with small losses.

If the field distribution around the line and the current flowing on the line are assumed to be the same as in the ideal transmission line, then the power loss in the conductors per unit length is

$$P_L = \tfrac{1}{2} R_m \oint_{S_1 + S_2} |\mathbf{J}_s|^2 \, dl \tag{23}$$

where $|\mathbf{J}_s|$ is the magnitude of the current density for the unperturbed mode. A resistance $R$ per unit length of line may be defined by the relation $\frac{1}{2} I_0^2 R = P_L$, or

$$R = \frac{R_m \oint_{S_1 + S_2} |\mathbf{J}_s|^2 \, dl}{\left( \oint_{S_2} |\mathbf{J}_s| \, dl \right)^2} \tag{24}$$

This expression can be evaluated when the field distribution, and, hence, $\mathbf{J}_s$, is known. The attenuation constant $\alpha$ can be found from the rate of power loss with distance along the line. If the propagation constant is $\gamma = j\beta + \alpha$, the power carried along the line relative to that at $z = 0$ is given by $P = P_0 e^{-2\alpha z}$, and the rate of decrease of power is $-dP/dz = 2\alpha P$. The rate of decrease of power is equal to the power loss per unit length, and, hence,

$$\alpha = \frac{P_L}{2P} \tag{25}$$

where $P_L$ includes both the series resistance loss $\frac{1}{2} R I_0^2$ and the shunt conductance loss $\frac{1}{2} V_0^2 G$ if present. Just as the shunt conductance can be associated with a complex value of dielectric constant, the series resistance can be associated with a complex value of permeability. The characteristic impedance of the line is thus given by

$$Z_c = \left[\frac{L(1 - jR/\omega L)}{C(1 - jG/\omega C)}\right]^{1/2} = \left(\frac{R + j\omega L}{G + j\omega C}\right)^{1/2} \approx \left(\frac{L}{C}\right)^{1/2} \tag{26}$$

since $G \ll \omega C$, $R \ll \omega L$ when the losses are small. Equation (12) for power flow along the line is still valid to a first approximation, and, hence, (25) gives

$$\alpha = \tfrac{1}{2}(RY_c + GZ_c) \tag{27a}$$

The propagation constant $\gamma$ is now given by

$$\gamma = j\beta + \alpha = j\omega(\mu\epsilon)^{1/2}\left(1 - \frac{jR}{\omega L}\right)^{1/2}\left(1 - \frac{jG}{\omega C}\right)^{1/2}$$
$$\approx j\omega(LC)^{1/2} + \tfrac{1}{2}(RY_c + GZ_c) \tag{27b}$$

The approximation of taking $Z_c$ real is inconsistent from an energy-conservation point of view but has negligible effect on the calculation of reflection coefficient and input impedance of a terminated line and is, therefore, frequently made in practice.

The above simple analysis of the transmission line with small losses gives results which are sufficiently accurate for most engineering purposes but is somewhat lacking in rigor. The variational principle to be given next will provide a somewhat firmer foundation for the above approach.

The starting point for developing a variational expression for the eigenvalues or propagation constants for a wave function is the *wave equation*. If it is assumed that the only currents on the conductors are in the $z$ direction, the vector potential $\mathbf{A}$ may be represented as follows:

$$\mathbf{A} = \mathbf{a}_z\psi(x,y)e^{-\gamma z} \tag{28}$$

where $\psi$ is a solution of the scalar Helmholtz equation

$$\nabla_t^2\psi + k_c^2\psi = 0 \tag{29}$$

and satisfies the nonhomogeneous boundary condition

$$g\frac{\partial\psi}{\partial n} - \psi = 0 \tag{30}$$

where $k_c^2 = k^2 + \gamma^2$ and $g = j\omega\epsilon Z_m/k_c^2$, and the normal $\mathbf{n}$ is assumed directed into the conductor rather than outward. Multiplying (29) by $\psi$ and integrating over the whole $xy$ plane gives

$$k_c^2 = -\frac{\iint \psi\nabla_t^2\psi\, da}{\iint \psi^2\, da} \tag{31}$$

To determine whether this expression gives a solution for $k_c^2$ whose first variation is zero when we perturb the functional form of $\psi$ by a small amount $\delta\psi$ we conduct the variation to obtain

$$2k_c\,\delta k_c \iint \psi^2\,da + 2k_c^2 \iint \psi\,\delta\psi\,da = -\iint (\delta\psi\,\nabla_t^2\psi + \psi\,\nabla_t^2\delta\psi)\,da$$

If Green's second identity† is used, which in the present case gives

$$\iint (\psi\,\nabla_t^2\delta\psi - \delta\psi\,\nabla_t^2\psi)\,da = \oint_C \left(\psi\,\frac{\partial\delta\psi}{\partial n} - \delta\psi\,\frac{\partial\psi}{\partial n}\right) dl$$

where $\mathbf{n}$ is directed outward from the closed contour $C$ as illustrated previously in Fig. 4.2, the variation in $k_c^2$ is found to be

$$2k_c\,\delta k_c \iint \psi^2\,da = -2 \iint \delta\psi(k_c^2\psi + \nabla_t^2\psi)\,da - \oint_C \left(\psi\,\frac{\partial\delta\psi}{\partial n} - \delta\psi\,\frac{\partial\psi}{\partial n}\right) dl \quad (32)$$

The surface integral vanishes since $\psi$ satisfies the scalar Helmholtz equation. If the contour integral would also vanish, then the variation $\delta k_c$ in $k_c^2$ would vanish as well. In this case (31) would provide a variational expression which would permit us to calculate a value for $k_c^2$ accurate to the second order for an assumed approximate form for $\psi$ accurate to the first order only. The logical approximation to use for $\psi$ would be the potential function for the same transmission line but with perfect conductors. To ensure that our variation $\delta\psi$ in $\psi$ would make the contour integral vanish is a difficult condition to meet in practice. We therefore attempt to find a suitable term to add to (31) such that the first variation in $k_c^2$ will vanish independently of what boundary conditions $\delta\psi$ satisfies. The term that we add to the numerator of (31) must be such that it does not affect the value of $k_c^2$ when the true eigenfunction $\psi$ is used in the resultant expression, and must also be such that its first variation will cancel the contour integral in (32). Any function multiplied by the expression $j\omega\epsilon Z_m(\partial\psi/\partial n) - k_c^2\psi$ from (30) will satisfy the first condition. After a few trial substitutions involving combinations of $\psi$ and $\partial\psi/\partial n$, we find that a suitable expression to add to the numerator in (31) is

$$\oint_C \left(\frac{j\omega\epsilon Z_m}{k_c^2}\,\frac{\partial\psi}{\partial n} - \psi\right)\frac{\partial\psi}{\partial n}\,dl$$

In place of (32), we now obtain

$$2\delta k_c \left[k_c \iint \psi^2\,da + \frac{g}{k_c}\oint_C \left(\frac{\partial\psi}{\partial n}\right)^2 dl\right] = -2\iint \delta\psi(k_c^2\psi + \nabla_t^2\psi)\,da + 2\oint_C \left(g\,\frac{\partial\psi}{\partial n} - \psi\right)\frac{\partial\delta\psi}{\partial n}\,dl = 0$$

† Section A.1c.

since $\psi$ is a solution of (29) and (30). Thus an appropriate variational integral or expression for $k_c^2$ is

$$k_c^2 = -\frac{\iint \psi \nabla_t^2 \psi \, da - \oint_C \left( \frac{j\omega\epsilon Z_m}{k_c^2} \frac{\partial \psi}{\partial n} - \psi \right) \frac{\partial \psi}{\partial n} \, dl}{\iint \psi^2 \, da} \tag{33}$$

The integral in the denominator serves as a normalization term, and, in practice, we may always choose our approximate eigenfunction, say $\psi_0$, such that $\iint \psi_0^2 \, da$ is equal to unity or some other convenient factor.

For the approximation to $\psi$ we choose the correct potential function $\psi_0$ for the ideal transmission line. This function is a solution of Laplace's equation $\nabla_t^2 \psi_0 = 0$ in the $xy$ plane. In terms of $\psi_0$, the fields are given by $H_x = \mu^{-1} \, \partial\psi_0/\partial y$, $H_y = -\mu^{-1} \, \partial\psi_0/\partial x$, $E_x = ZH_y$, and $E_y = -ZH_x$, where $Z = (\mu/\epsilon)^{1/2}$. In terms of the potential function $\phi$ introduced for the analysis of the ideal line, we had $H_x = Y \, \partial\phi/\partial y$ and $H_y = -Y \, \partial\phi/\partial x$, and, hence, apart from any normalization factor, we may identify $\psi_0$ with $\mu Y \phi$. The current density on the conductors is given by $\mu|\mathbf{J}_s| = |\partial\psi_0/\partial n|$. On the conductor $S_2$ we have $\psi_0 = \mu Y V_0/2$, while on $S_1$ we have

$$\psi_0 = \frac{-\mu Y V_0}{2}$$

For later convenience it will be assumed that $\psi_0$ has been normalized so that $\iint \psi_0^2 \, da = \mu\epsilon$, and thus $\iint \phi^2 \, da$ equals unity. Introducing the above values of $\psi_0$ and $\partial\psi_0/\partial n$ at the conductor surfaces, the expression for $k_c^2$ becomes

$$k_c^2 = -Z \frac{V_0}{2} \left( \oint_{S_1} |\mathbf{J}_s| \, dl + \oint_{S_2} |\mathbf{J}_s| \, dl \right) + \frac{jkZZ_m}{k_c^2} \oint_{S_1+S_2} |\mathbf{J}_s|^2 \, dl \tag{34}$$

since $\partial\psi_0/\partial n$ is positive on $S_2$ and negative on $S_1$. The surface integral vanishes since $\psi_0$ is a solution of Laplace's equation in the $xy$ plane. The first two integrals give $ZV_0I_0$, which is equal to $2ZP$, where $P$ is the power flowing along the line. In the last integral $R_m \oint_{S_1+S_2} |\mathbf{J}_s|^2 \, dl = 2P_L$, where $P_L$ is the power loss in the conductors. Since the conductivity is finite, the magnetic field penetrates into the conductor, and, hence, there is a net amount of magnetic energy stored internal to the conductor surface. At the surface the magnetic field is equal to $\mathbf{J}_s$ in magnitude and decays exponentially with the distance $u$ into the conductor according to the factor $e^{-u/\delta_s}$. Thus the internal magnetic energy is

$$W_{mi} = \tfrac{1}{4}\mu_0 \int_0^{u_0} \oint_{S_1+S_2} |\mathbf{J}_s|^2 e^{-2u/\delta_s} \, du \, dl \tag{35}$$

where the integral over $u$ need be taken only to some interior point $u_0$ where the field is negligible. From (35) we get

$$\begin{aligned} W_{mi} &= \tfrac{1}{4}\mu_0 \frac{\delta_s}{2} \oint_{S_1+S_2} |\mathbf{J}_s|^2 \, dl \\ &= \tfrac{1}{4}(\omega\sigma\delta_s)^{-1} \oint_{S_1+S_2} |\mathbf{J}_s|^2 \, dl \\ &= \frac{1}{2\omega} P_L \end{aligned}$$

when $(2/\omega\mu_0\sigma)^{1/2}$ is substituted for $\delta_s$. An internal inductance $L_i$ may be defined by the relation $\tfrac{1}{4}I_0{}^2L_i = W_{mi}$, and thus $L_i = R/\omega$, where $R$ is the series resistance of the line per unit length. With this result we are able to replace the last integral in (34) by $-2kZ(-jP_L + 2\omega W_{mi})/k_c{}^2$, and, hence, the equation for $k_c{}^2$ becomes

$$k_c{}^4 + 2ZPk_c{}^2 + 2kZ(-jP_L + 2\omega W_{mi}) = 0 \tag{36}$$

The solution for $k_c{}^2$ is

$$\begin{aligned} k_c{}^2 &= -ZP + [(ZP)^2 + 2kZ(jP_L - 2\omega W_{mi})]^{1/2} \\ &\approx \frac{k(jP_L - 2\omega W_{mi})}{P} \end{aligned} \tag{37}$$

since the second term in the brackets is small compared with the first, and the expression to the one-half power may be expanded according to the binomial expansion. Replacing $k_c{}^2$ by $k^2 + (j\beta + \alpha)^2$, we get

$$2\beta\alpha = \frac{kP_L}{P} \tag{38a}$$

$$k^2 - \beta^2 + \alpha^2 = -\frac{2W_{mi}}{P}\omega k \tag{38b}$$

Since the losses are small, we must have $\beta \approx k$, and, hence, (38*a*) gives essentially

$$\alpha = \frac{P_L}{2P} \tag{39a}$$

In (38*b*) we may replace $k^2 - \beta^2$ by $(k + \beta)(k - \beta) \approx 2k(k - \beta)$. Also, since the right-hand side of (38*b*) is of magnitude $2k\alpha$, we see that $\alpha^2$ is negligible in comparison, and, hence,

$$2k(k - \beta) = -\frac{2\omega k W_{mi}}{P} - \alpha^2 \approx -\frac{2\omega k W_{mi}}{P}$$

and thus

$$\beta = k + \frac{\omega W_{mi}}{P} \tag{39b}$$

The use of the variational principle has established that the simple approach employed to calculate the parameters of the line with losses is valid and gives results which are correct to the second order for an assumed field distribution correct to only the first order, provided the transverse currents may be neglected. It also shows that the change in $\beta$ from the unperturbed value $k$ is principally due to an increase in the magnetic energy in the region internal to the conductors. Equations (39) may also be written in the following form when the approximation $Z_c = (L/C)^{1/2}$ is made:

$$\alpha = \frac{I_0^2 R}{2I_0^2 Z_c} = \frac{R}{2Z_c} \tag{40a}$$

$$\beta = \omega(LC)^{1/2} + \frac{\omega L_i}{2Z_c} \approx \omega[C(L + L_i)]^{1/2} \tag{40b}$$

In practical lines at high frequencies the internal inductance $L_i$ is negligible in comparison with the external inductance $L$. When $\epsilon$ is made complex to account for the shunt conductance, we will obtain the complete expression for $\alpha$ corresponding to (27) given earlier. Since the internal inductance and series resistance arise because of the surface impedance $Z_m$ of the conductor, we may regard the series resistance as an imaginary part added to the inductance and, hence, also consider the effect of the series resistance as equivalent to making the permeability $\mu$ complex.

## 4.2. The Characteristic Impedance of Transmission Lines

We have seen that, if we can determine the capacitance $C$ of a transmission line, we can then obtain the characteristic impedance from the relation $Z_c = (\mu\epsilon)^{1/2}C^{-1}$. There are many methods which exist for the solution of Laplace's equation in two dimensions. We shall discuss only two of them here: the conformal transformation method and variational methods which yield upper and lower bounds on $Z_c$. Many configurations which occur in practice are extremely difficult, if not impossible, to obtain rigorous solutions for, so that an approximate solution such as can be obtained by variational methods has to suffice. The big advantage of a variational method is that the capacitance is correct to the second order for an approximate solution for the field distribution correct to only the first order.

### *Characteristic Impedance by Conformal Mapping*

Consider the solution of Laplace's equation $\nabla_t^2\phi$ for the two-dimensional electrostatic problem illustrated in Fig. 4.5, with the boundary conditions $\phi = \phi_1$ on $S_1$ and $\phi = \phi_2$ on $S_2$. The curves $S_1$ and $S_2$ can respresent the two conductors of a transmission line. In general, the solution to this problem is difficult unless $S_1$ and $S_2$ coincide with constant coordinate curves. For a more suitable choice of coordinates this may actually be the case, and we would then obviously use this set of coordinates in preference to the $x$, $y$ coordinate system. Consider, therefore, a transformation of coordinates, from the $x$, $y$ coordinates over to the general curvilinear coordinates, $u$, $v$, where

$$u = u(x,y) \tag{41a}$$
$$v = v(x,y) \tag{41b}$$

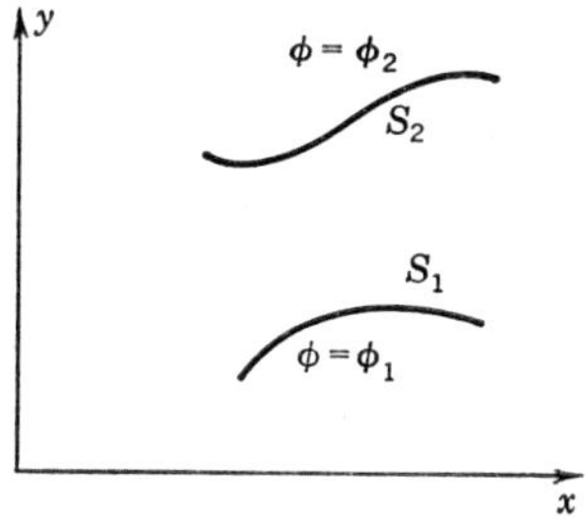

FIG. 4.5

We will take the unit vectors along the $u$ and $v$ coordinate curves as $\mathbf{a}_1$ and $\mathbf{a}_2$, respectively. We will restrict our coordinate transformations to those that can be effected through the complex function transformation

$$W = F(Z) = F(x + jy) = u + jv \tag{42}$$

The function $F(Z)$ is restricted to be an analytic function of $Z$; that is, $dW/dZ = dF/dZ$ must have a unique value at each point in the $xy$ plane. The condition that $F$ should be analytic is ensured if $u$ and $v$ satisfy the Cauchy-Riemann equations†

$$\frac{\partial u}{\partial x} = \frac{\partial v}{\partial y} \qquad \frac{\partial u}{\partial y} = -\frac{\partial v}{\partial x} \tag{43}$$

Conversely, if $F(Z)$ is analytic, the Cauchy-Riemann equations are satisfied. Since $\nabla_t u = \mathbf{a}_x\, \partial u/\partial x + \mathbf{a}_y\, \partial u/\partial y$ and

$$\nabla_t v = \mathbf{a}_x \frac{\partial v}{\partial x} + \mathbf{a}_y \frac{\partial v}{\partial y} = -\mathbf{a}_x \frac{\partial u}{\partial y} + \mathbf{a}_y \frac{\partial u}{\partial x}$$

from Eq. (43), we have the result that $\nabla_t u \cdot \nabla_t v = 0$ and, hence, $u$ and $v$ form an orthogonal coordinate system. The scale factors $h_1$ and $h_2$ are given by

$$\frac{1}{h_1^2} = \left(\frac{\partial u}{\partial x}\right)^2 + \left(\frac{\partial u}{\partial y}\right)^2 = \left(\frac{\partial v}{\partial y}\right)^2 + \left(\frac{\partial v}{\partial x}\right)^2 = \frac{1}{h_2^2} = \frac{1}{h^2} = \left|\frac{dW}{dZ}\right|^2 \tag{44}$$

and are equal by virtue of the Cauchy-Riemann equations. The coordinate transformation effected by the complex function transformation (42) is a very special type of coordinate transformation for which $h_1 = h_2$. Laplace's equation in the $uv$ coordinate system is

$$\frac{\partial}{\partial u}\frac{h_2}{h_1}\frac{\partial \phi}{\partial u} + \frac{\partial}{\partial v}\frac{h_1}{h_2}\frac{\partial \phi}{\partial v} = \frac{\partial^2 \phi}{\partial u^2} + \frac{\partial^2 \phi}{\partial v^2} = 0 \tag{45}$$

since $h_1 = h_2 = h$. The potential function $\phi$, therefore, satisfies the same equation in the $u$, $v$ coordinate systems that it does in the $x$, $y$ coordinate system. If in the $u$, $v$ coordinate system the curves $S_1$ and $S_2$ can

† Continuity of $u$ and $v$ and their first-order partial derivatives is also required.

be represented by constant coordinate curves as illustrated in Fig. 4.6, the solution to the equation $\partial^2\phi/\partial u^2 + \partial^2\phi/\partial v^2 = 0$ will be much simpler than the solution to $\partial^2\phi/\partial x^2 + \partial^2\phi/\partial y^2 = 0$.

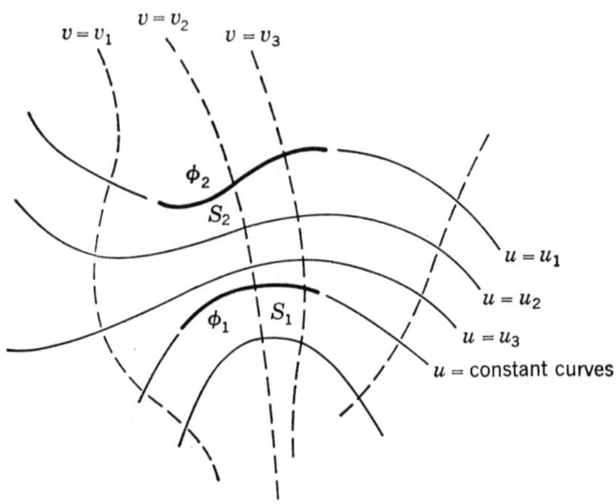


FIG. 4.6

The energy stored in the electrostatic field is

$$W_e = \tfrac{1}{2}\epsilon \iint\limits_{\substack{xy\\ \text{plane}}} \left[\left(\frac{\partial\phi}{\partial x}\right)^2 + \left(\frac{\partial\phi}{\partial y}\right)^2\right] dx\,dy = \tfrac{1}{2}\epsilon \iint\limits_{\substack{xy\\ \text{plane}}} |\nabla_t\phi|^2\,dx\,dy$$

$$= \tfrac{1}{2}\epsilon \iint\limits_{\substack{uv\\ \text{plane}}} \left[\frac{1}{h_1^2}\left(\frac{\partial\phi}{\partial u}\right)^2 + \frac{1}{h_2^2}\left(\frac{\partial\phi}{\partial v}\right)^2\right] h_1\,du\,h_2\,dv$$

$$= \tfrac{1}{2}\epsilon \iint\limits_{\substack{uv\\ \text{plane}}} \left[\left(\frac{\partial\phi}{\partial u}\right)^2 + \left(\frac{\partial\phi}{\partial v}\right)^2\right] du\,dv = \tfrac{1}{2}C(\phi_2 - \phi_1)^2 \qquad (46)$$

where $C$ is the capacitance of the line per unit length perpendicular to the $xy$ plane. Equations (45) and (46) show that we may treat $u$ and $v$ as rectangular coordinates, and, hence, we may replot the curves $S_1(x,y)$ and $S_2(x,y)$ onto a rectangular $uv$ grid by expressing the $x$ and $y$ values along $S_1$ and $S_2$ as $u$ and $v$ values by means of the coordinate transformation equations (41). This mapping of the original curves $S_1$ and $S_2$ into new curves in the $uv$ coordinate plane, i.e., in the complex $W$ plane, is called a conformal transformation. Equation (46) shows that the capacitance of the conductor configuration in the $uv$ plane is the same as that in the $xy$ plane. Thus, provided we can solve Laplace's equation for the conformal mapping of the original configuration, we can then transform back to the $x$, $y$ coordinates to obtain the field distribution for our actual physical problem. If we are interested only in the capacitance, we do not even need to transform back to our original $x$, $y$ coordinates.

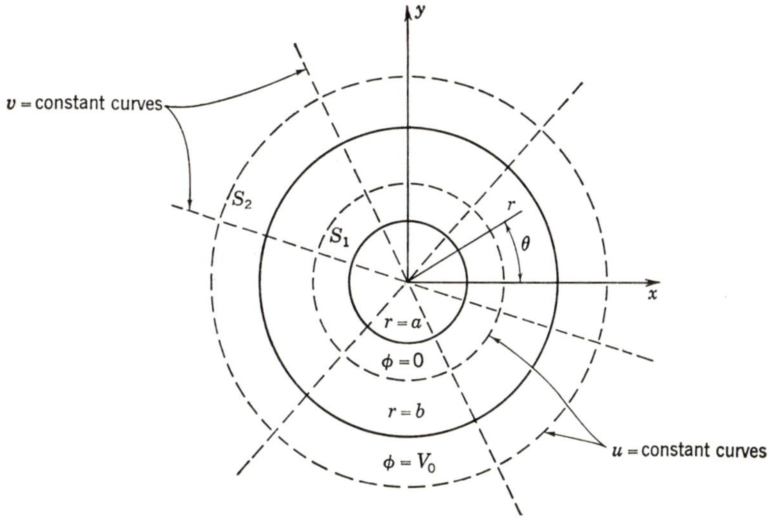


FIG. 4.7. A coaxial transmission line.

As a simple example, consider the problem of two coaxial cylinders as illustrated in Fig. 4.7. The solution to the equation

$$\frac{\partial^2 \phi}{\partial x^2} + \frac{\partial^2 \phi}{\partial y^2} = 0$$

is not simple for this configuration. Our natural choice of coordinates would be cylindrical coordinates. This coordinate transformation is obtained through the following complex function transformation $W = \ln Z = \ln re^{j\theta} = \ln r + j\theta$ and, hence, $u = \ln r$, $v = \theta$. A plot of the constant $uv$ coordinate curves is given in Fig. 4.7 also. The scale factor $h$ is equal to $r$. If we treat $u$ and $v$ as rectangular coordinates, the curves $S_1$ and $S_2$ defined by $r = a$, $r = b$, $0 < \theta < 2\pi$, map into the curves $u = \ln a$, $u = \ln b$, $0 < v < 2\pi$, as in Fig. 4.8.

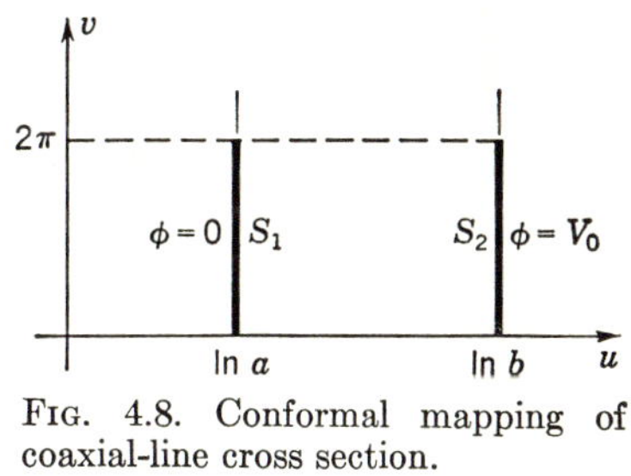


FIG. 4.8. Conformal mapping of coaxial-line cross section.

The region between the two coaxial cylinders maps into the region inside the rectangle bounded by $S_1$, $S_2$ and $v = 0$, $v = 2\pi$. As $\theta$ ranges from $-\infty$ to $+\infty$, $v$ does also, and we obtain a periodic mapping of the region between the two coaxial cylinders into the whole infinite strip bounded by the lines $u = \ln a$ and $u = \ln b$. The field in the region

$2n\pi < v < 2(n+1)\pi$ is the same as that in the region $0 < v < 2\pi$. The equivalent electrostatic problem in the complex $W$ plane is, therefore, just the simple problem of finding the potential distribution between the two infinite planes. The solution is obviously

$$\phi = \frac{u - \ln a}{\ln b - \ln a} V_0$$

The energy stored in the field is

$$W_e = \tfrac{1}{2}\epsilon \int_0^{2\pi} \int_a^b \left(\frac{\partial \phi}{\partial u}\right)^2 du = \frac{\pi \epsilon V_0^2}{\ln (b/a)} = \tfrac{1}{2} C V_0^2 \tag{47}$$

For the equivalent transmission-line problem with a simple-harmonic time variation, the time-average electric energy stored in the field is one-half the above value. The capacitance of the coaxial cylinders is $C = \dfrac{2\pi\epsilon}{\ln (b/a)}$ per unit length. When used as a transmission line, the characteristic impedance is $Z_c = \left(\dfrac{\mu\epsilon}{C}\right)^{1/2} = \dfrac{1}{2\pi}\left(\dfrac{\mu}{\epsilon}\right)^{1/2} \ln \dfrac{b}{a}$. For an air-filled line, $(\mu_0/\epsilon_0)^{1/2} = 120\pi$ and $Z_c = 60 \ln (b/a)$ ohms.

We may regard the mapping of the curves $S_1$ and $S_2$ into the complex $W$ plane as a distorted representation of our original $S_1$ and $S_2$ curves and the $uv$ coordinate curves. In this mapping, the curvilinear coordinates have been straightened out into a rectangular grid. Obviously any constant potential curve $\phi = K$ in the $xy$ plane is a constant potential curve $\phi = K$ in the $uv$ coordinate plane also, since we have not changed the physical problem. Similarly, any curve $\nabla_t \phi = 0$ goes over into a curve $\nabla_t \phi = 0$ in the $uv$ coordinate frame. At any point in the $xy$ plane, $\nabla_t \phi = \mathbf{K}$ goes over into $\dfrac{\mathbf{a}_1}{h}\dfrac{\partial \phi}{\partial u} + \dfrac{\mathbf{a}_2}{h}\dfrac{\partial \phi}{\partial v} = \mathbf{K}$ in the $u$, $v$ curvilinear coordinate system, and hence $\mathbf{a}_1\, \partial\phi/\partial u + \mathbf{a}_2\, \partial\phi/\partial v = h\mathbf{K}$. Therefore, the gradient of $\phi$, when calculated in the complex $W$ plane according to the relation $\mathbf{a}_1\, \partial\phi/\partial u + \mathbf{a}_2\, \partial\phi/\partial v$, must be divided by the scale factor $h$ to obtain the value of the gradient of $\phi$ at the corresponding point in the $xy$ plane.

If we differentiate Eq. (43) with respect to $x$ and $y$, we get

$$\frac{\partial^2 u}{\partial x^2} - \frac{\partial^2 v}{\partial x\, \partial y} = \frac{\partial^2 u}{\partial x^2} + \frac{\partial^2 u}{\partial y^2} = 0$$

and similarly $\nabla_t^2 v = 0$. Hence, if in our coordinate transformation the curves $S_1$ and $S_2$ map into constant coordinate curves, we then have the result that $u$ (or $v$) gives the potential distribution directly; that is, $\phi = u$ (or $\phi = v$) is the required solution. It is not necessary that the original $S_1$ and $S_2$ curves should map into constant-value coordinate curves, although the method is usually of only limited help if this does not happen. Several transformations in succession may be used in order to obtain the desired final configuration. In practice, the main difficulty is in obtaining the desired transformation. Problems which can be solved by conformal-mapping techniques can be solved by other methods also, although in many cases by considerably more labor and dexterity on the part of the individual.

### 4.3. The Schwarz-Christoffel Transformation

The Schwarz-Christoffel transformation is a conformal transformation which will map the real axis in the $Z$ plane into a general polygon in the $W$ plane with the upper half of the $Z$ plane mapping into the region

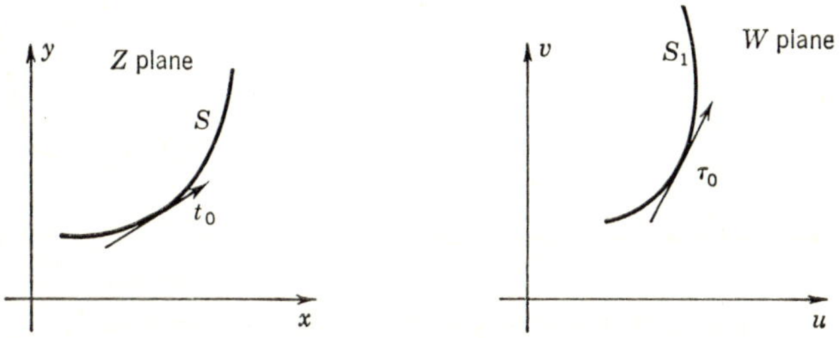


FIG. 4.9. Conformal mapping of $S$ into $S_1$.

interior to the polygon. To derive the basic transformation, we consider first a curve $S$ in the $Z$ plane and its conformal mapping $S_1$ in the $W$ plane as illustrated in Fig. 4.9.

Let the unit tangent to $S$ in the $Z$ plane be $t_0$ and the unit tangent to $S_1$ in the $W$ plane be $\tau_0$, where

$$t_0 = \lim_{\Delta Z \to 0} \frac{\Delta Z}{|\Delta Z|}$$

$$\tau_0 = \lim_{\Delta W \to 0} \frac{\Delta W}{|\Delta W|}$$

If the mapping function is $W = F(Z)$, we have

$$\frac{dW}{dZ} = F'(Z) = \lim_{\Delta Z \to 0} \frac{\Delta W}{\Delta Z} = \lim_{\Delta Z \to 0} \frac{\Delta W/|\Delta W|}{\Delta Z/|\Delta Z|} \left| \frac{\Delta W}{\Delta Z} \right| = \frac{\tau_0}{t_0} |F'(Z)|$$

and, hence,

$$\tau_0 = \frac{F'(Z)}{|F'(Z)|} t_0 \tag{48}$$

The angle that $\tau_0$ makes with the $u$ axis is given by

$$\angle \tau_0 = \angle t_0 + \angle F'(Z) \tag{49}$$

Consider next the following function:

$$F'(Z) = A(Z - x_1)^{-k_1}(Z - x_2)^{-k_2} \cdots (Z - x_N)^{-k_N}$$
$$= A \prod_{i=1}^{N} (Z - x_i)^{-k_i} \qquad (50)$$

where $A$ is an arbitrary constant, $k_i$ is a real number, and $x_1 < x_2 < \cdots < x_N$. If the $x$ axis is chosen as the curve $S$ in the $Z$ plane, the angle of

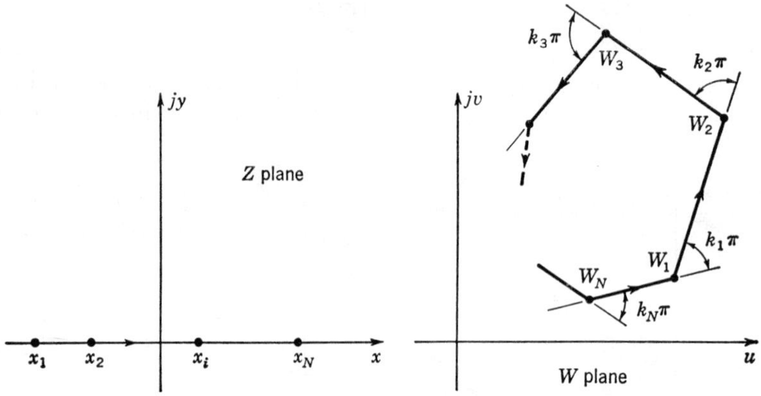


FIG. 4.10. The Schwarz-Christoffel transformation.

the unit tangent to the mapping of $S$, that is, to $S_1$, in the $W$ plane will be

$$\angle\tau_0 = \angle F' = \angle A - \sum_{i=1}^{N} k_i \angle(Z - x_i) \qquad (51)$$

For $x < x_i$, we have $\angle(x - x_i) = \pi$, and, for $x > x_i$, we have

$$\angle(x - x_i) = 0$$

Thus, as each point $x_i$ is passed in the $Z$ plane, the angle of $\tau_0$ changes discontinuously by an amount $k_i\pi$, and a polygon such as illustrated in Fig. 4.10 is traced out. The exterior angles to the polygon are $k_i\pi$, and these angles must add up to $2\pi$ if the polygon is to be closed; hence,

$$\sum_{i=1}^{N} k_i = 2 \qquad (52)$$

The constant $A$ serves to rotate and magnify the figure in the $W$ plane. The points $x_i$ map into the points $W_i = F(x_i)$. The mapping function is given by

$$W = \int^Z F'(Z)\,dZ + B = A \int^Z \prod_{i=1}^{N} (Z - x_i)^{-k_i}\,dZ + B \qquad (53)$$

where $B$ is an arbitrary constant which serves to translate the figure in the $W$ plane. The integration of (53) is usually not possible unless $|k_i| = 0, \frac{1}{2}, 1, \frac{3}{2}$, or 2. It should be noted that in this transformation there is a factor for each vertex corresponding to a finite value of $x$, but no terms for the points $x = \pm\infty$.

In practice, we would normally wish to map a given polygon in the $W$ plane into the $x$ axis. This requires the inverse mapping function

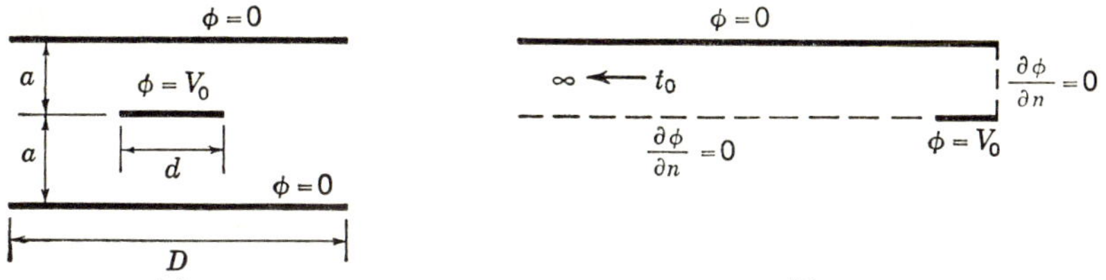


(a) (b)

FIG. 4.11. A symmetrical-strip transmission line.

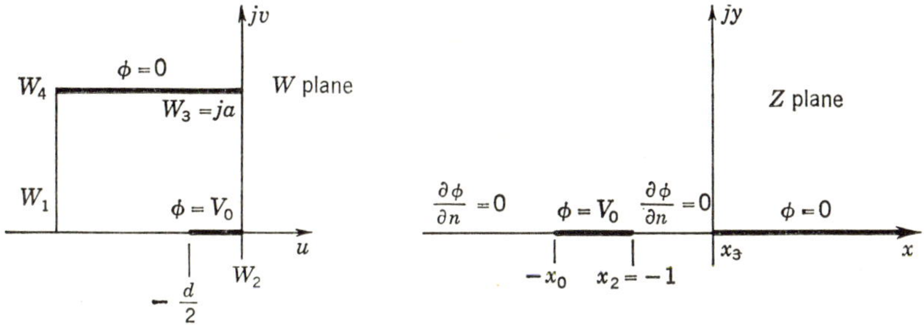


(a) (b)

FIG. 4.12. Conformal mapping for strip line.

giving $Z$ as a function of $W$. Generally, it is difficult to obtain this inverse transformation, and so we would usually proceed by trial and error to set up a transformation of the form (53) which will map the $x$ axis into the given polygon. In this procedure we are aided by the condition that three of the points $x_i$ may be arbitrarily chosen.

To illustrate the use of the Schwarz-Christoffel transformation, we will find the capacitance per unit length of the symmetrical-strip transmission line illustrated in Fig. 4.11$a$. If the spacing $a$ is small compared with the width $D$ of the two outer strips, the field is essentially zero at the edges of the outer strips, and, hence, we may simplify the problem by assuming that the outer strips are infinitely wide. The center strip is assumed to have negligible thickness and the medium between the ground planes to be free space with electrical parameters $\epsilon_0$, $\mu_0$.

From a consideration of symmetry we may reduce the electrostatic boundary-value problem to that illustrated in Fig. 4.11$b$. Along the dotted portion of the boundary, the normal gradient of the potential $\phi$ is zero. With reference to Fig. 4.12$a$, the capacitor configuration is seen to be obtained as tne points $W_1$ and $W_4$ approach $-\infty$. As a first step toward the solution of this boundary-value problem, we will map the boundary of the polygon in the $W$ plane into the real axis in the $Z$ plane, as in Fig. 4.12*b*. The external angles of the polygon are equal to $\pi/2$, and so all the parameters $k_i$ are equal to $+\frac{1}{2}$. The mapping function is of the following form:

$$
\begin{aligned}
W &= A' \int^Z (Z - x_1)^{-1/2}(Z - x_2)^{-1/2}(Z - x_3)^{-1/2}(Z - x_4)^{-1/2}\,dZ + B \\
&= A \int^Z \left[\left(1 - \frac{Z}{x_1}\right)\left(1 - \frac{Z}{x_4}\right)(Z - x_3)(Z - x_2)\right]^{-1/2} dZ + B
\end{aligned}
$$

where $A = A'x_1x_4$. In order to cover the whole $x$ axis, we let $x_1$ and $x_4$ tend to infinity, and the corresponding terms in the integral may be replaced by unity. If we also choose $x_2 = -1$ and $x_3 = 0$, our mapping function becomes

$$
\begin{aligned}
W &= A \int^Z \frac{dZ}{[Z(Z+1)]^{1/2}} + B \\
&= 2A \ln\,[Z^{1/2} + (Z+1)^{1/2}] + B
\end{aligned} \tag{54}
$$

When $W = 0$, $Z = -1$, and, when $W = ja$, $Z = 0$; so Eq. (54) gives

$$
\begin{aligned}
2A \ln\,(-1)^{1/2} &= j\pi A = -B \\
ja &= B
\end{aligned}
$$

and, hence, our final transformation becomes

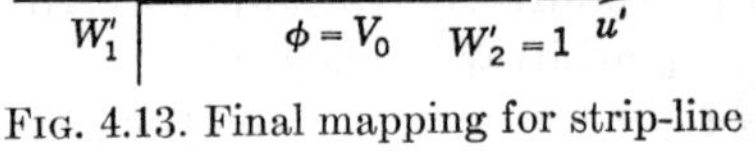


FIG. 4.13. Final mapping for strip-line cross section.

$$
W = -\frac{2a}{\pi} \ln\,[Z^{1/2} + (Z+1)^{1/2}] + ja \tag{55}
$$

When $W = -d/2$, let the corresponding value of $Z$ be $-x_0$. From Eq. (55) we then get

$$
x_0^{1/2} + (x_0 - 1)^{1/2} = e^{d\pi/4a}
$$

or

$$
x_0 = \cosh^2 \frac{\pi d}{4a} \tag{56}
$$

since $Z = -x_0$ is a negative number.

Up to this point in the analysis we are really no better off than we were at the beginning since the boundary-value problem in the $Z$ plane (Fig. 4.12*b*) is just as difficult to solve as the original one. We therefore will attempt to find another mapping function which will map the $x$ axis in the $Z$ plane into a rectangle in the $W'$ plane as illustrated in Fig. 4.13. If this can be done, we may determine the capacitance at once, since the final boundary-value problem in the $W'$ plane is an elementary one. The capacitance is obviously given by $C = \epsilon_0/v_0$, where $v_0$ is the plate spacing in the $W'$ plane. In the $W'$ plane we make $W'_1 = 0$ correspond to the point $Z = -x_0$, $W'_2 = 1$ correspond to $Z = -1$, and $W'_3 = 1 + jv_0$ correspond to $Z = 0$. Our required mapping function thus has the form

$$W' = A_1 \int^Z \frac{dZ}{[Z(Z + x_0)(Z + 1)]^{1/2}} + B_0 \tag{57}$$

This integral cannot be evaluated in terms of elementary functions. It does, however, represent an inverse elliptic function, and tables are available for its evaluation. To evaluate the constants $A_1$ and $B_0$, we substitute the values $Z = -x_0, -1, 0$ and the corresponding values of $W'$ into (57). Since the upper limit of integration is a negative number or zero, we make the following change in variables, $Z = -\lambda^2$, and (57) becomes

$$W' = \frac{2jA_1}{x_0^{1/2}} \int^{(-Z)^{1/2}} \frac{d\lambda}{[(1 - \lambda^2)(1 - \lambda^2/x_0)]^{1/2}} + B_0 \tag{58}$$

The inverse elliptic function $\text{sn}^{-1}(x,k)$ is defined by the following integral:

$$\text{sn}^{-1}(x,k) = \int_0^x \frac{d\lambda}{[(1 - \lambda^2)(1 - k^2\lambda^2)]^{1/2}} \tag{59}$$

where $x$ may be a complex variable, and $k$ is called the modulus of the elliptic function. Since $B_0$ is as yet arbitrary in (58), we may put in a lower limit of integration and, by introducing a new constant

$$A_0 = \frac{2jA_1}{x_0^{1/2}}$$

we get

$$W' = A_0 \,\text{sn}^{-1}[(-Z)^{1/2}, x_0^{-1/2}] + B_0 \tag{60}$$

since $x_0^{-1/2}$ corresponds to $k$ in (59). When $k^2 = 0$, that is, $x_0$ infinite, the integration in (59) is easily performed and gives

$$\int_0^{(-Z)^{1/2}} \frac{d\lambda}{(1 - \lambda^2)^{1/2}} = \sin^{-1}(-Z)^{1/2}$$

The particular elliptic function occurring in (60) is a generalization of the inverse sine function and the notation $\text{sn}^{-1}[(-Z)^{1/2},k]$ has been chosen for this reason.† The elliptic function $x = f = \text{sn}\, W$, where $W$ is given by the integral in (59), is a *doubly periodic function*, i.e., a function which

† For a brief treatment of the elliptic functions, see P. M. Morse and H. Feshbach, "Methods of Theoretical Physics," part I, chap. 4, McGraw-Hill Book Company, Inc., New York, 1953.

For a more complete treatment see H. Hancock, "Lectures on the Theory of Elliptic Functions," Dover Publications, New York, 1958.

is periodic along two different directions in the complex plane. In a unit cell or rectangle with sides corresponding to the two periods, the function $f = \text{sn}\, W$ takes on all its possible values as $W$ varies throughout the unit cell. Consequently, as $f$ takes on all its possible values, the inverse function $W = \text{sn}^{-1} f$ varies throughout the unit cell or rectangle. It is for these reasons that inverse elliptic functions always occur in the mapping of finite rectangular polygons. As one of the sides of the polygon or rectangle becomes infinite, the inverse elliptic functions degenerate into inverse trigonometric or hyperbolic functions.

Returning to our problem and putting in the boundary conditions $Z = -x_0$ for $W' = 0$, $Z = -1$ for $W' = 1$, and $Z = 0$ for $W' = 1 + jv_0$, we get

$$A_0 \,\text{sn}^{-1}\,(x_0^{1/2},x_0^{-1/2}) + B_0 = 0 \tag{61a}$$

$$A_0 \,\text{sn}^{-1}\,(1,x_0^{-1/2}) + B_0 = 1 \tag{61b}$$

$$B_0 = 1 + jv_0 \tag{61c}$$

Solving (61*a*) and (61*b*) for $B_0$ and substituting into (61*c*) gives

$$v_0 = \frac{-j\,\text{sn}^{-1}\,(1,x_0^{-1/2})}{\text{sn}^{-1}\,(x_0^{1/2},x_0^{-1/2}) - \text{sn}^{-1}\,(1,x_0^{-1/2})} \tag{62}$$

The function $\text{sn}\,(x + jy)$ is periodic in $x$ with period $4K$, where

$$K = \int_0^1 \frac{d\lambda}{[(1-\lambda^2)(1-k^2\lambda^2)]^{1/2}} \tag{63}$$

and periodic in $y$ with period $2K'$, where

$$K' = \int_0^1 \frac{d\lambda}{[(1-\lambda^2)(1-\lambda^2+k^2\lambda^2)]^{1/2}} = \int_1^{1/k} \frac{d\lambda}{[(\lambda^2-1)(1-k^2\lambda^2)]^{1/2}} \tag{64}$$

Also $\text{sn}\,K = 1$, and $\text{sn}\,(K - jK') = 1/k$; so $\text{sn}^{-1}\,(1,k) = K$, and $\text{sn}^{-1}\,(1/k,k) = K - jK'$. Hence, (62) gives

$$v_0 = \frac{K}{K'} = \frac{K(k)}{K(k')} \tag{65}$$

where $k' = (1 - k^2)^{1/2}$, and $K$ as given by (63) is called the complete elliptic integral of the first kind and is tabulated.† In Table 4.1 the values of $x_0$, $K$, $K'$, and $v_0$ are given for several typical values of the parameter $d/a$. The value of $x_0$ for a given value of $d/a$ is determined by (56). The capacitance per unit length of the strip line is four times that of the mapping of one quarter section in the $W'$ plane, that is, $C = 4\epsilon_0/v_0$. The characteristic impedance is given by $Z_c = 30\pi v_0$ for

† E. Jahnke and F. Emde, "Tables of Functions," 4th ed., Dover Publications, New York, 1945. This reference also contains several relief drawings which clearly illustrate the double-periodicity property of the Jacobi elliptic functions.

an air-filled line. Values of $Z_c$ are also given in the table as well as the approximate values $Z_{c,\text{approx}}$ determined by considering only the parallel-plate capacitance $2d\epsilon_0/a$. For $d/a > 4$, the error is much less than 1 per cent if $K$ is replaced by $\pi/2$ and $K'$ by $\ln 2 + \pi d/4a$. It is thus seen that, for $d/a$ sufficiently large, the ln 2 term is negligible, and the parallel-plate capacitance, neglecting fringing fields, is all that needs to be taken into account.

TABLE 4.1

| $d/a$ | $x_0$ | $K$ | $K'$ | $v_0$ | $Z_c$ | $Z_{c,\text{approx}}$ |
|---|---|---|---|---|---|---|
| 2 | 6.3 | 1.64 | 2.26 | 0.695 | 65.5 | 94.2 |
| 4 | 134 | 1.57 | 3.84 | 0.41 | 38.6 | 47.1 |
| 8 | $0.25e^{4\pi}$ | 1.57 | 6.98 | 0.225 | 21.2 | 23.6 |
| 10 | $0.25e^{5\pi}$ | 1.57 | 8.55 | 0.184 | 17.3 | 18.8 |
| 20 | $0.25e^{10\pi}$ | 1.57 | 16.4 | 0.0958 | 9.03 | 9.4 |

Once the required conformal transformations have been obtained, the capacitance per unit length and the characteristic impedance are quite straightforward to evaluate. Unfortunately, this is not true for the series resistance of the line since this depends on $(\partial\phi/\partial n)^2$. The charge density as obtained, in say the $W'$ plane, according to the relation $\epsilon_0(\partial\phi/\partial v')$ must be multiplied by $\left|\frac{dW'}{dZ}\right|\left|\frac{dZ}{dW}\right| = \left|\frac{dW'}{dW}\right|$ to obtain the true charge density corresponding to that on the actual conductors. The resultant expression is often very complicated to integrate in closed form. In practice, it is not realistic to deal with infinitely thin conducting strips with acute angles, since the charge density becomes infinite at such an edge and leads to values of attenuation and conductor losses which are too large and may even become infinite.† For the purpose of evaluating the capacitance, the resulting error introduced by replacing a thin strip by an infinitely thin strip is usually negligible. For evaluating the attenuation we may replace the infinitely thin strip by a strip which coincides with one of the original equipotential contours surrounding the thin strip. The equipotential contour has rounded corners and is often a better approximation to the conductor cross section than an ideal rectangular cross section. In addition, it has also the great advantage that the same mapping function used to map the infinitely thin strip also maps the desired equipotential contour, and the final mapping of the conductor cross section is normally a simple contour in the final mapped representation of the strip-line cross section. The above ideas will be clarified by an application to the strip line originally considered.

† See Prob. 4.6.

Figure 4.14 illustrates the original strip-line cross section and an equipotential contour $\phi = V_1$, which is now taken to be the cross section of the actual center conductor. The mapping of one quadrant of this equipotential contour onto the $W'$ plane is also illustrated in the same figure. If the separation of the two contours $\phi = V_0$ and $\phi = V_1$ is designated as $\Delta v$ in the $W'$ plane, the conductor elliptical cross section in the $uv$ plane is determined by finding the $u$ and $v$ values corresponding to $jv' = j\,\Delta v$, $0 < u' < 1$, by means of the two mapping functions (55)

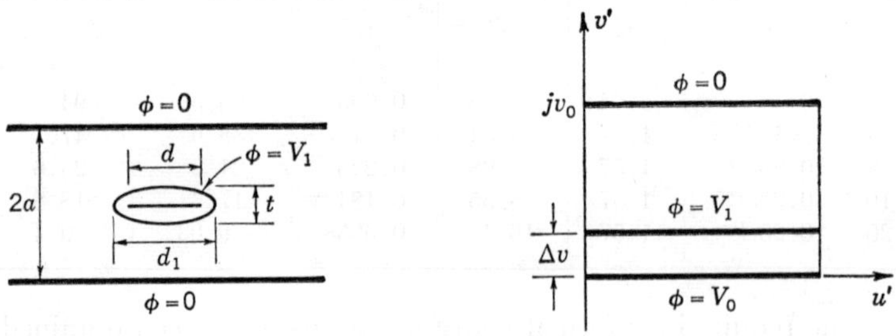


FIG. 4.14. Mapping of equipotential contour corresponding to conductor cross section of finite thickness.

and (60). The spacing $\Delta v$ in the $W'$ plane has to be chosen so as to give the correct thickness $t$ for the conductor cross section in the actual line. It should be noted that the conductor width has also been modified from its zero-thickness width $d$ to a new width $d_1$ which again depends on $\Delta v$. If $\Delta v \ll v_0$, the capacitance in the $W'$ plane is changed by a negligible amount; that is, $\Delta C = \epsilon_0\,\Delta v[v_0(v_0 - \Delta v)]^{-1}$, and the thickness $t$ may be assumed zero for the purpose of evaluating the capacitance.

From (23) the conductor loss is given by

$$P_L = \tfrac{1}{2}R_m \oint_W |\mathbf{J}_s|^2|dW| = \tfrac{1}{2}R_m Y_0^2 \oint_W \left|\frac{\partial\phi}{\partial n}\right|^2 |dW| \tag{66}$$

since

$$|\mathbf{J}_s| = Y_0 \left|\frac{\partial\phi}{\partial n}\right|$$

where $\partial\phi/\partial n$ is the normal gradient at the conductor surface, and the integral is taken around the conductor cross sections in the $W$ plane with $|dW| = dl$ being the element of path length. The value of $\partial\phi/\partial n$ is given by

$$\left|\frac{\partial\phi}{\partial n}\right| = \left|\frac{\partial\phi}{\partial v'}\right|\left|\frac{dW'}{dW}\right| = \frac{V_0}{v_0}\left|\frac{dW'}{dW}\right| \tag{67}$$

since

$$\frac{\partial\phi}{\partial v'} = \frac{\partial}{\partial v'}\frac{V_1 v'}{v_0 - \Delta v} = \frac{V_1}{v_0 - \Delta v} = \frac{V_0}{v_0}$$

The conductor losses are, therefore, given by

$$P_L = \tfrac{1}{2} R_m Y_0^2 \frac{V_0^2}{v_0^2} I \tag{68}$$

where

$$I = \oint_W \left| \frac{dW'}{dW} \right|^2 |dW|$$

The integral $I$ may also be evaluated in either the $Z$ or $W'$ plane. Since

$$\left| \frac{dW'}{dW} \right| |dW| = \left| \frac{dW'}{dW} \right| \left| \frac{dW}{dZ} \right| |dZ| = \left| \frac{dW'}{dZ} \right| |dZ| = |dW'|$$

we get

$$I = \oint_W \left| \frac{dW'}{dW} \right|^2 |dW| = 4 \oint_Z \left| \frac{dW'}{dW} \right| \left| \frac{dW'}{dZ} \right| |dZ| = 4 \oint_{W'} \left| \frac{dW'}{dW} \right| |dW'| \tag{69}$$

The factor 4 arises in the integration in the $Z$ and $W'$ planes because only one quadrant of the strip-line cross section has been mapped. The particular plane in which to carry out the integration is dependent on the complexity of the integrand. From (55), we get

$$\left| \frac{dW}{dZ} \right| = \frac{a}{\pi} |[Z(Z+1)]^{-\frac{1}{2}}|$$

while, from (60),

$$\left| \frac{dW'}{dZ} \right| = \frac{|A_0| x_0^{\frac{1}{2}}}{2} |[Z(Z+x_0)(Z+1)]^{-\frac{1}{2}}|$$

where $A_0 = [\text{sn}^{-1}(1,k) - \text{sn}^{-1}(k^{-1},k)]^{-1} = (jK')^{-1}$, with $k^2 = x_0^{-1}$. Substituting into (69) gives

$$\begin{aligned} I &= \frac{\pi}{a(kK')^2} \oint_Z \frac{|dZ|}{|(Z+x_0)[Z(Z+1)]^{\frac{1}{2}}|} = 4 \oint_{W'} \left| \frac{dW'}{dZ} \right| \left| \frac{dZ}{dW} \right| |dW'| \\ &= \frac{2\pi}{K'a} \oint_{W'} \frac{|dW'|}{|[1 - k^2 \, \text{sn}^2 (jK'W' - jK' + K, k)]^{\frac{1}{2}}|} \end{aligned} \tag{70}$$

since, from (60), $(-Z)^{\frac{1}{2}} = \text{sn}\,[(W' - B_0)/A_0] = \text{sn}\,(jK'W' - jK' + K, k)$. By means of the formulas and relations between the various elliptic functions tabulated in the book by Jahnke and Emde, we shall be able to perform the integration in the $W'$ plane along the contour $v' = v_0$, $0 < u' < 1$, corresponding to the ground planes in the strip line. By considering $\Delta v$ very small, we shall also be able to obtain an approximate value for the integral along $v' = \Delta v$, $0 < u' < 1$, corresponding to the center conductor.

For $W' = u' + jv_0 = u' + jK/K'$, we have

$$\begin{aligned} \text{sn}\,(+jK'W' - jK' + K, k) &= \text{sn}\,(+jK'u' - jK', k) \\ &= [k \, \text{sn}\,(jK'u',k)]^{-1} = [jk \, \text{tn}\,(K'u',k')]^{-1} \end{aligned}$$

where $k'$ is the complementary modulus $(1 - k^2)^{1/2}$. Along this contour the integrand becomes

$$[1 - k^2 \operatorname{sn}(jK'W' - jK' + K, k)]^{-1/2} = \operatorname{sn}(K'u',k')$$

upon using the relations $\operatorname{tn}(K'u',k') = \operatorname{sn}(K'u',k')/\operatorname{cn}(K'u',k')$ and $\operatorname{sn}^2(K'u',k') + \operatorname{cn}^2(K'u',k') = 1$. If we denote the part of $I$ corresponding to the integration along the $v' = v_0$ contour by $I_1$, we have

$$\begin{aligned} I_1 = \frac{2\pi}{aK'}\int_0^1 \operatorname{sn}(K'u',k')\,du' &= \frac{2\pi}{aK'^2}\int_0^{K'} \operatorname{sn}(K'u',k')\,d(K'u') \\ &= \frac{2\pi}{ak'K'^2}\ln\frac{\operatorname{dn}(K',k') - k'\operatorname{cn}(K',k')}{1-k'} \end{aligned} \tag{71}$$

where $\operatorname{dn}^2(K',k') = 1 - k'^2\operatorname{sn}^2(K',k')$. Now $\operatorname{dn}(K',k') = k$,

$$\operatorname{cn}(K',k') = 0$$

(71) becomes

$$I_1 = \frac{2\pi}{ak'K'^2}\ln\frac{k}{1-k'} \tag{72}$$

and the conductor loss arising from the ground planes is

$$P_{L,1} = \frac{\pi R_m Y_0^2 V_0^2}{aK^2k'}\ln\frac{k}{1-k'} \tag{73}$$

where $K/K'$ has been substituted for $v_0$. For $d \geq 4a$, this expression simplifies considerably since under these conditions we have, to an accuracy better than 1 per cent,

$$\begin{aligned} k &= x_0^{-1/2} = 2e^{-\pi d/4a} \\ k' &= 1 - 2e^{-\pi d/2a} \approx 1 \\ \frac{k}{1-k'} &= e^{\pi d/4a} \\ K &= \frac{\pi}{2} \end{aligned}$$

and (73) becomes

$$P_{L,1} = R_m Y_0^2 \left(\frac{V_0}{a}\right)^2 d \tag{74}$$

which is just the loss which would arise for a uniform field existing between the center strip and ground planes and a zero field outside this region.

For the second part of the integral, which we will denote as $I_2$, the contour is along $v' = \Delta v$, $0 < u' < 1$. In this case

$$\operatorname{sn}[jK'(u' + j\,\Delta v) - jK' + K, k] = k^{-1}\operatorname{dn}(K'u' + jK'\,\Delta v, k')$$

and, since $1 - \text{dn}^2\,(\;\;) = k'^2\,\text{sn}^2\,(\;\;)$, we get

$$I_2 = \frac{2\pi}{aK'k'} \int_0^1 \frac{du'}{|\text{sn}\,(K'u' + jK'\,\Delta v,\,k')|}$$
$$= \frac{2\pi}{ak'K'^2} \int_0^{K'} \frac{du}{|\text{sn}\,(u + jK'\,\Delta v,\,k')|} \tag{75}$$

If we choose $\Delta v \leq 0.1 v_0$ and also assume that $d \geq 4a$, then

$$K'\,\Delta v = K\,\frac{\Delta v}{v_0} \leq \frac{\pi}{20}$$

and $\text{sn}\,(u + jK'\,\Delta v,\,k') \approx \text{sn}\,(u,k') + jK'\,\Delta v\,\text{cn}\,(u,k')\,\text{dn}\,(u,k')$. The modulus of this term is

$$[\text{sn}^2\,(u,k') + (K'\,\Delta v)^2\,\text{cn}^2\,(u,k')\,\text{dn}^2\,(u,k')]^{1/2} \approx [\text{sn}^2\,(u,k') + (K'\,\Delta v)^2]^{1/2}$$

since, for $u$ near zero, $\text{cn}\,(u,k')$ and $\text{dn}\,(u,k')$ are nearly equal to unity. When $u = K'$, $\text{cn}\,(K',k')$ equals zero, but now $\text{sn}\,(K',k')$ equals unity, and so there is negligible error in still having the term $(K'\,\Delta v)^2$, which is less than 0.03, present. For values of $u$ up to $u_1$, where $u_1$ is about 0.4 or less, $\text{sn}\,(u,k') \approx u$. For values of $u > u_1$, we may neglect the term $(K'\,\Delta v)^2$ in comparison with $\text{sn}^2\,(u,k')$, and this also removes the error at the upper end of the range, where $u = K'$. The integral $I_2$ is thus evaluated in two parts, corresponding to the two ranges in $u$, where different approximations are used. This artifice permits us to carry out the integrations readily, and we get

$$\begin{aligned}
I_2 &= \frac{2\pi}{ak'K'^2}\left[\int_0^{u_1} \frac{du}{[u^2 + (K'\,\Delta v)^2]^{1/2}} + \int_{u_1}^{K'} \frac{du}{\text{sn}\,(u,k')}\right] \\
&= \frac{2\pi}{ak'K'^2}\left\{\ln \frac{u_1 + [u_1^2 + (K'\,\Delta v)^2]^{1/2}}{K'\,\Delta v} + \ln \frac{\text{cn}\,(u_1,k') + \text{dn}\,(u_1,k')}{k\,\text{sn}\,(u_1,k')}\right\} \\
&\approx \frac{2\pi}{ak'K'^2}\left(\ln \frac{2u_1}{K'\,\Delta v} + \ln \frac{2}{ku_1}\right) \\
&= \frac{2\pi}{ak'K'^2} \ln \frac{4}{kK'\,\Delta v}
\end{aligned} \tag{76}$$

since $u_1$ is small and $\ln kK'\,\Delta v$ is the dominant term. Before proceeding any further, it is necessary to establish the value of $\Delta v$ in terms of the thickness $t$ of the center conductor. Provided $t$ and $\Delta v$ are small, this may be done directly from the relation $\Delta W = (dW/dW')\,\Delta W'$, where $\Delta W = j(t/2)$ and $\Delta W' = j\,\Delta v$ for small displacements from the origin along the $v$ axis. Employing the same approximations as those used to arrive at (76), we get

$$K'\,\Delta v = \frac{\pi t}{4a} \tag{77}$$

The conductor loss associated with the center conductor is now readily evaluated and is given by

$$P_{L,2} = \frac{R_m Y_0^2 V_0^2}{\pi a}\left(4 \ln \frac{8a}{\pi t} + \frac{\pi d}{a}\right) \tag{78}$$

The second term in parentheses is the loss which would arise on the assumption of a uniform field between the center strip and the ground planes, while the logarithmic term gives the increase in the losses arising from the concentration of the current at the edges of the center strip.

The power flowing along the line is given by $\frac{1}{2}V_0^2/Z_c$, and, hence, by using (25), the attenuation factor arising from conductor losses is found to be

$$\alpha = \frac{R_m}{2Z_0 a}\,\frac{\pi d/2a + \ln (8a/\pi t)}{\ln 2 + \pi d/4a} \qquad \text{nepers/unit length} \tag{79}$$

a result valid for $d \geq 4a$ and $t \leq a/5$.

## 4.4. Characteristic Impedance by Variational Methods

*Variational Expression for Lower Bound on $Z_{c0}$*

The electrostatic energy per unit length stored in the field surrounding the two conducting surfaces represented by the curves $S_1$ and $S_2$ in Fig. 4.1 is given by

$$W_e = \tfrac{1}{2}\epsilon \iint \left[\left(\frac{\partial \phi}{\partial x}\right)^2 + \left(\frac{\partial \phi}{\partial y}\right)^2\right] dx\,dy = \tfrac{1}{2}C_0(\phi_2 - \phi_1)^2 = \tfrac{1}{2}C_0 V_0^2 \tag{80}$$

where $C_0$ is the capacitance of the line per unit length, and $V_0 = \phi_2 - \phi_1$ and is the potential difference between the two conductors. Without loss in generality, we may take $\phi_1 = 0$.

If we compute the first-order variation in $W_e$ due to a first-order variation in the functional form of $\phi$, subject to the condition that $\phi = 0$ on $S_1$ and $\phi = \phi_2$ on $S_2$, we find, by applying the Euler-Lagrange equations to the integrand, that this first-order variation vanishes, provided $\phi$ satisfies the equation†

$$\frac{\partial^2 \phi}{\partial x^2} + \frac{\partial^2 \phi}{\partial y^2} = \nabla_t^2 \phi = 0 \tag{81}$$

(The result also follows from Thomson's theorem given in Sec. 1.3.) Since we know that $\phi$ is a solution of Laplace's equation, we obtain at once the following variational expression for the capacitance and, hence, for the characteristic impedance $Z_{c0}$:

$$\frac{C_0}{(\epsilon\mu)^{1/2}} = \frac{1}{Z_{c0}} = \frac{\iint \nabla_t \phi \cdot \nabla_t \phi \, dx\,dy}{(\mu/\epsilon)^{1/2} V_0^2} \tag{82}$$

† Section A.4.

The integral is always positive, and, hence, the stationary value is an absolute minimum.† Therefore, if we substitute an approximate value for $\nabla_t\phi$ into the integral, the calculated value $C$ for the capacitance will always be too large; that is, $C > C_0$, $Z_c < Z_{c0}$. The calculated value of $Z_c$ is always too small, i.e., is a lower bound to the true value $Z_{c0}$. In practice, we choose a functional form for our approximate solution that has several parameters $\alpha_1, \alpha_2, \ldots, \alpha_N$, by which we can change the functional form of $\phi$. When we compute the integral, we get a value $C(\alpha_1,\alpha_2, \ldots ,\alpha_N)$ for the capacitance which is an ordinary function of the variational parameters $\alpha_i$. We obtain the best possible solution

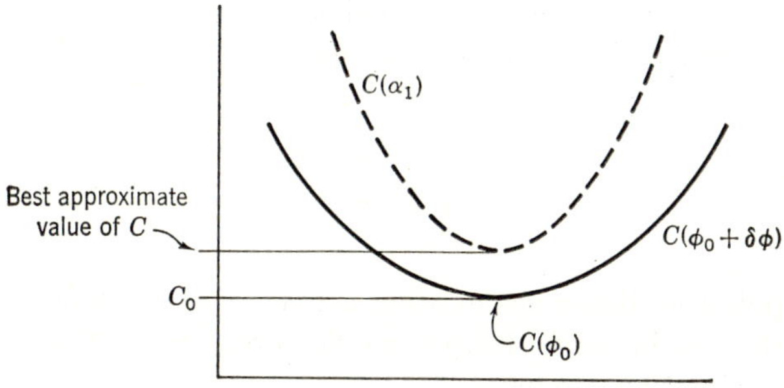


FIG. 4.15

for $C$ by choosing the minimum value of $C$ that varying the parameters $\alpha_i$ can produce. This required minimum is obtained by treating the $\alpha_i$ as independent variables and equating all $\partial C/\partial\alpha_i$ equal to zero; thus

$$\frac{\partial C}{\partial \alpha_i} = 0 \qquad i = 1, 2, \ldots, N \tag{83}$$

Equations (83) are a set of $N$ homogeneous equations which, together with the equation $\phi = V_0$ on $S_2$, allows us to solve for the $\alpha_i$ and, hence, compute the minimum value of $C$. Equations (83) only give the ratio of the $\alpha_i$ to any one, say $\alpha_1$, and, hence, the requirement of the additional condition $\phi = V_0$ on $S_2$. The variational parameters $\alpha_i$ may be the coefficients in a Fourier-series expansion of $\phi$.

We may illustrate what the above procedure does as follows. Let $\phi_0$ be the correct solution for the problem, and $\phi(\alpha_1)$, a function of one variational parameter only, be the approximate solution. Let $\delta\phi$ be a variation in the functional form of $\phi_0$. If we plot the value of $C$ as a function of $\phi_0 + \delta\phi$ for a range of functions $\delta\phi$, we get a curve as illustrated in Fig. 4.15. The true value $C_0$ is obtained when $\delta\phi = 0$. The dashed curve shows a plot of the calculated value of the capacitance as a function of the parameter $\alpha_1$ in our approximate solution. The best

† See Sec. 1.3, where it is shown that the change in $W_e$ is always positive.

possible solution is the minimum value as obtained by equating $dC(\alpha_1)/d\alpha_1$ to zero. If our function $\phi(\alpha_1)$ is sufficiently flexible, it will be able to approximate $\phi_0$ closely, and, since $C$ varies slowly in the vicinity of $\phi_0$ [$\delta C$ is of the order of $\iint|(\nabla_t\delta\phi)|^2\,dx\,dy$], the calculated value of $C$ will be an excellent approximation to $C_0$. The more variational parameters $\alpha_i$ we include in our approximation, the more nearly will $\phi$ be able to approximate $\phi_0$ and yield a closer approximation to $C_0$. Unfortunately, the numerical work involved increases rapidly with an increase in the number of parameters $\alpha_i$.

As a simple example, we will find the capacitance of the two coaxial cylinders illustrated in Fig. 4.7. For a first approximation, we take $\phi = \alpha_0 + \alpha_1 r$. Rather than have to choose our $\phi$ so that $\phi = 0$ on the inner conductor, and $\phi = V_0$ on the outer conductor, we can modify (82) to

$$C = \epsilon\frac{\iint|\nabla_t\phi|^2\,dx\,dy}{\left(\int_{S_1}^{S_2}\nabla_t\phi\cdot d\mathbf{l}\right)^2} \tag{84}$$

since the potential difference existing between the conductors $S_1$ and $S_2$ is simply the line integral of $\nabla_t\phi$ from $S_1$ to $S_2$. Since $\nabla_t\phi = \alpha_1$ for the above approximation, we have

$$C = \epsilon\frac{\int_0^{2\pi}\int_a^b \alpha_1{}^2 r\,dr\,d\theta}{\left(\int_a^b \alpha_1\,dr\right)^2} = \epsilon\pi\frac{b+a}{b-a} \tag{85}$$

The true value of capacitance is

$$C_0 = \frac{2\pi\epsilon}{\ln(b/a)}$$

For $b > a$, we have

$$\ln\frac{b}{a} = 2\left[\frac{b-a}{b+a} + \frac{1}{3}\left(\frac{b-a}{b+a}\right)^3 + \cdots\right] \approx 2\frac{b-a}{b+a}$$

when $b$ is not much greater than $a$. Thus we have

$$\frac{\pi\epsilon(b+a)}{b-a} = C \approx C_0$$

when $b - a$ is small.

If $b = 2a$, $C = 3\pi\epsilon$, $C_0 \approx 2.9\pi\epsilon$, and so the error is around 3 per cent. We note that $\alpha_0$ does not enter into the expression for $C$. The reason is that the potential is unique only to within an arbitrary constant. The only value of any significance is the difference in potential.

As a second approximation, take $\phi = \alpha_0 + \alpha_1 r + \alpha_2 r^2$; thus

$$C = \epsilon \frac{\int_0^{2\pi} \int_a^b (\alpha_1 + 2\alpha_2 r)^2 r\, dr\, d\theta}{\left[ \int_a^b (\alpha_1 + 2\alpha_2 r)\, dr \right]^2}$$

$$= \frac{2\pi\epsilon \left[ \dfrac{b^2 - a^2}{2} \alpha_1^2 + \dfrac{b^3 - a^3}{3} 4\alpha_1\alpha_2 + (b^4 - a^4)\alpha_2^2 \right]}{(b - a)^2\alpha_1^2 + (b^2 - a^2)(b - a)2\alpha_1\alpha_2 + (b^2 - a^2)^2\alpha_2^2} \tag{86a}$$

Hence, $$\frac{C}{\pi\epsilon} [(b - a)^2\alpha_1^2 + (b^2 - a^2)(b - a)2\alpha_1\alpha_2 + (b^2 - a^2)^2\alpha_2^2]$$
$$- [(b^2 - a^2)\alpha_1^2 + \tfrac{8}{3}(b^3 - a^3)\alpha_1\alpha_2 + 2(b^4 - a^4)\alpha_2^2] = 0 \tag{86b}$$

In order to obtain the best possible value of $C$ with this approximation, we equate $\partial/\partial\alpha_i$ to zero in Eq. (86*b*). This is equivalent to equating $\partial C/\partial\alpha_i$ to zero in (86*a*). Let $D$ be the denominator in (86*a*) and $N$ be the numerator; thus

$$\frac{\partial C}{\partial \alpha_i} = \frac{D(\partial N/\partial\alpha_i) - N(\partial D/\partial\alpha_i)}{D^2} = \frac{1}{D}\left( \frac{\partial N}{\partial \alpha_i} - \frac{N}{D}\frac{\partial D}{\partial \alpha_i} \right) = 0$$

and hence $\partial N/\partial\alpha_i - C\, \partial D/\partial\alpha_i = 0$, which is just the derivative of (86*b*). Carrying out the differentiation, we obtain the following two homogeneous equations in $\alpha_1$ and $\alpha_2$:

$$\alpha_1 \left[ \frac{2C}{\pi\epsilon} (b - a)^2 - 2(b^2 - a^2) \right] + \alpha_2 \left[ \frac{2C}{\pi\epsilon} (b^2 - a^2)(b - a) - \tfrac{8}{3}(b^3 - a^3) \right] = 0$$

$$\alpha_1 \left[ \frac{2C}{\pi\epsilon} (b^2 - a^2)(b - a) - \tfrac{8}{3}(b^3 - a^3) \right] + \alpha_2 \left[ \frac{2C}{\pi\epsilon} (b^2 - a^2)^2 - 4(b^4 - a^4) \right] = 0$$

For a solution for the $\alpha_i$ to exist, the determinant of the above equations must vanish, and this gives directly the value of $C$ without the necessity of evaluating the $\alpha_i$. By performing the following operations on this determinant:

1. Divide first row by $2(b - a)^2$
2. Divide second row by $2(b^2 - a^2)(b - a)$
3. Divide second column by $b + a$
4. Subtract first row from second

it reduces to

$$\begin{vmatrix} \dfrac{C}{\pi\epsilon} - \dfrac{b+a}{b-a} & \dfrac{C}{\pi\epsilon} - \dfrac{4}{3}\dfrac{b^3-a^3}{(b^2-a^2)(b-a)} \\ \dfrac{b+a}{b-a} - \dfrac{4}{3}\dfrac{b^3-a^3}{(b^2-a^2)(b-a)} & \dfrac{4}{3}\dfrac{b^3-a^3}{(b^2-a^2)(b-a)} - 2\dfrac{b^4-a^4}{(b^2-a^2)^2} \end{vmatrix} = 0$$

If we expand this determinant in terms of the elements of the first row, we get

$$\frac{C}{\pi\epsilon}\begin{vmatrix} 1 & 1 \\ \dfrac{b+a}{b-a} - \dfrac{4}{3}\dfrac{b^3-a^3}{(b^2-a^2)(b-a)} & \dfrac{\frac{4}{3}(b^3-a^3)}{(b^2-a^2)(b-a)} - 2\dfrac{b^4-a^4}{(b^2-a^2)^2} \end{vmatrix}$$
$$= \left(\frac{b+a}{b-a}\right)^2 \begin{vmatrix} 1 & \dfrac{4}{3}\dfrac{b^3-a^3}{(b^2-a^2)(b+a)} \\ 1 - \dfrac{4}{3}\dfrac{b^3-a^3}{(b^2-a^2)(b+a)} & \dfrac{\frac{4}{3}(b^3-a^3)}{(b^2-a^2)(b+a)} - \dfrac{2(b^4-a^4)}{(b^2-a^2)(b+a)^2} \end{vmatrix} \tag{87}$$

There is only one root for $C$. When $b/a = 2$, this second approximation gives $C = 2.889\pi\epsilon$. The true value is $C_0 = 2.886\pi\epsilon$, and so the error in $C$ is only 0.1 per cent. However, this accuracy has been obtained at the expense of considerably more labor.

*Variational Expression for Upper Bound on $Z_{c0}$*

The previous section presented a method whereby the upper bound on $C$ and, hence, a lower bound on $Z_{c0}$ could be obtained. By solving the electrostatic problem in terms of the unknown charge distribution, we shall be able to obtain a variational expression which will give a lower bound on the capacitance and, hence, an upper bound on the characteristic impedance $Z_{c0}$. The method utilizes the Green's function technique for solving boundary-value problems. By comparing the upper and lower bounds for $Z_{c0}$, we can find at once the maximum possible error in our approximate values.

Consider the two-dimensional electrostatic problem illustrated in Fig. 4.16. The two conductors $S_1$ and $S_2$ are of infinite length and parallel to the $z$ axis. We may take $S_1$ at zero potential and $S_2$ at a potential $V_0$. In order that $S_2$ should be at a potential $V_0$, there must exist a distribution of positive electric charge on its surface. The only way in which this conductor $S_2$ can make its presence known to the field surrounding it is by this charge distribution on its surface. The magnitude of the charge is equal to $\epsilon E_n = -\epsilon(\partial\phi/\partial n)$, where $E_n$ is the normal electric field at the surface of $S_2$ and is equal to the normal derivative of the potential function $\phi$ on the surface $S_2$. Provided we leave the charge distribution undisturbed, we may remove the conductor $S_2$ without disturbing the field. We are then left with the problem of finding the potential function $\phi$ which satisfies Poisson's equation

$$\nabla_t^2\phi = -\frac{1}{\epsilon}\rho(x',y') \tag{88}$$

and the boundary condition that it vanishes on $S_1$. The coordinates $x'$, $y'$ are the values of $x$ and $y$ which defined the original conductor surface $S_2$, and $\rho(x',y')$ is the charge distribution which existed on $S_2$. As a

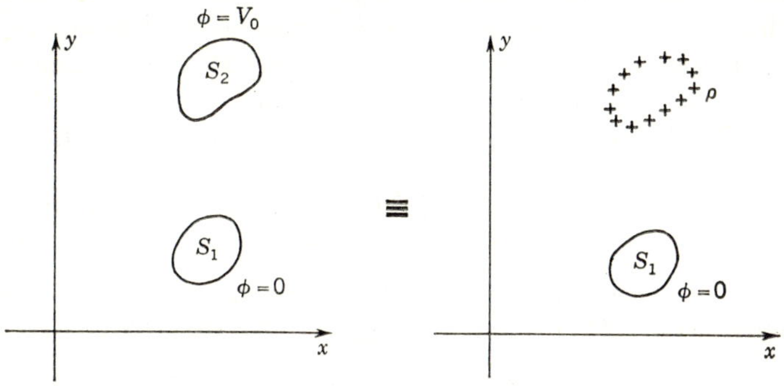


FIG. 4.16

first step toward the solution of Eq. (88), we will find the solution to the problem of a unit charge located at $(x',y')$ in the presence of $S_1$ and such that the potential vanishes on $S_1$. [Since the problem is a two-dimensional one, we mean by a unit charge at $(x',y')$ a unit line charge extending from $-\infty$ to $+\infty$ along the $z$ axis.] The most convenient way of representing a unit charge mathematically is by means of the delta function, which is zero for all values of $x$ and $y$ except $x'$, $y'$. At the point $(x',y')$, the function is such that, if we integrate it over a small interval including the point $(x',y')$, then the value of the integral is unity. We therefore require a solution of the following equation:

$$\nabla_t^2\phi(x,y) = -\frac{1}{\epsilon}\delta(x-x')\delta(y-y') \tag{89}$$

subject to the boundary condition that $\phi$ equals zero on $S_1$ and at infinity. The above equation defines the two-dimensional Green's function of the first kind for the given boundary.† Let the solution to (89) be represented by $G(x,y|x',y')$, and by superposition it follows that the solution to (88) is

$$\phi(x,y) = \oint_{S_2} G(x,y|x',y')\rho(x',y')\,dl' \tag{90}$$

On the surface $S_1$, the potential $\phi$ must reduce to zero, while, on $S_2$, it

† The Green's function for Poisson's equation may or may not include the factor $\epsilon$. We have included this factor here for later convenience, but this is not always done.

reduces to $V_0$. Hence, from (90), we have

$$\phi(x,y) = V_0 = \oint_{S_2} G(x,y|x',y')\rho(x',y')\,dl' \qquad x,y \text{ on } S_2 \tag{91}$$

Equation (91) is an integral equation involving the unknown charge distribution $\rho(x',y')$, whose solution would give $\rho(x',y')$ and, hence, would determine the potential function $\phi(x,y)$ everywhere by means of (90). This is the condition which must be imposed on the charge distribution in order that $\phi$ should reduce to $V_0$ on $S_2$. Since $G$ vanishes on $S_1$ and at infinity, the potential $\phi$ does also. Equation (91) is the basic equation from which we will now derive our variational expression for the capacitance $C_0$. Multiply (91) by $\rho(x,y)$, and integrate over the surface $S_2$; thus

$$V_0 \oint_{S_2} \rho(x,y)\,dl = V_0 Q = \oint\!\!\oint_{S_2} G(x,y|x',y')\rho(x,y)\rho(x',y')\,dl\,dl' \tag{92}$$

where $Q$ is the total charge on $S_2$. The capacitance $C_0$ is equal to $Q/V_0$, and, hence, $QV_0 = Q^2/C_0$, and we get from (92) the result

$$\frac{1}{C_0} = \frac{\oint\!\!\oint_{S_2} G(x,y|x',y')\rho(x,y)\rho(x',y')\,dl\,dl'}{\left[\oint_{S_2} \rho(x,y)\,dl\right]^2} \tag{93}$$

which is our required variational expression. To show that this expression is a stationary expression for $1/C_0$ for arbitrary first-order changes in the functional form of $\rho(x,y)$, let $\rho$ change by $\delta\rho$. Thus

$$\begin{aligned}\left[\oint_{S_2} \rho(x,y)\,dl\right]^2 \delta\left(\frac{1}{C_0}\right) + \frac{2}{C_0}\left[\oint_{S_2} \rho(x,y)\,dl \oint_{S_2} \delta\rho\,dl\right] &= \oint\!\!\oint_{S_2} G(x,y|x',y')\rho(x,y)\delta\rho(x',y')\,dl\,dl' \\ &\quad + \oint\!\!\oint_{S_2} G(x,y|x',y')\rho(x',y')\delta\rho(x,y)\,dl\,dl' \\ &= 2\oint\!\!\oint_{S_2} G(x,y|x',y')\rho(x',y')\delta\rho(x,y)\,dl\,dl'\end{aligned}$$

since $G$ is symmetrical in the variables $x$, $y$ and $x'$, $y'$, and similarly for $\rho$ and $\delta\rho$. Introducing the total charge $Q$, we may write the above equations as follows:

$$Q^2\delta\left(\frac{1}{C_0}\right) = 2\left\{\oint_{S_2} \delta\rho(x,y)\left[-\frac{Q}{C_0} + \oint_{S_2} G(x,y|x',y')\rho(x',y')\,dl'\right]dl\right\}$$

which shows at once that the variation in $1/C_0$ vanishes by virtue of (91), and that $Q = C_0V_0$. We have, therefore, proved that (93) is a stationary expression for $C_0^{-1}$, and, hence, it follows that, if we can find an approximate solution for $\rho(x',y')$, we will be able to compute a value for $C^{-1}$ which is correct to the second order. Since $G$ is a symmetrical function, the integral in (93) is always positive, and the stationary value must be an absolute minimum. Consequently, our approximate value of $C^{-1}$ is too large, and, hence, $C < C_0$, $Z_c > Z_{c0}$, and we have an upper bound for our characteristic impedance $Z_{c0}$.

In practice, the charge distribution would be approximated by a function containing one or more variational parameters, and the best possible solution for $C_0$ would be obtained by minimizing (93) with respect to the adjustable variational parameters. The procedure to be followed is the same as that used to treat Eq. (84).

## 4.5. Characteristic Impedance of a Strip Line by Variational Methods

*Lower Bound to* $Z_{c0}$

As an application of the variational methods given in the preceding section, the upper and lower bounds to the characteristic impedance of the shielded strip line with a rectangular center conductor will be obtained. The dimensions of the line are given in Fig. 4.17. In view

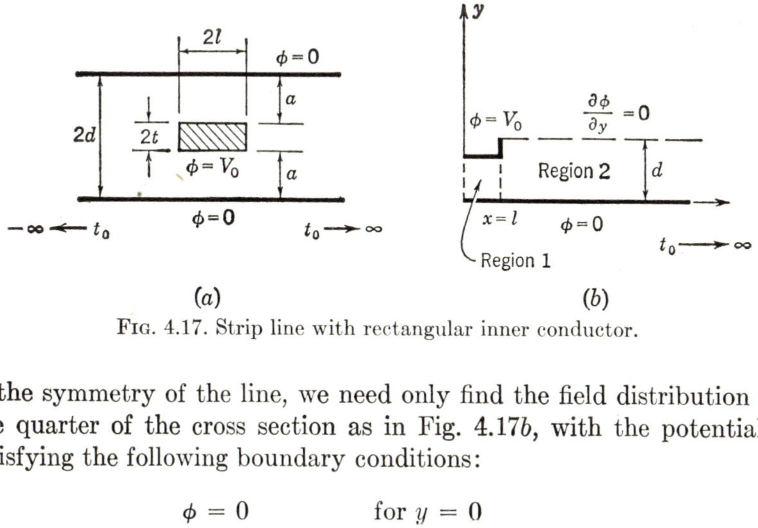


FIG. 4.17. Strip line with rectangular inner conductor.

of the symmetry of the line, we need only find the field distribution for one quarter of the cross section as in Fig. 4.17*b*, with the potential $\phi$ satisfying the following boundary conditions:

$$\begin{aligned} \phi &= 0 && \text{for } y = 0 \\ \phi &= V_0 && \text{on inner conductor} \\ -E_x &= \frac{\partial \phi}{\partial x} = 0 && x = 0 \qquad 0 \le y \le a \\ -E_y &= \frac{\partial \phi}{\partial y} = 0 && y = d \qquad l \le x \end{aligned}$$

In each region $\phi$ is a solution of Laplace's equation $\nabla_t^2\phi = 0$. In region 1, a suitable Fourier-series expansion of the potential is†

$$\phi = \frac{V_0 y}{a} + \sum_{n=1}^{\infty} a_n \cosh \frac{n\pi x}{a} \sin \frac{n\pi y}{a} \tag{94}$$

Each term $\cosh (n\pi x/a) \sin (n\pi y/a)$ satisfies Laplace's equation, and the derivative with respect to $x$ vanishes at $x = 0$. The term $V_0 y/a$ is included to satisfy the boundary condition at $y = a$, and the $a_n$ are unknown coefficients to be determined. In region 2, a suitable expansion of the potential is

$$\phi = \sum_{n=1,3,\ldots}^{\infty} b_n \sin \frac{n\pi y}{2d} e^{-n\pi x/2d} \tag{95}$$

since $\cos (n\pi y/2d)$ vanishes at $y = d$ for odd values of $n$, and, hence, $\partial\phi/\partial y$ does also. The potential vanishes as $x \to \infty$ because of the decaying exponential terms. The unknown coefficients $a_n$ and $b_n$ must be determined so that the electric field and potential are continuous at $x = l$, $0 \le y \le a$. Continuity of the potential at $x = l$ gives

$$V_0 \frac{y}{a} + \sum_{n=1}^{\infty} a_n \cosh \frac{n\pi l}{a} \sin \frac{n\pi y}{a} = \sum_{1,3,\ldots}^{\infty} b_n e^{-n\pi l/2d} \sin \frac{n\pi y}{2d} \tag{96}$$

Since we have made $\phi$ continuous at $x = l$, then $\partial\phi/\partial y$, and, hence, $E_y$, is also continuous, as may be seen at once by differentiating (96) with respect to $y$. The continuity of $E_x$ is ensured by making $\partial\phi/\partial x$ continuous, i.e.,

$$\sum_{1}^{\infty} \frac{n\pi}{a} a_n \sinh \frac{n\pi l}{a} \sin \frac{n\pi y}{a} = -\sum_{1,3,\ldots}^{\infty} \frac{n\pi}{2d} b_n e^{-n\pi l/2d} \sin \frac{n\pi y}{2d} \qquad 0 \le y \le a \tag{97}$$

A rigorous solution to Eqs. (96) and (97) is difficult to obtain, and, hence, we appeal to the stationary property of the electrostatic energy integral to obtain an approximate value for the capacitance of the line.

† This form of solution is obtained at once by using the finite Fourier sine transform. See R. V. Churchill, "Operational Mathematics," 2d ed., chap. 10, McGraw-Hill Book Company, Inc., New York, 1958.

The electrostatic energy stored in region 1 is

$$\tfrac{1}{2}\epsilon_0 \int_0^a \int_0^l \left[ \left( \sum_1^\infty \frac{n\pi}{a} a_n \sinh \frac{n\pi x}{a} \sin \frac{n\pi y}{a} \right)^2 + \left( \frac{V_0}{a} + \sum_1^\infty \frac{n\pi}{a} a_n \cosh \frac{n\pi x}{a} \cos \frac{n\pi y}{a} \right)^2 \right] dx\, dy = W_{e1}$$

Using the orthogonal properties of the sin $(n\pi y/a)$ and cos $(n\pi y/a)$ functions, i.e.,

$$\int_0^a \sin \frac{n\pi y}{a} \sin \frac{m\pi y}{a}\, dy = \begin{cases} 0 & n \neq m \\ \dfrac{a}{2} & n = m \end{cases}$$

and similarly for cos $(n\pi y/a)$, the integration is readily carried out to give

$$W_{e1} = \tfrac{1}{2}\epsilon_0 \left( \frac{V_0{}^2 l}{a} + \frac{\pi}{4} \sum_1^\infty a_n{}^2 n \sinh \frac{2n\pi l}{a} \right) \tag{98}$$

Similarly, the electrostatic energy stored in region 2 is found to be

$$W_{e2} = \tfrac{1}{2}\epsilon_0 \left( \frac{\pi}{4} \sum_{1,3,\ldots}^\infty b_n{}^2 n e^{-n\pi l/d} \right) \tag{99}$$

Let the unknown potential distribution at $x = l$ be

$$G(y) = \begin{cases} V_0 \left( \dfrac{y}{a} \right) + g(y) & 0 \le y \le a \\ V_0 & a \le y \le d \end{cases} \tag{100}$$

We may express the unknown coefficients $a_n$ and $b_n$ in terms of this potential distribution by Fourier-series analysis since both sides of (96) are equal to the potential distribution given by Eqs. (100); thus

$$a_n \cosh \frac{n\pi l}{a} = \frac{2}{a} \int_0^a g(y) \sin \frac{n\pi y}{a}\, dy \tag{101a}$$

$$b_n e^{-n\pi l/2d} = \frac{2}{d} \int_0^d G(y) \sin \frac{n\pi y}{d}\, dy \tag{101b}$$

Substituting into Eqs. (98) and (99), we get

$$W_e = W_{e1} + W_{e2} = \tfrac{1}{2}\epsilon_0 \frac{V_0^2 l}{a} + \frac{\pi\epsilon_0}{a^2} \sum_1^\infty n \tanh \frac{n\pi l}{a} \left[ \int_0^a g(y) \sin \frac{n\pi y}{a}\, dy \right]^2 + \frac{\pi\epsilon_0}{2d^2} \sum_{1,3,\ldots}^\infty n \left[ \int_0^d G(y) \sin \frac{n\pi y}{2d}\, dy \right]^2 \qquad (102)$$

where the identity $\sinh 2\theta = 2 \sinh \theta \cosh \theta$ was used in (98). We may readily show that (102) is stationary with respect to arbitrary first-order variations in the functional form $G(y)$, that is, in $g(y)$. The total energy stored in the space surrounding the strip line in Fig. 4.17 is $4W_e$ per unit length. Hence the capacitance per unit length is $C = 8W_e/V_0^2$. For a first approximation to $C$ we will take $g(y)$ as zero; thus, $a_n = 0$, and

$$b_n = \frac{2}{d} e^{n\pi l/2d} \left( \int_0^a \frac{V_0 y}{a} \sin \frac{n\pi y}{2d}\, dy + \int_a^d V_0 \sin \frac{n\pi y}{2d}\, dy \right) = \frac{8V_0 d}{n^2\pi^2 a} e^{-n\pi l/2d} \sin \frac{n\pi a}{2d}$$

Hence
$$W_e = \tfrac{1}{2}\epsilon_0 \frac{V_0^2 l}{a} + \frac{8\epsilon_0 V_0^2 d^2}{\pi^3 a^2} \sum_{1,3,\ldots}^\infty \frac{1}{n^3} \sin^2 \frac{n\pi a}{2d} \qquad (103)$$

The series may be summed by the methods discussed in the Mathematical Appendix and gives

$$\begin{aligned} S_0 &= \sum_{1,3,\ldots}^\infty \frac{1}{n^3} \sin^2 \frac{n\pi a}{2d} = \sum_{n=1,3,\ldots}^\infty \frac{1}{n^3} \cos^2 \frac{n\pi t}{2d} \\ &= \frac{1}{2}\left(\frac{\pi a}{2d}\right)^2 \left(1.5 - \ln \frac{\pi a}{2d}\right) - \frac{1}{36}\left(\frac{\pi a}{2d}\right)^4 - \frac{7}{2{,}700}\left(\frac{\pi a}{2d}\right)^6 - \cdots \\ &= \sum_{1,3,\ldots}^\infty \frac{1}{n^3} - \frac{1}{2}\left(\frac{\pi t}{2d}\right)^2 \left(1.5 - \ln \frac{\pi t}{2d}\right) + \frac{1}{36}\left(\frac{\pi t}{2d}\right)^4 \\ &\quad + \frac{7}{2{,}700}\left(\frac{\pi t}{2d}\right)^6 + \cdots \end{aligned} \qquad (104)$$

If we let $t = a = d/2$ and subtract the two expressions for the sum $S_0$, we get the interesting result

$$\sum_{1,3,\ldots}^\infty \frac{1}{n^3} \le \frac{\pi^2}{16} \left[ 1.5 + \ln \frac{4}{\pi} - \frac{\pi^2}{288} - \frac{14}{2{,}700}\left(\frac{\pi}{4}\right)^4 - \cdots \right] \qquad (105)$$

If the first $N$ terms in (105) are summed directly, a comparison with the right-hand side of (105) gives at once the maximum error involved since the sum of $N + P$ terms is greater than the sum of $N$ terms but yet is less than the right-hand side of (105) because all the higher-order terms are preceded by minus signs. The value of the sum is 1.052 to four figures.† It is convenient to have the two forms for the sum $S_0$, depending on whether $t$ is much smaller than $a$ or vice versa. The capacitance $C$ is given by $C = 4\epsilon_0 l/a + (64\epsilon_0 d^2/\pi^3 a^2)S_0$, and, hence, the characteristic impedance is given by

$$Z_c = Z_0\left(\frac{4l}{a} + \frac{64d^2S_0}{\pi^3a^2}\right)^{-1} \tag{106}$$

and is a lower bound to the true value $Z_{c0}$. The term $4\epsilon_0 l/a$ is just the ordinary parallel-plate capacitance, while the term involving $S_0$ is the increase in capacitance due to the field existing past the edges of the center strip. Higher-order approximations to $C$ may be obtained by taking

$$g(y) = \sum_{n=1}^{N} c_n \sin\frac{n\pi y}{a}$$

and making the energy integral stationary with respect to the $N$ unknown variational parameters (coefficients) $c_n$. Needless to say, the numerical work increases rapidly, and so we will not pursue the problem any further here.

In practice, the upper and lower plates are of finite width. The field decays as $e^{-n\pi x/2d}$, and so, provided the plates are wide enough so that the first mode $e^{-\pi x/2d}$ has decayed to a negligible value, the perturbing effect of finite-width plates is small. A value of $x$ equal to $l + 3d$ makes the exponential term about 1 per cent of what its value at $x = l$ is. Thus a total width of $2l + 6d$ is sufficient to make the effect of finite-width plates negligible. The upper and lower plates should extend past the center strip by approximately one and a half times the spacing of the upper and lower strips.

### *Upper Bound for $Z_{c0}$*

If we do not disturb the charge distribution which exists on the center conductor, we may remove the center conductor, leaving the charge behind. The electrostatic problem is thus to find a suitable solution to Poisson's equation. As a first step, we require the potential (Green's function) due to a unit line charge at $x'$, $y'$ and subject to the following

† The sum may also be evaluated in terms of the Riemann zeta function. See Jahnke and Emde, *op. cit.*

boundary conditions (see Fig. 4.18):

$$\phi = 0 \qquad y = 0, 2d$$
$$\phi \to 0 \qquad |x| \to \infty$$

The Green's function satisfies the following equation:

$$\frac{\partial^2 G}{\partial x^2} + \frac{\partial^2 G}{\partial y^2} = -\frac{1}{\epsilon_0}\,\delta(x - x')\delta(y - y') \tag{107}$$

A suitable solution may be obtained by the method given in Sec. 2.3.

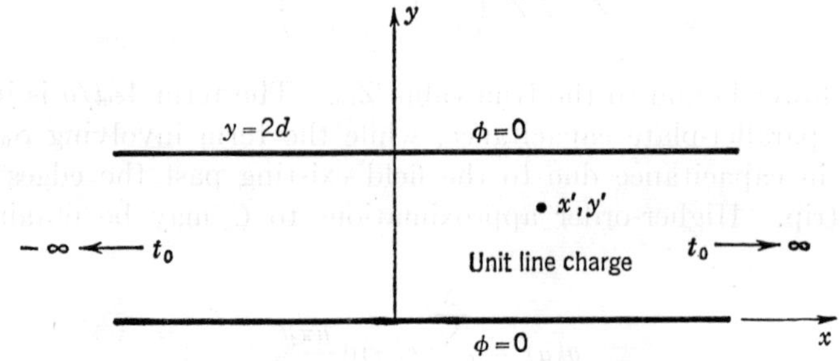


FIG. 4.18. Line charge between two infinite parallel planes.

However, it will be instructive to find the required solution by a different method, and this will be done. At all points except $(x',y')$, $G$ satisfies Laplace's equation. Since Laplace's equation is separable in rectangular coordinates, we will have a solution of the form $\sum_n f_n(x)f_n(x')g_n(y)g_n(y')$, since we know that $G$ is symmetrical in the variables $x$, $y$ and $x'$, $y'$. At $(x',y')$, $G$ is continuous since it is a potential distribution (which is, of course, continuous; i.e., the potential cannot have two different values at the same point). We may expand $\delta(y - y')$ into a Fourier series. Let

$$\delta(y - y') = \sum_1^\infty a_n \sin \frac{n\pi y}{2d}$$

Multiply both sides by $\sin (n\pi y/2d)$, and integrate from 0 to $2d$ to obtain $a_n = (1/d) \sin (n\pi y'/2d)$. Thus we have

$$\delta(y - y') = \frac{1}{d}\sum_1^\infty \sin \frac{n\pi y'}{2d} \sin \frac{n\pi y}{2d}$$

as the required Fourier-series expansion of the delta function. A Fourier-series expansion of $\sum_n g_n(y')g_n(y)$ which satisfies the boundary conditions at $y = 0$, $2d$ is

$$\sum_1^\infty \sin\frac{n\pi y}{2d}\sin\frac{n\pi y'}{2d} = \sum_1^\infty g_n(y)g_n(y')$$

Substituting into the differential equation (107), we get

$$\sum_1^\infty \left(\frac{\partial^2}{\partial x^2} - \frac{n^2\pi^2}{4d^2}\right) f_n(x)f_n(x')\sin\frac{n\pi y}{2d}\sin\frac{n\pi y'}{2d}$$
$$= -\frac{1}{\epsilon_0 d}\,\delta(x - x')\sin\frac{n\pi y}{2d}\sin\frac{n\pi y'}{2d}$$

This equation must hold for all values of $n$, and thus $f_n(x)$ must satisfy the following differential equation:

$$f_n(x')\left[\frac{d^2f_n(x)}{dx^2} - \frac{n^2\pi^2}{4d^2}f_n(x)\right] = -\frac{1}{\epsilon_0 d}\,\delta(x - x') \tag{108}$$

If we integrate this equation from $x = x' - \Delta$ to $x' + \Delta$, and let $\Delta$ approach zero, we obtain

$$\lim_{\Delta\to 0} f_n(x')\left[\frac{df_n}{dx}\bigg|_{x'-\Delta}^{x'+\Delta} - \frac{n^2\pi^2}{4d^2}\int_{x'-\Delta}^{x'+\Delta} f_n(x)\,dx\right] = -\frac{1}{\epsilon_0 d} = f_n(x')\,\frac{df_n(x)}{dx}\bigg|_{x'-}^{x'+}$$

since $f_n$ is a continuous function at $x'$, and the interval of integration vanishes. Thus the derivative of $f_n(x)$ is discontinuous at $x = x'$ by the amount $-1/[\epsilon_0\, df_n(x')]$. The solution to (108) for all values of $x \neq x'$ which vanishes as $|x|$ tends to infinity is

$$f_n(x) = \begin{cases} a_n e^{n\pi x/2d} & x < x' \\ b_n e^{-n\pi x/2d} & x > x' \end{cases}$$

where the unknown coefficients $a_n$ and $b_n$ will depend on $x'$. When $x = x'$, $f_n$ is continuous; hence $a_n e^{n\pi x'/2d} = b_n e^{-n\pi x'/2d}$. The discontinuity condition on the derivative gives

$$a_n e^{n\pi x'/2d}\left(\frac{-n\pi}{2d}\,b_n e^{-n\pi x'/2d} - \frac{n\pi}{2d}\,a_n e^{n\pi x'/2d}\right) = -\frac{1}{\epsilon_0 d}$$

and solving these latter two equations for $a_n$ gives

$$\frac{n\pi}{d}\,a_n{}^2 e^{n\pi x'/d} = \frac{1}{\epsilon_0 d} \qquad \text{or} \qquad a_n = \frac{1}{(n\pi\epsilon_0)^{1/2}}\,e^{-n\pi x'/2d}$$

The required solution for the Green's function is

$$G(x,y|x',y') = \frac{1}{\epsilon_0\pi}\begin{cases}\displaystyle\sum_1^\infty \frac{1}{n}\sin\frac{n\pi y}{2d}\sin\frac{n\pi y'}{2d}\,e^{-(n\pi/2d)(x'-x)} & x \le x' \\ \displaystyle\sum_1^\infty \frac{1}{n}\sin\frac{n\pi y}{2d}\sin\frac{n\pi y'}{2d}\,e^{(n\pi/2d)(x'-x)} & x \ge x'\end{cases}$$

$$= \frac{1}{\epsilon_0\pi}\sum_1^\infty \frac{1}{n}\sin\frac{n\pi y}{2d}\sin\frac{n\pi y'}{2d}\,e^{-(n\pi/2d)|x-x'|} \tag{109}$$

The reason why the derivative of $G$ is discontinuous at $(x',y')$ is that the derivative is equal to the negative electric field component which terminates on the unit line charge. Whenever the electric field terminates on a charge, it is discontinuous by the amount $1/\epsilon_0$ times the charge density.

From Eq. (93), the capacity of the strip line is given by

$$\frac{1}{C} = \frac{\displaystyle\oint\oint_{S_2} G(x,y|x',y')\rho(x,y)\rho(x',y')\,dl\,dl'}{\left[\displaystyle\oint_{S_2}\rho(x,y)\,dl\right]^2} \tag{110}$$

where $\rho(x,y)$ is the charge distribution on the center conductor, and the integration is carried out around the center conductor. The simplest approximation to $\rho$ which we can make is to assume that it is constant. We may take the magnitude of $\rho$ as unity since (110) does not depend on the absolute magnitude of $\rho$. It depends only on the functional form of $\rho$. Thus the denominator in (110) is simply $16(l + t)^2$. We will integrate the numerator over the variables $x'$, $y'$ first. The integral from $-l$ to $l$ is the sum of two terms, one term for $y' = d - t$ and the other for $y' = d + t$. We may carry out the integration for the $n$th term in the Green's function and then sum this typical term over all values of $n$. Bearing in mind that the Green's function has two different forms, depending on whether $x' > x$ or $x' < x$ [see Eq. (109)], we obtain

$$\sin\frac{n\pi y}{2d}\int_{-l}^{x}\left(\sin n\pi\frac{d+t}{2d} + \sin n\pi\frac{d-t}{2d}\right)e^{-n\pi x/2d}e^{n\pi x'/2d}\,dx'$$
$$+ \sin n\pi\frac{y}{2d}\int_{x}^{l}\left(\sin n\pi\frac{d+t}{2d} + \sin n\pi\frac{d-t}{2d}\right)e^{n\pi x/2d}e^{-n\pi x'/2d}\,dx'$$
$$+ \int_{d-t}^{d+t}\sin n\pi\frac{y}{2d}\sin n\pi\frac{y'}{2d}\,e^{-n\pi l/2d}\left(e^{-n\pi x/2d} + e^{n\pi x/2d}\right)dy'$$

where the Green's function for $x > x'$ is used when $x' = -l$, and the

Green's function for $x < x'$ is used when $x' = l$ in the integral over $y'$. The integration gives the following result:

$$\frac{8d}{n\pi} \sin n\pi \frac{y}{2d} \sin \frac{n\pi}{2} \left[ \cos n\pi \frac{t}{2d} \left( 1 - e^{-n\pi l/2d} \cosh n\pi \frac{x}{2d} \right) + \sin n\pi \frac{t}{2d} e^{-n\pi l/2d} \cosh n\pi \frac{x}{2d} \right]$$

Owing to the factor $\sin (n\pi/2)$ all the even terms vanish. Integrating over the variables $x$ and $y$, we obtain the following result for the typical $n$th term:

$$\frac{32d^2}{n^2\pi^2} \left[ \frac{n\pi l}{2d} \cos^2 n\pi \frac{t}{2d} + \sin n\pi \frac{t}{d} - \cos n\pi \frac{t}{d} + e^{-n\pi l/d} \left( 1 - \sin n\pi \frac{t}{d} \right) \right]$$

Multiplying by $1/\epsilon_0 n\pi$ and summing over $n$, we obtain the following final result for the upper bound to $Z_{c0}$:

$$Z_c = \frac{(\mu_0\epsilon_0)^{1/2}}{C} = \frac{2d^2 Z_0}{\pi^3 (l + t)^2} (S_1 + S_2 - S_3 + S_4) \tag{111}$$

where

$$S_1 = \frac{\pi l}{d} \sum_{1,3,\ldots}^{\infty} \frac{\cos^2 n\pi(t/2d)}{n^2}$$

$$S_2 = \sum_{1,3,\ldots}^{\infty} \frac{\sin n\pi(t/d)}{n^3} = \sum_{1,3,\ldots}^{\infty} \frac{\sin n\pi(a/d)}{n^3}$$

$$S_3 = \sum_{1,3,\ldots}^{\infty} \frac{\cos n\pi(t/d)}{n^3} = - \sum_{1,3,\ldots}^{\infty} \frac{\cos n\pi(a/d)}{n^3}$$

$$S_4 = \sum_{1,3,\ldots}^{\infty} \frac{e^{-n\pi l/d}}{n^3} \left( 1 - \sin n\pi \frac{t}{d} \right)$$

The series may be summed by the methods described in the Mathematical Appendix and yield the following results:

$$S_1 = \frac{\pi^3}{8} \frac{l}{d} \left( 1 - \frac{t}{d} \right) \qquad S_2 = \frac{\pi^3}{8} \frac{t}{d} \left( 1 - \frac{t}{d} \right)$$

$$S_3 = \sum_{1,3,\ldots}^{\infty} \frac{1}{n^3} + \frac{1}{4} \left[ \left( \frac{\pi t}{d} \right)^2 \ln \frac{\pi t}{2d} - \frac{3}{2} \left( \frac{\pi t}{d} \right)^2 + \frac{1}{72} \left( \frac{\pi t}{d} \right)^4 + \cdots \right]$$

$$= - \sum_{1,3,\ldots}^{\infty} \frac{1}{n^3} - \frac{1}{4} \left[ \left( \frac{\pi a}{d} \right)^2 \ln \frac{\pi a}{2d} - \frac{3}{2} \left( \frac{\pi a}{d} \right)^2 + \frac{1}{72} \left( \frac{\pi a}{d} \right)^4 + \frac{7}{21{,}600} \left( \frac{\pi a}{d} \right)^6 + \cdots \right]$$

The series $S_4$ is readily summed term by term since it converges very rapidly because of the decaying exponential term.

In Fig. 4.19 the upper and lower bounds, as well as the average value of characteristic impedance, are plotted as a function of $l/d$ for a fixed

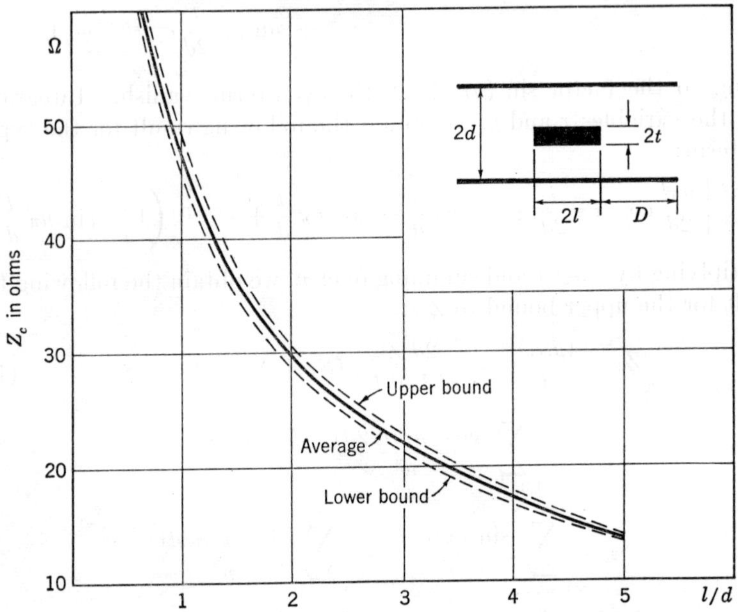


FIG. 4.19. Characteristic impedance of strip line for $t = 0.2d$, $\epsilon = \epsilon_0$, $D > 3d$.

value of $t/d$ equal to 0.2. For all values of $l/d$ greater than 0.5 the maximum error in the average value of characteristic impedance is less than 3 per cent, decreasing as $l/d$ becomes large.

## REFERENCES

*General theory of transmission lines*

4.1. S. Ramo and J. R. Whinnery: "Fields and Waves in Modern Radio," 2d ed., John Wiley & Sons, Inc., New York, 1953.

4.2. S. A. Schelkunoff: "Electromagnetic Waves," D. Van Nostrand Company, Inc., Princeton, N.J., 1943.

4.3. R. W. P. King: "Transmission Line Theory," McGraw-Hill Book Company, Inc., New York, 1955.

*Theory of conformal mapping*

4.4. R. V. Churchill: "Introduction to Complex Variables and Applications," McGraw-Hill Book Company, Inc., New York, 1948.

4.5. H. Kober: "Dictionary of Conformal Representation," 2d ed., Dover Publications, New York, 1957.

*Application of conformal mapping to strip lines*

4.6. Symposium on Microwave Strip Circuits, *IRE Trans.*, vol. MTT-3, March, 1955.

4.7. S. B. Cohn: Shielded Coupled Strip Transmission Line, *IRE Trans.*, vol. MTT-3, pp. 29–38, October, 1955.

4.8. B. A. Dahlman: A Double Ground Plane Strip Line System for Microwaves, *IRE Trans.*, vol. MTT-3, pp. 52–57, October, 1955. [This paper treats the same problem as in Sec. 4.3 but approximates center conductor by a semi-infinite strip and solves for the fringing field. Equation (21) is in error and, when corrected, gives the same value for the attenuation constant as obtained here in Sec. 4.3.]

4.9. D. Park: Planar Transmission Lines, *IRE Trans.*, vol. MTT-3, part I, pp. 8–12, April, 1955; part II, pp. 7–11, October, 1955. (The formula for attenuation for a single strip line appears to be in error.)

4.10. W. H. Hayt, Jr.: Potential Solution of a Homogeneous Strip Line of Finite Width, *IRE Trans.*, vol. MTT-3, pp. 16–18, July, 1955. (Gives solution to same problem as treated in Sec. 4.3 but does not assume that ground planes are infinite.)

*Application of variational methods to transmission lines*

4.11. R. E. Collin: Characteristic Impedance of a Slotted Coaxial Line, *IRE Trans.*, vol. MTT-4, pp. 4–8, January, 1956.

4.12. R. M. Chisholm: The Characteristic Impedance of Trough and Slab Lines, *IRE Trans.*, vol. MTT-4, pp. 166–172, July, 1956.

*Alternative methods of analysis*

4.13. J. J. Skiles and T. J. Higgins: Determination of the Characteristic Impedance of UHF Coaxial Rectangular Transmission Lines, *Proc. Natl. Electronics Conf.*, vol. 10, pp. 97–108, 1954.

4.14. W. B. Swift and T. J. Higgins: Determination of the Operating Performance of Waveguides of Arbitrary Cross Section by Hypercircle Analysis, *Proc. Natl. Electronics Conf.*, vol. 10, pp. 83–96, 1954.

## PROBLEMS

**4.1.** Find the characteristic impedance and attenuation constant arising from conductor loss for a coaxial line with inner radius $a$ and outer radius $b$.

*Answer*

$$Z_c = \frac{Z_0}{2\pi}\ln\frac{b}{a} \qquad \alpha = \frac{R_m}{4\pi Z_c}\left(\frac{1}{a}+\frac{1}{b}\right)$$

**4.2.** Show that the conformal transformation $W = \ln\,[(Z + a)/(Z - a)]$ maps the cross section of a two-wire line into the strips $u = \pm u_1$, $0 \le v \le 2\pi$, in the $W$ plane. The conductor radius and spacing are given by $R = a(\sinh u_1)^{-1}$, $D = 2a \coth u_1$. Find the characteristic impedance of the line. By a suitable integration along the conductor strips in the $W$ plane, determine the power loss in the conductors and, hence, the attenuation constant for the line.

*Answer*

$$Z_c = \frac{Z_0}{\pi}\cosh^{-1}\frac{D}{2R} \qquad \alpha = \frac{R_m}{\pi Z_c 2R}\frac{D/2R}{[(D/2R)^2 - 1]^{1/2}}$$

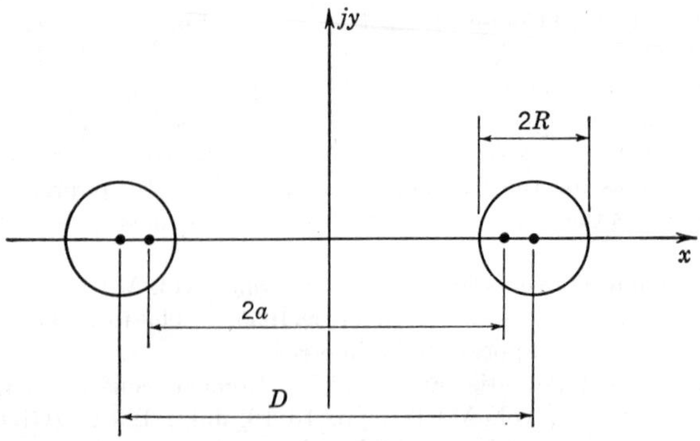


FIG. P 4.2

**4.3.** For a constant value of outer radius $b$, show that the optimum ratio of outer to inner conductor radius, i.e., $b/a$, which will make the attenuation constant of a coaxial line a minimum satisfies the equation $x \ln x = 1 + x$, where $x = b/a$. Find the corresponding value of characteristic impedance.

**4.4.** By a suitable conformal transformation, determine the characteristic impedance of the strip line illustrated. Assume that the ground planes are infinitely wide and that the thickness of the center conductor is negligible.

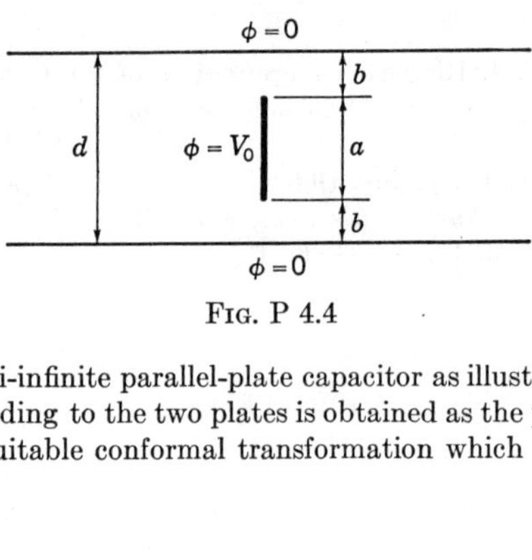


FIG. P 4.4

**4.5.** Consider a semi-infinite parallel-plate capacitor as illustrated. In the $W$ plane the polygon corresponding to the two plates is obtained as the points $W_1$ and $W_3$ tend to infinity. Find a suitable conformal transformation which will map the capacitor

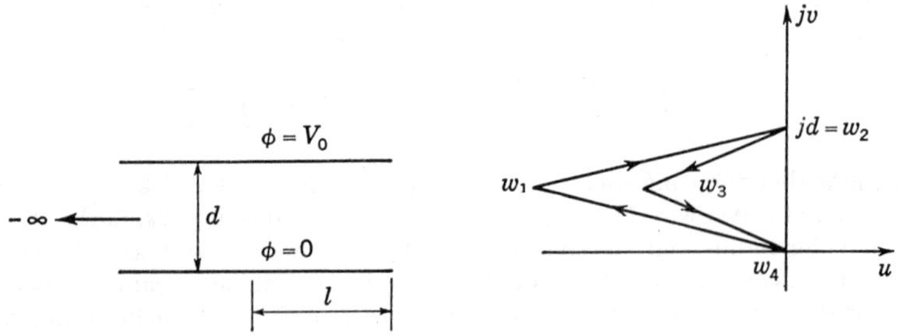


FIG. P 4.5

plates into the real axis in the $Z$ plane. Integrate the normal gradient of the potential along the $x$ axis over that region which corresponds to the inner and outer surfaces of one plate extending in a distance $l$ from the edge. Find the capacitance $C(l)$ as a function of $l$. Use this result to determine the approximate characteristic impedance of a parallel-plate transmission line with plate spacing $d$ and total width $2l$.

*Answer.* A suitable mapping function is $W = \frac{d}{\pi}\left(\frac{1 - Z^2}{2} + \ln Z\right)$. Charge $Q(l) = \frac{\epsilon_0 V_0}{\pi} \ln\left(\frac{x_2}{x_1}\right)$, where $x_2$ and $x_1$ are determined by $\frac{-l\pi}{d} = \frac{1 - x_i^2}{2} + \ln x_i$.

For $l \gg d$,

$$\ln x_1 \approx \frac{-l\pi}{d} - \frac{1}{2}$$

$$x_2^2 \approx \frac{2l\pi}{d} + 1$$

$$\text{Capacitance } C(l) \approx \frac{\epsilon_0 l}{d}\left[1 + \frac{d}{2\pi l}\ln\left(1 + \frac{2\pi l}{d}\right)\right] \qquad Z_c = \frac{(\mu_0\epsilon_0)^{1/2}}{2C(l)}$$

**4.6.** Consider a conducting wedge of internal angle $\alpha$ and a line charge located at $Z = 1$ as illustrated. By means of the transformation $W = Z^\beta$, where $\beta = \pi/(2\pi - \alpha)$, map this into a plane and a line charge in the $W$ plane. This latter problem is readily

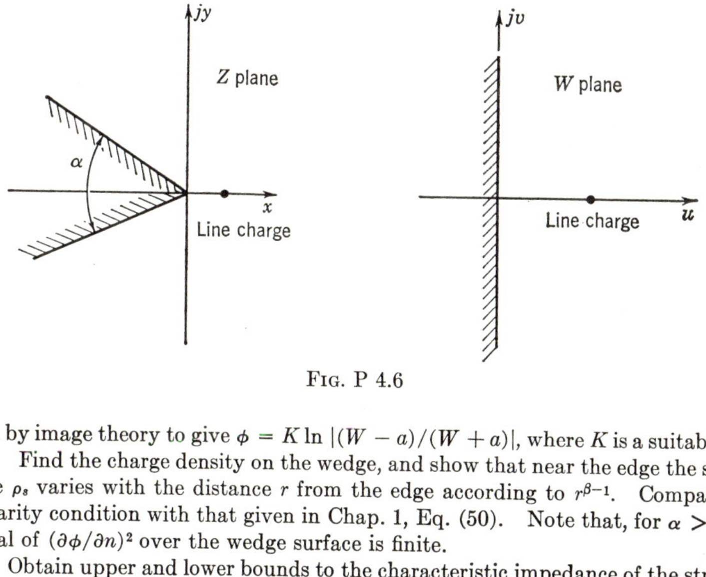


FIG. P 4.6

solved by image theory to give $\phi = K \ln |(W - a)/(W + a)|$, where $K$ is a suitable constant. Find the charge density on the wedge, and show that near the edge the surface charge $\rho_s$ varies with the distance $r$ from the edge according to $r^{\beta-1}$. Compare this singularity condition with that given in Chap. 1, Eq. (50). Note that, for $\alpha > 0$, the integral of $(\partial\phi/\partial n)^2$ over the wedge surface is finite.

**4.7.** Obtain upper and lower bounds to the characteristic impedance of the strip line illustrated in Prob. 4.4. In the variational expressions use the approximations that the potential varies linearly from the center conductor to the ground plane and that the charge density is constant on the center strip. Compare the value of $Z_c$ obtained for $a = 0.5d$ with the correct value obtained in Prob. 4.4.

**4.8.** Consider a series of three two-conductor transmission lines with cross sections $S, S_1$; $S, S_2$; and $S, S_3$, where the cross section $S_1$ completely overlaps $S_2$, and $S_2$ completely overlaps $S_3$ as illustrated. Let the capacitance per unit length for the three cross sections be $C_1$, $C_2$, and $C_3$, respectively. By means of the variational expression for capacitance show that $C_1 > C_2 > C_3$.

HINT: For an approximation to the potential $\phi_3$ for the cross section $S$, $S_3$, use the correct expression $\phi_2$ for the potential appropriate for the cross section $S$, $S_2$ in the region external to $S_2$, and zero in the region between $S_2$ and $S_3$. The variational

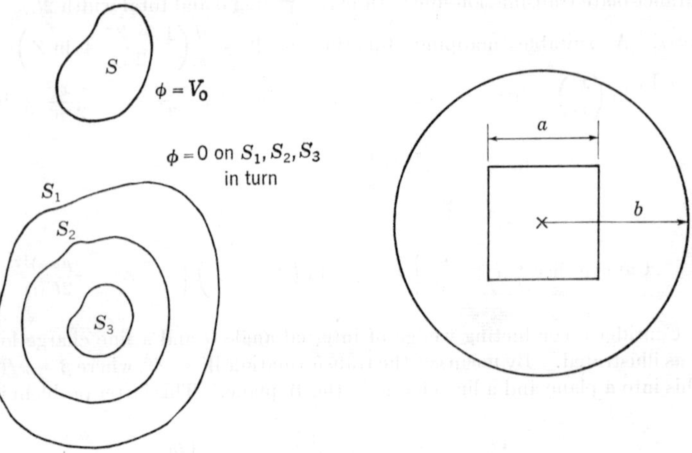


FIG. P 4.8

integral now gives $C_2$ as the upper bound on $C_3$, and hence $C_3 < C_2$. A similar procedure may be used to show $C_2 < C_1$.

Use this result to obtain upper and lower bounds for the characteristic impedance of a coaxial line with a square center conductor. What is the maximum error in the average value for $Z_c$ when $b = 3a$?

**4.9.** Apply Taylor's expansion

$$F = F_0 + \sum_i \Delta x_i \left.\frac{\partial F}{\partial x_i}\right|_0 + \frac{1}{2}\sum_i \sum_j \Delta x_i \, \Delta x_j \left.\frac{\partial^2 F}{\partial x_i \, \partial x_j}\right|_0 + \cdots$$

to the expression (98) plus (99) to show that the first variation in $W_e = W_{e1} + W_{e2}$ is zero, the second variation is positive, and all higher-order variations are zero. Let the variation of each coefficient $a_n$ and $b_n$ about the correct value be $\Delta a_n$ and $\Delta b_n$.

**4.10.** A coaxial line is terminated by a thin resistive disk having a resistance $R_0$ ohms per unit square and a perfect short circuit a distance $l$ away. Determine the required values of $R_0$ and $l$ in order to provide a matched termination, i.e., to prevent a reflected wave from the termination.

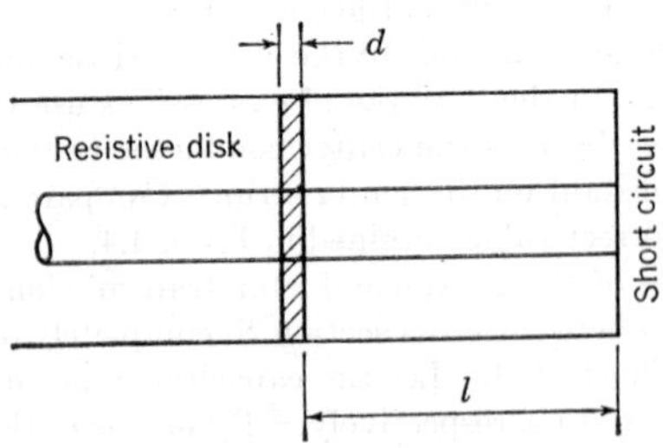


FIG. P 4.10

HINT: Consider a thin disk of thickness $d$ with parameters $\mu_2$, $\sigma$. When $l = \lambda_0/4$, show that the condition for a match is $Z = Z_0 \tanh kd$, where $Z = (j\omega\mu_2/\sigma)^{1/2}$ and $k = (j\omega\mu_2\sigma)^{1/2}$.

*Answer*

$$R_0 = \left(\frac{\mu_0}{\epsilon_0}\right)^{1/2} \qquad l = \frac{\lambda_0}{4}$$

**4.11.** From Eq. (35) it was found that $\omega L_i = R$. Equation (35) shows that the internal inductance $L_i$ arises because the magnetic field penetrates an effective distance $0.5\delta_s$ into the conductor. Thus an alternative way to evaluate $L_i$ is to find the change in the external inductance $L$ when all conductor surfaces are pushed in by an amount $0.5\delta_s$, that is, $L_i = (\delta_s/2)\ \partial L/\partial n$, where $n$ is the inward normal to the conductor surface. This is the incremental inductance rule proposed by Wheeler.† Use this rule to find the series resistance of the coaxial line and the two-wire line in Probs. 4.1 and 4.2. For the coaxial line, differentiate $L$ with respect to $-a$ and $+b$. For the two-wire line, differentiate $L$ with respect to $-R$. Note that $v_cL = Z_c$ and that $\alpha = R/2Z_c$. Comment on the limitations of this rule.

† *Proc. IRE*, vol. 30, p. 412, 1942.

CHAPTER 5

# PROPAGATION IN CYLINDRICAL WAVEGUIDES

In this chapter we shall be concerned with the fundamental properties of electromagnetic waves in the interior of uniform cylindrical waveguides. The type of waveguide considered is a hollow conducting cylindrical tube with a cross section which is uniform along the direction of propagation. The axis of the guide will be taken as the $z$ axis. Initially it will be assumed that the walls have infinite conductivity. Also it is assumed that the guide is either empty or else filled with a homogeneous isotropic medium with electrical parameters $\epsilon$, $\mu$. The guide will be taken as infinitely long and the sources for the electromagnetic field located at infinity. Thus, no consideration will be given to the type of field a particular kind of source generates. This aspect of the problem will be investigated in Chap. 7.

In the interior of such a waveguide it is found that the solutions to Maxwell's equations can be divided into two basic sets of solutions or modes. For one set of modes, no longitudinal or axial magnetic field component exists. These modes have an axial component of electric field, however, and are called "electric type," that is, $E$ modes or "transverse magnetic," that is, TM modes. The field components for these modes may be derived from an electric-type Hertzian potential having a single component along the axis of the guide. The other basic set of modes have an axial magnetic field but no axial component of electric field and are, therefore, referred to as "magnetic type," that is, $H$ or "transverse electric," that is, TE modes. These modes may be derived from a magnetic-type Hertzian potential again having only an axial component. In a hollow cylindrical waveguide, a TEM mode of propagation is not possible since the assumption of transverse fields leaves neither an axial current or axial components of electric displacement nor magnetic flux to generate the transverse field components. From another point of view, we can say that a solution to Laplace's equation in the transverse plane in the interior of a closed boundary on which the potential is constant does not exist (apart from perhaps a constant) unless a singularity is located within the boundary, and, hence, a potential function from which the TEM mode can be derived does not exist.

The division into $E$ and $H$ modes is not necessary, and in some cases it is more convenient to choose a different set of basic modes. This latter set can, however, be represented as a linear combination of the $E$ and $H$ modes because the $E$ and $H$ modes form a complete set in terms of which an arbitrary field can be represented. The question of mode completeness is taken up in some detail in the latter part of the chapter.

## 5.1. General Properties of Cylindrical Waveguides

All cylindrical waveguides have many properties in common. For this reason we will begin our discussion by considering a guide with an arbitrary cross section as illustrated in Fig. 5.1. The frame of reference

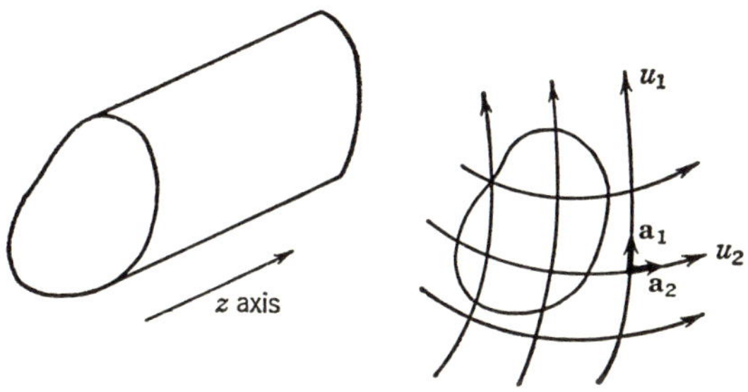


FIG. 5.1. A general cylindrical waveguide.

will be an orthogonal curvilinear cylindrical coordinate system $u_1$, $u_2$, $z$ with unit vectors $\mathbf{a}_1$, $\mathbf{a}_2$, and $\mathbf{a}_z$ directed along the coordinate curves. The interior of the guide is assumed to be free space and the time variation according to $e^{j\omega t}$.

*Transverse Electric Modes*

The TE modes may be derived from a magnetic Hertzian potential $\mathbf{\Pi}_h = \mathbf{a}_z\Pi_h$ by means of the following equations:

$$\mathbf{E} = -j\omega\mu_0 \nabla \times \mathbf{\Pi}_h \tag{1a}$$

$$\mathbf{H} = k_0^2\mathbf{\Pi}_h + \nabla\nabla \cdot \mathbf{\Pi}_h = \nabla \times \nabla \times \mathbf{\Pi}_h \tag{1b}$$

where $\mathbf{\Pi}_h$ is a solution of the following equation:

$$\nabla^2\mathbf{\Pi}_h + k_0^2\mathbf{\Pi}_h = \nabla\nabla \cdot \mathbf{\Pi}_h - \nabla \times \nabla \times \mathbf{\Pi}_h + k_0^2\mathbf{\Pi}_h = 0 \tag{1c}$$

and $k_0^2 = \omega^2\mu_0\epsilon_0 = 4\pi^2/\lambda_0^2$, with $\lambda_0$ being the free-space wavelength for TEM waves. Since we are looking for propagating waves, a solution for $\mathbf{\Pi}_h$ of the form

$$\mathbf{\Pi}_h = \mathbf{a}_z\psi_h(u_1,u_2)e^{\pm\Gamma z} \tag{2}$$

will be assumed. The function $\psi_h$ satisfies the following two-dimensional scalar Helmholtz equation:

$$\nabla_t^2\psi_h + k_c^2\psi_h = 0 \tag{3}$$

where $k_c^2 = k_0^2 + \Gamma^2$, and $\nabla_t^2$ is the transverse part of the $\nabla^2$ operator. In the $u_1u_2z$ coordinate frame with scale factors $h_1$, $h_2$, and unity, (3) becomes†

$$\frac{1}{h_1h_2}\frac{\partial}{\partial u_1}\frac{h_2}{h_1}\frac{\partial \psi_h}{\partial u_1} + \frac{1}{h_1h_2}\frac{\partial}{\partial u_2}\frac{h_1}{h_2}\frac{\partial \psi_h}{\partial u_2} + k_c^2\psi_h = 0 \tag{4}$$

In general, the solution to this equation is difficult to obtain unless it is separable so that a solution of the form $\psi_h = U_1(u_1)U_2(u_2)$ may be found. If such a solution for $\psi_h$ exists, it will still be a difficult task to make $\psi_h$ satisfy the required boundary conditions unless the waveguide boundary coincides with the constant coordinate curves. The allowed values of $k_c^2$ and, hence, the propagation constants $\Gamma^2$ are determined by the boundary conditions on $\psi_h$. An infinite number of discrete values for $k_c^2$ exist and, hence, an infinite number of modes. Only a finite number of the propagation constants are imaginary, and thus only a finite number of the modes will propagate freely along the guide.

By using the expressions for the curl and divergence in orthogonal curvilinear coordinates, the following equations for the field components are obtained from (1):

$$H_{u_1} = \frac{\pm\Gamma}{h_1}\frac{\partial \psi_h}{\partial u_1}e^{\pm\Gamma z} \tag{5a}$$

$$H_{u_2} = \pm\frac{\Gamma}{h_2}\frac{\partial \psi_h}{\partial u_2}e^{\pm\Gamma z} \tag{5b}$$

$$H_z = -\frac{1}{h_1h_2}\left(\frac{\partial}{\partial u_1}\frac{h_2}{h_1}\frac{\partial \psi_h}{\partial u_1} + \frac{\partial}{\partial u_2}\frac{h_1}{h_2}\frac{\partial \psi_h}{\partial u_2}\right)e^{\pm\Gamma z} \tag{5c}$$

$$E_{u_1} = -j\omega\mu_0\frac{1}{h_2}\frac{\partial \psi_h}{\partial u_2}e^{\pm\Gamma z} \tag{5d}$$

$$E_{u_2} = j\omega\mu_0\frac{1}{h_1}\frac{\partial \psi_h}{\partial u_1}e^{\pm\Gamma z} \tag{5e}$$

An examination of these equations shows that they may be more compactly written as follows:

$$\mathbf{H}_t = \pm\Gamma e^{\pm\Gamma z}\nabla_t\psi_h \tag{6a}$$

$$H_z = -\nabla_t^2\psi_h e^{\pm\Gamma z} = k_c^2\psi_h e^{\pm\Gamma z} \tag{6b}$$

$$\mathbf{E}_t = \mp\bar{\mathbf{Z}}_h\cdot\mathbf{H}_t = \pm Z_h\mathbf{a}_z\times\mathbf{H}_t \tag{6c}$$

where the subscript $t$ means the transverse part, $\bar{\mathbf{Z}}_h$ is the dyadic wave impedance given by

$$\bar{\mathbf{Z}}_h = Z_h(\mathbf{a}_1\mathbf{a}_2 - \mathbf{a}_2\mathbf{a}_1) \tag{7}$$

and $Z_h = j\omega\mu_0/\Gamma = (jk_0/\Gamma)Z_0$ is the scalar wave impedance, with $Z_0 = (\mu_0/\epsilon_0)^{1/2}$ being the intrinsic impedance of free space. For a propa-

† Section A.1e.

gating mode, $\Gamma$ is imaginary and will be taken as $j\beta$. The wave impedance $Z_h$ is now real and equal to $Z_0k_0/\beta$. For a nonpropagating mode, the wave impedance is imaginary and inductive in nature.

For a waveguide with perfectly conducting walls, $\psi_h$ must satisfy the boundary condition $\partial\psi_h/\partial n = 0$, in order to make the tangential electric field and normal magnetic field vanish on the boundary. This boundary condition determines the allowed values of $k_c^2$ and $\Gamma^2$. The parameter $k_c$ is called the cutoff wave number, since it determines the wavelength at which free propagation ceases, i.e., for which $\Gamma$ becomes equal to zero. This occurs when $\lambda_0$ is chosen so that $k_c = k_0$ for a given mode, and, hence, the cutoff wavelength is given by

$$\lambda_c = \frac{2\pi}{k_c} \tag{7a}$$

For a guide filled with a medium with parameters $\epsilon$, $\mu$, cutoff occurs when $k_c = \omega(\mu\epsilon)^{1/2} = k$, and, hence, in general the free-space cutoff wavelength is given by

$$\lambda_c = \frac{2\pi}{k_c}\left(\frac{\mu\epsilon}{\mu_0\epsilon_0}\right)^{1/2} \tag{7b}$$

When the cutoff wavelength (which is a function of the guide cross section and $\epsilon\mu$ only) has been determined for a particular propagating mode, the value of $\beta$ and the guide wavelength may be found. The guide wavelength $\lambda_g$ is the distance the wave must propagate to undergo a phase change of $2\pi$ radians, and hence

$$\lambda_g = \frac{2\pi}{\beta} \tag{8}$$

Using the relation $k_0^2 - \beta^2 = k_c^2$, we may solve for $\lambda_g$ to get

$$\lambda_g = \frac{\lambda_0}{[1 - (\lambda_0/\lambda_c)^2]^{1/2}} \tag{9a}$$

or, for a guide filled with material with parameters $\epsilon$, $\mu$,

$$\lambda_g = \frac{\lambda_0(\mu_0\epsilon_0/\mu\epsilon)^{1/2}}{[1 - (\lambda_0/\lambda_c)^2]^{1/2}} \tag{9b}$$

where $\lambda_c$ is now given by (7*b*), and $\lambda_0$ is still the free-space wavelength. For all values of $\lambda_0 < \lambda_c$, free propagation of the mode occurs. As $\lambda_0$ approaches $\lambda_c$, the guide wavelength approaches infinity until, at $\lambda_0 = \lambda_c$, free propagation stops. For $\lambda_0 > \lambda_c$ the mode is exponentially damped with distance along the guide and is commonly referred to as an evanescent mode.

*Transverse Magnetic Modes*

The TM or $E$ modes may be derived from an electric-type Hertzian potential $\mathbf{\Pi}_e = \mathbf{a}_z\Pi_e$ by means of the following equations:

$$\mathbf{E} = k_0^2\mathbf{\Pi}_e + \nabla\nabla \cdot \mathbf{\Pi}_e = \nabla \times \nabla \times \mathbf{\Pi}_e \tag{10a}$$
$$\mathbf{H} = j\omega\epsilon_0 \nabla \times \mathbf{\Pi}_e \tag{10b}$$

where $\mathbf{\Pi}_e$ is a solution of

$$\nabla^2\mathbf{\Pi}_e + k_0^2\mathbf{\Pi}_e = \nabla\nabla \cdot \mathbf{\Pi}_e - \nabla \times \nabla \times \mathbf{\Pi}_e + k_0^2\mathbf{\Pi}_e = 0 \tag{10c}$$

If a solution for $\mathbf{\Pi}_e$ of the form $\mathbf{a}_z\psi_e(u_1,u_2)e^{\pm\Gamma z}$ is assumed, a set of equations similar to (6) for the field components is obtained. These are

$$\mathbf{E}_t = \pm\Gamma \nabla_t\psi_e e^{\pm\Gamma z} \tag{11a}$$
$$E_z = -\nabla_t^2\psi_e e^{\pm\Gamma z} = k_c^2\psi_e e^{\pm\Gamma z} \tag{11b}$$
$$\mathbf{H}_t = \mp\bar{\mathbf{Y}}_e \cdot \mathbf{E}_t = \mp Y_e\mathbf{a}_z \times \mathbf{E}_t \tag{11c}$$

where $\bar{\mathbf{Y}}_e$ is the dyadic wave admittance given by

$$\bar{\mathbf{Y}}_e = Y_e(\mathbf{a}_2\mathbf{a}_1 - \mathbf{a}_1\mathbf{a}_2) \tag{11d}$$

In (11*d*), $Y_e$ is the scalar wave admittance equal to $jk_0Y_0/\Gamma$, and $Y_0$ is the intrinsic admittance of free space given by $(\epsilon_0/\mu_0)^{1/2}$. For perfectly conducting walls, $E_z$ must vanish on the boundary, and, hence, $\psi_e$ must vanish on the boundary also. Again an infinite number of values for $k_c^2$ exist. Also a cutoff wavelength $\lambda_c$ for each mode may be found, and the guide wavelength is given by (9). For $\Gamma = j\beta$, the wave admittance $Y_e$ is real and equal to $Y_0k_0/\beta$, while, for $\Gamma$ real, i.e., for a nonpropagating mode, it is capacitive.

## 5.2. Orthogonal Properties of the Modes

In a waveguide with perfectly conducting walls, the $E$ and $H$ modes have several interesting and useful orthogonal properties. Let $\psi_i$ and $\psi_j$ be the solutions for the $i$th and $j$th $E$ modes or $H$ modes with corresponding propagation constants $\Gamma_i$ and $\Gamma_j$. Multiplying the equation satisfied by $\psi_i$ by $\psi_j$ gives

$$\psi_j \nabla_t^2\psi_i + (k_0^2 + \Gamma_i^2)\psi_i\psi_j = 0 \tag{12a}$$

Similarly, multiplying the equation for $\psi_j$ by $\psi_i$ gives

$$\psi_i \nabla_t^2\psi_j + (k_0^2 + \Gamma_j^2)\psi_i\psi_j = 0 \tag{12b}$$

Subtracting these two equations and substituting into the two-dimensional form of Green's second identity gives†

$$(\Gamma_i^2 - \Gamma_j^2) \iint_S \psi_i \psi_j \, da = \iint_S (\psi_i \nabla_t^2 \psi_j - \psi_j \nabla_t^2 \psi_i) \, da$$
$$= \oint_C \left( \psi_i \frac{\partial \psi_j}{\partial n} - \psi_j \frac{\partial \psi_i}{\partial n} \right) dl \tag{13}$$

where $S$ denotes the cross-sectional surface of the guide, and $C$ denotes the guide boundary. For $E$ modes, $\psi_i$ and $\psi_j$ vanish on the boundary, while, for $H$ modes, $\partial \psi_i / \partial n$ and $\partial \psi_j / \partial n$ vanish on the boundary. Therefore, in either case the contour integral vanishes, and, provided $\Gamma_i^2 \neq \Gamma_j^2$, we obtain the following orthogonal property for both the $E$ and $H$ modes:

$$\iint_S \psi_i \psi_j \, da = 0 \qquad i \neq j \tag{14}$$

Thus the axial components of the field for two different modes are orthogonal.

If the two modes are degenerate such that $\Gamma_i = \Gamma_j$, the proof just given breaks down. However, for degenerate modes we may choose a suitable linear combination of the degenerate modes such that this subset of modes is an orthogonal set. For example, if $\psi_1$ and $\psi_2$ are two degenerate modes with eigenvalue $\Gamma^2$, we may choose a new subset of modes as follows: $\psi_1' = \psi_1$, $\psi_2' = \psi_2 + \alpha \psi_1$, where $\alpha$ is a constant to be determined so that $\iint_S \psi_1' \psi_2' \, da = 0$. If we let $\iint_S \psi_j \psi_i \, da = P_{ij}$, $i, j = 1, 2$, we find that $\alpha = -P_{12}/P_{11}$. With this value of $\alpha$, the two modes $\psi_1'$ and $\psi_2'$ are orthogonal. This procedure may be generalized so that any subset of $n$ degenerate modes can be converted into a new subset of $n$ mutually orthogonal modes. Thus we may, in a general discussion, assume that $\Gamma_i^2 \neq \Gamma_j^2$ without invalidating the final result.

The transverse electric fields for two different $E$ modes or for two different $H$ modes, as well as the transverse electric field for one $E$ mode and one $H$ mode, are mutually orthogonal. The same is true for the transverse magnetic field for any two ($E$, $H$, or $E$ and $H$) modes. The proof for the case of the transverse electric fields will be given; that for the transverse magnetic fields may be constructed along similar lines.

For two different $E$ modes, the transverse electric fields are orthogonal, provided the following integral vanishes:

$$I = \iint_S \nabla_t \psi_i \cdot \nabla_t \psi_j \, da \tag{15}$$

† Section A.1c.

where $\psi_i$ and $\psi_j$ are two different eigenfunctions from which the $E$ modes may be derived by means of (11). Using the two-dimensional form of Green's first identity, the surface integral becomes

$$\iint_S \nabla_t\psi_i \cdot \nabla_t\psi_j \, da = -\iint_S \psi_i \nabla_t^2\psi_j \, da + \oint_C \psi_i \frac{\partial\psi_j}{\partial n} \, dl \tag{16}$$

The contour integral vanishes since $\psi_i = 0$ on the boundary. In the surface integral on the right $\nabla_t^2\psi_j$ may be replaced by $-k_c^2\psi_j$, and, using the result that $\psi_i$ and $\psi_j$ are orthogonal, it is seen to vanish also. Consequently Eq. (15) vanishes, and the stated orthogonal property is proved.

For two different $H$ modes the transverse electric fields are orthogonal, provided

$$\iint_S (\mathbf{a}_z \times \nabla_t\psi_i) \cdot (\mathbf{a}_z \times \nabla_t\psi_j) \, da \tag{17}$$

vanishes. In expression (17), $\psi_i$ and $\psi_j$ are two eigenfunctions, giving rise to two $H$ modes, and the expressions for the transverse electric fields have been obtained from (6). The integrand in (17) may be written as $\mathbf{a}_z \cdot [\nabla_t\psi_i \times (\mathbf{a}_z \times \nabla_t\psi_j)]$ and expanded to give

$$\mathbf{a}_z \cdot [(\nabla_t\psi_i \cdot \nabla_t\psi_j)\mathbf{a}_z - (\nabla_t\psi_i \cdot \mathbf{a}_z)\nabla_t\psi_j] = \nabla_t\psi_i \cdot \nabla_t\psi_j$$

since $\mathbf{a}_z \cdot \nabla_t\psi_i = 0$, because these two vectors are perpendicular. Replacing the integrand by $\nabla_t\psi_i \cdot \nabla_t\psi_j$ and using Green's first identity gives a result similar to (16). For $H$ modes, $\partial\psi_i/\partial n$ vanishes on the boundary, and, since $\psi_i$ and $\psi_j$ are orthogonal, the vanishing of the integral (17) follows as a consequence.

Finally, for the transverse electric fields for one $E$ mode and one $H$ mode we must show that

$$\iint_S (\mathbf{a}_z \times \nabla_t\psi_{hi}) \cdot \nabla_t\psi_{ej} \, da \tag{18}$$

vanishes. To avoid confusion between the eigenfunction for the $E$ and $H$ modes in (18), the subscripts $h$ and $e$ have been introduced again. Consider the following expression:

$$\begin{aligned} \nabla_t \cdot (\mathbf{a}_z\psi_{hi} \times \nabla_t\psi_{ej}) &= (\nabla_t \times \mathbf{a}_z\psi_{hi}) \cdot \nabla_t\psi_{ej} - \mathbf{a}_z\psi_{hi} \cdot \nabla_t \times \nabla_t\psi_{ej} \\ &= -(\mathbf{a}_z \times \nabla_t\psi_{hi}) \cdot \nabla_t\psi_{ej} \end{aligned}$$

since $\nabla_t \times \nabla_t\psi_{ej}$ is identically zero, and $\nabla_t \times \mathbf{a}_z\psi_{hi} = -\mathbf{a}_z \times \nabla_t\psi_{hi}$. Using this result, the integral in (18) may be replaced by

$$-\iint_S \nabla_t \cdot (\mathbf{a}_z\psi_{hi} \times \nabla_t\psi_{ej}) \, da = +\oint_C (\mathbf{a}_z \times \nabla_t\psi_{ej}) \cdot \mathbf{n}\psi_{hi} \, dl$$

where $\mathbf{n}$ is the unit normal to the waveguide wall and directed inward. In the contour integral, $\nabla_t \psi_{ej}$ is proportional to the transverse electric field and, hence, has only a normal component at the guide wall. Thus $\mathbf{a}_z \times \nabla_t \psi_{ej}$ is a factor which is tangential to the guide wall (it is a vector proportional to the transverse magnetic field), and, hence, the dot product with $\mathbf{n}$ and also the integrand vanish. This establishes the orthogonality of the transverse electric fields corresponding to one $E$ mode and one $H$ mode.

The above orthogonality properties are useful in practice because they enable a given arbitrary field to be expanded into a series of $E$ and $H$ modes. These orthogonal properties are not, however, completely general. For a guide with finite conducting walls, the proofs given above are not valid because the eigenfunctions $\psi_e$ and $\psi_h$ no longer satisfy the assumed boundary conditions; that is, $\psi_e = 0$ and $\partial\psi_h/\partial n = 0$ no longer hold on the guide boundary. The presence of finite conductivity results in a cross coupling between the various $E$ and $H$ modes. This phenomenon is considered later in the section on attenuation. In inhomogeneously filled waveguides it is found, in many instances, that the normal modes of propagation are not either $E$ or $H$ modes but linear combinations of these modes. In order to expand an arbitrary field into a series of these latter normal modes, a more general orthogonal principle must be resorted to. This more general orthogonal relation follows as a result of the Lorentz reciprocity principle and will be demonstrated below.

Let $\mathbf{H}_{tn}$, $\mathbf{E}_{tn}$ and $\mathbf{H}_{tm}$, $\mathbf{E}_{tm}$ be the transverse fields for two linearly independent solutions to Maxwell's equations. The curl equations for the electric field give $\nabla \times \mathbf{E}_n = -j\omega\mu\mathbf{H}_n$, $\nabla \times \mathbf{E}_m = -j\omega\mu\mathbf{H}_m$, where $\mathbf{E}_n$, $\mathbf{H}_n$ and $\mathbf{E}_m$, $\mathbf{H}_m$ are the total fields, i.e., both the transverse and axial components. Scalar-multiplying the above equations by $\mathbf{H}_m$ and $\mathbf{H}_n$, respectively, and subtracting gives $\mathbf{H}_m \cdot \nabla \times \mathbf{E}_n - \mathbf{H}_n \cdot \nabla \times \mathbf{E}_m = 0$. Scalar-multiplying the curl equations for $\mathbf{H}_n$, $\mathbf{H}_m$ by $\mathbf{E}_m$, $\mathbf{E}_n$ and subtracting gives a similar result, but with the role of $\mathbf{E}$ and $\mathbf{H}$ interchanged; thus $\mathbf{E}_m \cdot \nabla \times \mathbf{H}_n - \mathbf{E}_n \cdot \nabla \times \mathbf{H}_m = 0$. Adding these equations gives $\nabla \cdot (\mathbf{E}_n \times \mathbf{H}_m - \mathbf{E}_m \times \mathbf{H}_n) = 0$, since

$$\mathbf{H}_m \cdot \nabla \times \mathbf{E}_n - \mathbf{E}_n \cdot \nabla \times \mathbf{H}_m = \nabla \cdot \mathbf{E}_n \times \mathbf{H}_m \text{ etc.}$$

If $\mathbf{E}_n$, $\mathbf{H}_n$ depend on the coordinate $z$ only as $e^{-\Gamma_n z}$ and $\mathbf{E}_m$, $\mathbf{H}_m$ as $e^{-\Gamma_m z}$, this last relation reduces to

$$\begin{aligned}
\nabla \cdot (\mathbf{E}_n \times \mathbf{H}_m - \mathbf{E}_m \times \mathbf{H}_n) &= \nabla_t \cdot (\mathbf{E}_n \times \mathbf{H}_m - \mathbf{E}_m \times \mathbf{H}_n) \\
&\quad + \mathbf{a}_z \frac{\partial}{\partial z} \cdot (\mathbf{E}_n \times \mathbf{H}_m - \mathbf{E}_m \times \mathbf{H}_n) \\
&= -(\Gamma_n + \Gamma_m)\mathbf{a}_z \cdot (\mathbf{E}_n \times \mathbf{H}_m - \mathbf{E}_m \times \mathbf{H}_n) \\
&= -(\Gamma_n + \Gamma_m)\mathbf{a}_z \cdot (\mathbf{E}_{tn} \times \mathbf{H}_{tm} - \mathbf{E}_{tm} \times \mathbf{H}_{tn})
\end{aligned}$$

since the terms involving the axial components of the fields vanish upon taking the scalar product with $\mathbf{a}_z$. Using the two-dimensional form of the divergence theorem now gives

$$\iint_S \nabla_t \cdot (\mathbf{E}_n \times \mathbf{H}_m - \mathbf{E}_m \times \mathbf{H}_n)\, da = \oint_C \mathbf{n} \cdot (\mathbf{E}_n \times \mathbf{H}_m - \mathbf{E}_m \times \mathbf{H}_n)\, dl$$
$$= (\Gamma_n + \Gamma_m) \iint_S \mathbf{a}_z \cdot (\mathbf{E}_{tn} \times \mathbf{H}_{tm} - \mathbf{E}_{tm} \times \mathbf{H}_{tn})\, da \quad (19)$$

The contour integral vanishes since $\mathbf{n} \times \mathbf{E}_n$ and $\mathbf{n} \times \mathbf{E}_m$ vanish on the perfectly conducting guide walls. Thus we obtain

$$(\Gamma_n + \Gamma_m) \iint_S \mathbf{a}_z \cdot (\mathbf{E}_{tn} \times \mathbf{H}_{tm} - \mathbf{E}_{tm} \times \mathbf{H}_{tn})\, da = 0 \quad (20)$$

It is convenient to write the expression for the transverse fields in the following form:

$$\mathbf{H}_{tn} = \mathbf{h}_n(u_1,u_2)e^{-\Gamma_n z} \quad (21a)$$
$$\mathbf{E}_{tn} = \mathbf{e}_n(u_1,u_2)e^{-\Gamma_n z} \quad (21b)$$

where $\mathbf{h}_n$ and $\mathbf{e}_n$ are transverse vector functions of the transverse coordinates $u_1$, $u_2$. Introducing these, (20) becomes

$$(\Gamma_n + \Gamma_m) \iint_S \mathbf{a}_z \cdot (\mathbf{e}_n \times \mathbf{h}_m - \mathbf{e}_m \times \mathbf{h}_n)\, da = 0 \quad (22)$$

since the common exponential terms can be canceled. To show that each term vanishes separately, we consider the two solutions $\mathbf{E}_n$, $\mathbf{H}_n$ and $\mathbf{E}'_m$, $\mathbf{H}'_m$, where $\mathbf{E}'_m$, $\mathbf{H}'_m$ is the same mode as considered previously, but depending on $z$ according to $e^{\Gamma_m z}$ rather than $e^{-\Gamma_m z}$. This corresponds to a reversal in the direction of propagation, and, consequently, the direction of the transverse magnetic field is reversed; that is, $\mathbf{E}'_{tm} = \mathbf{e}_m e^{\Gamma_m z}$, $\mathbf{H}'_{tm} = -\mathbf{h}_m e^{\Gamma_m z}$. The equation corresponding to (22) is

$$(\Gamma_n - \Gamma_m) \iint_S \mathbf{a}_z \cdot (-\mathbf{e}_n \times \mathbf{h}_m - \mathbf{e}_m \times \mathbf{h}_n)\, da = 0 \quad (23)$$

Addition and subtraction of (22) and (23) gives

$$\iint_S \mathbf{e}_n \times \mathbf{h}_m \cdot \mathbf{a}_z\, da = 0 \quad (24a)$$
$$\iint_S \mathbf{e}_m \times \mathbf{h}_n \cdot \mathbf{a}_z\, da = 0 \quad (24b)$$

which is the desired orthogonality relation. When there are no losses present, a similar derivation shows that

$$\iint_S \mathbf{e}_n \times \mathbf{h}_m^* \cdot \mathbf{a}_z\, da = 0 \quad (25)$$

where $\mathbf{h}_m^*$ is the complex conjugate of $\mathbf{h}_m$. This latter relation shows that the power flow in a lossless guide is the sum of the power carried by each mode individually. These results will be used in the derivation of Green's functions for general cylindrical waveguides.

### 5.3. Power, Energy, and Attenuation

In a waveguide with perfectly conducting walls, each mode propagates power along the guide independently of the presence of other modes. That this should be so is apparent from the orthogonal properties of the modes, in particular the orthogonal property expressed by (25). The power flow along the guide is given by the real part of the integral of the complex Poynting vector over the guide cross section.

For a single propagating $H$ mode, the time-average power flow is given by

$$P = \tfrac{1}{2}\,\mathrm{Re}\iint_S \mathbf{E}\times\mathbf{H}^*\cdot\mathbf{a}_z\,da = \tfrac{1}{2}\,\mathrm{Re}\iint_S \mathbf{E}_t\times\mathbf{H}_t^*\cdot\mathbf{a}_z\,da$$
$$= -\tfrac{1}{2}\,\mathrm{Re}\iint_S Z_h\beta^2[(\mathbf{a}_z\times\nabla_t\psi_h)\times\nabla_t\psi_h]\cdot\mathbf{a}_z\,da$$

by using (6) for the transverse fields and where $\Gamma$ has been replaced by $j\beta$. The integrand may be expanded to give

$$(\mathbf{a}_z\times\nabla_t\psi_h)\times\nabla_t\psi_h = -(\nabla_t\psi_h\cdot\nabla_t\psi_h)\mathbf{a}_z + (\nabla_t\psi_h\cdot\mathbf{a}_z)\,\nabla_t\psi_h$$

which when scalar-multiplied by $\mathbf{a}_z$ gives $-\nabla_t\psi_h\cdot\nabla_t\psi_h$. Using Green's first identity, the expression for power flow becomes

$$\tfrac{1}{2}Z_h\beta^2\iint_S \nabla_t\psi_h\cdot\nabla_t\psi_h\,da = -\tfrac{1}{2}Z_h\beta^2\iint_S \psi_h\nabla_t^2\psi_h\,da + \tfrac{1}{2}Z_h\beta^2\oint_C \psi_h\frac{\partial\psi_h}{\partial n}\,dl$$

On the guide boundary, $\partial\psi_h/\partial n$ vanishes, and, replacing $\nabla_t^2\psi_h$ by $-k_c^2\psi_h$, and $Z_h$ by $(k_0/\beta)Z_0$, we finally get

$$P = \tfrac{1}{2}Z_0k_0\beta k_c^2\iint_S \psi_h^2\,da \tag{26}$$

for the time-average power flow in a propagating $H$ mode.

For a single $E$ mode propagating along the guide, the time-average power flow is given by

$$P = \tfrac{1}{2}\iint_S Y_e\beta^2[\nabla_{te}\psi\times(\mathbf{a}_z\times\nabla_t\psi_e)]\cdot\mathbf{a}_z\,da$$
$$= \tfrac{1}{2}Y_0k_0\beta k_c^2\iint_S \psi_e^2\,da \tag{27}$$

Equation (27) is obtained by expanding the integrand and using Green's first identity again and the boundary condition $\psi_e = 0$ on the guide walls.

For a propagating mode, the time-average electric and magnetic energies associated with the mode are equal. This is readily demonstrated by integrating the complex Poynting vector over a closed surface consisting of two cross-sectional planes separated by an arbitrary distance and the guide walls. Since the walls are assumed perfectly conducting, the only contribution to the surface integral is from the two cross-sectional planes. From Chap. 1, Eq. (25), we have

$$\tfrac{1}{2} \iint_S \mathbf{E} \times \mathbf{H}^* \cdot d\mathbf{a} = 2j\omega(W_m - W_e) + P_L = 0$$

since $\mathbf{E} \times \mathbf{H}^* \cdot \mathbf{a}_z$ is real, and just as much power enters at one cross-sectional plane as leaves at the other cross-sectional plane for a lossless guide. In the volume bounded by these cross-sectional planes, it is thus necessary that $W_m = W_e$ and $P_L = 0$.

For an $H$ mode, the time-average electric energy per unit length of guide is given by

$$\begin{aligned} W_e &= \tfrac{1}{4}\epsilon_0 \iint_S \mathbf{E}_t \cdot \mathbf{E}_t^* \, da \\ &= \tfrac{1}{4}\epsilon_0 \iint_S (Z_h\beta)^2(\mathbf{a}_z \times \nabla_t\psi_h)^2 \, da \\ &= \tfrac{1}{4}\epsilon_0(Z_0k_0)^2k_c^2 \iint_S \psi_h^2 \, da = W_m \end{aligned} \tag{28}$$

this latter result being obtained by expanding the integrand and using Green's first identity. In a similar fashion, we find that the time-average magnetic energy per unit length of guide in a propagating $E$ mode is given by

$$W_m = \tfrac{1}{4}\mu_0(Y_0k_0)^2k_c^2 \iint_S \psi_e^2 \, da = W_e \tag{29}$$

The power flowing along the guide is equal to the product of the total energy per unit length and the velocity $v_g$ of energy propagation. Using (26) and (28) and the relation $P = 2W_ev_g$, we may solve for the velocity of energy propagation to get

$$v_g = \frac{\beta}{k_0} v_c = \frac{\lambda_0}{\lambda_g} v_c \tag{30}$$

where $v_c = (\mu_0\epsilon_0)^{-1/2}$ is the velocity of TEM waves in free space. The phase velocity of a given mode is that velocity with which an observer would have to move to keep $\beta z - \omega t =$ constant. Differentiating gives

$$\frac{dz}{dt} = v_p = \frac{\omega}{\beta} = \frac{\omega(\mu_0\epsilon_0)^{1/2}}{\beta} v_c = \frac{k_0}{\beta} v_c \tag{31}$$

where $v_p$ is the phase velocity. Comparing (30) and (31) shows that

$$v_p v_g = v_c^2 \tag{32}$$

The velocity of energy propagation is never greater than the velocity of light $v_c$. In Chap. 7 it is shown that $v_g$ is also equal to the group velocity of the signal, and the subscript $g$ was chosen for this reason. The use of (27) and (29) shows that the velocity of energy propagation for $E$ modes is also given by (30). Equations (31) and (32) are valid for $E$ modes as well.

*Energy in Evanescent Modes*

We now turn to a consideration of the energy relations for nonpropagating $E$ and $H$ modes. Consider a section of waveguide bounded by two planes $S_1$ and $S_2$ a distance $l$ apart, as in Fig. 5.2. We wish to determine the time-average net amount of energy, electric or magnetic, which is associated with a nonpropagating mode in the volume bounded by the two cross-sectional surfaces $S_1$ and $S_2$. The volume $V$ is assumed to contain no lossy material. If the time-average complex Poynting vector is integrated over the closed surface surrounding the volume $V$, we get

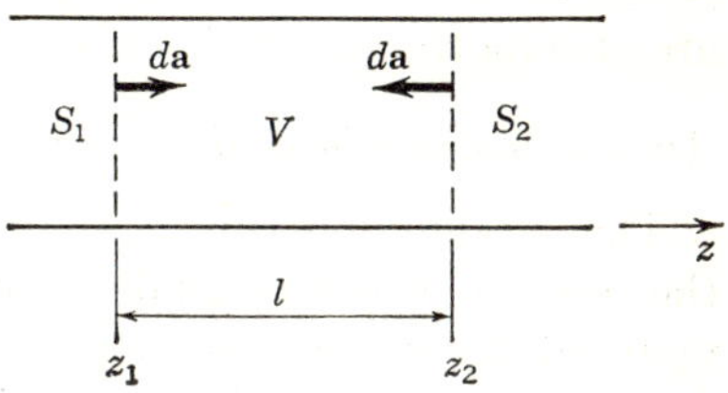

FIG. 5.2

$$\frac{1}{2} \iint_{S_1} \mathbf{E}_t \times \mathbf{H}_t^* \cdot \mathbf{a}_z \, da - \frac{1}{2} \iint_{S_2} \mathbf{E}_t \times \mathbf{H}_t^* \cdot \mathbf{a}_z \, da = 2j\omega(W_m - W_e) \tag{33}$$

since $\mathbf{E}_t \times \mathbf{H}_t^*$ is zero on the guide walls, and the vector element of area on $S_2$ is directed in the negative $z$ direction. The integrals over $S_1$ and $S_2$ are similar to those in (26) and (27) with the exception that $\Gamma$ is now real and, consequently, an exponential term $e^{-2\Gamma z}$ is present and $Z_h$ and $Y_e$ are imaginary. An analysis similar to that used to obtain (26) and (27) shows that, for $H$ modes, (33) becomes

$$\frac{1}{2} j Z_0 k_0 \Gamma k_c^2 (e^{-2\Gamma z_1} - e^{-2\Gamma z_2}) \iint_{S_1} \psi_h^2 \, da = 2j\omega(W_m - W_e)$$

and, hence, $$W_m - W_e = \frac{Z_0(\mu_0\epsilon_0)^{1/2}\Gamma k_c^2}{4} e^{-2\Gamma z_1}(1 - e^{-2\Gamma l}) \iint_{S_1} \psi_h^2 \, da \tag{34}$$

This result shows that, for nonpropagating $H$ modes, the time-average magnetic energy stored in a given length of guide is greater than the time-average electric energy stored in the same length of guide. For $E$ modes, we find that there is a net amount of electric energy stored in a given length of guide for a nonpropagating mode; that is, $W_e > W_m$. For $E$ modes, the expression corresponding to (34) is

$$W_e - W_m = \frac{Y_0(\mu_0\epsilon_0)^{1/2}k_c^2}{4} e^{-2\Gamma z_1}(1 - e^{-2\Gamma l}) \iint_{S_1} \psi_e^2 \, da \qquad (35)$$

The method of derivation is the same as for (34). The predominance of magnetic energy in the case of nonpropagating $H$ modes and electric energy in the case of nonpropagating $E$ modes is, of course, correlated with the inductive and capacitive nature of the wave impedances for these nonpropagating modes. In the analysis of discontinuities in waveguides, we find that the inductive or capacitive nature of a given obstacle depends on the net amount of energy present in the nonpropagating modes in the vicinity of the obstacle. The above energy relations are of importance for this reason.

*Attenuation in Waveguides*

The attenuation constant for electromagnetic waves propagating in the interior of a waveguide filled with homogeneous isotropic dielectric material having a loss tangent $\tan \delta_l$ is easily evaluated. The only modification required in the previously derived formulas is replacing $\epsilon_0$ by $\epsilon(1 - j \tan \delta_l)$. The equation for the propagation constant $\Gamma = j\beta + \alpha$ becomes

$$\Gamma^2 = (j\beta + \alpha)^2 = k_c^2 - \omega^2\mu_0\epsilon(1 - j \tan \delta_l) \qquad (36)$$

where $\beta$ is the phase constant, and $\alpha$ the attenuation constant. When the losses are small, $\delta_l \ll 1$, and (36) may be approximated by

$$-\beta^2 + 2j\alpha\beta = k_c^2 - \omega^2\mu_0\epsilon + j\omega^2\mu_0\epsilon \tan \delta_l$$

for frequencies sufficiently above the cutoff frequency so that $\alpha \ll \beta$. Hence,

$$\beta^2 = \omega^2\mu_0\epsilon - k_c^2 \qquad (37a)$$

$$\alpha = \frac{k_0^2\kappa \tan \delta_l}{2\beta} \qquad (37b)$$

where $\kappa = \epsilon/\epsilon_0$ is the relative dielectric constant.

A first approximation to the attenuation arising from the finite conductivity of the guide walls may be obtained in the same way as for transmission lines with small losses. In this method it is assumed that the currents flowing on the guide wall for a given mode of propagation in the presence of finite conductivity are essentially the same as when the conductivity is infinite. This permits us to evaluate the losses in the walls from the known current distribution of the unperturbed mode, and use of the relation $2\alpha P = P_L$ then leads to a value for the attenuation constant. In the above relation, $P$ is the power flow along the guide for a given mode, and $P_L$ is the associated power loss in the walls. The surface current density on the walls is equal to the tangential magnetic field. Thus, for $H$ modes we obtain, by integrating around the guide boundary $C$, the following result for the losses:

$$P_L = \tfrac{1}{2}R_m \oint_C [(\beta)^2|\mathbf{n} \times \nabla_t\psi_h|^2 + k_c^4\psi_h^2]\, dl$$

per unit length of guide, where $R_m = (\omega\mu/2\sigma)^{1/2}$ is the skin-effect surface resistance of a metal with permeability $\mu$. The attenuation constant for $H$ modes is hence given by

$$\alpha = \frac{R_m \oint_C [(\beta)^2|\mathbf{n} \times \nabla_t\psi_h|^2 + k_c^4\psi_h^2]\, dl}{2Z_0k_0\beta k_c^2 \iint_S \psi_h^2\, da} \tag{38}$$

upon using (26) for the power flow along the guide. The analogous expression for $E$ modes is

$$\alpha = \frac{R_m k_0 Y_0}{2\beta k_c^2} \frac{\oint_C |\nabla_t\psi_e|^2\, dl}{\iint_S \psi_e^2\, da} \tag{39}$$

In (38), $\mathbf{n}$ is the unit normal to the surface of the waveguide wall. The expressions for $\alpha$ may be evaluated for a given guide when the eigenfunctions $\psi_m$ and $\psi_e$ have been found. This simple theory is adequate for computing the attenuation of practical waveguides, provided the frequency is not too close to the cutoff value, and the skin depth is considerably greater than the average surface irregularities. In practice, it is found that, at the shorter wavelengths, normal surfaces cannot be considered as smooth surfaces, and the attenuation is greater than the theoretical formulas predict. This is attributed directly to the effects of surface roughness.†

At the cutoff frequency, $\beta = 0$ and (38) and (39) predict infinite values of attenuation. On the basis of the simple theory, what has happened is that propagation of power along the guide has ceased, while, at the same time, the losses in the walls remain finite. A more careful

† J. Allison and F. A. Benson, Surface Roughness and Attenuation of Precision Drawn, Chemically Polished, Electropolished, Electroplated and Electroformed Waveguides, *Proc. IEE (London)*, vol. 102, part B, pp. 251–259, 1955.

analysis shows that infinite values of $\alpha$ do not occur. The presence of finite conductivity results in a coupling between the $E$ and $H$ modes, and propagation along the guide does not stop at a particular frequency. Rather there is a range of frequencies, over which transition from a propagating mode to a more and more highly attenuated mode takes place, and, only as the frequency tends to zero, does propagation cease. In the particular case of a lossless guide filled with lossy dielectric material, this phenomenon is readily understood. Equation (36) may be written as

$$\Gamma = j\beta + \alpha = [(k_c^2 - \omega^2\mu_0\epsilon)^2 + (\omega^2\mu_0\epsilon \tan \delta_l)^2]^{1/4} e^{j\theta/2} \tag{40}$$

where
$$\theta = \tan^{-1} \frac{\omega^2\mu_0\epsilon \tan \delta_l}{k_c^2 - \omega^2\mu_0\epsilon}$$

For frequencies well above the no-loss cutoff frequency, i.e., for $\omega^2\mu_0\epsilon \gg k_c^2$, the angle $\theta$ is close to $\pi$, and (40) is very nearly a pure-imaginary quantity; hence $\beta$ is large, and $\alpha$ is small. At the no-loss cutoff frequency, $k_c^2 = \omega^2\mu_0\epsilon$, and $\theta = \pi/2$, and from (40) it is seen that

$$\beta = \alpha = 0.707\omega^2\mu_0\epsilon \tan \delta_l$$

Finally, for frequencies below the no-loss cutoff frequency, $\theta$ becomes progressively smaller, and thus $\alpha$ becomes progressively larger than $\beta$, until, when $\theta = 0$, $\beta = 0$ also. It is apparent that a sharp cutoff frequency does not exist as long as $\tan \delta_l$ does not vanish. For most practical purposes, the attenuation becomes so large for frequencies below the no-loss cutoff frequency that this latter frequency may be regarded as the practical lower-frequency bound for useful propagation of a given mode.

At high frequencies, where the skin depth $\delta_s$ is extremely small, we may replace a guide with lossy walls by a guide with perfectly conducting walls, lined with a resistance sheet of thickness a few times greater than $\delta_s$, and having a volume conductivity $\sigma$. In the resulting inhomogeneously filled guide it is found, in most cases, that a pure $E$ or $H$ mode of propagation is not possible. One exception is the circular cylindrical guide where a circularly symmetric $E$ or $H$ mode solution may be found.† For noncircular symmetry, each allowed mode of propagation is a combination of an $E$ and an $H$ mode. Also for the circular guide the problem of propagation in the case of finite conducting walls may be rigorously solved. This is not true for the rectangular guide and other general shapes, and so a perturbation technique must be employed. Even for the circular guide a perturbation technique is preferable, since it leads

† J. A. Stratton, "Electromagnetic Theory," secs. 9.15–9.18, McGraw-Hill Book Company, Inc., New York, 1941.

to a simpler analysis. The perturbation solution given below follows quite closely that given by Papadopoulos.†

In a guide with lossy walls, the modes of propagation in most cases are a combination of an $E$ and an $H$ mode and, consequently, cannot be derived from a scalar potential function, i.e., from a vector potential having only a single component. However, the solutions to Maxwell's equations in a waveguide may be derived from the transverse electric (or transverse magnetic) field component only. Let the electric field in the guide be

$$\mathbf{E} = \mathbf{e}(u_1,u_2)e^{-\gamma z} + \mathbf{e}_z(u_1,u_2)e^{-\gamma z} \tag{41a}$$
$$\mathbf{H} = \mathbf{h}(u_1,u_2)e^{-\gamma z} + \mathbf{h}_z(u_1,u_2)e^{-\gamma z} \tag{41b}$$

where $\mathbf{e}$ and $\mathbf{h}$ are transverse vector functions of the transverse coordinates, $\mathbf{e}_z$ and $\mathbf{h}_z$ are vector functions directed along the $z$ axis only, and $\gamma$ is the propagation constant. Our task is to express all the field components in terms of $\mathbf{e}$. Since $\nabla \cdot \mathbf{E} = 0$, we get $\nabla \cdot \mathbf{e}e^{-\gamma z} + \nabla \cdot \mathbf{e}_z e^{-\gamma z} = 0$ or

$$\gamma \mathbf{e}_z = \mathbf{a}_z \nabla_t \cdot \mathbf{e} \tag{42}$$

where $\nabla_t$ is the transverse part of the vector operator. From the curl equation for $\mathbf{E}$, we get

$$\begin{aligned}\nabla \times \mathbf{E} &= \nabla_t \times \mathbf{e}e^{-\gamma z} + \nabla_t \times \mathbf{e}_z e^{-\gamma z} + \mathbf{a}_z \frac{\partial}{\partial z} \times \mathbf{e}e^{-\gamma z} + \mathbf{a}_z \frac{\partial}{\partial z} \times \mathbf{e}_z e^{-\gamma z} \\ &= \nabla_t \times \mathbf{e}e^{-\gamma z} - \mathbf{a}_z \times \nabla_t e_z e^{-\gamma z} - \gamma \mathbf{a}_z \times \mathbf{e}e^{-\gamma z} = -j\omega\mu_0(\mathbf{h}e^{-\gamma z} + \mathbf{h}_z e^{-\gamma z})\end{aligned}$$

Equating the transverse and axial components now gives

$$-j\omega\mu_0 \mathbf{h}_z = \nabla_t \times \mathbf{e} \tag{43a}$$
$$\begin{aligned} j\omega\mu_0 \mathbf{h} &= \mathbf{a}_z \times \nabla_t e_z + \gamma \mathbf{a}_z \times \mathbf{e} \\ &= \frac{1}{\gamma} \mathbf{a}_z \times \nabla_t \nabla_t \cdot \mathbf{e} + \gamma \mathbf{a}_z \times \mathbf{e} \end{aligned} \tag{43b}$$

where $e_z = \mathbf{a}_z \cdot \mathbf{e}_z$. The field $\mathbf{E}$ is a solution of the vector Helmholtz equation

$$\nabla^2 \mathbf{E} + k_0^2 \mathbf{E} = 0 \tag{44}$$

In general orthogonal curvilinear coordinates, the operator $\nabla^2$ is to be interpreted as the $\nabla\nabla \cdot - \nabla \times \nabla \times$ operator. In a cylindrical coordinate system, (44) separates into two parts given by

$$\nabla_t^2 \mathbf{e}_z + (k_0^2 + \gamma^2)\mathbf{e}_z = 0 \tag{45a}$$
$$\nabla_t^2 \mathbf{e} + (k_0^2 + \gamma^2)\mathbf{e} = 0 \tag{45b}$$

In (45$a$), $\nabla_t^2$ is the transverse part of the Laplacian operator, but, in

† V. M. Papadopoulos, Propagation of Electromagnetic Waves in Cylindrical Waveguides with Imperfectly Conducting Walls, *Quart. J. Mech. and Appl. Math.*, vol. 7, pp. 325–334, September, 1954.

(45*b*), it must be interpreted as the $\nabla_t\nabla_t \cdot - \nabla_t \times \nabla_t \times$ operator for general orthogonal curvilinear cylindrical coordinates.

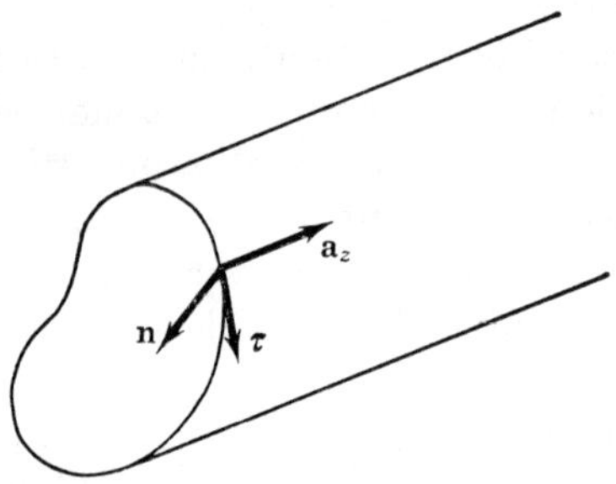

FIG. 5.3

On the guide wall, the field **E** must satisfy the boundary condition

$$\mathbf{E}_{\tan} = Z_m \mathbf{J} e^{-\gamma z}$$

where $Z_m$ is the surface impedance of the metal, $\mathbf{E}_{\tan}$ is the tangential electric field, and $\mathbf{J}e^{-\gamma z}$ is the surface current density. Let $\boldsymbol{\tau}$ be a unit vector tangential to the guide wall, and let **n** be the unit inward normal chosen so that $\boldsymbol{\tau}$, **n**, $\mathbf{a}_z$ form a right-handed system as in Fig. 5.3. The boundary condition for **e** may be written as

$$\begin{aligned} \boldsymbol{\tau} \cdot \mathbf{e} &= Z_m \boldsymbol{\tau} \cdot \mathbf{J}_t = \boldsymbol{\tau} \cdot (\mathbf{n} \times \mathbf{h}_z) Z_m = Z_m h_z \\ &= -\frac{Z_m}{j\omega\mu_0} \mathbf{a}_z \cdot \nabla_t \times \mathbf{e} \end{aligned} \tag{46a}$$

since the transverse current $\mathbf{J}_t$ is given by $\mathbf{n} \times \mathbf{h}_z$. The boundary condition on $\mathbf{e}_z$ is

$$\begin{aligned} \mathbf{a}_z \cdot \mathbf{e}_z = e_z = \nabla_t \cdot \frac{\mathbf{e}}{\gamma} &= Z_m J_z = Z_m \mathbf{a}_z \cdot \mathbf{n} \times \mathbf{h} \\ &= \frac{Z_m}{j\omega\mu_0\gamma} \mathbf{a}_z \cdot \mathbf{n} \times (\mathbf{a}_z \times \nabla_t\nabla_t \cdot \mathbf{e} + \gamma^2 \mathbf{a}_z \times \mathbf{e}) \\ &= \frac{Z_m}{j\omega\mu_0\gamma} \mathbf{n} \cdot (\nabla_t\nabla_t \cdot \mathbf{e} + \gamma^2 \mathbf{e}) \end{aligned}$$

which may be written, by substituting from the wave equation (45*b*),

$$\nabla_t \cdot \mathbf{e} = \frac{Z_m}{j\omega\mu_0} (\mathbf{n} \cdot \nabla_t \times \nabla_t \times \mathbf{e} - k_0^2 \mathbf{n} \cdot \mathbf{e}) \tag{46b}$$

The boundary conditions on **E** have now been written in terms of boundary conditions on the transverse field **e** alone.

The next step in the analysis is to expand **e** in terms of the normal mode functions for the same guide with perfectly conducting walls. Let

$$\mathbf{e} = \sum_s a_s \mathbf{e}_s \tag{47}$$

where the sum is extended over all the *E* and *H* modes, $\mathbf{e}_s$ is the transverse field for a given mode in the lossless guide, and $a_s$ is an unknown amplitude coefficient. Corresponding to the transverse electric field $\mathbf{e}_s$ is a transverse magnetic field $\mathbf{h}_s$, longitudinal fields $\mathbf{e}_{zs}$, $\mathbf{h}_{zs}$, and a propagation constant $\Gamma_s$. For *E* modes, $\mathbf{h}_{zs}$ is zero, while, for *H* modes, $\mathbf{e}_{zs}$ is zero. The transverse-mode function $\mathbf{e}_s$ is a solution of (45*b*) with $\gamma^2$ replaced by $\Gamma_s^2$, and all other field components may be derived from $\mathbf{e}_s$ by equations similar to (42) and (43). On the guide walls, the surface currents associated with the mode $\mathbf{e}_s$ will be denoted as $\mathbf{J}_{ts}$ and $\mathbf{J}_{zs}$. Also the tangential field components $\boldsymbol{\tau} \cdot \mathbf{e}_s$ and $\mathbf{e}_{zs}$ are zero on the walls. It will be assumed that the mode functions $\mathbf{e}_s$ have been normalized so that $\iint_S \mathbf{e}_s \cdot \mathbf{e}_s \, da = 1$. In view of the orthogonal properties of these modes, we then have

$$\iint_S \mathbf{e}_s \cdot \mathbf{e}_n \, da = \begin{cases} 1 & n = s \\ 0 & n \neq s \end{cases} \tag{48}$$

and the mode functions are called an orthonormal set, i.e., a normalized orthogonal set. Using these properties of the functions $\mathbf{e}_s$, it is readily found that the coefficients $a_s$ are given by

$$a_s = \iint_S \mathbf{e} \cdot \mathbf{e}_s \, da \tag{49}$$

where $\mathbf{e}$ is the actual transverse electric field in the lossy guide.

A system of equations for determining the coefficients $a_s$ and propagation constant $\gamma$ may be obtained as follows. Scalar-multiply (45$b$) by $\mathbf{e}_n$ and the corresponding equation for $\mathbf{e}_n$ by $\mathbf{e}$; subtracting the two resulting equations and integrating over the guide cross section gives

$$(\gamma^2 - \Gamma_n^2) \iint_S \mathbf{e} \cdot \mathbf{e}_n \, da = \iint_S (\mathbf{e} \cdot \nabla_t^2 \mathbf{e}_n - \mathbf{e}_n \cdot \nabla_t^2 \mathbf{e}) \, da$$

The right-hand side may be converted to a contour integral by using the two-dimensional vector form of the relation (63$b$) given in Sec. 2.6. We find at once that

$$(\gamma^2 - \Gamma_n^2)a_n = \oint_C \{(\mathbf{n} \times \mathbf{e}_n) \cdot (\nabla_t \times \mathbf{e}) + [\mathbf{n} \times (\nabla_t \times \mathbf{e}_n) \cdot \mathbf{e}] + \mathbf{n} \cdot \mathbf{e}_n \nabla_t \cdot \mathbf{e} - \mathbf{n} \cdot \mathbf{e} \nabla_t \cdot \mathbf{e}_n\} \, dl \tag{50}$$

On the guide walls denoted by $C$, we have $\mathbf{n} \times \mathbf{e}_n = 0$, $\nabla_t \cdot \mathbf{e}_n = \Gamma_n e_{zn} = 0$, and also

$$[\mathbf{n} \times (\nabla_t \times \mathbf{e}_n)] \cdot \mathbf{e} = -j\omega\mu_0 \mathbf{n} \times \mathbf{a}_z h_{zn} \cdot \mathbf{e} = -j\omega\mu_0 h_{zn} \boldsymbol{\tau} \cdot \mathbf{e}$$

Replacing $\boldsymbol{\tau} \cdot \mathbf{e}$ by its value on the surface as given by Eq. (46$a$) and using Eq. (43$a$), this latter term becomes $-(Z_m/j\omega\mu_0)(\nabla_t \times \mathbf{e}_n) \cdot (\nabla_t \times \mathbf{e})$. Using the boundary condition (46$b$) for $\nabla_t \cdot \mathbf{e}$, (50) now becomes

$$(\gamma^2 - \Gamma_n^2)a_n = -\frac{Z_m}{j\omega\mu_0} \oint_C [(\nabla_t \times \mathbf{e}_n) \cdot (\nabla_t \times \mathbf{e}) - \mathbf{n} \cdot \mathbf{e}_n(\mathbf{n} \cdot \nabla_t \times \nabla_t \times \mathbf{e} - k_0^2 \mathbf{n} \cdot \mathbf{e})] \, dl \tag{51a}$$

As a last step, we replace $\mathbf{e}$ by the expansion (47) to get

$$(\gamma^2 - \Gamma_n{}^2)a_n = \frac{-Z_m}{j\omega\mu_0} \sum_s a_s \oint_C [(\nabla_t \times \mathbf{e}_n) \cdot (\nabla_t \times \mathbf{e}_s)$$
$$- \mathbf{n} \cdot \mathbf{e}_n(\mathbf{n} \cdot \nabla_t \times \nabla_t \times \mathbf{e}_s - k_0{}^2\mathbf{n} \cdot \mathbf{e}_s)]\, dl \qquad n = 1, 2, 3, \ldots \quad (51b)$$

The above represents an infinite set of linear equations for the unknown coefficients $a_s$. A solution for the $a_s$ exists only if the determinant vanishes. The vanishing of the determinant results in an infinite number of roots for $\gamma^2$, corresponding to the infinite number of modes which may exist in the lossy guide. For each root for $\gamma^2$, a set of coefficients $a_s$ is obtained, and these determine the field configuration for the mode. In practice, of course, a finite expansion must be used in order to obtain a finite number of equations. Examination of (51$b$) shows that all the coefficients $a_n$ are small except those for which $\Gamma_n{}^2 \approx \gamma^2$, since $Z_m$ is very small. Thus, when a solution for the perturbed mode with propagation constant $\gamma = \gamma_n \approx \Gamma_n$ is desired, the first approximation to (51$b$) is obtained by including only the mode $\mathbf{e}_n$. If degeneracy exists among the modes in the lossless guide, all modes with propagation constant $\Gamma_n$ must be included.

A clearer interpretation of (51$b$) may be obtained by noting that the normal mode functions $\mathbf{e}_n$ may be chosen as real vector functions, and, hence,

$$(\nabla_t \times \mathbf{e}_n) \cdot (\nabla_t \times \mathbf{e}_s) = j\omega\mu_0\mathbf{h}_{zn} \cdot j\omega\mu_0\mathbf{h}_{zs} = \omega^2\mu_0{}^2\mathbf{J}_{ts} \cdot \mathbf{J}_{tn}^*$$

where $\mathbf{J}_{tn}^*$ is the complex conjugate of $\mathbf{J}_{tn}$, since, when $\mathbf{e}_n$ is real, $\mathbf{J}_{tn}$ is imaginary. Also we have

$$\mathbf{n} \cdot \nabla_t \times \nabla_t \times \mathbf{e}_s = -j\omega\mu_0\mathbf{n} \cdot \nabla_t \times \mathbf{h}_{zs} = -j\omega\mu_0(j\omega\epsilon_0\mathbf{n} \cdot \mathbf{e}_s + \Gamma_s\mathbf{n} \cdot \mathbf{a}_z \times \mathbf{h}_s)$$
$$= k_0{}^2\mathbf{n} \cdot \mathbf{e}_s + j\omega\mu_0\Gamma_sJ_{zs}$$

and one further result that $\mathbf{n} \cdot \mathbf{e}_n = -Z_n\boldsymbol{\tau} \cdot \mathbf{h}_n = Z_nJ_{zn}$, where $J_{zn}$ and $Z_n$ are real for propagating modes, and $Z_n$ is the scalar wave impedance for the mode. Substituting these expressions into (51$b$), we obtain a relation in terms of the surface currents only:

$$(\gamma^2 - \Gamma_n{}^2)a_n = Z_m \sum_s a_s \oint_C (j\omega\mu_0\mathbf{J}_{ts} \cdot \mathbf{J}_{tn}^* + \Gamma_sZ_n\mathbf{J}_{zs} \cdot \mathbf{J}_{zn})\, dl$$
$$n = 1,2, \ldots \quad (52)$$

where, for $H$ modes, $Z_n = j\omega\mu_0/\Gamma_n$, and, for $E$ modes, $Z_n = -j\omega\mu_0\Gamma_n/k_0{}^2$. This result shows at once that modes with current distributions on the walls which are orthogonal are not coupled, even if such modes are degenerate. For nondegenerate modes a single-mode approximation in (52) leads to results for the attenuation constant $\alpha$ which is the same as given by the power-loss method, i.e., Eqs. (38) and (39) for frequencies above the cutoff frequency. In addition, (52) gives the first-order change in the phase constant $\beta$. An application of (52) to the rectangular waveguide is made in the next section.

## 5.4. The Rectangular Waveguide

The most commonly used waveguide is the rectangular guide. Some of the reasons for this are a good bandwidth of operation for single-mode propagation, reasonably low attenuation, and good mode stability for the fundamental mode of propagation. The dimensions of the common

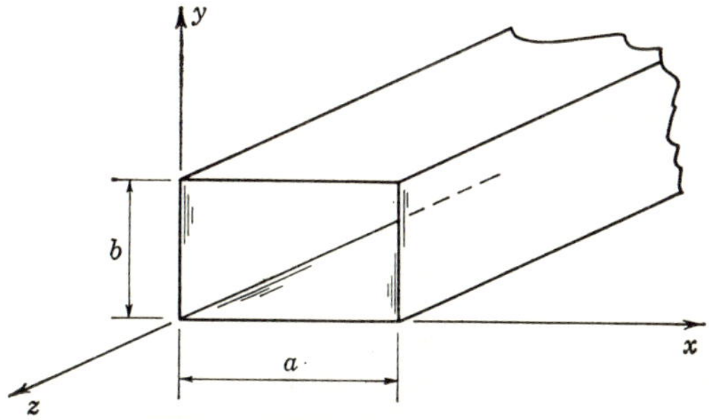


FIG. 5.4. The rectangular guide.

rectangular guide are such that the width $a$ is approximately equal to twice the height $b$, as illustrated in Fig. 5.4.

*Transverse Electric or H Modes*

The transverse electric or $H$ modes may be derived from the scalar function

$$\psi_{h,nm} = \cos\frac{n\pi x}{a}\cos\frac{m\pi y}{b} \qquad \begin{array}{l} n = 0, 1, 2, \ldots \\ m = 0, 1, 2, \ldots \\ m \neq n = 0 \end{array} \tag{53}$$

by means of (6). It is readily seen that $\psi_h$ as given is a solution to the scalar Helmholtz equation (3) and satisfies the boundary condition $\partial\psi_h/\partial n = 0$, provided the propagation constant is chosen as

$$\Gamma_{nm}^2 = \left(\frac{n\pi}{a}\right)^2 + \left(\frac{m\pi}{b}\right)^2 - k_0^2 = k_c^2 - k_0^2 \tag{54}$$

To each set of integers $n$, $m$, a solution or mode exists, and these modes are designated as the $\text{TE}_{nm}$ or $H_{nm}$ modes. The cutoff wavelength $\lambda_{c,nm}$ for the $nm$th mode is given by

$$\lambda_{c,nm} = \frac{2\pi}{k_c} = \frac{2ab}{(n^2b^2 + m^2a^2)^{1/2}} \tag{55}$$

The dominant mode of propagation (mode with the largest cutoff wavelength) is the $H_{10}$ mode when $a > b$, and its cutoff wavelength is $2a$. If $a = 2b$, the next mode to propagate is the $H_{01}$ mode, which has a cutoff wavelength equal to $a$.

The properties of $H$ modes which are common to all guides are summarized in Table 5.1, while those properties of $H$ modes which are valid for the rectangular guide only are summarized in Table 5.2. The expression given for the attenuation constant in this table is that computed by the power-loss method, i.e., Eq. (38).

TABLE 5.1. COMMON PROPERTIES OF PROPAGATING MODES IN EMPTY CYLINDRICAL WAVEGUIDES

| Property | $H$ modes | $E$ modes |
|---|---|---|
| Wave impedance | $Z_h = (k_0/\beta)Z_0$ | $Z_e = (\beta/k_0)Z_0$ |
| Propagation constant | $\Gamma = j\beta = (k_c^2 - k_0^2)^{1/2}$ | Same |
| Cutoff wavelength | $\lambda_c = 2\pi/k_c$ | Same |
| Guide wavelength | $\lambda_g = \lambda_0/(1 - \lambda_0^2/\lambda_c^2)^{1/2}$ | Same |
| Group velocity | $v_g = v_c\lambda_0/\lambda_g$ | Same |
| Phase velocity | $v_p = v_c\lambda_g/\lambda_0$ | Same |

$\lambda_0$ = free-space wavelength

$Z_0 = \left(\dfrac{\mu_0}{\epsilon_0}\right)^{1/2}$ = intrinsic impedance of free space

$v_c = (\mu_0\epsilon_0)^{-1/2}$ = velocity of light in free space

*Transverse Magnetic or E Modes*

The appropriate solution for $\psi_e$ which satisfies the boundary condition $\psi_e = 0$ on the guide walls is

$$\psi_{e,nm} = \sin\frac{n\pi x}{a}\sin\frac{m\pi y}{b} \qquad \begin{array}{l} n = 1, 2, \ldots \\ m = 1, 2, \ldots \end{array} \tag{56}$$

and $\Gamma_{nm}^2 = (n\pi/a)^2 + (m\pi/b)^2 - k_0^2$. The fields may be derived from $\psi_{e,nm}$ by means of (11). The modes are designated as $\mathrm{TM}_{nm}$ or $E_{nm}$ modes. The dominant $E$ mode is the $E_{11}$ mode with a cutoff wavelength $2ab/(a^2 + b^2)^{1/2}$, which is equal to $2a/5^{1/2}$ for $a = 2b$. The properties of $E$ modes in rectangular guides are summarized in Tables 5.1 and 5.2. The attenuation constant given is based on Eq. (39), that is, that determined by the power-loss method. It should be noted that the $H_{nm}$ and $E_{nm}$ modes are degenerate and have the same propagation constant $\Gamma_{nm}$.

*Coupling of Modes in Lossy Rectangular Guides*

The normal mode functions $\mathbf{e}_n$ introduced in Sec. 5.3 to represent the transverse electric field distribution will be written as $\mathbf{e}_{nm}$ for the $H_{nm}$

TABLE 5.2. PROPERTIES OF MODES IN EMPTY RECTANGULAR WAVEGUIDES

| Property | $H_{nm}$ modes | $E_{nm}$ modes |
|---|---|---|
| Generating function | $\psi_{h,nm} = \cos\frac{n\pi x}{a}\cos\frac{m\pi y}{b}$ | $\psi_{e,nm} = \sin\frac{n\pi x}{a}\sin\frac{m\pi y}{b}$ |
| Cutoff wave number | $k_{c,nm}^2 = \left(\frac{n\pi}{a}\right)^2 + \left(\frac{m\pi}{b}\right)^2$ | Same |
| Propagation constant | $\Gamma_{nm}^2 = k_{c,nm}^2 - k_0^2$ | Same |
| Dyadic wave impedance | $\bar{\mathbf{Z}}_h = \frac{jk_0}{\Gamma_{nm}} Z_0(\mathbf{a}_x\mathbf{a}_y - \mathbf{a}_y\mathbf{a}_x)$ | $\bar{\mathbf{Z}}_e = \frac{\Gamma_{nm}}{jk_0} Z_0(\mathbf{a}_x\mathbf{a}_y - \mathbf{a}_y\mathbf{a}_x)$ |
| Dyadic wave admittance | $\bar{\mathbf{Y}}_h = \frac{\Gamma_{nm}}{jk_0} Y_0(\mathbf{a}_y\mathbf{a}_x - \mathbf{a}_x\mathbf{a}_y)$ | $\bar{\mathbf{Y}}_e = \frac{jk_0}{\Gamma_{nm}} Y_0(\mathbf{a}_y\mathbf{a}_x - \mathbf{a}_x\mathbf{a}_y)$ |
| Axial magnetic field | $k_{c,nm}^2\psi_{h,nm}e^{\pm\Gamma_{nm}z}$ | Zero |
| Axial electric field | Zero | $k_{c,nm}^2\psi_{e,nm}e^{\pm\Gamma_{nm}z}$ |
| Transverse magnetic field | $\mathbf{H}_t = \pm\Gamma_{nm}e^{\pm\Gamma_{nm}z}\nabla_t\psi_{h,nm}$ | $\mp\bar{\mathbf{Y}}_e\cdot\mathbf{E}_t$ |
| Transverse electric field | $\mp\bar{\mathbf{Z}}_h\cdot\mathbf{H}_t$ | $\mathbf{E}_t = \pm\Gamma_{nm}e^{\pm\Gamma_{nm}z}\nabla_t\psi_{e,nm}$ |
| Power flow, Eqs. (26), (27) $\Gamma_{nm} = j\beta_{nm}$ | $\frac{abZ_0k_0\beta_{nm}k_{c,nm}^2}{2\epsilon_{on}\epsilon_{om}}$ | $\frac{abY_0k_0\beta_{nm}k_{c,nm}^2}{2\epsilon_{on}\epsilon_{om}}$ |
| Attenuation, Eqs. (38), (39) $R_m = (\omega\mu/2\sigma)^{1/2}$, $\alpha$ in nepers per meter | $\frac{2R_m}{bZ_0(1 - k_{c,nm}^2/k_0^2)^{1/2}}\left[\left(1+\frac{b}{a}\right)\frac{k_{c,nm}^2}{k_0^2} + \frac{b}{a}\left(\frac{\epsilon_{om}}{2} - \frac{k_{c,nm}^2}{k_0^2}\right)\left(\frac{n^2ab + m^2a^2}{n^2b^2 + m^2a^2}\right)\right]$ <br> $\epsilon_{on} = \begin{cases} 1 & n = 0 \\ 2 & n > 0 \end{cases}$ | $\frac{2R_m}{bZ_0(1 - k_{c,nm}^2/k_0^2)^{1/2}}\left(\frac{n^2b^3 + m^2a^3}{n^2b^2a + m^2a^3}\right)$ |

modes and as $\mathbf{e}'_{nm}$ for the $E_{nm}$ modes. The corresponding transverse and longitudinal magnetic field components will be written as $\mathbf{h}_{nm}$, $\mathbf{h}_{znm}$ for the $H_{nm}$ modes and as $\mathbf{h}'_{nm}$ for the $E_{nm}$ modes. The surface currents on the guide wall will be designated by $\mathbf{J}_{tnm}$, $\mathbf{J}_{znm}$, and $\mathbf{J}'_{znm}$, where the prime again corresponds to the $E_{nm}$ modes. The surface currents are given by

$$\begin{aligned} \mathbf{J}_{tnm} &= \mathbf{n} \times \mathbf{h}_{znm} \\ \mathbf{J}_{znm} &= \mathbf{n} \times \mathbf{h}_{nm} \end{aligned} \tag{57}$$

where $\mathbf{n}$ is the unit inward normal to the guide surface. A similar set of equations hold for the $E_{nm}$ modes with all quantities replaced by primed quantities. Note that $\mathbf{e}_{znm}$, $\mathbf{h}'_{znm}$, $\mathbf{J}'_{tnm}$ are all zero, and that the propagation factor $e^{-\Gamma_{nm}z}$ is not included in the above.

The normal mode functions are readily obtained from Table 5.2 in terms of $\psi_{h,nm}$ and $\psi_{e,nm}$ and, when normalized so that (48) holds, are given by the following equations:

For $H_{nm}$ modes

$$\mathbf{e}_{nm} = \left[\left(\frac{m\pi}{b}\right)^2 + \left(\frac{n\pi}{a}\right)^2\right]^{-\frac{1}{2}} \left(\frac{\epsilon_{on}\epsilon_{om}}{ab}\right)^{\frac{1}{2}} \left(\mathbf{a}_x \frac{m\pi}{b} \cos\frac{n\pi x}{a} \sin\frac{m\pi y}{b} - \mathbf{a}_y \frac{n\pi}{a} \sin\frac{n\pi x}{a} \cos\frac{m\pi y}{b}\right) \tag{58a}$$

For $E_{nm}$ modes

$$\mathbf{e}'_{nm} = \left[\left(\frac{m\pi}{b}\right)^2 + \left(\frac{n\pi}{a}\right)^2\right]^{-\frac{1}{2}} \left(\frac{\epsilon_{on}\epsilon_{om}}{ab}\right)^{\frac{1}{2}} \left(\mathbf{a}_x \frac{n\pi}{a} \cos\frac{n\pi x}{a} \sin\frac{m\pi y}{b} + \mathbf{a}_y \frac{m\pi}{b} \sin\frac{n\pi x}{a} \cos\frac{m\pi y}{b}\right) \tag{58b}$$

The magnetic field may be obtained by means of (43*a*) and (43*b*), and then the surface currents are found by means of (57). For the $H_{nm}$ modes the surface currents are

$$\mathbf{J}_{tnm} = \frac{\left[\left(\frac{n\pi}{a}\right)^2 + \left(\frac{m\pi}{b}\right)^2\right]^{\frac{1}{2}} \left(\frac{\epsilon_{on}\epsilon_{om}}{ab}\right)^{\frac{1}{2}}}{j\omega\mu_0} \begin{cases} -\mathbf{a}_y \cos\frac{m\pi y}{b} & x = 0 \\ \mathbf{a}_y \cos n\pi \cos\frac{m\pi y}{b} & x = a \\ \mathbf{a}_x \cos\frac{n\pi x}{a} & y = 0 \\ -\mathbf{a}_x \cos m\pi \cos\frac{n\pi x}{a} & y = b \end{cases} \tag{59a}$$

$$J_{znm} = \frac{\beta_{nm}\left(\dfrac{\epsilon_{on}\epsilon_{om}}{ab}\right)^{1/2}}{\omega\mu_0\left[\left(\dfrac{n\pi}{a}\right)^2 + \left(\dfrac{m\pi}{b}\right)^2\right]^{1/2}} \begin{cases} \dfrac{m\pi}{b}\sin\dfrac{m\pi y}{b} & x = 0 \\ \dfrac{-m\pi}{b}\cos n\pi \sin\dfrac{m\pi y}{b} & x = a \\ \dfrac{-n\pi}{a}\sin\dfrac{n\pi x}{a} & y = 0 \\ \dfrac{n\pi}{a}\cos m\pi \sin\dfrac{n\pi x}{a} & y = b \end{cases} \tag{59b}$$

where $j\beta_{nm} = \Gamma_{nm}$, and $\epsilon_{on}$ is the Neumann factor, and equals unity for $n = 0$ and is equal to 2 otherwise. For the $E_{nm}$ modes, there are no transverse currents, and the longitudinal current is given by

$$J'_{znm} = \frac{\left[\left(\dfrac{n\pi}{a}\right)^2 + \left(\dfrac{m\pi}{b}\right)^2\right]^{-1/2}\left(\dfrac{\epsilon_{on}\epsilon_{om}}{ab}\right)^{1/2} k_0^2}{\omega\mu_0\beta_{nm}} \begin{cases} \dfrac{n\pi}{a}\sin m\pi\dfrac{y}{b} & x = 0 \\ \dfrac{-n\pi}{a}\cos n\pi \sin\dfrac{m\pi y}{b} & x = a \\ \dfrac{m\pi}{b}\sin\dfrac{n\pi x}{a} & y = 0 \\ \dfrac{-m\pi}{b}\cos m\pi \sin\dfrac{n\pi x}{a} & y = b \end{cases} \tag{59c}$$

An examination of the above expressions for the surface currents shows that, for $H_{nm}$ and $E_{rs}$ modes, with $r \neq n$ and $s \neq m$, the currents are orthogonal, and, hence, these modes are not coupled. The coupling between $E$ and $H$ modes is brought about by the axial currents only. From (52) a first-order solution for the propagation constant $\gamma$ may be obtained by including only the $E_{nm}$ and $H_{nm}$ modes since $\gamma \approx \Gamma_{nm}$, and all the coefficients in the expansion of the transverse electric field are small, except the coefficients of the two modes $\mathbf{e}_{nm}$ and $\mathbf{e}'_{nm}$. The coefficients for these two modes will be designated as $a_{nm}$ and $a'_{nm}$. Substituting (59) into (52) and performing the integrations gives the following two homogeneous equations:

$$\left\{\frac{(\gamma^2 + \beta_{nm}^2)j\omega\mu_0 ab}{Z_m\epsilon_{on}\epsilon_{om}} + k_{c,nm}^2\left(\frac{2b}{\epsilon_{om}} + \frac{2a}{\epsilon_{on}}\right) + \frac{\beta_{nm}^2}{k_{c,nm}^2}\left[b\left(\frac{m\pi}{b}\right)^2 + a\left(\frac{n\pi}{a}\right)^2\right]\right\} a_{nm} + \frac{k_0^2}{k_{c,nm}^2}\frac{mn\pi^2}{ab}(b-a)a'_{nm} = 0 \tag{60a}$$

$$\frac{\beta_{nm}^2}{k_{c,nm}^2}\frac{mn\pi^2}{ab}(b-a)a_{nm} + \left\{\frac{(\gamma^2 + \beta_{nm}^2)j\mu\omega_0 ab}{Z_m\epsilon_{on}\epsilon_{om}} + \frac{k_0^2}{k_{c,nm}^2}\left[\left(\frac{n\pi}{a}\right)^2 b + \left(\frac{m\pi}{b}\right)^2 a\right]\right\} a'_{nm} = 0 \qquad n = m \neq 0 \tag{60b}$$

One interesting result immediately apparent from (60) is the absence of coupling between $E$ and $H$ modes in a square guide for which $a = b$ and the off-diagonal terms in (60) vanish. For $a \neq b$, (60) determines a

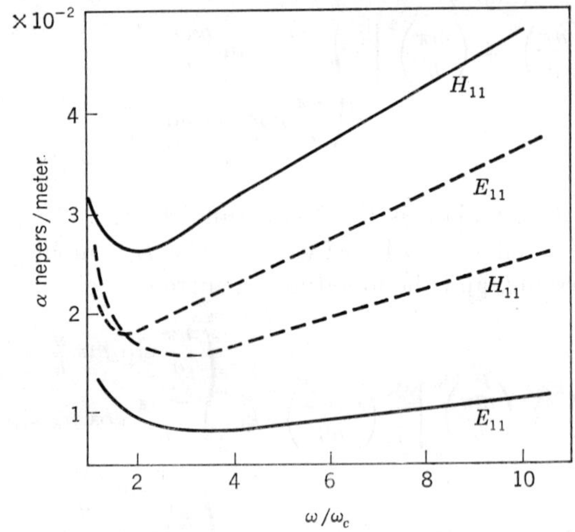


FIG. 5.5. Attenuation in a copper waveguide, $a = 2b = 1$ inch. Solid line, perturbation solution; dashed line, power-loss-method solution.

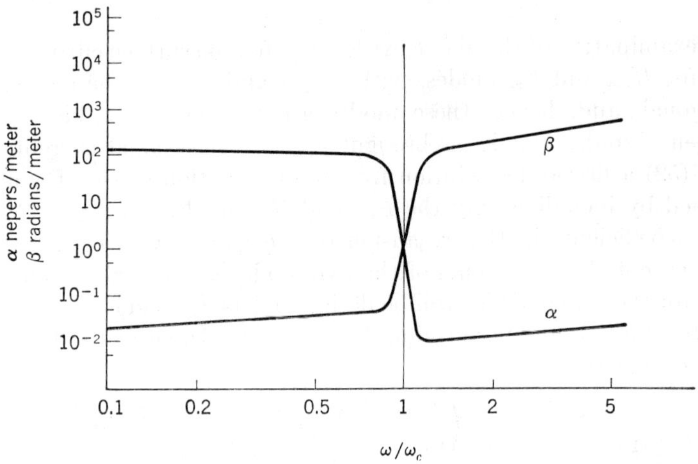


FIG. 5-6. Attenuation and phase constant for $H_{10}$ mode in a copper rectangular guide with $a = 2b = 1$ inch.

solution for $a_{nm}$ and $a'_{nm}$, provided the determinant vanishes. The vanishing of the determinant leads to two roots for $\gamma^2$. For one root $a_{nm} > a'_{nm}$, and this is the propagation constant for a mode corresponding to a perturbed $H_{nm}$ mode. The other root corresponds to a perturbed $E_{nm}$ mode, for which $a'_{nm} > a_{nm}$. In Fig. 5.5 the attenuation constants for the perturbed $E_{11}$ and $H_{11}$ modes are plotted as a function of frequency for $a = 2b$ and a conductivity corresponding to that of copper ($5.8 \times 10^7$ mhos per meter). The dashed curves in the same figure give the attenuation constants for the same two modes as computed by the power-loss method. There is an appreciable difference between the results as obtained by the two methods. We can conclude from this example that the power-loss method does not give accurate results in those cases where degenerate $E$ and $H$ modes are coupled together.

If $m = 0$ and $n = 1$, Eq. (60$a$) reduces to a single term and gives the phase constant and attenuation constant for the $H_{10}$ mode. A plot of $\alpha$ and $\beta$ versus frequency is given in Fig. 5.6 for $a = 2b = 1$ inch and a copper waveguide. For this mode, the power-loss method gives essentially the same results except in the immediate vicinity of the no-loss cutoff frequency $\omega_c/2\pi$.

### 5.5. Circular Cylindrical Waveguides

In a circular cylindrical waveguide of radius $a$, the generating functions $\psi_h$ and $\psi_e$ for the $H$ and $E$ modes are solutions of the following equation:

$$\frac{\partial^2\psi_i}{\partial r^2} + \frac{1}{r}\frac{\partial\psi_i}{\partial r} + \frac{1}{r^2}\frac{\partial^2\psi_i}{\partial\theta^2} + k_c{}^2\psi_i = 0 \qquad i = h, e \tag{61}$$

where $k_c{}^2 = \Gamma^2 + k_0{}^2$ and $\psi_i = 0$ at $r = a$ for $E$ modes, and $\partial\psi_i/\partial r = 0$ at $r = a$ for $H$ modes. Figure 5.7 illustrates a circular guide and the coordinates $r$, $\theta$, $z$. The solution to (61), which is finite at the origin and single-valued throughout the interior, is

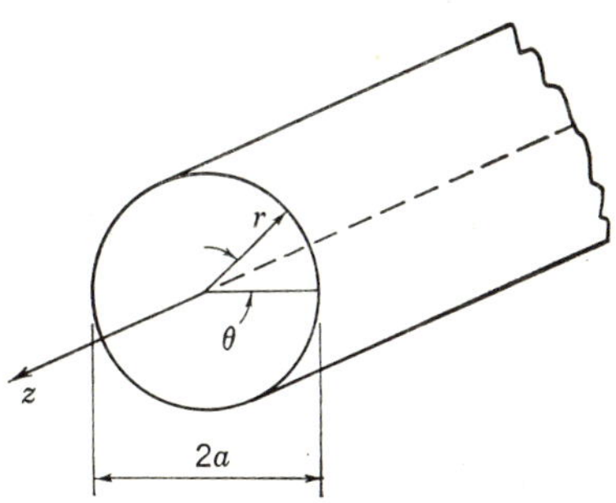

FIG. 5.7. A circular cylindrical waveguide.

$$\psi_i = J_n(k_c r)\begin{cases}\sin n\theta \\ \cos n\theta\end{cases} \qquad n = 0, 1, 2, \ldots \tag{62}$$

For $H$ modes, $dJ_n(k_c r)/dr = 0$ at $r = a$. This equation determines an infinite number of roots for $k_c a$ which will be designated by $p'_{nm}$. The corresponding $H$ mode will be labeled $H_{nm}$ or $\text{TE}_{nm}$, where the first subscript refers to the number of cyclic variations with $\theta$, and the second subscript refers to the $m$th root of the Bessel function. A number of the roots $p'_{nm}$ are listed in Table 5.3. The dominant, or first, mode to propagate in a circular guide is the $\text{TE}_{11}$ mode.

For $E$ modes, $J_n(k_c a) = 0$ determines the allowed values of $k_c$ and, hence, the propagation constants. The roots of this equation are

TABLE 5.3. ROOTS OF $\left.\frac{dJ_n(k_c r)}{dr}\right|_{r=a} = 0$

| $n$ | $p'_{n1}$ | $p'_{n2}$ | $p'_{n3}$ | $p'_{n4}$ |
|---|---|---|---|---|
| 0 | 3.832 | 7.016 | 10.174 | 13.324 |
| 1 | 1.841 | 5.331 | 8.536 | 11.706 |
| 2 | 3.054 | 6.706 | 9.970 | 13.170 |

designated by $p_{nm}$, and the corresponding modes labeled $E_{nm}$ or $TM_{nm}$. Table 5.4 gives the first few roots of the above equation.

TABLE 5.4. ROOTS OF $J_n(k_c a) = 0$

| $n$ | $p_{n1}$ | $p_{n2}$ | $p_{n3}$ | $p_{n4}$ |
|---|---|---|---|---|
| 0 | 2.405 | 5.520 | 8.654 | 11.792 |
| 1 | 3.832 | 7.016 | 10.174 | 13.324 |
| 2 | 5.135 | 8.417 | 11.620 | 14.796 |

The properties of $E$ and $H$ modes in circular waveguides are summarized in Table 5.5. In circular cylindrical coordinates the transverse operator $\nabla_t$ is given by

$$\nabla_t = \mathbf{a}_r \frac{\partial}{\partial r} + \frac{\mathbf{a}_\theta}{r} \frac{\partial}{\partial \theta} \tag{63}$$

The derivatives of the Bessel function $J_n(k_c r)$ are given by the following relation:

$$\frac{1}{k_c} \frac{d}{dr} J_n(k_c r) = \frac{-n}{k_c r} J_n(k_c r) + J_{n-1}(k_c r) = \frac{n}{k_c r} J_n(k_c r) - J_{n+1}(k_c r) \tag{64}$$

The integral of the product of two Bessel functions of the same order $n$ and of different argument is given by

$$\int_o^a J_n(k_1 r) J_n(k_2 r) r\, dr = \frac{a}{k_1^2 - k_2^2} [k_2 J_n(k_1 a) J_{n-1}(k_2 a) - k_1 J_{n-1}(k_1 a) J_n(k_2 a)] \tag{65a}$$

and, when $k_1 = k_2$, the value of the integral is

$$\int_o^a r[J_n(k_1 r)]^2\, dr = \frac{a^2}{2} \left\{ \left[ \frac{1}{k_1} \frac{dJ_n(k_1 a)}{dr} \right]^2 + \left( 1 - \frac{n^2}{k_1^2 a^2} \right) J_n^2(k_1 a) \right\} \tag{65b}$$

If $k_1 a$ and $k_2 a$ correspond to two of the roots given in Table 5.3 or Table 5.4, the value of the integral (65a) is zero. This result follows in view of the orthogonality properties of the modes in a cylindrical waveguide.

TABLE 5.5. PROPERTIES OF MODES IN EMPTY CIRCULAR WAVEGUIDES

| Property | $H_{nm}$ modes | $E_{nm}$ modes |
|---|---|---|
| Generating function | $\psi_{h,nm} = J_n(k_{c,nm}r) \begin{cases} \sin n\theta \\ \cos n\theta \end{cases}$ | $\psi_{e,nm} = J_n(k_{c,nm}r) \begin{cases} \sin n\theta \\ \cos n\theta \end{cases}$ |
| Cutoff wave number | $k_{c,nm} = \dfrac{p'_{nm}}{a}$ | $k_{c,nm} = \dfrac{p_{nm}}{a}$ |
| Propagation constant | $\Gamma_{nm}^2 = \left(\dfrac{p'_{nm}}{a}\right)^2 - k_0{}^2$ | $\Gamma_{nm}^2 = \left(\dfrac{p_{nm}}{a}\right)^2 - k_0{}^2$ |
| Dyadic wave impedance | $\bar{\mathbf{Z}}_h = \dfrac{jk_0}{\Gamma_{nm}} Z_0(\mathbf{a}_r\mathbf{a}_\theta - \mathbf{a}_\theta\mathbf{a}_r)$ | $\bar{\mathbf{Z}}_e = \dfrac{\Gamma_{nm}}{jk_0} Z_0(\mathbf{a}_r\mathbf{a}_\theta - \mathbf{a}_\theta\mathbf{a}_r)$ |
| Dyadic wave admittance | $\bar{\mathbf{Y}}_h = \dfrac{\Gamma_{nm}}{jk_0} Y_0(\mathbf{a}_\theta\mathbf{a}_r - \mathbf{a}_r\mathbf{a}_\theta)$ | $\bar{\mathbf{Y}}_e = \dfrac{jk_0}{\Gamma_{nm}} Y_0(\mathbf{a}_\theta\mathbf{a}_r - \mathbf{a}_r\mathbf{a}_\theta)$ |
| Axial magnetic field........ | $k_{c,nm}^2\psi_{h,nm}e^{\pm\Gamma_{nm}z}$ | Zero |
| Axial electric field........ | Zero | $k_{c,nm}^2\psi_{e,nm}e^{\pm\Gamma_{nm}z}$ |
| Transverse magnetic field........ | $\mathbf{H}_t = \pm\Gamma_{nm}e^{\pm\Gamma_{nm}z}\nabla_t\psi_{h,nm}$ | $\mp\bar{\mathbf{Y}}_e\cdot\mathbf{E}_t$ |
| Transverse electric field | $\mp\bar{\mathbf{Z}}_h\cdot\mathbf{H}_t$ | $\mathbf{E}_t = \pm\Gamma_{nm}e^{\pm\Gamma_{nm}z}\nabla_t\psi_{e,nm}$ |
| Power flow, Eqs. (26), (27); $\Gamma_{nm} = j\beta_{nm}$ | $\dfrac{Z_0k_0\beta_{nm}\pi}{4}(p'^2_{nm} - n^2)J_n{}^2(p'_{nm})$ | $\dfrac{Y_0k_0\beta_{nm}\pi}{4}p_{nm}^2\left[\dfrac{dJ_n(k_cr)}{dk_cr}\Bigg\vert_a\right]^2$ |
| Attenuation in nepers per meter, Eqs. (38), (39); $R_m = (\omega\mu/2\sigma)^{1/2}$ | $\dfrac{R_m}{aZ_0}\left(1 - \dfrac{k_{c,nm}^2}{k_0{}^2}\right)^{-1/2}\left(\dfrac{k_{c,nm}^2}{k_0{}^2} + \dfrac{n^2}{p'^2_{nm} - n^2}\right)$ | $\dfrac{R_m}{aZ_0}\left(1 - \dfrac{k_{c,nm}^2}{k_0{}^2}\right)^{-1/2}$ |

The result may be proved directly from Bessel's differential equation by the methods given in Sec. 5.2.

An examination of the formula for the attenuation constant for $H_{nm}$ modes shows that the $H_{01}$ mode has the unique property that its attenuation decreases as the square root of the frequency. It is for this reason that considerable work has been done on techniques and components for the utilization of the $H_{01}$ mode in low-loss long-distance communication links.†

## 5.6. Green's Functions and Mode Completeness

The general dyadic Green's function for a cylindrical guide is a solution of one of the following two equations:

$$\nabla \times \nabla \times \bar{\mathbf{G}} - k^2\bar{\mathbf{G}} = \bar{\mathbf{I}}\delta(x - x')\delta(y - y')\delta(z - z') \tag{66a}$$
$$\nabla^2\bar{\mathbf{G}} + k^2\bar{\mathbf{G}} = \bar{\mathbf{I}}\delta(x - x')\delta(y - y')\delta(z - z') \tag{66b}$$

subject to certain boundary conditions on the guide walls. Very often, in practice, the field can be derived from a single scalar function, and then a scalar Green's function is all that is required. For arbitrary known current distributions it is usually easier to compute the radiated field directly than to first obtain the solution for the appropriate dyadic Green's function. In this section we will consider first of all the scalar Green's function for $H_{n0}$ modes in rectangular waveguides, and then a general method for the derivation of the field radiated by arbitrary current filaments in cylindrical guides. The vector field radiated by a small current element or loop is often called a Green's function, but it should be noted that this is not, in general, a dyadic Green's function which is a solution of (66). The dyadic Green's function is most appropriately viewed as an operator which will serve to generate the field from a given vector source function. In some cases it cannot be readily interpreted in a physical sense to be the field radiated by a unit current source. One exception is the dyadic Green's function corresponding to the vector potential arising from a unit vector current source (see Prob. 5.13 at end of chapter). Since the vast majority of waveguide discontinuity problems can be readily handled by the techniques to be presented here, a detailed consideration of dyadic Green's functions is not undertaken in this section. More on dyadic Green's functions is given in the problems at the end of this chapter.

### *Green's Function for $H_{n0}$ Modes in a Rectangular Guide*

Consider a uniform unit line current extending across the rectangular guide, parallel with the $y$ axis and located at $(x',z')$, as in Fig. 5.8. The current is uniform along the $y$ coordinate, and, hence, the field will not vary with $y$. We will use the vector potential **A** from which to derive

† See, for example:

S. E. Miller, Waveguide as a Communication Medium, *Bell System Tech. J.*, vol. 33, pp. 1209–1265, November, 1954.

H. E. M. Barlow and E. G. Effemey, Propagation Characteristics of Low Loss Tubular Waveguides, *Proc. IEE (London)*, vol. 104, part B, pp. 254–260, May, 1957.

our field components, where $\mathbf{A} = \mathbf{a}_y A_y$, and

$$\frac{\partial^2 A_y}{\partial z^2} + \frac{\partial^2 A_y}{\partial x^2} + k_0^2 A_y = -\mu_0 \delta(x - x')\delta(z - z') \tag{67}$$

since the current has only a $y$ component, and, hence, $\mathbf{A}$ has only a $y$ component. The unit current is represented by the delta function $\delta(x - x')\delta(z - z')$. Now $\nabla \cdot \mathbf{J} = \partial J/\partial y = 0$, and, hence, we may

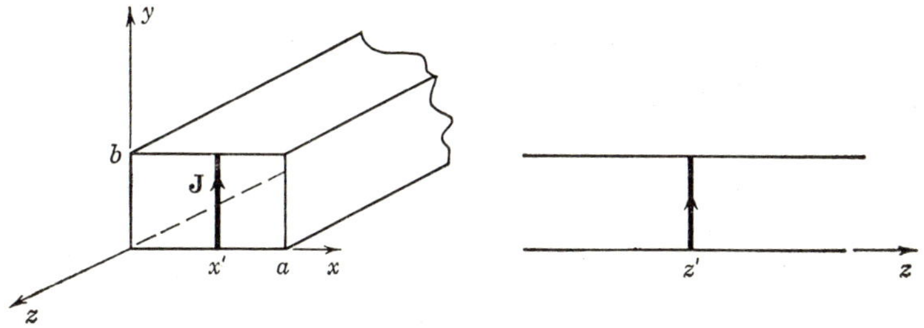


FIG. 5.8. Unit current source for $H_{n0}$ modes.

choose the gauge where $\nabla \cdot \mathbf{A} = 0$, and thus the fields are given by the following relations:

$$\mathbf{E} = -j\omega\mathbf{A} \tag{68a}$$

$$\mathbf{B} = \nabla \times \mathbf{A} \tag{68b}$$

There is only a $y$ component of electric field, and, hence, the only possible modes excited are $H_{n0}$ modes since these are the only modes which exhibit only a $y$ component of electric field and no variation with $y$. The field must also represent waves propagating away from the source at $(x',z')$. The wave equation satisfied by $E_y$ is

$$\frac{\partial^2 E_y}{\partial z^2} + \frac{\partial^2 E_y}{\partial x^2} + k_0^2 E_y = j\omega\mu_0 \delta(x - x')\delta(z - z') \tag{69}$$

since $E_y = -j\omega A_y$. Taking the finite Fourier sine transform of this equation, i.e., multiplying by $\sin(n\pi x/a)$ and integrating from $0 \le x \le a$ gives

$$\left(k_0^2 - \frac{n^2\pi^2}{a^2} + \frac{\partial^2}{\partial z^2}\right) \int_0^a E_y \sin\frac{n\pi x}{a}\, dx = j\omega\mu_0 \sin\frac{n\pi x'}{a}\, \delta(z - z')$$

since $E_y$ vanishes at $x = 0$, $a$. Let

$$\int_0^a E_y(x,z) \sin\frac{n\pi x}{a}\, dx = \mathcal{E}_n(z)$$

Thus $\mathcal{E}_n$ must satisfy the following equation:

$$\left(\frac{\partial^2}{\partial z^2} - \Gamma_n^2\right) \mathcal{E}_n(z) = j\omega\mu_0 \sin\frac{n\pi x'}{a}\, \delta(z - z') \tag{70}$$

where $\Gamma_n^2 = n^2\pi^2/a^2 - k_0^2$. For $z \neq z'$, the solution which represents waves propagating away from the source at $z'$ is $\mathcal{E}_n = b_n e^{\Gamma_n z}$, $z < z'$; $\mathcal{E}_n = c_n e^{-\Gamma_n z}$, $z > z'$, where $b_n$ and $c_n$ depend on $z'$. At $z = z'$, $\mathcal{E}_n$ is continuous, and, hence, $b_n e^{\Gamma_n z'} = c_n e^{-\Gamma_n z'}$. Integrate (70) from $z' - \Delta$ to $z' + \Delta$, and let $\Delta$ approach zero; thus

$$\left.\frac{\partial \mathcal{E}_n}{\partial z}\right|_{z'-}^{z'+} = j\omega\mu_0 \sin\frac{n\pi x'}{a}$$

which shows that the derivative of $\mathcal{E}_n$ with respect to $z$ is discontinuous at $z'$. Substituting for $\mathcal{E}_n$, we can solve for $b_n$ and $c_n$ and thus obtain

$$b_n = \frac{-j\omega\mu_0 e^{-\Gamma_n z'}}{2\Gamma_n}\sin\frac{n\pi x'}{a} \qquad c_n = b_n e^{2\Gamma_n z'}$$

There is a solution for $\mathcal{E}_n$ for each value of $n$, and this solution is

$$\mathcal{E}_n(z) = \frac{-j\omega\mu_0}{2\Gamma_n}\sin\frac{n\pi x'}{a}\begin{cases} e^{-\Gamma_n(z'-z)} & z \leq z' \\ e^{\Gamma_n(z'-z)} & z \geq z' \end{cases}$$

and is the finite Fourier sine transform of $E_y$. The inversion formula gives

$$E_y = G(x,z|x',z') = \frac{-j\omega\mu_0}{a}\sum_{n=1}^{\infty}\frac{1}{\Gamma_n}\sin\frac{n\pi x}{a}\sin\frac{n\pi x'}{a}e^{-\Gamma_n|z-z'|} \tag{71}$$

and this is the Green's function for a uniform current at $(x',z')$. This Green's function will be of use to us later in the study of waveguide discontinuities. If the unit vectors $\mathbf{a}_y\mathbf{a}_y$ are associated with $G$, we may, in a restricted sense, consider this as a dyadic Green's function.

*Green's Functions for General Cylindrical Guides*

The method to be used to find the field radiated by an arbitrary current filament in a cylindrical guide is to expand the radiated field in terms of a suitable set of normal modes, and to determine the amplitude coefficients in this expansion by an application of the Lorentz reciprocity theorem. With reference to Fig. 5.9, let $C$ represent an arbitrary infinitely thin unit current element, with the current flowing in the direction of the unit vector $\boldsymbol{\tau}$ along $C$. Such a current filament must be main-

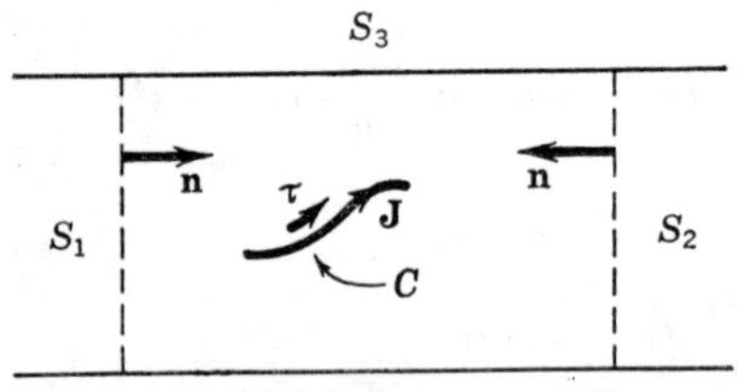

FIG. 5.9. An arbitrary current filament in a cylindrical guide.

tained by some external field or source, but in the evaluation of the Green's function we are concerned only with the radiated fields, and consequently the field or source which maintains the specified current does not enter into the picture here. In calculating the fields radiated by probes and loops in waveguides, we must take into account the source which gives rise to the currents on the probe or loop antenna. The Green's functions derived here will be found useful in the analysis of such problems.

An arbitrary field in a waveguide can be represented as an infinite series of the normal modes for the guide. These modes may be chosen as the $E$ and $H$ modes discussed earlier or any other convenient set of modes. The field for any one mode may be expressed in terms of the transverse electric field $\mathbf{e}_n e^{\pm\Gamma_n z}$ by means of Eqs. (42) and (43). For propagation in the positive $z$ direction, let the fields for the $n$th mode be represented as follows:

$$\mathbf{E}_n^+ = (\mathbf{e}_n + \mathbf{e}_{zn})e^{-\Gamma_n z} \tag{72a}$$

$$\mathbf{H}_n^+ = (\mathbf{h}_n + \mathbf{h}_{zn})e^{-\Gamma_n z} \tag{72b}$$

where for $E$ modes $\mathbf{h}_{zn}$ is zero, and for $H$ modes $\mathbf{e}_{zn}$ is zero. For propagation along the negative $z$ direction, the fields for the $n$th mode are then given by

$$\mathbf{E}_n^- = (\mathbf{e}_n - \mathbf{e}_{zn})e^{\Gamma_n z} \tag{73a}$$

$$\mathbf{H}_n^- = (-\mathbf{h}_n + \mathbf{h}_{zn})e^{\Gamma_n z} \tag{73b}$$

In (72), $\mathbf{e}_n$ and $\mathbf{h}_n$ are transverse vector functions of the transverse coordinates, while $\mathbf{e}_{zn}$ and $\mathbf{h}_{zn}$ are axial vector functions of the transverse coordinates. It will be assumed that the normal mode functions have been normalized, so that

$$\iint_S \mathbf{e}_n \times \mathbf{h}_n \cdot \mathbf{a}_z \, da = 1 \tag{74a}$$

The modes are orthogonal, and so we also have

$$\iint_S \mathbf{e}_n \times \mathbf{h}_m \cdot \mathbf{a}_z \, da = 0 \qquad n \neq m \tag{74b}$$

In (74), the integration is over the guide cross section $S$. An examination of (22) and (23) shows that the above orthogonal property is valid for degenerate modes also.

Let the field radiated in the positive $z$ direction by the current filament be represented by

$$\mathbf{E}^+ = \sum_n a_n \mathbf{E}_n^+ \tag{75a}$$

$$\mathbf{H}^+ = \sum_n a_n \mathbf{H}_n^+ \tag{75b}$$

and the field radiated in the negative $z$ direction by

$$\mathbf{E}^- = \sum_n b_n \mathbf{E}_n^- \tag{76a}$$

$$\mathbf{H}^- = \sum_n b_n \mathbf{H}_n^- \tag{76b}$$

In particular, the expansion (75) gives the radiated field on the cross-sectional plane $S_2$ while (76) gives the radiated field on $S_1$.

To determine the expansion coefficients $a_n$ and $b_n$, we need the Lorentz reciprocity principle, which is valid for a region containing a current source. Let $\mathbf{E}_1$ and $\mathbf{H}_1$ be the fields radiated by the current $\mathbf{J}$. These fields are solutions of the following equations: $\nabla \times \mathbf{E}_1 = -j\omega\mu_0\mathbf{H}_1$, $\nabla \times \mathbf{H}_1 = j\omega\epsilon_0\mathbf{E}_1 + \mathbf{J}$. The normal mode function $\mathbf{E}_n^\pm$, $\mathbf{H}_n^\pm$ is a solution of the source-free equations. Consider the expansion

$$\begin{aligned}\nabla \cdot (\mathbf{E}_n^\pm \times \mathbf{H}_1 - \mathbf{E}_1 \times \mathbf{H}_n^\pm) &= \mathbf{H}_1 \cdot \nabla \times \mathbf{E}_n^\pm - \mathbf{E}_n^\pm \cdot \nabla \times \mathbf{H}_1 \\ &\quad - \mathbf{H}_n^\pm \cdot \nabla \times \mathbf{E}_1 + \mathbf{E}_1 \cdot \nabla \times \mathbf{H}_n^\pm = -\mathbf{J} \cdot \mathbf{E}_n^\pm\end{aligned}$$

when the terms involving the curl are replaced by their equivalents as obtained from Maxwell's equations. Integrating over a volume bounded by a closed surface $S$, and converting the volume integral of the divergence to a surface integral, we get the desired form of the Lorentz reciprocity principle:

$$\oiint_S (\mathbf{E}_n^\pm \times \mathbf{H}_1 - \mathbf{E}_1 \times \mathbf{H}_n^\pm) \cdot \mathbf{n}\, da = \iiint \mathbf{J} \cdot \mathbf{E}_n^\pm\, dV \tag{77}$$

where $\mathbf{n}$ has been chosen as the inward directed normal to $S$.

We now choose as our volume that bounded by the perfectly conducting guide walls and the two cross-sectional planes $S_1$ and $S_2$ as in Fig. 5.9. There is no contribution to the surface integral arising from the guide walls since $\mathbf{n} \times \mathbf{E}_1$ and $\mathbf{n} \times \mathbf{E}_n^\pm$ are zero on the guide wall. For the normal mode function $\mathbf{E}_n^+$, $\mathbf{H}_n^+$, it is readily found that

$$\iint_{S_1} (\mathbf{E}_n^+ \times \mathbf{H}_1 - \mathbf{E}_1 \times \mathbf{H}_n^+) \cdot \mathbf{a}_z\, da = -2b_n \tag{78a}$$

$$-\iint_{S_1} (\mathbf{E}_n^+ \times \mathbf{H}_1 - \mathbf{E}_1 \times \mathbf{H}_n^+) \cdot \mathbf{a}_z\, da = a_n e^{-2\Gamma_n z_2} - a_n e^{-2\Gamma_n z_2} = 0 \tag{78b}$$

when the expansion (75) and (76) for $\mathbf{H}_1$, $\mathbf{E}_1$ and the orthogonal property (74*b*) are used. Consequently, (77) gives

$$2b_n = -\int_C \boldsymbol{\tau} \cdot \mathbf{E}_n^+\, dl \tag{79}$$

since it is assumed that the total current is of unit magnitude and concentrated in an infinitely thin filament. The integral in (79) may be regarded as the voltage impressed along $C$ by the normal mode field $\mathbf{E}_n^+$, and, because of the reciprocity principle which holds for the field, the current generates a mode with this same amplitude. If we begin with the normal mode function $\mathbf{E}_n^-$, $\mathbf{H}_n^-$ instead, we find in a similar manner that the coefficient $a_n$ is given by

$$2a_n = -\int_C \boldsymbol{\tau} \cdot \mathbf{E}_n^- \, dl \tag{80}$$

If the current is a function of the distance along $C$, the integrands in (79) and (80) must be replaced by $I(l)\boldsymbol{\tau} \cdot \mathbf{E}_n^{\pm}$, where $I(l)$ is the total current along $C$.

If $\boldsymbol{\tau}$ lies in a transverse plane (79) and (80) show that $a_n = b_n$, and, consequently, the transverse electric field (analogous to voltage) is continuous at the current filament, and the current filament behaves as a shunt source. If $\boldsymbol{\tau}$ is along the $z$ direction, $a_n = -b_n$, the transverse magnetic field (analogous to current) is continuous, and the current element acts as a series source.

If $C$ is a small closed loop, we may modify (79) and (80) as follows:

$$\begin{aligned} 2b_n &= -\oint_C \mathbf{E}_n^+ \cdot \boldsymbol{\tau} \, dl = -\iint_{S_0} \nabla \times \mathbf{E}_n^+ \cdot d\mathbf{a} \\ &= j\omega\mu_0 \iint_{S_0} \mathbf{H}_n^+ \cdot d\mathbf{a} \end{aligned} \tag{81a}$$

$$2a_n = j\omega\mu_0 \iint_{S_0} \mathbf{H}_n^- \cdot d\mathbf{a} \tag{81b}$$

where $S_0$ is the surface spanning the closed contour $C$. The integrals in (81) represent the negative induced voltage in the loop $C$ arising from the rate of change of the magnetic flux of the $n$th normal mode through $C$. Let the induced voltages be

$$V_n^+ = -j\omega \iint_{S_0} \mu_0 \mathbf{H}_n^+ \cdot d\mathbf{a} \tag{82a}$$

and

$$V_n^- = -j\omega \iint_{S_0} \mu_0 \mathbf{H}_n^- \cdot d\mathbf{a} \tag{82b}$$

and we may rewrite (81) as follows:

$$2b_n = -V_n^+ \tag{83a}$$

$$2a_n = -V_n^- \tag{83b}$$

For an infinitesimal loop having a vector area $\mathbf{A}$ and carrying a current $I$, the magnetic moment $\mathbf{m}$ is given by

$$\mathbf{m} = I\mathbf{A} \tag{84}$$

Introducing this concept into (81), we find that we may also write

$$2b_n = j\omega\mu_0\mathbf{H}_n^+ \cdot \mathbf{m} \tag{85a}$$
$$2a_n = j\omega\mu_0\mathbf{H}_n^- \cdot \mathbf{m} \tag{85b}$$

If the field $\mathbf{H}_n^\pm$ cannot be considered constant over the area of the loop, (85) must be replaced by an integral as in (82). These latter results are useful for finding the Green's functions from magnetic dipole sources. Using (72) and (73), we get

$$2b_n = j\omega\mu_0(\mathbf{h}_n + \mathbf{h}_{zn}) \cdot \mathbf{m}e^{-\Gamma_n z} \tag{86a}$$
$$2a_n = j\omega\mu_0(-\mathbf{h}_n + \mathbf{h}_{zn}) \cdot \mathbf{m}e^{\Gamma_n z} \tag{86b}$$

for an infinitesimal current loop of moment $\mathbf{m}$. When $\mathbf{m}$ is in the $z$ direction, the loop appears as a shunt source, but, if $\mathbf{m}$ lies in a transverse plane, it appears as a series source. The Green's function is the expansion (75) and (76), with the coefficients $a_n$ and $b_n$ given by the appropriate expression above. In an empty guide or a homogeneously filled one, a current element in the $z$ direction excites only $E$ modes, while a magnetic dipole in the $z$ direction (a current loop in the transverse plane) excites only $H$ modes.

The above discussion has been restricted to the case of an empty waveguide. The extension to the case of a homogeneously or inhomogeneously filled guide is obtained by using the appropriate normal mode functions for such a guide. It should also be kept in mind that, even though a given current filament may be in a single direction, say along the $z$ axis, it may be necessary to use a vector potential with both a $z$ component and a transverse component, in order to obtain a field which will have the proper continuity across surfaces along which the electrical parameters of the medium change discontinuously. When the method of expansion in terms of normal mode functions is used, such discontinuity surfaces are taken care of by the normal mode functions which satisfy the appropriate continuity conditions.

### *Mode Completeness*

An implicit assumption in the preceding derivation of the fields radiated by current elements in cylindrical waveguides is that the normal mode functions are capable of representing the field radiated by an arbitrary current source in the guide. Fortunately, this is true, and the total set of normal modes for the guide is said to form a complete set of eigenfunctions for this reason. A more specific statement of the completeness property is the following. Let $\mathbf{E}_t$ be an arbitrary transverse electric field in a given guide. The normal mode functions $\mathbf{e}_n$ form a complete set, provided the integrated mean-square error in representing or approximating $\mathbf{E}_t$ by a finite series of $N$ normal mode functions tends to zero as $N$ tends to infinity. If we approximate $\mathbf{E}_t$ by $\sum_{n=1}^{N} c_n\mathbf{e}_n$, where $c_n$ is given by $\iint_S \mathbf{E}_t \cdot \mathbf{e}_n \, da$, we must have

$$\lim_{N\to\infty} \iint_S \left(\mathbf{E}_t - \sum_{n=1}^{N} c_n\mathbf{e}_n\right) \cdot \left(\mathbf{E}_t - \sum_{n=1}^{N} c_n\mathbf{e}_n\right) da = 0 \tag{87}$$

in order for the $\mathbf{e}_n$ to form a complete set.† It is assumed here that the functions $\mathbf{e}_n$ form an orthonormal set such that

$$\iint_S \mathbf{e}_n \cdot \mathbf{e}_m \, da = \begin{cases} 0 & n \neq m \\ 1 & n = m \end{cases} \tag{88}$$

where $S$ denotes the guide cross section. Expanding (87) gives

$$\iint_S \mathbf{E}_t \cdot \mathbf{E}_t \, da - 2 \iint_S \sum_{n=1}^{N} c_n\mathbf{e}_n \cdot \mathbf{E}_t \, da + \sum_{n=1}^{N} c_n^2$$

and, using the expansion for $\mathbf{E}_t$ given above, we find that (87) is equivalent to

$$\lim_{N\to\infty} \left( \iint_S \mathbf{E}_t \cdot \mathbf{E}_t \, da - \sum_{n=1}^{N} c_n^2 \right) = 0 \tag{89}$$

The field $\mathbf{E}_t$ must be restricted to one which is at least piecewise-continuous if (89) is to be valid. Physical fields are, of course, sufficiently well behaved to satisfy these requirements. For the rectangular and circular guide, the completeness relation (89) is readily proved from the known properties of the functions involved. For example, if $\mathbf{E}_t$ is a field directed along the $y$ axis in a rectangular guide, and having an arbitrary variation with $x$ but no variation with $y$, then the only normal mode functions required to expand $\mathbf{E}_t$ are the $H_{n0}$ modes, for which

$$\mathbf{e}_n = \left(\frac{2}{ab}\right)^{1/2} \mathbf{a}_y \sin \frac{n\pi x}{a}$$

The expansion is just a Fourier sine series, which is known to be able to

† When this relation is true, it can be shown that a transverse vector function $\mathbf{f}$ having the property $\iint_S \mathbf{f} \cdot \mathbf{e}_n \, da = 0$ does not exist, and, hence, no other functions orthogonal to the $\mathbf{e}_n$ can be found. This is known as the closure property of the function set $\mathbf{e}_n$. It should be noted that the words closure and completeness are often interchanged.

represent an arbitrary piecewise-continuous function of $x$ over the interval $0 \le x \le a$. For fields that are restricted to the above form, the $H_{n0}$ modes form a complete set.

For those waveguide systems for which the normal modes may be derived from the functions $\psi_h$ and $\psi_e$ introduced in Sec. 5.1, the completeness relation may be proved from a variational principle which the functions $\psi_h$ and $\psi_e$ satisfy.† Thus, the $E$ modes are capable of representing an arbitrary field which has no $z$ component of magnetic intensity, while the $H$ modes are capable of representing an arbitrary field with no axial electrical component. A general field can, in turn, be represented by a combination of these two sets of modes.

An equivalent statement of the completeness property (89) is the following:

$$\begin{aligned}\sum_n \mathbf{e}_n(x',y')\mathbf{e}_n(x,y) &= (\mathbf{a}_x\mathbf{a}_x + \mathbf{a}_y\mathbf{a}_y)\delta(x - x')\delta(y - y') \\ &= \bar{\mathbf{I}}\delta(x - x')\delta(y - y')\end{aligned} \tag{90}$$

where $\bar{\mathbf{I}}$ is the two-dimensional unit dyadic. Equation (90) is seen to be the expansion of the unit source on the right-hand side by scalar-post-multiplying both sides by $\mathbf{e}_m(x,y)$ and integrating over the guide cross section to get

$$\begin{aligned}\sum_n \mathbf{e}_n(x',y') \iint_S \mathbf{e}_n(x,y) \cdot \mathbf{e}_m(x,y)\, da &= \iint_S \bar{\mathbf{I}} \cdot \mathbf{e}_m(x,y)\delta(x-x')\delta(y - y')\, da \\ &= \mathbf{e}_m(x',y')\end{aligned}$$

by virtue of the orthogonal properties of the $\mathbf{e}_n$ and the property of the delta function. From (90), the expansion for an arbitrary field $\mathbf{E}_t$ follows at once by superposition, i.e.,

$$\begin{aligned}\mathbf{E}_t(x,y) &= \iint_S \mathbf{E}_t(x',y') \cdot \bar{\mathbf{I}}\delta(x - x')\delta(y - y')\, dx'\, dy' \\ &= \sum_n \iint_S \mathbf{E}_t(x',y') \cdot \mathbf{e}_n(x',y')\mathbf{e}_n(x,y)\, dx'\, dy' \\ &= \sum_n c_n\mathbf{e}_n(x,y)\end{aligned} \tag{91}$$

In a closed-boundary system such as a waveguide with perfectly conducting walls, the eigenfunctions form an infinite discrete set with an infinite number of discrete eigenvalues (propagation constants). For open structures, such as a dielectric rod, we find that only a finite number of discrete eigenfunctions exist. In addition, we find as part of the

† P. M. Morse and H. Feshbach, "Methods of Theoretical Physics," part I, sec. 6.3, McGraw-Hill Book Company, Inc., New York, 1953.

solution a continuously varying eigenfunction with a continuous range of eigenvalues. The completeness relation (90) must be replaced by a sum over the discrete eigenfunctions plus an integral over the continuous part of the eigenvalue spectrum in such cases. These ideas may be clarified by noting the analogy with a function of time which has a periodic part and a nonperiodic part. The frequency spectrum of such a function consists of a discrete part corresponding to the fundamental and harmonics of the periodic part of the time function and a continuous spectrum arising from the nonperiodic part. The periodic part may be represented by a Fourier series, but the nonperiodic part must be represented by a Fourier integral.

In some instances the solution for a Green's function may be obtained by using Fourier or bilateral Laplace transforms. In such cases the completeness relation (90) is automatically satisfied, since the completeness property is contained within the integral transform itself, e.g., for the bilateral Laplace transform

$$\frac{1}{2\pi j} \int_C e^{w(\lambda-\lambda')}\, dw = \delta(\lambda - \lambda')$$

where $C$ is a suitable inversion contour. Consequently, a function $f(\lambda)$, defined over $-\infty \le \lambda \le \infty$, may be represented by

$$f(\lambda) = \frac{1}{2\pi j} \int_C g(w) e^{w\lambda}\, dw$$

where

$$g(w) = \int_{-\infty}^{\infty} f(\lambda) e^{-w\lambda}\, d\lambda$$

The function $f$ is, of course, restricted to one for which its Laplace transform exists.

As an example of the use of these properties of integral transforms, we will rederive the Green's function for $H_{n0}$ modes in terms of a bilateral Laplace transform. With reference to Fig. 5.8, the electric field $E_y$ is a solution of the equation

$$\frac{\partial^2 \psi}{\partial z^2} + \frac{\partial^2 \psi}{\partial x^2} + k_0^2 \psi = \delta(x - x')\delta(z - z')$$

where $\psi = E_y/j\omega\mu_0$. If we multiply this by $e^{-wz}$ and integrate over $-\infty \le x \le \infty$, we get

$$\int_{-\infty}^{\infty} e^{-wz} \frac{\partial^2 \psi}{\partial z^2}\, dz + \left(\frac{\partial^2}{\partial x^2} + k_0^2\right) \int_{-\infty}^{\infty} e^{-wz} \psi\, dz = e^{-wz'} \delta(x - x')$$

If we consider the medium in the guide to have a small loss, then we may integrate the first term by parts, and discard the integrated parts which vanish at infinity for Re $w = 0$, that is,

$$\int_{-\infty}^{\infty} e^{-wz} \frac{\partial^2 \psi}{\partial z^2} dz = e^{-wz} \frac{\partial \psi}{\partial z}\Big|_{-\infty}^{\infty} + we^{-wz}\psi\Big|_{-\infty}^{\infty} + w^2 \int_{-\infty}^{\infty} e^{-wz}\psi \, dz$$
$$= w^2 \int_{-\infty}^{\infty} e^{-wz}\psi \, dz = w^2 g(w,x)$$

where $g(w,x)$ is the bilateral Laplace transform of $\psi(z,x)$. The equation satisfied by $g$ is

$$\frac{d^2 g}{dx^2} + (w^2 + k_0^2)g = e^{-wz'}\delta(x - x') \tag{92}$$

where $g(w,0) = g(w,a) = 0$. The solution for $g$ is readily found by the method given in Sec. 2.3. One solution for the homogeneous equation corresponding to (92) is $g_1 = \sin ux$, while another solution which vanishes at $x = a$ is $g_2 = \sin u(a - x)$, where $u = (w^2 + k_0^2)^{1/2}$. The Wronskian is equal to $W = g_1 \, dg_2/dx - g_2 \, dg_1/dx = -u \sin ua$, and does not vanish, and, hence, by (28*b*) of Chap. 2, the solution for $g$ is

$$g(w,x) = \frac{e^{-wz'}}{W}\left[g_2 \int^x g_1\delta(x - x') \, dx + g_1 \int_x g_2\delta(x - x') \, dx\right]$$

To make $g$ satisfy the required boundary conditions at $x = 0$, $a$, we choose $x > x'$ in the first integral and less than $x'$ in the second integral. Thus we get

$$g(w,x) = \frac{-e^{-wz'}}{u \sin ua}\begin{cases} \sin ux \sin u(a - x') & x \le x' \\ \sin u(a - x) \sin ux' & x \ge x' \end{cases} \tag{93}$$

The inverse transform now gives the following solution for our field $\psi$:

$$\psi(x,z|x',z') = \frac{-1}{2\pi j}\int_C \frac{e^{w(z-z')}}{u \sin ua}\begin{Bmatrix} \sin ux \sin u(a - x') & x \le x' \\ \sin ux' \sin u(a - x) & x \ge x' \end{Bmatrix} dw \tag{94}$$

and this may be regarded as the Green's function for the problem. If the integral is converted to a contour integral, it may be evaluated in terms of the poles which occur at $\sin ua = 0$ or $w^2 = n\pi^2/a - k_0^2 = \Gamma_n^2$. This evaluation in terms of the residues at the poles leads to the eigenfunction expansion (71) given earlier. The demonstration of the equivalence between the two forms follows the same general pattern as given in Sec. 2.4. Since (94) leads to the eigenfunction expansion (71), it ensures us of the completeness of this set of eigenfunctions.

In open-boundary problems it is generally found that the inversion integral contains one or more branch points. In converting the integral to a contour integral, we are then left with integrals along certain branch cuts or lines. These latter integrals yield that portion of the solution which corresponds to the continuous eigenvalue spectrum, since $w$ varies continuously along the branch cuts. The poles of the integrand yield the part of the solution corresponding to the discrete eigenvalues (discrete values of $w$). Examples of this type of behavior are given in the chapter on surface waves.

## 5.7. Analogy with Transmission Lines

The similarity between the normal mode solutions in a waveguide and the TEM mode on a transmission line permits us to set up a system of equations representing the fields in a waveguide which is formally identical with the current and voltage relations on a system of transmission lines.

With reference to Fig. 5.10, let the region between the two arbitrarily chosen reference planes $S_1$ and $S_2$ be a region in which a localized discontinuity (such as a post or diaphragm) exists. If it is assumed (this assumption is not necessary and is made only to simplify the development of the basic ideas involved) that the input and output guides are the same, the following expansion for the transverse fields in the guide may be written.

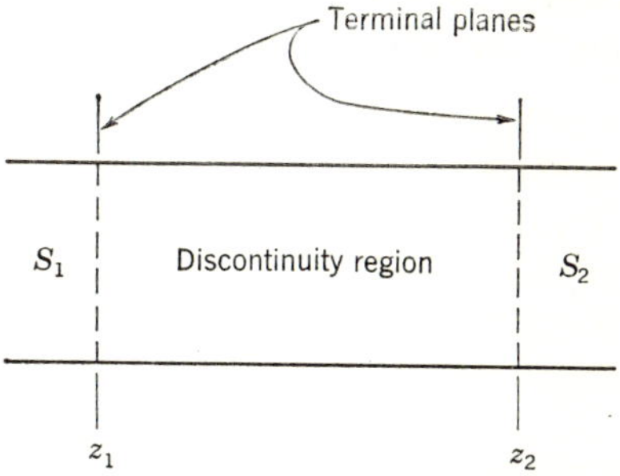

FIG. 5.10. Terminal planes in a guide containing a discontinuity.

For $z \leq z_1$, the transverse electric and magnetic fields are

$$\sum_{n=1}^{N} a_n \mathbf{e}_n e^{-\Gamma_n(z-z_1)} + \sum_{n=1}^{\infty} b_n \mathbf{e}_n e^{\Gamma_n(z-z_1)} \tag{95a}$$

$$\sum_{n=1}^{N} a_n \mathbf{h}_n e^{-\Gamma_n(z-z_1)} - \sum_{n=1}^{\infty} b_n \mathbf{h}_n e^{\Gamma_n(z-z_1)} \tag{95b}$$

while, for $z \geq z_2$, the corresponding fields are

$$\sum_{n=1}^{M} c_n \mathbf{e}_n e^{+\Gamma_n(z-z_2)} + \sum_{n=1}^{\infty} d_n \mathbf{e}_n e^{-\Gamma_n(z-z_2)} \tag{96a}$$

$$-\sum_{n=1}^{M} c_n \mathbf{h}_n e^{\Gamma_n(z-z_2)} + \sum_{n=1}^{\infty} d_n \mathbf{h}_n e^{-\Gamma_n(z-z_2)} \tag{96b}$$

In the above expressions, the finite sums represent incident modes, while the infinite sums represent reflected modes, which are set up by the discontinuity, or transmitted past the discontinuity because of modes incident from the other side. The normal mode functions $\mathbf{e}_n$ and $\mathbf{h}_n$ are transverse vector functions of the transverse coordinates. These normal mode functions may be $E$ and $H$ modes or suitable combinations of $E$ and $H$ modes. The normal mode functions $\mathbf{e}_n$ and $\mathbf{h}_n$ are related by means of the dyadic wave impedance $\bar{\mathbf{Z}}_n$ or dyadic wave admittance $\bar{\mathbf{Y}}_n$ for the mode as follows:

$$\mathbf{e}_n = \bar{\mathbf{Z}}_n \cdot \mathbf{h}_n \tag{97a}$$
$$\mathbf{h}_n = \bar{\mathbf{Y}}_n \cdot \mathbf{e}_n \tag{97b}$$

where

$$\bar{\mathbf{Z}}_n = Z_n(\mathbf{a}_x\mathbf{a}_y - \mathbf{a}_y\mathbf{a}_x)$$
$$\bar{\mathbf{Y}}_n = Y_n(\mathbf{a}_y\mathbf{a}_x - \mathbf{a}_x\mathbf{a}_y)$$

and $Z_n$ and $Y_n$ are the scalar wave impedance and admittance, respectively. If the following equivalent transmission-line voltages and currents are introduced;

$$\left.\begin{aligned} V_{n1}^+ &= a_n \\ V_{n1}^- &= b_n \\ I_{n1}^+ &= Y_n a_n = Y_n V_{n1}^+ \\ I_{n1}^- &= Y_n b_n = Y_n V_{n1}^- \end{aligned}\right\} \quad z \le z_1 \tag{98a}$$

$$\left.\begin{aligned} V_{n2}^+ &= c_n \\ V_{n2}^- &= d_n \\ I_{n2}^+ &= -Y_n c_n = -Y_n V_{n2}^+ \\ I_{n2}^- &= -Y_n d_n = -Y_n V_{n2}^- \end{aligned}\right\} \quad z \ge z_2 \tag{98b}$$

the following set of relations may be written in place of those given in (95) and (96):

$$\sum_{n=1}^{N} V_{n1}^+ e^{-\Gamma_n(z-z_1)} + \sum_{n=1}^{\infty} V_{n1}^- e^{\Gamma_n(z-z_1)} \tag{99a}$$

$$\sum_{n=1}^{N} I_{n1}^+ e^{-\Gamma_n(z-z_1)} - \sum_{n=1}^{\infty} I_{n1}^- e^{\Gamma_n(z-z_1)} \tag{99b}$$

$$\sum_{n=1}^{M} V_{n2}^+ e^{\Gamma_n(z-z_2)} + \sum_{n=1}^{\infty} V_{n2}^- e^{-\Gamma_n(z-z_2)} \tag{99c}$$

$$\sum_{n=1}^{M} I_{n2}^+ e^{\Gamma_n(z-z_2)} - \sum_{n=1}^{\infty} I_{n2}^- e^{-\Gamma_n(z-z_2)} \tag{99d}$$

When the currents and voltages are known, we may obtain the expression for the fields at the terminal planes by means of (95) to (98). The above equations are formally the same as those which would be used to describe the behavior of a junction of transmission lines (infinite in number), with waves incident on $N$ lines for $z < z_1$, and on $M$ lines for $z > z_2$. The currents and voltages on one side may be related to those on the other side by a suitable impedance, admittance, scattering, or some other matrix. This impedance matrix, etc., may be used to define an equivalent lumped-parameter network to represent the discontinuity.

The usual situation, in practice, is the one where the frequency of operation is chosen so that only the dominant mode propagates. It is then convenient to choose terminal planes sufficiently far removed from the discontinuity so that all evanescent modes have decayed to a negligible value at these planes. Even when several modes propagate, it is convenient to do this, since it is then possible to describe the effect of the discontinuity on the propagating modes by an equivalent transmission-line circuit, which couples together the various transmission lines used to represent each propagating mode. It should also be noted that other definitions of the equivalent voltages and currents may be chosen in place of those given by (98). It is not necessary that $\frac{1}{2}V_n^+I_n^{+*}$ should represent the power carried by the mode. However, it is necessary that the constant of proportionality between the power carried by each mode and that carried by its equivalent transmission line should be the same for all modes, if there is to be conservation of power and energy in the equivalent circuit.

In the discussion to follow, it will be assumed that only one mode propagates, and that the terminal planes are chosen far enough from the discontinuity so that all evanescent modes are of negligible amplitude at these planes. In place of (99), we now have

$$\begin{aligned} &V_1^+e^{-\Gamma(z-z_1)} + V_1^-e^{\Gamma(z-z_1)} \\ &I_1^+e^{-\Gamma(z-z_1)} - I_1^-e^{\Gamma(z-z_1)} \end{aligned} \qquad z \le z_1 \tag{100a}$$

$$\begin{aligned} &V_2^+e^{\Gamma(z-z_2)} + V_2^-e^{-\Gamma(z-z_2)} \\ &I_2^+e^{\Gamma(z-z_2)} - I_2^-e^{-\Gamma(z-z_2)} \end{aligned} \qquad z \ge z_2 \tag{100b}$$

where the subscript $n$ has been dropped (if $n = 1$ is the propagating mode, this would be a common subscript on all quantities and may be dropped for convenience). In view of the linearity of Maxwell's equations, the relationship between the equivalent voltages and currents at the terminal planes on the input and output sides is a linear one. Thus, we may write

$$V_1 = V_1^+ + V_1^- = I_1Z_{11} + I_2Z_{12} \tag{101a}$$
$$V_2 = V_2^+ + V_2^- = I_1Z_{21} + I_2Z_{22} \tag{101b}$$

where $V_1$ is the total voltage at $z = z_1$, which is the location of the terminal plane in the input side, $V_2$ is the total voltage at the terminal plane located at $z = z_2$ on the output side, $I_1$ is the total current $I_1^+ - I_1^-$, and $I_2$ is the total current $I_2^+ - I_2^-$, again measured at the respective terminal planes. In (101), the parameters $Z_{11}$, $Z_{12}$, $Z_{21}$, $Z_{22}$ may be identified with an equivalent-T-network representation of the discontinuity. For a reciprocal structure, $Z_{12} = Z_{21}$, and the equivalent circuit is as illustrated in Fig. 5.11.

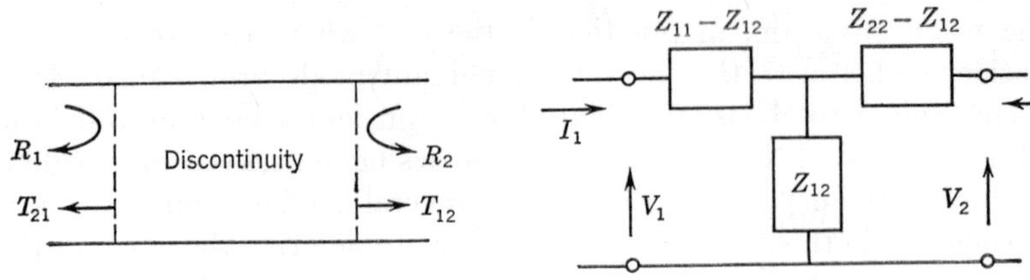


FIG. 5.11. Equivalent circuit for a waveguide discontinuity.

If the following reflection and transmission coefficients are introduced,

$$R_1 = \left.\frac{V_1^-}{V_1^+}\right|_{V_2^+=0} \tag{102a}$$

$$R_2 = \left.\frac{V_2^-}{V_2^+}\right|_{V_1^+=0} \tag{102b}$$

$$T_{12} = \left.\frac{V_2^-}{V_1^+}\right|_{V_2^+=0} \tag{102c}$$

$$T_{21} = \left.\frac{V_1^-}{V_2^+}\right|_{V_1^+=0} \tag{102d}$$

we may write in place of (101) the equations below, which are valid when $V_2^+ = 0$,

$$1 + R_1 = Y_1(1 - R_1)Z_{11} - Y_1T_{12}Z_{12} \tag{103a}$$

$$T_{12} = Y_1(1 - R_1)Z_{12} - Y_1T_{12}Z_{22} \tag{103b}$$

since $I_1^\pm = Y_1V_1^\pm$ and $I_2^- = -Y_1V_2^-$. Solving these equations for $R_1$ and $T_{12}$ gives

$$R_1 = \frac{Y_1(Z_{11}Z_{22} - Z_{12}^2) + Z_{11} - (Z_1 + Z_{22})}{Y_1(Z_{11}Z_{22} - Z_{12}^2) + Z_{11} + (Z_1 + Z_{22})} \tag{104a}$$

$$T_{12} = \frac{2Z_{12}}{Y_1(Z_{11}Z_{22} - Z_{12}^2) + Z_{11} + (Z_1 + Z_{22})} \tag{104b}$$

Expressions for $R_2$ and $T_{21}$ may be obtained by considering the special case $V_1^+ = 0$ and $V_2^+ \neq 0$, and are

$$R_2 = \frac{Y_1(Z_{11}Z_{22} - Z_{12}^2) + Z_{22} - (Z_1 + Z_{11})}{Y_1(Z_{11}Z_{22} - Z_{12}^2) + Z_{22} + (Z_1 + Z_{11})} \tag{105a}$$

$$T_{21} = T_{12} \tag{105b}$$

The equality of $T_{12}$ and $T_{21}$ follows because $Z_{12} = Z_{21}$, and the wave impedance for the mode is the same on both sides of the discontinuity. If the wave impedance on the input side was $Z_1$, and on the output side $Z_2$, then the equation we would obtain in place of (105*b*) is

$$T_{12}' = T_{12}\left(\frac{Z_1}{Z_2}\right)^{1/2} = T_{21}' = T_{21}\left(\frac{Z_2}{Z_1}\right)^{1/2} \tag{106}$$

where $T'_{12}$ and $T'_{21}$ are normalized transmission coefficients. The reflection coefficients are referred to their respective terminal planes, while the transmission coefficients represent transmission from one terminal plane to the other. The expressions for $R_1$, $R_2$, and $T_{12}$ must be modified when the input and output transmission lines do not have the same characteristic impedance.

The input impedance at the terminal plane at $z = z_1$ is

$$Z_{\text{in}} = \frac{V_1}{I_1} = \frac{1 + R_1}{1 - R_1} Z_1 = \frac{(Z_{11}Z_{22} - Z_{12}^2) + Z_1 Z_{11}}{Z_1 + Z_{22}} \tag{107}$$

a result which may be derived from (104$a$) or from the equivalent circuit in Fig. 5.11 by the usual methods of circuit analysis. If the discontinuity is lossless, no power is absorbed in the circuit elements $Z_{ij}$, and, hence, they must be pure-imaginary elements. Also, for a lossless discontinuity, $|R_1| = |R_2|$ and $1 - |R_1|^2 = |T_{12}|^2$. From (104$a$) and (105$b$), we get $(R_1 - R_2)/T_{12} = (Z_{11} - Z_{22})/Z_{12}$, and thus, when $R_1 = R_2$, the equivalent network is symmetrical; that is, $Z_{11} = Z_{22}$. For a lossless network, the right-hand side of the above relation is real, and, hence, the left-hand side must be real also. Let $R_1 = Re^{j\theta_1}$, $R_2 = Re^{j\theta_2}$, and $T_{12} = (1 - R^2)^{1/2}e^{j\phi}$. It is necessary for $\sin(\theta_2 - \phi)$ to equal $\sin(\theta_1 - \phi)$ if $(R_1 - R_2)/T_{12}$ is to be real. Thus we get $\theta_2 - \phi = \theta_1 - \phi + 2n\pi$ or $\theta_2 - \phi = -(\theta_1 - \phi) + (2n + 1)\pi$, and, hence, $\theta_2 = \theta_1 + 2n\pi$, or $\theta_2 = -\theta_1 + 2\phi + (2n + 1)\pi$. Since $R_i$ is invariant to a change in its phase by $2n\pi$ radians, the solution $\theta_1 = \theta_2$ applies to a symmetrical structure only. For a nonsymmetrical lossless structure,

$$R_2 = Re^{j\theta_2} = -R_1 e^{j2(\phi - \theta_1)} \tag{108a}$$

$$2\angle T_{12} = \angle R_1 + \angle R_2 \pm \pi \tag{108b}$$

The above theory may be extended to cover the case of the junction of more than two waveguides as well. If there are two or more propagating modes, a transmission line is introduced for each mode, and the discontinuity is now represented by a multiterminal network coupling together the various transmission lines. The equivalent-circuit representation of a discontinuity neglects all knowledge of the detailed structure of the field in the vicinity of the discontinuity, and contents itself with a knowledge of the fields at the terminal planes only. On the other hand, the parameters of the equivalent circuit can be determined only by a detailed solution of Maxwell's field equations at the discontinuity, or by measurements of the fields at the terminal planes. The analytical approach will be covered in later chapters.

A further insight into the relationship between the field and the equivalent circuit may be obtained by an application of the complex

Poynting vector result given in Sec. 1.3. If the complex Poynting vector is integrated over the closed surface consisting of the two terminal planes $S_1$, $S_2$ and the perfectly conducting guide wall, we get

$$\tfrac{1}{2}a_1a_1^*(1 + R_1)(1 - R_1^*) \iint_{S_1} \mathbf{e}_1 \times \mathbf{h}_1^* \cdot d\mathbf{a} = 2j\omega(W_m - W_e) + P_L + \tfrac{1}{2}a_1a_1^*T_{12}T_{12}^* \iint_{S_2} \mathbf{e}_1 \times \mathbf{h}_1^* \cdot d\mathbf{a} \quad (109)$$

where it is assumed that we have a guide in which only one mode propagates; that the only field incident on the discontinuity is the $n = 1$ mode from the region $z < z_1$; and that $W_m$, $W_e$, $P_L$ are respectively the time-average magnetic energy, electric energy, and power loss associated with the fields in the vicinity of the discontinuity, i.e., between the terminal planes. The last integrals in (109) are equal to the power $P_t$ transmitted past the discontinuity. The first integrals may be modified as follows:

$$\begin{aligned}
&\tfrac{1}{2}a_1a_1^*(1 + R_1)(1 - R_1^*) \iint_{S_1} \mathbf{e}_1 \times \mathbf{h}_1^* \cdot d\mathbf{a} \\
&= \frac{1 + R_1}{1 - R_1}\frac{1}{2}\, a_1a_1^*(1 - R_1)(1 - R_1^*) \iint_{S_1} \mathbf{e}_1 \times \mathbf{h}_1^* \cdot d\mathbf{a} \\
&= \frac{1 + R_1}{1 - R_1} Z_1 \left\{ \tfrac{1}{2}a_1a_1^*(1 - R_1)(1 - R_1^*) \iint_{S_1} [(\mathbf{a}_x\mathbf{a}_y - \mathbf{a}_y\mathbf{a}_x) \cdot \mathbf{h}_1 \times \mathbf{h}_1^*] \cdot d\mathbf{a} \right\} \\
&= \frac{1 + R_1}{1 - R_1} Z_1 P_i \tfrac{1}{2} I_1 I_1^*
\end{aligned}$$

where $P_i$ may also be written as $P_i = \iint_{S_1} \mathbf{h}_1 \cdot \mathbf{h}_1^*\, da$, and $I_1$ is the total equivalent transmission-line current at the terminal plane $S_1$. Substituting into (109), the following result for the input impedance at the terminal plane $S_1$ is obtained:

$$Z_{\text{in}} = \frac{1 + R_1}{1 - R_1} Z_1 = \frac{4j\omega(W_m - W_e)}{P_i I_1 I_1^*} + 2\frac{P_L + P_t}{P_i I_1 I_1^*} \quad (110)$$

In this equation, $P_i$ serves as a normalization factor. It is seen that the reactive part of the input impedance is determined by the difference in the time-average magnetic and electric energy stored in the field around the discontinuity, while the resistive part arises from the power loss in the discontinuity and the power transmitted past the discontinuity. Equation (110) is formally the same as that derived in Sec. 1.3 for the input impedance of a simple *RLC* series network.

## 5.8. The Tangent Method for the Experimental Determination of the Equivalent-circuit Parameters

In this section a brief treatment of the tangent method for determining the equivalent-circuit parameters of a lossless discontinuity will be given. This method is discussed here because, in Chap. 8, an analytical method, which is equivalent to this experimental method, will be given. The tangent method, or nodal-shift method, as it is sometimes called, was developed independently by Weissfloch and Feenberg.†

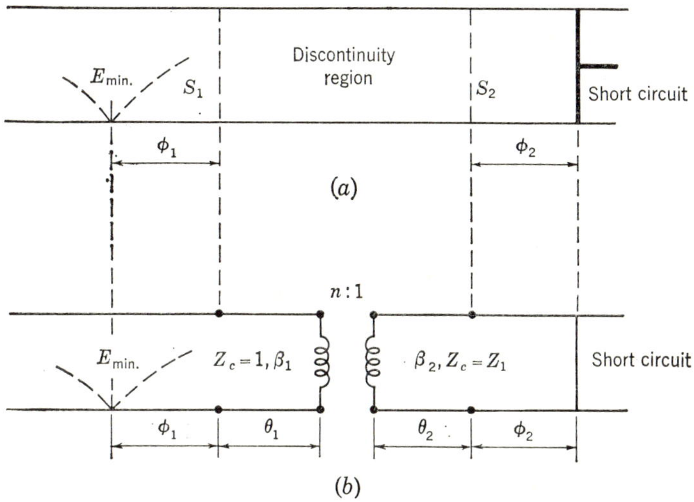


FIG. 5.12. Equivalent circuit for a lossless discontinuity.

Consider a guide containing a lossless discontinuity, located between the terminal planes $S_1$ and $S_2$. This section of guide and the discontinuity may be represented by an equivalent circuit containing three independent parameters. A convenient circuit to choose is one consisting of two lengths of transmission line of electrical length $\theta_1$ and $\theta_2$, connected together by an ideal transformer of turns ratio $n:1$, as in Fig. 5.12. It will be assumed for generality that the transmission line on the output side has a characteristic impedance of $Z_1$, and that the input line has a characteristic impedance of unity. The propagation constants of the input and output guide are $j\beta_1$ and $j\beta_2$, respectively.

† A. Weissfloch, *Hochfrequenz und Elektroakustik*, vol. 60, p. 67, September, 1942.
E. Feenberg, *J. Appl. Phys.*, vol. 17, p. 530, 1946.
Good descriptions of the method may also be found in:
E. L. Ginzton, "Microwave Measurements," sec. 5.8, McGraw-Hill Book Company, Inc., New York, 1957.
N. Marcuvitz, "Waveguide Handbook," vol. 10 of MIT Rad. Lab. Series, sec. 3.4, McGraw-Hill Book Company, Inc., New York, 1951.

In practice, the measurements must be made at points sufficiently far away from the discontinuity so that only the dominant modes are present. In theoretical work it is convenient to shift the terminal planes used in the experimental measurements back toward the discontinuity by integral multiples of a half guide wavelength so as to coincide more nearly with the actual discontinuity. The equivalent-circuit parameters are invariant under such a shift in the terminal-plane positions.

A plot of the electrical field null position $\phi_1$ as a function of the short-circuit position $\phi_2$ yields a curve of the form illustrated in Fig. 5.13. In

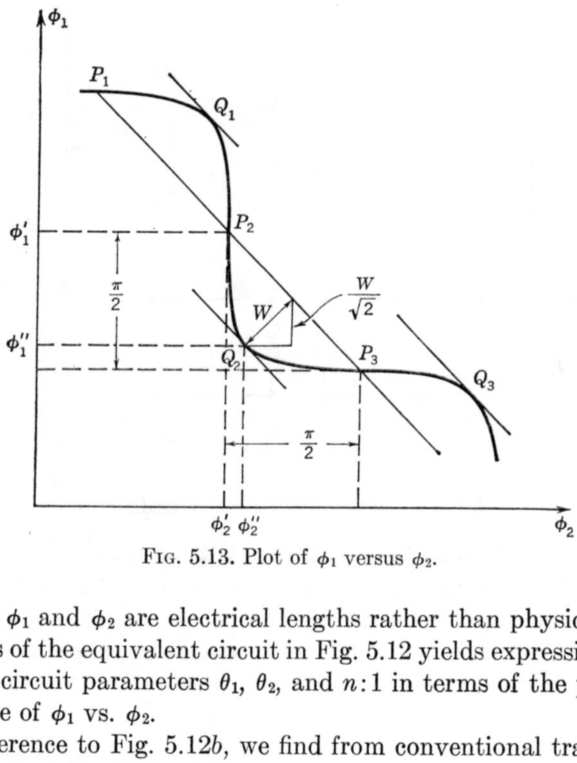


FIG. 5.13. Plot of $\phi_1$ versus $\phi_2$.

this figure, $\phi_1$ and $\phi_2$ are electrical lengths rather than physical lengths. An analysis of the equivalent circuit in Fig. 5.12 yields expressions for the equivalent-circuit parameters $\theta_1$, $\theta_2$, and $n{:}1$ in terms of the parameters of the curve of $\phi_1$ vs. $\phi_2$.

With reference to Fig. 5.12*b*, we find from conventional transmission-line theory that the field minimum position $\phi_1$ (this corresponds to the point where the input impedance is zero) is given by

$$\tan(\phi_1 + \theta_1) = -N^2 \tan(\phi_2 + \theta_2) \tag{111}$$

where $N^2 = n^2 Z_1$. A plot of $\phi_1$ vs. $\phi_2$ as determined by this relation results in an oscillatory curve symmetrical about a line of slope $-1$ and with a period $\pi$ for both variables as in Fig. 5.13. From (111) it is readily found that, if $N < 1$, the slope of the curve at $P_1$ is $-N^2$, and the slope at $P_2$ is $-1/N^2$. If $N > 1$, the reverse is true. The slope of the curve is given by

$$\frac{d\phi_1}{d\phi_2} = -N^2 \frac{\sec^2(\phi_2 + \theta_2)}{\sec^2(\phi_1 + \theta_1)} \tag{112}$$

At the points $Q_1$, $Q_2$, $Q_3$, the slope is equal to $-1$. Whenever the curve crosses the straight line, we have $\theta_1 + \phi_1 = s\pi/2$, $\theta_2 + \phi_2 = n\pi/2$, where $s$ and $n$ are suitable integers. When $s$ and $n$ are even, Eq. (112) shows that the slope is $-N^2$, and, hence, if $N > 1$, the coordinates of the point $P_2$ are $\phi_1' = n_1\pi - \theta_1$, $\phi_2' = n_2\pi - \theta_2$, where $n_1$ and $n_2$ are suitable integers. If $N < 1$, the coordinates of the point $P_2$ are

$$\phi_1' = n_1'\pi + \frac{\pi}{2} - \theta_1 \qquad \text{and} \qquad \phi_2' = n_2'\pi + \frac{\pi}{2} - \theta_2$$

where $n_1'$ and $n_2'$ are again a suitable set of integers. Equating (112) to $-1$ and solving this equation and (111) simultaneously for $N$ gives

$$N = \pm\cot(\phi_2'' + \theta_2) \tag{113}$$

where $\phi_2''$ is the value of $\phi_2$ where the slope equals $-1$, say at $Q_2$. If the positive root is chosen, the value of $\phi_2 + \theta_2$ must be equal to an even multiple of $\pi/2$ at $P_2$ since the cotangent function must be positive, and this is possible only if the angle increases from some multiple of $\pi$. In this case, the value of $N$ obtained is greater than unity, as the preceding discussion shows. If the negative root is chosen, the corresponding value of $\phi_2 + \theta_2$ at $P_2$ must be equal to an odd multiple of $\pi/2$, and $N$ is less than unity. Either root may be chosen and leads to a correct equivalent circuit. Along the straight line, and in particular at $P_2$, we have $\phi_1 + \theta_1 + \phi_2 + \theta_2 = n\pi$, where $n$ is a suitable integer. At the point $Q_2$ we have $\phi_1'' + \theta_1 + \phi_2'' + \theta_2 = -2^{1/2}W + n\pi$, where $W$ is the peak amplitude of the curve in radians, measured from the straight line. Expanding the tangent of this angle gives

$$\begin{aligned} \tan 2^{1/2}W &= -\frac{\tan(\phi_1'' + \theta_1) + \tan(\phi_2'' + \theta_1)}{1 - \tan(\phi_1'' + \theta_1)\tan(\phi_2'' + \theta_2)} \\ &= \pm\frac{N^2 - 1}{2N} = \pm\frac{1 - 1/N^2}{2/N} \end{aligned} \tag{114}$$

when (111) and (113) are used. Now

$$\tan 2^{1/2}W = \cot\left(\frac{\pi}{2} - 2^{1/2}W\right) = \frac{1 - \tan^2(\pi/4 - 2^{1/2}W/2)}{2\tan(\pi/4 - 2^{1/2}W/2)}$$

and, hence, upon comparison with Eq. (114), we get for one solution

$$N = \cot\left(\frac{\pi}{4} - 2^{-1/2}W\right) \tag{115}$$

Equation (115) determines the value of $N$ in terms of the amplitude $W$ of the curve. Furthermore, the value of $N$ given by (115) is greater than unity, and, therefore, the point $P_2$ has coordinates $\phi_1' = n_1\pi - \theta_1$, $\phi_2' = n_2\pi - \theta_2$. From the measured values of $\phi_1'$ and $\phi_2'$, the equivalent-circuit parameters $\theta_1$ and $\theta_2$ are now readily found to within a multiple of $\pi$.

If the negative sign in (114) is chosen, we get in place of (115) the solution

$$N = \tan\left(\frac{\pi}{4} - 2^{-1/2}W\right) \tag{116}$$

and, hence, $N$ is less than unity, and the coordinates of the point $P_2$ are $\phi_1' = n_1'\pi + \pi/2 - \theta_1$, $\phi_2' = n_2'\pi + \pi/2 - \theta_2$. Either solution may be chosen; however, it is important to note that the coordinates of the point $P_2$ are different, and, hence, the values of $\theta_1$ and $\theta_2$ are different, depending on whether $N$ is chosen greater or smaller than unity.

In analytical work we generally obtain a bilinear relation between $\phi_1$ and $\phi_2$ of the following form:†

$$\tan\phi_1 = \frac{A + B\tan\phi_2}{C + D\tan\phi_2} \tag{117}$$

where $A$, $B$, $C$, and $D$ are known constants. It is now convenient to be able to determine the equivalent-circuit parameters in terms of the known constants, without having to construct the curve of $\phi_1$ vs. $\phi_2$ and analyzing this curve by the previous method. Equating the slope as obtained from (117) to $-1$, and using (117) to eliminate $\tan\phi_1$, we get

$$(D^2 + B^2 - AD + BC)\tan^2\phi_2 + (2CD + 2AB)\tan\phi_2 + (C^2 + A^2 + BC - AD) = 0 \tag{118}$$

This equation may be solved for $\phi_2$, and determines the points $Q_1$, $Q_2$, $Q_3$, etc. The solution is of the form

$$\tan\phi_2 = x \pm y \tag{119}$$

If the two solutions for $\phi_2$ are averaged, either the point $P_2$ or $P_3$ is determined. If the two solutions for $\phi_2$ given by (119) are designated $\phi_{21}$ and $\phi_{22}$, we get

$$\tan(\phi_{21} + \phi_{22}) = \frac{\tan\phi_{21} + \tan\phi_{22}}{1 - \tan\phi_{21}\tan\phi_{22}} = \frac{2x}{1 - x^2 + y^2}$$

When $x$ and $y$ are obtained from (118), we get

$$\phi_2' = \frac{\phi_{21} + \phi_{22}}{2} = \tfrac{1}{2}\tan^{-1}\frac{2(AB + CD)}{A^2 + C^2 - B^2 - D^2} \tag{120}$$

† R. E. Collin, Determination of Equivalent Circuit Parameters, *IRE Trans.*, vol. MTT-5, pp. 266–267, October, 1957.

The corresponding value of $\phi_1'$ may be found from (117) and is

$$\phi_1' = \tan^{-1} \frac{A + B \tan \phi_2'}{C + D \tan \phi_2'} \tag{121}$$

If the derivative $d\phi_1/d\phi_2$ is equated to $-N^2$ at the point $(\phi_1',\phi_2')$, we get

$$N^2 = n^2 Z_1 = \frac{(AD - BC)(1 + \tan^2 \phi_2')}{(C + D \tan \phi_2')^2 (1 + \tan^2 \phi_1')} \tag{122}$$

If (112) determines $N$ as greater than unity, then we have

$$\theta_1 = \phi_1' + n_1\pi \tag{123a}$$
$$\theta_2 = -\phi_2' + n_2\pi \tag{123b}$$

whereas if $N$ is found to be less than unity, $\pm\pi/2$ must be added to the parameters $\theta_1$ and $\theta_2$ as given by (123).

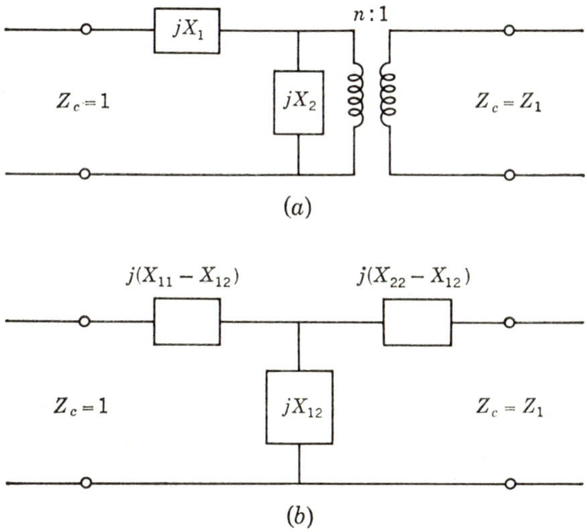


FIG. 5.14. Alternative equivalent circuits.

The parameters of alternative equivalent circuits as illustrated in Fig. 5.14 may also be obtained in terms of the constants $A$, $B$, $C$, and $D$. For Fig. 5.14$a$, we get

$$X_1 = \frac{-A}{C} \tag{124a}$$

$$X_2 = \frac{A}{C} - \frac{B}{D} \tag{124b}$$

$$n^2 Z_1 = \frac{DA}{C^2} - \frac{B}{C} \tag{124c}$$

while, for Fig. 5.14*b*,

$$X_{11} = \frac{-B}{D} \tag{125a}$$

$$X_{22} = \frac{CZ_1}{D} \tag{125b}$$

$$X_{12} = \pm\left(\frac{AZ_1}{D} - \frac{BCZ_1}{D^2}\right)^{1/2} \tag{125c}$$

## REFERENCES

*General references on waveguides*

5.1. S. Ramo and J. R. Whinnery: "Fields and Waves in Modern Radio," 2d ed., John Wiley & Sons, Inc., New York, 1953.
5.2. G. Southworth: "Principles and Application of Waveguide Transmission," D. Van Nostrand Company, Inc., Princeton, N.J., 1950.

The following papers are largely of historical interest now but nevertheless contain a good deal of the fundamental theory and principles of waveguides:

5.3. Lord Rayleigh: On the Passage of Electric Waves through Tubes, or the Vibration of Dielectric Cylinders, *Phil. Mag.*, vol. 43, pp. 125–132, February, 1897.
5.4. J. R. Carson, S. P. Mead, and S. A. Schelkunoff: Hyper-frequency Waveguides: Mathematical Theory, *Bell System Tech. J.*, vol. 15, pp. 310–333, April, 1936.
5.5. G. Southworth: Hyper-frequency Waveguides: General Considerations and Experimental Results, *Bell System Tech. J.*, vol. 15, pp. 284–309, April, 1936.
5.6. W. L. Barrow: Transmission of Electromagnetic Waves in Hollow Tubes of Metal, *Proc. IRE*, vol. 24, pp. 1298–1328, October, 1936.

*Attenuation in waveguides*

5.7. S. Kuhn: Calculation of Attenuation in Waveguides, *Proc. IEE (London)*, vol. 93, part IIIA, pp. 663–678, 1946.
5.8. A. E. Karbowiak: Theory of Imperfect Waveguides, the Effect of Wall Impedance, *Proc. IEE (London)*, vol. 102, part B, pp. 698–707, 1955.
5.9. P. N. Butcher: A New Treatment of Lossy Periodic Waveguides, *Proc. IEE (London)*, vol. 103, part B, pp. 301–306, 1956.

*Mode completeness and Green's functions.*

5.10. G. M. Roe: Normal Modes in the Theory of Waveguides, *Phys. Rev.*, vol. 69, p. 255, March, 1946.
5.11. J. Van Bladel: Expandability of a Waveguide Field in Terms of Normal Modes, *J. Appl. Phys.*, vol. 22, pp. 68–69, Janury, 1951.
5.12. J. Van Bladel: Field Expandability in Normal Modes for a Multilayered Rectangular or Circular Waveguide, *J. Franklin Inst.*, vol. 253, pp. 313–321, 1952.
5.13. J. J. Freeman: The Field Generated by an Arbitrary Current within a Waveguide, *J. Research Natl. Bur. Standards*, vol. 44, pp. 193–198, February, 1950.

*Transmission-line analogy and equivalent circuits*

5.14. N. Marcuvitz: "Waveguide Handbook," vol. 10 of MIT Rad. Lab. Series, McGraw-Hill Book Company, Inc., New York, 1951.
5.15. C. G. Montgomery, R. H. Dicke, and E. M. Purcell: "Principles of Microwave Circuits," vol. 8 of MIT Rad. Lab. Series, McGraw-Hill Book Company, Inc., New York, 1948.

*Variational solution*

5.16. R. M. Soria and T. J. Higgins: A Critical Study of Variational and Finite Difference Methods for Calculating the Operating Characteristics of Wave Guides and Other Electromagnetic Devices, *Proc. Natl. Electronic Conf.*, vol. 3, pp. 670–679, 1947.

## PROBLEMS

**5.1.** Show that the following integral expression is a stationary expression for the cutoff wave number $k_c$; that is, show that the first variation in $k_c$ is zero for arbitrary first-order variations in $\psi$, subject to the boundary constraint that the variation in $\psi$ vanishes on the perfectly conducting guide wall for $E$ modes, and that the variation in $\partial\psi/\partial n$ vanishes on the boundary for $H$ modes.

$$k_c^2 = \frac{\iint_S \nabla_t\psi \cdot \nabla_t\psi \, da}{\iint_S \psi^2 \, da}$$

The integration is over the guide cross section.

**5.2.** Show that, by a conformal transformation of the guide cross section, say $W = u + jv = F(x + jy)$, the wave equation for $\psi$ becomes

$$\frac{\partial^2\psi}{\partial u^2} + \frac{\partial^2\psi}{\partial v^2} + h^2k_c^2\psi = 0 \qquad \text{where } h = \left|\frac{dF}{dZ}\right|^{-1}$$

The transformation $W = \ln Z$ maps a circular guide in the $Z$ plane into an infinitely wide rectangular guide in the $uv$ plane.

**5.3.** Derive the orthogonality relation $\iint_S \mathbf{E}_{tn} \times \mathbf{H}_{tm}^* \cdot d\mathbf{a} = 0$ given in Eq. (25).

HINT: Begin with the expression $\mathbf{E}_n \cdot \nabla \times \mathbf{H}_m^* + \mathbf{E}_m^* \cdot \nabla \times \mathbf{H}_n$ and the similar one with the role of $\mathbf{E}$ and $\mathbf{H}$ interchanged.

**5.4.** Show that, in a rectangular guide lined with a thin resistance sheet at the wall $x = a$, the propagation constants for the $H_{n0}$ modes are given by a solution of the following equations:

$$\tan k_c a = -k_c(j\omega\mu_0\sigma)^{1/2} \qquad \text{and} \qquad \Gamma^2 = k_c^2 - k_0^2$$

where $\sigma$ is the conductivity of the resistance sheet.

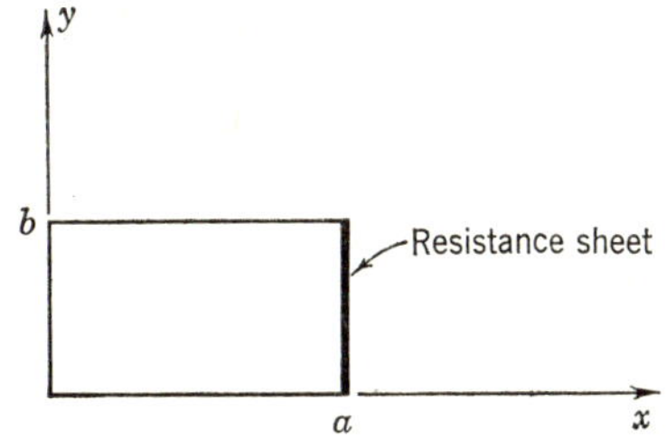


FIG. P 5.4

**5.5.** Show that Eq. (52) gives Eqs. (38) and (39) for the attenuation of nondegenerate $E$ and $H$ modes. Note that

$$P = \frac{1}{2}\iint_S \mathbf{e}_n \times h_n^* \cdot da = \frac{1}{2Z_n}\iint_S \mathbf{e}_n \cdot \mathbf{e}_n \, da = \frac{1}{2Z_n} \qquad \text{where } \mathbf{a}_z \times \mathbf{e}_n = Z_n \mathbf{h}_n$$

**5.6.** Derive the expressions for attenuation given in Tables 5.2 and 5.5.

**5.7.** Use the method of expansion in normal modes to derive the Green's function for $H_{n0}$ modes [Eq. (71)] in a rectangular guide.

**5.8.** Find the Green's function representing the $E$ modes excited by a current element $\mathbf{a}_z\delta(x - x')\delta(y - y')\delta(z - z')$ in a rectangular guide. This is a scalar Green's function proportional to the axial field component $E_z$ from which the total field may be derived.

**5.9.** Find the position and orientation of two small loops in a rectangular guide which will radiate in one direction only. Assume that only the $H_{10}$ mode can propagate. The two loops are to be located in the same transverse plane. The loop currents are $I$ and $Ie^{j\theta}$. Specify the required value of the phase angle $\theta$ also.

**5.10.** By a suitable contour integration show the equivalence between (71) and (94). Note that (94) must be multiplied by $j\omega\mu_0$ to give the electric field.

**5.11.** Find the reflection and transmission coefficients for the transmission-line circuit illustrated. Prove the relation $T_{12}Z_1 = T_{21}Z_2$.

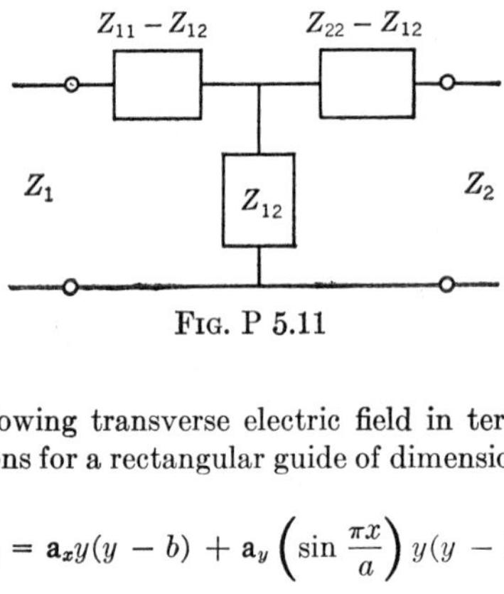


FIG. P 5.11

**5.12.** Expand the following transverse electric field in terms of the normal mode ($E$ and $H$ modes) functions for a rectangular guide of dimensions $a$ by $b$:

$$\mathbf{E}_t = \mathbf{a}_x y(y - b) + \mathbf{a}_y \left(\sin \frac{\pi x}{a}\right) y(y - b)$$

**5.13.** Find the dyadic Green's function (corresponding to the vector potential from a unit current source) which is a solution of the equation

$$\nabla^2\bar{\mathbf{G}} + k^2\bar{\mathbf{G}} = -\bar{\mathbf{I}}\delta(x - x')\delta(y - y')\delta(z - z')$$

and satisfies the boundary conditions $\mathbf{n} \cdot (\nabla \times \bar{\mathbf{G}}) = 0$, $\nabla \cdot \bar{\mathbf{G}} = 0$, in a rectangular guide. The solution may be constructed from the three components of the vector potential $\mathbf{A}$, where $\nabla^2 \mathbf{A}_i + k^2\mathbf{A}_i = -\mu \mathbf{a}_i \delta(x - x')\delta(y - y')\delta(z - z')$, $i = x, y, z$. The required Green's dyadic function is $\mu\bar{\mathbf{G}} = \mathbf{a}_x\mathbf{a}_x A_x + \mathbf{a}_y\mathbf{a}_y A_y + \mathbf{a}_z\mathbf{a}_z A_z$. Note that $\bar{\mathbf{G}}$ is a symmetric dyadic, so that $\bar{\mathbf{G}} \cdot \mathbf{J} = \mathbf{J} \cdot \bar{\mathbf{G}}$ and also

$$\bar{\mathbf{G}}(x, \ldots |x', \ldots) = \bar{\mathbf{G}}(x', \ldots |x, \ldots)$$

*Answer*

$$G = \sum_{n=0}^{\infty} \sum_{m=0}^{\infty} \frac{\epsilon_{on}\epsilon_{om}}{2ab\Gamma_{nm}} e^{-\Gamma_{nm}|z-z'|} \left[ \mathbf{a}_x\mathbf{a}_x \left( \cos\frac{n\pi x}{a} \cos\frac{n\pi x'}{a} \sin\frac{m\pi y}{b} \sin\frac{m\pi y'}{b} \right) + \mathbf{a}_y\mathbf{a}_y \right.$$
$$\left. \left( \sin\frac{n\pi x}{a} \sin\frac{n\pi x'}{a} \cos\frac{m\pi y}{b} \cos\frac{m\pi y'}{b} \right) + \mathbf{a}_z\mathbf{a}_z \left( \sin\frac{n\pi x}{a} \sin\frac{n\pi x'}{a} \sin\frac{m\pi y}{b} \sin\frac{m\pi y'}{b} \right) \right]$$

**5.14.** Since $\mu\bar{\mathbf{G}}$ in Prob. 5.13 corresponds to a vector potential, the electric field $\mathbf{E}$ from an arbitrary current souce is given by

$$\mathbf{E}(x,y,z) = \frac{-j\omega\mu}{k^2} \iiint_V (k^2 + \nabla\nabla\cdot)\mathbf{G}(x,y,z|x',y',z') \cdot J(x',y',z')\, dV'$$
$$= -j\omega\mu \iiint_V \bar{\mathbf{G}}_1(x,y,z|x',y',z') \cdot J(x',y',z')\, dV'$$

where
$$\bar{\mathbf{G}}_1 = \left(1 + \frac{1}{k^2}\nabla\nabla\cdot\right)\bar{\mathbf{G}}(x,y,z|x',y',z')$$

and is the dyadic Green's function which satisfies the same equation as $\mathbf{E}$. Show that $\bar{\mathbf{G}}_1$ is a solution of the equation

$$\nabla \times \nabla \times \bar{\mathbf{G}}_1 - k^2\bar{\mathbf{G}}_1 = \bar{\mathbf{I}}\delta(x - x')\delta(y - y')\delta(z - z')$$

by using the vector identity $\nabla \times \nabla \times = \nabla\nabla\cdot - \nabla^2$. Also show that $\bar{\mathbf{G}}_1$ satisfies the boundary condition $\mathbf{n} \times \bar{\mathbf{G}}_1 = 0$ on the guide wall. By direct differentiation of the result given in Prob. 5.13, show that

$$\nabla\nabla \cdot \bar{\mathbf{G}}(x,y,z|x',y',z') = \bar{\mathbf{g}}(x,y,z|x',y',z')$$

where $\bar{\mathbf{g}}$ is a dyadic with the property

$$\bar{\mathbf{g}}(x,y,z|x',y',z') = \bar{\mathbf{g}}_t(x',y',z'|x,y,z)$$

in which the subscript $t$ means the transposed dyadic; i.e., replace $g_{xy}$ by $g_{yx}$, etc. Thus the expression for the electric field may also be written as

$$\mathbf{E}(x,y,z) = -j\omega\mu \iiint_V \mathbf{J}(x', \ldots) \cdot \bar{\mathbf{G}}_1(x',y',z'|x,y,z)\, dV'$$

but not as

$$-j\omega\mu \iiint_V \mathbf{J}(x', \ldots) \cdot \bar{\mathbf{G}}_1(x,y,z|x',y',z')\, dV'$$

in general.

# CHAPTER 6

# INHOMOGENEOUSLY FILLED WAVEGUIDES

Waveguides partially filled with dielectric slabs find application in a variety of waveguide components such as phase changers, matching transformers, quarter-wave plates, etc. The first few sections of this chapter will be devoted to an examination of some of the basic properties of inhomogeneously filled waveguides. It will be assumed that the dielectric loss is very small and may be neglected. When this assumption is not valid, the solution may be obtained by replacing the permittivity $\epsilon$ by $\epsilon(1 - j \tan \delta_l)$ in the solutions for the lossless case. Some of the properties of guides loaded with ferrite material are discussed in the latter part of the chapter.

## 6.1. Dielectric-slab-loaded Rectangular Guides

Figure 6.1 illustrates the cross-sectional view of several typical slab-loaded rectangular waveguides. For the configurations lettered $a$, $b$, $c$,

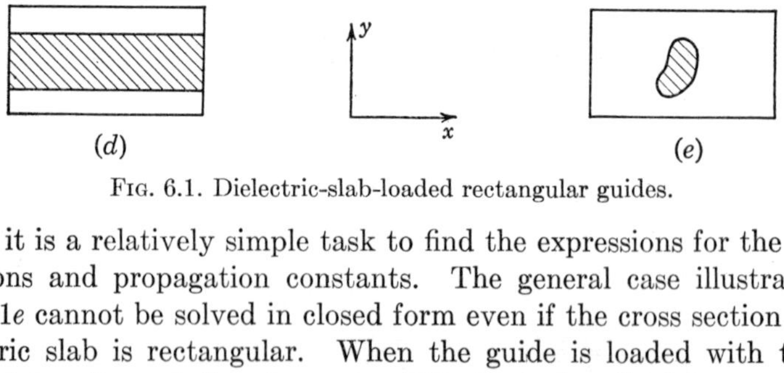


Fig. 6.1. Dielectric-slab-loaded rectangular guides.

and $d$, it is a relatively simple task to find the expressions for the eigenfunctions and propagation constants. The general case illustrated in Fig. 6.1$e$ cannot be solved in closed form even if the cross section of the dielectric slab is rectangular. When the guide is loaded with two or more slabs, the solution becomes difficult to handle because of the involved transcendental equations which occur for the determination of the eigenvalues. In such cases a perturbation or variational technique may be employed to advantage. One such technique, the Rayleigh-Ritz method, will be discussed in a later section.

The normal modes of propagation for guides loaded with dielectric slabs as in Fig. 6.1 are not, in general, either $E$ or $H$ modes, but a combination of an $E$ and an $H$ mode. An exception is the case of $H_{n0}$ modes with the electric field parallel to the slab and no variation of the fields along the dielectric-air interface. For a rectangular guide terminated

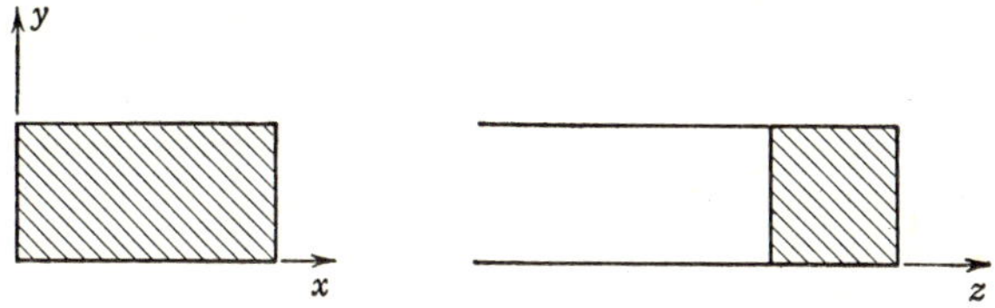


FIG. 6.2

by a dielectric plug as in Fig. 6.2, it is readily found that the boundary conditions at the air-dielectric interface can be satisfied by $E$ modes alone or $H$ modes alone. The configurations illustrated in Fig. 6.1 are essentially the same, except that the air-dielectric interface lies in the $xz$ or $yz$ plane, instead of in the $xy$ plane as in Fig. 6.2. This suggests that the basic modes of propagation for slab-loaded rectangular guides may be derived from magnetic and electric types of Hertzian potential functions, having single components directed normal to the air-dielectric interface. This is indeed the case, and the resultant modes may be classified as $E$ or $H$ modes with respect to the interface normal ($x$ or $y$ axis). From the magnetic Hertzian potential, we obtain the solution for a mode which has no component of electric field normal to the interface. The electric field thus lies in the longitudinal interface plane, and the mode is referred to as a longitudinal-section electric (LSE) mode. From the electric Hertzian potential, a mode with no magnetic field component normal to the interface is obtained. This mode is called a longitudinal-section magnetic (LSM) mode.

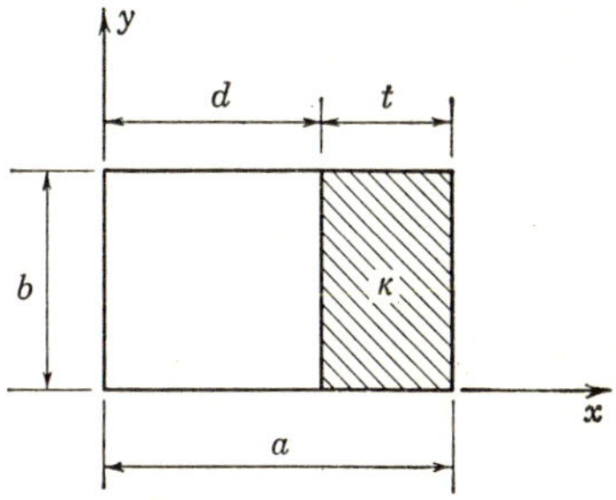


FIG. 6.3. Asymmetrical dielectric-slab-loaded guide.

As a first example we will consider the solution for an asymmetrically loaded guide as illustrated in Fig. 6.3. The thickness of the slab is $t$, and its relative dielectric constant is $\kappa$.

*LSE Modes*

For a magnetic-type Hertzian potential $\mathbf{\Pi}_h = \mathbf{a}_x\psi_h(x,y)e^{-\gamma z}$, the electric and magnetic fields are given by

$$\mathbf{E} = -j\omega\mu_0\nabla \times \mathbf{\Pi}_h \tag{1a}$$

$$\mathbf{H} = \nabla \times \nabla \times \mathbf{\Pi}_h = \kappa(x)k_0^2\mathbf{\Pi}_h + \nabla\nabla \cdot \mathbf{\Pi}_h \tag{1b}$$

and

$$\nabla_t^2\psi_h + [\gamma^2 + \kappa(x)k_0^2]\psi_h = 0 \tag{1c}$$

where

$$\kappa(x) = \begin{cases} 1 & 0 \le x \le d \\ \kappa & d \le x \le a \end{cases}$$

Propagation along the guide may be assumed to be according to the exponential factor $e^{-\gamma z}$ (or $e^{\gamma z}$) and must be the same in the empty portion and dielectric-filled portion of the guide, if the boundary conditions at the interface are to be satisfied for all arbitrary values of $z$. A solution for $\psi_h$ which satisfies (1$c$) in the two regions of interest, and also gives an electric field whose tangential components vanish on the perfectly conducting guide walls, is readily found to be

$$\psi_h = \begin{cases} A \sin hx \cos \dfrac{m\pi y}{b} & 0 \le x \le d \\ B \sin l(a - x) \cos \dfrac{m\pi y}{b} & d \le x \le a \end{cases} \tag{2}$$

where

$$\gamma^2 = l^2 + \left(\frac{m\pi}{b}\right)^2 - \kappa k_0^2 = h^2 + \left(\frac{m\pi}{b}\right)^2 - k_0^2 \tag{3}$$

and $A$ and $B$ are amplitude constants. The variation with $y$ must be the same in both regions, in order to satisfy the boundary conditions for all $y$ along the interface. The solution given in (2) is valid for all integer values of $m$. The ratio of the coefficients and the wave numbers $l$, $h$ and propagation constant $\gamma$ will be determined by the boundary conditions which must hold at $x = d$. From (1$a$), it is found that $E_z = j\omega\mu_0 e^{-\gamma z}\ \partial\psi_h/\partial y$, while $H_y = (\partial^2\psi_h/\partial y\ \partial x)e^{-\gamma z}$. In order for $E_z$ and $H_y$ to be continuous at the interface, the following conditions are imposed on the solution given by (2),

$$A \sin hd = B \sin lt \tag{4a}$$

$$Ah \cos hd = -Bl \cos lt \tag{4b}$$

Dividing (4$a$) by (4$b$), the followingtran scendental equation is obtained:

$$h \tan lt = -l \tan hd \tag{5a}$$

which together with the relation

$$l^2 = h^2 + (\kappa - 1)k_0^2 \tag{5b}$$

derivable from (3) determines an infinite number of solutions for the wave numbers $l$ and $h$. An examination of the expressions for the remaining field components shows that they satisfy the proper boundary conditions at the interface when $E_z$ and $H_y$ do. When $l$ and $h$ become large, they approach equality, and the $n$th solution to (5$a$) approaches

$n\pi/a$, and hence $l_n$ and $h_n$ approach $n\pi/a$ for large $n$. By eliminating the coefficient $B$ by means of (4$a$), the solution for the $nm$th LSE mode may be written as

$$\psi_{h,nm}e^{-\gamma_{nm}z} = A_{nm}e^{-\gamma_{nm}z}\begin{cases} \sin h_n x \cos \dfrac{m\pi y}{b} & 0 \le x \le d \\ \dfrac{\sin h_n d}{\sin l_n t} \sin l_n(a-x) \cos \dfrac{m\pi y}{b} & d \le x \le a \end{cases} \tag{6}$$

where $A_{nm}$ may be chosen arbitrarily. When $m = 0$, both $E_z$ and $H_y$ are zero, and the only field components present are $H_x$, $E_y$, and $H_z$. For this special case, the LSE mode degenerates into an $H_{n0}$ mode.

The wave impedance measured in the $x$ direction is given by

$$-\frac{E_z}{H_y} = \frac{E_y}{H_z} = -\frac{j\omega\mu_0\psi_h}{\partial\psi_h/\partial x} \tag{7}$$

The eigenvalue equation (5$a$) may be obtained directly by considering the equivalent circuit of the loaded guide in the transverse plane as

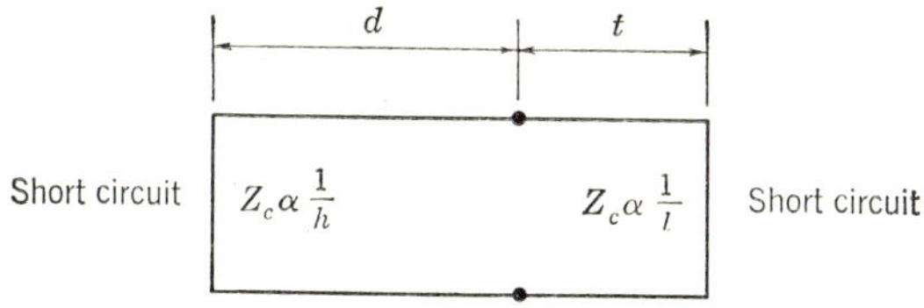


FIG. 6.4. Transverse equivalent transmission-line circuit.

illustrated in Fig. 6.4. By considering propagation to take place along the $x$ direction, the equivalent transmission-line circuit is a junction of two lines with characteristic impedance proportional to the wave impedances (inversely proportional to the wave numbers) in the two sections and short-circuited at the ends. The wave numbers $l$ and $h$ are determined by the condition that the zero impedance load (short circuit) at $x = a$ must transform into a zero impedance at $x = 0$. By conventional transmission-line theory we find that the input impedance at $x = d$ is $jl^{-1} \tan lt$, and at $x = 0$ is given by

$$Z_{\text{in}} = \frac{1}{h}\frac{jl^{-1}\tan lt + jh^{-1}\tan hd}{h^{-1} - l^{-1}\tan lt \tan hd}$$

The condition that $Z_{\text{in}}$ should equal zero is found at once to be that given by (5$a$). This transverse-resonance procedure is useful for finding the eigenvalue equation for guides which are loaded with several dielectric slabs.

The eigenvalue equation for LSE modes for an inhomogeneously filled guide as illustrated in Fig. 6.5 will be determined by this transverse-resonance method. At $x = x_2$, the input impedance is $jh^{-1} \tan hd_2$. At $x = x_1$, the input impedance looking to the right must be equal to the negative value of the input impedance seen at this point looking to the left. Hence, we get

$$-jh^{-1} \tan hd_1 = \frac{1}{l} \frac{jh^{-1} \tan hd_2 + jl^{-1} \tan lt}{l^{-1} - h^{-1} \tan hd_2 \tan lt}$$

which may be simplified to

$$h^2 \tan lt + hl \tan hd_2 + lh \tan hd_1 - l^2 \tan lt \tan hd_1 \tan hd_2 = 0 \quad (8)$$

This equation together with (3) determines the allowed solutions for $l$, $h$, and $\gamma$. The solution to (8) is obviously not a pleasant one to obtain.

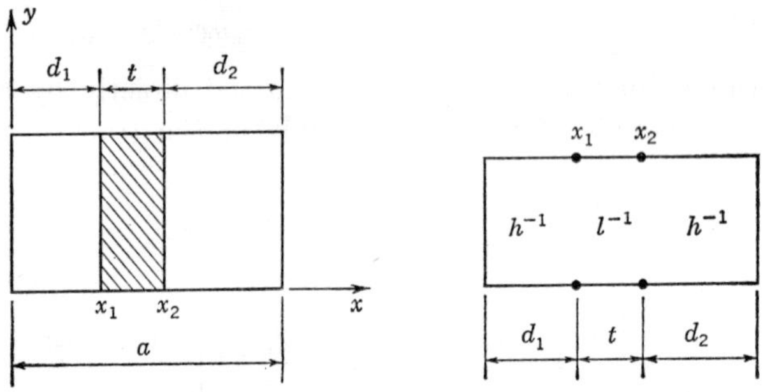


FIG. 6.5

For the special case of a symmetrically placed slab, $d_1 = d_2 = d$, and our transverse-resonance condition is equivalent to the condition that a short circuit should appear at the center of the slab for asymmetrical modes, and that an open circuit should appear at $x = a/2$ for symmetrical modes. The corresponding eigenvalue equations are, respectively,

$$-l \tan hd = h \tan \frac{lt}{2} \quad (9a)$$

$$h \cot hd = l \tan \frac{lt}{2} \quad (9b)$$

These equations may be obtained from (8), but are more readily determined directly by using the appropriate transverse-resonance conditions given above. The symmetrical modes are those which have a symmetrical variation of $E_y$ with respect to $x$ about the point $x = a/2$.

*LSM Modes*

The LSM modes may be derived from an electric type of Hertzian potential of the following form, $\mathbf{\Pi}_e = \mathbf{a}_x\psi_e(x,y)e^{-\gamma z}$, for the case of a guide inhomogeneously filled as in Fig. 6.3. The equations determining the fields are

$$\mathbf{H} = j\omega\epsilon_0\kappa(x)\nabla \times \mathbf{\Pi}_e \tag{10a}$$

$$\mathbf{E} = \nabla \times \nabla \times \mathbf{\Pi}_e = \kappa(x)k_0^2\mathbf{\Pi}_e + \nabla\nabla \cdot \mathbf{\Pi}_e \tag{10b}$$

$$\nabla_t^2\psi_e + [\gamma^2 + \kappa(x)k_0^2]\psi_e = 0 \tag{10c}$$

These equations are valid in each region over which $\kappa(x)$ is constant, but do not hold at the interface across which $\kappa(x)$ changes discontinuously. However, we may solve for the fields in each section and match the tangential components at the interface to obtain the solution valid for all $x$.

Appropriate solutions for $\psi_e$ in each region, such that the tangential electric field will vanish on the guide boundary, are

$$\psi_e = \begin{cases} A \cos hx \sin \dfrac{m\pi y}{b} & 0 \le x \le d \\ B \cos l(a - x) \sin \dfrac{m\pi y}{b} & d \le x \le a \end{cases} \tag{11}$$

The field component $E_z$ is proportional to $\partial\psi_e/\partial x$, while $H_y$ is proportional to $\kappa(x)\psi_e$ in both regions. Continuity of these tangential field components at $x = d$ imposes the following restriction on the solution given by (11):

$$Ah \sin hd = -Bl \sin lt \tag{12a}$$

$$A \cos hd = \kappa B \cos lt \tag{12b}$$

Dividing (12$a$) by (12$b$) yields the eigenvalue equation

$$\kappa h \tan hd = -l \tan lt \tag{13}$$

which, together with the relation $\gamma^2 - (m\pi/b)^2 = h^2 - k_0^2 = l^2 - \kappa k_0^2$, determines the allowed values of $h$, $l$, and $\gamma$. Again an infinite number of discrete solutions exist, and both $h_n$ and $l_n$ approach $n\pi/a$ for large $n$. From (12), the ratio of the coefficients $A$ and $B$ may be found, and, when substituted into (11), completes the solution.

For more general slab arrangements, the eigenvalue equation may be determined by the transverse-resonance method. It should be noted that the wave impedances in the $x$ direction are proportional to $h$ and $l\kappa^{-1}$ in the empty and dielectric-loaded sections, respectively, for LSM modes. For rectangular guides loaded with dielectric slabs which fill the guide uniformly along the $x$ direction, but discontinuously along the $y$ direction, the field may be found from Hertzian potentials directed along the $y$ axis.

*Orthogonality of Modes*

The various orthogonality relations derived for the $E$ and $H$ modes in an empty guide in the preceding chapter are not all valid for the LSM and

LSE modes in an inhomogeneously filled guide. However, the following orthogonality relation does hold:

$$\iint_S \mathbf{E}_{tn} \times \mathbf{H}_{tm} \cdot \mathbf{a}_z \, da = 0 \qquad n \neq m$$

where $\mathbf{E}_{tn}$ is the transverse electric field for the $n$th mode and $\mathbf{H}_{tm}$ is the transverse magnetic field for the $m$th mode. If the two fields involved are for one LSE and one LSM mode, the above orthogonality property holds for $n = m$ also, since these latter fields are independent nondegenerate solutions of Maxwell's equations. The above relation will be proved as a special case of a more general orthogonality property.

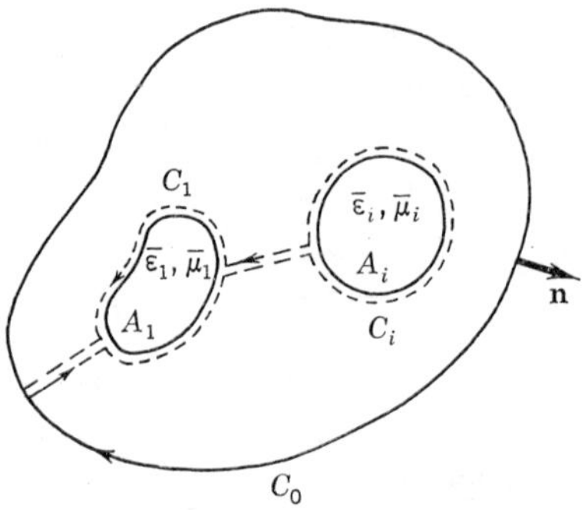


FIG. 6.6. Cross section of a general inhomogeneously filled guide.

Consider a general cylindrical guide, with perfectly conducting walls, and inhomogeneously filled, as in Fig. 6.6. Over each cross section $A_i$, the medium filling that section is characterized by a general permittivity tensor $\bar{\boldsymbol{\varepsilon}}_i$ and permeability tensor $\bar{\boldsymbol{\mu}}_i$. In general, the modes cannot be classified as $E$, $H$, LSM, or LSE modes, but nevertheless an infinite number of discrete modes may exist.

Let $\mathbf{E}_n$, $\mathbf{H}_n$ be the solution for the $n$th mode, and $\mathbf{E}'_m$, $\mathbf{H}'_m$ be the solution for the $m$th mode, with $\bar{\boldsymbol{\mu}}_i$ and $\bar{\boldsymbol{\varepsilon}}_i$ replaced by their transpose dyadics $(\bar{\boldsymbol{\mu}}_i)_t$ and $(\bar{\boldsymbol{\varepsilon}}_i)_t$. Over each cross section, the curl equations for the fields are

$$\nabla \times \mathbf{E}_n = -j\omega\bar{\boldsymbol{\mu}}_i \cdot \mathbf{H}_n \qquad \nabla \times \mathbf{H}_n = j\omega\bar{\boldsymbol{\varepsilon}}_i \cdot \mathbf{E}_n$$
$$\nabla \times \mathbf{E}'_m = -j\omega(\bar{\boldsymbol{\mu}}_i)_t \cdot \mathbf{H}'_m \qquad \nabla \times \mathbf{H}'_m = j\omega(\bar{\boldsymbol{\varepsilon}}_i)_t \cdot \mathbf{E}'_m$$

By forming the dot products indicated below, we get

$$j\omega[\mathbf{E}'_m \cdot \bar{\boldsymbol{\varepsilon}}_i \cdot \mathbf{E}_n - \mathbf{E}_n \cdot (\bar{\boldsymbol{\varepsilon}}_i)_t \cdot \mathbf{E}'_m - \mathbf{H}'_m \cdot \bar{\boldsymbol{\mu}}_i \cdot \mathbf{H}_n + \mathbf{H}_n \cdot (\bar{\boldsymbol{\mu}}_i)_t \cdot \mathbf{H}'_m] = \nabla \cdot (\mathbf{E}_n \times \mathbf{H}'_m - \mathbf{E}'_m \times \mathbf{H}_n) = 0$$

since $\mathbf{E}'_m \cdot \bar{\boldsymbol{\varepsilon}}_i \cdot \mathbf{E}_n = \mathbf{E}_n \cdot (\bar{\boldsymbol{\varepsilon}}_i)_t \cdot \mathbf{E}'_m$, etc. Let the $z$ dependence of the fields $\mathbf{E}_n$, $\mathbf{H}_n$ and $\mathbf{E}'_m$, $\mathbf{H}'_m$ be according to $e^{-\gamma_n z}$ and $e^{-\gamma_m' z}$, respectively. The divergence relation now gives

$$\nabla_t \cdot (\mathbf{E}_n \times \mathbf{H}'_m - \mathbf{E}'_m \times \mathbf{H}_n) = (\gamma_n + \gamma'_m)\mathbf{a}_z \cdot (\mathbf{E}_n \times \mathbf{H}'_m - \mathbf{E}'_m \times \mathbf{H}_n)$$

Integrating over the cross section $A_i$ and converting the two-dimensional divergence integral to a contour integral gives

$$\oint_{C_i} (\mathbf{n} \times \mathbf{E}_n \cdot \mathbf{H}'_m - \mathbf{n} \times \mathbf{E}'_m \cdot \mathbf{H}_n) \, dl = (\gamma_n + \gamma'_m) \iint_{A_i} \mathbf{a}_z \cdot (\mathbf{E}_n \times \mathbf{H}'_m - \mathbf{E}'_m \times \mathbf{H}_n) \, da$$

The cross product $\mathbf{n} \times \mathbf{E}_n$ selects the component of $\mathbf{E}_n$ tangential to the bounding surface of section $A_i$. The dot product with $\mathbf{H}'_m$ subsequently selects only the tangential component of $\mathbf{H}'_m$. The tangential components are continuous, and hence we may sum over all cross-sectional areas, and the contribution from all the interior contours $C_i$ will vanish since they are traversed twice, but in opposite senses. On the exterior contour, $\mathbf{n} \times \mathbf{E}_n$ and $\mathbf{n} \times \mathbf{E}'_m$ vanish, and hence the total contour integral is zero. Thus we obtain the final result

$$(\gamma_n + \gamma'_m) \iint_S \mathbf{a}_z \cdot (\mathbf{E}_n \times \mathbf{H}'_m - \mathbf{E}'_m \times \mathbf{H}_n)\, da = 0 \tag{14}$$

This result is the general statement of the reciprocity principle for the electromagnetic field in a source-free guide bounded by perfectly conducting walls. For general media such as ferrites, which are characterized by nonsymmetrical tensors, the reciprocity principle involves fields which are solutions to the related problem, with media characterized by the transposed permittivity and permeability tensors.

If a guide exhibits reflection symmetry with respect to the $z$ coordinate, then corresponding to the solutions $\mathbf{E}_n$, $\mathbf{H}_n$; $\mathbf{E}'_m$, $\mathbf{H}'_m$ with transverse and longitudinal components $\mathbf{E}_{tn}$, $\mathbf{E}_{zn}$; $\mathbf{H}_{tn}$, $\mathbf{H}_{zn}$ and $\mathbf{E}'_{tm}$, $\mathbf{E}'_{zm}$; $\mathbf{H}'_{tm}$, $\mathbf{H}'_{zm}$, respectively, there exist solutions $\mathbf{E}_{tn}$, $-\mathbf{E}_{zn}$; $-\mathbf{H}_{tn}$, $\mathbf{H}_{zn}$ and $\mathbf{E}'_{tm}$, $-\mathbf{E}'_{zm}$; $-\mathbf{H}'_{tm}$, $\mathbf{H}'_{zm}$ with eigenvalues $-\gamma_n$ and $-\gamma'_m$, respectively. This is just the definition of reflection symmetry.

For guides of the above type, we obtain from (14) the result

$$(\gamma_n - \gamma'_m) \iint_S \mathbf{a}_z \cdot (\mathbf{E}_n \times \mathbf{H}'_m + \mathbf{E}'_m \times \mathbf{H}_n)\, da = 0 \tag{15}$$

upon replacing the solution $(\mathbf{H}'_{tm},\mathbf{H}'_{zm},\gamma'_m)$ by $(-\mathbf{H}'_{tm},\mathbf{H}'_{zm},-\gamma'_m)$. The combination of (14) and (15) gives

$$\iint_S \mathbf{a}_z \cdot \mathbf{E}_{tn} \times \mathbf{H}'_{tm}\, da = 0 \tag{16}$$

after canceling the factors $\gamma_n + \gamma'_m$, which is permissible when $\gamma_n \neq \pm\gamma'_m$. Only the transverse field components enter into the triple scalar product in (14) and (15), the terms involving the longitudinal fields being orthogonal to $\mathbf{a}_z$.

All uniform guides inhomogeneously filled with isotropic media have reflection symmetry. The same is true of guides filled with media which have transverse anisotropy only; for example, $\mu_{xz} = \mu_{zx} = \mu_{zy} = \mu_{yz} = 0$. When the medium filling the guide is lossless, the following orthogonality relation is also valid:

$$\iint_S \mathbf{a}_z \cdot \mathbf{E}_{tn} \times \mathbf{H}^*_{tm}\, da = 0 \tag{17}$$

where * denotes the complex conjugate value.† In the case of the inhomogeneously filled rectangular guide, the transverse electric fields for two different LSE modes are orthogonal. The same is true for the transverse magnetic field and the longitudinal field components. For the LSM modes, none of these orthogonal properties hold. In place of these we have orthogonality with respect to a suitable weighting function, a topic to be considered in the next section.

## 6.2. The Rayleigh-Ritz Method

The Rayleigh-Ritz method is a systematic scheme for determining a finite set of approximate eigenfunctions and eigenvalues to a given differential equation and its boundary conditions. These are obtained by means of a variational integral, whose stationary values correspond to the true eigenvalues, when the true eigenfunctions are employed in the integrand. The details of this method will be developed in this section.

As noted in the previous section, the equations for the LSM modes do not hold at the interface where $\kappa(x)$ changes discontinuously. For the LSE modes, both $\psi_h$ and $\partial\psi_h/\partial x$ are continuous, and so the equations given for these modes are valid at the interface. To avoid the complications associated with a piecewise-continuous function $\kappa(x)$, we will assume that $\kappa(x)$ is continuous and specialize to the case of a discontinuous dielectric constant after the general equations have been derived. The theory will be given for LSM modes, and only the results for LSE modes.

As a first step, we need to derive the equation for $\mathbf{\Pi}_e = \mathbf{a}_x\psi_e e^{-\gamma z}$ for a rectangular guide filled with a dielectric whose dielectric constant $\kappa = \kappa(x)$ is a continuous function of $x$. Maxwell's equations in a source-free region are

$$\nabla \times \mathbf{H} = j\omega\epsilon_0\kappa\mathbf{E} \tag{18a}$$
$$\nabla \times \mathbf{E} = -j\omega\mu_0\mathbf{H} \tag{18b}$$
$$\nabla \cdot \mathbf{B} = 0 \tag{18c}$$
$$\nabla \cdot \mathbf{D} = 0 = \nabla \cdot \kappa\mathbf{E} \tag{18d}$$

Since $\mu_0$ is a constant, $\nabla \cdot \mathbf{H} = 0$, we may take

$$\mathbf{H} = j\omega\epsilon_0 \nabla \times \mathbf{\Pi}_e \tag{19}$$

Note that this equation differs from (10*a*) by the absence of the factor $\kappa(x)$. From (18*b*) we get $\nabla \times \mathbf{E} = k_0^2 \nabla \times \mathbf{\Pi}_e$, which integrates to

$$\mathbf{E} = k_0^2\mathbf{\Pi}_e + \nabla\phi$$

† A much more complete discussion of general orthogonality properties of waveguides will be found in the reports by Bresler and Marcuvitz listed at the end of this chapter.

where $\phi$ is as yet an arbitrary scalar function. Taking the curl of (19) [the factor $\kappa(x)$ was omitted from this equation so that the curl could be readily taken], we get

$$\nabla \times \mathbf{H} = j\omega\epsilon_0 \nabla \times \nabla \times \mathbf{\Pi}_e = j\omega\epsilon_0\kappa(x)\mathbf{E}$$

Expanding and substituting for $\mathbf{E}$ gives

$$\begin{aligned}\nabla\nabla \cdot \mathbf{\Pi}_e - \nabla^2\mathbf{\Pi}_e &= \kappa(x)k_0^2\mathbf{\Pi}_e + \kappa(x)\,\nabla\phi \\ &= \kappa k_0^2\mathbf{\Pi}_e + \nabla\kappa\phi - \phi\,\nabla\kappa\end{aligned}$$

Since $\nabla \cdot \mathbf{\Pi}_e$ and $\phi$ are arbitrary up to this point, we choose the following relation between them:

$$\nabla\nabla \cdot \mathbf{\Pi}_e = \nabla\kappa\phi$$

which apart from an irrelevant constant integrates to $\kappa\phi = \nabla \cdot \mathbf{\Pi}_e$. Thus, the following equation for $\mathbf{\Pi}_e$ is obtained:

$$\nabla^2\mathbf{\Pi}_e - \kappa^{-1}(\nabla\kappa)\,\nabla \cdot \mathbf{\Pi}_e + \kappa k_0^2\mathbf{\Pi}_e = 0 \tag{20}$$

The equation for $\mathbf{E}$ becomes

$$\begin{aligned}\mathbf{E} &= k_0^2\mathbf{\Pi}_e + \nabla(\kappa^{-1}\,\nabla \cdot \mathbf{\Pi}_e) \\ &= k_0^2\mathbf{\Pi}_e + \nabla \cdot \mathbf{\Pi}_e\,\nabla\kappa^{-1} + \kappa^{-1}\,\nabla\nabla \cdot \mathbf{\Pi}_e \\ &= \kappa^{-1}\,\nabla \times \nabla \times \mathbf{\Pi}_e\end{aligned} \tag{21}$$

when substitution from (20) is made. From (21) it is readily seen that $\nabla \cdot \kappa\mathbf{E} = 0$, since the divergence of the curl of any vector is identically zero. If now it is assumed that $\mathbf{\Pi}_e$ is of the form

$$\mathbf{\Pi}_e = \mathbf{a}_x\phi_e(x)\sin\frac{m\pi y}{b}\,e^{-\gamma z}$$

we find that $\phi_e$ must satisfy the following equation:

$$\frac{d^2\phi_e}{dx^2} - \kappa^{-1}\frac{d\kappa}{dx}\frac{d\phi_e}{dx} + (\kappa k_0^2 + \gamma^2 - h^2)\phi_e = 0$$

or, equivalently,

$$\frac{d}{dx}\frac{1}{\kappa}\frac{d\phi_e}{dx} + \frac{1}{\kappa}(\kappa k_0^2 + \gamma^2 - h^2)\phi_e = 0 \tag{22}$$

where $h^2$ has been written for $(m\pi/b)^2$. In order that the tangential electric field derivable from $\mathbf{\Pi}_e$ shall vanish on the guide boundary at $x = 0, a$, the function $\phi_e$ must satisfy the boundary conditions

$$\frac{d\phi_e}{dx} = 0 \qquad x = 0, a \tag{23}$$

The differential equation (22), together with the boundary conditions (23), constitutes what is known as a Sturm-Liouville system. Such a system has, among others, the following important properties:

1. An infinite number of eigenfunctions $\phi_{en}$ with discrete eigenvalues $\gamma_n^2$ exists.
2. The eigenfunctions $\phi_{en}$ form an orthogonal set over the closed interval $0 \le x \le a$, with respect to the weighting function $\kappa^{-1}$.
3. The eigenfunctions form a complete set, in terms of which a piecewise-continuous function of $x$ defined over the interval $0 \le x \le a$ may be expanded.
4. The solution to the system may be formulated equivalently as the problem of finding the function which will render a certain integral stationary.

The orthogonality of the eigenfunctions is readily proved as follows. Multiply the equation for $\phi_{en}$ by $\phi_{em}$, and the equation for $\phi_{em}$ by $\phi_{en}$; subtracting the two resultant equations and integrating over the interval $0 \le x \le a$, we get

$$\int_0^a \left(\phi_{em} \frac{d}{dx} \frac{1}{\kappa} \frac{d\phi_{en}}{dx} - \phi_{en} \frac{d}{dx} \frac{1}{\kappa} \frac{d\phi_{em}}{dx}\right) dx$$
$$= (\gamma_m^2 - \gamma_n^2) \int_0^a \kappa^{-1} \phi_{em} \phi_{en}\, dx = \left(\frac{\phi_{em}}{\kappa} \frac{d\phi_{en}}{dx} - \frac{\phi_{en}}{\kappa} \frac{d\phi_{em}}{dx}\right)\Bigg|_0^a$$
$$- \int_0^a \kappa^{-1} \left(\frac{d\phi_{em}}{dx} \frac{d\phi_{en}}{dx} - \frac{d\phi_{en}}{dx} \frac{d\phi_{em}}{dx}\right) dx = 0 \qquad (24)$$

upon integrating by parts once and using the boundary conditions (23). The eigenfunctions are thus seen to be orthogonal with respect to the weighting function $\kappa^{-1}$, since $\gamma_n^2 \neq \gamma_m^2$. The eigenfunctions may be normalized so that they form an orthonormal set, i.e.,

$$\int_0^a \kappa^{-1} \phi_{en} \phi_{em}\, dx = \delta_{nm} \qquad (25)$$

where $\delta_{nm}$ is the Kronecker delta and equals unity for $n = m$, and zero otherwise. An arbitrary piecewise-continuous function of $x$, say $g(x)$, may be expanded into an infinite series of these eigenfunctions as follows:

$$g(x) = \sum_{n=0}^{\infty} a_n \phi_{en} \qquad (26a)$$

where $a_n$ is determined by multiplying both sides by $\kappa^{-1}\phi_{en}$, integrating over $0 \le x \le a$, and using (25) to get

$$a_n = \int_0^a \kappa^{-1} g \phi_{en}\, dx \qquad (26b)$$

*Minimum Characterization of the Eigenvalues*

If (22) is multiplied by $\phi_e$ and integrated over the interval $0 \le x \le a$, we get

$$\gamma^2 \int_0^a \kappa^{-1}\phi_e^2\,dx = \int_0^a \kappa^{-1}\left[\left(\frac{d\phi_e}{dx}\right)^2 - (\kappa k_0^2 - h^2)\phi_e^2\right] dx \tag{27}$$

by integrating by parts once. The integrated term vanishes by virtue of the boundary conditions on $d\phi_e/dx$ at $x = 0, a$. Equation (27) is a variational expression for the eigenvalue $\gamma^2$. The condition that (27) should be stationary may be obtained by imposing the condition that the first-order change or variation in $\gamma^2$ should vanish for arbitrary first-order variations $\delta\theta_e$ in $\phi_e$. This, of course, is equivalent to making the integrand satisfy the Euler-Lagrange equation from the calculus of variations, and imposes on $\phi_e$ the condition that it satisfy the differential equation (22).†

An extremization of (27) by that class of functions $\phi(x)$ which are continuous, with at least a piecewise-continuous first derivative which vanishes at $x = 0, a$, and are orthogonal to the first $K$ true eigenfunctions $\phi_{en}$, yields an upper bound on the true eigenvalue $\gamma_K^2$. By means of the expansion theorem for a complete set of functions, i.e., Eq. (26), we may write

$$\phi = \sum_{n=0}^{\infty} a_n\phi_{en} \tag{28a}$$

The orthogonality conditions imposed on $\phi$ give

$$a_n = 0 \qquad n = 0, 1, 2, \ldots, K-1 \tag{28b}$$

Note that the $K$th eigenfunction is the one labeled $K - 1$. Substituting into (27) gives

$$\begin{aligned}\gamma^2 \sum_{n=K}^{\infty} a_n^2 &= \int_0^a \kappa^{-1} \sum_{s=K}^{\infty}\sum_{n=K}^{\infty}\left[\frac{d\phi_{en}}{dx}\frac{d\phi_{es}}{dx} - (\kappa k_0^2 - h^2)\phi_{en}\phi_{es}\right] \\ a_n a_s\,dx &= \sum_{s=K}^{\infty}\sum_{n=K}^{\infty} a_n a_s \left\{\kappa^{-1}\phi_{es}\frac{d\phi_{en}}{dx}\Big|_0^a \right. \\ &\quad \left. - \int_0^a \left[\phi_{es}\frac{d}{dx}\frac{1}{\kappa}\frac{d\phi_{en}}{dx} + \kappa^{-1}(\kappa k_0^2 - h^2)\phi_{es}\phi_{en}\right] dx\right\}\end{aligned} \tag{29}$$

The integrated term vanishes, and, substituting into (29) from the

† Section A.1*d*.

differential equation (22), we get

$$\gamma^2 \sum_{n=K}^{\infty} a_n^2 = \sum_{n=K}^{\infty} a_n^2 \gamma_n^2 \tag{30}$$

If it is assumed that the eigenvalues have been arranged into a monotonically increasing sequence, so that $\gamma_0^2 < \gamma_1^2 < \gamma_2^2 < \gamma_3^2 \cdots < \gamma_K^2$, the result (30) may be written as

$$\gamma^2 = \frac{\sum_{n=K}^{\infty} a_n^2 \gamma_n^2}{\sum_{n=K}^{\infty} a_n^2} = \gamma_K^2 + \frac{\sum_{n=K+1}^{\infty} a_n^2(\gamma_n^2 - \gamma_K^2)}{\sum_{n=K}^{\infty} a_n^2} \geq \gamma_K^2 \tag{31}$$

since $\gamma_n^2 > \gamma_K^2$ for all $n > K$. Only when all $a_n = 0$ for $n > K$, that is, $\phi \equiv \phi_{eK}$, will $\gamma^2 = \gamma_K^2$. In general, the approximate eigenvalue is too large, that is, $\gamma^2$ is an upper bound on $\gamma_K^2$. A suitable set of functions to use for the extremization of (27) are the corresponding eigenfunctions for the empty guide. For the LSM modes these are

$$f_n(x) = \left(\frac{\epsilon_{on}}{a}\right)^{1/2} \cos \frac{n\pi x}{a} \qquad n = 0, 1, 2, \ldots \tag{32}$$

where $\epsilon_{on} = 1$ for $n = 0$, and 2 otherwise.

*The Approximate Eigenfunctions*

In many practical cases the true eigenfunctions are very complex, and there is considerable advantage in working with a finite set of approximate eigenfunctions instead. In constructing this set of eigenfunctions, we do not have available any of the true eigenfunctions to make the $n$th approximate eigenfunction orthogonal too. However, it turns out that the $(K + 1)$st extremization of the integral by that class of functions which are orthogonal to the first $K$ approximate eigenfunctions again yields an upper bound on the true eigenvalue $\gamma_K^2$. Also, if the eigenfunctions given by (32) for the empty guide are used for the approximation, the approximate eigenfunctions are automatically orthogonal, and, hence, the orthogonality conditions originally imposed on the function used for the $(K + 1)$st extremization may be dispensed with. Both these statements will be proved, but first we accept the results and construct a set of $N + 1$ approximate eigenfunctions. The approximate eigenfunctions and eigenvalues will be written as $\bar{\phi}_{en}$ and $\bar{\gamma}_n^2$.

For the $(K + 1)$st approximate eigenfunction, we choose the following series:

$$\bar{\phi}_{eK} = \sum_{n=0}^{N} a_{nK} f_n(x) \tag{33}$$

where the $a_{nK}$ are a set of coefficients to be determined, subject to the normalization condition

$$\int_0^a \kappa^{-1} \bar{\phi}_{eK}^2 \, dx = \sum_{n=0}^{N} \sum_{s=0}^{N} a_{nK} a_{sK} P_{sn} = 1 \tag{34}$$

where
$$P_{sn} = \int_0^a \kappa^{-1} f_n f_s \, dx = P_{ns} \tag{35a}$$

Substituting into (27) gives

$$\sum_{n=0}^{N} \sum_{s=0}^{N} \int_0^a \kappa^{-1} \left[ \frac{df_n}{dx} \frac{df_s}{dx} - (\kappa k_0^2 - h^2 + \bar{\gamma}_K^2) f_n f_s \right] a_{nK} a_{sK} \, dx$$
$$= \text{a stationary quantity} \tag{35b}$$

Let
$$T_{sn} = T_{ns} = \int_0^a \kappa^{-1} \left[ \frac{df_n}{dx} \frac{df_s}{dx} - (\kappa k_0^2 - h^2) f_n f_s \right] dx \tag{36}$$

and (35$b$) may be written as

$$\sum_{n=0}^{N} \sum_{s=0}^{N} a_{nK} a_{sK} (T_{sn} - \bar{\gamma}_K^2 P_{sn}) = \text{a stationary quantity} \tag{37}$$

To render (37) stationary, all the partial derivatives $\partial/\partial a_{nK}$ for $n = 0, 1, \ldots, N$ are equated to zero, and the following set of homogeneous equations are obtained:

$$\sum_{n=0}^{N} a_{nK} (T_{sn} - \bar{\gamma}_K^2 P_{sn}) = 0 \qquad s = 0, 1, 2, \ldots, N \tag{38}$$

This set of equations constitutes a matrix eigenvalue problem of the type discussed in Sec. A.3. Because of the symmetry of the $T_{sn}$ and $P_{sn}$ matrices, all the eigenvalues $\bar{\gamma}_K^2$ are real, and the eigenvectors, whose components are the coefficients $a_{nK}$, form an orthogonal set with respect to the weighting factors $P_{sn}$. The proofs of these properties are given in the Mathematical Appendix. For (38) to hold, the determinant of the coefficients must vanish. The vanishing of this determinant determines $N + 1$ roots for $\bar{\gamma}_K^2$, which are the first $N + 1$ eigenvalues. For each root, say $\bar{\gamma}_K^2$, a set of coefficients (eigenvector) $a_{nK}$ may be determined. This set of coefficients is unique when subjected to the normalization condition (34). Since the totality of eigenvectors so determined forms an orthonormal set, it is clear that there is no need to impose any orthogonality restriction on the functions used to extremize (27). The orthogonality relations which hold are

$$\sum_{n=0}^{N} \sum_{s=0}^{N} a_{nK} a_{sL} P_{sn} = \delta_{KL} \tag{39}$$

Having found the coefficients $a_{nK}$, the approximate eigenfunction is given by (33). If $N$ was chosen as infinite, the eigenfunctions determined by the above method would be identical with the set of true eigenfunctions, since the set of functions $f_n$ is complete, and hence able to represent the set $\phi_{en}$.

To prove that $\bar{\gamma}_K{}^2 \geq \gamma_K{}^2$ requires introduction of the concept of a class of functions. The complete set of true eigenfunctions $\phi_{e0}, \phi_{e1}, \ldots, \phi_{en}, \ldots$, with eigenvalues $\gamma_0{}^2, \gamma_1{}^2, \ldots, \gamma_n{}^2, \ldots$, where $n$ runs from zero to infinity, will be called the class $C$. The partial set $C_K$ consists of the functions $\phi_{en}$, with $n$ running from 0 to $K$. The set of approximate eigenfunctions $\bar{\phi}_{e0}, \ldots, \bar{\phi}_{eN}$, with eigenvalues $\bar{\gamma}_0{}^2, \ldots, \bar{\gamma}_N{}^2$, will be called the class $\bar{C}_N$. The partial set consisting of the first $K + 1$ approximate eigenfunctions will be called the class $\bar{C}_K$. We also construct one further class of functions by removing from the class $\bar{C}_N$ those functions which are orthogonal to the first $K$ true eigenfunctions $\phi_{e0}, \ldots, \phi_{e(K-1)}$, and call this the class $\bar{C}_N - C_{K-1}$.

The variational integral for $\gamma^2$ may be written as follows:

$$\gamma^2(g) = \frac{\displaystyle\int_0^a \kappa^{-1}\left[\left(\frac{dg}{dx}\right)^2 - (\kappa k_0{}^2 - h^2)g^2\right] dx}{\displaystyle\int_0^a \kappa^{-1} g^2\, dx} \tag{40}$$

Consider a function $g(x)$ in the class $\bar{C}_K - C_{K-1}$. This function may be written as

$$g = \sum_{n=0}^{K} b_n \bar{\phi}_{en} \tag{41a}$$

and must satisfy the $K$ orthogonality conditions

$$\begin{aligned} \int_0^a \kappa^{-1} g \phi_{es}\, dx &= \sum_{n=0}^{K} b_n \int_0^a \kappa^{-1} \bar{\phi}_{en} \phi_{es}\, dx \\ &= \sum_{n=0}^{K} b_n Q_{sn} = 0 \qquad s = 0, 1, \ldots, K-1 \end{aligned} \tag{41b}$$

where

$$Q_{sn} = \int_0^a \kappa^{-1} \bar{\phi}_{en} \phi_{es}\, dx$$

These orthogonality conditions ensure that $g$ is in the class $\bar{C}_K - C_{K-1}$.

For convenience, it will also be assumed that $g$ has been normalized so that

$$\int_0^a \kappa^{-1}g^2\,dx = \sum_{n=0}^{K} b_n^2 = 1 \tag{42}$$

Equations (41) and (42) are a set of $K + 1$ equations with $K + 1$ unknowns and may be solved for the coefficients $b_n$. Substituting into (40), integrating, and making use of (36), (38), and (39), we readily find that

$$\begin{aligned}\gamma^2(\bar{C}_K - C_{K-1}) = \gamma^2(g) &= \sum_{n=0}^{K} b_n^2\bar{\gamma}_n^2 \\ &= \sum_{n=0}^{K} b_n^2\bar{\gamma}_K^2 + \sum_{n=0}^{K} b_n^2(\bar{\gamma}_n^2 - \bar{\gamma}_K^2) \\ &= \bar{\gamma}_K^2 - \sum_{n=0}^{K} b_n^2(\bar{\gamma}_K^2 - \bar{\gamma}_n^2) \le \bar{\gamma}_K^2\end{aligned} \tag{43}$$

We know that any function in the class $C - C_{K-1}$, that is, any function $\sum_{n=K}^{\infty} b_n\phi_{en}$ with $b_n$ arbitrary, yields $\gamma^2 \ge \gamma_K^2$. Any function in the class $\bar{C}_K - C_{K-1}$ does not have so many degrees of freedom, i.e., is not so flexible, as a function in the class $C - C_{K-1}$, and hence cannot yield a value for $\gamma^2$ less than $\gamma_K^2$. Thus we are able to state that

$$\gamma^2(\bar{C}_K - C_{K-1}) \ge \gamma^2(C - C_{K-1})$$

and, hence, by using (43), we have

$$\gamma^2(C - C_{K-1}) \le \gamma^2(\bar{C}_K - C_{K-1}) \le \bar{\gamma}_K^2 \tag{44}$$

If the function in the class $C - C_{K-1}$ is chosen as the true eigenfunction $\phi_{eK}$, (44) gives at once $\gamma_K^2 \le \bar{\gamma}_K^2$, and, therefore, the approximate eigenvalues are approximations from above, i.e., upper bounds on the true eigenvalues.

*Proof of Completeness*

Since the machinery for proving that the set of true eigenfunctions is a complete set is now available, it seems worthwhile to give this proof here. For an arbitrary function $G(x)$ such that $dG/dx = 0$, $x = 0$, $a$, the set of eigenfunctions $\phi_{en}$ is complete provided

$$\lim_{N\to\infty} \int_0^a \kappa^{-1}\left[G(x) - \sum_{n=0}^{N} a_n\phi_{en}\right]^2 dx = 0$$

where

$$a_n = \int_0^a \kappa^{-1}G\phi_{en}\,dx$$

Since some of the first few eigenvalues $\gamma_n^2$ may be negative, we remove from $G(x)$ the part which has a component along the eigenfunctions with negative eigenvalues. (The reason for doing this will be pointed out later.) A value of $M$ may be chosen so that all $\gamma_n^2 < 0$ for $n < M$ and all $\gamma_n^2 > 0$ for $n \geq M$. Thus, consider the modified function

$$g(x) = G(x) - \sum_{n=0}^{M-1} a_n \phi_{en}$$

Also let

$$g_N(x) = \frac{1}{\Delta_N}\left(g - \sum_{n=M}^{N} a_n \phi_{en}\right)$$

where $\Delta_N$ is a normalization factor such that

$$\Delta_N^2 = \int_0^a \left(g - \sum_{n=M}^{N} a_n \phi_{en}\right)^2 \kappa^{-1}\, dx$$

From the definition of $\Delta_N$ it is seen that, provided we can prove $\Delta_N$ approaches zero as $N$ goes to infinity, then the completeness property will have been proved. Substitution of the function $g_N$ into (40) gives

$$\gamma^2(g_N) = \frac{1}{\Delta_N^2}\left[\gamma^2(g)\int_0^a \kappa^{-1} g^2\, dx - \sum_{n=M}^{N} a_n^2 \gamma_n^2\right] \tag{45}$$

after integration by parts and using the series expansion of $g$ to arrive at the second term. The function $g_N$ is eligible for the $(N + 2)$nd minimization of the integral (40), and hence $\gamma^2(g_N) \geq \gamma_{N+1}^2$. Since all $\gamma_n^2 > 0$ for $n \geq M$, we obtain from (45) the result

$$\gamma^2(g)\int_0^a \kappa^{-1} g^2\, dx \geq \Delta_N^2 \gamma_{N+1}^2 + \sum_{n=M}^{N} a_n^2 \gamma_n^2$$

The sum is a positive term, and hence

$$\Delta_N^2 \leq \frac{\gamma^2(g)\int_0^a \kappa^{-1} g^2\, dx}{\gamma_{N+1}^2} \tag{46}$$

The terms with negative eigenvalues were removed in order to leave a positive term to permit the step leading to expression (46). The numerator is independent of $N$ and $\gamma_{N+1}^2$ approaches infinity, and consequently $\Delta_N^2$ vanishes, as $N$ becomes infinite.

Since the functions $\sin(m\pi y/b)$ form a complete set also, we are able to state that, for a rectangular guide inhomogeneously filled in the manner considered here, the LSM modes form a complete set. Any arbitrary physical field having no $x$ component of magnetic field may be expanded as a series of LSM modes.

*Equations for LSE Modes*

The equations given in Sec. 6.1 for the LSE modes are valid when $\kappa$ is any arbitrary function of $x$, since these modes do not have an $x$ component of electric field. The generating function $\psi_h$ may be taken as

$$\psi_h(x,y) = \phi_h(x) \cos \frac{m\pi y}{b} \tag{47}$$

The eigenfunction $\phi_h$ is a solution of

$$\frac{d^2\phi_h}{dx^2} + [\gamma^2 - h^2 + \kappa(x)k_0^2]\phi_h = 0 \tag{48}$$

and satisfies the boundary conditions $\phi_h = 0$, $x = 0, a$. The variational integral for $\gamma^2$ is

$$\gamma^2 \int_0^a \phi_h^2\, dx = \int_0^a \left[\left(\frac{d\phi_h}{dx}\right)^2 - (\kappa k_0^2 - h^2)\phi_h^2\right] dx \tag{49}$$

The approximate eigenfunctions may be found by substituting into (49) a series of the corresponding eigenfunctions for the empty guide, i.e., a series of the functions

$$f_n(x) = \left(\frac{2}{a}\right)^{1/2} \sin \frac{n\pi x}{a} \qquad n = 1, 2, \ldots \tag{50}$$

Thus, if we take

$$\bar{\phi}_{hK} = \sum_{n=1}^{N} a_{nK} f_n \tag{51a}$$

subject to the normalization condition

$$\int_0^a \bar{\phi}_{hK}^2\, dx = \sum_{n=1}^{N} a_{nK}^2 = 1 \tag{51b}$$

the following set of homogeneous equations is obtained:

$$\sum_{n=1}^{N} a_{nK}(T_{sn} - \bar{\gamma}_K^2\, \delta_{sn}) = 0 \qquad s = 1, 2, \ldots, N \tag{52}$$

where $$T_{sn} = T_{ns} = \int_0^a \left[\frac{df_n}{dx}\frac{df_s}{dx} - (\kappa k_0^2 - h^2) f_n f_s\right] dx$$

The approximate eigenfunctions form an orthonormal set, i.e.,

$$\sum_{n=1}^{N} a_{nK} a_{nL} = \delta_{KL} \tag{53}$$

In the preceding equations, $h$ is used to represent the quantity $m\pi/b$.

*Application to Dielectric-slab-loaded Guides*

In deriving the variational expression (27) for the propagation constant $\gamma^2$, the term $\phi_e \dfrac{d}{dx}\dfrac{1}{\kappa}\dfrac{d\phi_e}{dx}$ was integrated by parts, and the integrated term $\phi_e \dfrac{1}{\kappa}\dfrac{d\phi_e}{dx}$ vanished at $x = 0$, $a$. When $\kappa$ is a discontinuous function of $x$, the boundary conditions on $\phi_e$ require that both $\phi_e$ and $\dfrac{1}{\kappa}\dfrac{d\phi_e}{dx}$ be continuous for all values of $x$, and that $d\phi_e/dx = 0$ at $x = 0$, $a$. Consequently, the integrated term will also vanish when $\kappa$ is a discontinuous

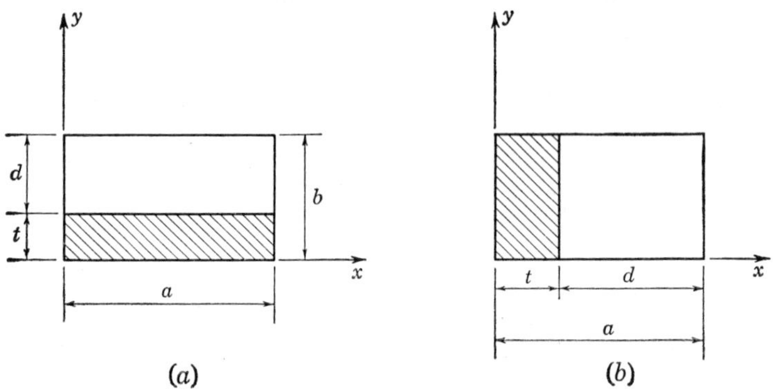


FIG. 6.7

function of $x$, and, therefore, (27) is a valid variational expression for $\gamma^2$ for this case as well. All the previously derived formulas for both LSM and LSE modes may be applied to find the approximate eigenvalues and eigenfunctions for dielectric-slab-loaded rectangular guides.

Very often, in practice, all we want to know is the propagation constant for the dominant mode. Expressions for the propagation constant for the dominant mode for rectangular guides asymmetrically filled as in Fig. 6.7 will be given below. By using a two-mode approximation, results for $\gamma$ accurate to within a few per cent are obtained in most cases.

For Fig. 6.7*a*, with variation along the $x$ direction according to $\sin(\pi x/a)$ the propagation constant $\gamma_0^2$ is given by (38) and is

$$\gamma_0^2 = \frac{Q}{2|P|} - \left(\frac{Q^2}{4|P|^2} - \frac{|T|}{|P|}\right)^{1/2} \tag{54}$$

where

$$Q = T_{11}P_{00} + T_{00}P_{11} - 2T_{01}P_{01}$$
$$|P| = P_{11}P_{00} - P_{01}^2$$
$$|T| = T_{11}T_{00} - T_{01}^2$$

The matrix elements $P_{ij}$ are given by (34) and are

$$P_{00} = \frac{1}{b}\int_0^b \kappa^{-1}\,dy = \frac{1}{b}\left(\int_0^b dy - \frac{\kappa - 1}{\kappa}\int_0^t dy\right) = 1 - \frac{\kappa - 1}{\kappa}\frac{t}{b} \tag{55a}$$

$$P_{01} = \frac{2}{b}\left(\int_0^b \cos\frac{\pi y}{b}\,dy - \frac{\kappa - 1}{\kappa}\int_0^t \cos\frac{\pi y}{b}\,dy\right) = -\frac{\kappa - 1}{\kappa}\frac{\sqrt{2}}{\pi}\sin\frac{\pi t}{b} \tag{55b}$$

$$P_{11} = \frac{2}{b}\left(\int_0^b \cos^2\frac{\pi y}{b}\,dy - \frac{\kappa - 1}{\kappa}\int_0^t \cos^2\frac{\pi y}{b}\,dy\right) = P_{00} - \frac{\kappa - 1}{2\pi\kappa}\sin\frac{2\pi t}{b} \tag{55c}$$

From (36) the elements $T_{ij}$ are found to be

$$T_{00} = \left(\frac{\pi}{a}\right)^2 P_{00} - k_0^2 \tag{56a}$$

$$T_{01} = \left(\frac{\pi}{a}\right)^2 P_{01} \tag{56b}$$

$$T_{11} = \left(\frac{\pi}{a}\right)^2 P_{11} - k_0^2 + \left(\frac{\pi}{b}\right)^2 (2P_{00} - P_{11}) \tag{56c}$$

Because of the symmetry of the dominant mode, it follows by image

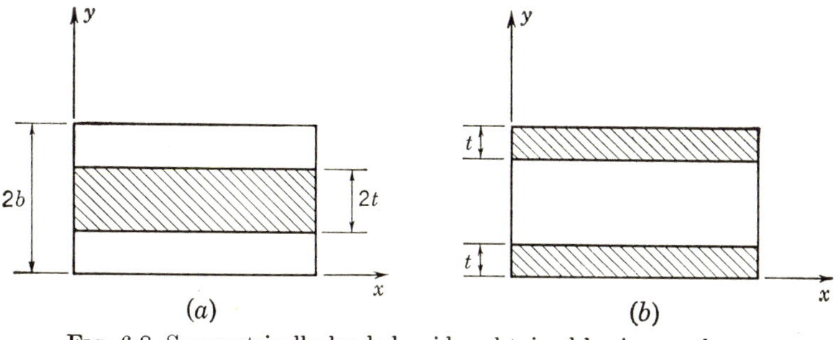


FIG. 6.8. Symmetrically loaded guides obtained by image theory.

theory that the above expression for $\gamma_0^2$ also gives the propagation constant for the dominant mode in a guide inhomogeneously filled as in Fig. 6.8$a$ and $b$.

For Fig. 6.7$b$ the propagation constant for the dominant mode ($H_{10}$ mode) is given by (52) as follows:

$$\gamma_1^2 = \frac{T_{11} + T_{22}}{2} - \left[\frac{(T_{11} + T_{22})^2}{4} - |T|\right]^{1/2} \tag{57}$$

where $|T| = T_{11}T_{22} - T_{12}^2$. The matrix elements are

$$T_{11} = \left(\frac{\pi}{a}\right)^2 - k_0{}^2 - (\kappa - 1)k_0{}^2\left(\frac{t}{a} - \frac{1}{2\pi}\sin\frac{2\pi t}{a}\right) \tag{58a}$$

$$T_{12} = (\kappa - 1)k_0{}^2\left(\frac{1}{3\pi}\sin\frac{3\pi t}{a} - \frac{1}{\pi}\sin\frac{\pi t}{a}\right) \tag{58b}$$

$$T_{22} = 4\left(\frac{\pi}{a}\right)^2 - k_0{}^2 - (\kappa - 1)k_0{}^2\left(\frac{t}{a} - \frac{1}{4\pi}\sin\frac{4\pi t}{a}\right) \tag{58c}$$

## 6.3. A Dielectric Step Discontinuity

In this section an example of how the Rayleigh-Ritz method may be employed to obtain an approximate solution for the equivalent circuit of the junction of an empty and inhomogeneously filled rectangular guide will be given. The problem to be treated is illustrated in Fig. 6.9. For

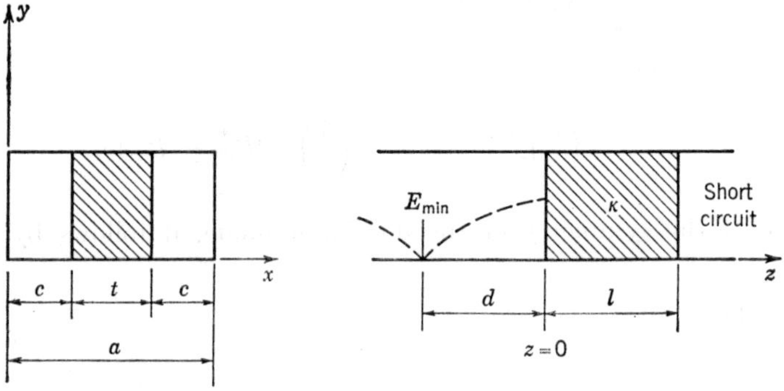


FIG. 6.9. A dielectric step discontinuity.

$z > 0$, the guide is symmetrically loaded with a dielectric slab of thickness $t$ and dielectric constant $\kappa$. An $H_{10}$ mode is assumed incident on the step from the region $z < 0$. From symmetry considerations it is apparent that only those modes which are symmetrical about $x = a/2$ will be excited by the discontinuity. It will be assumed that only the dominant mode propagates. The only modes present are the $H_{n0}$ modes (degenerate LSE modes) with $n$ odd. The field components present are $E_y$, $H_x$, and $H_z$. The electric field $E_y$ is proportional to the eigenfunctions $\phi_h(x)e^{-\gamma z}$, while $H_x$ is proportional to $\partial E_y/\partial z$. In order to make $E_y$ and $H_x$ continuous at $z = 0$, it is thus necessary to make $E_y$ and $\partial E_y/\partial z$ continuous at $z = 0$.

As a first step, we have to determine a set of approximate eigenfunctions for the region $z > 0$. Only two modes will be taken into account, in order to keep the analysis relatively simple. The approximate eigenfunctions are taken as

$$\phi_1 e^{\pm\gamma_1 z} = e^{\pm\gamma_1 z}(a_{11}f_1 + a_{31}f_3) \tag{59a}$$
$$\phi_3 e^{\pm\gamma_3 z} = e^{\pm\gamma_3 z}(a_{13}f_1 + a_{33}f_3) \tag{59b}$$

where $f_n = (2/a)^{1/2} \sin (n\pi x/a)$. The eigenvector components $a_{ij}$ and propagation constants $\gamma_j$ are determined by a solution of the following matrix eigenvalue system:

$$\sum_{n=1,3} a_{nj}(T_{nj} - \gamma^2\delta_{nj}) = 0 \qquad j = 1, 3 \tag{60}$$

subject to the normalization conditions

$$\sum_{1,3} a_{nj}^2 = a_{1j}^2 + a_{3j}^2 = 1 \qquad j = 1, 3$$

The matrix elements $T_{nj}$ are given by

$$T_{11} = \left(\frac{\pi}{a}\right)^2 - k_0^2 - (\kappa - 1)k_0^2\left(\frac{t}{a} + \frac{1}{\pi}\sin\frac{\pi t}{a}\right) \tag{61a}$$
$$T_{13} = (\kappa - 1)k_0^2\left(\frac{1}{\pi}\sin\frac{\pi t}{a} + \frac{1}{2\pi}\sin\frac{2\pi t}{a}\right) \tag{61b}$$
$$T_{33} = 9\left(\frac{\pi}{a}\right)^2 - k_0^2 - (\kappa - 1)k_0^2\left(\frac{t}{a} + \frac{1}{3\pi}\sin\frac{3\pi t}{a}\right) \tag{61c}$$

The propagation constants $\gamma_j^2$ are the roots of the determinant of the coefficients $a_{nj}$ in (60).

The two modes for the region $z < 0$ corresponding to (59) are

$$f_1 e^{\pm\Gamma_1 z} \tag{62a}$$
$$f_3 e^{\pm\Gamma_3 z} \tag{62b}$$

where $\Gamma_n^2 = (n\pi/a)^2 - k_0^2$. The electric field in the two regions $z > 0$ and $z < 0$ may be written as a series expansion in terms of the functions given in (59) and (62). If the output guide is assumed to extend to infinity, the amplitude coefficients and reflection and transmission coefficients are complex. To avoid this complication in the subsequent numerical work, it is preferable to terminate the output guide at $z = l$ by a short circuit. All amplitude coefficients now turn out to be real quantities.

A suitable expansion of the electric field in the two regions $z < 0$ and $z > 0$ is

$$E_y = \begin{cases} A_1 \sin |\Gamma_1|(z + d)f_1(x) + A_3 f_3(x)e^{\Gamma_3 z} & z < 0 \quad (63a) \\ B_1 \sin |\gamma_1|(z - l)\phi_1(x) + B_3\phi_3(x)e^{-\gamma_3 z} & z > 0 \quad (63b) \end{cases}$$

where $d$ is seen to correspond to a position where the electric field is zero in the region $z < 0$. It is assumed that $l$ is chosen large enough so that the mode $\phi_3 e^{-\gamma_3 z}$ is negligible at $z = l$. If this is not the case, $e^{-\gamma_3 z}$ must be replaced by $\sinh \gamma_3(z - l)$. Equations (63) are, of course, only an approximation, since, in general, an infinite number of modes are present. However, this simple two-term approximation does yield useful results in practical cases. To determine the amplitude coefficients $A_i$ and $B_i$, $E_y$ and $\partial E_y/\partial z$ are made continuous at $z = 0$. These boundary conditions yield the two following equations:

$$\begin{aligned} A_1 f_1 \sin |\Gamma_1| d + A_3 f_3 &= -B_1 \phi_1 \sin |\gamma_1| l + B_3 \phi_3 \\ |\Gamma_1| A_1 f_1 \cos |\Gamma_1| d + A_3 \Gamma_3 f_3 &= |\gamma_1| B_1 \phi_1 \cos |\gamma_1| l - \gamma_3 B_3 \phi_3 \end{aligned}$$

By substituting for $\phi_i$ from (59), and letting $A_1 \cos |\Gamma_1| d = A_1'$, and $B_1 \cos |\gamma_1| l = B_1'$, the coefficients of $f_1$ and $f_3$ may be equated, and we get

$$A_1' \tan |\Gamma_1| d + B_1' a_{11} \tan |\gamma_1| l - a_{13} B_3 = 0 \tag{64a}$$
$$A_3 + B_1' a_{31} \tan |\gamma_1| l - a_{33} B_3 = 0 \tag{64b}$$
$$|\Gamma_1| A_1' - |\gamma_1| a_{11} B_1' + \gamma_3 a_{13} B_3 = 0 \tag{64c}$$
$$\Gamma_3 A_3 - |\gamma_1| a_{31} B_1' + \gamma_3 a_{33} B_3 = 0 \tag{64d}$$

For a solution, the determinant of the coefficients must vanish, and this gives

$$\tan |\Gamma_1| d = \frac{A + B \tan |\gamma_1| l}{C + D \tan |\gamma_1| l} \tag{65}$$

where

$$\begin{aligned} A &= -|\Gamma_1 \gamma_1| a_{13} a_{31} \\ B &= |\Gamma_1| [\gamma_3 a_{11} a_{33} + \Gamma_3 (a_{11} a_{33} - a_{13} a_{31})] \\ C &= |\gamma_1| [\gamma_3 (a_{31} a_{13} - a_{11} a_{33}) - \Gamma_3 a_{11} a_{33}] \\ D &= \Gamma_3 \gamma_3 a_{13} a_{31} \end{aligned}$$

An examination of these latter equations shows that (65) is independent of any normalization condition on the $a_{ij}$, and thus we may take $a_{11} = a_{33} = 1$. Since the approximate eigenfunctions are orthogonal, $a_{13} = -a_{31}$, and hence the following simplified equations are obtained:

$$A = |\Gamma_1 \gamma_1| a_{13}^2 \tag{66a}$$
$$B = |\Gamma_1| [\gamma_3 + \Gamma_3 (1 + a_{13}^2)] \tag{66b}$$
$$C = -|\gamma_1| [\gamma_3 (1 + a_{13}^2) + \Gamma_3] \tag{66c}$$
$$D = -\Gamma_3 \gamma_3 a_{13}^2 \tag{66d}$$

From the coefficients $A$, $B$, $C$, $D$, a variety of equivalent circuits may be determined. Some of these are given at the end of Chap. 5. For the problem considered here, an equivalent circuit of the form given in Fig. 6.10 is chosen. The normalized characteristic impedance of the output line is chosen equal to $\Gamma_1/\gamma_1$, which is equal to the ratio of the wave impedances of the dominant modes on the input and output sides. From Chap.

5, the equivalent-circuit parameters are given by

$$X_1 = -\frac{A}{C}$$

$$X_2 = \frac{A}{C} - \frac{B}{D}$$

$$n^2 Z_1 = \frac{AD}{C^2} - \frac{B}{C}$$

The equivalent-circuit parameters have been computed for $k_0^2 = 4$,

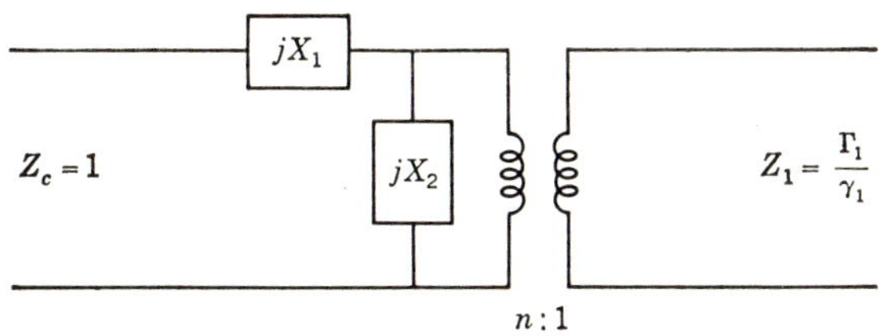


FIG. 6.10. Equivalent circuit for dielectric step.

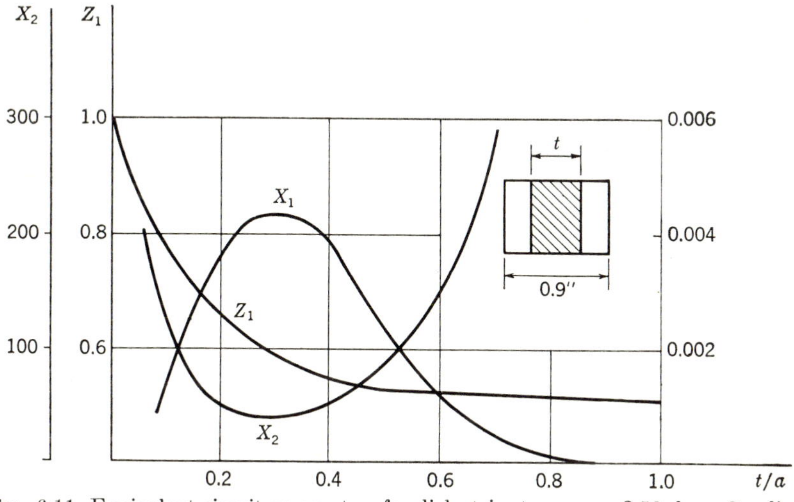


FIG. 6.11. Equivalent-circuit parameters for dielectric step. $\kappa = 2.56$, $k_0 = 2$ radians per centimeter.

$\kappa = 2.56$, $a = 0.90$ inch, and $t$ ranging from 0 to $a$, and are plotted as a function of $t/a$ in Fig. 6.11. The turns ratio $n:1$ is found to be equal to unity for all values of $t$. The series reactance $X_1$ is very small, while the shunt reactance $X_2$ is large. For all practical purposes, the equivalent circuit is just a junction of two transmission lines, with characteristic impedances proportional to the wave impedances of the $H_{10}$ mode in the two regions.

## 6.4. Ferrite Slabs in Rectangular Guides

The general theory of wave propagation in waveguides inhomogeneously filled with anisotropic ferrite material is very complex, and no attempt will be made to present such a theory here. There are special cases, however, which have considerable practical importance, and at the same time are quite readily analyzed. One particular case which has received considerable study is illustrated in Fig. 6.12. The static magnetizing field $B_0$ is applied along the $y$ axis, and hence the permeability dyadic is of the following form:

$$\bar{\bar{\mathbf{u}}} = \mathbf{a}_x\mathbf{a}_x\mu - \mathbf{a}_x\mathbf{a}_z j\kappa + \mathbf{a}_y\mathbf{a}_y\mu_0 + \mathbf{a}_z\mathbf{a}_x j\kappa + \mathbf{a}_z\mathbf{a}_z\mu \tag{67}$$

The dielectric permittivity of the ferrite is $\epsilon$, and will be written as $\epsilon$ throughout this section. The Greek letter $\kappa$ is reserved for representing the off-diagonal components in the permeability dyadic. Any losses which are present can be taken into account by making $\mu$, $\kappa$, and $\epsilon$ complex.

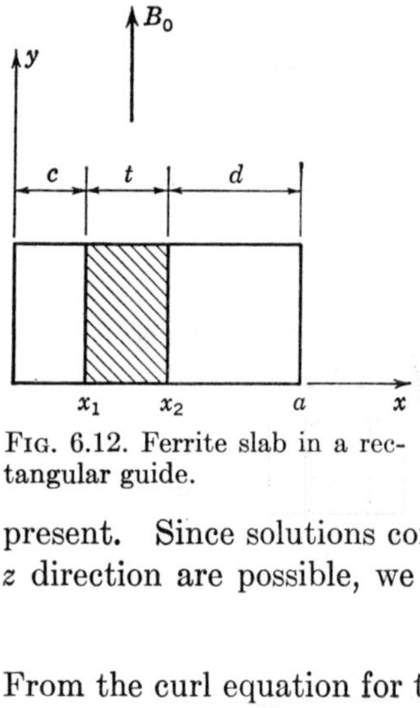


FIG. 6.12. Ferrite slab in a rectangular guide.

The off-diagonal terms in $\bar{\bar{\mathbf{u}}}$ result in a coupling between the field components $H_x$ and $H_z$. This coupling is independent of $y$, and the structure is uniform along the $y$ direction, and, consequently, a solution for modes of the $H_{n0}$ type may be found. These modes may be derived from the single electric field component $E_y$ which is present. Since solutions corresponding to waves propagating along the $z$ direction are possible, we may take

$$E_y = \psi(x)e^{-\gamma z}$$

From the curl equation for the electric field, it is then found that, in the ferrite slab,

$$\mathbf{a}_y \times \nabla[\psi(x)e^{-\gamma z}] = j\omega\bar{\bar{\mathbf{u}}}\cdot\mathbf{H} \tag{68}$$

From this equation we obtain

$$j\omega\mu H_x + \kappa\omega H_z = -\gamma\psi e^{-\gamma z} \tag{69a}$$

$$-\kappa\omega H_x + j\omega\mu H_z = -\frac{d\psi}{dx}e^{-\gamma z} \tag{69b}$$

Solving for $H_x$ and $H_z$ gives

$$H_x = \frac{j\omega\mu\gamma\psi - \kappa\omega(d\psi/dx)}{\omega^2(\mu^2 - \kappa^2)}e^{-\gamma z} \tag{70a}$$

$$H_z = \frac{\gamma\kappa\omega\psi + j\omega\mu(d\psi/dx)}{\omega^2(\mu^2 - \kappa^2)}e^{-\gamma z} \tag{70b}$$

To obtain the equation satisfied by $\psi$ in the ferrite medium, scalar-multiply (68) by the transposed dyadic $\bar{\mathbf{\mu}}_t$ to get

$$\bar{\mathbf{\mu}}_t \cdot \mathbf{a}_y \times \nabla(\psi e^{-\gamma z}) = j\omega\bar{\mathbf{\mu}}_t \cdot \bar{\mathbf{\mu}} \cdot \mathbf{H} = j\omega(\mu^2 - \kappa^2)\mathbf{H}$$

since $\mathbf{H}$ has no $y$ component. Taking the curl of both sides and substituting $\mathbf{a}_y j\omega\epsilon\psi e^{-\gamma z}$ for $\nabla \times \mathbf{H}$ gives

$$\nabla \times [\bar{\mathbf{\mu}}_t \cdot \mathbf{a}_y \times \nabla(\psi e^{-\gamma z})] = -\mathbf{a}_y\omega^2\epsilon(\mu^2 - \kappa^2)\psi e^{-\gamma z}$$

Expansion of the left-hand side of this equation yields the final result, which is

$$\frac{d^2\psi}{dx^2} + \left(\gamma^2 + \frac{\mu^2 - \kappa^2}{\mu}\omega^2\epsilon\right)\psi = 0 \tag{71}$$

The parameter $(\mu^2 - \kappa^2)/\mu$ will be called the effective permeability, and will be written as $\mu_e$. In the empty part of the guide, the solution is obtained by replacing $\mu$, $\epsilon$ by $\mu_0$, $\epsilon_0$ and putting $\kappa = 0$ in (70) and (71).

A general solution for $\psi$, which vanishes at $x = 0, a$, is readily found to be

$$\psi = \begin{cases} D \sin hx & 0 \le x \le x_1 \\ A \sin l(x - x_1) + B \sin l(x - x_2) & x_1 \le x \le x_2 \\ C \sin h(a - x) & x_2 \le x \le a \end{cases} \tag{72}$$

The particular form for $\psi$ in the region $x_1 \le x \le x_2$ was chosen to facilitate matching the tangential fields at $x = x_1, x_2$. In order for (72) to be a solution of (71) and the corresponding equation for the empty guide, the following restrictions on the transverse wave numbers $h$ and $l$ must be imposed:

$$\gamma^2 = h^2 - k_0{}^2 = l^2 - k^2 \tag{73}$$

where $k^2 = \omega^2\mu_e\epsilon$.

At $x = x_1, x_2, E_y$ and hence $\psi$ must be continuous; therefore

$$D \sin hc = -B \sin lt \tag{74a}$$

$$C \sin hd = A \sin lt \tag{74b}$$

Also, at $x = x_1, x_2$, the tangential magnetic field component $H_z$ must be continuous. Using (70$b$) and (72) the two following equations are obtained:

$$-Dh \cos hc = -\frac{\mu_0}{\mu^2 - \kappa^2}(j\gamma\kappa B \sin lt + \mu lB \cos lt + \mu lA) \tag{75a}$$

$$-hC \cos hd = \frac{\mu_0}{\mu^2 - \kappa^2}(-j\gamma\kappa A \sin lt + \mu lA \cos lt + \mu lB) \tag{75b}$$

A solution for the coefficients $A, B, C, D$ exists only if the determinant of the set of equations (74) and (75) vanishes. The resulting equation is the eigenvalue equation which, together with (73), determines the wave numbers $l$, $h$ and the propagation factor $\gamma$. The eigenvalue equation is easily set up by substituting for $B$ and $C$ from (74) into (75), and is

$$h \cot hc \left( \frac{\mu_0}{\mu_e} l \cot lt - \frac{j\gamma\kappa\mu_0}{\mu\mu_e} \right) + h \cot hd \left( \frac{\mu_0}{\mu_e} l \cot lt + \frac{\mu_0}{\mu\mu_e} j\gamma\kappa \right)$$
$$+ h^2 \cot hc \cot hd + \left( \frac{\mu_0}{\mu_e} \right)^2 \left( \frac{\gamma^2\kappa^2}{\mu^2} - l^2 \right) = 0 \quad (76)$$

For the special case when $c = 0$, (76) reduces to

$$\frac{\mu_0}{\mu_e} l \cot lt - \frac{j\gamma\kappa\mu_0}{\mu\mu_e} + h \cot hd = 0 \tag{77a}$$

and, when $d = 0$, it becomes

$$\frac{\mu_0}{\mu_e} l \cot lt + \frac{j\gamma\kappa\mu_0}{\mu\mu_e} + h \cot hc = 0 \tag{77b}$$

These two eigenvalue equations are not the same and show that the propagation factor in the positive $z$ direction will be different for a ferrite slab placed along the guide wall at $x = 0$, as compared with the propagation factor obtained with the slab against the guide wall at $x = a$. For propagation in the negative $z$ direction, $\gamma$ is replaced by $-\gamma$ in (76) and (77). This has the effect of interchanging the two equations (77$a$) and (77$b$). This nonreciprocal propagation behavior may be understood in a qualitative way from a consideration of the properties of the magnetic field for an $H_{10}$ mode in an empty guide. In the empty guide the magnetic field for the $H_{10}$ mode is given by

$$j\omega\mu_0 H_x = \pm\Gamma \sin \frac{\pi x}{a} e^{\pm\Gamma z}$$

$$j\omega\mu_0 H_z = -\frac{\pi}{a} \cos \frac{\pi x}{a} e^{\pm\Gamma z}$$

where $$\Gamma^2 = \left(\frac{\pi}{a}\right)^2 - k_0^2$$

Since $\Gamma$ is imaginary, the magnetic field is circularly polarized, but of opposite sense, at the two positions where $|H_x| = |H_z|$. These positions are determined by a solution of the following equation:

$$\sin \frac{\pi x}{a} = \frac{\lambda_0}{2a}$$

Since the transverse field $H_x$ changes sign when the direction of propagation is reversed, the positions where the total magnetic field is positive and negative circularly polarized are interchanged. If a ferrite slab is placed in a circularly polarized magnetic field, its effective permeability is different for the two circular polarizations, and hence the perturbation of the field by the ferrite slab depends on the direction of propagation.

These effects exist when the magnetic field is elliptically polarized as well. The nonreciprocal propagation is brought about because of the reversal in sign of $H_x$, but not $H_z$, when the direction of propagation is changed. A reversal in direction of the applied static field $B_0$ has the same effect as changing the direction of propagation or changing the position of the ferrite slab to the other side of the center position $x = a/2$. A guide symmetrically loaded with ferrite slabs, and with the applied static field in the same direction for all slabs, will not exhibit a nonreciprocal propagation property.

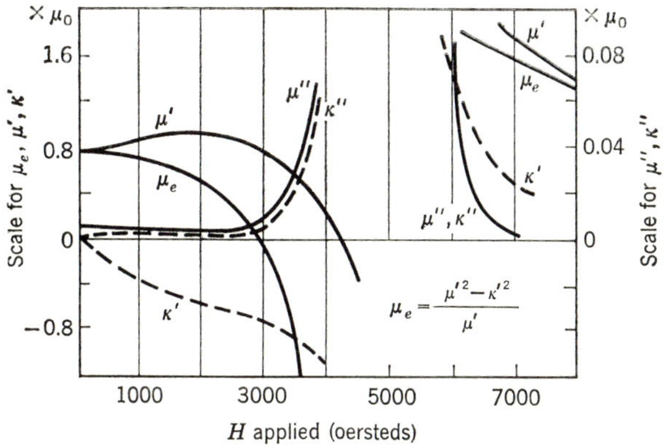


FIG. 6.13. Permeability components for Ferramic $R_1$ at 9,000 megacycles, $\epsilon = (11.9 - j4 \times 10^{-4})\epsilon_0$; the double-primed parameters arise because of the losses in the material. (*Adopted by permission from C. E. Fay, Ferrite Tuned Cavities, Proc. IRE, vol. 44, p. 1446, fig. 1, October, 1956.*)

The particular case of a ferrite slab placed along one wall, i.e., $c = 0$, will now be examined in greater detail. In order to obtain numerical values for the various parameters of interest, a knowledge of the intrinsic parameters $\mu$, $\kappa$, and $\epsilon$ for the ferrite is required. Typical measured values are illustrated in Fig. 6.13. For a frequency well above the resonant frequency of the magnetization (weak applied static field), the losses are negligible. We may choose a value of applied static field such that the effective permeability $\mu_e$ is positive, zero, or even negative. Thus the product $\mu_e\epsilon$ may be chosen greater or smaller than the free-space value $\mu_0\epsilon_0$.

The wave numbers $l$ and $h$ are solutions of the two equations (73) and (77$a$). For numerical evaluation of these parameters it is convenient to introduce the two following functions:

$$f_1(\beta) = l \cot lt + \frac{\mu_e}{\mu_0} h \cot hd \tag{78a}$$

$$f_2(\beta) = \frac{-\beta\kappa}{\mu} \tag{78b}$$

where $\gamma$ has been placed equal to $j\beta$. When $f_1 = f_2$, (77a) is satisfied. By choosing a value of $\beta$, both $l$ and $h$ are determined, and $f_1$ and $f_2$ may be evaluated. A plot of $f_1$ and $f_2$ as a function of $\beta$ for two values of $t$ is given in Fig. 6.14. The assumed parameters for this example are $\mu' = \mu = 0.8\mu_0$, $\kappa' = \kappa = 0.6\mu_0$, $\epsilon = 10\epsilon_0$, $k_0 = 2$, $a = 2.286$ centimeters, and the losses are neglected so that $\mu'' = \kappa'' = 0$.

The intersection of the curves $f_1 = \pm f_2$ determines the propagation constants for the forward- and backward-propagating dominant modes.

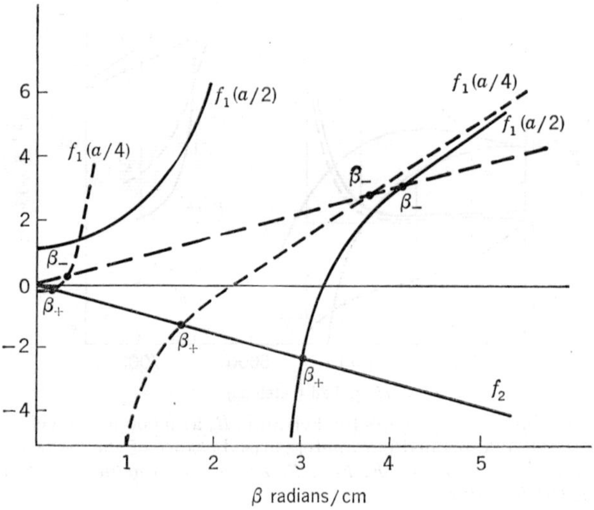


FIG. 6.14. Plot of $f_1$ and $f_2$ for finding the propagation constants for a ferrite-loaded rectangular guide.

These are designated as $\beta+$ and $\beta-$ in Fig. 6.14. For large values of $t$, only a single forward-and-backward-propagating mode exists. For small values of $t$, we find that two forward-and-backward-propagating modes may exist. The modes for $\beta > 3.72$ have both $l$ and $h$ imaginary, and hence have transverse field components which are described by the hyperbolic functions in both the empty and the filled portions of the guide. These modes bear a close resemblance to the surface waves, which may be propagated along the surface of a thin ferrite sheet backed by a conducting plane and located in free space. The modes for $\beta < 3.72$ are the normal perturbed waveguide modes.

Equation (77a) does not have a solution for real values of $\gamma$. Consequently, all the higher-order modes have complex propagation constants. When $\gamma$ is complex, the transverse wave numbers $l$ and $h$ are also complex, and this makes the numerical evaluation of the eigenvalues very laborious to carry out.

*A Variational Formulation*

A general study of a rectangular guide partially filled with a transversely magnetized slab is facilitated if it is assumed that the parameters $\mu$, $\kappa$, and $\epsilon$ are all continuous functions of $x$. The resulting differential equation for the transverse electric field is considerably more complicated in this case, since it must take into account the behavior of the field at the air-ferrite interface, if such an interface exists. The derivation of the required equation may proceed directly from the field equations provided we take $\mu$, $\kappa$, and $\epsilon$ as functions of $x$. The following alternative procedure leads to the desired end result in a somewhat simpler manner.

For any propagating mode in a lossless guide, the time-average electric and magnetic energy densities are equal. By minimizing the integral of the energy function $W_m - W_e$ over the guide cross section, we are led to a variational integral formulation for the problem. The condition imposed on the minimizing function, in order that it shall render the integral of $W_m - W_e$ stationary, is that it be a solution of the required differential equation that we are seeking.

Let the transverse electric field in the guide be represented by $E_y = \psi(x)e^{-\gamma z}$. The magnetic field and magnetic flux density components are given by

$$j\omega B_x = -\gamma\psi e^{-\gamma z} \qquad j\omega B_z = -\psi' e^{-\gamma z}$$
$$\omega^2(\mu^2 - \kappa^2)H_x = (j\omega\mu\gamma\psi - \kappa\omega\psi')e^{-\gamma z}$$
$$\omega^2(\mu^2 - \kappa^2)H_z = (\gamma\kappa\omega\psi + j\omega\mu\psi')e^{-\gamma z}$$

where $\psi' = d\psi/dx$. The time-average magnetic energy density is

$$W_m = \tfrac{1}{4}\mathbf{B} \cdot \mathbf{H}^* = \frac{\beta^2\mu\psi^2 + 2\kappa\beta\psi\psi' + \mu(\psi')^2}{4\omega^2(\mu^2 - \kappa^2)}$$

where $\gamma$ has been replaced by $j\beta$. The time-average electric energy density is $W_e = (\epsilon/4)\psi^2$. Forming the difference function $4(W_m - W_e)$, introducing the effective permeability $\mu_e = (\mu^2 - \kappa^2)/\mu$, and integrating over $x$, we get

$$I = \int_0^a (\omega^2\mu_e)^{-1} \left[\beta^2\psi^2 + \frac{2\kappa}{\mu}\beta\psi\psi' + (\psi')^2 - \omega^2\mu_e\epsilon\psi^2\right] dx \qquad (79)$$

We will now impose on the function $\psi$ the condition that it shall render the integral $I$ stationary.

Let $\psi$ vary by a small amount $\delta\psi$. The variation $\delta I$ in $I$ becomes

$$\delta I = 2\int_0^a (\omega^2\mu_e)^{-1} \left[(\beta^2 - \omega^2\mu_e\epsilon)\psi\,\delta\psi + \psi'\,\delta\psi' + \frac{\kappa}{\mu}\beta\psi\,\delta\psi' + \frac{\kappa}{\mu}\beta\psi'\,\delta\psi\right] dx$$

Consider the term

$$2\int_0^a (\omega^2\mu_e)^{-1}\left(\psi' + \frac{\kappa}{\mu}\beta\psi\right)\delta\psi'\,dx = 2(\omega^2\mu_e)^{-1}\left(\psi' + \frac{\kappa}{\mu}\beta\psi\right)\delta\psi\Big|_0^a - 2\int_0^a\left[\frac{d}{dx}\frac{\psi' + (\kappa/\mu)\beta\psi}{\omega^2\mu_e}\right]\delta\psi\,dx$$

upon integrating by parts once. The integrated term is proportional to $H_z$ times $\delta\psi$. Since $H_z$ is continuous, this term will vanish, provided $\delta\psi$ is continuous and vanishes at $x = 0, a$. Using this latter result in the expression for $\delta I$ gives

$$\delta I = 2\int_0^a \delta\psi\left[\frac{\kappa}{\omega^2\mu\mu_e}\beta\psi' - \frac{d}{dx}\frac{\psi' + \beta\psi\kappa/\mu}{\omega^2\mu_e} + \frac{(\beta^2 - \omega^2\mu_e\epsilon)\psi}{\omega^2\mu_e}\right]dx$$

Since $\delta\psi$ is arbitrary, the variation $\delta I$ will vanish provided

$$\frac{d}{dx}\left(\frac{\psi'}{\mu_e} + \frac{\kappa}{\mu\mu_e}\beta\psi\right) - \frac{\kappa}{\mu\mu_e}\beta\psi' + \frac{k^2 - \beta^2}{\mu_e}\psi = 0 \tag{80}$$

where $k^2 = \omega^2\mu_e\epsilon$. Equation (80) is the required equation for the function $\psi$. In the derivation we considered only propagating modes for which $\gamma\gamma^* = +\beta^2$. However, (80) is perfectly general and holds for all the modes, and $\beta$ in this equation may be replaced by $-j\gamma$. For nonpropagating modes both $\gamma$ and $\psi$ are complex. Replacing $\beta$ by $-j\gamma$, we obtain

$$\frac{d}{dx}\left(\frac{\psi'}{\mu_e} - j\gamma\frac{\kappa}{\mu\mu_e}\psi\right) + \frac{j\gamma\kappa}{\mu\mu_e}\psi' + \frac{\gamma^2 + k^2}{\mu_e}\psi = 0 \tag{81}$$

The above equation clearly shows that, if $\psi(x)$ with an eigenvalue $\gamma$ is a solution, then $\psi(x)$ with an eigenvalue $-\gamma$ is a solution to the problem with a medium characterized by the transposed permeability dyadic. A transposition of the permeability dyadic is equivalent to changing the sign of $\kappa$ (reversing the d-c magnetizing field changes the sign of $\kappa$), and, if the sign of $\gamma$ is changed at the same time, the differential equation remains invariant, since $\gamma$ and $\kappa$ occur as a product in the differential equation. We may also readily see from the form of the equation that a solution for $\gamma$ pure-real cannot exist.

Let $L$ be the differential operator occurring in (80), such that $L\psi = 0$ represents the differential equation (80). The adjoint differential equation corresponding to the adjoint differential operator $L_a$ operating on a function $\phi$ may be defined according to the following relation:

$$\int_0^a \phi L\psi\,dx = \int_0^a \psi L_a\phi\,dx \tag{82}$$

Multiplying (80) by $\phi$ and integrating by parts once gives

$$\phi\left(\frac{\psi'}{\mu_e} + \sigma\beta\psi\right)\Big|_0^a - \int_0^a \left[\phi'\frac{\psi'}{\mu_e} + \sigma\beta(\psi\phi' + \phi\psi') - \frac{k^2 - \beta^2}{\mu_e}\psi\phi\right] dx = 0$$

where $\kappa/\mu\mu_e$ has been denoted by $\sigma$. A further integration by parts to eliminate $\psi'$ yields

$$\frac{\phi\psi' - \psi\phi'}{\mu_e}\Big|_0^a + \int_0^a \psi\left[\frac{d}{dx}\left(\frac{\phi'}{\mu_e} + \sigma\beta\phi\right) - \sigma\beta\phi' + \frac{k^2 - \beta^2}{\mu_e}\phi\right] dx = 0 \quad (83)$$

Adding and subtracting $\beta\sigma\phi\psi$ from the integrated part gives

$$\left[\phi\left(\frac{\psi'}{\mu_e} + \sigma\beta\psi\right) - \psi\left(\frac{\phi'}{\mu_e} + \sigma\beta\phi\right)\right]\Big|_0^a$$

These terms will vanish provided $\phi$ and $\phi'/\mu_e + \beta\sigma\phi$ are continuous functions of $x$ and $\phi(0) = \phi(a) = 0$. This is true for the function $\psi$, since $\psi$ is proportional to the electric field $E_y$, and $\psi'/\mu_e + \beta\sigma\psi$ is proportional to the magnetic field $H_z$. Thus we find by comparing (83) and (82) that the adjoint operator $L_a$ is equal to the operator $L$. Since $\phi$ and $\psi$ also satisfy identical boundary conditions, the differential equation (80) is said to be self-adjoint.

Because of the self-adjoint property of (80) we may obtain an orthogonality relation for the eigenfunctions $\psi_m$ of this equation. If the equation was not self-adjoint, the orthogonality relations would involve the eigenfunctions of the adjoint operator $L_a$ as well. Let $\psi_n$, $\psi_m$ be two eigenfunctions of (80) with nonidentical eigenvalues $\beta_n$, $\beta_m$. Multiplying the equation for $\psi_n$ by $\psi_m$, the equation for $\psi_m$ by $\psi_n$, subtracting, and performing an integration by parts, we get

$$\left[\psi_m\left(\frac{\psi_n'}{\mu_e} + \sigma\beta_n\psi_n\right) - \psi_n\left(\frac{\psi_m'}{\mu_e} + \sigma\beta_m\psi_m\right)\right]\Big|_0^a$$
$$- \int_0^a \left[\psi_m'\left(\frac{\psi_n'}{\mu_e} + \sigma\beta_n\psi_n\right) - \psi_n'\left(\frac{\psi_m'}{\mu_e} + \sigma\beta_m\psi_m\right) + \sigma\beta_n\psi_m\psi_n'\right.$$
$$\left. - \sigma\beta_m\psi_n\psi_m' + \frac{\beta_n^2 - \beta_m^2}{\mu_e}\psi_n\psi_m\right] dx = 0$$

The integrated part vanishes and the remaining terms may be combined to give

$$(\beta_n - \beta_m)\int_0^a \left[\sigma(\psi_n\psi_m' + \psi_m\psi_n') + (\beta_n + \beta_m)\frac{\psi_n\psi_m}{\mu_e}\right] dx = 0$$

Rearranging the terms, we may write this latter result as follows:

$$\int_0^a \left[\psi_n\left(\sigma\psi_m' + \frac{\beta_m\psi_m}{\mu_e}\right) + \psi_m\left(\sigma\psi_n' + \frac{\beta_n\psi_n}{\mu_e}\right)\right] dx = 0 \quad (84)$$

Referring to (70$a$) shows that $\sigma\psi_n' + \beta_n\psi_n/\mu_e$ is proportional to the transverse magnetic field component $H_x$ for the $n$th mode, and similarly for the other term in parentheses. If we note that the solution for the $n$th mode when the permeability dyadic is transposed is $\psi_n$, with an eigenvalue $-\beta_n$, and a transverse magnetic field proportional to $-\sigma\psi_n' - \beta_n\psi_n/\mu_e$, then (84) will be seen to be the same orthogonality property which was deduced directly from Maxwell's equations, i.e., equivalent to (15). The rather involved orthogonality relation given by (84) is a consequence of the fact that the differential equation (80) is a more general form than the Sturm-Liouville equation discussed in Sec. 6.2. It is indicative of the complex nature of electromagnetic-wave propagation in guides filled with anisotropic media.

## REFERENCES

*Guides partially filled with dielectric media*

6.1. L. Pincherle: Electromagnetic Waves in Metal Tubes Filled Longitudinally with Two Dielectrics, *Phys. Rev.*, vol. 66, pp. 118–130, September, 1944.

6.2. L. G. Chambers: Propagation in Waveguides Filled Longitudinally with Two or More Dielectrics, *Brit. J. Appl. Phys.*, vol. 4, pp. 39–45, February, 1953. (This is a review article containing 16 references.)

*Variational techniques and the Rayleigh-Ritz method*

6.3. L. G. Chambers: An Approximate Method for the Calculation of Propagation Constants for Inhomogeneously Filled Waveguides, *Quart. J. Mech. and Appl. Math.*, vol. 7, pp. 299–316, September, 1954.

6.4. R. E. Collin and R. M. Vaillancourt: Application of Rayleigh-Ritz Method to Dielectric Steps in Waveguides, *IRE Trans.*, vol. MTT-5, pp. 177–184, July, 1957.

6.5. A. D. Berk: Variational Principles for Electromagnetic Resonators and Waveguides, *IRE Trans.*, vol. AP-4, pp. 104–111, April, 1956.

The literature on propagation in ferrite-filled guides is extensive. A representative series of papers are listed below:

6.6. A. A. Th. M. Van Trier: Guided Electromagnetic Waves in Anisotropic Media, *Appl. Sci. Research*, vol. 3B, pp. 305–371, 1953.

6.7. H. Suhl and L. R. Walker: Topics in Guided Wave Propagation through Gyromagnetic Media, *Bell System Tech. J.*, vol. 33, pp. 579–660, 939–986, 1133–1144, May, July, September, 1954.

6.8. M. L. Kales: Modes in Waveguides Containing Ferrites, *J. Appl. Phys.*, vol. 24, pp. 604–608, May, 1953.

6.9. Ferrites Issue, *Proc. IRE*, vol. 44, October, 1956.

6.10. A. D. Bresler and N. Marcuvitz: Operator Methods in Electromagnetic Field Theory, part I, *Polytech. Inst. Brooklyn Research Rept. R*-495-56, May, 1956; part II, *Polytech. Inst. Brooklyn Research Rept. R*-565-57, March, 1957. (These reports present a very general treatment of guided waves in anisotropic media based on an abstract operator formalism.)

## PROBLEMS

**6.1.** Show that the LSM and LSE modes are a linear combination of the $E$ and $H$ modes.

**6.2.** Derive the eigenvalue equations for the symmetrical and unsymmetrical modes (both LSM and LSE modes) for rectangular guides inhomogeneously filled with lossless dielectric material as illustrated.

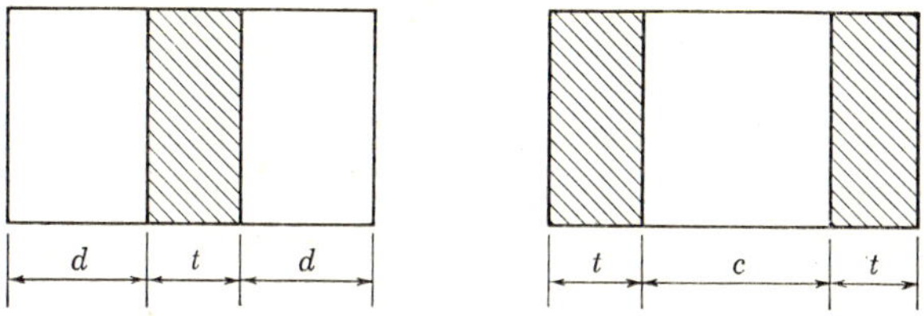


FIG. P 6.2

**6.3.** Use a two-term approximation in the Rayleigh-Ritz method to compute an approximate value for the propagation constant $\gamma_1$ for the dominant mode in a guide loaded with a centered dielectric slab as in Prob. 6.2. Compare the approximate eigenvalue with the true eigenvalue. Note that, since $\gamma_1$ is imaginary, the absolute value of the approximate eigenvalue will be too small. Evaluate the approximate normalized eigenfunction, and compare it with the true eigenfunction by plotting them as functions of $x$. Carry out the computation only for the LSE mode and take $a = 0.9$ inch, $t = 0.3a$, $\kappa = 2.56$, $\lambda_0 = 3.14$ centimeters.

**6.4.** Show that Eq. (27) is a stationary expression for $\gamma^2$ by evaluating the variation $\delta\gamma$ and also by applying the Euler-Lagrange equations from the calculus of variations.

**6.5.** Give a detailed derivation of Eqs. (43) and (45).

**6.6.** Show that the variational expressions for the eigenvalues $\gamma^2$ for the LSE and LSM modes is equivalent to minimizing the integral over the guide cross section of the energy function $4(W_m - W_e) = \mu_0\mathbf{H} \cdot \mathbf{H}^* - \kappa x \epsilon_0 \mathbf{E} \cdot \mathbf{E}^*$.

HINT: Express $W_m - W_e$ in terms of the Hertzian potential function.

**6.7.** Use a two-mode approximation in the Rayleigh-Ritz method to compute the equivalent-circuit parameters of the asymmetrical dielectric step in a rectangular guide. Assume that an $H_{10}$ mode is incident and that $a = 0.9$ inch, $t = 0.3a$, $\kappa = 2.56$, $\lambda_0 = 3.14$ centimeter, and find $X_1$, $X_2$, $n^2Z_1$.

**6.8.** A circular guide of radius $b$ is inhomogeneously filled with a dielectric cylinder of radius $t$, where $t < b$. The relative dielectric constant is $\kappa$. Obtain the solutions for circularly symmetric $E$ and $H$ modes.

**6.9.** For the structure of Prob. 6.8, show that all modes which have angular variation are not pure $E$ or $H$ modes but a linear combination of the two. Obtain the eigenvalue equation for these hybrid modes.

**6.10.** Derive suitable variational expressions for the propagation constants of the circularly symmetric $E$ and $H$ modes for the inhomogeneously filled circular guide of Prob. 6.8.

**6.11.** A rectangular guide is completely filled with a ferrite medium. The d-c magnetizing field is applied along the longitudinal axis. Determine the possible modes of propagation in the guide.

CHAPTER 7

# EXCITATION OF WAVEGUIDES

In any wave-guiding structure, the field must be generated or excited by a suitable source. In a cylindrical guide the field is generally excited by means of an antenna (probe or loop type), which in turn is driven by a generator. Coupling of fields from one guide to another may be accomplished by means of small antennas also, but is more often done by means of small radiating apertures located at appropriate positions in the common wall separating the two guides. A completely rigorous solution of the antenna boundary-value problem in a waveguide is just as difficult to obtain as the solution for antennas located in free space. Nevertheless, by making suitable approximations, solutions which shed considerable light on the behavior of antennas in waveguides, as well as providing useful engineering results, can be obtained for many practical cases.

In this chapter attention will be confined to the small coaxial-line probe and loop antennas in a rectangular guide. In addition, the problem of coupling by small apertures will be examined. Finally, in the last part of the chapter, the important problem of the transient behavior of a guide will be considered briefly.

## 7.1. Probe Antenna

The type of coaxial-line probe antenna to be analyzed is illustrated in Fig. 7.1. It consists of a small coaxial line, terminated in the center of

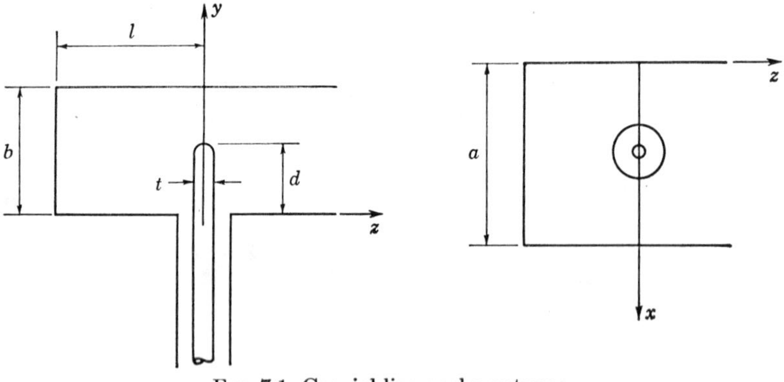


FIG. 7.1. Coaxial-line probe antenna.

the broad face of a rectangular guide, with its inner conductor extending a distance $d$ into the guide. In order to have the antenna radiate in one direction only, a short-circuiting plunger is placed at a distance $l$ to the left of the probe. The analysis to be presented will show that, by proper adjustment of the parameters $l$ and $d$, substantially all the power incident along the coaxial line can be coupled into the waveguide.

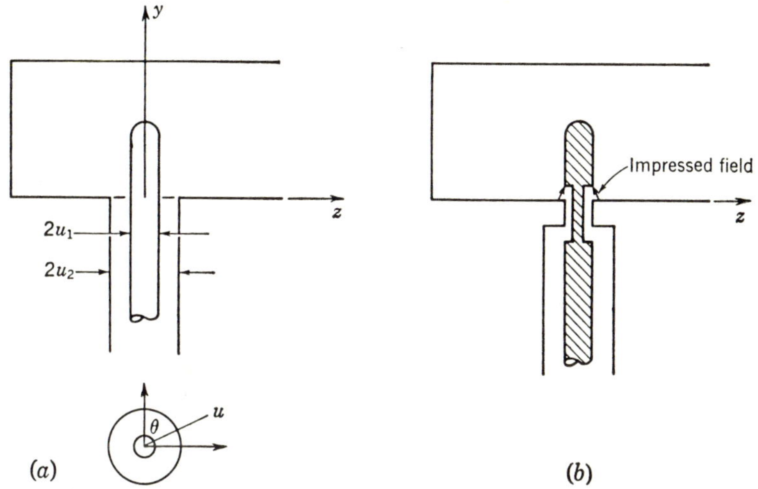


FIG. 7.2. (a) Coaxial-line probe, and (b) its approximate equivalent.

According to Love's field equivalence principle,† the total field in the waveguide may be found from an equivalent magnetic current $\mathbf{J}_m = -\mathbf{n} \times \mathbf{E}_a$ and an equivalent electric current $\mathbf{J} = \mathbf{n} \times \mathbf{H}_a$ located in the coaxial-line aperture $S_a$, where $\mathbf{E}_a$ and $\mathbf{H}_a$ are the total tangential fields in the aperture. Unfortunately, these aperture fields are unknown, but, if the coaxial-line opening is small, we may, to a first approximation, assume that the higher-order coaxial-line modes which are excited are negligible. The field in the aperture will then be that associated with the incident and reflected TEM modes in the coaxial line. If $V$ is the total voltage (incident plus reflected voltage waves) at the aperture, and $I_t$ is the total probe current at this plane the aperture fields are given by

$$\mathbf{E}_a = \frac{V}{\ln (u_2/u_1)} \frac{\mathbf{a}_u}{u}$$

$$\mathbf{H}_a = \frac{-I_t}{2\pi u} \mathbf{a}_\theta$$

where $u$ is the radial coordinate in the aperture, and $u_1$, $u_2$ are the inner and outer coaxial-line radii as in Fig. 7.2a. The current $I_t$ flows outward on the probe.

† Section 1.7.

Having specified approximate aperture sources, the problem of finding the field radiated by these sources, such that the tangential electric field vanishes on the probe and is equal to $\mathbf{E}_a$ in the aperture, is a difficult one. An approximate technique which is not too difficult to carry out will be used instead. A variational integral for the input impedance to an antenna in free space has been given by Storer, and this same formulation may be adapted to the problem considered here.†

The electric field in the guide may be determined from the electric currents that flow on the probe and in the aperture surface. Let $\bar{\mathbf{G}}(\mathbf{r}|\mathbf{r}')$ be the dyadic Green's function corresponding to the electric field radiated by a unit dyadic source, where $\mathbf{r}$ designates the field point $(x,y,z)$ and $\mathbf{r}'$ the source point $(x',y',z')$. The electric field radiated by the currents on the probe and aperture surface is given by

$$\mathbf{E}(\mathbf{r}) = \iint_S \bar{\mathbf{G}}(\mathbf{r}|\mathbf{r}') \cdot \mathbf{J}(\mathbf{r}')\, da'$$

where $S$ is the total surface $S_0 + S_a$. This field satisfies the following boundary conditions:

$$\mathbf{n} \times \mathbf{E}(\mathbf{r}) = \mathbf{n} \times \iint_S \bar{\mathbf{G}}(\mathbf{r}|\mathbf{r}') \cdot \mathbf{J}(\mathbf{r}')\, da' = \begin{cases} \mathbf{n} \times \mathbf{E}_a & \text{on } S_a \\ 0 & \text{on } S_0 \end{cases} \tag{1a}$$

This is an integral equation whose solution would determine the current $\mathbf{J}$ on $S_0$. Let $\boldsymbol{\tau}$ be a unit vector in the direction of $\mathbf{J}$. In place of (1a), we may write

$$\boldsymbol{\tau} \cdot \mathbf{E}(\mathbf{r}) = \begin{cases} \boldsymbol{\tau} \cdot \mathbf{E}_a & \text{on } S_a \\ 0 & \text{on } S_0 \end{cases} \tag{1b}$$

If, instead of dot-multiplying by $\boldsymbol{\tau}$, we dot-multiply by $\mathbf{J}(\mathbf{r})$, and integrate over the total surface $S$, we get

$$\iint_{S_a} \mathbf{J}(\mathbf{r}) \cdot \mathbf{E}_a\, da = \iint_S \iint_S \mathbf{J}(\mathbf{r}) \cdot \bar{\mathbf{G}}(\mathbf{r}|\mathbf{r}') \cdot \mathbf{J}(\mathbf{r}')\, da\, da'$$

The integral on the left is readily evaluated since both $\mathbf{J}$ and $\mathbf{E}_a$ are known on the aperture surface $S_a$. We have

$$\mathbf{J} \cdot \mathbf{E}_a = \frac{-VI_t}{2\pi \ln (u_2/u_1)} \frac{(\mathbf{a}_y \times \mathbf{a}_\theta) \cdot \mathbf{a}_u}{u^2} = \frac{-VI_t}{2\pi \ln (u_2/u_1)} \frac{1}{u^2}$$

The integral over $S_a$ becomes

$$\frac{-VI_t}{2\pi \ln (u_2/u_1)} \int_0^{2\pi} \int_{u_1}^{u_2} \frac{u\, du\, d\theta}{u^2} = -VI_t = -Z_{\text{in}} I_t^2$$

† An account of Storer's method is given in R. W. P. King, "The Theory of Linear Antennas," sec. II.39, Harvard University Press, Cambridge, Mass., 1956.

since the input impedance of the antenna at the aperture plane is equal to $V/I_t$. Hence, we get

$$Z_{\text{in}} = -\frac{1}{I_t^2} \iint\limits_S \iint\limits_S \mathbf{J}(\mathbf{r}) \cdot \bar{\mathbf{G}}(\mathbf{r}|\mathbf{r}') \cdot \mathbf{J}(\mathbf{r}')\, da\, da' \tag{2}$$

which we will show is a variational expression for the input impedance. In practice, a first-order approximation for $\mathbf{J}$ used in (2) gives a second-order approximation to $Z_{\text{in}}$. Let the current change by an amount $\delta\mathbf{J}$. The change in $Z_{\text{in}}$ is given by

$$I_t^2\, \delta Z_{\text{in}} + 2Z_{\text{in}} I_t\, \delta I_t = \iint\limits_S \iint\limits_S \delta\mathbf{J}(\mathbf{r}) \cdot \bar{\mathbf{G}}(\mathbf{r}|\mathbf{r}') \cdot \mathbf{J}(\mathbf{r}')\, da'\, da + \iint\limits_S \iint\limits_S \mathbf{J}(\mathbf{r}) \cdot \bar{\mathbf{G}}(\mathbf{r}|\mathbf{r}') \cdot \delta\mathbf{J}(\mathbf{r}')\, da\, da'$$

The Green's function obeys the symmetry relation

$$\mathbf{J}(\mathbf{r}) \cdot \bar{\mathbf{G}}(\mathbf{r}|\mathbf{r}') \cdot \delta\mathbf{J}(\mathbf{r}') = \delta\mathbf{J}(\mathbf{r}') \cdot \bar{\mathbf{G}}(\mathbf{r}'|\mathbf{r}) \cdot \mathbf{J}(\mathbf{r})$$

and for the purpose of carrying out the integration we may interchange $\mathbf{r}$ and $\mathbf{r}'$, and hence

$$I_t^2\, \delta Z_{\text{in}} = -2\left(Z_{\text{in}} I_t\, \delta I_t - \iint\limits_S \iint\limits_S \delta\mathbf{J} \cdot \bar{\mathbf{G}} \cdot \mathbf{J}\, da\, da'\right) \tag{3}$$

We now note that

$$\iint\limits_{S_a} \mathbf{E}_a \cdot \delta\mathbf{J}\, da = V\, \delta I_t = Z_{\text{in}} I_t\, \delta I_t = \iint\limits_S \mathbf{E}_a \cdot \delta\mathbf{J}\, da$$

since $\mathbf{E}_a$ is zero on the probe; i.e., it is the aperture field only. We now get

$$I_t^2\, \delta Z_{\text{in}} = -2 \iint\limits_S \delta\mathbf{J}(\mathbf{r}) \cdot \left[\mathbf{E}_a - \iint\limits_S \bar{\mathbf{G}}(\mathbf{r}|\mathbf{r}') \cdot \mathbf{J}(\mathbf{r}')\, da'\right] da = 0 \tag{4}$$

since the dot product with $\delta\mathbf{J}$ selects only the tangential component which, by virtue of the boundary condition (1*b*), vanishes on $S$. Thus (2) gives us a method of finding $Z_{\text{in}}$ correct to the second order for an assumed current distribution correct to only the first order.

In view of the similarity of the present antenna to a stub antenna above a conducting plane in free space, the current is approximately sinusoidal. A current distribution of the form $I_1 \sin k_0(d - y) + I_2[1 - \cos k_0(d - y)]$ would be suitable to use in (2). After evaluating the integrals, we get $Z_{\text{in}}(I_1,I_2)$ as a function of the two parameters $I_1$, $I_2$. The next step is to equate $\partial Z_{\text{in}}/\partial I_1$ and $\partial Z_{\text{in}}/\partial I_2$ to zero, and solve for the values of $I_1$ and $I_2$ which render the integral expression for $Z_{\text{in}}$ stationary.

The corresponding value of $Z_{in}$ is the best approximation that the assumed current distribution can give to the true value of $Z_{in}$.

We will carry out a first-order analysis only in order to keep the details of the steps involved in the solution clear. First, however, we will replace the probe antenna by the approximate equivalent antenna illustrated in Fig. 7.2*b*. This allows us to assume a $y$-directed current only, from which the field $\mathbf{E}(\mathbf{r})$ may be found. The justification for assuming an excitation of the form in Fig. 7.2*b* can only be obtained by comparison of the theory with experimental results. The work of R. W. P. King and his group at the Cruft Laboratories has shown that the effect of changing the coaxial-line gap width on the input impedance to an antenna is negligible, provided the gap radii $u_2$, $u_1$ are small compared with the wavelength.† Thus we may consider the probe antenna as effectively driven by a $y$-directed localized field at the base, with negligible error in most cases of practical interest.

The current has a $y$ component only, and, hence, the field may be found in terms of a vector potential function $\mathbf{A} = \mathbf{a}_y A_y$ having only a $y$ component by means of the following equations:

$$\mathbf{E} = -j\omega\mathbf{A} + \frac{\nabla\nabla \cdot \mathbf{A}}{j\omega\epsilon_0\mu_0} \tag{5a}$$

$$\mathbf{B} = \nabla \times \mathbf{A} \tag{5b}$$

These equations give rise to the longitudinal-section $H$ modes. As a first step toward finding $\mathbf{A}$, we will find a dyadic Green's function $\bar{\mathbf{G}} = \mathbf{a}_y G \mathbf{a}_y$, which is a solution of the following inhomogeneous equation:

$$\nabla^2\bar{\mathbf{G}} + k_0^2\bar{\mathbf{G}} = \mathbf{a}_y\mathbf{a}_y\,\delta(x - x')\,\delta(z - z')\delta(y - y') \tag{6}$$

and satisfies the boundary conditions

$$\bar{\mathbf{G}} = 0 \qquad \text{at } x = 0, a;\ z = -l$$

$$\frac{\partial\bar{\mathbf{G}}}{\partial y} = 0 \qquad \text{at } y = 0, b$$

FIG. 7.3. Unit source in a short-circuited guide.

This Green's function corresponds to the vector potential from a unit $y$-directed current element, and the boundary conditions are dictated by the condition that the tangential electric field, to be derived from $\bar{\mathbf{G}}$, must vanish on the guide boundary.

With reference to Fig. 7.3 it is seen that appropriate expansions for $\bar{\mathbf{G}}$, which are solutions of the homogeneous Helmholtz equation in the two

† *Ibid.*, sec. II.38.

regions of interest, are

$$G(x,y,z|x',y',z') = \sum_{n=1}^{\infty}\sum_{m=0}^{\infty} C_{nm} \sin\frac{n\pi x}{a}\cos\frac{m\pi y}{b} \times \sinh \Gamma_{nm}(z+l) \qquad z \le z' \tag{7a}$$

$$G(x,y,z|x',y',z') = \sum_{n=1}^{\infty}\sum_{m=0}^{\infty} D_{nm} \sin\frac{n\pi x}{a}\cos\frac{m\pi y}{b}\, e^{-\Gamma_{nm}z} \qquad z \ge z' \tag{7b}$$

where $\Gamma_{nm}^2 = (n\pi/a)^2 + (m\pi/b)^2 - k_0{}^2$, and $C_{nm}$ and $D_{nm}$ are unknown coefficients. At $z = z'$, $E_y$ and $E_x$ are continuous, and, hence, $G$ is continuous, and, therefore,

$$C_{nm} \sinh \Gamma_{nm}(z'+l) = D_{nm}e^{-\Gamma_{nm}z'} \tag{8}$$

If (6) is multiplied by $\sin (n\pi x/a) \cos (m\pi y/b)$ and integrated over the guide cross section, we get

$$\left[\frac{d^2}{dz^2} + k_0{}^2 - \left(\frac{n\pi}{a}\right)^2 - \left(\frac{m\pi}{b}\right)^2\right] g_{nm}(z|z') = \sin\frac{n\pi x'}{a}\cos\frac{m\pi y'}{b}\,\delta(z-z') \tag{9}$$

where $$g_{nm}(z|z') = \int_0^a \int_0^b G \sin\frac{n\pi x}{a}\cos\frac{m\pi y}{b}\,dx\,dy$$

The terms involving the partial derivatives with respect to $x$ and $y$ were integrated by parts twice, and the integrated terms vanished by virtue of the boundary conditions on $G$. Equation (9) is the inhomogeneous equation for the finite double Fourier sine and cosine transform of $G$. Using the condition that $g_{nm}$ is continuous at $z = z'$, (9) may be integrated with respect to $z$ once over a small vanishing interval including the point $z'$ to get

$$\left.\frac{dg_{nm}}{dz}\right|_{z_-'}^{z_+'} = \sin\frac{n\pi x'}{a}\cos\frac{m\pi y'}{b} \tag{10}$$

a result which determines the discontinuity in the derivative of $g_{nm}$. In terms of the expansion (7), the transform of $G$ is found to be

$$g_{nm} = \begin{cases} \dfrac{ab}{2\epsilon_{0m}} C_{nm} \sinh \Gamma_{nm}(z+l) & z \le z' \\[2ex] \dfrac{ab}{2\epsilon_{0m}} D_{nm}e^{-\Gamma_{nm}z} & z \ge z' \end{cases} \tag{11}$$

where $\epsilon_{0m} = 1$ for $m = 0$, and 2 otherwise. Using (8) and (10), the coefficients $C_{nm}$ and $D_{nm}$ are readily found and are

$$C_{nm} = -\frac{2\epsilon_{0m}}{ab\Gamma_{nm}} e^{-\Gamma_{nm}l} \sin\frac{n\pi x'}{a} \cos\frac{m\pi y'}{b} e^{-\Gamma_{nm}z'} \tag{12a}$$

$$D_{nm} = -\frac{2\epsilon_{0m}}{ab\Gamma_{nm}} e^{-\Gamma_{nm}l} \sin\frac{n\pi x'}{a} \cos\frac{m\pi y'}{b} \sinh\Gamma_{nm}(l+z') \tag{12b}$$

Substituting back into (7), the complete solution for $\bar{\mathbf{G}}$ is obtained and is

$$\bar{\mathbf{G}}(x,y,z|x',y',z') = -\mathbf{a}_y\mathbf{a}_y \frac{2}{ab} \sum_{n=1}^{\infty}\sum_{m=0}^{\infty} \frac{\epsilon_{0m}}{\Gamma_{nm}} e^{-\Gamma_{nm}l}$$

$$\sin\frac{n\pi x}{a}\cos\frac{m\pi y}{b}\sin\frac{n\pi x'}{a}\cos\frac{m\pi y'}{b}\begin{cases} e^{-\Gamma_{nm}z'}\sinh\Gamma_{nm}(z+l) & z \le z' \\ e^{-\Gamma_{nm}z}\sinh\Gamma_{nm}(z'+l) & z \ge z' \end{cases} \tag{13}$$

For a uniform current distribution $\mathbf{J} = \mathbf{a}_y J_o \sin k_0(d - y')$ on the probe, the vector potential $\mathbf{A}$ is given by

$$\mathbf{A}(x,y,z) = -\mu_0 \iint_{S_0} \bar{\mathbf{G}}(x,y,z|x',y',z') \cdot \mathbf{J}(y')\, da' \tag{14}$$

since $\mathbf{A}$ is a solution of the same equation as $\bar{\mathbf{G}}$ with a source function $-\mu_0\mathbf{J}$. In (14), $da'$ is an element of area on the surface of the probe. Introducing a local cylindrical coordinate system as in Fig. 7.4, it is seen that the element of surface area $da' = (t/2)\, d\theta'\, dy'$. The $y$ component of electric field is obtained from the vector potential function $\mathbf{A}$ by means of (5a).

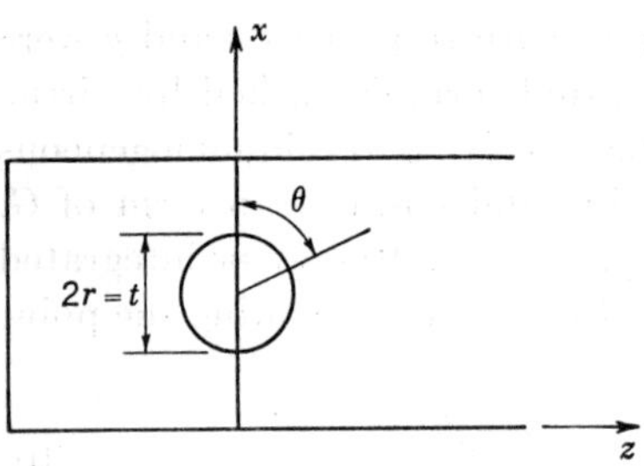


FIG. 7.4. Coordinate system for probe surface.

At this point it is instructive to note that, if the complex Poynting vector is integrated over the coaxial-line opening, we get

$$\tfrac{1}{2}VI_0^* \sin k_0 d = \tfrac{1}{2}VI_0 \sin k_0 d = \tfrac{1}{2}Z_{\text{in}}I_0^2 \sin^2 k_0 d$$

since the total current at the base is real and equal to $I_0 \sin k_0 d$. This must be equal to the total scattered power $P_s$ plus $2j\omega$ times the time-average net reactive energy $W_m - W_e$ in the waveguide. For the present approximation we are able, therefore, to interpret the integral $-\frac{1}{2}\iint \mathbf{J}\cdot\bar{\mathbf{G}}\cdot\mathbf{J}\, da\, da'$ as giving the complex power $P_s + 2j\omega(W_m - W_e)$

radiated into the guide. This interpretation would not be true for a different current distribution which was complex in general. For the rest of the analysis presented here, this interpretation will be retained, since it gives a convenient way of thinking about the various terms which contribute to the input impedance of the antenna.

To determine the power flow and reactive energy in the field, the following integral must be evaluated over the probe surface, $-\frac{1}{2}\iint\limits_{S_0} E_y J_y \, da$. Substituting for $E_y$ in terms of $\mathbf{A}$ as given by (14) leads to the following expression:

$$P_s + 2j\omega(W_m - W_e) = \frac{1}{2j\omega\epsilon_0}\iint\limits_{S_0}\iint\limits_{S_0} \mathbf{J}(y)\cdot\left[\left(k_0^2 + \frac{\partial^2}{\partial y^2}\right)\bar{\mathbf{G}}\right] \cdot \mathbf{J}(y')\,\frac{t^2\,d\theta\,d\theta'}{4}\,dy\,dy' \qquad (15)$$

which gives the radiated power and net time-average reactive energy $W_m - W_e$ stored in the field around the antenna. Since it is assumed that the frequency of operation has been chosen so that only the $H_{10}$ mode propagates, then all $\Gamma_{nm}$ except $\Gamma_{10}$ are real. Thus the radiated power $P_s$ arises only from the $n = 1$, $m = 0$ term.

The integration of (15) over the $y$ and $y'$ coordinates is easily performed, a typical term being

$$\int_0^d \cos\frac{m\pi y}{b}\sin k_0(d - y)\,dy = \frac{k_0[\cos k_0 d - \cos m\pi(d/b)]}{(m\pi/b)^2 - k_0^2} \qquad (16)$$

For a small probe diameter $t$, the variation of the first few terms ($n$ and $m$ small) of the series expansion of $\bar{\mathbf{G}}$ with the coordinates $\theta$ and $\theta'$ may be neglected. In this case, the integration over the angle coordinate gives only a factor $2\pi$. This approximation cannot be used for the higher-order terms, however, since it would lead to a divergent series for the reactive energy. This is to be expected since the approximation is equivalent to assuming that the antenna is infinitely thin, and such an antenna is known to have an infinite reactance. For the $n = 1$, $m = 0$ term, we replace $x$ and $x'$ by $a/2$, and put $z = z' = 0$ (the coordinates of the center of the probe). Also, noting that $2\pi(t/2)J_0 = I_0$, the total current on the probe, the $n = 1$, $m = 0$ term from (15) becomes

$$-\frac{I_0^2}{j\omega\epsilon_0 ab\Gamma_{10}}(\cos k_0 d - 1)^2 e^{-\Gamma_{10}l}\sinh\Gamma_{10}l = \frac{4I_0^2 Z_0\sin^4(k_0 d/2)}{ab\beta_{10}k_0}\left(\sin^2\beta_{10}l + j\,\frac{\sin 2\beta_{10}l}{2}\right)$$

where $\beta_{10} = |\Gamma_{10}|$, and $Z_0 = (\mu_0/\epsilon_0)^{1/2}$. The real part of this term represents the power $P_s$ radiated into the guide and propagating away from the antenna as an $H_{10}$ mode. The imaginary term represents $2j\omega(W_m - W_e)_{10}$, where $(W_m - W_e)_{10}$ is the time-average reactive energy in the $H_{10}$ standing wave between the antenna and the short-circuited end of the guide. The input impedance is given by [see Chap. 1, Eq. (25)]

$$Z_{\text{in}} = R + jX = \frac{P_s + 2j\omega(W_m - W_e)}{\frac{1}{2}I_0^2 \sin^2 k_0 d} \tag{17}$$

Using the result already obtained for $P_s$, the radiation resistance $R$ of the probe antenna is found to be given by

$$R = \frac{2Z_0}{ab\beta_{10}k_0} \sin^2 \beta_{10}l \tan^2 k_0 \frac{d}{2} \tag{18}$$

The contribution $X_{10}$ of the $H_{10}$ mode to the total input reactance is similarly found to be

$$X_{10} = \frac{Z_0}{ab\beta_{10}k_0} \sin 2\beta_{10}l \tan^2 k_0 \frac{d}{2} \tag{19}$$

At this point in the analysis it is apparent that, by a proper choice of the probe length $d$ and short-circuit position $l$, the radiation resistance $R$ can be made equal to the characteristic impedance $Z_c$ of the coaxial line, and $X_{10}$ can be adjusted to cancel the part of the input reactance contributed by all the higher-order evanescent modes. With such a choice of parameters, $Z_{\text{in}} = Z_c$, and all the incident power is coupled into the guide.

The evaluation of the remainder of the reactive term is very laborious unless a number of simplifying approximations are made. Since the probe is symmetrically located in the guide, all the terms with $n = 2, 4, 6, \ldots$, in (14) integrate to zero. The double integration over $y$ results in a factor $[(m\pi/b)^2 - k_0^2]^2$ in the denominator. One of these factors is canceled when the operation $(k_0^2 + \partial^2/\partial y^2)G$ is carried out as indicated in (15). As a function of $m$, the propagation factor $\Gamma_{nm}$ falls off as $m^{-1}$, and hence the terms in (15) decay as $m^{-3}$. The significant contribution to the input reactance will, therefore, come from the terms with $m$ small, in particular from the terms $m = 0, 1; n = 3, 5, 7, \ldots$.

The vector potential $A_y$ arising from a uniform current distribution on a small-diameter probe does not differ much from that arising from a concentrated current filament located at the center of the probe. Thus, in the integration over $\theta'$ in (14) and (15), we may put $z' = 0$ and $x' = a/2$, and the integration then yields a factor $2\pi$ only. For the integration over $x$ and $z$, that is, over $\theta$, we will find an approximate sum for the terms that depend on $n$ first. In the Green's function (13), the terms that depend on $z$, $z'$, and $l$ may be written as

$$\tfrac{1}{2}e^{-\Gamma_{nm}(l+z')}\left(e^{\Gamma_{nm}(l+z)} - e^{-\Gamma_{nm}(l+z)}\right) = \tfrac{1}{2}\left(e^{-\Gamma_{nm}(z'-z)} - e^{-2\Gamma_{nm}l-\Gamma_{nm}(z'+z)}\right) \qquad z \le z'$$

or as

$$\tfrac{1}{2}\left(e^{-\Gamma_{nm}|z'-z|} - e^{-2\Gamma_{nm}l-\Gamma_{nm}(z'+z)}\right)$$

which is valid for all $z$ and $z'$. The term multiplied by $e^{-2\Gamma_{nm}l}$ is very small, since $\Gamma_{nm}$ is real for the reactive terms, and, hence, may be dropped with negligible error. These terms arise because of the interaction of the localized field between the antenna and its image located at $z = -2l$. This interaction is negligible in practice, since $l$ is usually a quarter wavelength or more. When $x$ is replaced by $a/2 + r \cos \theta$, where $r = t/2$ is the probe radius, the term $\sin (n\pi x/a)$ becomes

$$\sin \frac{n\pi}{2} \cos \left(\frac{n\pi r}{a} \cos \theta\right)$$

since $n$ is odd. Replacing $x'$ by $a/2$ and putting $z' = 0$ and $z = r \sin \theta$, we are finally able to replace (15) by

$$-\frac{jI_0{}^2Z_0k_0}{4ab\pi} \sum_{n=1}^{\infty}{}' \sum_{m=0}^{\infty}{}' \frac{\epsilon_{0m}}{\Gamma_{nm}} \frac{[\cos k_0d - \cos m\pi(d/b)]^2}{(m\pi/b)^2 - k_0{}^2} \times \int_0^{2\pi} \cos\left(\frac{n\pi r}{a}\cos\theta\right) e^{-\Gamma_{nm}r|\sin\theta|}\, d\theta \qquad (20)$$

where the primes on the summation mean omission of the term $n = 1$, $m = 0$, which has already been evaluated.

First we consider the terms for $m = 0$, $n = 3, 5, \ldots,$ of which $\cos [(n\pi r/a) \cos \theta]e^{-\Gamma_{n0}r \sin \theta}$ is a typical one. The series to be summed over $n$ may be written as follows:

$$\operatorname{Re} \sum_{n=3,5,\ldots}^{\infty} \frac{e^{j(n\pi r/a)\cos\theta - \Gamma_{n0}r\sin\theta}}{\Gamma_{n0}} = \operatorname{Re} \sum_{n=1,3,\ldots}^{\infty} \frac{a}{n\pi} e^{j(n\pi r/a)(\cos\theta + j\sin\theta)}$$
$$- \operatorname{Re} \frac{a}{\pi} e^{j(\pi r/a)(\cos\theta+j\sin\theta)} + \operatorname{Re} \sum_{n=3,5,\ldots}^{\infty} \left(\frac{e^{-\Gamma_{n0}r\sin\theta}}{\Gamma_{n0}} - \frac{a}{n\pi} e^{-(n\pi r/a)\sin\theta}\right) \times e^{j(\pi r/a)\cos\theta}$$

The last series is a rapidly converging series since $\Gamma_{n0}$ approaches $n\pi/a$ rapidly. For $n$ small, the exponential factors may be expanded according to $e^x = 1 + x$, and each term is then seen to be equal to $\Gamma_{n0}^{-1} - a/n\pi$. This form may be retained for large $n$ as well, since these terms in the correction series are negligible. The first series is of the type that is summed in the Mathematical Appendix, and its sum is given by

$$\operatorname{Re}\frac{a}{\pi}\sum_{n=1,3,\ldots}^{\infty}\frac{1}{n}e^{j(n\pi r/a)e^{j\theta}} = -\operatorname{Re}\frac{a}{2\pi}\ln j\tan\frac{\pi r}{2a}e^{j\theta} \approx -\frac{a}{2\pi}\ln\frac{\pi r}{2a} \tag{21}$$

since the argument of the tangent function is small.

The $n = 1$ term which was subtracted away may be replaced by $(a/\pi)e^{-(\pi r/a)\sin\theta} \approx (a/\pi)[1 - (\pi r/a)\sin\theta]$. When this term is integrated over $\theta$ from 0 to $\pi$ and multiplied by 2, it becomes $2a(1 - 2r/a)$. Integration of the two series over $\theta$ yields a factor $2\pi$ only. An approximate closed-form expression for the correction series may be obtained as follows. Expanding $\Gamma_{n0}^{-1}$, we get

$$\Gamma_{n0}^{-1} = \left[\left(\frac{n\pi}{a}\right)^2 - k_0^2\right]^{-\frac{1}{2}} \approx \frac{a}{n\pi}\left(1 + \frac{k_0^2a^2}{2n^2\pi^2}\right) = \frac{a}{n\pi} + \frac{k_0^2a^3}{2n^3\pi^3}$$

Hence, the correction series becomes

$$\frac{k_0^2a^3}{2\pi^3}\sum_{n=3,5,\ldots}^{\infty}\frac{1}{n^3} = \frac{0.0259k_0^2a^3}{\pi^3} \tag{22}$$

Collecting together our results, we find that the contribution $\sum_n X_{n0}$ to the total input reactance is given by

$$\sum_n X_{n0} = \frac{Z_0\tan^2(k_0d/2)}{2\pi k_0 b}\left[\ln\frac{2a}{\pi r} + \frac{0.0518k_0^2a^2}{\pi^2} - 2\left(1 - \frac{2r}{a}\right)\right] \tag{23}$$

when (20) is divided by $\frac{1}{2}I_0^2\sin^2 k_0d$, and the terms $m = 0$, $n = 3, 5, \ldots$ separated out. The above contribution is inductive, which of course it should be since, for $m = 0$, $H_{n0}$ modes are involved, and these store a net time average of magnetic energy in the evanescent field.

For the terms $m > 0$, the Poisson summation formula† will be used to convert the sum over $n$ into a more rapidly converging series. The following Fourier transform is required in order to carry out the summation:

$$\int_{-\infty}^{\infty}\frac{\exp[jw(z + z') - \alpha(w^2 + \beta^2)^{\frac{1}{2}}]}{(w^2 + \beta^2)^{\frac{1}{2}}}\,dw = 2K_0\{\beta[\alpha^2 + (z + z')^2]^{\frac{1}{2}}\} \tag{24}$$

where $K_0(x)$ is the modified Bessel function of the second kind.‡ The

† Section A.6.

‡ N. W. McLachlan, "Bessel Functions for Engineers," 2d ed., Oxford University Press, New York, 1946.

series to be summed over $n$ is

$$\sum_{n=1,3,\ldots}^{\infty} \frac{\cos[(n\pi r/a)\cos\theta]\exp\{-r\sin\theta[(n\pi/a)^2+k_m^2]^{1/2}\}}{[(n\pi/a)^2+k_m^2]^{1/2}}$$

where $k_m^2 = (m\pi/b)^2 - k_0^2$. First we will sum the following series:

$$\sum_{-\infty}^{\infty} S\left(n\frac{\pi}{a}\right) = \sum_{-\infty}^{\infty} \frac{\exp\{j(n\pi r/a)\cos\theta - r\sin\theta[(n\pi/a)^2+k_m^2]^{1/2}\}}{[(n\pi/a)^2+k_m^2]^{1/2}}$$

According to the Poisson summation formula, the sum is given by

$$\sum_{-\infty}^{\infty} S\left(n\frac{\pi}{a}\right) = \frac{a}{\pi}\sum_{n=-\infty}^{\infty} f(2na)$$

where $f(z)$ is the Fourier transform of $S(w)$. Using (24) with $w = n\pi/a$, $z' = r\cos\theta$, $\alpha = r\sin\theta$, $\beta = k_m$, the result is

$$\sum_{-\infty}^{\infty} S\left(n\frac{\pi}{a}\right) = \frac{2a}{\pi}\sum_{n=-\infty}^{\infty} K_0\{k_m[r^2+(2na)^2+4nar\cos\theta]^{1/2}\}$$

Since we want only the odd terms, we subtract from this result the same series with $a$ replaced by $a/2$ (this series is the same as the original series summed over the even values of $n$). Hence we get

$$\begin{aligned}\sum_{n=1,3,\ldots}^{\infty} \cos\left(\frac{n\pi r}{a}\cos\theta\right)\frac{1}{\Gamma_{nm}}e^{-\Gamma_{nm}r\sin\theta} &= \frac{1}{2}\sum_{n=\pm1,\pm3,\ldots}^{\infty} e^{j(n\pi/a)r\cos\theta}\frac{1}{\Gamma_{nm}}e^{-\Gamma_{nm}r\sin\theta} \\ &= \frac{a}{2\pi}\sum_{-\infty}^{\infty}(2K_0\{k_m[r^2+(2na)^2+4nar\cos\theta]^{1/2}\} \\ &\qquad - K_0\{k_m[r^2+(na)^2+2nar\cos\theta]^{1/2}\}) \qquad (25)\end{aligned}$$

The modified Bessel function of the second kind decays rapidly so that the only significant terms in (25) are those for $n = 0$, and, hence, for all practical purposes the original series has been summed. For small values of $x$, we have $K_0(x) \approx -[\gamma + \ln(x/2)]$, where $\gamma = 0.5772$, whereas, for large $x$, $K_0(x) \approx (\pi/2x)^{1/2}e^{-x}(1 - 1/8x + \cdots)$. The function $K_0(x)$ is equal to $j(\pi/2)H_0^1(jx)$, where $H_0^1(jx)$ is the Hankel function of the first kind with imaginary argument. This latter function is tabulated.† A graph of $K_0(x)$ is given in Fig. 7.5. For $n = 0$, (25) gives $(a/2\pi)K_0(k_m r)$ as the required sum. For small values of $k_m r$, we have approximately $-(a/2\pi)[\gamma + \ln (k_m r/2)]$. Since this result is independent of $\theta$, the integration over $\theta$ gives a factor $2\pi$. Extracting from (20) the

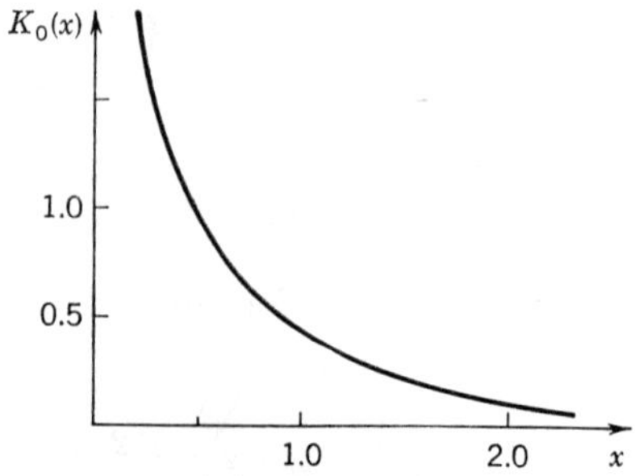


FIG. 7.5. Modified Bessel function of the second kind.

terms for $m > 0$ and dividing by $\frac{1}{2}I_0^2 \sin^2 k_0 d$, we finally obtain the remainder of the expression for the antenna reactance. This latter contribution is

$$-\frac{Z_0 k_0}{b\pi} \tan^2 k_0 \frac{d}{2} \sum_{m=1}^{\infty} \left[1 - \frac{\sin^2 (m\pi d/2b)}{\sin^2 (k_0 d/2)}\right] \frac{K_0(k_m r)}{k_m^2} \tag{26}$$

The series converges very rapidly, and usually only three or four terms are required. This contribution is seen to be capacitive and will partially or completely cancel the inductive part.

Combining all our results, the following expression is obtained for the total input reactance:

$$X = \frac{Z_0 \tan^2 (k_0 d/2)}{2\pi k_0 b} \left\{ \ln \frac{2a}{\pi r} + \frac{0.0518 k_0^2 a^2}{\pi^2} + \frac{2\pi}{\beta_{10} a} \sin 2\beta_{10} l \right.$$
$$\left. - 2\left(1 - \frac{2r}{a}\right) - 2k_0^2 \sum_{m=1}^{\infty} \left[1 - \frac{\sin^2 (m\pi d/2b)}{\sin^2 (k_0 d/2)}\right]^2 \frac{K_0(k_m r)}{k_m^2} \right\} \tag{27}$$

where $k_m^2 = (m\pi/b)^2 - k_0^2$, $r$ is the probe radius, and $l$ is the short-circuit position.

Results similar to these were obtained by Infeld and coworkers.‡ They treated the problem from the image viewpoint, and their final result differs essentially only in the form for the inductive part of the reactance. If the $m = 0$, $n = 3, 5, \ldots$ terms were summed using the

† E. Jahnke and F. Emde, "Tables of Functions," 4th ed., Dover Publications, New York, 1945.

‡ L. Infeld, A. F. Stevenson, and J. L. Synge, Contributions to the Theory of Waveguides, *Can. J. Research*, vol. 27, pp. 68–129, July, 1949.

Poisson summation formula, a slowly converging infinite series of Bessel functions of the second kind would result. This is the form obtained by Infeld et al.

The required values of $l$ and $d$ to make $X = 0$ and $R = Z_c$, so that all the incident power is coupled into the guide, may be found graphically from a plot of the $X = 0$ and $R =$ constant contours in the $k_0d - \beta_{10}l$ plane. Such a plot is given in Fig. 7.6 for a guide of dimensions 0.9 by

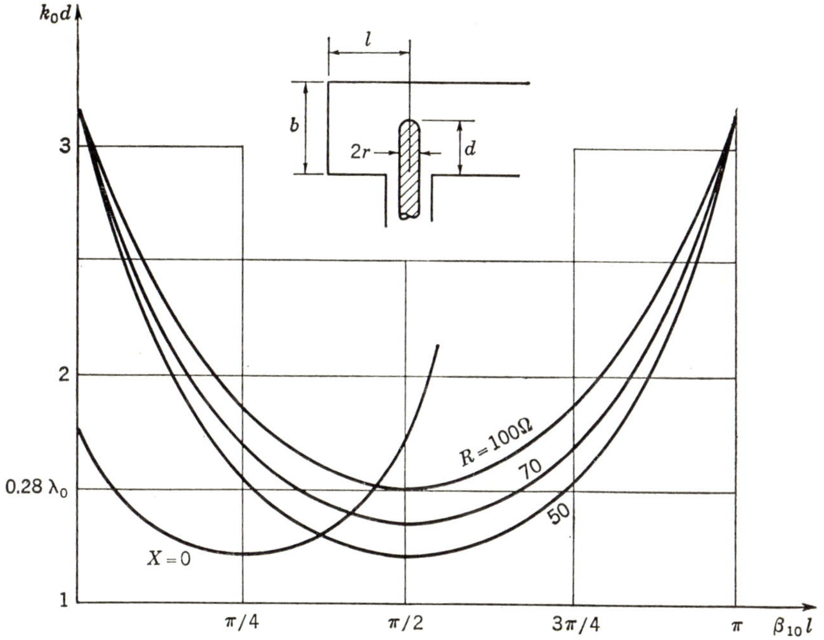


FIG. 7.6. $X = 0$ and $R =$ constant contours in $k_0d - \beta_{10}l$ plane for $a = 0.9$ inch, $b = 0.4$ inch, $r = 0.05a$, $\lambda_0 = 3.14$ centimeters.

0.4 inch and a probe diameter equal to 0.09 inch at a wavelength of 3.14 centimeters. The intersection of the two curves determines the required parameters. From the figure it is seen that the self-resonant length of the antenna, corresponding to $\beta_{10}l = 0$ or $\pi/2$, is $0.28\lambda_0$. Also it is seen that, for $k_0d$ less than 1.22 ($d < 0.61$ centimeter), a value of $\beta_{10}l$ which will make $X = 0$ does not exist. The reason is that the $H_{10}$ mode standing wave between the antenna and the short-circuiting plunger cannot provide enough inductive reactance to counterbalance the high capacitive reactance of a short probe.

### 7.2. The Loop Antenna

As an example of a loop antenna, we consider a coaxial line terminated in the narrow wall of an infinitely long rectangular guide and with its center conductor bent into a semicircular loop of radius $d$, as in Fig. 7.7. The center of the loop is located midway between the top and bottom walls of the guide and coincides with the origin of the coordinate system to be used. The plane of the loop is parallel to the $xy$ plane, and hence only those modes ($H_{nm}$ modes) with an axial magnetic field component are excited. Provided the size of the loop is small compared with a wavelength, say $d$ less than $0.1\lambda_0$, the current in the loop may be assumed constant.

For the purpose of computing the power radiated into the guide, the current may be considered concentrated in a thin filament along the center

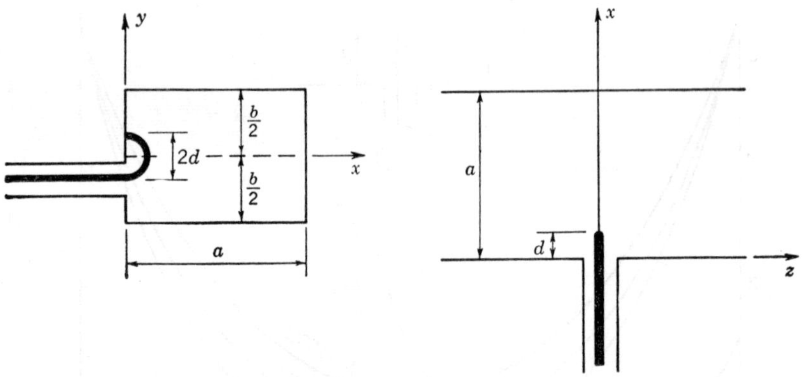


FIG. 7.7. Coaxial-line loop antenna.

of the conductor. The evaluation of the radiated power leads directly to an expression for the input resistance (radiation resistance) of the antenna. The radiated power is readily evaluated by finding the coupling coefficient between the propagating $H_{10}$ mode (all other modes are assumed to be evanescent) and the loop antenna, according to the methods given in Sec. 5.6. To evaluate the input reactance to the antenna, the finite radius $r$ of the conductor must be taken into account, or else we run into the embarrassing situation of an infinite reactance. To evaluate the reactance term, the antenna and guide will be replaced by a double infinite array of image antennas. Our calculation then reduces to finding the self-inductance of a single loop in free space and the mutual inductance between the loop at the origin and all the image loops. In this formulation it is quite clear what approximations can be made with negligible error. An analysis based entirely on an expansion in terms of the normal waveguide modes could also be carried out, but the approximations which can be made are not so apparent in this case.

The normalized $H_{nm}$ modes may be derived from the scalar function

$$\psi_{nm}(x,y) = \left(\frac{\epsilon_{0n}\epsilon_{0m}}{abjk_0Z_0\Gamma_{nm}k_{c,nm}^2}\right)^{1/2} \cos\frac{n\pi x}{a}\cos m\pi\frac{2y-b}{b} \tag{28}$$

by means of the following equations:

$$\mathbf{h}_{znm} = \mathbf{a}_z k_{c,nm}^2 \psi_{nm} \tag{29a}$$

$$\mathbf{h}_{nm} = -\Gamma_{nm} \nabla_t \psi_{nm} \tag{29b}$$

$$\mathbf{e}_{nm} = \frac{jk_0 Z_0}{\Gamma_{nm}} (\mathbf{a}_x\mathbf{a}_y - \mathbf{a}_y\mathbf{a}_x) \cdot \mathbf{h}_{nm} \tag{29c}$$

$$\mathbf{E}_{nm}^{\pm} = \mathbf{e}_{nm} e^{\mp\Gamma_{nm}z} \tag{29d}$$

$$\mathbf{H}_{nm}^{\pm} = (\pm\mathbf{h}_{nm} + \mathbf{h}_{znm}) e^{\mp\Gamma_{nm}z} \tag{29e}$$

where $k_{c,nm}^2 = (m\pi/b)^2 + (n\pi/a)^2 - k_0^2$, $\mathbf{h}_{nm}$ and $\mathbf{e}_{nm}$ are the transverse magnetic and electric normal mode fields, and $\Gamma_{nm}^2 = k_{c,nm}^2 - k_0^2$. The normalization in (28) has been chosen so that

$$\int_0^a \int_0^b \mathbf{e}_{nm} \times \mathbf{h}_{nm} \cdot \mathbf{a}_z \, dx \, dy = 1 \tag{30}$$

The axial magnetic field in the two regions $z < 0$ and $z > 0$ may be represented as an expansion in terms of the above normal mode functions as follows:

$$\mathbf{H}_z = \begin{cases} \sum_{n=0}^{\infty} \sum_{m=0}^{\infty} C_{nm} \mathbf{h}_{znm} e^{-\Gamma_{nm}z} & z > 0 \\ \sum_{n=0}^{\infty} \sum_{m=0}^{\infty} D_{nm} \mathbf{h}_{znm} e^{\Gamma_{nm}z} & z < 0 \end{cases} \tag{31}$$

From Chap. 5, Eqs. (81), the coefficients are given by

$$C_{nm} = D_{nm} = \frac{j\omega\mu_0 I_0}{2} \iint_{S_0} \mathbf{h}_{znm} \cdot \mathbf{a}_z \, dx \, dy \tag{32}$$

where $S_0$ is the surface spanned by the loop, and $I_0$ is the total loop current assumed concentrated in a thin filament. For the $H_{10}$ mode, the

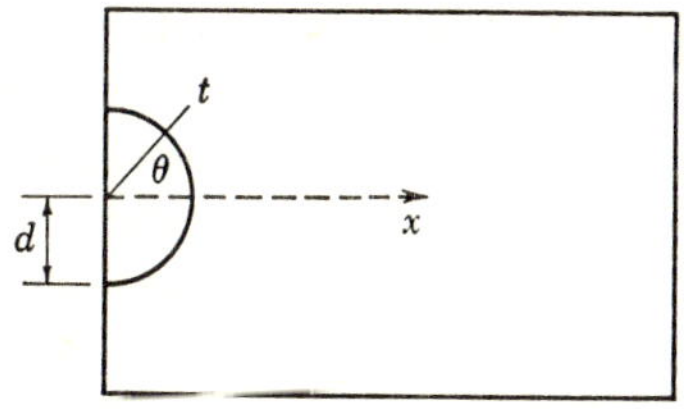


FIG. 7.8

coupling coefficient $C_{10}$ is given by

$$C_{10} = \frac{j\omega\mu_0 I_0}{2} \left(\frac{\pi}{a}\right)^2 \left(\frac{2a}{jk_0 Z_0 \Gamma_{10} \pi^2 b}\right)^{1/2} \iint_{S_0} \cos\frac{\pi x}{a} \, dx \tag{33}$$

Introducing a local cylindrical coordinate system $t\theta$ as in Fig. 7.8 and expanding $\cos(\pi x/a)$ since $\pi d/a$ is small, the integral in Eq. (33) becomes

$$\int_0^d \int_0^\pi \left(1 - \frac{1}{2}\frac{\pi^2}{a^2} t^2 \cos^2\theta\right) t\,dt\,d\theta = \frac{\pi d^2}{2}\left(1 - \frac{\pi^2 d^2}{8a^2}\right) \approx \frac{\pi d^2}{2}$$

The power radiated into the guide is twice that radiated in one direction and, hence, is given by

$$P = \int_0^a \int_0^b C_{10}C_{10}^* \mathbf{e}_{10} \times \mathbf{h}_{10}^* \cdot \mathbf{a}_z\,dx\,dy = C_{10}C_{10}^*$$

since, from the definition of the normal mode functions and the normalization condition (30), we have

$$\int_0^a \int_0^b \mathbf{e}_{nm} \times \mathbf{h}_{nm}^* \cdot \mathbf{a}_z\,dx\,dy = \begin{cases} 1 & \Gamma_{nm} \text{ imaginary} \\ j & \Gamma_{nm} \text{ real} \end{cases}$$

Substituting for $C_{10}$, we get

$$P = \frac{k_0 Z_0}{2\beta_{10}ab}\left(\frac{\pi}{a}\right)^2 \left(\frac{\pi d^2}{2}\right)^2 I_0^2 \tag{34}$$

where $\beta_{10}^2 = k_0^2 - (\pi/a)^2$. If the higher-order modes in the coaxial line are neglected, the integral of the complex Poynting vector over the coaxial-line opening gives $\frac{1}{2}Z_{\text{in}}I_0^2$, since $I_0$ may be assumed real. The input impedance is given by

$$Z_{\text{in}} = \frac{P + 2j\omega(W_m - W_e)}{\frac{1}{2}I_0^2}$$

and, hence, the input resistance, i.e., radiation resistance, is

$$R = \frac{k_0 Z_0}{ab\beta_{10}}\left(\frac{\pi}{a}\right)^2 \left(\frac{\pi d^2}{2}\right)^2 \tag{35}$$

The radiation resistance is seen to be proportional to the square of the loop area and for a small loop is quite small; typical values range from 10 to 30 ohms.

To evaluate the input reactance, we replace the semicircular loop and the guide by a lattice of loop antennas as illustrated in Fig. 7.9. After we have found the field radiated by a single circular loop, it will be seen that the system of loop antennas in Fig. 7.9 radiates a total field which has a zero tangential electric field on the guide boundary. The semicircular loop and its image in the $x = 0$ guide wall are illustrated in Fig. 7.10 and are equivalent to a circular loop driven by a voltage generator having twice the voltage required to maintain the current in the semicircular loop.

To evaluate the input reactance to the antenna, we must compute the total flux linking the loop at the origin due to all of the image antennas, plus the self-flux linkage arising from the magnetic field set up by its own current. The time rate of change of the flux linking the antenna results in a back emf being induced. The applied voltage must be equal and opposite to this induced emf, in order to maintain the current $I_0$ in the loop. The flux linking the antenna and in phase with the current $I_0$

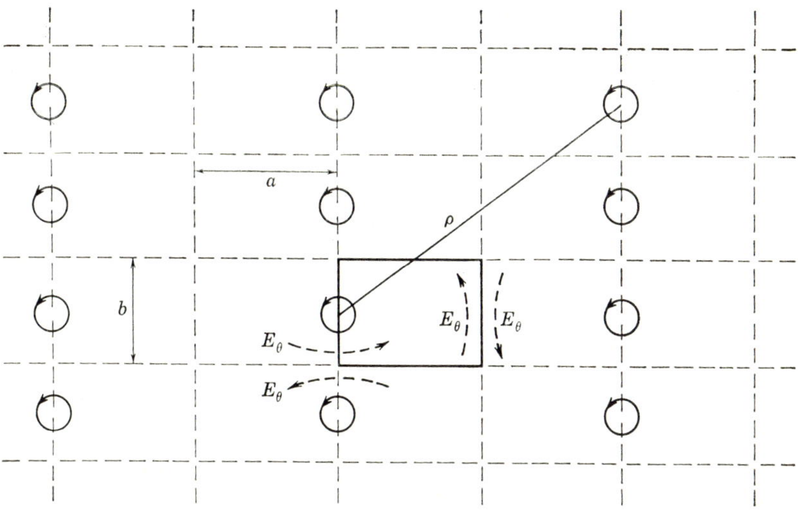


FIG. 7.9. The loop antenna and its images.

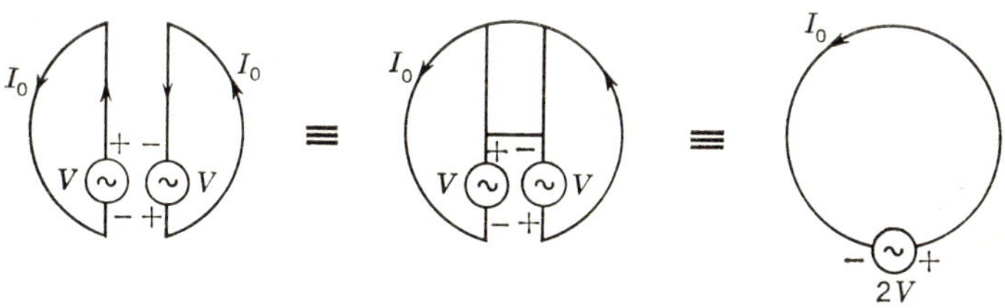


FIG. 7.10. Semicircular loop, its image, and their equivalent loop configuration.

leads to an induced emf in phase quadrature with the current. It is this flux which gives rise to the inductive reactance term. The flux linking the antenna and in phase quadrature with the current gives rise to an emf in phase with the current, and hence contributes to the radiation resistance. Since we have already obtained an expression for the radiation resistance, we are interested only in that part of the flux linkage which is in phase with the current $I_0$.

The field from the image antennas at the position of the loop located at the origin may be found with negligible error by assuming the current $I_0$ to be concentrated in a thin filament. Each image loop may be replaced by a $z$-directed magnetic dipole of moment $M = I_0\pi d^2$. To find the field radiated by this array of magnetic dipoles it is convenient to first find a magnetic-type Hertzian potential $\Pi_z$ having only a $z$ component. The relevant equations are

$$\nabla^2\Pi_z + k_0{}^2\Pi_z = -M$$

$$H_z = k_0{}^2\Pi_z + \frac{\partial^2\Pi_z}{\partial z^2}$$

For a single loop antenna located at the origin as in Fig. 7.11 the solution is

$$\Pi = \frac{M}{4\pi\rho}\, e^{-jk_0\rho}$$

where $\rho = (x^2 + y^2 + z^2)^{1/2}$. For a row of dipoles located at $y = mb$,

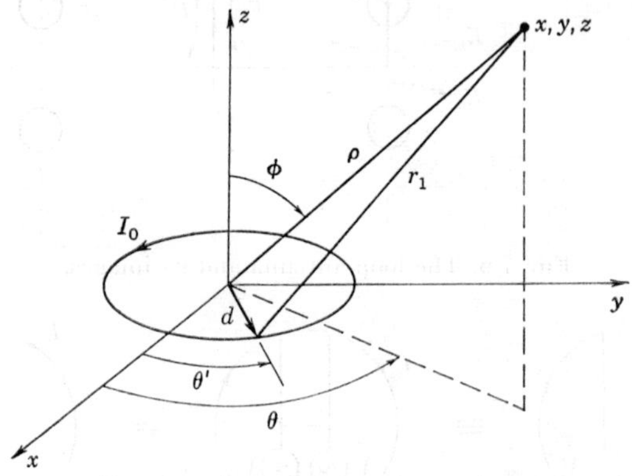


FIG. 7.11. Loop antenna in $xy$ plane.

$m = 0, \pm 1, \pm 2, \ldots$, the potential is given by

$$\Pi_1 = \frac{M}{4\pi} \sum_{m=-\infty}^{\infty} \frac{\exp\{-jk_0[x^2 + z^2 + (y - mb)^2]^{1/2}\}}{[x^2 + z^2 + (y - mb)^2]^{1/2}} \tag{36}$$

This series is readily converted to a more rapidly converging one by using the Poisson summation formula. The required Fourier transform is

$$\int_{-\infty}^{\infty} e^{jwu} \frac{\exp\{-jk_0[x^2 + z^2 + (y - u)^2]^{1/2}\}}{[x^2 + z^2 + (y - u)^2]^{1/2}}\, du = 2e^{jwy}K_0[(x^2 + z^2)(w^2 - k_0{}^2)]^{1/2} \tag{37}$$

where $K_0$ is the modified Bessel function of the second kind. Using the

Poisson summation formula now converts (36) into the following:

$$\frac{M}{4\pi}\sum_{m=-\infty}^{\infty}\frac{\exp\{-jk_0[x^2+z^2+(y-mb)^2]^{1/2}\}}{[x^2+z^2+(y-mb)^2]^{1/2}} = \frac{M}{2\pi b}\sum_{m=-\infty}^{\infty}e^{j2m\pi y/b}$$
$$\times K_0[\Gamma_{0m}(x^2+z^2)^{1/2}] = \frac{M}{\pi b}\left\{\tfrac{1}{2}K_0[jk_0(x^2+z^2)^{1/2}] + \sum_{m=1}^{\infty}\cos\frac{2m\pi y}{b}\right.$$
$$\left.\times K_0[\Gamma_{0m}(x^2+z^2)^{1/2}]\right\} \tag{38}$$

where $\Gamma_{0m} = [(2m\pi/b)^2 - k_0{}^2]^{1/2}$.

The potential arising from all the image dipoles is obtained by replacing $x$ by $x - 2na$ in (38) and summing over all $n$ but excluding the contribution from the dipole at the origin (the driven loop). The flux linkage will be computed by evaluating the magnetic field at the center of the loop and multiplying by the loop area. The evaluation of $H_z$ involves derivatives with respect to $z$ only, and so we may place $x$ and $y$ equal to zero at this point. For the rows of dipoles located at $x = 2na$, $n = \pm 1$, $\pm 2$, . . . , we get, from (38),

$$\Pi_2 = \frac{M}{2\pi b}\sum_{n=-\infty}^{\infty}{}' K_0[jk_0(z^2+4n^2a^2)^{1/2}]$$
$$+ \frac{2M}{\pi b}\sum_{n=1}^{\infty}\sum_{m=1}^{\infty}K_0[\Gamma_{0m}(z^2+4n^2a^2)^{1/2}] \tag{39a}$$

To the above we must add the contribution from dipoles located at $x = z = 0$, $y = mb$, $m = \pm 1, \pm 2, \ldots.$ This latter contribution is, from (36),

$$\frac{M}{2\pi}\sum_{m=1}^{\infty}\frac{\exp[-jk_0(z^2+m^2b^2)^{1/2}]}{(z^2+m^2b^2)^{1/2}} \tag{39b}$$

In (39$a$) the double series may be neglected, since, for the range of parameters involved here, $K_0[\Gamma_{0m}(z^2 + 4n^2a^2)^{1/2}]$ is extremely small. The first series, however, converges very slowly because the argument is imaginary. By using the Poisson summation formula again, this series may be transformed as follows [the required transform may be obtained by inverting (37)]:

$$\frac{M}{2\pi b}\sum_{n=-\infty}^{\infty}K_0[jk_0(z^2+4n^2a^2)^{1/2}] - \frac{M}{2\pi b}K_0(jk_0z)$$
$$= \frac{M}{4ab}\sum_{n=-\infty}^{\infty}\frac{e^{-\Gamma_{n0}z}}{\Gamma_{n0}} - \frac{M}{2\pi b}K_0(jk_0z)$$

where $\Gamma_{n0} = [(n\pi/a)^2 - k_0^2]^{1/2}$. Our final expression for the total Hertzian potential now becomes

$$\Pi_z = \frac{M}{4ab} \sum_{n=-\infty}^{\infty} \frac{e^{-\Gamma_{n0}z}}{\Gamma_{n0}} - \frac{M}{2\pi b} K_0(jk_0z) + \frac{M}{2\pi} \sum_{m=1}^{\infty} \frac{\exp[-jk_0(z^2 + m^2b^2)^{1/2}]}{(z^2 + m^2b^2)^{1/2}} \qquad (40)$$

The field $H_z$, at the origin, is given by $k_0^2\Pi_z + \partial^2\Pi_z/\partial z^2$ evaluated for $z = 0$. The first term gives a contribution

$$\frac{k_0^2M}{4ab} \sum_{n=-\infty}^{\infty} \frac{e^{-\Gamma_{n0}z}}{\Gamma_{n0}} - \frac{k_0^2M}{2\pi b} K_0(jk_0z) + \frac{k_0^2M}{2\pi} \sum_{m=1}^{\infty} \frac{e^{-jk_0mb}}{mb}$$

where the limit as $z$ tends to zero is to be taken. We have $\Gamma_{n0}^{-1} \approx a/n\pi$, and hence the first series is a dominant series

$$\frac{k_0^2M}{2ab} \sum_{m=1}^{\infty} \frac{ae^{-n\pi z/a}}{n\pi}$$

together with the correction series

$$\frac{k_0^2M}{2ab} \left[ \sum_{n=1}^{\infty} \left( \frac{1}{\Gamma_{n0}} - \frac{a}{n\pi} \right) - \frac{j}{2k_0} \right]$$

The dominant series is readily summed by the methods given in the Mathematical Appendix and gives

$$-\frac{k_0^2aM}{2\pi ab} \left( \ln 2 \sinh \frac{\pi z}{2a} - \frac{\pi z}{2a} \right)$$

The logarithmic singularity occurring here is canceled by that arising from the term $K_0(jk_0z)$ since for $z$ small we have $K_0(jk_0z) \to -[\gamma + \ln(jk_0z/2)]$, where $\gamma = 0.577$ and is Euler's constant. If the series in $m$ is also summed and the real part of all terms taken, we finally obtain for the part of $k_0^2\Pi_z$ which is in phase with the current $I_0$ the result

$$\text{Re}\,(k_0^2\Pi_z) = \frac{k_0^2M}{2\pi b} \left[ \gamma + \ln \frac{k_0a}{2\pi} - \ln 2 \sin \frac{k_0b}{2} + \frac{\pi}{a} \sum_{n=2}^{\infty} \left( \frac{1}{\Gamma_{n0}} - \frac{a}{n\pi} \right) - 1 \right] \qquad (41)$$

To find the second part of $H_z$, we first evaluate $\partial^2\Pi_z/\partial z^2$ to get

$$\frac{\partial^2\Pi_z}{\partial z^2} = \frac{M}{4ab}\sum_{n=-\infty}^{\infty}\Gamma_{n0}e^{-\Gamma_{n0}z} + \frac{Mk_0^2}{2\pi b}\left[K_0(jk_0z) + \frac{K_1(jk_0z)}{jk_0z}\right]$$
$$+ \frac{M}{2\pi}\sum_{m=1}^{\infty}\frac{e^{-jk_0r_m}}{r_m}\left[\frac{jk_0z}{r_m^2} + \frac{2z}{r_m^3} - \frac{k_0^2z}{r_m} + \frac{jk_0z}{r_m^2} - \left(jk_0 + \frac{1}{r_m}\right)\left(\frac{1}{r_m} - \frac{z^2}{r_m^3}\right)\right]$$

where $r_m = (z^2 + m^2b^2)^{1/2}$.

The first series consists of a dominant part

$$\frac{M}{2ab}\left[\sum_{n=1}^{\infty}e^{-n\pi z/a}\left(\frac{n\pi}{a} - \frac{k_0^2a}{2\pi n}\right)\right]$$

plus a correction series

$$\frac{M}{2ab}\left[\sum_{n=1}^{\infty}\left(\Gamma_{n0} - \frac{n\pi}{a} + \frac{k_0^2a}{2\pi n}\right) + \frac{jk_0}{2}\right]$$

since $\Gamma_{n0} \approx n\pi/a - k_0^2a/2\pi n$. The dominant part of the series sums to

$$\frac{M}{2ab}\left[\frac{\pi}{4a\sinh^2(\pi z/2a)} + \frac{k_0^2a}{2\pi}\left(\ln 2\sinh\frac{\pi z}{2a} - \frac{\pi z}{2a}\right)\right]$$

As $z$ approaches zero, the singularities occurring in the above expression are canceled by those arising from the Bessel functions $K_0$ and $K_1$. As $x$ approaches zero, we have $K_0(x) \to -[\gamma + \ln(x/2)]$, and $K_1(x) \to x^{-1} + [\ln(x/2) + \gamma - \frac{1}{2}]\,x/2$, and these expansions may be used to evaluate the contribution from the terms $K_0(jk_0z)$ and $K_1(jk_0z)$.

In the series over $m$, we may place $z$ equal to zero, and the resultant series are then readily summed. The required summation formulas are tabulated in the Mathematical Appendix, and so we will write down only the final results that we require. After performing the indicated steps, we obtain for the real part of $\partial^2\Pi_z/\partial z^2$ the following result:

$$\operatorname{Re}\frac{\partial^2\Pi_z}{\partial z^2} = \frac{M}{2ab}\left[\sum_{n=2}^{\infty}\left(\Gamma_{n0} - \frac{n\pi}{a} + \frac{k_0^2a}{2\pi n}\right) - \frac{13}{12}\frac{\pi}{a} + \frac{k_0^2a}{4\pi} - \frac{\gamma k_0^2a}{2\pi}\right.$$
$$\left. + \frac{k_0^2a}{2\pi}\ln\frac{2\pi}{k_0a}\right] - \frac{M}{2\pi b^3}\left(1.2 + \frac{k_0^2b^2}{4} + \frac{3k_0^4b^4}{288} - \frac{k_0^2b^2}{2}\ln k_0b\right) \quad (42)$$

The total inphase flux linking the loop antenna at the origin is obtained by multiplying the real part of the field $H_z$ from the image antennas by $\mu_0\pi d^2$, that is, multiplying the sum of (41) and (42) by $\mu_0\pi d^2$. Thus, the inphase flux linkage arising from the mutual coupling with the image antennas is given by

$$\sum_{n,m}{}' F_{nm} = \mu_0 \pi d^2 I_0 \pi d^2 \left[\frac{k_0^2}{4\pi b} \ln \frac{k_0^2 ab}{4\pi(1-\cos k_0 b)} - \frac{k_0^2}{4\pi b}(2-\gamma) - \frac{13\pi}{24a^2 b} - \frac{0.6}{\pi b^3} - \frac{3k_0^4 b}{576\pi} + \frac{1}{2ab}\sum_{n=2}^{\infty}\left(\frac{k_0^2}{\Gamma_{n0}} + \Gamma_{n0} - \frac{n\pi}{a} - \frac{k_0^2 a}{2n\pi}\right)\right] \quad (43)$$

where $M$ has been replaced by $I_0 \pi d^2$.

The only flux linkage left to be evaluated is that due to the current in

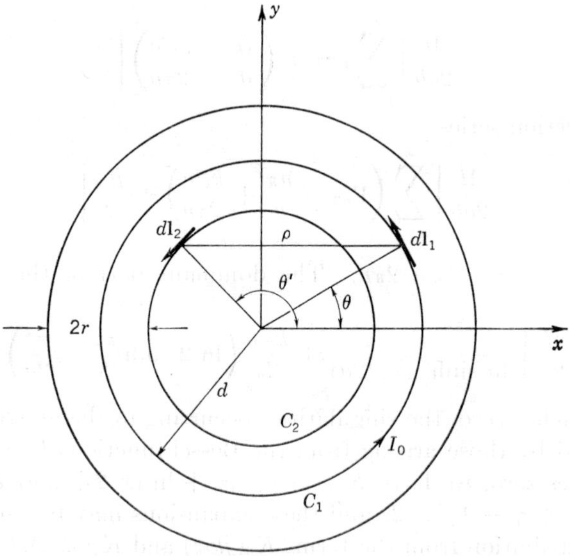


FIG. 7.12. Coordinate system for single loop antenna.

the loop at the origin, i.e., the flux giving rise to the self-inductance of the loop. The vector potential is given by

$$\mathbf{A} = \frac{\mu_0}{4\pi} \iint_{S_0} \mathbf{J}_0 \frac{e^{-jk_0\rho}}{\rho}\, da$$

where $\mathbf{J}_0$ is the uniform circumferential current density on the loop-antenna surface $S_0$, and $\rho$ is the radial distance from a current element to the point in space where $\mathbf{A}$ is computed. For a small loop with $\rho \le 2d$, the exponential term does not vary significantly across the domain of the loop. To be consistent with our assumption of a uniform loop current, we therefore place $e^{-jk_0\rho}$ equal to unity, and we are then left with the static field problem of evaluating the self-inductance of a small circular loop of radius $d$ with a conductor radius $r$. The real part of the flux linkage will be correct to order $(k_0 d)^2$, since the second term in the expansion of $e^{-jk_0\rho}$ is $-jk_0\rho$, and contributes to the radiation resistance only. With reference to Fig. 7.12, the field at the surface of the conductor is $H = J_0 = I_0/2\pi r$, and is the same as that produced by a current filament $I_0$ at the center, i.e., located along the contour $C_1$. The magnetic field does not penetrate into the conductor, and, hence, we only require the flux linking the contour $C_2$ which coincides with the inner surface of the conductor. This flux linkage is given by

$$F_{00} = \iint \mathbf{B} \cdot d\mathbf{a} = \iint \nabla \times \mathbf{A} \cdot d\mathbf{a} = \oint_{C_2} \mathbf{A} \cdot d\mathbf{l}_2$$

Substituting for **A** gives

$$F_{00} = \frac{\mu_0 I_0}{4\pi} \oint_{C_2} \oint_{C_1} \frac{d\mathbf{l}_1 \cdot d\mathbf{l}_2}{\rho} \tag{44}$$

From the figure it is seen that

$$\rho = [d^2 + (d - r)^2 - 2d(d - r)\cos(\theta' - \theta)]^{1/2}$$

and $d\mathbf{l}_1 \cdot d\mathbf{l}_2 = d(d - r)\cos(\theta' - \theta)\, d\theta\, d\theta'$. If we integrate over $\theta$ first, we must get a result independent of $\theta'$ in view of the symmetry involved. We may, therefore, place $\theta'$ equal to zero, and replace the integration over $\theta'$ by a factor $2\pi$. Hence we get

$$\begin{aligned} F_{00} &= \frac{\mu_0 I_0}{2} \int_0^{2\pi} \frac{d(d - r)\cos\theta\, d\theta}{[d^2 + (d - r)^2 - 2d(d - r)\cos\theta]^{1/2}} \\ &= -\mu_0 I_0 d(d - r) \int_0^{\pi} \frac{\cos\theta\, d\theta}{[d^2 + (d - r)^2 + 2d(d - r)\cos\theta]^{1/2}} \end{aligned}$$

This integral may be reduced to the standard form for elliptic integrals with the following substitutions, $\theta = 2\phi$, $\cos\theta = 1 - 2\sin^2\phi$, and

$$k^2 = \frac{4d(d - r)}{(2d - r)^2}$$

We now get

$$\begin{aligned} -\frac{2d - r}{2}\mu_0 I_0 k^2 \int_0^{\pi/2} \frac{1 - 2\sin^2\phi}{(1 - k^2\sin^2\phi)^{1/2}}\, d\phi &= (2d - r)\mu_0 I_0 \left[\left(1 - \frac{k^2}{2}\right) \int_0^{\pi/2} \frac{d\phi}{(1 - k^2\sin^2\phi)^{1/2}} - \int_0^{\pi/2} \frac{1 - k^2\sin^2\phi}{(1 - k^2\sin^2\phi)^{1/2}}\, d\phi\right] \\ &= (2d - r)\mu_0 I_0 \left[\left(1 - \frac{k^2}{2}\right) K(k) - E(k)\right] \end{aligned} \tag{45}$$

where $K$ and $E$ are complete elliptic integrals of modulus† $k$. Referring back to the definition of $k$, it is seen that $k \approx 1$ and, hence, $K(k) \approx \ln [4/(1 - k^2)^{1/2}]$, $E(k) \approx 1$. Replacing $(1 - k^2)^{1/2}$ by $r/2d$, we get

$$F_{00} = \mu_0 I_0 d\left(\ln \frac{8d}{r} - 2\right) \tag{46}$$

† Jahnke and Emde, *op. cit.*

If $F$ is the total flux linkage, the total induced back emf will be $j\omega F$ and must equal the applied voltage $2V$. Hence we have

$$2V = j\omega F$$
$$2VI_0 = 2I_0^2 Z_{\text{in}} = j\omega F I_0$$
$$Z_{\text{in}} = R + jX = \frac{j\omega F}{2I_0}$$

and
$$X = \operatorname{Re}\frac{\omega F}{2I_0}$$

Adding together all the contributions to $F$ gives the following expression for the input reactance to the small-loop antenna:

$$X = \frac{Z_0 k_0 d}{2}\left(\ln\frac{8d}{r} - 2\right) - \frac{\pi^2 d^4 k_0 Z_0}{2}\left[\frac{k_0^2}{4\pi b}(2-\gamma) + \frac{13\pi}{24a^2 b} + \frac{0.6}{\pi b^3} + \frac{3k_0^4 b}{576\pi} - \frac{k_0^2}{4\pi b}\ln\frac{k_0^2 ab}{4\pi(1-\cos k_0 b)} - \frac{1}{2ab}\sum_{n=2}^{\infty}\left(\frac{k_0^2}{\Gamma_{n0}} + \Gamma_{n0} - \frac{n\pi}{a} - \frac{k_0^2 a}{2n\pi}\right)\right] \tag{47}$$

For a small loop and a thin conductor ($r$ small), the self-inductance term predominates.

For the purpose of obtaining an antenna radiating in one direction only, a short-circuiting plunger may be placed a distance $l$ away from the loop. This has negligible effect on the evanescent modes excited by the antenna, since these will have decayed to a negligible value at the short-circuit position for $l > \lambda_0/4$. The effect on the dominant mode is the same as superimposing the $H_{10}$ mode radiated by an image antenna carrying a current $-I_0$ and located at $z = -2l$. The total $H_{10}$ mode axial magnetic field in the guide for $z > 0$ becomes

$$C_{10}\mathbf{h}_{z10}(e^{-\Gamma_{10}z} - e^{-\Gamma_{10}z - 2\Gamma_{10}l}) = 2jC_{10}\mathbf{h}_{z10}e^{-\Gamma_{10}(z+l)}\sin\beta_{10}l$$

where $\beta_{10} = |\Gamma_{10}|$. The amplitude of the wave propagating in the positive $z$ direction is increased by a factor $2\sin\beta_{10}l$, and the power flow is increased by a factor $4\sin^2\beta_{10}l$ in the positive $z$ direction and reduced to zero in the negative $z$ direction. The total radiated power, and, hence, the radiation resistance, becomes

$$R_0 = \frac{k_0 Z_0}{ab\beta_{10}}\left(\frac{\pi}{a}\right)^2 \frac{(\pi d^2)^2}{2}\sin^2\beta_{10}l \tag{48}$$

The reactive energy associated with the $H_{10}$ mode standing wave between the antenna and short-circuiting plunger is given by

$$2j\omega(W_m - W_e)_{10} = \tfrac{1}{2}\int_0^a\int_0^b E_y H_x^* \, dx\, dy$$

where, for $z = 0$, we have

$$\mathbf{a}_y E_y = C_{10}(1 - e^{-j2\beta_{10}l})\mathbf{e}_{10}$$
$$\mathbf{a}_x H_x = -C_{10}(1 + e^{-j2\beta_{10}l})\mathbf{h}_{10}$$

Since
$$\int_0^a \int_0^b \mathbf{e}_{10} \times \mathbf{h}_{10}^* \cdot \mathbf{a}_z \, dx \, dy = 1$$

we get, after substituting for $C_{10}$,

$$2j\omega(W_m - W_e)_{10} = jC_{10}C_{10}^* \sin 2\beta_{10}l = jP \sin 2\beta_{10}l$$

where $P$ is given by (34). The additional contribution to the input

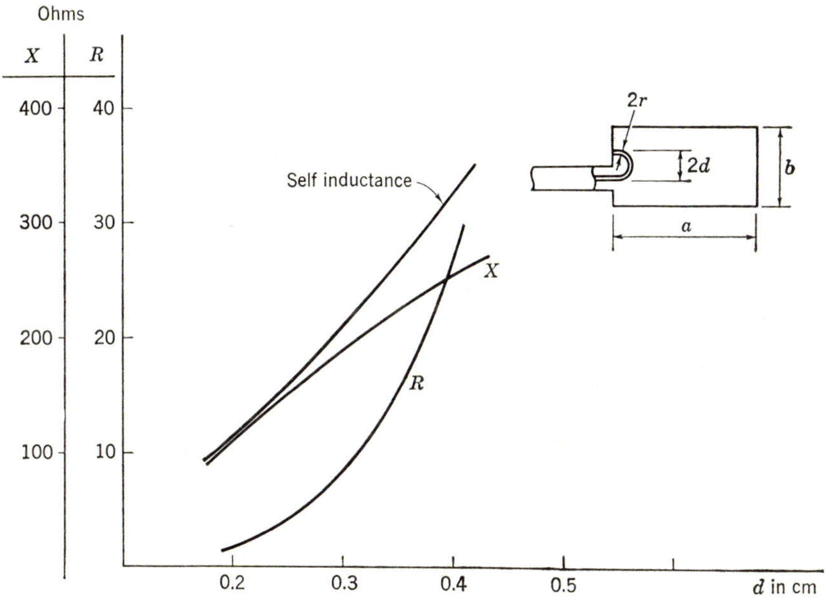


FIG. 7.13. Radiation resistance and input reactance of a small-loop antenna in an infinite guide for $a = 0.9$ inch, $b = 0.4$ inch, $r = 0.5$ millimeter, $\lambda_0 = 3.14$ centimeters.

reactance is obtained by dividing by $\frac{1}{2}I_0^2$, and the total antenna reactance now becomes

$$X_0 = \frac{k_0 Z_0}{ab\beta_{10}} \left(\frac{\pi}{a}\right)^2 \left(\frac{\pi d^2}{2}\right)^2 \sin 2\beta_{10}l + X \tag{49}$$

where $X$ is given by (47). To partially cancel the inductive term $X$, the value of $l$ should be between $\lambda_g/4$ and $\lambda_g/2$, where $\lambda_g$ is the guide wavelength.

In Fig. 7.13 the input reactance $X$ and radiation resistance $R$ of a small loop in an infinite guide are plotted as a function of loop radius $d$ for the following parameters, $k_0 = 2$, $a = 0.9$ inch, $b = 0.4$ inch, $r = 0.5$ millimeter. The self-inductance part is very large (for $d = 0.4$ centimeter, it is equal to 324 ohms), and, consequently, placing a short-circuiting plunger in the guide will not produce enough capacitive loading on the loop to tune out the inductive reactance. The maximum capacitive reactance which can be obtained is equal to the radiation resistance $R$ as plotted, although, with one end of the guide short-circuited, the radiation resistance $R$ must be multiplied by $2 \sin^2 \beta_{10} l$ to obtain the radiation resistance $R_0$ which is applicable now. The inductive reactance may, of course, be tuned out by a shunt stub located at an appropriate position along the coaxial line. The input resistance may then also be made equal to the characteristic impedance of the coaxial line. The use of a larger loop will result in a decrease in the self-inductance because of the variation in phase of the field across the loop. The analysis presented here does not cover this latter situation. The following qualitative argument provides some indication of what may be expected to happen for a larger loop.

On a straight cylindrical wire, various modes of propagation are possible.† All these modes are very highly damped except the circularly symmetric TM mode, which for practical purposes differs by a negligible amount from a TEM mode. To a first approximation we may think of the coaxial line as launching a TEM wave on the loop, which subsequently propagates along the loop a distance $\pi d$, and is reflected at the guide wall. The properties of the loop as a transmission line may be considered to be those of an equivalent coaxial line with an outer conductor radius equal to the loop radius $d$. Neglecting radiation effects, the input impedance seen at the coaxial-line aperture is

$$\begin{aligned} Z_{\text{in}} &= jZ_c \tan k_0 \pi d \\ &= j \frac{Z_0}{2\pi} \ln \frac{d}{r} \tan k_0 \pi d \\ &\approx j \frac{Z_0 k_0 d}{2} \ln \frac{d}{r} \\ &= j \frac{\omega \mu_0 d}{2} \ln \frac{d}{r} \qquad k_0 \pi d \leq 0.8 \end{aligned}$$

For $d/r = 8$ this differs by less than 5 per cent from the static formula for the self-inductive reactance of a semicircular loop in front of a conducting plane, that is, $(j\omega\mu_0/2)d[\ln (8d/r) - 2]$.

Because of radiation there will not be a complete current standing wave on the loop. The effect of a reflected current wave which partially cancels the reflected wave existing in the absence of any radiation may be accounted for by placing an equivalent load $Z_L$ at the end of the loop.

† J. A. Stratton, "Electromagnetic Theory," sec. 9.15, McGraw-Hill Book Company, Inc., New York, 1941.

The input impedance now becomes

$$\begin{aligned} Z_{\text{in}} &= Z_c \frac{Z_L + jZ_c \tan k_0\pi d}{Z_c + jZ_L \tan k_0\pi d} \\ &\approx Z_L + jZ_c\pi k_0 d \qquad |Z_L| \ll Z_c \qquad \pi k_0 d \leq 0.8 \end{aligned}$$

where $Z_L$ has a real part corresponding to the radiation resistance, and an imaginary part corresponding to the radiation reactance or mutual coupling with the image antennas. These latter two terms do not depend critically on the current distribution. For the numerical example presented in Fig. 7.13, $Z_c = (Z_0/2\pi) \ln (d/r)$ is more than 10 times greater than $|Z_L|$ for $d < 0.35$ centimeter, and, hence, the current distribution is almost a complete standing wave. From these considerations we may expect the small-loop analysis to break down when $k_0\pi d$ approaches $\pi/2$ or for a half-circumferential length $\pi d$ approaching one-quarter wavelength. In this length region, resonance may be expected to occur, the input reactance to the antenna will vanish, and a relatively large radiation resistance will occur. Experimental measurements that support this transmission-line model of the loop have been reported.† It is found that the standing-wave minimum in the coaxial line occurs at a position very nearly that computed by assuming that the line is effectively $\pi d$ greater in length and terminates in a short circuit at this point.

## 7.3. Coupling by Small Apertures

Electromagnetic energy may be coupled from one waveguide into another guide or into a cavity resonator by a small aperture located at a suitable position in the common wall. For apertures whose linear dimensions are small compared with the wavelength, an approximate theory is available which states that the aperture is equivalent to a combination of radiating electric and magnetic dipoles, whose dipole moments are respectively proportional to the normal electric field and tangential magnetic field of the incident wave. This theory was originally developed by H. A. Bethe‡ and will be presented here. The approach to be used here differs from that of Bethe, and is chosen because it leads to the final result in a somewhat more direct manner.

As a first step we present two field equivalence principles, due to Schelkunoff, which provide a method of formulating the problem.§

† J. C. Slater, "Microwave Electronics," sec. 7.4, D. Van Nostrand Company, Inc., Princeton, N.J., 1950.

‡ H. A. Bethe, Theory of Diffraction by Small Holes, *Phys. Rev.*, vol. 66, pp. 163–182, 1944.

§ S. A. Schelkunoff, Field Equivalence Theorems, *Comm. on Pure and Appl. Math*, vol. 4, pp. 43–59, June, 1951.

*Theorem 1*

With reference to Fig. 7.14, which illustrates a rectangular guide partially closed by a metal septum or diaphragm at $z = 0$, let $\mathbf{E}_0$, $\mathbf{H}_0$ be the field produced by given sources located in the left half guide when the aperture surface $S_a$ is closed by a perfectly conducting wall. The electric current flowing on this wall is $-\mathbf{n} \times \mathbf{H}_0$. With the wall removed, the given sources radiate a field $\mathbf{E}_2$, $\mathbf{H}_2$ into the right half guide which is identical with the field radiated by an equivalent current $\mathbf{J} = \mathbf{n} \times \mathbf{H}_0$ located on the aperture surface. This equivalent current radiates a field

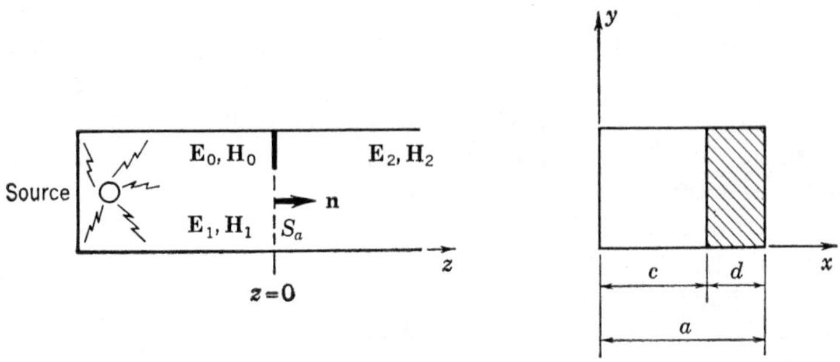


FIG. 7.14. Illustration for field equivalence principles.

$\mathbf{E}_1$, $\mathbf{H}_1$ into the left half guide. The total field in the latter region is the sum of $\mathbf{E}_1$, $\mathbf{H}_1$ and $\mathbf{E}_0$, $\mathbf{H}_0$.

The proof is readily established by noting that $\mathbf{E}_1$, $\mathbf{H}_1$ are source-free in the left-hand region, while $\mathbf{E}_0$, $\mathbf{H}_0$ provide the proper discontinuity or singularities at the source point, so that the total field $\mathbf{E}_1 + \mathbf{E}_0$, $\mathbf{H}_1 + \mathbf{H}_0$ satisfies all required boundary conditions in this region. At the aperture surface, the total tangential electric field is continuous since $\mathbf{n} \times \mathbf{E}_0 = 0$ on $S_a$, and $\mathbf{E}_1$, $\mathbf{E}_2$ have been determined so that $\mathbf{n} \times \mathbf{E}_1 = \mathbf{n} \times \mathbf{E}_2$ on $S_a$. The tangential magnetic field has been determined so that

$$\mathbf{n} \times \mathbf{H}_2 - \mathbf{n} \times \mathbf{H}_1 = \mathbf{n} \times \mathbf{H}_0$$

and, hence, the total tangential magnetic field $\mathbf{n} \times (\mathbf{H}_1 + \mathbf{H}_0)$ is equal to $\mathbf{n} \times \mathbf{H}_2$ in the aperture. Therefore, the specified field satisfies all the required boundary conditions and, by the uniqueness theorem, is the unique field radiated by the given sources in the open guide.

The following example will clarify some of the concepts involved. Let the sources be located at $z = -\infty$ and radiate an $H_{10}$ mode. With the aperture closed, the tangential fields in the region $z < 0$ are

$$\begin{aligned} E_{0y} &= A_1\phi_1(x)(e^{-\Gamma_1 z} - e^{\Gamma_1 z}) \\ j\omega\mu_0 H_{0x} &= -A_1\Gamma_1\phi_1(e^{-\Gamma_1 z} + e^{\Gamma_1 z}) \end{aligned}$$

and the aperture current is

$$-J_y = -\frac{2A_1\Gamma_1\phi_1}{j\omega\mu_0} \qquad 0 \le x \le c$$

where $\phi_n = \sin(n\pi x/a)$ and $\Gamma_n^2 = (n\pi/a)^2 - k_0^2$. With the aperture open, we must determine the field radiated by the known current $J_y$ such that the tangential electric field is continuous across $S_a$ and vanishes on the diaphragm, and such that the tangential magnetic field has a discontinuity equal to $J_y$ at $z = 0$ in the aperture. To solve the problem, we assume an unknown current $J_0$ on the diaphragm. Using the Green's function for $H_{n0}$ modes as derived in Chap. 5, the electric field radiated by the two current distributions is

$$E_y = -\frac{j\omega\mu_0}{a}\sum_{n=1}^{\infty}\frac{1}{\Gamma_n}e^{-\Gamma_n|z|}\phi_n(x)\left[\int_0^c \frac{-2A_1\Gamma_1\phi_1(x')}{j\omega\mu_0}\phi_n(x')\,dx' + \int_c^a J_0(x')\phi_n(x')\,dx'\right] \tag{50}$$

where $J_0$ must be found such that $E_y(x,0) = 0$ for $c \le x \le a$. From (50) it is easily seen that $E_y$ is continuous at $z = 0$, and, hence, the total tangential electric field is also continuous at $z = 0$ since $E_{0y}(x,0) = 0$. The total tangential magnetic field at $z = 0$ in the two regions must be continuous on $S_a$; hence,

$$\begin{aligned} j\omega\mu_0(H_{1x} + H_{0x}) &= -2\Gamma_1A_1\phi_1 - \frac{1}{a}\sum_{n=1}^{\infty}\phi_n\int_0^c(-2A_1\Gamma_1\phi_1\phi_n)\,dx' \\ &\quad - \frac{j\omega\mu_0}{a}\sum_{n=1}^{\infty}\phi_n\int_c^a J_0\phi_n\,dx' = \frac{1}{a}\sum_{n=1}^{\infty}\phi_n\int_0^c(-2A_1\Gamma_1\phi_1\phi_n)\,dx' \\ &\quad + \frac{j\omega\mu_0}{a}\sum_{n=1}^{\infty}\phi_n\int_c^a J_0\phi_n\,dx' = j\omega\mu_0H_{2x} \qquad \text{on } S_a \end{aligned} \tag{51}$$

The Fourier-series expansion of the current function $-2A_1\Gamma_1\phi_1/j\omega\mu_0$, $0 \le x \le c$, and zero otherwise, is

$$\sum_{n=1}^{\infty} a_n\phi_n \qquad \text{where } a_n = \frac{2}{a}\int_0^c \frac{-2A_1\Gamma_1\phi_1\phi_n}{j\omega\mu_0}\,dx'$$

Using this result (51) reduces to

$$-\Gamma_1 A_1 \phi_1 - \frac{j\omega\mu_0}{a} \sum_{n=1}^{\infty} \phi_n \int_c^a J_0 \phi_n \, dx' = -\Gamma_1 A_1 \phi_1 + \frac{j\omega\mu_0}{a} \sum_{n=1}^{\infty} \phi_n \int_0^a J_0 \phi_n \, dx' \qquad \text{on } S_a \quad (52)$$

The infinite series are Fourier series for a current function $J_0$ which is zero in the aperture, and, hence, the field is continuous in the aperture $S_a$. On the diaphragm, the tangential magnetic field is discontinuous by an amount equal to the current $J_0$ flowing on the diaphragm. This example clearly shows that the field equivalence principle leads to a unique solution.

For aperture problems the following equivalence principle is generally more useful.

*Theorem 2*

The sources are assumed located in the region $z < 0$. Let these sources radiate a field $\mathbf{E}_0$, $\mathbf{H}_0$ when the aperture is closed by a perfect magnetic

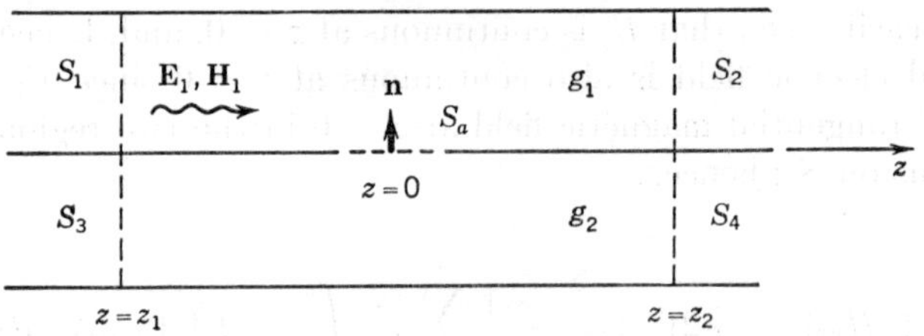


FIG. 7.15. Two guides coupled by a small aperture.

wall. With the aperture open, the total field for $z > 0$ may be found from an equivalent magnetic current $-\mathbf{n} \times \mathbf{E}_0 = \mathbf{J}_m$ on $S_a$. In the region $z < 0$ the given sources produce a field identical with the sum of the fields $\mathbf{E}_0$, $\mathbf{H}_0$ and the field radiated by $\mathbf{J}_m$. The proof is similar to that for theorem 1.

Consider two infinite guides $g_1$ and $g_2$ with a small aperture $S_a$ in the common wall, as illustrated in Fig. 7.15. If we decide to use an equivalent electric current distribution in the aperture, in order to compute the scattered field when a normal mode is incident in guide 1, we run into convergence difficulties. Each normal mode has a vanishing electric field over the aperture surface, and, hence, we are not able to find the coupling coefficient for each mode individually by the methods given in Sec. 5.6, even though we know the current in the aperture. The total tangential electric field in the aperture will differ from zero, and an infinite sum of normal mode functions will be required to represent this finite value of tangential electric field on $S_a$. The situation is much like that of expanding a square wave, which is not zero at the origin, into a Fourier sine series. The expansion can be found, but it will be discontinuous at the origin.

On the other hand, if we use an equivalent magnetic current in the aperture, the coupling with the normal modes arises from the tangential magnetic field, which is not equal to zero on the guide wall for the normal mode functions. With the aperture closed by a magnetic wall, let a normal mode $\mathbf{E}_1$, $\mathbf{H}_1$ be incident from the left. Because of the magnetic wall discontinuity, a scattered field $\mathbf{E}_s$, $\mathbf{H}_s$ will be excited such that the total field satisfies the following boundary conditions on the magnetic wall in the aperture $S_a$,

$$\mathbf{n} \times \mathbf{H}_s = -\mathbf{n} \times \mathbf{H}_1 \quad \text{or} \quad \mathbf{n} \times (\mathbf{H}_s + \mathbf{H}_1) = 0 \tag{53a}$$

$$\mathbf{n} \cdot \mathbf{E}_s = -\mathbf{n} \cdot \mathbf{E}_1 \quad \text{or} \quad \mathbf{n} \cdot (\mathbf{E}_s + \mathbf{E}_1) = 0 \tag{53b}$$

The normal magnetic field and the tangential electric field are not equal to zero on $S_a$, and, hence, a magnetic charge and a magnetic current distribution given by

$$\rho_m = \mu_0 \mathbf{n} \cdot \mathbf{H}_s \tag{54a}$$

$$\mathbf{J}_m = -\mathbf{n} \times \mathbf{E}_s \tag{54b}$$

will exist on $S_a$. For the incident mode, we have $\mathbf{n} \cdot \mathbf{H}_1 = \mathbf{n} \times \mathbf{E}_1 = 0$ on $S_a$. The scattered field may be expanded in terms of the normal waveguide modes as follows (see Sec. 5.6):

$$\mathbf{E}_s = \Sigma a_n \mathbf{E}_n^+ \qquad z > 0 \tag{55a}$$

$$\mathbf{H}_s = \Sigma a_n \mathbf{H}_n^+ \qquad z > 0 \tag{55b}$$

$$\mathbf{E}_s = \Sigma b_n \mathbf{E}_n^- \qquad z < 0 \tag{55c}$$

$$\mathbf{H}_s = \Sigma b_n \mathbf{H}_n^- \qquad z < 0 \tag{55d}$$

The scattered field is a solution of the equations

$$\nabla \times \mathbf{E}_s = -j\omega\mu_0 \mathbf{H}_s - \mathbf{J}_m$$

$\nabla \times \mathbf{H}_s = j\omega\epsilon_0 \mathbf{E}_s$, while each normal mode function $\mathbf{E}_n$, $\mathbf{H}_n$ is a solution of the source-free equations. If we apply the Lorentz reciprocity principle along the same lines as was done in Sec. 5.6, we find that the expansion coefficients $a_n$, $b_n$ are given by

$$2b_n = \iint_{S_a} \mathbf{H}_n^+ \cdot \mathbf{J}_m \, da \tag{56a}$$

$$2a_n = \iint_{S_a} \mathbf{H}_n^- \cdot \mathbf{J}_m \, da \tag{56b}$$

For a small aperture we will show that Eqs. (56) represent coupling to an electric and magnetic dipole plus a magnetic quadrupole.

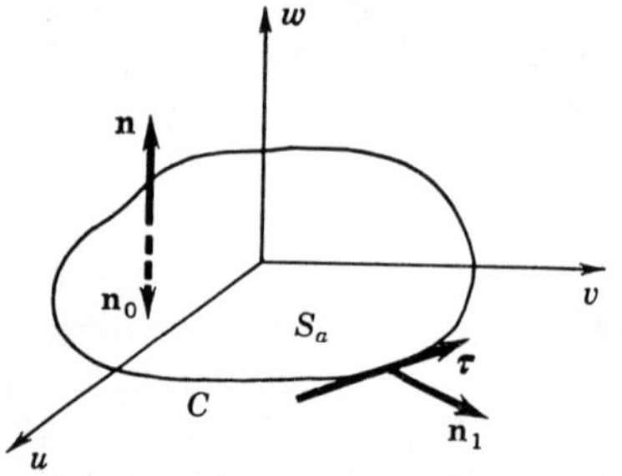


FIG. 7.16. Aperture coordinates.

With reference to Fig. 7.16 the following parameters are introduced:

1. $\boldsymbol{\tau}$ a unit vector tangent to the aperture contour $C$
2. $\mathbf{n}_1$ a unit vector normal to $\boldsymbol{\tau}$ and in the plane of the aperture
3. $\mathbf{n}$ a unit vector perpendicular to $S_a$ and directed into guide $g_1$
4. $u$, $v$, $w$ a localized rectangular coordinate system with the origin at the center of the aperture and $w$ directed along $\mathbf{n}$
5. $\mathbf{n}_0$ a unit vector normal to $S_a$ and directed from guide $g_1$ into guide $g_2$

The scattered field satisfies the boundary condition $\boldsymbol{\tau} \cdot \mathbf{E}_s = 0$, and, hence, $\mathbf{n}_1 \cdot \mathbf{J}_m$ equals zero on $C$; that is, there is no normal component of magnetic current at the boundary of the aperture. Since $\rho_m = \mu_0 \mathbf{n} \cdot \mathbf{H}_s$, we have

$$\mu_0^{-1} \iint_{S_a} \rho_m \, da = \iint_{S_a} \mathbf{H}_s \cdot \mathbf{n} \, da = \iint_{S_a} \frac{\nabla \times \mathbf{E}_s \cdot \mathbf{n} \, da}{-j\omega\mu_0}$$
$$= -\frac{1}{j\omega\mu_0} \oint_C \mathbf{E}_s \cdot \boldsymbol{\tau} \, dl = 0$$

which shows that there is no net magnetic charge on $S_a$. For a small aperture we may expand $\mathbf{H}_n$ in a Taylor series about the origin to get

$$\mathbf{H}_n(u,v) = \mathbf{H}_n(0) + \frac{u \, \partial \mathbf{H}_n}{\partial u}\bigg|_0 + \frac{v \, \partial \mathbf{H}_n}{\partial v}\bigg|_0 + \cdots$$
$$\approx \mathbf{H}_n(0) + \mathbf{r} \cdot \nabla \mathbf{H}_n$$

where $\mathbf{r} = \mathbf{a}_u u + \mathbf{a}_v v$, and $\nabla \mathbf{H}_n$ is to be evaluated at the origin. The coupling coefficient $b_n$ is given by

$$2b_n = \iint_{S_a} \mathbf{H}_n^+ \cdot \mathbf{J}_m \, da = \mathbf{H}_n^+(0) \cdot \iint_{S_a} \mathbf{J}_m \, da + \iint_{S_a} (\mathbf{r} \cdot \nabla \mathbf{H}_n^+) \cdot \mathbf{J}_m \, da \quad (57)$$

Consider the following integral:

$$\iint_{S_a} \nabla \cdot \phi \mathbf{J}_m \, da = \iint_{S_a} (\phi \nabla \cdot \mathbf{J}_m + \mathbf{J}_m \cdot \nabla \phi) \, da$$
$$= \oint_C \phi \mathbf{J}_m \cdot \mathbf{n}_1 \, dl = 0$$

since $\mathbf{J}_m \cdot \mathbf{n}_1 = 0$, and $\phi$ is an arbitrary scalar function. Let $\phi = u$, and the above result gives

$$\mathbf{a}_u \iint\limits_{S_a} \mathbf{J}_m \cdot \nabla u \, da = \mathbf{a}_u \iint\limits_{S_a} \mathbf{J}_m \cdot \mathbf{a}_u \, da = -\mathbf{a}_u \iint\limits_{S_a} u \nabla \cdot \mathbf{J}_m \, da \qquad (58)$$

A similar result is obtained if we let $\phi = v$. Combining the two gives

$$\begin{aligned} \iint\limits_{S_a} \mathbf{J}_m \, da &= -\iint\limits_{S_a} \mathbf{r} \nabla \cdot \mathbf{J}_m \, da \\ &= j\omega \iint\limits_{S_a} \mathbf{r}\rho_m \, da \\ &= j\omega\mu_0 \mathbf{M} \end{aligned} \qquad (59)$$

since $\nabla \cdot \mathbf{J}_m = -j\omega\rho_m$ from the continuity equation relating current and charge, and the integral $\frac{1}{\mu_0} \iint\limits_{S_a} \mathbf{r}\rho_m \, da$ defines an equivalent magnetic dipole moment $\mathbf{M}$ in direct analogy with the electric dipole moment arising from electric charge $\rho$. The first term in (57) is thus seen to represent coupling with a magnetic dipole $\mathbf{M}$.

The second integral in (57) is readily evaluated by writing the integrand in component form. We have

$$\begin{aligned} \mathbf{r} \cdot \nabla \mathbf{H}_n^+ \cdot \mathbf{J}_m &= (\mathbf{a}_u u + \mathbf{a}_v v) \cdot \left( \mathbf{a}_u \mathbf{a}_u \frac{\partial H_{nu}^+}{\partial u} + \mathbf{a}_u \mathbf{a}_v \frac{\partial H_{nv}^+}{\partial u} + \mathbf{a}_v \mathbf{a}_u \frac{\partial H_{nu}^+}{\partial v} \right. \\ &\qquad \left. + \mathbf{a}_v \mathbf{a}_v \frac{\partial H_{nv}^+}{\partial v} \right) \cdot (\mathbf{a}_u J_{mu} + \mathbf{a}_v J_{mv}) \\ &= u \frac{\partial H_{nu}^+}{\partial u} J_{mu} + u \frac{\partial H_{nv}^+}{\partial u} J_{mv} + v \frac{\partial H_{nu}^+}{\partial v} J_{mu} + v \frac{\partial H_{nv}^+}{\partial v} J_{mv} \end{aligned}$$

Subtracting and adding similar terms, this may be rewritten as

$$\begin{aligned} &\frac{u}{2} J_{mv} \left( \frac{\partial H_{nv}^+}{\partial u} - \frac{\partial H_{nu}^+}{\partial v} \right) - \frac{v J_{mu}}{2} \left( \frac{\partial H_{nv}^+}{\partial u} - \frac{\partial H_{nu}^+}{\partial v} \right) + \frac{u J_{mv}}{2} \frac{\partial H_{nu}^+}{\partial v} \\ &+ \frac{v J_{mu}}{2} \frac{\partial H_{nv}^+}{\partial u} + u J_{mu} \frac{\partial H_{nu}^+}{\partial u} + v J_{mv} \frac{\partial H_{nv}^+}{\partial v} + \frac{u J_{mv}}{2} \frac{\partial H_{nv}^+}{\partial u} + \frac{v J_{mu}}{2} \frac{\partial H_{nu}^+}{\partial v} \end{aligned} \qquad (60)$$

The first two terms are readily recognized as being equal to $j\omega\epsilon_0 \mathbf{E}_n^+(0) \cdot (\mathbf{r} \times \mathbf{J}_m)/2$. The magnetic dipole moment of an electric current distribution is defined by the integral $\iiint \frac{\mathbf{r} \times \mathbf{J}}{2} \, dv$. By analogy, we now see that

$$-j\omega\epsilon_0 \mathbf{E}_n^+(0) \cdot \iint\limits_{S_a} \frac{-\mathbf{r} \times \mathbf{J}_m}{2} \, da = -j\omega \mathbf{E}_n^+(0) \cdot \mathbf{P}$$

where **P** is the equivalent electric dipole moment of the circulating magnetic current. This dipole is directed normal to the aperture.

If we let $\phi = u^2/2,\ v^2/2$, and $uv$ in turn, then in (58) we get the results

$$\iint_{S_a} uJ_{mu}\,da = -\iint_{S_a} \frac{u^2}{2}\nabla\cdot\mathbf{J}_m\,da = \frac{j\omega}{2}\iint_{S_a} u^2\rho_m\,da$$

$$\iint_{S_a} vJ_{mv}\,da = \frac{j\omega}{2}\iint_{S_a} v^2\rho_m\,da$$

$$\iint_{S_a} (vJ_{mu} + uJ_{mv})\,da = j\omega\iint_{S_a} uv\rho_m\,da$$

The components of a dyadic magnetic quadrupole $\bar{\mathbf{Q}}$ are given by

$$Q_{uu} = \frac{1}{\mu_0}\iint_{S_a} u^2\rho_m\,da$$

$$Q_{vu} = Q_{uv} = \frac{1}{\mu_0}\iint_{S_a} uv\rho_m\,da \qquad \text{etc.}$$

and, hence, the remaining six terms in (60) may be combined by using the above relations to give finally

$$\frac{j\omega\mu_0}{2}\nabla\mathbf{H}_n^+ : \bar{\mathbf{Q}}$$

where the double-dot product is taken between the two dyadics $\nabla\mathbf{H}_n^+$ and $\bar{\mathbf{Q}}$.

The expansion coefficients $a_n$ and $b_n$ are thus seen to be given by

$$2a_n = j\omega\left(\mu_0\mathbf{H}_n^- \cdot \mathbf{M} - \mathbf{E}_n^- \cdot \mathbf{P} + \frac{\mu_0}{2}\nabla\mathbf{H}_n^- : \bar{\mathbf{Q}}\right) \tag{61a}$$

$$2b_n = j\omega\left(\mu_0\mathbf{H}_n^+ \cdot \mathbf{M} - \mathbf{E}_n^+ \cdot \mathbf{P} + \frac{\mu_0}{2}\nabla\mathbf{H}_n^+ : \bar{\mathbf{Q}}\right) \tag{61b}$$

In the definition of the dipole moments, the origin for the radius vector **r** is arbitrary, since there is no net charge in the aperture. The magnetic dipole moments have been defined to be dimensionally the same as **H** times length cubed. In most applications the quadrupole term can be neglected, since it represents a small quantity depending on the square of the aperture dimension.

Although we have written the expansion coefficients in terms of equivalent dipole moments, we do not know these before we find a solution for the current $\mathbf{J}_m$ and charge $\rho_m$ in the aperture. This solution is difficult to obtain in general. However, for a small aperture we may find a static field solution for the dipole moments of elliptic- and circular-shaped apertures quite readily. Bouwkamp has shown that this does not lead to a solution for $\mathbf{J}_m$ correct to the first order in the parameter $k_0d$, where $d$ is the radius for a circular aperture.† In the practical application of the theory, it is nevertheless found that the static field solution gives results of acceptable accuracy for small apertures. To further enhance the applicability of the theory, Cohn has described an electrolytic-tank method for measuring polarizabilities of arbitrary-shaped apertures.‡

Before presenting the static field solution for the dipole moments, the rest of the theory required to obtain the field coupled through the aperture into guide $g_2$ will be given. When the aperture is open, the field in $g_2$ is that radiated by an equivalent current $-\mathbf{J}_m$ on $S_a$. The field in $g_1$ is the sum of the field radiated by $-\mathbf{J}_m$ in the aperture and the field $\mathbf{E}_1 + \mathbf{E}_s$, $\mathbf{H}_1 + \mathbf{H}_s$ existing in $g_1$ with the aperture closed by a magnetic wall. Let the current $-\mathbf{J}_m$ radiate a field $\mathbf{E}_{s1}$, $\mathbf{H}_{s1}$ into $g_1$, and $\mathbf{E}_{s2}$, $\mathbf{H}_{s2}$ into $g_2$. At the aperture, the total tangential magnetic field must be continuous, and, hence, $\mathbf{n}_0 \times \mathbf{H}_{s2} = \mathbf{n}_0 \times (\mathbf{H}_{s1} + \mathbf{H}_s + \mathbf{H}_1) = \mathbf{n}_0 \times \mathbf{H}_{s1}$ from (53$a$). The total tangential electric field must also be continuous, and so we have $\mathbf{n}_0 \times \mathbf{E}_{s2} = \mathbf{n}_0 \times (\mathbf{E}_{s1} + \mathbf{E}_s + \mathbf{E}_1)$, or, by using (54$b$), $\mathbf{n}_0 \times (\mathbf{E}_{s1} - \mathbf{E}_{s2}) = -\mathbf{J}_m$. The field in $g_2$ may be expanded in terms of the normal mode functions for this guide as follows:

$$\mathbf{E}_{s2} = \Sigma c_n \mathbf{E}_{n2}^+ \qquad z > 0 \tag{62a}$$
$$\mathbf{H}_{s2} = \Sigma c_n \mathbf{H}_{n2}^+ \qquad z > 0 \tag{62b}$$
$$\mathbf{E}_{s2} = \Sigma d_n \mathbf{E}_{n2}^- \qquad z < 0 \tag{62c}$$
$$\mathbf{H}_{s2} = \Sigma d_n \mathbf{H}_{n2}^- \qquad z < 0 \tag{62d}$$

where the additional subscript 2 is used to distinguish between the normal modes for the two guides. If the two guides are identical, then the current $-\mathbf{J}_m$ will radiate identical fields into the two guides. The expansion coefficients may be found in the same way as for the field $\mathbf{E}_s$, $\mathbf{H}_s$, but it must be borne in mind that the tangential electric field and normal magnetic field in one guide will undergo only one-half the discontinuous change across the source region that the field $\mathbf{E}_s$, $\mathbf{H}_s$ did. The effective dipole moments associated with the aperture current $-\mathbf{J}_m$ and charge $-\rho_m$ for radiation into guide $g_2$ will be $-\mathbf{M}/2$ and $-\mathbf{P}/2$. One-half the required discontinuity across the source is provided by $\mathbf{E}_{s2}$, $\mathbf{H}_{s2}$; the other half by $\mathbf{E}_{s1}$, $\mathbf{H}_{s1}$. The total field in $g_1$ is now seen to be equivalent to the sum of the incident fields $\mathbf{E}_1$, $\mathbf{H}_1$ and the field radiated by dipoles of strength $\mathbf{M}/2$ and $\mathbf{P}/2$ since the field radiated by $-\mathbf{M}/2$, $-\mathbf{P}/2$ cancels one-half of the field $\mathbf{E}_s$, $\mathbf{H}_s$ radiated by $\mathbf{M}$ and $\mathbf{P}$.

† C. J. Bouwkamp, Diffraction Theory, *Repts. Progr. in Phys.*, vol. 17, pp. 75–100, 1954.

‡ S. B. Cohn, Determination of Aperture Parameters by Electrolytic Tank Measurements, *Proc. IRE*, vol. 39, pp. 1416–1421, November, 1951.

In the more general case, when the two guides are not identical, this symmetry argument does not apply any more. However, if the aperture is located in a plane section of the guide wall, and not too close to a corner or bend, we may approximately consider the problem as an aperture in a plane conducting screen for which the above symmetry conditions do hold. With these reservations in mind, we are thus able to state that the field radiated in $g_2$ through a small aperture $S_a$ when a mode is incident in $g_1$ is equivalent to that radiated by a magnetic dipole $-\mathbf{M}/2$ and an electric dipole $-\mathbf{P}/2$ located in the aperture. The scattered field in $g_1$ is that produced by dipoles $\mathbf{M}/2$ and $\mathbf{P}/2$ located in the aperture. Furthermore, the field radiated into each guide is to be computed from the respective aperture dipoles with the other guide considered absent; i.e., the proper continuity of the field at the aperture has been assured by this choice of dipole moments.

*Dipole Moments of Elliptic Apertures*

The dipole moments of an elliptic aperture may be obtained from the static dipole moments of general ellipsoidal dielectric and permeable magnetic bodies, placed in uniform static electric and magnetic fields, by suitable limiting processes. Consider a dielectric ellipsoid with semi-axes $l_1$, $l_2$, $l_3$ placed in a uniform electric field $E_0$ directed along the $l_3$ axis as in Fig. 7.17.

The problem of the ellipsoid in a uniform field is readily solved in ellipsoidal coordinates.† We find that the total field in the interior is uniform and parallel to the applied field if the latter is directed along one of the axes of the ellipsoid. The dipole polarization per unit volume, therefore, has zero divergence, and the equivalent polarization volume charge in the interior is zero. On the surface, an equivalent surface-polarization charge density equal to the discontinuity in the normal component of the polarization vector exists. The total dipole moment of the ellipsoid may be calculated by a volume integral either in terms of the polarization density or in terms of the equivalent polarization surface charge. The field outside the ellipsoid is the sum of the applied field plus a dipole field arising from an elementary dipole at the origin and having a moment equal to the dipole moment of the ellipsoid. According to Stratton, the dipole moment $P_3$, when a uniform field $E_0$ is applied along the $l_3$ axis, is

$$P_3 = \frac{\epsilon_0 E_0 V}{L_3 + \epsilon_0/(\epsilon - \epsilon_0)} \tag{63}$$

where $$L_3 = \frac{l_1 l_2 l_3}{2} \int_0^\infty \frac{ds}{(s + l_3^2)[(s + l_1^2)(s + l_2^2)(s + l_3^2)]^{1/2}}$$

† J. A. Stratton, "Electromagnetic Theory," sec. 3.27, McGraw-Hill Book Company, Inc., New York, 1941.

and $V = \frac{4}{3}\pi l_1 l_2 l_3$ is the volume of the ellipsoid. If we let $\epsilon$ be equal to zero, we obtain the dual of a perfect diamagnetic material. For $\epsilon = 0$, the internal polarization sets up a field equal and opposite to the applied field, so that the net internal displacement flux $D$ is zero. The surface polarization charge now gives rise to an external field, which cancels the normal component of the applied field $E_0$ at the surface of the ellipsoid. This is precisely the boundary condition that the electric field

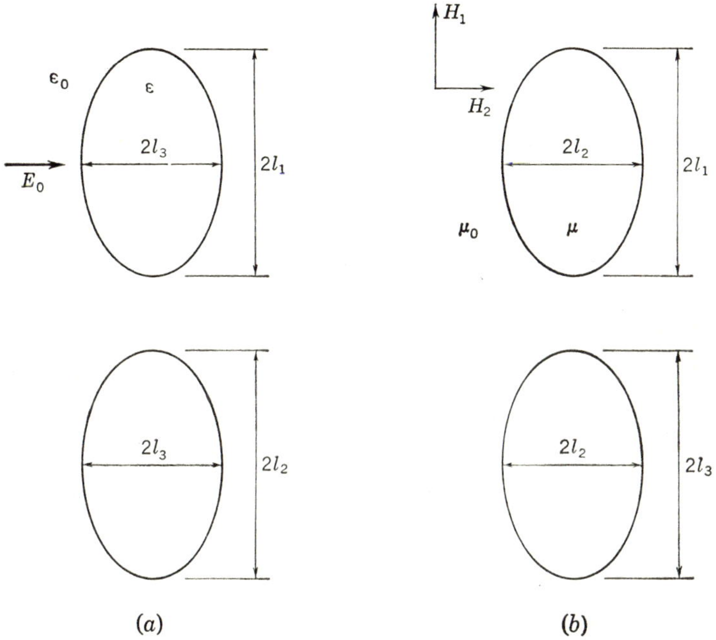


FIG. 7.17. Dielectric and permeable magnetic ellipsoids.

must satisfy at a magnetic wall. Hence, a dielectric body with $\epsilon = 0$ is equivalent to a body with a magnetic wall surface. If we now let $l_3$ approach zero, (63) gives the dipole moment of an elliptic disk which is made from a perfect "magnetic conductor." The dipole moment of the aperture is only one-half that of a disk, since only one side of the aperture is under consideration while the disk has two sides with polarization charge on both. Finally, the effective radiating dipole moment in the aperture is one-half of the moment of an aperture closed by a magnetic wall, and hence equals one-quarter of the moment of a complete disk. Before evaluating the integral for $L_3$, we will give the expressions for the magnetic dipole moments of an ellipsoid.

Figure 7.17*b* illustrates a permeable ellipsoid placed in a uniform magnetic field. There are two dipole moments to consider: $M_1$ due to a field $H_1$ along the $l_1$ axis, and $M_2$ due to a field $H_2$ along the $l_2$ axis. Each case may be treated separately, and the solution is essentially the same as for the dielectric case. We find that

$$M_1 = \frac{VH_1}{L_1 + \mu_0/(\mu - \mu_0)} \tag{64a}$$

where $$L_1 = \frac{l_1 l_2 l_3}{2} \int_0^\infty \frac{ds}{(s + l_1^2)[(s + l_1^2)(s + l_2^2)(s + l_3^2)]^{1/2}}$$

and $$M_2 = \frac{VH_2}{L_2 + \mu_0/(\mu - \mu_0)} \tag{64b}$$

where $$L_2 = \frac{l_1 l_2 l_3}{2} \int_0^\infty \frac{ds}{(s + l_2^2)[(s + l_2^2)(s + l_3^2)(s + l_1)^2]^{1/2}}$$

If we let $\mu$ approach infinity, the internal magnetic field vanishes, since $\mathbf{n} \cdot \mu_0 \mathbf{H}_e$ equals $\mathbf{n} \cdot \mu \mathbf{H}_i$, where $\mathbf{H}_e$ is the external field, and $\mathbf{H}_i$ is the internal field, and, hence, the product $\mu \mathbf{H}_i$ is finite and $\mathbf{H}_i$ tends to zero as $\mu$ tends to infinity. For an external field applied along a principal axis, the internal field is uniform and parallel with the applied field, and, since it vanishes at the surface, it follows that the induced external field just cancels the tangential component of the applied field at the surface. Again we have boundary conditions corresponding to those at a magnetic wall, so that, by letting $l_3$ approach zero and $\mu$ tend to infinity, we obtain the magnetic dipole moments of an elliptical disk having perfect magnetic conductivity.

The integrals for $L_1$, $L_2$, and $L_3$ may be evaluated in terms of elliptic integrals. First we show that $L_1 + L_2 + L_3 = 1$, and so we need evaluate only two of the integrals. If a new variable

$$u^2 = (s + l_1^2)(s + l_2^2)(s + l_3^2)$$

is introduced, we get

$$L_1 + L_2 + L_3$$
$$= \frac{l_1 l_2 l_3}{2} \int_0^\infty \frac{(s + l_2^2)(s + l_3^2) + (s + l_3^2)(s + l_1^2) + (s + l_1^2)(s + l_2^2)}{u^3}\, ds$$

Using the familiar rule for the differential of a product, it is seen at once that the numerator is equal to $2u\,du$. Putting in the proper limits of integration for the variable $u$, the integral becomes

$$\frac{l_1 l_2 l_3}{2} \int_{l_1 l_2 l_3}^\infty \frac{2\,du}{u^2} = 1$$

The integrals for $L_1$ and $L_2$ are somewhat easier to evaluate than that for $L_3$. The procedure used will be given only for the case of $L_2$, the evaluation of $L_1$ being similar. We may take $l_1 > l_2 > l_3$ without any loss in generality. In the integral for $L_2$, we introduce a new variable $t$ according to the relation $s + l_2^2 = t^2$ and get

$$L_2 = l_1 l_2 l_3 \int_{l_2}^{\infty} \frac{dt}{t^2(t^2 + l_1^2 - l_2^2)^{1/2}(t^2 - l_2^2 + l_3^2)^{1/2}}$$

Multiplying the numerator by $[(t^2 + l_1^2 - l_2^2) - t^2]/(l_1^2 - l_2^2)$ the integral may be split into two parts as follows:

$$\frac{l_1 l_2 l_3}{l_1^2 - l_2^2}\left[\int_{l_2}^{\infty} \frac{(t^2 + l_1^2 - l_2^2)^{1/2}}{t^2(t^2 - l_2^2 + l_3^2)^{1/2}}\,dt - \int_{l_2}^{\infty} \frac{dt}{(t^2 + l_1^2 - l_2^2)^{1/2}(t^2 - l_2^2 + l_3^2)^{1/2}}\right]$$

Apart from the limits of integration, the integrals occurring here are of the form that Jahnke and Emde have tabulated the solutions for.† In order to use the solutions presented, the integrals here are rewritten as the sum of two integrals with limits chosen as indicated below:

$$\int_{l_2}^{x} = \int_{(l_2^2 - l_3^2)^{1/2}}^{x} - \int_{(l_2^2 - l_3^2)^{1/2}}^{l_2}$$

For our application we will set $l_3$ equal to zero eventually, and so the second integral will vanish and, therefore, will not be evaluated. The value of the required integral is

$$\frac{l_1 l_2 l_3}{l_1^2 - l_3^2}\left[\frac{(l_1^2 - l_3^2)^{1/2}}{l_2^2 - l_3^2}\,\mathcal{E}(k,\phi) - \frac{1}{(l_1^2 - l_3^2)^{1/2}}\,F(k,\phi)\right]$$

where $$k = \left(\frac{l_1^2 - l_2^2}{l_1^2 - l_3^2}\right)^{1/2} \qquad \cos\phi = \frac{(l_2^2 - l_3^2)^{1/2}}{x}$$

and $F$ and $\mathcal{E}$ are the incomplete elliptic integrals of the first and second kind. As $x$ tends to infinity, $\phi$ becomes equal to $\pi/2$, and, when $l_3$ is placed equal to zero, we get

$$\frac{L_2}{V} = \frac{3}{4\pi l_1^3 e^2}\left[(1 - e^2)^{-1}E(e) - K(e)\right] \tag{65}$$

where $K$ and $E$ are complete elliptic integrals of the first and second kind with modulus $e$ equal to the eccentricity $(1 - l_2^2/l_1^2)^{1/2}$ of the ellipse.

In a similar fashion we find that

$$\frac{L_1}{V} = \frac{3}{4\pi l_1^3 e^2}\left[K(e) - E(e)\right] \tag{66}$$

$$\frac{L_3 - 1}{V} = -\frac{L_1 + L_2}{V} = \frac{-3E(e)}{4\pi l_1^3(1 - e^2)} \tag{67}$$

† E. Jahnke and F. Emde, "Tables of Functions," 4th ed., p. 58, Dover Publications, New York, 1945. Fifth formula pair for evaluation of $L_2$, and last pair for evaluation of $L_1$.

Substituting into (63) and (64), the dipole moments are obtained:

$$P_3 = -\frac{4\pi l_1^3(1-e^2)}{3E(e)}\epsilon_0 E_0 \tag{68a}$$

$$M_1 = \frac{4\pi l_1^3 e^2}{3[K(e)-E(e)]}H_1 \tag{68b}$$

$$M_2 = \frac{4\pi l_1^3 e^2(1-e^2)}{3[E(e)-(1-e^2)K(e)]}H_2 \tag{68c}$$

Returning to our waveguide-coupling problem, let $\mathbf{E}_1$, $\mathbf{H}_1$ be the incident mode in guide $g_1$. The scattered field in this guide is that radiated by an electric dipole $\mathbf{P}_0 = \frac{1}{2}\mathbf{P}$ and a magnetic dipole $\mathbf{M}_0 = \frac{1}{2}\mathbf{M}$, where

$$\mathbf{P}_0 = \epsilon_0\alpha_e\mathbf{nn}\cdot\mathbf{E}_1 \tag{69a}$$

$$\mathbf{M}_0 = \bar{\boldsymbol{\alpha}}_m\cdot\mathbf{H}_1 \tag{69b}$$

where from (68) we have, after dividing by 4,

$$\alpha_e = -\frac{\pi l_1^3(1-e^2)}{3E(e)} \tag{70a}$$

$$\bar{\boldsymbol{\alpha}}_m = \mathbf{a}_u\mathbf{a}_u\frac{\pi l_1^3 e^2}{3[K(e)-E(e)]} + \mathbf{a}_v\mathbf{a}_v\frac{\pi l_1^3 e^2(1-e^2)}{3[E(e)-(1-e^2)K(e)]} \tag{70b}$$

In (70*a*), $\alpha_e$ is the electric polarizability of the aperture, while, in (70*b*), $\bar{\boldsymbol{\alpha}}_m$ is the dyadic magnetic polarizability of the aperture. The aperture coordinates $u$ and $v$ are to be oriented with $u$ along the major axis of the ellipse. For small values of $e$, the following formulas may be used to evaluate the elliptic integrals:

$$E = \frac{\pi}{2}\left(1-\frac{e^2}{4}\right)$$

$$K = \frac{\pi}{2}\left(1+\frac{e^2}{4}\right)$$

while, for values of $e$ approaching unity,

$$E = 1$$

$$K = \ln 4\frac{l_1}{l_2}$$

For $e$ equal to zero, we obtain a circular aperture with polarizabilities

$$\alpha_e = -\tfrac{2}{3}l^3 \tag{71}$$

$$\bar{\boldsymbol{\alpha}}_m = \tfrac{4}{3}l^3(\mathbf{a}_u\mathbf{a}_u + \mathbf{a}_v\mathbf{a}_v) \tag{72}$$

where $l$ is the radius of the aperture. The field radiated into guide $g_2$ is that due to radiating dipoles $-\mathbf{P}_0$ and $-\mathbf{M}_0$ in the aperture. The procedure to be followed in evaluating the scattered fields will be illustrated by specific examples pertaining to the rectangular waveguide.

*Aperture Coupling in Rectangular Guides*

Consider an elliptic aperture in the common side wall of two rectangular guides as in Fig. 7.18. The coordinates $x_1$, $y_1$, $z_1$ pertain to guide $g_1$, and $x_2$, $y_2$, $z_2$ to guide $g_2$. The center of the aperture is at $y_1 = b/2$, $z_1 = 0$. The major and minor semiaxes are $l_1$ and $l_2$. An $H_{10}$ mode is incident from the left in guide $g_1$. We will assume that only the $H_{10}$

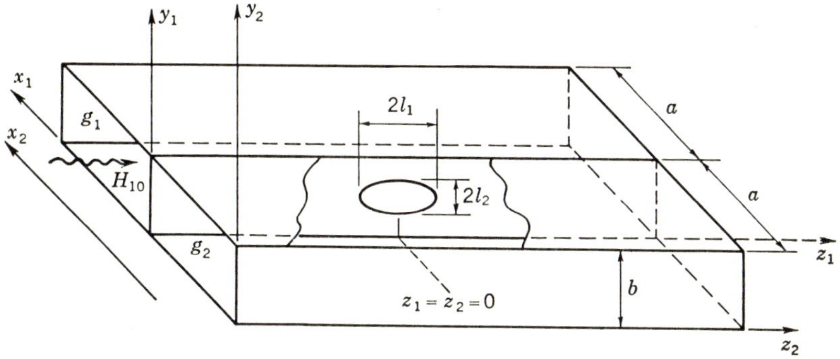


FIG. 7.18. Aperture-coupled rectangular guide.

mode propagates, and hence we will evaluate only the amplitudes of the scattered $H_{10}$ modes. The normalized mode functions for the $H_{10}$ mode are

$$
\begin{aligned}
\mathbf{E}_{10}^{+} &= \mathbf{e}_{10} e^{-\Gamma_{10} z} \\
\mathbf{E}_{10}^{-} &= \mathbf{e}_{10} e^{\Gamma_{10} z} \\
\mathbf{H}_{10}^{+} &= (\mathbf{h}_{10} + \mathbf{h}_{z10}) e^{-\Gamma_{10} z} \\
\mathbf{H}_{10}^{-} &= (-\mathbf{h}_{10} + \mathbf{h}_{z10}) e^{\Gamma_{10} z} \\
\mathbf{e}_{10} &= -jk_0 Z_0 \left( \frac{2}{jabk_0 Z_0 \Gamma_{10}} \right)^{1/2} \sin \frac{\pi x}{a} \mathbf{a}_y \\
\mathbf{h}_{10} &= \Gamma_{10} \left( \frac{2}{jabk_0 Z_0 \Gamma_{10}} \right)^{1/2} \sin \frac{\pi x}{a} \mathbf{a}_x \\
\mathbf{h}_{z10} &= \left( \frac{2}{jabk_0 Z_0 \Gamma_{10}} \right)^{1/2} \frac{\pi}{a} \cos \frac{\pi x}{a} \mathbf{a}_z
\end{aligned}
$$

Let the incident mode be $A_1\mathbf{E}_{10}^{+}$, $A_1\mathbf{H}_{10}^{+}$, where $A_1$ is the amplitude constant. The incident mode does not have a component of electric field normal to the aperture, and so there will not be an induced electric dipole $\mathbf{P}_0$. The incident magnetic field has a tangential component along the major axis of the ellipse only, and so the induced magnetic dipole moment is

$$
\begin{aligned}
\mathbf{M}_0 &= \bar{\boldsymbol{\alpha}}_m \cdot A_1 \mathbf{H}_{10}^{+} \\
&= \frac{\pi l_1^3 e^2}{3[K(e) - E(e)]} A_1 \mathbf{h}_{z10} \\
&= \frac{\pi l_1^3 e^2}{3[K(e) - E(e)]} A_1 \frac{\pi}{a} \left( \frac{2}{jabk_0 Z_0 \Gamma_{10}} \right)^{1/2} \mathbf{a}_z
\end{aligned} \tag{73}
$$

Let the scattered field in $g_1$ be

$$\begin{matrix} a_1\mathbf{E}_{10}^+ & a_1\mathbf{H}_{10}^+ & z_1 > 0 \\ b_1\mathbf{E}_{10}^- & b_1\mathbf{H}_{10}^- & z_1 < 0 \end{matrix}$$

Equations (61) are applicable in the present case, and hence

$$2b_1 = j\omega\mu_0\mathbf{H}_{10}^+ \cdot \mathbf{M}_0 = j\omega\mu_0\mathbf{h}_{z10} \cdot \mathbf{M}_0 \tag{74a}$$

$$2a_1 = j\omega\mu_0\mathbf{H}_{10}^- \cdot \mathbf{M}_0 = j\omega\mu_0\mathbf{h}_{z10} \cdot \mathbf{M}_0 = 2b_1 \tag{74b}$$

The total field in $g_1$ is the sum of the incident field plus the scattered field. If a reflection coefficient $R$ is defined by the relation $b_1 = RA_1$, we find that

$$R = -\frac{j\omega\mu_0}{abk_0Z_0\beta_{10}}\left(\frac{\pi}{a}\right)^2 \frac{\pi l_1^3e^2}{3[K(e) - E(e)]} \tag{75}$$

where $\beta_{10} = |\Gamma_{10}|$. A small shunt susceptance of normalized value $jB$ across a transmission line produces a reflection coefficient

$$\frac{-jB}{2 + jB} \approx \frac{-jB}{2}$$

and, hence, for the incident wave the aperture appears as a shunt susceptance $jB = -2R$.

In guide $g_2$ let the scattered field be

$$\begin{matrix} c_1\mathbf{E}_{10}^+ & c_1\mathbf{H}_{10}^+ & z_2 > 0 \\ d_1\mathbf{E}_{10}^- & d_1\mathbf{H}_{10}^- & z_2 < 0 \end{matrix}$$

If we had a current element $\mathbf{J}$ in $g_2$, an application of the Lorentz reciprocity theorem would give

$$2c_1 = -\iiint \mathbf{E}_{10}^- \cdot \mathbf{J}\,dv$$

If $\mathbf{J}$ consists of a small linear element $\mathbf{J}_1$ and a circulating element $J_2\boldsymbol{\tau}$, where $\boldsymbol{\tau}$ is a unit vector tangent to the contour around which $J_2$ flows, we have

$$2c_1 = -\int \mathbf{E}_{10}^- \cdot \mathbf{J}_1\,dl - \oint J_2\mathbf{E}_{10}^- \cdot \boldsymbol{\tau}\,dl$$

The second integral is equal to

$$-J_2\iint \nabla \times \mathbf{E}_{10}^- \cdot d\mathbf{a} = j\omega\mu_0 J_2 \iint \mathbf{H}_{10}^- \cdot d\mathbf{a}$$

or, for a very small loop, equal to

$$j\omega\mu_0\mathbf{H}_{10}^- \cdot \iint J_2\,d\mathbf{a}$$

This latter integral defines a magnetic dipole $\mathbf{M}$, and, since $\int \mathbf{J}_1\,dl$ is equivalent to an electric dipole $j\omega\mathbf{P}$, we get

$$2c_1 = -j\omega\mathbf{E}_{10}^- \cdot \mathbf{P} + j\omega\mu_0\mathbf{H}_{10}^- \cdot \mathbf{M}$$

For radiation into $g_2$ the effective aperture dipoles are $-\mathbf{P}_0$, $-\mathbf{M}_0$ in general. In the present case $\mathbf{P}_0 = 0$, and so we have

$$2c_1 = -j\omega\mu_0\mathbf{H}_{10}^- \cdot \mathbf{M}_0 = -j\omega\mu_0\mathbf{h}_{z10} \cdot \mathbf{M}_0$$
$$2d_1 = -j\omega\mu_0\mathbf{H}_{10}^+ \cdot \mathbf{M}_0 = 2c_1$$

In the coordinate system used for $g_2$, we have, at $z_2 = 0$, $x_2 = a$,

$$\mathbf{h}_{z10} = -\left(\frac{2}{jabk_0Z_0\Gamma_{10}}\right)^{1/2}\frac{\pi}{a}\mathbf{a}_z$$

and hence $c_1 = d_1 = a_1 = b_1$. Note that the tangential magnetic fields in the scattered waves are of opposite sign at the aperture. This must be the case since their difference must give the nonzero value equal to the tangential component of the incident magnetic field.

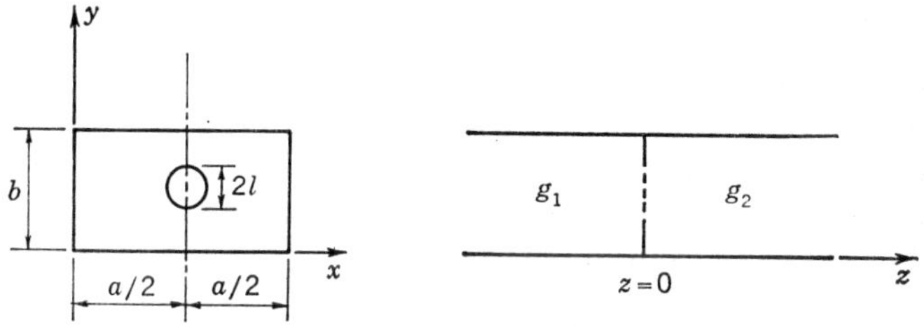


FIG. 7.19. Aperture in a transverse wall.

As a second example, consider a rectangular guide with a transverse wall containing a small central circular aperture as in Fig. 7.19. The radius of the aperture is $l$. The incident transverse fields in $g_1$ are $A_1\mathbf{e}_{10}(e^{-\Gamma_{10}z} - e^{\Gamma_{10}z})$, $A_1\mathbf{h}_{10}(e^{-\Gamma_{10}z} + e^{\Gamma_{10}z})$. There is only a tangential magnetic field giving rise to a dipole moment. The magnetic dipole moment is

$$\begin{aligned}\mathbf{M}_0 &= \frac{4}{3}l^3\mathbf{a}_x\mathbf{a}_x \cdot 2A_1\mathbf{h}_{10}\\ &= \frac{8}{3}l^3A_1\Gamma_{10}\left(\frac{2}{jabk_0Z_0\Gamma_{10}}\right)^{1/2}\mathbf{a}_x\end{aligned} \tag{76}$$

In $g_1$ the dominant-mode scattered field is $b_1\mathbf{E}_{10}^-$, $b_1\mathbf{H}_{10}^-$, while in $g_2$ it is $c_1\mathbf{E}_{10}^+$, $c_1\mathbf{H}_{10}^+$. The Lorentz reciprocity theorem gives (see Probs. 7.12 and 7.14)

$$\begin{aligned}b_1 &= j\omega\mu_0\mathbf{h}_{10} \cdot \mathbf{M}_0\\ &= \frac{16}{3}\frac{l^3\beta_{10}j\omega\mu_0}{abk_0Z_0}A_1\end{aligned} \tag{77}$$

The total reflected wave is $(b_1 - A_1)\mathbf{E}_{10}^-$ and the equivalent reflection coefficient may be defined so that $b_1 - A_1 = RA_1$ or $(1 + R)A_1 = b_1$. A susceptance $jB$ gives a reflection coefficient $R = -jB/(2 + jB)$, and, when $B$ is very large, $1 + R \approx 2/jB$. Thus the aperture appears as an inductive susceptance

$$B = -\frac{3ab}{8l^3\beta_{10}} \tag{78}$$

For $z > 0$, we get $c_1 = -j\omega\mu_0\mathbf{h}_{10} \cdot \mathbf{M}_0 = b_1$, as would be expected, since the transmission coefficient for any size aperture in a plane transverse wall is equal to $1 + R$.

In the case of the rectangular guide more accurate expressions for the aperture dipole strengths may be obtained by taking into account the interaction field, due to the dipole images in the guide wall, which also acts to polarize the aperture. The procedure to be used is similar to that given in Chap. 12 in connection with artificial dielectric media.†

## 7.4. Transients in Waveguides

In any waveguide, the phase velocity $v_p$ is a function of frequency. As a consequence, all signals having a finite frequency spectrum will undergo dispersion when transmitted through a length of guide. The phase relationship between the frequency components of the original signal at the feeding point continually changes as the signal progresses along the guide. The analysis of this phenomenon belongs in the domain of transient analysis, and may be handled by the conventional techniques utilizing Laplace and Fourier transforms. Any realistic analysis should take into account the frequency characteristics of the antenna or aperture that couples the signal into the guide as well as the characteristics of the circuit elements used to extract the signal at the receiving end. In this section we shall consider only the properties of the guide itself. The effect of losses and their variation will also be neglected. In practice, this does not lead to significant errors, because in most cases the frequency bandwidth of the signal is relatively narrow, and the mid-band frequency of operation is usually chosen far enough above the cutoff frequency so that the attenuation curve is approximately constant throughout the band.

† For circular apertures W. Eggimann has recently evaluated the polarizabilities $\alpha_e$ and $\bar{\alpha}_m$ correct to terms up to and including $(k_0l)^2$ (work to be published). The results are

$$\alpha_e = -\tfrac{2}{3}l^3\left[1 - \frac{(k_0l)^2}{10}(2 + \sin^2\theta)\right]$$

$$\bar{\alpha}_m = (\mathbf{a}_u\mathbf{a}_u + \mathbf{a}_v\mathbf{a}_v)\tfrac{4}{3}l^3\left[1 + \frac{(k_0l)^2}{15}(8 - 1.5\sin^2\theta)\right]$$

and these expressions should be used for larger apertures. The angle $\theta$ is the angle of incidence for a TEM wave, with respect to the aperture normal. For an aperture in the transverse wall of a rectangular guide, $\sin\theta = \lambda_0/2a$.

Before some time, which we choose as the time origin $t = 0$, the disturbance in the guide is zero. When a current element is introduced into the guide, a disturbance or signal is generated. This signal is a solution of the time-dependent field equations. Let $\mathcal{E}(\mathbf{r},t)$, $\mathcal{H}(\mathbf{r},t)$ be the time-dependent field vectors. For a unit impulse current element $\delta(t - t')$ applied at time $t = t'$ and located at the point $(x',y',z')$, the field vectors are a solution of

$$\nabla \times \mathcal{E} = -\mu_0 \frac{\partial \mathcal{H}}{\partial t} \tag{79a}$$

$$\nabla \times \mathcal{H} = \epsilon_0 \frac{\partial \mathcal{E}}{\partial t} + \boldsymbol{\tau}\delta(t - t')\delta(\mathbf{r} - \mathbf{r}') \tag{79b}$$

where $\boldsymbol{\tau}$ is a unit vector giving the direction of the current element, and $\mathbf{r}$ designates the field point $(x,y,z)$, while $\mathbf{r}'$ designates the source point or location. If the Laplace transform of these equations is taken, we get

$$\nabla \times \mathbf{E}(\mathbf{r},p) = -\mu_0 p\mathbf{H}(\mathbf{r},p) \tag{80a}$$

$$\nabla \times \mathbf{H}(\mathbf{r},p) = \epsilon_0 p\mathbf{E}(\mathbf{r},p) + \boldsymbol{\tau}e^{-pt'}\delta(\mathbf{r} - \mathbf{r}') \tag{80b}$$

where
$$\mathbf{E}(\mathbf{r},p) = \int_0^\infty \mathcal{E}(\mathbf{r},t)e^{-pt}\, dt$$

with a similar definition for $\mathbf{H}(\mathbf{r},p)$. These equations are formally the same as those obtained by assuming a time dependence $e^{j\omega t}$, and the solution may be obtained by the methods we have previously discussed. All our previous solutions may be converted into solutions of (80) by replacing $j\omega$ by $p$ and $e^{j\omega t}$ by $e^{-pt'}$. The Laplace transform has the effect of suppressing the time variable. The solution to (80) constitutes the Laplace transform of the time-dependent Green's function. Inverting the transform yields the time-dependent Green's function. For an arbitrary spatial and time variation, the solution may be obtained by a superposition integral.

If we restrict ourselves to a line current extending across the narrow dimension of a rectangular guide, the appropriate Green's function is obtained from Chap. 5, Eq. (71). We have

$$E_y(\mathbf{r},p) = G(\mathbf{r}|\mathbf{r}',p) = \frac{-\mu_0}{a} \sum_{n=1}^{\infty} \phi_n(x)\phi_n(x') \frac{p \exp\{-pt' - [(n\pi/a)^2 + \mu_0\epsilon_0 p^2]^{1/2}|z - z'|\}}{[(n\pi/a)^2 + \mu_0\epsilon_0 p^2]^{1/2}} \tag{81}$$

where $\phi_n(x) = \sin(n\pi x/a)$. The inversion of this expression gives us the time-dependent Green's function corresponding to the disturbance set up in the waveguide by an impulse line current located at $x'$, $z'$.

The Laplace transform of the time derivative of a function $f(t)$ is

$$\mathcal{L}\frac{df}{dt} = \int_0^\infty e^{-pt}\frac{df}{dt}\,dt = fe^{-pt}\Big|_0^\infty + p\int_0^\infty e^{-pt}f\,dt$$

If $f$ vanishes at $t = 0$, we have $\mathcal{L}df/dt = p\mathcal{L}f$, where $\mathcal{L}$ indicates the operation of taking the Laplace transform of the function. Using this result, we may invert each term in (81), disregarding the factor $p$ in the numerator, and then differentiate the result with respect to $t$. A typical term from (81) gives

$$\frac{1}{2\pi j}\int_C \frac{\exp\{p(t-t') - [(n\pi/a)^2 + \mu_0\epsilon_0 p^2]^{1/2}|z-z'|\}}{[(n\pi/a)^2 + \mu_0\epsilon_0 p^2]^{1/2}}\,dp$$
$$= \begin{cases} v_c J_0\left\{\dfrac{n\pi}{a}[v_c^2(t-t')^2 - (z-z')^2]^{1/2}\right\} & 0 < t - t' < \dfrac{|z-z'|}{v_c} \\ 0 & \text{otherwise} \end{cases}$$

where $J_0$ is the Bessel function of the first kind, and $v_c = (\mu_0\epsilon_0)^{-1/2}$ is the velocity of light in vacuum.† At any given distance $|z - z'|$ from the source, the disturbance is zero until a time $t = t' + |z - z'|/v_c$ is reached when the presence of the signal first becomes known to the observer at this position. No information reaches the observer in a time interval less than the time required to propagate a disturbance with the velocity of light. The velocity of light is, therefore, the wavefront velocity.

The time derivative of the above function is

$$\frac{-v_c^3(n\pi/a)(t-t')}{[v_c^2(t-t')^2 - (z-z')^2]^{1/2}} J_1\left\{\frac{n\pi}{a}[v_c^2(t-t')^2 - (z-z')^2]^{1/2}\right\}$$

where $J_1$ is the Bessel function of the first kind and order 1. The solution for the time-dependent Green's function which is equal to $\mathcal{E}_y(\mathbf{r},t)$ becomes

$$\mathcal{G}(\mathbf{r}|\mathbf{r}', t-t') = \frac{Z_0 v_c^2}{a}\sum_{n=1}^{\infty}\frac{n\pi}{a}\phi_n(x)\phi_n(x') \\ (t-t')\frac{J_1\{(n\pi/a)[v_c^2(t-t')^2 - (z-z')^2]^{1/2}\}}{[v_c^2(t-t')^2 - (z-z')^2]^{1/2}} \tag{82}$$

for $t' < t < t' + |z - z'|/v_c$. Outside this range for $t$ the Green's function is zero. In the interpretation of (82) it should be noted that

$$\lim_{u\to 0}\frac{J_1(ku)}{u} = \lim_{u\to 0}\frac{dJ_1}{du} = \lim_{u\to 0}\frac{k}{2}[J_0(ku) - J_2(ku)] = \frac{k}{2}$$

† This transform is listed by R. V. Churchill, "Operational Mathematics," 2d ed., p. 329, McGraw-Hill Book Company, Inc., New York, 1958.

Alternatively, the recurrence relation $J_1(ku) = (ku/2)[J_2(ku) + J_0(ku)]$ may be used to eliminate the factor $[v_c^2(t - t')^2 - (z - z')^2]^{1/2}$ in the denominator. The field contributed by each mode first makes its appearance as a discontinuous step change, and thereafter oscillates similar to a damped sinusoidal wave. Thus what was originally a pulse localized in both time and space has become a disturbance distributed throughout both space and time.

For an impressed current $\mathbf{J}(\mathbf{r},t)$ which varies arbitrarily with time for $t > 0$, the response is obtained by a superposition integral, i.e.,

$$\iiint_0^t \mathcal{G}(\mathbf{r}|\mathbf{r}', t - t')J(\mathbf{r}',t')\,dt'\,da' \tag{83}$$

when $J(t) = 0$ for $t < 0$. For a step input, the solution is readily found since the Laplace transform of a unit step applied at $t = t'$ is $p^{-1}e^{-pt'}$. This corresponds to (81) with the factor $p$ deleted from the numerator. The terms that arise are those obtained in arriving at (82) before the derivative with respect to time was taken.

Impulse and step inputs are not typical of the signal we wish to transmit along a waveguide. More realistic signals are those consisting of a narrow spectrum of frequencies, centered around some high frequency $\omega_0$, such as is encountered in amplitude-modulated sinusoids. For an analysis of these we turn to an application of Fourier transforms.

Let the input current at $x = a/2$, $z = 0$ be of the form of an amplitude-modulated signal

$$I = I_0[1 + f(t)] \cos \omega_0 t \tag{84}$$

where $f(t)$ is an arbitrary function of time subject to the restriction that it contains frequencies in the range $0 \le \omega \le \omega_1$ with $\omega_1 \ll \omega_0$. Thus the resultant signal has frequency components in the range $\omega_0 - \omega_1 \le \omega \le \omega_0 + \omega_1$, and $\omega_0$ corresponds to the carrier frequency. Furthermore, let $\omega_1$ be chosen so that $(\pi/a)v_c < \omega < (2\pi/a)v_c$. This latter restriction ensures us that only the $H_{10}$ mode in a rectangular guide of width $a$ will propagate.

The current input is the real part of $(1 + f)e^{j\omega_0 t}$. The function $f(t)$ has a frequency spectrum given by the Fourier transform of $f(t)$:

$$g(\omega) = \int_{-\infty}^{\infty} e^{-j\omega t} f(t)\,dt \tag{85}$$

The function $f(t)$ may be recovered by the inverse transform relation

$$\begin{aligned} f(t) &= \frac{1}{2\pi}\int_{-\infty}^{\infty} e^{j\omega t} g(\omega)\,d\omega \\ &= \frac{1}{2\pi}\int_{-\omega_1}^{\omega_1} e^{j\omega t} g(\omega)\,d\omega \end{aligned} \tag{86}$$

The transform of $e^{j\omega_0 t}$ is $2\pi\delta(\omega - \omega_0)$, and, hence, for the function $(1 + f)e^{j\omega_0 t}$ the frequency spectrum is given by $2\pi\delta(\omega - \omega_0) + g(\omega - \omega_0)$. Each frequency component propagates along the guide in a manner described by the steady-state solution corresponding to an impressed current varying with time according to $e^{j\omega t}$. For the $H_{10}$ mode, the $y$ component of electric field, at a distance $z$ from the origin, arising from a single frequency component is

$$-\frac{j\omega\mu_0}{a}\phi_1(x)\frac{\exp\{j\omega t - j|z|[(\omega/v_c)^2 - (\pi/a)^2]^{1/2}\}}{[(\omega/v_c)^2 - (\pi/a)^2]^{1/2}} = -\frac{j\omega\mu_0}{a}\phi_1(x)\frac{e^{j\omega t - j\beta(\omega)|z|}}{\beta(\omega)}$$

where $\beta(\omega) = [(\omega/v_c)^2 - (\pi/a)^2]^{1/2}$. This expression without the factor $e^{j\omega t}$ may be considered as the transfer function, in the frequency domain, relating the response at $|z|$ to the input at $z = 0$. With the specified input current, the total response is

$$\begin{aligned}\mathcal{E}_y &= -\operatorname{Re}\frac{\mu_0 I_0}{a2\pi}\phi_1\int_{-\infty}^{\infty}\frac{j\omega e^{-j\beta(\omega)|z|}}{\beta(\omega)}[2\pi\delta(\omega - \omega_0) + g(\omega - \omega_0)]e^{j\omega t}\,d\omega \\ &= \frac{-\mu_0 I_0}{2\pi a}\phi_1\operatorname{Re}\left[\frac{j2\pi\omega_0 e^{j\omega_0 t - j\beta_0|z|}}{\beta_0} + \int_{-\infty}^{\infty}\frac{j\omega e^{j\omega t - j\beta(\omega)|z|}}{\beta(\omega)}g(\omega - \omega_0)\,d\omega\right] \qquad (87)\end{aligned}$$

where $\beta_0 = \beta(\omega_0)$. The first term in (87) corresponds to the unmodulated carrier and does not contain any information. The second term is the signal containing the original modulation-frequency components. In this term, the integration need be extended only over the range $\omega_0 - \omega_1$ to $\omega_0 + \omega_1$, since $g(\omega - \omega_0)$ is zero outside this range.

To evaluate the integral, we expand $\beta(\omega)$ in a Taylor series about the point $\omega_0$ to get

$$\beta(\omega) = \beta_0 + \beta_0'(\omega - \omega_0) + \frac{1}{2!}\beta_0''(\omega - \omega_0)^2 + \cdots \qquad (88)$$

where $\beta_0' = \left.\frac{d\beta}{d\omega}\right|_{\omega=\omega_0}$, etc. For a signal containing only a narrow band of frequencies, we retain only the first two terms in (88) and approximate the factor $\omega/\beta$ by $\omega_0/\beta_0$. For this case we then get

$$\begin{aligned}\frac{1}{2\pi}\int_{-\infty}^{\infty}\frac{j\omega e^{j\omega t - j\beta|z|}}{\beta}g(\omega - \omega_0)\,d\omega \\ &\approx \frac{j\omega_0}{2\pi\beta_0}e^{j\omega_0 t - j\beta_0|z|}\int_{\omega_0-\omega_1}^{\omega_0+\omega_1}e^{j(\omega-\omega_0)(t-\beta_0'|z|)}g(\omega - \omega_0)\,d\omega \\ &= \frac{j\omega_0}{\beta_0}e^{j\omega_0 t - j\beta_0|z|}f(t - \beta_0'|z|) \qquad (89)\end{aligned}$$

The latter result follows by making the change of variables $\omega - \omega_0 = \omega'$

and comparing with (86). Combining the two terms and taking the real part, we obtain the total response at $|z|$:

$$\mathcal{E}_y = \frac{\mu_0 I_0}{a} \phi_1(x) \frac{\omega_0}{\beta_1} [1 + f(t - \beta_0'|z|)] \sin \omega_0 \left(t - \frac{\beta_0}{\omega_0}|z|\right) \tag{90}$$

To this order of approximation the original modulation is reproduced without distortion but delayed in time by an amount $\beta_0'|z|$. The velocity with which the signal propagates is equal to the distance $|z|$ divided by the time delay, and hence is given by

$$\begin{aligned} v_g &= \frac{1}{\beta_0'} = \left[\frac{d\beta(\omega)}{d\omega}\bigg|_{\omega_0}\right]^{-1} = \frac{v_c{}^2\beta_0}{\omega_0} \\ &= \frac{v_c{}^2}{v_p} \end{aligned} \tag{91}$$

where the carrier phase velocity $v_p$ is equal to $\omega_0/\beta_0 = (k_0/\beta_0)v_c$. This particular definition of signal velocity is called the group velocity, since it corresponds to the velocity with which a narrow band or group of frequency components are propagated. It is also equal to the velocity of energy propagation introduced in Chap. 5 and is always less than the velocity of light for a uniform hollow waveguide.

If the band of frequencies involved is too large for only two terms of the Taylor series expansion of $\beta(\omega)$ to give a good representation of $\beta(\omega)$ throughout the band, then additional terms must be included. These higher-order terms always lead to distortion of the signal. In general, these terms are difficult to evaluate unless $f(t)$ happens to be of such a form that the inverse transforms can be found. Forrer has presented an analysis for the propagation of a pulse of the form $e^{-(t-t_0)^2}$. In this case terms up to and including $\frac{1}{2}\beta_0''(\omega - \omega_0)^2$ in the expansion of $\beta$ may be evaluated, since the transforms involved are of the type that may be readily inverted.† Several other papers which treat various types of transients in waveguides have also been published. Several of these papers are listed in the general references below.

## REFERENCES

*Antennas and apertures in waveguides*

7.1. H. Motz: "Electromagnetic Problems of Microwave Theory," Methuen Monograph, Methuen & Co., Ltd., London, 1951.

7.2. L. Lewin: A Contribution to the Theory of Probes in Waveguides, *Proc. IEE (London)*, vol. 105, part C, pp. 109–116, March, 1958.

7.3. L. B. Felson and N. Marcuvitz: Slot Coupling of Rectangular and Spherical Waveguides, *J. Appl. Phys.*, vol. 24, pp. 755–770, June, 1953.

† M. P. Forrer, Analysis of Millimicrosecond RF Pulse Transmission, *Proc. IRE*, vol. 46, pp. 1830–1835, November, 1958.

7.4. S. B. Cohn: The Electric Polarizability of Apertures of Arbitrary Shape, *Proc. IRE*, vol. 40, pp. 1069–1071, September, 1952.
7.5. S. B. Cohn: Microwave Coupling by Large Apertures, *Proc. IRE*, vol. 40, pp. 696–699, 1952.

*Waveguide transients*

7.6. M. Cerrillo: Transient Phenomena in Waveguides, *MIT Research Lab. Electronics Tech. Rept.* 33, January, 1948.
7.7. G. I. Cohn: Electromagnetic Transients in Waveguides, *Proc. Natl. Electronics Conf.*, vol. 8, pp. 284–295, 1952.
7.8. M. Cotte: Propagation of a Pulse in a Waveguide, *Ondé élec.*, vol. 34, pp. 143–146, February, 1954.
7.9. R. S. Elliott: Pulse Waveform Degradation Due to Dispersion in Waveguides, *IRE Trans.*, vol. MTT-5, pp. 245–257, October, 1957.
7.10. A. E. Karbowiak: Propagation of Transients in Waveguides, *Proc. IEE (London)*, vol. 104, part C, pp. 339–349, February, 1957.

## PROBLEMS

**7.1.** A small probe antenna is located in the center of the end wall of a rectangular guide. The guide dimensions are such that the only $E$ mode which will propagate is the $E_{11}$ mode. Obtain an expression for the radiation resistance of the antenna.

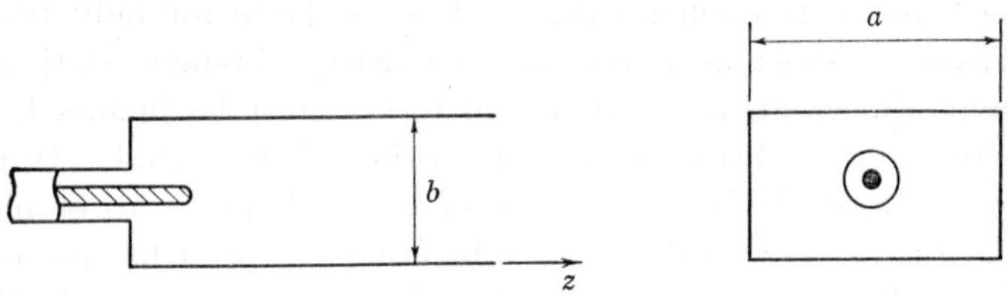


FIG. P 7.1

**7.2.** Consider a small circular loop antenna of radius $d$, carrying a uniform current $I_0$, and located in free space. Expand $e^{-jk_0\rho}$ to get $1 - jk_0\rho - \frac{1}{2}k_0^2\rho^2 + (jk_0^3/6)\rho^3 + \cdots$. Evaluate the self-flux linkage through the loop contributed by the terms in phase quadrature with the current, and show that the term with the factor $jk_0\rho$ does not contribute while the term with the factor $(jk_0^3/6)\rho^3$ gives $(-j\mu_0\pi I_0/6k_0)(k_0 d)^4$. Evaluate the induced emf around the loop due to this portion of the flux linkage, and show that it leads to an input resistance $Z_0(k_0 d)^4\pi/6$. Compute the power radiated by the loop by integrating the complex Poynting vector over the surface of a large sphere. If $P$ is the radiated power, use the relation $\frac{1}{2}RI_0^2 = P$ to define a radiation resistance, and show that this is equal to the input resistance obtained from a consideration of the flux linkages.

**7.3.** Perform the summations leading to Eq. (38).

**7.4.** Sum the following series directly, and compare with the sum obtained by using the Poisson summation formula.

$$\sum_{n=1}^{\infty} \frac{e^{-(n^2+1)^{1/2}}}{(n^2+1)^{1/2}}$$

**7.5.** Use the Poisson summation formula to sum the following series:

$$\sum_{n=1}^{\infty} e^{-\alpha(n^2+\beta^2)^{1/2}} \qquad \sum_{n=1}^{\infty} e^{-\pi n^2/1{,}000}$$

Note the large difference in the rate of convergence of the second series and its transformed equivalent. Consult a table of Fourier transforms for the required transforms.

**7.6.** Modify the small-aperture theory for waveguides to obtain an approximate theory for the reflected and transmitted fields from a small aperture in an infinite-plane conducting screen. Assume a parallel-polarized TEM wave incident at an angle $\theta_i$ with respect to the normal.

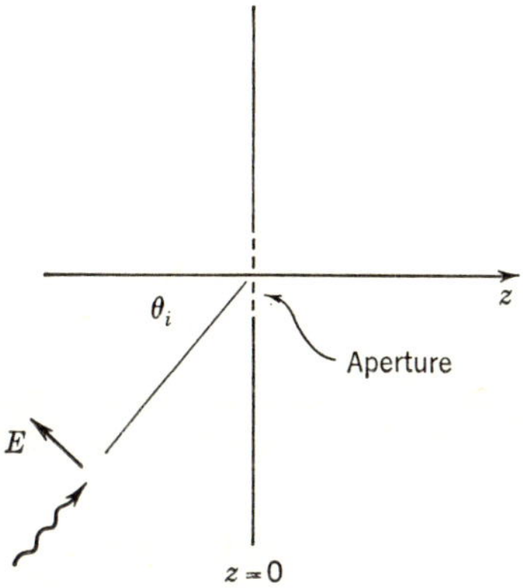


FIG. P 7.6

**7.7.** Develop a similar theory as in Prob. 7.6 for the scattered field from a small conducting disk when a perpendicular-polarized TEM wave is incident at an angle $\theta_i$. (See Stratton† for an expansion of the vector potential which gives the radiation from

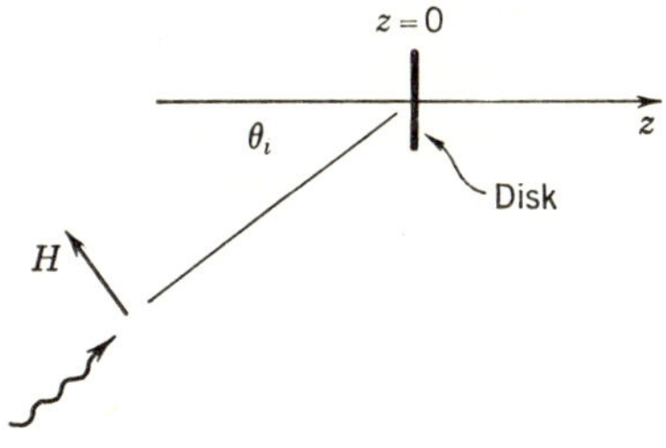


FIG. P 7.7

a small-volume distribution of current as radiation from an electric and magnetic dipole and other higher-order multipoles.) Use Babinet's principle to obtain the solution for a small aperture in a conducting screen from the solution of the disk problem. Show that this leads to the same solution as that obtained in Prob. 7.6.

† J. A. Stratton, "Electromagnetic Theory," sec. 8.4, McGraw-Hill Book Company, Inc., New York, 1941.

**7.8.** Show that an elliptic aperture in a transverse wall in a rectangular guide as illustrated is equivalent to a shunt inductive susceptance

$$B = \left(\frac{-2\beta_{10}}{ab}\sin^2\frac{\pi x_0}{a}\,\bar{\boldsymbol{\alpha}}_m : \mathbf{a}_x\mathbf{a}_x\right)^{-1}$$

where $\bar{\boldsymbol{\alpha}}_m : \mathbf{a}_x\mathbf{a}_x$ is the double-dot product of $\bar{\boldsymbol{\alpha}}_m$ with the unit vectors $\mathbf{a}_x$ and $x_0$ is the coordinate of the center of the aperture.

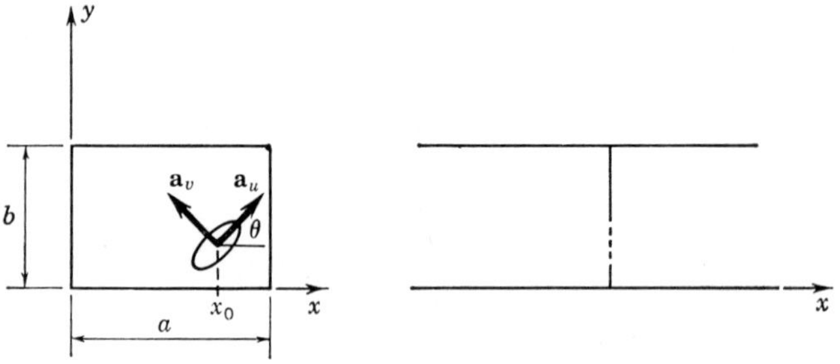


FIG. P 7.8

**7.9.** A small circular aperture of radius $l$ is located in the center of the transverse wall in a circular guide of radius $a$. Show that, for a $TE_{11}$ mode, the aperture is equivalent to an inductive susceptance

$$B = \frac{-0.955a^2\lambda_g}{4\alpha_m}$$

where $\lambda_g$ is the guide wavelength, and $\alpha_m = \frac{4}{3}l^3$. For a $TM_{01}$ mode, show that the aperture is equivalent to a shunt capacitive susceptance

$$B = \frac{0.92a^4}{|\alpha_e|\lambda_g} \qquad |\alpha_e| = \tfrac{2}{3}l^3$$

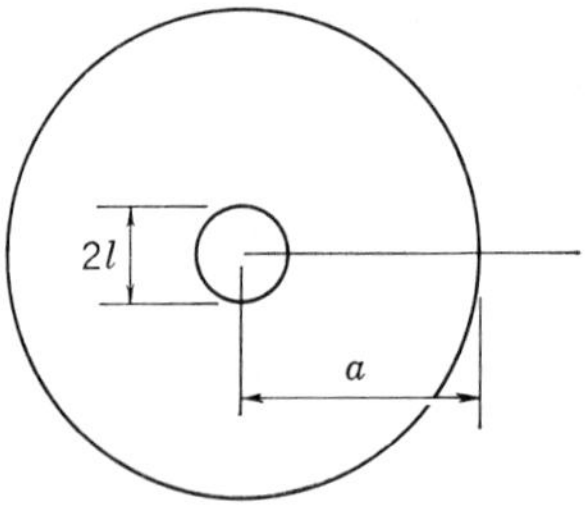


FIG. P 7.9

**7.10.** The Bethe hole directional coupler consists of two rectangular guides with their axe soriented with an angle $\theta$ between them and coupled together through a small circular aperture in the center of the broad wall. Find the required angle $\theta$ such that the combination of the electric and magnetic dipole coupling results in a wave propagating in the direction of arrow $A_2$ only when a wave is incident along the direction of arrow $A_1$.

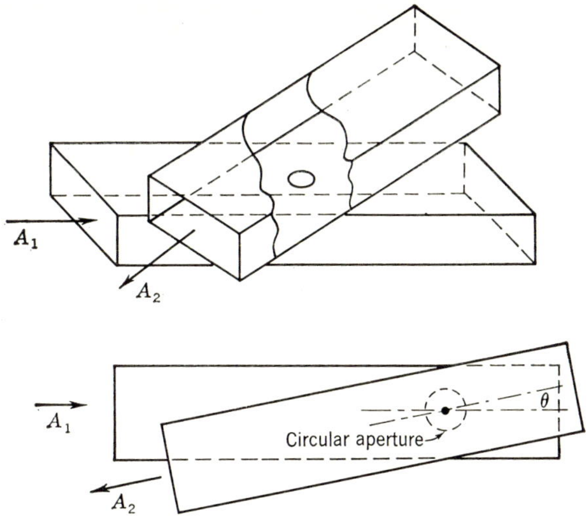


FIG. P 7.10

**7.11.** If it is desirable to maintain the angle $\theta$ between the two guide axes equal to zero in Prob. 7.10, a similar type of coupler may be built, using a properly oriented

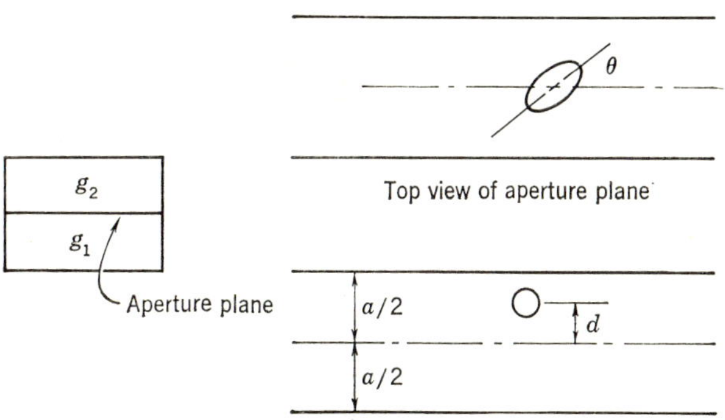


FIG. P 7.11

elliptical aperture or an offset circular aperture as illustrated above. Determine the orientation angle $\theta$ of the elliptical aperture and the offset $d$ of the circular aperture in order to obtain a directional coupler.

*Answer*

$$\alpha_{mu} \sin^2 \theta + \alpha_{mv} \cos^2 \theta = \frac{k_0^2}{\beta_0^2} |\alpha_e|$$

**7.12.** Consider a rectangular guide closed at $z = 0$ by an electric wall in which is located a small magnetic disk carrying a linear magnetic current of density $\mathbf{J}_m$. Let the field radiated by the magnetic current in the region $z > 0$ be

$$\mathbf{E}_s = \Sigma c_n \mathbf{E}_n e^{-\Gamma_n z}$$
$$\mathbf{H}_s = \Sigma c_n \mathbf{H}_n e^{-\Gamma_n z}$$

In the volume bounded by the guide walls and transverse planes at $z = 0$ and $z = z_1$, the field $\mathbf{E}_s$, $\mathbf{H}_s$ is source-free. On the electric wall at $z = 0$ we have $\mathbf{n} \times \mathbf{E}_s = 0$,

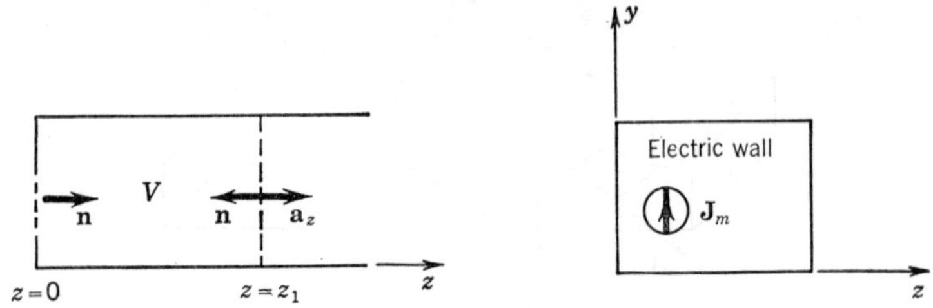


FIG. P 7.12

while on the surface of the magnetic disk $\mathbf{n} \times \mathbf{E}_s = -\mathbf{J}_m$. Let $\mathbf{E}$, $\mathbf{H}$ be a normal-mode standing wave:

$$\mathbf{E} = \mathbf{e}_m(e^{-\Gamma_m z} - e^{\Gamma_m z}) + \mathbf{e}_{zm}(e^{-\Gamma_m z} + e^{\Gamma_m z})$$
$$\mathbf{H} = \mathbf{h}_m(e^{-\Gamma_m z} + e^{\Gamma_m z}) + \mathbf{h}_{zm}(e^{-\Gamma_m z} - e^{\Gamma_m z})$$

This field is also source-free in $V$. The Lorentz reciprocity principle gives

$$\oint\!\!\oint_S \mathbf{n} \cdot (\mathbf{E}_s \times \mathbf{H} - \mathbf{E} \times \mathbf{H}_s)\, da = 0$$

Show that, over the plane $z = z_1$, the integral gives $-2c_m \int\!\!\int \mathbf{e}_m \times \mathbf{h}_m \cdot \mathbf{a}_z\, da$ while, over the $z = 0$ plane, the result is $-\int\!\!\int \mathbf{J}_m \cdot \mathbf{H}\, da = -2\int\!\!\int \mathbf{J}_m \cdot \mathbf{h}_m\, da$. A linear magnetic current element is equivalent to a magnetic dipole, and, if $\mathbf{h}_m$ is assumed constant over the magnetic disk, the coefficient $c_m$ is found to be given by

$$2c_m \int\!\!\int \mathbf{e}_m \times \mathbf{h}_m \cdot \mathbf{a}_z\, da = -j\omega\mu_0 2\mathbf{M}_0 \cdot \mathbf{h}_m$$

where $\mathbf{M}_0$ is the magnetic dipole moment associated with the current $\mathbf{J}_m$. This result was used to obtain Eq. (77).

**7.13.** Consider a $z$-directed magnetic dipole $M\mathbf{a}_z$ located at the origin between two infinite conducting planes placed parallel with the $xz$ plane at $y = \pm b/2$. Obtain the solution for the magnetic Hertzian potential $\Pi_z$ by solving the scalar Helmholtz equation

$$\nabla^2\Pi_z + k_0^2\Pi_z = -M\,\delta(x)\delta(y)\delta(z)$$

in cylindrical coordinates. The solution is the same as that given by (38).

HINT: Expand $\Pi_z$ in a Fourier series of the form

$$\Pi_z = \sum_{m=0}^{\infty} a_m \cos\frac{2m\pi y}{b} K_0(\Gamma_m r) \qquad \Gamma_m^2 = \left(\frac{2m\pi}{b}\right)^2 - k_0^2$$

Next use Fourier analysis to obtain

$$\left(\frac{1}{r}\frac{\partial}{\partial r} r \frac{\partial}{\partial r} - \Gamma_m^2\right) a_m K_0(\Gamma_m r) = -\frac{M\epsilon_{0m}}{b}\delta(r)$$

Each term must have a logarithmic singularity $-(M\epsilon_{0m}/2b\pi)\ln r$, and hence $a_m = M\epsilon_{0m}/2\pi b$, $\epsilon_{0m} = 1$, $m = 0$; $\epsilon_{0m} = 2$, $m > 0$.

**7.14.** Consider a transverse conducting plane at $z = 0$ in a rectangular guide and containing a small aperture. For odd excitation with incident $H_{10}$ modes having electric fields

$$\mathbf{E}_y = \begin{cases} -A_1\mathbf{e}_1 e^{-\Gamma_1 z} & z < 0 \\ A_1\mathbf{e}_1 e^{\Gamma_1 z} & z > 0 \end{cases}$$

show that the aperture may be closed by an electric wall and hence the reflection coefficient $R_0 = -1$. For the case of even excitation with incident electric fields

$$\mathbf{E}_y = A_1\mathbf{e}_1 e^{\Gamma_1|z|} \qquad \text{all } z$$

show that the aperture may be closed by a magnetic wall. Let the reflection coefficient in this case be $R_e$. By superimposing the two solutions show that the reflection coefficient $R$ for a wave incident from $z > 0$ only is given by $R = (R_0 + R_e)/2 = (R_e - 1)/2$. Thus the problem of an aperture in a transverse wall is reduced to that of a magnetic disk located in an electric wall. Since $R_e$ is nearly equal to $-1$, it is convenient to assume that the dominant-mode transverse fields in the guide are

$$\begin{gathered} A_1\mathbf{e}_1(e^{\Gamma_1 z} - e^{-\Gamma_1 z}) + c_1\mathbf{e}_1 e^{-\Gamma_1 z} \\ -A_1\mathbf{h}_1(e^{\Gamma_1 z} + e^{-\Gamma_1 z}) + c_1\mathbf{h}_1 e^{-\Gamma_1 z} \end{gathered}$$

where $c_1$ is small and the portion of the field multiplied by $c_1$ is due to the radiation from the magnetic current on the disk. The standing-wave dominant field satisfies the boundary conditions on the electric wall but not on the magnetic disk. On the magnetic disk the polarizing magnetic field is, therefore, $-2A_1\mathbf{h}_1$, and hence the induced magnetic dipole moment is

$$\mathbf{M} = -2\bar{\alpha}_m \cdot 2A_1\mathbf{h}_1 = -2\mathbf{M}_0 A_1 \qquad \text{where } \mathbf{M}_0 = 2\bar{\alpha}_m \cdot \mathbf{h}_1$$

since one-half of the disk is involved. The evaluation of $c_1$ is given in Prob. 7.12. Thus show that $R_e = -1 + j\omega\mu_0 2\mathbf{M}_0 \cdot \mathbf{h}_1$ and $R = -1 + j\omega\mu_0\mathbf{M}_0 \cdot \mathbf{h}_1$ [compare with Eqs. (77) and (78)].

The above results show that, for an incident wave on one side only, the problem of an aperture in a transverse wall may be analyzed by assuming a standing-wave mode for the incident field, using the Lorentz principle as in Prob. 7.12, and taking the aperture polarizabilities equal to $\alpha_e$ and $\bar{\alpha}_m$. This is the method used in the text and parallels that used for coupling by an aperture in a side wall common to two guides.

CHAPTER 8

# VARIATIONAL METHODS FOR WAVEGUIDE DISCONTINUITIES

When waveguides are used in practice, it is often necessary to introduce discontinuities such as diaphragms or irises, metallic posts, dielectric slabs, and so forth, for the purposes of matching the waveguide to a given load or termination, to obtain a phase shift in the transmitted wave, or for a variety of other reasons. The presence of a discontinuity gives rise to a reflected wave and a storage of reactive energy in the vicinity of the discontinuity because of excitation of higher-order waveguide modes which are evanescent, i.e., decay exponentially with distance from the discontinuity. Under most conditions, in practice, only the dominant mode can propagate in the guide. Under these conditions three parameters are the maximum number required to describe the effect of the discontinuity on the propagating mode if the discontinuity is lossless. These parameters are the modulus $\rho$ and phase angle $\theta$ of the reflection coefficient $R$ and the phase angle $\alpha$ of the transmission coefficient $T$. The modulus of $T$ is given by $TT^* = 1 - \rho^2$, a result which follows at once from the conservation-of-energy principle. Alternatively, we may describe the discontinuity by an equivalent transmission-line circuit which would give rise to a reflected and transmitted wave of magnitude proportional to $R$ and $T$, respectively. This latter description is most commonly used.

As a general rule, the rigorous solution of Maxwell's equations for the field distribution existing near a discontinuity is very difficult, if not impossible, to obtain. However, since we are interested only in the effect produced on the dominant mode, the detailed nature of the diffraction field around the discontinuity is not required. The higher-order modes excited act as a perturbation, giving rise to a phase shift in the reflected and transmitted waves, and an impedance transformation at the discontinuity. When the magnitude of these higher-order modes is small, approximate methods may be used to determine the parameters of the equivalent circuit for the discontinuity. A very useful method capable of handling a large variety of problems was introduced by Schwinger during the period 1940 to 1945.† This method, referred to as the variational method, has a variety of forms, some of which are preferable in certain cases to others. Some of the various formulations possible will be illustrated by application to a specific problem.

## 8.1. Outline of Variational Methods

The problem of evaluating the equivalent-circuit parameters for the junction between an empty rectangular waveguide and one inhomogeneously filled with an asymmetrical dielectric slab will serve to illustrate the various variational techniques which may be used. The output guide for $z > 0$ is inhomogeneously filled with a dielectric slab of

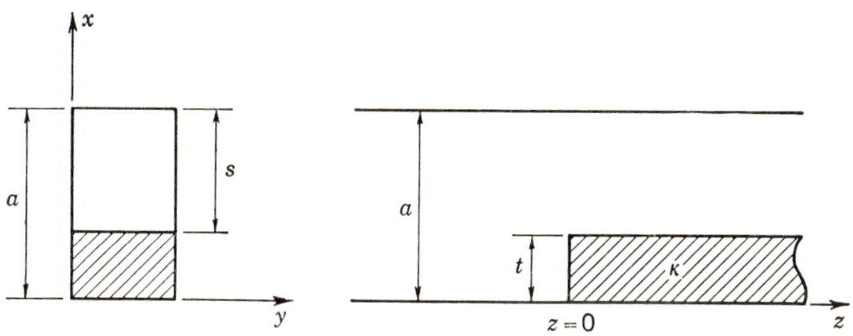


FIG. 8.1. Junction between an empty and inhomogeneously filled rectangular guide.

thickness $t$ and relative dielectric constant $\kappa$, as illustrated in Fig. 8.1. The junction is assumed lossless, which at microwave frequencies is usually a very good approximation.

The incident wave from the left is an $H_{10}$ mode, and, since the discontinuity is uniform along the $y$ axis, the only types of higher-order modes which are excited at the junction are $H_{n0}$ modes. It is therefore possible to derive the various field components from the one scalar component $E_y$ by means of the following equations:

$$j\omega\mu_0 H_z = -\frac{\partial E_y}{\partial x} \qquad j\omega\mu_0 H_x = \frac{\partial E_y}{\partial z} \tag{1}$$

where $E_y$ satisfies the scalar Helmholtz equation

$$\frac{\partial^2 E_y}{\partial x^2} + \frac{\partial^2 E_y}{\partial z^2} + k_0^2 \kappa E_y = 0 \tag{2}$$

and

$$k_0^2 = \omega^2 \mu_0 \epsilon_0 = \frac{2\pi}{\lambda_0}$$

† J. Schwinger, "Discontinuities in Waveguides," MIT lecture notes (edited by D. S. Saxon).

The solutions to these equations in the empty guide are

$$E_y = \sin \frac{n\pi x}{a} e^{\mp\Gamma_n z} \tag{3a}$$

$$H_x = \mp \frac{\Gamma_n}{jZ_0k_0} \sin \frac{n\pi x}{a} e^{\mp\Gamma_n z} \tag{3b}$$

$$H_z = \frac{jn\pi}{aZ_0k_0} \cos \frac{n\pi x}{a} e^{\mp\Gamma_n z} \tag{3c}$$

where $\Gamma_n^2 = (n\pi/a)^2 - k_0^2$ and $Z_0 = (\mu_0/\epsilon_0)^{1/2}$ is the wave impedance of free space. The term $\Gamma_n/jZ_0k_0$ has the dimensions of an admittance, and is the wave admittance of the $n$th mode in the empty guide. It is equal to $\pm H_x/E_y$, and we see that the transverse magnetic and electric field components play the role of a current and a voltage, respectively, on a fictitious transmission line. When the mode propagates, $\Gamma_n$ is imaginary, and the wave admittance $Y_n$ is real. When the mode is evanescent, $Y_n$ is imaginary. The following convention will be adopted in regard to the sign of $Y_n$. For modes that propagate, $Y_n$ will be taken as a positive-real quantity. For evanescent $H_{n0}$ modes, $Y_n$ will be taken as negative-imaginary, since evanescent $H_{n0}$ modes store more magnetic energy than electric energy, corresponding to a net inductive impedance.

Let the amplitude of the wave incident from the left on the junction be $a_1$. Owing to the discontinuity, a reflected wave of amplitude $R_1a_1$ and an infinite number of higher-order modes of amplitudes $a_n$ are excited. Thus, in the region $z \leq 0$, we may write the following expressions for the total transverse fields:

$$E_y = a_1(e^{-\Gamma_1 z} + R_1e^{\Gamma_1 z})\phi_1 + \sum_2^\infty a_n\phi_n e^{\Gamma_n z} \tag{4a}$$

$$H_x = -Y_1a_1(e^{-\Gamma_1 z} - R_1e^{\Gamma_1 z})\phi_1 + \sum_2^\infty a_nY_n\phi_n e^{\Gamma_n z} \tag{4b}$$

where $\phi_n = (2/a)^{1/2} \sin (n\pi x/a)$. The factor $(2/a)^{1/2}$ is introduced for normalization purposes only. The analogy between waves propagating on a transmission line is readily seen when the following definitions for the equivalent transmission-line currents and voltages are made:

$$V_{11}^+ = a_1 \qquad V_{11}^- = a_1R_1 \qquad V_{n1}^- = a_n \tag{5a}$$

$$I_{11}^+ = a_1Y_1 \qquad I_{11}^- = -a_1R_1Y_1 \qquad I_{n1}^- = -a_nY_n \tag{5b}$$

$$V_1 = V_{11}^+e^{-\Gamma_1 z} + V_{11}^-e^{\Gamma_1 z} \tag{5c}$$

$$I_1 = I_{11}^+e^{-\Gamma_1 z} + I_{11}^-e^{\Gamma_1 z} \tag{5d}$$

The power flow along the guide for the $n$th mode is

$$\operatorname{Re} \frac{1}{2} \int_0^a \int_0^b a_n a_n^* Y_n^* \phi_n{}^2 \, dx \, dy = \frac{b}{2} \operatorname{Re} (a_n a_n^* Y_n^*)$$
$$= \operatorname{Re} \left( \frac{b}{2} V_n I_n^* \right) \tag{6}$$

For an evanescent mode the real power flow is zero.

At a distance $d$ in front of the junction, an input impedance may be defined by the following relation:

$$Z_{\text{in}}(d) = \frac{V_1(d)}{I_1(d)} = \frac{e^{\Gamma_1 d} + R_1 e^{-\Gamma_1 d}}{e^{\Gamma_1 d} - R_1 e^{-\Gamma_1 d}} Z_1$$
$$= Z_1 \frac{Z_L + jZ_1 \tan \beta_1 d}{Z_1 + jZ_L \tan \beta_1 d} \tag{7}$$

where $\beta_1 = |\Gamma_1|$, and $Z_L$ is the equivalent input impedance referred to the junction plane $z = 0$, that is,

$$Z_L = \frac{1 + R_1}{1 - R_1} Z_1 \tag{8}$$

A description in terms of an equivalent transmission-line circuit is not necessary, but does provide a very convenient terminology for the description of a junction. Usually it is the practice to describe a junction in terms of normalized impedance and reactive elements. In the present case, the normalized input impedance of the junction is

$$\frac{Z_L}{Z_1} = \frac{1 + R_1}{1 - R_1}$$

A knowledge of the normalized input impedance permits the physically meaningful reflection coefficient to be computed at once.

So far we have written down expansions for the transverse fields in the input guide only. In the output guide an infinite number of $H_{n0}$ modes are excited. The method of solution for these has been covered in Chap. 6. For the purpose of discussion here, all we need to know is that an infinite number of $H_{n0}$ modes exist; that these modes form a complete set, in terms of which an arbitrary electromagnetic field having only a $y$ component of electric field, and no variation with $y$, may be expanded; and, finally, that these functions may be normalized so that the eigenfunction set forms an orthonormal system. We will let $\psi_m(x)$ be the $m$th eigenfunction for the inhomogeneously filled guide. These functions satisfy the following orthogonal relations:

$$\int_0^a \psi_m(x) \psi_n(x) \, dx = \delta_{nm} \tag{9}$$

where the Kronecker delta $\delta_{nm}$ equals unity for $n = m$, and zero otherwise.

The transverse fields for $z \geq 0$ are given by

$$E_y = \sum_{m=1}^{\infty} b_m \psi_m e^{-\gamma_m z} \tag{10a}$$

$$H_x = -\sum_{m=1}^{\infty} b_m Y_{0m} \psi_m e^{-\gamma_m z} \tag{10b}$$

where $b_m$ are amplitude coefficients, $Y_{0m}$ is the mode admittance $\gamma_m/jk_0Z_0$, and $\gamma_m$ is the propagation constant. The propagation constants are determined by the following equations:

$$\frac{\tan l_m t}{l_m} = \frac{-\tan k_m s}{k_m} \tag{11a}$$

$$\gamma_m{}^2 = l_m{}^2 - \kappa k_0{}^2 = k_m{}^2 - k_0{}^2 \tag{11b}$$

It will be assumed that the guide dimensions and frequency have been chosen so that only the $m = 1$ mode propagates. Thus all $\Gamma_n$ and $\gamma_n$ are real except $\Gamma_1$ and $\gamma_1$, which are pure-imaginary. For the output guide, the equivalent transmission-line voltages and currents will be defined by

$$V_{n2}^+ = b_n$$

$$I_{n2}^+ = Y_{0n} b_n$$

At the junction $z = 0$, the transverse fields must be continuous. This condition leads to the two following equations:

$$a_1(1 + R_1)\phi_1 + \sum_{n=2}^{\infty} a_n \phi_n = \sum_{m=1}^{\infty} b_m \psi_m \tag{12a}$$

$$a_1(1 - R_1)Y_1\phi_1 - \sum_{n=2}^{\infty} a_n Y_n \phi_n = \sum_{m=1}^{\infty} b_m Y_{0m} \psi_m \tag{12b}$$

A simultaneous solution of these two equations would give us a complete description of the field. However, it is quite apparent that such a solution would be extremely difficult to obtain, and recourse to an approximate method must be made. The variational method is capable of giving an accurate approximate solution with a minimum of computational labor, and so we will now proceed to develop this method.

*Method* 1

The two fundamental continuity equations, i.e., Eqs. (12), will be repeated here for convenience with the amplitude of the incident wave assumed to be unity. Thus

$$(1 + R_1)\phi_1 + \sum_{2}^{\infty} a_n\phi_n = \sum_{1}^{\infty} b_m\psi_m \tag{13a}$$

$$(1 - R_1)Y_1\phi_1 = \sum_{2}^{\infty} a_nY_n\phi_n + \sum_{1}^{\infty} b_mY_{0m}\psi_m \tag{13b}$$

In the aperture plane $z = 0$, let the electric field distribution be $\mathcal{E}(x)$. Since $\phi_n$ and $\psi_m$ form complete sets of orthonormal functions, we may develop $\mathcal{E}(x)$ into a Fourier series in terms of them, i.e.,

$$\mathcal{E}(x) = (1 + R_1)\phi_1 + \sum_{2}^{\infty} a_n\phi_n = \sum_{1}^{\infty} b_m\psi_m$$

so that

$$1 + R_1 = \int_0^a \mathcal{E}\phi_1\,dx$$

$$a_n = \int_0^a \mathcal{E}\phi_n\,dx$$

$$b_m = \int_0^a \mathcal{E}\psi_m\,dx$$

The coefficients $a_n$ and $b_m$ are thus expressible in terms of the aperture electric field distribution. The continuity of the electric field in the aperture is automatically ensured since both sides of the first equation in (13) has been equated to $\mathcal{E}(x)$. If these values of the coefficients are substituted into the second equation, we obtain

$$(1 - R_1)Y_1\phi_1\frac{\int_0^a \mathcal{E}\phi_1\,dx}{1 + R_1} = \sum_{2}^{\infty} Y_n\phi_n\int_0^a \mathcal{E}\phi_n\,dx + \sum_{1}^{\infty} Y_{0m}\psi_m\int_0^a \mathcal{E}\psi_m\,dx \tag{14a}$$

Interchanging the order of integration and summation gives

$$\frac{1 - R_1}{1 + R_1}Y_1\phi_1\int_0^a \mathcal{E}\phi_1\,dx = \int_0^a \mathcal{E}(x')G_1(x|x')\,dx' \tag{14b}$$

where

$$G_1(x|x') = \sum_{2}^{\infty} Y_n\phi_n(x)\phi_n(x') + \sum_{1}^{\infty} Y_{0m}\psi_m(x)\psi_m(x')$$

The function $G_1(x|x')$ is symmetrical in the variables $x$ and $x'$, and for convenience it will be called a Green's function, although, strictly speaking, it is not a true Green's function in the normal sense. Equation (14$b$) is an identity, and we may, therefore, multiply both sides by an arbitrary function of $x$ and integrate from $0 \le x \le a$. Let this arbitrary function of $x$ be $\mathcal{E}_1(x)$; thus

$$Y_{\text{in}}Y_1 \iint_0^a \mathcal{E}_1(x)\mathcal{E}(x')\phi_1(x)\phi_1(x')\,dx\,dx' = \iint_0^a \mathcal{E}_1(x)\mathcal{E}(x')G_1(x|x')\,dx\,dx'$$

where $Y_{\text{in}}$ has been written for $(1 - R_1)/(1 + R_1)$, since it is the normalized input admittance at the plane $z = 0$. Let us calculate the variation in $Y_{\text{in}}$ due to a small variation in the functional form of the function $\mathcal{E}_1(x)$. We have

$$Y_{\text{in}}Y_1 \int_0^a \delta\mathcal{E}_1\phi_1\,dx \int_0^a \mathcal{E}\phi_1\,dx + \delta Y_{\text{in}}Y_1 \int_0^a \mathcal{E}_1\phi_1\,dx \int_0^a \mathcal{E}\phi_1\,dx = \int_0^a \delta\mathcal{E}_1(x) \int_0^a G_1(x|x')\mathcal{E}(x')\,dx'\,dx$$

which may be written as

$$\int_0^a \delta\mathcal{E}_1 \left[ Y_{\text{in}}Y_1\phi_1 \int_0^a \mathcal{E}\phi_1\,dx - \int_0^a G_1(x|x')\mathcal{E}(x')\,dx' \right] dx + \delta Y_{\text{in}}Y_1 \int_0^a \mathcal{E}_1\phi_1\,dx \int_0^a \mathcal{E}\phi_1\,dx = 0$$

By virtue of (14*b*), the integrand vanishes, and, hence, $\delta Y_{\text{in}}$ is zero for a variation in $\mathcal{E}_1$. Since we want to make $\delta Y_{\text{in}}$ zero for arbitrary small variation in $\mathcal{E}$ also, we conduct the variation with respect to $\mathcal{E}$ to obtain

$$\int_0^a \delta\mathcal{E} \left[ Y_{\text{in}}Y_1\phi_1 \int_0^a \mathcal{E}_1\phi_1\,dx - \int_0^a G_1(x|x')\mathcal{E}_1(x)\,dx \right] dx' + \delta Y_{\text{in}}Y_1 \int_0^a \mathcal{E}_1\phi_1\,dx \int_0^a \mathcal{E}\phi\,dx = 0$$

as the condition to be imposed on $\mathcal{E}_1$. Examination of this equation shows that $\delta Y_{\text{in}}$ will equal zero if $\mathcal{E}_1 = \mathcal{E}$.

We, therefore, have the following variational expression for $Y_{\text{in}}$:

$$Y_{\text{in}} = \frac{\displaystyle\iint_0^a G_1(x|x')\mathcal{E}(x)\mathcal{E}(x')\,dx\,dx'}{Y_1 \left[ \displaystyle\int_0^a \mathcal{E}(x)\phi_1\,dx \right]^2} \tag{15}$$

which has the property that $Y_{\text{in}}$ is stationary for small arbitrary variations in the electric field distribution about its correct value. It, therefore, follows that a first-order approximation to the electric field distribution will yield a second-order approximation to the input admittance. Also the expression for $Y_{\text{in}}$ is homogeneous in the sense that it does not depend on the amplitude of the electric field distribution $\mathcal{E}(x)$, but only on its functional form.

Alternative to the above formulation, we may derive a similar expression for the normalized input impedance $(1 + R_1)/(1 - R_1)$. If we let the transverse magnetic field in the aperture plane be $\mathcal{H}(x)$, then

$$1 - R_1 = Z_1 \int_0^a \mathcal{H}\phi_1\,dx$$

$$a_n = -Z_n \int_0^a \mathcal{H}\phi_n\,dx$$

$$b_m = Z_{0m} \int_0^a \mathcal{H}\psi_m\,dx$$

Using the same procedure as before, we find that

$$Z_{\text{in}} = \frac{\displaystyle\iint_0^a G_2(x|x')\mathcal{H}(x)\mathcal{H}(x')\,dx\,dx'}{Z_1\left[\displaystyle\int_0^a \mathcal{H}(x)\phi_1\,dx\right]^2} \tag{16}$$

where $$G_2(x|x') = \sum_2^\infty Z_n\phi_n(x)\phi_n(x') + \sum_1^\infty Z_{0m}\psi_m(x)\psi_m(x')$$

We will now show how, in practice, we may obtain a suitable approximation to the aperture fields and, hence, to $Y_{\text{in}}$ and $Z_{\text{in}}$. First it will be necessary to consider some of the properties of the expansion of one set or complete orthonormal functions in terms of another set. We may expand each function $\psi_m$ in a Fourier series consisting of the functions $\phi_n$ as follows:

$$\psi_m = \sum_{n=1}^\infty P_{nm}\phi_n \tag{17}$$

where the coefficients $P_{nm}$ are given by $\int_0^a \phi_n\psi_m\,dx = P_{nm}$. Similarly, we may expand $\phi_n$ in terms of $\psi_m$ as follows:

$$\phi_n = \sum_{m=1}^\infty P_{nm}\psi_m \tag{18}$$

where $$P_{nm} = \int_0^a \phi_n\psi_m\,dx$$

Substituting the expansion for $\psi_m$ into (18) gives

$$\phi_n = \sum_{m=1}^\infty P_{nm}\sum_{s=1}^\infty P_{sm}\phi_s = \sum_{s=1}^\infty \phi_s \sum_{m=1}^\infty P_{nm}P_{sm} \tag{19}$$

from which we conclude that

$$\sum_{m=1}^\infty P_{nm}P_{sm} = \delta_{sn}$$

and hence

$$\sum_{m=1}^\infty P_{nm}^2 = 1 \qquad \text{and} \qquad \sum_{m=1}^\infty P_{nm}P_{sm} = 0 \qquad s \neq n$$

Similarly we find that

$$\sum_{n=1}^\infty P_{nm}P_{nr} = \delta_{mr}$$

The transformation from one set of orthonormal functions to another set is analogous to the transformation from one orthogonal coordinate system to another one. The matrix of the transformation is composed of the elements $P_{nm}$. The product of the matrix $[P_{nm}]$ with its transpose $[P_{mn}]$ is equal to the unit matrix so that the matrix $[P_{nm}]$ is unitary. Since the elements are all real, the matrix of the transformation is also orthogonal.

For a suitable approximation to the aperture electric field $\mathcal{E}(x)$, we may take a finite series of the functions $\phi_n(x)$, say

$$\sum_1^N a_n \phi_n$$

Since $Y_{\text{in}}$ depends only on the functional form of $\mathcal{E}$, we may take $a_1$ equal to unity without any loss in generality. The variational expression for $Y_{\text{in}}$ involves the integral of $G_1(x|x')$ times the functions $\phi_n(x)\phi_n(x')$ and is computed as follows:

$$\begin{aligned}
\sum_{s=1}^N \sum_{r=1}^N \iint_0^a G_1(x|x')\phi_s(x)\phi_r(x')a_r a_s \, dx\, dx'
&= \sum_{s=1}^N \sum_{r=1}^N a_s a_r \iint_0^a \phi_r(x')\phi_s(x) \Big[ \sum_{n=2}^\infty Y_n \phi_n(x)\phi_n(x') \\
&\quad + \sum_{m=1}^\infty Y_{0m}\psi_m(x)\psi_m(x') \Big] dx\, dx' \\
&= \sum_{n=2}^N Y_n a_n{}^2 + \sum_{s=1}^N \sum_{r=1}^N a_r a_s \sum_{m=1}^\infty Y_{0m} P_{rm} P_{sm} \\
&= \sum_{s=1}^N \sum_{r=1}^N a_s a_r \left( \sum_{n=2}^\infty Y_n \delta_{ns}\delta_{nr} + \sum_{m=1}^\infty Y_{0m} P_{sm} P_{rm} \right)
\end{aligned} \tag{20}$$

For brevity, we will denote the quantity

$$\sum_{n=2}^\infty Y_n \delta_{ns}\delta_{nr} + \sum_{m=1}^\infty Y_{0m} P_{sm} P_{rm}$$

by $g_{sr}$. It should be noted that $g_{sr} = g_{rs}$. Thus, (15) for $Y_{\text{in}}$ becomes

$$Y_{\text{in}} = \frac{1}{Y_1} \sum_{s=1}^N \sum_{r=1}^N a_s a_r g_{sr} \tag{21}$$

Since $P_{rm}$ and $P_{sm}$ approach zero as $m$ becomes large when $r$ and $s$ are different from $m$, the summation over $m$ in the evaluation of $g_{sr}$ may be terminated at a given finite value of $m$ consistent with the required accuracy. We must now impose on (21) the condition that $Y_{\text{in}}$ should be stationary for arbitrary variations in $\mathcal{E}$, that is, in $a_s$. This condition is obtained by equating the partial derivatives of the unknown coefficients $a_s$ to zero and gives the best possible solution for $Y_{\text{in}}$ that can be obtained with the assumed approximation for $\mathcal{E}$. From (21) we get

$$\sum_{r=1}^{N} a_r g_{sr} = 0 \qquad s = 2, 3, 4, \ldots, N \tag{22}$$

since $g_{sr}$ is symmetrical in the indices $s$ and $r$. The coefficient $a_1$ has been taken equal to unity, and hence $\partial Y_{\text{in}}/\partial a_1$ is automatically zero. Equation (22) is a set of $N - 1$ equations which may be solved for the $N - 1$ unknowns $a_r$ in terms of $g_{11}a_1$ or $g_{11}$. These values of $a_r$ are now substituted back into (21), and $Y_{\text{in}}$ is evaluated. We may handle (16) for $Z_{\text{in}}$ in an entirely analogous manner. The following points are of interest:

1. Since $Y_{01}$ is real while all $Y_n$ and $Y_{0m}$ for $n$ and $m$ greater than one are imaginary, the factors $g_{sr}$ are complex. Consequently, the coefficients $a_r$ obtained from (22) are complex also.
2. Since the $a_r$ are complex, $Y_{\text{in}}$ is complex, as, of course, it must be, since it represents a transmission of power past the discontinuity plus a storage of reactive energy at the junction.
3. Because of the complex nature of $Y_{\text{in}}$, the stationary value of $Y_{\text{in}}$ is neither a maximum nor a minimum. The same is true of the stationary value of $Z_{\text{in}}$. Therefore, a comparison of the calculated values of $Y_{\text{in}}$ and $Z_{\text{in}}^{-1}$ will not give upper and lower bounds to the true value of $Y_{\text{in}}$.
4. The calculation of $Y_{\text{in}}$ gives only two of the three parameters required to specify the junction completely. To compute the third parameter, the above analysis has to be repeated for a wave incident on the junction from the inhomogeneously filled region.
5. The necessity of repeating the analysis twice, and the fact that the coefficients $a_r$ are complex, makes the numerical computation laborious and lengthy.

We may organize the numerical work in a more compact fashion by treating (15) for $Y_{\text{in}}$ [similarly with (16) for $Z_{\text{in}}$] in a somewhat different manner. As before, we approximate $\mathcal{E}(x)$ by a finite series of the $\phi_n$, but we do not equate $a_1$ to unity; rather we leave it as arbitrary for the present. In place of (21), we then get

$$Y_{\text{in}} Y_1 a_1^2 - \sum_{s=1}^{N} \sum_{r=1}^{N} a_s a_r g_{sr} = 0 \tag{23}$$

When we equate the partial derivatives with respect to $a_s$ equal to zero, we get

$$2Y_{\text{in}}Y_1a_1 - 2\sum_{r=1}^{N} a_r g_{1r} = 0 \tag{24a}$$

$$-2\sum_{r=1}^{N} a_r g_{sr} = 0 \qquad s = 2, 3, \ldots, N \tag{24b}$$

This is a set of $N$ homogeneous equations for $N$ unknowns, which has a solution only if the determinant vanishes. The vanishing of the determinant gives the solution for $Y_{\text{in}}$ without the necessity of computing the values of the coefficients. We have

$$\begin{vmatrix} Y_{\text{in}}Y_1 - g_{11} & -g_{12} & \cdots & -g_{1N} \\ -g_{21} & \cdots & \cdots & \cdots \\ \cdots & \cdots & \cdots & \cdots \\ -g_{N1} & -g_{N2} & \cdots & -g_{NN} \end{vmatrix} = 0$$

This determinant may be factored into the sum of two determinants as follows:

$$\begin{vmatrix} Y_{\text{in}}Y_1 & -g_{12} & \cdots & -g_{1N} \\ 0 & -g_{22} & \cdots & \cdots \\ \cdots & \cdots & \cdots & \cdots \\ 0 & -g_{N2} & \cdots & -g_{NN} \end{vmatrix} + (-1)^N \begin{vmatrix} g_{11} & \cdots & g_{1N} \\ g_{21} & \cdots & \cdots \\ \cdots & \cdots & \cdots \\ g_{N1} & \cdots & g_{NN} \end{vmatrix}$$

and, hence,

$$Y_{\text{in}} = \frac{1}{Y_1} \frac{\begin{vmatrix} g_{11} & \cdots & g_{1N} \\ \cdots & \cdots & \cdots \\ g_{N1} & \cdots & g_{NN} \end{vmatrix}}{\begin{vmatrix} g_{22} & \cdots & g_{2N} \\ \cdots & \cdots & \cdots \\ g_{N2} & \cdots & g_{NN} \end{vmatrix}} \tag{25}$$

Thus we have reduced the computation of $Y_{\text{in}}$ to the evaluation of two determinants of order $N$ and $N - 1$.

*Method* 2

The main objections to the previously derived variational solution are those listed in the previous section as (3), (4), and (5). All these objections can be removed by a modification of the method. If we terminate the output guide in a short circuit located at $z = l$, we have a pure reactive termination since all of the incident power is reflected. Under these conditions $Y_{\text{in}}$ is pure-imaginary, and the electric field is everywhere in time quadrature with the magnetic field. We also find that the field distributions in the aperture plane, apart from a constant multiplier which may be complex, may be represented by a real function. Since the variational expressions depend only on the functional form of the aperture fields, this constant multiplier may be dropped. We will assume that only the dominant mode propagates in the output guide, and the short circuit will be placed at $z = l + n\lambda_{0g}/2$, where $n$ is chosen large enough so that all evanescent modes have decayed to a negligible value at the short circuit. The guide wavelength in the inhomogeneously filled guide is denoted by $\lambda_{0g}$.

In order for the electric field to vanish at the short circuit, the field of the dominant mode in the output guide must be

$$b_1 \sin \beta_{01}(z - l)\psi_1(x) \tag{26}$$

where $\beta_{01} = |\gamma_1|$ is the phase constant for the output guide. From (1), the transverse magnetic field is

$$-jY_{01}b_1\psi_1 \cos \beta_{01}(z - l) \tag{27}$$

If the aperture electric field is $\mathcal{E}(x)$, the coefficient $b_1$ is given by

$$b_1 = -\frac{1}{\sin \beta_{01}l}\int_0^a \mathcal{E}\psi_1 \, dx$$

Since all the incident energy is reflected, the modulus of the reflection coefficient $R_1$ is unity. We may therefore put $R_1$ equal to $e^{j\theta}$, and, hence,

$$Y_{\text{in}} = \frac{1 - e^{j\theta}}{1 + e^{j\theta}} = -j \tan \frac{\theta}{2}$$

In the input guide, the dominant-mode electric field will have a series of maxima and minima as a function of $z$. The first minimum occurs when the incident and reflected fields are out of phase, i.e., when

$$e^{-j\beta_1 z} + e^{j(\theta + \beta_1 z)} = 0$$

Let the corresponding value of $|z|$ be $d$, and we get $\theta = \pi + 2\beta_1 d$, and, hence,

$$Y_{\text{in}} = -j \tan\left(\frac{\pi}{2} + \beta_1 d\right) = j \cot \beta_1 d \tag{28}$$

The input admittance at the junction is given by (28) in terms of the location or position $d$ of the electric field minimum from the junction. We may now carry through the same analysis as before, and, in place of (15), we get

$$\cot \beta_1 d = -\frac{\iint_0^a G_1'(x|x')\mathcal{E}(x)\mathcal{E}(x')\, dx\, dx'}{Y_1\left(\int_0^a \mathcal{E}\phi_1\, dx\right)^2} \tag{29}$$

where

$$G_1'(x|x') = \sum_{n=2}^{\infty} j[Y_n\phi_n(x)\phi_n(x')] + \sum_{m=2}^{\infty} j[Y_{0m}\psi_m(x)\psi_m(x')] + Y_{01} \cos \beta_{01}l\, \psi_1(x)\psi_1(x')$$

We may handle Eq. (29) in the same manner as Eq. (15). If we define $g'_{sr}$ as

$$g'_{sr} = \iint_0^a G'_1(x|x')\phi_s(x)\phi_r(x')\,dx\,dx' = j\left(\sum_{n=2}^{\infty} Y_n\delta_{ns}\delta_{nr} + \sum_{m=2}^{\infty} Y_{0m}P_{sm}P_{rm}\right) + Y_{01}P_{s1}P_{r1}\cot\beta_{01}l$$

we get, in place of (21),

$$\cot\beta_1 d = -\frac{1}{Y_1 a_1^2}\sum_{s=1}^{N}\sum_{r=1}^{N} a_s a_r g'_{sr} \tag{30}$$

As before, we may take $a_1$ equal to unity, and then determine the other $N - 1$ coefficients so that $\cot\beta_1 d$ is stationary. Let us compute the second-order variation in $\cot\beta_1 d$ due to arbitrary but small variations in the coefficients $a_s$.

Taylor's expansion of a function of $N$ variables may be written symbolically as

$$F(x_1,x_2, \ldots ,x_N) = \left(\exp\sum_{s=1}^{N}\Delta x_s\frac{\partial}{\partial x_s}\right)F(x_1, \ldots ,x_N)\Big|_{x_i=x_{i0}}$$
$$= \left(1 + \sum_{s=1}^{N}\Delta x_s\frac{\partial}{\partial x_s} + \frac{1}{2!}\sum_{s=1}^{N}\sum_{r=1}^{N}\Delta x_s\,\Delta x_r\frac{\partial}{\partial x_s}\frac{\partial}{\partial x_r} + \cdots +\right) \times F(x_1, \ldots ,x_N)\Big|_{x_i=x_{i0}}$$

where all the partial derivatives are to be evaluated at the point $x_i = x_{i0}$, $i = 1, 2, \ldots, N$, with $x_{i0}$ being the initial values of the variables. This expansion gives the variation or change in the function $F$ correct to the second order if we retain the first three terms in the expansion of the exponential. Applying this result to $\cot\beta_1 d$, we get

$$Y_1\left[\cot\beta_1 d - \cot\beta_1 d\Big|_{a_s=a_{s0}}\right] = Y_1\delta^2\cot\beta_1 d$$
$$= -\left(\sum_{i=2}^{N}\Delta a_i\frac{\partial}{\partial a_i} + \frac{1}{2}\sum_{i=2}^{N}\sum_{k=2}^{N}\Delta a_i\,\Delta a_k\frac{\partial}{\partial a_i}\frac{\partial}{\partial a_k}\right)\sum_{s=1}^{N}\sum_{r=1}^{N}a_s a_r g'_{sr}\Big|_{a_i=a_{i0}}$$

The operation $\sum_{i=2}^{N}\Delta a_i\frac{\partial}{\partial a_i}$ gives the first variation of $\cot\beta_1 d$, and, since it is assumed that we have chosen the $a_s$ so that $\cot\beta_1 d$ is stationary, the first variation vanishes, and we are left with

$$Y_1\delta^2 \cot \beta_1 d = -\frac{1}{2}\left(\sum_{i=2}^{N}\sum_{k=2}^{N} \Delta a_i\, \Delta a_k \frac{\partial}{\partial a_i}\frac{\partial}{\partial a_k}\right)\sum_{s=1}^{N}\sum_{r=1}^{N} a_s a_r g'_{sr}\Bigg|_{a_i=a_{i0}}$$

$$= -\sum_{i=2}^{N} \Delta a_i \frac{\partial}{\partial a_i}\sum_{k=2}^{N}\Delta a_k \sum_{s=1}^{N} a_s g'_{ks}\Bigg|_{a_i=a_{i0}}$$

$$= -\sum_{k=2}^{N}\sum_{i=2}^{N} \Delta a_i\, \Delta a_k g'_{ki} \qquad (31)$$

This is, in fact, the total variation in $\cot \beta_1 d$, since there are now no coefficients $a_s$ left on which higher-order operators in the Taylor-series expansion can operate.

In (31) the second variation in $\cot \beta_1 d$ may be written as

$$-\sum_{k=2}^{N}\sum_{i=2}^{N} \Delta a_k\, \Delta a_i \left[\sum_{n=2}^{\infty} (jY_n\delta_{nk}\delta_{ni} + jY_{0n}P_{kn}P_{in}) + Y_{01}P_{k1}P_{i1}\cot \beta_{01} l\right]$$

$$= -\sum_{n=2}^{\infty}\left[jY_n\left(\sum_{k=2}^{N}\Delta a_k\, \delta_{nk}\right)^2 + jY_{0n}\left(\sum_{k=2}^{N}\Delta a_k\, P_{kn}\right)^2\right]$$

$$- Y_{01}\cot \beta_{01} l\left(\sum_{k=2}^{N}\Delta a_k\, P_{k1}\right)^2$$

When $\cot \beta_{01} l$ is positive, the second variation is always a negative real quantity since $jY_n$, $jY_{0n}$, and $Y_{01}$ are real and positive. Apart from the minus sign, the second variation is a positive definite quadratic form. The same is true for Eq. (30). Therefore, any variation in the coefficients $a_s$ from their true value will decrease the value of $\cot \beta_1 d$. Thus our approximation to $\mathcal{E}(x)$ will give a value for $\cot \beta_1 d$ which is always algebraically too small, except for that particular choice of trial functions that gives a correct functional form for $\mathcal{E}(x)$.

We may solve our problem in a similar fashion for the input impedance involving the transverse magnetic aperture field $\mathcal{H}(x)$. We would then get

$$\tan \beta_1 d = -\frac{1}{Z_1 a_1^2}\sum_{s=1}^{N}\sum_{r=1}^{N} a_s a_r h'_{sr} \qquad (32)$$

where

$$h'_{sr} = -j\left(\sum_{n=2}^{\infty} Z_n\delta_{ns}\delta_{nr} + \sum_{m=2}^{\infty} Z_{0m}P_{rm}P_{sm}\right) + Z_{01}\tan \beta_{01} l\, P_{s1}P_{r1}$$

This expression gives a lower algebraic bound on $\tan \beta_1 d$ and, consequently, an upper bound on $(\tan \beta_1 d)^{-1} = \cot \beta_1 d$. We thus have available two expressions giving upper and lower bounds on $\cot \beta_1 d$ and, therefore, are able to find the maximum error in our approximate solution.

Equations (30) and (32) are the analytic equivalent of the experimental tangent method for finding the equivalent-circuit parameters of a lossless junction. The solutions given by (30) and (32) are always of the form

$$\tan \beta_1 d = \frac{A + B \tan \beta_{01} l}{C + D \tan \beta_{01} l}$$

and the equivalent-circuit parameters may be found in terms of the coefficients $A$, $B$, $C$, and $D$ by the methods given in Sec. 5.8.

As in the first method, it is not necessary to solve for the unknown coefficients. We may obtain the value of $\cot \beta_1 d$ and $\tan \beta_1 d$ from the ratio of two determinants. Carrying out the analysis gives

$$\cot \beta_1 d \begin{vmatrix} g'_{22} & \cdots & g'_{2N} \\ \cdots & \cdots & \cdots \\ g'_{N2} & \cdots & g'_{NN} \end{vmatrix} = -Z_1 \begin{vmatrix} g'_{11} & \cdots & g'_{1N} \\ \cdots & \cdots & \cdots \\ g'_{N1} & \cdots & g'_{NN} \end{vmatrix} \tag{33a}$$

$$\tan \beta_1 d \begin{vmatrix} h'_{22} & \cdots & h'_{2N} \\ \cdots & \cdots & \cdots \\ h'_{N2} & \cdots & h'_{NN} \end{vmatrix} = -Y_1 \begin{vmatrix} h'_{11} & \cdots & h'_{1N} \\ \cdots & \cdots & \cdots \\ h'_{N1} & \cdots & h'_{NN} \end{vmatrix} \tag{33b}$$

The advantages of this second method over the first method are:

1. Upper and lower bounds for the equivalent-circuit parameters are obtained and, hence, a knowledge of the maximum error involved.
2. All the functions needed in the computation are real, and this simplifies the numerical work.
3. The method gives a convenient solution for all three parameters required for a complete description of the junction.

*Method* 3

The third method which we shall illustrate gives a direct solution for the impedance elements in the equivalent T-network representation of the junction or, alternatively, the admittance elements in the $\pi$-network representation.

If we assume an incident dominant mode of amplitude $b_1$ from the region $z > 0$, as well as an incident mode from the region $z < 0$, the two continuity equations at the aperture plane $z = 0$ become

$$[(1 + R_1)a_1 + T_{21}b_1]\phi_1 + \sum_{2}^{\infty} a_n\phi_n = [(1 + R_2)b_1 + T_{12}a_1]\psi_1 + \sum_{2}^{\infty} b_m\psi_m \tag{34a}$$

$$[(1 - R_1)a_1 - T_{21}b_1]Y_1\phi_1 - \sum_{2}^{\infty} a_n Y_n \phi_n$$
$$= -[(1 - R_2)b_1 - T_{12}a_1]Y_{01}\psi_1 + \sum_{2}^{\infty} b_m Y_{0m}\psi_m \quad (34b)$$

where $R_2$ is the reflection coefficient in the inhomogeneously filled guide, $T_{12}$ is the transmission coefficient from the empty guide into the partially filled guide, and $T_{21}$ is the transmission coefficient in the reverse direction. Equations (34) have been written so as to show explicitly that the dominant mode propagating away from the junction on each side depends linearly on the amplitudes of the incident modes on both sides of the junction. The following equivalent transmission-line voltages and currents are now introduced:

$$V_1 = (1 + R_1)a_1 + T_{21}b_1 \qquad V_2 = (1 + R_2)b_1 + T_{12}a_1$$
$$I_1 = (1 - R_1)Y_1a_1 - T_{21}Y_1b_1 \qquad I_2 = -(1 - R_2)Y_{01}b_1 + T_{12}Y_{01}a_1$$

The amplitudes $a_1$, $b_1$ may be chosen so that $I_2$ equals zero. The aperture electric field, which is proportional to $a_1$ and $b_1$ and, hence, proportional to $I_1$ and $I_2$ in general, in this case is proportional to $I_1$ only. If incident amplitudes are chosen so that $I_1$ equals zero, the aperture electric field will be proportional to $I_2$ only. For arbitrary amplitudes for the incident modes the aperture electric field is just a linear superposition of the fields which are proportional to $I_1$ and $I_2$, and may be written as

$$\mathcal{E}(x) = I_1\mathcal{E}_1(x) - I_2\mathcal{E}_2(x) \quad (35)$$

By the usual method of Fourier analysis, the equivalent voltages $V_1$, $V_2$ and the amplitude coefficients $a_n$, $b_m$ are found to be given by

$$V_1 = \int_0^a (I_1\mathcal{E}_1 - I_2\mathcal{E}_2)\phi_1\,dx \quad (36a)$$
$$V_2 = \int_0^a (I_1\mathcal{E}_1 - I_2\mathcal{E}_2)\psi_1\,dx \quad (36b)$$
$$a_n = \int_0^a (I_1\mathcal{E}_1 - I_2\mathcal{E}_2)\phi_n\,dx \quad (36c)$$
$$b_m = \int_0^a (I_1\mathcal{E}_1 - I_2\mathcal{E}_2)\psi_m\,dx \quad (36d)$$

The first two equations show that the voltages and currents are related by linear relations of the form

$$V_1 = Z_{11}I_1 - Z_{12}I_2 \quad (37a)$$
$$V_2 = Z_{21}I_1 - Z_{22}I_2 \quad (37b)$$

It will be shown that $Z_{12} = Z_{21}$, and hence (37) are a system of equations which may be interpreted as those describing an equivalent T network for the junction as illustrated in Fig. 8.2. The impedance elements are given by

$$Z_{11} = \int_0^a \mathcal{E}_1\phi_1\,dx \tag{38a}$$

$$Z_{12} = \int_0^a \mathcal{E}_2\phi_1\,dx \tag{38b}$$

$$Z_{21} = \int_0^a \mathcal{E}_1\psi_1\,dx \tag{38c}$$

$$Z_{22} = \int_0^a \mathcal{E}_2\psi_1\,dx \tag{38d}$$

Substituting for $a_n$, $b_m$ from (36) into (34b) gives

$$I_1\phi_1 - I_2\psi_1 = -j\int_0^a [I_1\mathcal{E}_1(x') - I_2\mathcal{E}_2(x')]G_1(x|x')\,dx' \tag{39}$$

where, for convenience, we have defined

$$-jG_1 = \sum_2^\infty [Y_n\phi_n(x)\phi_n(x') + Y_{0n}\psi_n(x)\psi_n(x')]$$

$$= -j\sum_2^\infty [|Y_n|\phi_n(x)\phi_n(x') + |Y_{0n}|\psi_n(x)\psi_n(x')]$$

It should be noted that this definition of $G_1$ differs from that used in method 1. For $n \geq 2$, the mode admittances $Y_n$ and $Y_{0n}$ are negative-imaginary, and hence $G_1$ is a real function. Since $I_1$ and $I_2$ have been

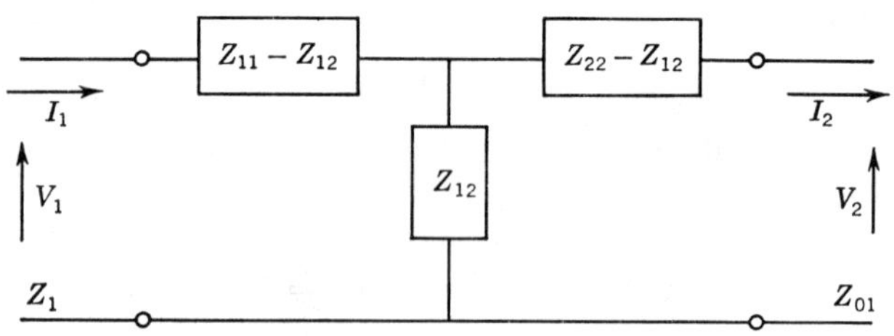


FIG. 8.2. Equivalent T network for junction.

chosen as the linearly independent variables, (39) is equivalent to the two separate equations

$$\phi_1 = -j\int_0^a \mathcal{E}_1(x')G_1(x|x')\,dx' \tag{40a}$$

$$\psi_1 = -j\int_0^a \mathcal{E}_2(x')G_1(x|x')\,dx' \tag{40b}$$

The required variational expressions for the impedance elements are obtained as follows. Multiplying (40a) by $\mathcal{E}_1(x)$ and integrating over $x$, dividing both sides by $\left(\int_0^a \mathcal{E}_1\phi_1\,dx\right)^2$, and comparing the result with (38a), we get

$$\frac{1}{\int_0^a \mathcal{E}_1\phi_1\,dx} = \frac{1}{Z_{11}} = -j\,\frac{\iint\limits_0^a G_1(x|x')\mathcal{E}_1(x)\mathcal{E}_1(x')\,dx\,dx'}{\left(\int_0^a \mathcal{E}_1\phi_1\,dx\right)^2} \tag{41a}$$

In a similar fashion we get from (40$b$)

$$\frac{1}{Z_{22}} = -j\,\frac{\iint\limits_0^a G_1(x|x')\mathcal{E}_2(x)\mathcal{E}_2(x')\,dx\,dx'}{\left(\int_0^a \mathcal{E}_2\psi_1\,dx\right)^2} \tag{41b}$$

For $Z_{21}$ we multiply (40$a$) by $\mathcal{E}_2(x)$, integrate over $x$, and divide by $\int_0^a \mathcal{E}_2\phi_1\,dx \int_0^a \mathcal{E}_1\psi_1\,dx$ to get

$$\frac{1}{Z_{21}} = -j\,\frac{\iint\limits_0^a G_1(x|x')\mathcal{E}_2(x)\mathcal{E}_1(x')\,dx\,dx'}{\iint\limits_0^a \mathcal{E}_1(x)\mathcal{E}_2(x')\phi_1(x')\psi_1(x)\,dx\,dx'} \tag{41c}$$

If we begin with (40$b$), we obtain an expression for $Z_{12}$ which is

$$\frac{1}{Z_{12}} = -j\,\frac{\iint\limits_0^a G_1(x|x')\mathcal{E}_2(x')\mathcal{E}_1(x)\,dx\,dx'}{\iint\limits_0^a \mathcal{E}_1(x)\mathcal{E}_2(x')\phi_1(x')\psi_1(x)\,dx\,dx'} = \frac{1}{Z_{21}} \tag{41d}$$

and is identical with (41$c$), since $G_1(x|x') = G_1(x'|x)$; that is, it is symmetrical in the variables $x$, $x'$. For later convenience a new set of real impedance elements will be defined by the following relations, $Z'_{ij} = -jZ_{ij}$; $i, j = 1, 2$.

The stationary property of the integral expressions for the impedance elements with respect to arbitrary first-order variations in the aperture field functions $\mathcal{E}_1$ and $\mathcal{E}_2$ is readily proved. For example, consider (41$a$), which gives the value of $Z_{11}$. If we vary the function $\mathcal{E}_1$ from its true value, we get

$$\left(\int_0^a \mathcal{E}_1\phi_1\,dx\right)^2 \delta\frac{1}{Z_{11}} + \frac{2}{Z_{11}} \iint\limits_0^a \mathcal{E}_1(x')\,\delta\mathcal{E}_1(x)\phi_1(x)\phi_1(x')\,dx\,dx'$$
$$= -2j \iint\limits_0^a G_1(x|x')\mathcal{E}_1(x')\,\delta\mathcal{E}_1(x)\,dx\,dx$$

or

$$\left(\int_0^a \mathcal{E}_1\phi_1\,dx\right)^2 \delta\frac{1}{Z_{11}} = 2\int_0^a \delta\mathcal{E}_1(x')\left[-j\int_0^a G_1(x|x')\mathcal{E}_1(x)\,dx - \phi_1(x')\right]dx'$$

after canceling $\int_0^a \mathcal{E}_1\phi_1\,dx$ with $Z_{11}$. An examination of the integrand on the right-hand side shows that it is identically zero, by virtue of (40$a$). Therefore, we again have the property that a first-order approximation to the aperture fields will yield a second-order approximation to the impedance elements $Z_{ij}$.

The expressions (41) are homogeneous in the sense that they are independent of the magnitudes of the aperture fields $\mathcal{E}_1$ and $\mathcal{E}_2$. The impedance elements depend only on the functional form of $\mathcal{E}_1$ and $\mathcal{E}_2$. Since $Z'_{11}$, $Z'_{22}$, and $Z'_{12}$ are real quantities, and $G_1$ is real, the functions $\mathcal{E}_1$ and $\mathcal{E}_2$ are also real, with the possible exception of a constant complex multiplier, which cancels out and does not affect the magnitude of $Z'_{ij}$. We may hence show that the expressions for $(Z'_{11})^{-1}$ and $(Z'_{22})^{-1}$ are absolute minima for the correct field distributions. Thus an approximate expression for $\mathcal{E}_1$ and $\mathcal{E}_2$ will yield values for $(Z'_{11})^{-1}$ and $(Z'_{22})^{-1}$ which are numerically too large, i.e., values for $Z'_{11}$ and $Z'_{22}$ which are too small. The proof is as follows. Since $Z'_{11}$ depends only on the functional form of $\mathcal{E}_1$, we may choose $\mathcal{E}_1$ so that $\int_0^a \mathcal{E}_1\phi_1\,dx = 1$. We have therefore placed a constraint on $\mathcal{E}_1$ so that, when we vary $\mathcal{E}_1$, the variation will be orthogonal to $\phi_1$, in order to satisfy the condition $\int_0^a \mathcal{E}_1\phi_1\,dx = 1$. Let $(Z'_{11})_0$ be the value of $Z'_{11}$ for the correct field distribution. Thus we have

$$\begin{aligned}\delta\frac{1}{Z'_{11}} = \frac{1}{Z'_{11}} - \frac{1}{(Z'_{11})_0} &= \frac{(Z'_{11})_0 - Z'_{11}}{Z'_{11}(Z'_{11})_0} \\ &= \iint_0^a G_1(x|x')[\mathcal{E}_1(x) + \delta\mathcal{E}_1(x)][\mathcal{E}_1(x') + \delta\mathcal{E}_1(x')]\,dx\,dx' \\ &\qquad - \iint_0^a G_1(x|x')\mathcal{E}_1(x)\mathcal{E}_1(x')\,dx\,dx' \\ &= \iint_0^a G_1(x|x')[\mathcal{E}_1(x)\delta\mathcal{E}_1(x') + \mathcal{E}_1(x')\delta\mathcal{E}_1(x)]\,dx\,dx' \\ &\qquad + \iint_0^a G_1(x|x')\delta\mathcal{E}_1(x)\delta\mathcal{E}_1(x')\,dx\,dx'\end{aligned}$$

The first integral on the right is equal to $j\int_0^a \delta\mathcal{E}_1\phi_1\,dx + j\int_0^a \delta\mathcal{E}_1\phi_1\,dx$ in view of (40$a$) and vanishes since we have constrained $\delta\mathcal{E}_1$ to be orthogonal to $\phi_1$. Thus our result reduces to

$$\frac{(Z'_{11})_0 - Z'_{11}}{Z'_{11}(Z'_{11})_0} = \iint_0^a G_1(x|x')\,\delta\mathcal{E}_1(x')\,\delta\mathcal{E}_1(x)\,dx\,dx'$$

which is a positive-definite quadratic form. Hence $(Z'_{11})_0 - Z'_{11}$ is positive, and $Z'_{11} \leq (Z'_{11})_0$, which proves the stated minimum property. The proof for $Z'_{12}$ does not hold since the quantity

$$\iint_0^a G_1(x|x')\,\delta\mathcal{E}_1(x)\,\delta\mathcal{E}_2(x')\,dx\,dx'$$

is not a positive-definite quadratic form; i.e., it may be either positive or negative, as, for example, if $\delta\mathcal{E}_1 = -\delta\mathcal{E}_2$, it will be negative.

The evaluation of the impedance elements may proceed along lines similar to that used in the previous two variational methods. Suitable approximations for the aperture fields $\mathcal{E}_1$ and $\mathcal{E}_2$ which lead to a straightforward computational procedure are

$$\mathcal{E}_1 = \sum_1^N a_n\phi_n \qquad \mathcal{E}_2 = \sum_1^N b_m\psi_m$$

For the impedance elements $Z'_{11}$ and $Z'_{22}$ we get

$$Z'_{11}\begin{vmatrix} g_{11} & \cdots & g_{1N} \\ \cdots & \cdots & \cdots \\ g_{N1} & \cdots & g_{NN} \end{vmatrix} = \begin{vmatrix} g_{22} & \cdots & g_{2N} \\ \cdots & \cdots & \cdots \\ g_{N2} & \cdots & g_{NN} \end{vmatrix} \tag{42a}$$

$$Z'_{22}\begin{vmatrix} t_{11} & \cdots & t_{1N} \\ \cdots & \cdots & \cdots \\ t_{N1} & \cdots & t_{NN} \end{vmatrix} = \begin{vmatrix} t_{22} & \cdots & t_{2N} \\ \cdots & \cdots & \cdots \\ t_{N2} & \cdots & t_{NN} \end{vmatrix} \tag{42b}$$

where

$$g_{sr} = \sum_{n=2}^{\infty} (|Y_n|\delta_{sn}\delta_{rn} + |Y_{0n}|P_{rn}P_{sn})$$

$$t_{sr} = \sum_{n=2}^{\infty} (|Y_n|P_{ns}P_{nr} + |Y_{0n}|\delta_{sn}\delta_{rn})$$

and $P_{sn}$ is given by $\int_0^a \phi\,\psi_n\,dx$.

The computation of $Z'_{12}$ is not so straightforward, and so we will work it out in more detail. From (41*d*), we have

$$(Z'_{12})^{-1} \int_0^a \sum_{s=1}^N a_s\phi_s\psi_1\,dx \int_0^a \sum_{r=1}^N b_r\psi_r\phi_1\,dx$$
$$= \frac{1}{Z'_{12}} \sum_{s=1}^N \sum_{r=1}^N a_s b_r P_{s1} P_{1r} = \iint_0^a \sum_{n=2}^\infty [|Y_n|\phi_n(x)\phi_n(x') + |Y_{0n}|$$
$$\times \psi_n(x)\psi_n(x')] \sum_{s=1}^N a_s\phi_s(x') \sum_{r=1}^N b_r\psi_r(x)\,dx\,dx' = \sum_{s=1}^N \sum_{r=1}^N a_s b_r$$
$$\left[\sum_{n=2}^\infty (|Y_n|\delta_{ns}P_{nr} + |Y_{0n}|\delta_{nr}P_{sn})\right] = \sum_{s=1}^N \sum_{r=1}^N [a_s b_r P_{sr}(|Y_s| + |Y_{0r}|)]$$

and, hence,

$$\sum_{s=1}^N \sum_{r=1}^N a_s b_r P_{s1} P_{1r} \left[\frac{1}{Z'_{12}} - \frac{P_{sr}}{P_{s1}P_{1r}}(|Y_s| + |Y_{0r}|)\right]$$
$$= \sum_{s=1}^N \sum_{r=1}^N a_s b_r P_{s1} P_{1r} \left(\frac{1}{Z'_{12}} - Q_{sr}\right) = 0$$

where $Q_{sr} = (P_{sr}/P_{s1}P_{1r})(|Y_s| + |Y_{0r}|)$. It should be noted that $Q_{sr}$ is not symmetrical in the indices $s$ and $r$ since $Q_{sr} \neq Q_{rs}$. When we equate the partial derivatives with respect to $a_s$ and $b_r$ to zero, we obtain a homogeneous set of $2N$ equations with the following determinant:

$$\begin{vmatrix} \frac{1}{Z'_{12}} - Q_{11} & \cdots & \frac{1}{Z'_{12}} - Q_{N1} & 0 & \cdots & 0 \\ \cdots & \cdots & \cdots & \cdots & \cdots & \cdots \\ \frac{1}{Z'_{12}} - Q_{1N} & \cdots & \frac{1}{Z'_{12}} - Q_{NN} & 0 & \cdots & 0 \\ 0 & \cdots & 0 & \frac{1}{Z'_{12}} - Q_{11} & \cdots & \frac{1}{Z'_{12}} - Q_{1N} \\ \cdots & \cdots & \cdots & \cdots & \cdots & \cdots \\ 0 & \cdots & 0 & \frac{1}{Z'_{12}} - Q_{N1} & \cdots & \frac{1}{Z'_{12}} - Q_{NN} \end{vmatrix}$$

The vanishing of this determinant gives the solution for $Z'_{12}$. The above $2N \times 2N$ determinant is equal to the product of the two determinants in the upper left-hand corner and the lower right-hand corner. We may interchange the rows and columns of the determinant in the lower right-hand corner without changing its value. This transposition of the rows and columns makes the two determinants equal, and thus $Z'_{12}$ is given by the condition that the following determinant shall vanish:

$$\begin{vmatrix} \frac{1}{Z'_{12}} - Q_{11} & \cdots & \frac{1}{Z'_{12}} - Q_{N1} \\ \frac{1}{Z'_{12}} - Q_{1N} & \cdots & \frac{1}{Z'_{12}} - Q_{NN} \end{vmatrix} = 0$$

If the first row is subtracted from all of the other rows, the resultant determinant may be factored into the sum of two determinants which gives the solution for $Z'_{12}$ in explicit form as follows:

$$Z'_{12}\begin{vmatrix} Q_{11} & Q_{21} & \cdots & Q_{N1} \\ Q_{11} - Q_{12} & Q_{21} - Q_{22} & \cdots & Q_{N1} - Q_{N2} \\ \cdots & \cdots & \cdots & \cdots \\ Q_{11} - Q_{1N} & \cdots & \cdots & Q_{N1} - Q_{NN} \end{vmatrix}$$

$$= \begin{vmatrix} 1 & 1 & \cdots & 1 \\ Q_{11} - Q_{12} & Q_{21} - Q_{22} & \cdots & Q_{N1} - Q_{N2} \\ \cdots & \cdots & \cdots & \cdots \\ Q_{11} - Q_{1N} & \cdots & \cdots & Q_{N1} - Q_{NN} \end{vmatrix} \tag{43}$$

A similar analysis may be carried out to obtain variational expressions for the admittance elements in the equivalent $\pi$-network representation of the junction. Our point of departure is again the set of equations (34) expressing the continuity of the tangential fields in the aperture plane. As before, a set of equivalent transmission-line voltages and currents is introduced. The incident mode amplitudes may be chosen so that the voltages $V_1$ and $V_2$ are the independent variables. The tangential magnetic field in the aperture plane may be chosen as the linear superposition of two partial fields proportional to $V_1$ and $V_2$, and thus

$$\mathcal{H}(x) = V_1\mathcal{H}_1(x) - V_2\mathcal{H}_2(x)$$

The unknown coefficients are then given by Fourier analysis as follows:

$$I_1 = \int_0^a (V_1\mathcal{H}_1 - V_2\mathcal{H}_2)\phi_1\,dx \tag{44a}$$

$$I_2 = \int_0^a (V_1\mathcal{H}_1 - V_2\mathcal{H}_2)\psi_1\,dx \tag{44b}$$

$$a_n = -Z_n \int_0^a (V_1\mathcal{H}_1 - V_2\mathcal{H}_2)\phi_n\,dx \tag{44c}$$

$$b_n = Z_{0n} \int_0^a (V_1\mathcal{H}_1 - V_2\mathcal{H}_2)\psi_n\,dx \tag{44d}$$

The first two equations give a linear relationship between the currents and voltages of the form

$$I_1 = Y_{11}V_1 - Y_{12}V_2 \tag{45a}$$

$$I_2 = Y_{21}V_1 - Y_{22}V_2 \tag{45b}$$

where the admittance elements are given by

$$Y_{11} = \int_0^a \mathfrak{H}_1\phi_1\,dx \tag{46a}$$

$$Y_{12} = \int_0^a \mathfrak{H}_2\phi_1\,dx \tag{46b}$$

$$Y_{21} = \int_0^a \mathfrak{H}_1\psi_1\,dx \tag{46c}$$

$$Y_{22} = \int_0^a \mathfrak{H}_2\psi_1\,dx \tag{46d}$$

Further along in the analysis we will find that $Y_{12} = Y_{21}$, and hence (45) are equations which describe the behavior of the equivalent $\pi$ network illustrated in Fig. 8.3. If we invert the set of Eqs. (37), we get

$$I_1 = \frac{Z_{22}}{\Delta}V_1 - \frac{Z_{12}}{\Delta}V_2$$

$$I_2 = \frac{Z_{12}}{\Delta}V_1 - \frac{Z_{11}}{\Delta}V_2$$

where $\Delta = Z_{11}Z_{22} - Z_{12}^2$. Thus we see that $Y_{11} = Z_{22}/\Delta$, $Y_{12} = Z_{12}/\Delta$, $Y_{22} = Z_{11}/\Delta$. The admittance elements $Y_{ij}$ are pure-imaginary for a

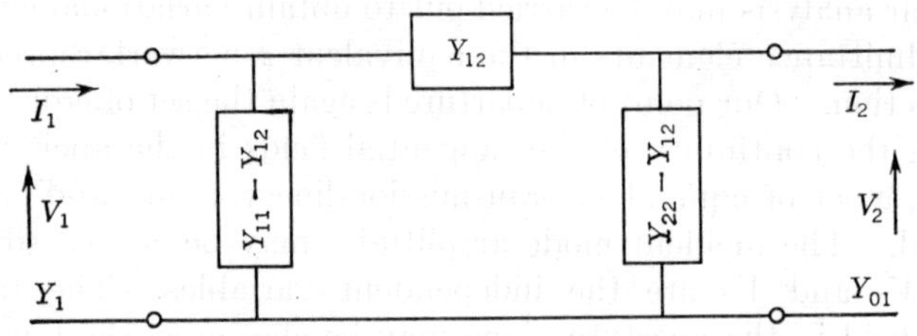


FIG. 8.3. Equivalent $\pi$ network for junction.

lossless function, and, for convenience, we will define the admittance elements $Y'_{ij}$ as $jY_{ij}$. The admittance elements $Y'_{ij}$ as defined are all real. Substituting for the unknown coefficients from (44) into (34a) gives

$$V_1\phi_1 - V_2\psi_1 = j\int_0^a (V_1\mathfrak{H}_1 - V_2\mathfrak{H}_2)G_2(x|x')\,dx'$$

where $$G_2(x|x') = \sum_{n=2}^{\infty} [|Z_n|\phi_n(x)\phi_n(x') + |Z_{0n}|\psi_n(x)\psi_n(x')]$$

Since $V_1$ and $V_2$ are linearly independent, we must have

$$\phi_1 = j\int_0^a \mathfrak{H}_1G_2\,dx' \qquad \psi_1 = j\int_0^a \mathfrak{H}_2G_2\,dx'$$

These latter two equations may be used to construct the stationary expressions for the admittance elements in the same manner as was done for the impedance elements. The results are

$$\frac{1}{Y'_{11}} = -\frac{j}{Y_{11}} = \frac{\iint_0^a G_2(x|x')\mathfrak{H}_1(x)\mathfrak{H}_1(x')\,dx\,dx'}{\left(\int_0^a \mathfrak{H}_1\phi_1\,dx\right)^2} \tag{47a}$$

$$\frac{1}{Y'_{22}} = -\frac{j}{Y_{22}} = \frac{\iint_0^a G_2(x|x')\mathfrak{H}_2(x)\mathfrak{H}_2(x')\,dx\,dx'}{\left(\int_0^a \mathfrak{H}_2\psi_1\,dx\right)^2} \tag{47b}$$

$$\frac{1}{Y'_{12}} = -\frac{j}{Y_{12}} = -\frac{j}{Y_{21}} = \frac{\iint_0^a G_2(x|x')\mathfrak{H}_1(x)\mathfrak{H}_2(x')\,dx\,dx'}{\int_0^a \mathfrak{H}_1\psi_1\,dx \int_0^a \mathfrak{H}_2\phi_1\,dx} \tag{47c}$$

The evaluation of $Y'_{ij}$ proceeds in an identical manner with that for $Z'_{ij}$. As before, we may show that Eqs. (47$a$) and (47$b$) are absolute minima for the correct field distributions, and hence give lower bounds on $Y'_{11}$ and $Y'_{22}$, respectively, when the approximate field distributions are used in the integrands. The functions $\mathfrak{H}_1$ and $\mathfrak{H}_2$ may be taken as real.

Although we can obtain lower bounds on $Y'_{11}$ and $Y'_{22}$, and thus upper bounds on $-Z'_{22}/\Delta$ and $-Z'_{11}/\Delta$, we do not obtain upper and lower bounds on the impedance elements $Z'_{11}$ and $Z'_{22}$ by comparing the lower bounds obtained for $Z'_{11}$ and $Z'_{22}$ with the upper bounds on $-Z'_{22}/\Delta$ and $-Z'_{11}/\Delta$, since we do not know if $\Delta$ is too large or too small. The bounds on $\Delta$ are not known, since the error in $Z_{12}$ may be either positive or negative.

If we have a plane discontinuity, i.e., a discontinuity localized in a plane such as $z = 0$, and the input and output guides are identical, then $\phi_n = \psi_n$, $Z_n = Z_{0n}$, and the functions $\mathcal{E}_1$, $\mathfrak{H}_1$ are identical with $\mathcal{E}_2$ and $\mathfrak{H}_2$, respectively. Thus $Z_{11} = Z_{12} = Z_{22} = Z_s$, and $Y_{11} = Y_{12} = Y_{22} = Y_s$, and the equivalent circuit reduces to a pure shunt reactive element. Furthermore $Y_s = 1/Z_s$, and we thus can obtain upper and lower bounds on the shunt impedance element, and, consequently, we know the maximum error involved in our calculations.

The main objection to the variational method for the T- or $\pi$-network parameters is that it does not give an estimate of the error involved in the calculations. Also three separate calculations must be made to determine all three elements. When the equivalent circuit reduces to a shunt element, only two calculations are required to obtain upper and lower bounds on the shunt impedance element, and this variational technique is then a convenient one to use.

The above three methods are by no means exhaustive, but they are perhaps the most useful ones. Under some conditions of symmetry, when a discontinuity can be represented by two parameters only, it is convenient to modify method 3 slightly. We will illustrate this modified technique in connection with the analysis of the thick inductive window. All the above methods may be extended in a straightforward manner to the vector problem, i.e., to the case where the functions $\mathcal{E}$ and $\mathcal{H}$ are vector functions rather than scalar functions. Under these conditions, the Green's function becomes a dyadic quantity. As a general rule, for the vector problem positive-definite quadratic forms for the equivalent-circuit parameters are not obtained because a mixture of $E$ and $H$ modes is excited. For such cases we obtain stationary expressions whose stationary values are not either absolute minima or absolute maxima. Positive-definite quadratic forms are obtained only when the modes excited all store a net amount of either electric energy or magnetic energy.

## 8.2. Capacitive Diaphragm

As a first example of a discontinuity problem for which a variational solution will be obtained, we consider an asymmetrical capacitive diaphragm in an infinitely wide parallel-plate transmission line as in Fig. 8.4.

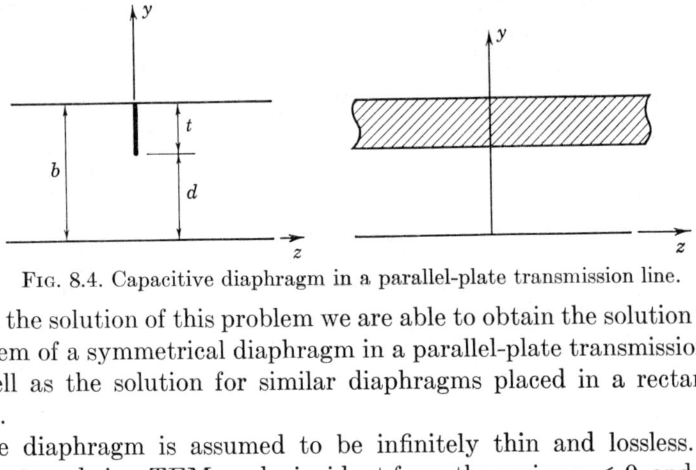


FIG. 8.4. Capacitive diaphragm in a parallel-plate transmission line.

From the solution of this problem we are able to obtain the solution to the problem of a symmetrical diaphragm in a parallel-plate transmission line, as well as the solution for similar diaphragms placed in a rectangular guide.

The diaphragm is assumed to be infinitely thin and lossless. The incident mode is a TEM mode, incident from the region $z < 0$, and polarized with the electric vector along the $y$ direction. The incident field is independent of the $x$ coordinate, and, because of the uniformity of the discontinuity, all the higher-order modes which are excited are also independent of $x$. The only higher-order modes which are independent of $x$ and can be coupled with the incident mode in the aperture plane are the TM or $E$ modes. If the frequency of operation is chosen so that $\lambda_0 > 2b$, none of the higher-order modes will propagate. The excited evanescent modes store a net amount of reactive electric energy in the vicinity of the diaphragm, and, since the total electric field of the dominant mode is continuous across the diaphragm, the discontinuity is equivalent to a shunt capacitive reactance. Since the discontinuity requires only a single parameter to describe it, the general variational formulation presented in the preceding section is not required. It will be just as expedient to proceed from basic principles in setting up the solution for the shunt susceptance.

For a capacitive diaphragm in a rectangular guide, the higher-order modes which are excited are the longitudinal-section electric (LSE) modes. When there is no variation with $x$, these modes reduce to the $E$ modes. For generality we will, therefore, derive the required modes from a magnetic-type Hertzian potential having a single component in the $x$ direction rather than from a $z$-directed electric-type Hertzian potential which is required for $E$ modes. The relevant equations are

$$\mathbf{E} = -j\omega\mu_0\nabla \times \mathbf{\Pi}_h \tag{48a}$$
$$\mathbf{H} = k_0^2\mathbf{\Pi}_h + \nabla\nabla\cdot\mathbf{\Pi}_h \tag{48b}$$
$$\nabla^2\mathbf{\Pi}_h + k_0^2\mathbf{\Pi}_h = 0 \tag{48c}$$

For $\mathbf{\Pi}_h$ we may choose a form such as $\mathbf{a}_x\psi(y)e^{\pm\Gamma z}$. A solution for $\psi$ which results in an electric field with a vanishing tangential component on the parallel plates is

$$\psi_n = \cos\frac{n\pi y}{b}\,e^{\pm\Gamma_n z} \qquad n = 0, 1, 2, \ldots$$

where $\Gamma_n^2 = (n\pi/b)^2 - k_0^2$. From these solutions the following expansion for the transverse electric and magnetic fields are readily established:

$$E_y = \begin{cases} a_0(e^{-jk_0z} + Re^{jk_0z}) + \sum_1^\infty a_n \cos\frac{n\pi y}{b}\,e^{\Gamma_n z} & z < 0 \quad (49a) \\ Ta_0e^{-jk_0z} + \sum_1^\infty b_n \cos\frac{n\pi y}{b}\,e^{-\Gamma_n z} & z > 0 \quad (49b) \end{cases}$$

$$H_x = \begin{cases} -a_0Y_0(e^{-jk_0z} - Re^{jk_0z}) + \sum_1^\infty a_nY_n \cos\frac{n\pi y}{b}\,e^{\Gamma_n z} & z < 0 \quad (49c) \\ -Ta_0Y_0e^{-jk_0z} - \sum_1^\infty b_nY_n \cos\frac{n\pi y}{b}\,e^{-\Gamma_n z} & z > 0 \quad (49d) \end{cases}$$

where $R$ is the reflection coefficient, and $T$ is the transmission coefficient. In (49) the factor $-j\omega\mu_0\Gamma_n$, which arises when the electric field is derived from the Hertzian potential, has been absorbed in the amplitude coefficients $a_n$, $b_n$. The mode admittances $Y_n$ are given by

$$Y_n = \frac{jk_0}{\Gamma_n} Y_0 = \frac{jk_0}{\Gamma_n}\left(\frac{\epsilon_0}{\mu_0}\right)^{1/2} = j|Y_n| \qquad n > 0 \tag{50}$$

and are positive-imaginary for $n > 0$. For $n = 0$, $\Gamma_n = jk_0$, and $Y_n = Y_0$, the characteristic admittance of free space.

The equations expressing the continuity of the transverse fields in the aperture plane $z = 0$ are

$$(1 + R)a_0 + \sum_1^\infty a_n \cos \frac{n\pi y}{b} = Ta_0 + \sum_1^\infty b_n \cos \frac{n\pi y}{b} \qquad 0 \le y \le b \tag{51a}$$

$$(1 - R)a_0 Y_0 - \sum_1^\infty a_n Y_n \cos \frac{n\pi y}{b} = Ta_0 Y_0 + \sum_1^\infty b_n Y_n \cos \frac{n\pi y}{b} \qquad 0 \le y \le d \tag{51b}$$

The total transverse electric field vanishes on the diaphragm, and hence (51*a*) is valid over the total range of $y$. The tangential magnetic field is discontinuous across the diaphragm by an amount equal to the total current on the diaphragm, and hence (51*b*) holds only in the aperture opening. Let the aperture electric field be $\mathcal{E}(y)$, where $\mathcal{E}(y)$ vanishes on the diaphragm. A Fourier analysis of (51*a*), holding for $0 \le y \le b$, now gives

$$a_0(1 + R) = Ta_0 = \frac{1}{b}\int_0^b \mathcal{E}(y')\,dy' \tag{52a}$$

$$a_n = b_n = \frac{2}{b}\int_0^b \mathcal{E}(y') \cos \frac{n\pi y'}{b}\,dy' \tag{52b}$$

Replacing $T$ by $1 + R$ and substituting the integral expressions for the unknown coefficients into (51*b*) gives

$$-\frac{2R}{1 + R} Y_0 \int_0^b \mathcal{E}(y')\,dy' = 4 \sum_1^\infty Y_n \cos \frac{n\pi y}{b} \int_0^b \mathcal{E}(y') \cos \frac{n\pi y'}{b}\,dy' \qquad 0 \le y \le d \tag{53}$$

A shunt susceptance $jB$ across a transmission line of unit characteristic impedance gives rise to a reflection coefficient (see Fig. 8.5):

$$R = \frac{1 - Y_{\text{in}}}{1 + Y_{\text{in}}} = -\frac{jB}{2 + jB}$$

Solving for $jB$ gives $jB = -2R/(1 + R)$. Thus (53) represents an integral equation whose solution would give the normalized susceptance $B$.

A variational integral for $B$ may be obtained by multiplying (53) by $\mathcal{E}(y)$ and integrating over the aperture region $0 \le y \le d$. Since $\mathcal{E}(y)$ vanishes on the diaphragm, the integrals for the unknown amplitude coefficients need be extended over the aperture opening only. Thus we get

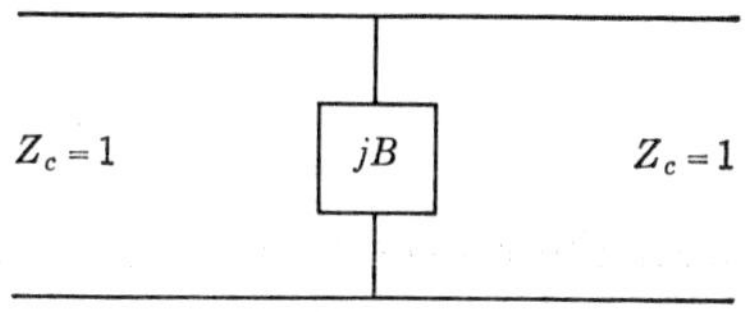


FIG. 8.5

$$B = 4k_0 \frac{\sum_{n=1}^{\infty} \frac{1}{\Gamma_n} \iint_0^d \mathcal{E}(y)\mathcal{E}(y') \cos \frac{n\pi y}{b} \cos \frac{n\pi y'}{b} \, dy \, dy'}{\left[ \int_0^d \mathcal{E}(y) \, dy \right]^2} \tag{54}$$

after replacing $-2R/(1 + R)$ by $jB$ and $Y_n/Y_0$ by $jk_0/\Gamma_n$. It is easily demonstrated by the usual method that (54) has the required stationary property.

As it stands, (54) is not directly amenable to a simple solution for $B$. By a rather ingenious transformation of variables, Schwinger was able to transform (54) into a dominant term consisting of a series of functions orthogonal over the range of integration plus a small correction series. As a first step, the order of integration and summation is interchanged. The series may be written as a dominant series plus a rapidly converging series, as follows:

$$\frac{b}{\pi} \sum_1^{\infty} \frac{1}{n} \cos \frac{n\pi y}{b} \cos \frac{n\pi y'}{b} + \sum_1^{\infty} \left( \frac{1}{\Gamma_n} - \frac{b}{n\pi} \right) \cos \frac{n\pi y}{b} \cos \frac{n\pi y'}{b}$$

The product of the two cosine terms in the first series is equal to

$$\frac{1}{2} \left[ \cos \frac{n\pi}{b} (y - y') + \cos \frac{n\pi}{b} (y + y') \right]$$

With this substitution this series may be summed according to the methods given in the Mathematical Appendix, and we get

$$\begin{aligned} \frac{b}{\pi} \sum_1^{\infty} \frac{1}{n} \cos \frac{n\pi y}{b} \cos \frac{n\pi y'}{b} &= -\frac{b}{2\pi} \ln 4 \left[ \sin \frac{\pi}{2b} (y - y') \sin \frac{\pi}{2b} (y + y') \right] \\ &= -\frac{b}{2\pi} \ln 2 \left( \cos \frac{\pi y}{b} - \cos \frac{\pi y'}{b} \right) \end{aligned} \tag{55}$$

As a next step, the following change of variables is made:

$$\cos\frac{\pi y}{b} = \alpha_1 + \alpha_2 \cos\theta \tag{56a}$$

$$\cos\frac{\pi y'}{b} = \alpha_1 + \alpha_2 \cos\theta' \tag{56b}$$

with the constants $\alpha_1$ and $\alpha_2$ chosen so that $\theta = 0, \pi$, when $y = 0, d$, respectively. Substitution of these particular values of $\theta$ and $y$ into (56) gives the following solutions for $\alpha_1$ and $\alpha_2$:

$$\alpha_1 = \frac{1}{2}\left(1 + \cos\frac{\pi d}{b}\right) = \cos^2\frac{\pi d}{2b} \tag{57a}$$

$$\alpha_2 = 1 - \alpha_1 = \sin^2\frac{\pi d}{2b} \tag{57b}$$

Substituting into Eq. (55) now gives

$$\begin{aligned}\frac{b}{\pi}\sum_1^\infty \frac{1}{n}\cos\frac{n\pi y'}{b}\cos\frac{n\pi y}{b} &= -\frac{b}{2\pi}[\ln\alpha_2 + \ln 2(\cos\theta - \cos\theta')] \\ &= -\frac{b}{\pi}\ln\sin\frac{\pi d}{2b} + \frac{b}{\pi}\sum_1^\infty \frac{1}{n}\cos n\theta\cos n\theta' \end{aligned}\tag{58}$$

where the logarithmic term involving the factor $2(\cos\theta - \cos\theta')$ has been expanded into an infinite series by direct analogy with the original series which was summed. The beauty of this result is that it gives a series which is orthogonal over the range of integration, i.e., orthogonal in the aperture domain.

The correction series cannot be transformed in this manner; however, this is not of great consequence, since it converges rapidly, and each term may be transformed individually. For example, we have

$$\begin{aligned}\cos\frac{2\pi y}{b} &= 2\cos^2\frac{\pi y}{b} - 1 = 2(\alpha_1 + \alpha_2\cos\theta)^2 - 1 \\ &= \alpha_2^2\cos 2\theta + 4\alpha_1\alpha_2\cos\theta + 2\alpha_1^2 + \alpha_2^2 - 1\end{aligned}$$

To complete the change in variables it should be noted that

$$-\frac{\pi}{b}\sin\frac{\pi y}{b}\,dy = -\alpha_2\sin\theta\,d\theta$$

and hence $$dy = \frac{b}{\pi}\csc\frac{\pi y}{b}\sin\theta\,d\theta = \frac{dy}{d\theta}\,d\theta$$

Thus $\mathcal{E}(y)\,dy$ becomes $\mathcal{E}(y)(dy/d\theta)\,d\theta = \mathcal{F}(\theta)\,d\theta$, where $\mathcal{F}(\theta)$ is equal to $\mathcal{E}(y)\,dy/d\theta$ expressed as a function of $\theta$. Each term $\cos(n\pi y/b)$ may be expressed as a series of the terms cos $m\theta$, with $m$ running from 0 up to $n$. For convenience we will write

$$\cos \frac{n\pi y}{b} = \sum_{m=0}^{n} P_{nm} \cos m\theta \tag{59}$$

where the $P_{nm}$ are suitably determined coefficients.

When all these substitutions are made in Eq. (54), we get

$$B = \frac{4k_0 b}{\pi} \frac{\displaystyle\iint_0^\pi \mathfrak{F}(\theta)\mathfrak{F}(\theta') \left[ \ln \csc \frac{\pi d}{2b} + \sum_1^\infty \frac{1}{n} \cos n\theta \cos n\theta' + \sum_{n=1}^\infty \left( \frac{\pi}{b\Gamma_n} - \frac{1}{n} \right) \sum_{m=0}^n \sum_{s=0}^n P_{nm} P_{ns} \cos m\theta \cos s\theta' \right]}{\left[ \int_0^\pi \mathfrak{F}(\theta)\, d\theta \right]^2} \tag{60}$$

Since a completely rigorous solution of (60) is difficult to obtain, we appeal to its stationary property and approximate $\mathfrak{F}(\theta)$ by a constant which may be chosen equal to unity. It turns out that this yields a surprisingly accurate result for $B$ for the usual range of parameters occurring in practice. The only terms in (60) which do not integrate to zero are the constant terms, and hence we get

$$B = \frac{4k_0 b}{\pi} \left[ \ln \csc \frac{\pi d}{2b} + \sum_{n=1}^\infty \left( \frac{\pi}{b\Gamma_n} - \frac{1}{n} \right) P_{n0}^2 \right] \tag{61}$$

It may be shown that all $P_{n0}^2$ are less than unity, and, hence, terms beyond $n$ equal to 3 or 4 are usually entirely negligible since the correction series converges as $n^{-3}$. The coefficients $P_{n0}$ may be obtained with reasonable facility by using available tabulated expansions of the Tchebysheff polynomials.† We have

$$\cos \frac{n\pi y}{b} = T_n \left( \cos \frac{\pi y}{b} \right) = T_n(\alpha_1 + \alpha_2 \cos \theta)$$

The expansion of the Tchebysheff polynomial gives a power series in cos $\theta$. Each term $(\cos \theta)^m$ may, in turn, be expressed as a series of Tchebysheff polynomials from which the coefficients $P_{nm}$ may be readily deduced. The required expansions for $n \leq 6$ are listed in Table 8.1.

† "Tables of Chebyshëv Polynomials," National Bureau of Standards, 1952. (Several spellings for the name Tchebysheff are in current use.)

With the aid of these relations the first four coefficients $P_{n0}$ are found to be

$$P_{10} = \alpha_1 \qquad P_{20} = 2\alpha_1^2 + \alpha_2^2 - 1$$
$$P_{30} = 4\alpha_1^3 + 6\alpha_1\alpha_2^2 - 3\alpha_1^2$$
$$P_{40} = 8\alpha_1^4 + 3\alpha_2^4 + 24\alpha_1^2\alpha_2^2 - 8\alpha_1^2 - 4\alpha_2^2 + 1$$

Equation (61) gives an upper bound on the true value of the susceptance. A lower bound may be obtained by solving the problem in

TABLE 8.1. TCHEBYSHEFF POLYNOMIALS

| $u = \cos x$ | $T_n(u) = T_n(\cos x) = \cos nx$ |
|---|---|
| $T_1(u)$ | $u$ |
| $T_2(u)$ | $2u^2 - 1$ |
| $T_3(u)$ | $4u^3 - 3u$ |
| $T_4(u)$ | $8u^4 - 8u^2 + 1$ |
| $T_5(u)$ | $16u^5 - 20u^3 + 5u$ |
| $T_6(u)$ | $32u^6 - 48u^4 + 18u^2 - 1$ |
| $T_n(u)$ | $2uT_{n-1} - T_{n-2}$ |
| | |
| $u$ | $T_1(u)$ |
| $u^2$ | $\frac{1}{2}T_2 + \frac{1}{2}$ |
| $u^3$ | $\frac{1}{4}T_3 + \frac{3}{4}T_1$ |
| $u^4$ | $\frac{1}{8}T_3 + \frac{1}{2}T_2 + \frac{3}{8}$ |
| $u^5$ | $\frac{1}{16}T_5 + \frac{5}{16}T_3 + \frac{5}{8}T_1$ |
| $u^6$ | $\frac{1}{32}T_6 + \frac{3}{16}T_4 + \frac{15}{32}T_2 + \frac{5}{16}$ |

terms of the magnetic field in the aperture, i.e., in terms of the current on the diaphragm. For this particular example, we may also obtain a lower bound by omitting the higher-order terms in the correction series, since the correction series always results in a positive contribution.

The function $\mathfrak{F}(\theta)$ may be completely represented by the infinite expansion $\sum_{m=0}^{\infty} C_m \cos m\theta$. If in (60) we neglect all the terms in the correction series, we get

$$B = \frac{4k_0 b}{\pi C_0^2}\left(C_0^2 \ln \csc \frac{\pi d}{2b} + \frac{1}{4}\sum_{m=1}^{\infty} \frac{C_m^2}{m}\right)$$

Equating all the partial derivatives $\partial/\partial C_m$ to zero shows that $C_m = 0$, $m > 0$, and hence

$$B = \frac{4k_0 b}{\pi} \ln \csc \frac{\pi d}{2b} \tag{62}$$

This result is, therefore, a rigorous solution for the low-frequency limit since, as the frequency approaches zero, the correction series vanishes. For the range of parameters encountered in practice (62) is not very accurate, although it is a lower bound. A significant improvement is obtained by retaining the first term in the correction series. We now get

$$B = \frac{4k_0b}{\pi C_0{}^2}\left[C_0{}^2 \ln \csc \frac{\pi d}{2b} + \frac{1}{4}\sum_{m=1}^{\infty}\frac{C_m{}^2}{m} + \left(\frac{\pi}{b\Gamma_1} - 1\right)\left(\alpha_1 C_0 + \frac{\alpha_2 C_1}{2}\right)^2\right]$$

Again, equating the partial derivatives with respect to $C_m$ equal to zero, we find that $C_m = 0$, $m > 1$. The first two partial derivatives, when equated to zero, yield two homogeneous equations for the coefficients $C_0$, $C_1$. The solution for these exists only if the determinant vanishes. Equating the determinant to zero gives the following solution for $B$, which is a lower bound on the true value,

$$B = \frac{4k_0b}{\pi}\left[\ln \csc \frac{\pi d}{2b} + \frac{(\pi/b\Gamma_1 - 1)\alpha_1{}^2}{1 + (\pi/b\Gamma_1 - 1)\alpha_2{}^2}\right] \tag{63}$$

That (63) does indeed present a lower bound on the susceptance will now be proved. Let $B_0$ be the true value of susceptance, and

$$\sum_{m=0}^{\infty} C_{0m} \cos m\theta$$

be the correct aperture field distribution. The susceptance $B_0$ is given by

$$B_0 = \frac{4k_0b}{\pi C_{00}^2}\left[C_{00}^2 \ln \csc \frac{\pi d}{2b} + \frac{1}{4}\sum_{m=1}^{\infty}\frac{C_{om}^2}{m} + \frac{1}{4}\sum_{n=1}^{\infty}\left(\frac{\pi}{b\Gamma_n} - \frac{1}{n}\right)\left(\sum_{m=0}^{n}\frac{2}{\epsilon_{0m}}P_{nm}C_{0m}\right)^2\right]$$

This expression for $B_0$ is an absolute minimum, so that any variation in the coefficients $C_{0m}$ will increase $B_0$. If we drop all the terms in the correction series for $n > N$, then clearly

$$B_1 = \frac{4k_0b}{\pi C_{00}^2}\left[C_{00}^2 \ln \csc \frac{\pi d}{2b} + \frac{1}{4}\sum_{m=1}^{\infty}\frac{C_{om}^2}{m} + \frac{1}{4}\sum_{n=1}^{N}\left(\frac{\pi}{b\Gamma_n} - \frac{1}{n}\right)\left(\sum_{m=0}^{n}\frac{2}{\epsilon_{0m}}P_{nm}C_{0m}\right)^2\right] < B_0$$

Now $\partial B_1/\partial C_{0m}$ will not, in general, equal zero, since the higher-order correction terms have been dropped, and therefore $B_1$ is not a minimum. However, if we compute a new set of values for $C_{0m}$, say $C_m$, so that $\partial B_1/\partial C_m = 0$, we will minimize the expression for $B_1$ to get a new susceptance $B < B_1 < B_0$. This latter procedure is, in fact, that used to obtain (63) and, therefore, demonstrates that (63) yields a lower bound on the true susceptance.

As an indication of the accuracy of (61) and (63), the upper and lower bounds on $B$ have been computed as a function of $d/b$ for $b = 0.4\lambda_0$ and are plotted in Fig. 8.6. For $d/b < 0.5$, the two bounds on $B$ differ by less than 4 per cent, and hence the average value is in error by less than

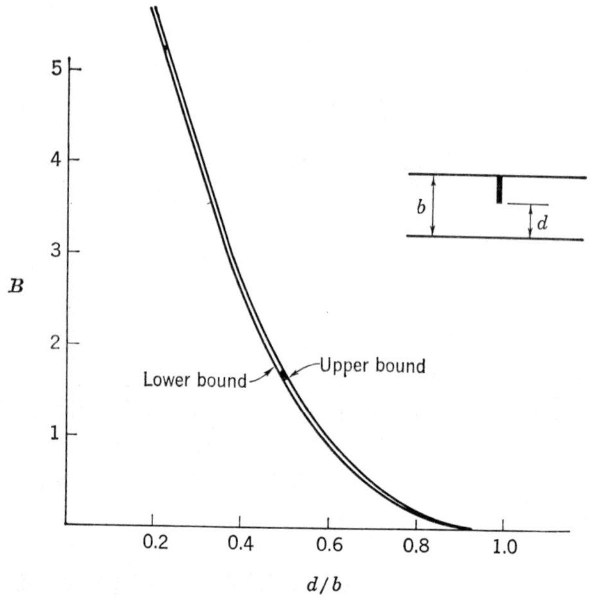


FIG. 8.6. Normalized capacitive susceptance $B$ for $b = 0.4\lambda_0$.

2 per cent. For $d/b = 0.8$, the difference amounts to about 10 per cent, but this occurs in a range for $d/b$ which is not commonly encountered in practice.

*Symmetrical Capacitive Diaphragms*

Figure 8.7*a* and *b* illustrate symmetrical capacitive diaphragms in parallel-plate transmission lines. An application of image theory reduces the solutions of both these problems to that for the asymmetrical diaphragm when a TEM mode is incident from the left. The symmetry of the structure results only in the excitation of those higher-order modes having a zero component of longitudinal electric field on the symmetry plane. A conducting plane may, therefore, be placed along the symmetry plane, and the problem is thereby reduced to that of the asymmetrical diaphragm. The normalized shunt capacitive susceptance is given by (61) and (63), with $d$ and $b$ replaced by $d/2$ and $b/2$, respectively.

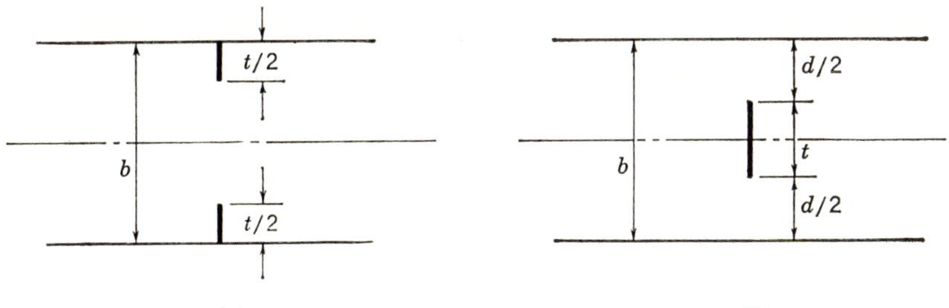


FIG. 8.7. Symmetrical capacitive diaphragms in a parallel-plate transmission line.

*Capacitive Diaphragms in a Rectangular Waveguide*

Figure 8.8*a*, *b*, and *c* illustrates capacitive diaphragms in a rectangular guide of height $b$ and width $a$. The solution to these discontinuity problems may be obtained from the corresponding solutions for similar diaphragms in a parallel-plate transmission line of height $b$ and with $a$ infinite.

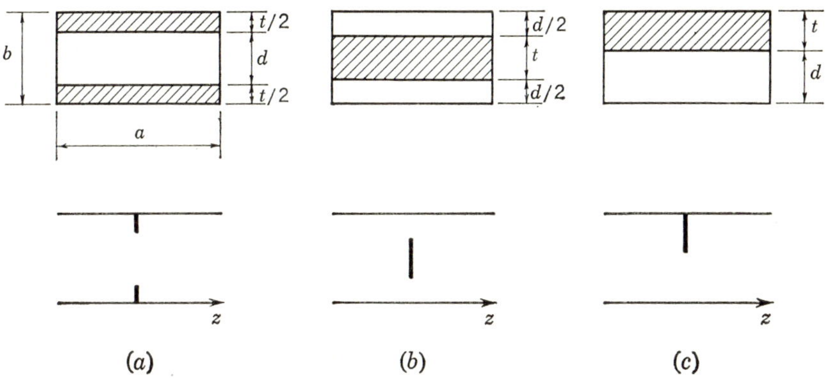


FIG. 8.8. Capacitive diaphragms in a rectangular guide.

In the parallel-plate transmission-line problem ($a$ infinite), let the total Hertzian potential function, from which all the field components may be derived, be $\mathbf{\Pi}_h(y,z;k_0^2)$. For the equivalent-waveguide problem, the solution is given by the following Hertzian potential function when the incident field is an $H_{10}$ mode:

$$\mathbf{\Pi}_h' = \sin\frac{\pi x}{a}\,\mathbf{\Pi}_h\left(y,z;\,k_0^2 - \frac{\pi^2}{a^2}\right)$$

Clearly $\mathbf{\Pi}_h(y,z;\,k_0^2 - \pi^2/a^2)$ is a solution of

$$\left(\frac{\partial^2}{\partial y^2} + \frac{\partial^2}{\partial z^2} + k_0^2 - \frac{\pi^2}{a^2}\right)\mathbf{\Pi}_h\left(y,z;\,k_0^2 - \frac{\pi^2}{a^2}\right) = 0$$

and hence $\mathbf{\Pi}_h'$ is a solution of the required Helmholtz equation

$$\left(\frac{\partial^2}{\partial x^2} + \frac{\partial^2}{\partial y^2} + \frac{\partial^2}{\partial z^2} + k_0^2\right)\mathbf{\Pi}_h' = 0$$

since the second partial derivative with respect to $x$ introduces a term $-(\pi/a)^2$. The function $\mathbf{\Pi}_h(y,z;k_0^2)$ has been determined so that $-j\omega\mu\ \partial\mathbf{\Pi}_h(y,z;k_0^2)/\partial z$, corresponding to the $y$ component of electric field, is continuous in the aperture and vanishes on the diaphragm. Also $k_0^2\mathbf{\Pi}_h(y,z;k_0^2)$, corresponding to the $x$ component of magnetic field, is continuous in the aperture and is discontinuous with respect to $z$ across the diaphragm. These boundary conditions hold for all values of $k_0^2$, and in particular when $k_0^2$ is replaced by $k_0^2 - (\pi/a)^2$. Multiplying the function by $\sin(\pi x/a)$ does not change the boundary conditions which are satisfied in the aperture plane. In the waveguide problem, the only new field components entering into the problem are $y$ and $z$ components of magnetic field, and these are easily shown to satisfy the proper boundary conditions since they are derivable from the curl of the electric field. Hence the potential function $\mathbf{\Pi}_h'$, as defined, gives a solution which satisfies all the required boundary conditions for the waveguide problem. Therefore the normalized capacitive susceptance for a diaphragm in a rectangular guide may be obtained by replacing $k_0^2$ in the formulas (61) and (63) by $\beta_{10}^2 = k_0^2 - (\pi/a)^2$. For the symmetrical diaphragms, $d$ and $b$ must also be replaced by $d/2$ and $b/2$ for the same reason that these changes were made to obtain formulas applicable to symmetrical diaphragms in a parallel-plate transmission line. The propagation constants $\Gamma_n^2$ in the correction series become $(n\pi/b)^2 + (\pi/a)^2 - k_0^2$ in the waveguide case. This modification results in a more rapidly converging series.

## 8.3. Thin Inductive Diaphragm in a Rectangular Guide

Initially we will consider a general infinitely thin inductive diaphragm as illustrated in Fig. 8.9. The solutions for a symmetrical or asymmetrical diaphragm are just special cases of the general solution.

With an $H_{10}$ mode incident from $z < 0$, only higher-order $H_{n0}$ modes are excited, since the discontinuity is uniform along the $y$ direction. The problem may be formulated in the same manner as that for the dielectric step considered in Sec. 8.1 in connection with the general discussion on variational methods. The particular problem under consideration here is simpler since the waveguides on both sides of the diaphragm are the same. Equations (41$a$) to (41$d$) are all applicable and, in actual fact, reduce to one single equation because of the identical input and output guides. The equivalent T network reduces to a single shunt element since $Z_{11} = Z_{22} = Z_{12}$. The normalized shunt inductive susceptance of the diaphragm is given by

$$B = \frac{jZ_1}{Z_{12}} = \frac{2 \displaystyle\iint\limits_{x_1}^{x_2} \sum_{n=2}^{\infty} \frac{|Y_n|}{Y_1} \sin \frac{n\pi x}{a} \sin \frac{n\pi x'}{a} \, \mathcal{E}(x)\mathcal{E}(x') \, dx \, dx'}{\left[ \displaystyle\int_{x_1}^{x_2} \mathcal{E}(x) \sin \frac{\pi x}{a} \, dx \right]^2} \tag{64}$$

where $|Y_n|/Y_1 = \Gamma_n/\beta_1$, $\Gamma_n^2 = (n\pi/a)^2 - k_0^2$, and $\beta_1 = |\Gamma_1|$. Equation (64) may also be readily set up from fundamental principles in much

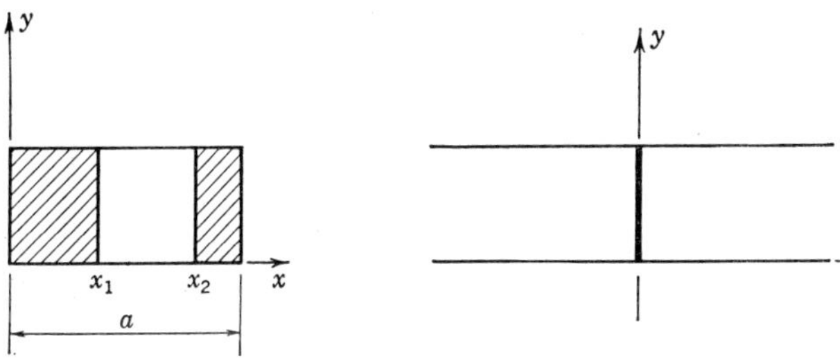


FIG. 8.9. Inductive diaphragm.

the same way as the capacitive-diaphragm problem was handled. The series in (64) may be written as a dominant series plus a correction series. The dominant series $\sum\limits_{2}^{\infty} n \sin \frac{n\pi x}{a} \sin \frac{n\pi x'}{a}$ may be summed, and then a change of variables may be introduced in order to obtain a series of functions which are orthogonal over the range of integration. It is somewhat more convenient to integrate each term by parts once first and thereby obtain a series similar to that occurring in the capacitive-diaphragm problem. We have

$$\int_{x_1}^{x_2} \mathcal{E}(x) \sin \frac{n\pi x}{a} \, dx = - \frac{a}{n\pi} \cos \frac{n\pi x}{a} \, \mathcal{E}(x) \Big|_{x_1}^{x_2} + \frac{a}{n\pi} \int_{x_1}^{x_2} \frac{d\mathcal{E}(x)}{dx} \cos \frac{n\pi x}{a} \, dx$$

Since the tangential electric field is zero at the edge of a thin conducting strip, the integrated term vanishes. Thus (64) is transformed over to

$$B = \frac{2}{\beta_1} \frac{\displaystyle\iint\limits_{x_1}^{x_2} \sum_{n=2}^{\infty} \frac{\Gamma_n}{n^2} \cos \frac{n\pi x}{a} \cos \frac{n\pi x'}{a} \, \mathcal{E}'(x)\mathcal{E}'(x') \, dx' \, dx}{\left[ \displaystyle\int_{x_1}^{x_2} \mathcal{E}'(x) \cos \frac{\pi x}{a} \, dx \right]^2} \tag{65}$$

where $\mathcal{E}'(x) = d\mathcal{E}(x)/dx$. The series may be written as

$$\frac{\pi}{a}\sum_{n=1}^{\infty}\frac{1}{n}\cos\frac{n\pi x}{a}\cos\frac{n\pi x'}{a} - \frac{\pi}{a}\cos\frac{\pi x}{a}\cos\frac{\pi x'}{a} + \sum_{n=2}^{\infty}\frac{\Gamma_n - n(\pi/a)}{n^2}\cos\frac{n\pi x}{a}\cos\frac{n\pi x'}{a}$$

where the last series converges rapidly since $\Gamma_n$ approaches $n\pi/a$ for large $n$. The first series sums to $(-\pi/2a)\ln 2[\cos(\pi x/a) - \cos(\pi x'/a)]$. The following change in variables is now introduced:

$$\cos\frac{\pi x}{a} = \alpha_1 + \alpha_2\cos\theta$$

$$\cos\frac{\pi x'}{a} = \alpha_1 + \alpha_2\cos\theta'$$

with $\alpha_1$ and $\alpha_2$ chosen so that $\theta = 0, \pi$, when $x = x_1, x_2$, respectively. Substituting these particular values into the transformation and solving for $\alpha_1$ and $\alpha_2$ gives

$$\alpha_1 = \frac{1}{2}\left(\cos\frac{\pi x_1}{a} + \cos\frac{\pi x_2}{a}\right) = \cos\pi\frac{x_2 - x_1}{2a}\cos\pi\frac{x_2 + x_1}{2a} = \cos\frac{\pi d}{2a}\cos\pi\frac{2x_1 + d}{2a} \tag{66a}$$

$$\alpha_2 = \frac{1}{2}\left(\cos\frac{\pi x_1}{a} - \cos\frac{\pi x_2}{a}\right) = \sin\frac{\pi d}{2a}\sin\pi\frac{2x_1 + d}{2a} \tag{66b}$$

The summed series transforms to

$$-\frac{\pi}{2a}\ln 2\alpha_2(\cos\theta - \cos\theta') = -\frac{\pi}{2a}\ln\alpha_2 + \frac{\pi}{a}\sum_{n=1}^{\infty}\frac{1}{n}\cos n\theta\cos n\theta'$$

after the logarithmic term is expanded. With the introduction of the new variables, $\mathcal{E}'(x)\,dx$ is replaced by $\mathcal{E}'(x)(dx/d\theta)\,d\theta = \mathcal{F}(\theta)\,d\theta$. Also each term $\cos(n\pi x/a)$ in the correction series is replaced by $\sum_{m=0}^{n} P_{nm}\cos m\theta$, where the $P_{nm}$ are suitably determined coefficients. With all these modifications substituted into (65), the variational expression for $B$ becomes

$$B = \frac{2\pi}{\beta_1 a} \frac{\displaystyle\iint_0^\pi \mathfrak{F}(\theta)\mathfrak{F}(\theta') \left[ -\tfrac{1}{2} \ln \alpha_2 + \sum_{n=1}^{\infty} \frac{1}{n} \cos n\theta \cos n\theta' - (\alpha_1 + \alpha_2 \cos\theta)(\alpha_1 + \alpha_2 \cos\theta') + \sum_{n=2}^{\infty} \frac{(a/\pi)\Gamma_n - n}{n^2} \sum_{m=0}^{n} \sum_{s=0}^{n} P_{nm} P_{ns} \cos m\theta \cos s\theta' \right] d\theta\, d\theta'}{\left[ \int_0^\pi (\alpha_1 + \alpha_2 \cos\theta)\mathfrak{F}(\theta)\, d\theta \right]^2} \tag{67}$$

Since (67) is a stationary expression for $B$, we may obtain a good approximation to $B$ (upper bound) by using a finite series in $\cos m\theta$ for $\mathfrak{F}(\theta)$. In order to simplify the details, we will restrict the analysis at this point to that applicable for a symmetrical diaphragm. For this case $2x_1 + d = a$ and $\alpha_1 = 0$, $\alpha_2 = \sin(\pi d/2a)$. If we substitute a general series $\sum_{m=0}^{N} C_m \cos m\theta$ for $\mathfrak{F}(\theta)$ into (67), we find that, when the minimization is carried out, all the even coefficients are zero. This we could anticipate in view of the symmetry. The electric field $\mathcal{E}(x)$ is an even function about $x = a/2$; hence $d\mathcal{E}/dx$ is an odd function about $x = a/2$, and therefore $\mathcal{E}'(x)\, dx/d\theta = \mathfrak{F}(\theta)$ will contain only the terms $\cos m\theta$ with $m$ odd. In the correction series, $\cos(n\pi x/a)$ becomes a series in $\cos m\theta$ with $m$ only odd or even, depending on whether $n$ is odd or even, respectively. This would not be true, in general, if $\alpha_1$ were not equal to zero.

If we approximate $\mathfrak{F}(\theta)$ by a single term, that is, $\cos\theta$, then (67) gives the following result for the symmetrical diaphragm with an aperture opening $d$:

$$B = \frac{2\pi}{\beta_1 a} \left( \csc \frac{\pi d}{2a} \right)^2 \left[ \cos^2 \frac{\pi d}{2a} + \sum_{n=3,5,\ldots}^{\infty} \frac{(a/\pi)\Gamma_n - n}{n^2} P_{n1}^2 \right] \tag{68}$$

The coefficients $P_{n1}$ may be found by using the properties of the Tchebysheff polynomials listed in Table 8.1. For example,

$$\cos \frac{3\pi x}{a} = T_3\left( \cos \frac{\pi x}{a} \right) = T_3(\alpha_2 \cos\theta) = 4(\alpha_2 \cos\theta)^3 - 3\alpha_2 \cos\theta$$

Now $(\cos\theta)^3 = \frac{1}{4} T_3(\cos\theta) + \frac{3}{4} T_1(\cos\theta) = \frac{1}{4}\cos 3\theta + \frac{3}{4}\cos\theta$, and hence $\cos(3\pi x/a)$ is transformed over to $\alpha_2^3 \cos 3\theta + (3\alpha_2^3 - 3\alpha_2)\cos\theta$. The coefficient $P_{31}$ is therefore given by

$$P_{31} = 3\alpha_2(\alpha_2^2 - 1) \tag{69a}$$

Similarly we find

$$P_{51} = 5\alpha_2(2\alpha_2^4 - 3\alpha_2^2 - 1) \tag{69b}$$

Terms beyond $n = 5$ are usually negligible, and the correction series may be terminated at this point.

The correction series for the inductive diaphragm gives a negative contribution since $\Gamma_n < n\pi/a$, and, hence, neglecting this series does not result in a lower bound to the susceptance $B$ as was the case for the capacitive diaphragm. A lower bound may, however, be obtained by solving the problem in terms of the current on the diaphragm. This solution is left for a problem at the end of this chapter.

## 8.4. Thick Inductive Window

The thick inductive window illustrated in Fig. 8.10 is an example of a symmetrical discontinuity which requires two parameters to completely

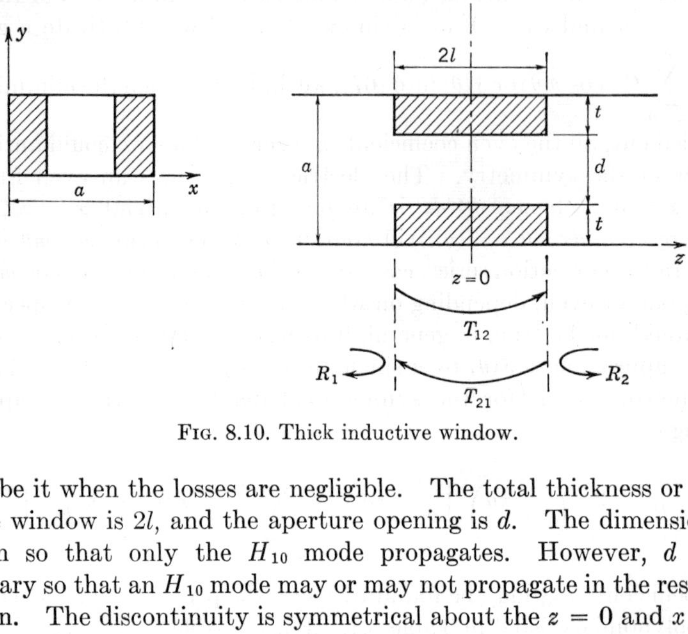


FIG. 8.10. Thick inductive window.

describe it when the losses are negligible. The total thickness or length of the window is $2l$, and the aperture opening is $d$. The dimension $a$ is chosen so that only the $H_{10}$ mode propagates. However, $d$ is left arbitrary so that an $H_{10}$ mode may or may not propagate in the restricted section. The discontinuity is symmetrical about the $z = 0$ and $x = a/2$ planes. If we choose terminal reference planes at $z = \pm l$, the equivalent circuit for the window is a symmetrical T network as illustrated in Fig. 8.11.

Let an $H_{10}$ mode of amplitude $a_1$ be incident on the diaphragm from the left and right simultaneously. For this particular choice of incident modes we have $V_1 = (1 + R_1)a_1 + T_{21}a_1 = (1 + R_2)a_1 + T_{12}a_1 = V_2$, since, from the symmetry of the structure, $R_1 = R_2$ and $T_{12} = T_{21}$. In view of the symmetrical excitation, the electric field $E_y$ must be symmetrical about $z = 0$. Since $H_x$ is proportional to $\partial E_y/\partial z$, it is antisymmetrical about $z = 0$ and, in particular, equal to zero on the symmetry plane. A magnetic wall may, therefore, be placed at $z = 0$. The circuit equations for the T network are

$$V_1 = Z_{11}I_1 - Z_{12}I_2$$
$$V_2 = Z_{12}I_1 - Z_{11}I_2$$

and, when $V_1 = V_2$, we have $I_1 = -I_2$. The input impedance to the network with an open circuit (magnetic wall) in the symmetry plane is

$$(Z_{\text{in}})_{oc} = \frac{V_1}{I_1} = Z_{11} + Z_{12} \tag{70}$$

If we choose incident waves with amplitudes $a_1$ and $-a_1$, we will have $E_y$ equal to zero on the symmetry plane because of the antisymmetrical

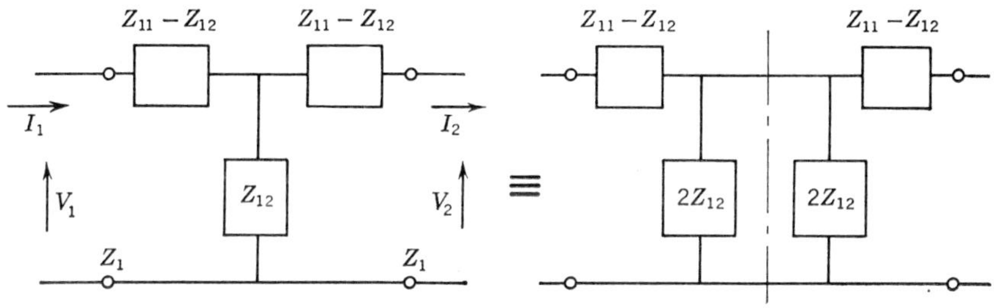


FIG. 8.11. Equivalent circuit for a thick inductive window.

method of excitation. In this case the input impedance is given by

$$(Z_{\text{in}})_{sc} = Z_{11} - Z_{12} \tag{71}$$

Hence, in order to find $Z_{11} + Z_{12}$, we solve the problem with a magnetic wall in the symmetry plane while, in order to find $Z_{11} - Z_{12}$, we solve the problem with a short circuit in the symmetry plane. From the two solutions we may then obtain $Z_{11}$ and $Z_{12}$ by addition and subtraction. The reduced problem is illustrated in Fig. 8.12. The field in the region $z \le -l$ will be given by an expansion in terms of an infinite set of $H_{n0}$ modes as follows:

$$E_y = a_1\phi_1 e^{-\Gamma_1(z+l)} + R_1 a_1 \phi_1 e^{\Gamma_1(z+l)} + \sum_{n=3,5,\ldots}^{\infty} a_n \phi_n e^{+\Gamma_n(z+l)} \tag{72a}$$

$$H_x = -Y_1 a_1 \phi_1 e^{-\Gamma_1(z+l)} + R_1 a_1 Y_1 \phi_1 e^{\Gamma_1(z+l)} + \sum_{n=3,5,\ldots}^{\infty} a_n Y_n \phi_n e^{+\Gamma_n(z+l)} \tag{72b}$$

where $\phi_n = \sin(n\pi x/a)$, $\Gamma_n{}^2 = n^2\pi^2/a^2 - k_0{}^2$, $Y_n = (-j\Gamma_n/k_0)Y_0$, and the origin for the waves is taken at $z = -l$. In the region $-l \le z \le 0$, we have a combination of forward- and backward-traveling waves chosen

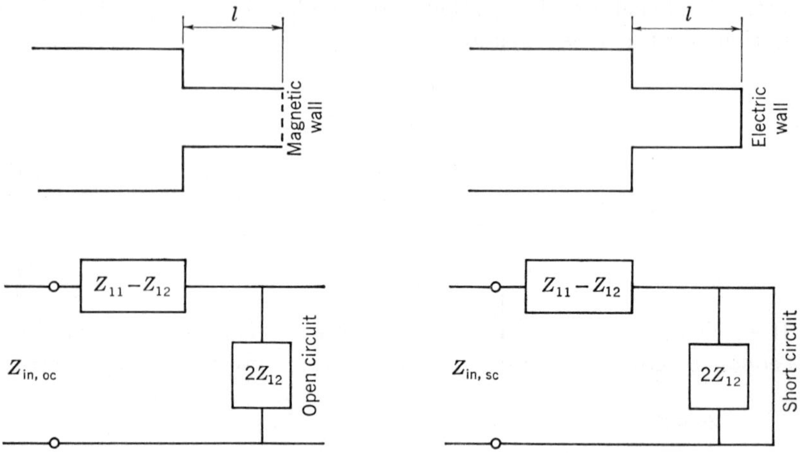


FIG. 8.12. Illustration of open-circuit and short-circuit problems.

so that, at $z = 0$, the transverse magnetic field vanishes. For the $n$th mode, we take $E_y$ equal to

$$b_n\psi_n \frac{e^{-\gamma_n z} + e^{\gamma_n z}}{2} = b_n\psi_n \cosh \gamma_n z$$

The transverse magnetic field is given by $\dfrac{-j}{\omega\mu_0}\dfrac{\partial E_y}{\partial z}$. Thus we may write the following expansion for the fields in the region $-l \le z \le 0$:

$$E_y = \sum_{n=1,3,\ldots}^{\infty} b_n\psi_n(x) \cosh \gamma_n z \tag{73a}$$

$$H_x = \sum_{n=1,3,\ldots}^{\infty} b_n Y_{0n}\psi_n(x) \sinh \gamma_n z \tag{73b}$$

where $\psi_n(x) = \sin (n\pi/d)(x - t)$, $\gamma_n{}^2 = n^2\pi^2/d^2 - k_0{}^2$, and

$$Y_{0n} = \frac{-j\gamma_n}{k_0} Y_0$$

At $z = -l$, the transverse fields must be continuous, and so we get

$$(1 + R_1)a_1\phi_1 + \sum_{n=3,5,\ldots}^{\infty} a_n\phi_n = \sum_{n=1,3,\ldots}^{\infty} b_n\psi_n \cosh \gamma_n l \tag{74a}$$

$$(1 - R_1)Y_1 a_1\phi_1 - \sum_{n=3,5,\ldots}^{\infty} a_n Y_n\phi_n = \sum_{n=1,3,\ldots}^{\infty} b_n Y_{0n}\psi_n \sinh \gamma_n l \tag{74b}$$

where $(1 + R_1)a_1 = V_1$ and $(1 - R_1)Y_1 a_1 = I_1 = (Z_{11} + Z_{12})^{-1}V_1$.

Let the transverse electric field in the aperture at $z = -l$ be $\mathcal{E}(x)$. By Fourier analysis, we get

$$V_1 = \frac{2}{a} \int_t^{t+d} \mathcal{E}(x')\phi_1(x')\,dx' \tag{75a}$$

$$a_n = \frac{2}{a} \int_t^{t+d} \mathcal{E}(x')\phi_n(x')\,dx' \tag{75b}$$

$$b_n = \frac{2}{d \cosh \gamma_n l} \int_t^{t+d} \mathcal{E}(x')\psi_n(x')\,dx' \tag{75c}$$

A variational expression for $(Z_{11} + Z_{12})^{-1}$ may be obtained by substituting the integral expressions for the amplitude coefficients into (74*b*), multiplying (74*b*) by $\mathcal{E}(x)$, integrating over the aperture, and, finally, dividing both sides by $\left(\int_t^{t+d} \mathcal{E}\phi_1\,dx\right)^2$. By this procedure, we obtain

$$(Z_{11} + Z_{12})^{-1} = -\frac{j \displaystyle\iint_t^{t+d} \mathcal{E}(x)\mathcal{E}(x')G_2(x|x')\,dx\,dx'}{\left[\displaystyle\int_t^{t+d} \mathcal{E}(x)\phi_1(x)\,dx\right]^2} \tag{76}$$

where

$$-jG_2(x|x') = \sum_{n=3,5,\ldots}^{\infty} Y_n\phi_n(x)\phi_n(x') + \frac{a}{d} \sum_{n=1,3,\ldots}^{\infty} Y_{0n}\psi_n(x)\psi_n(x') \tanh \gamma_n l$$

Equation (76) leads to a positive definite quadratic form if all $\gamma_n$ and hence $Y_{0n}$ are real. If $\gamma_1$ is imaginary, the term $Y_{01} \tanh \gamma_1 l$ is real; and as long as $l$ is chosen so that $\tan |\gamma_1 l|$ is positive, then (76) is still a positive definite quadratic form. When this is the case, an approximate solution for $\mathcal{E}(x)$ used in (76) gives a value for $Z_{11} + Z_{12}$ which is too small.

In the case of an electric wall at $z = 0$, the transverse electric field for the $n$th mode in the region $-l \le z \le 0$ is taken proportional to $\psi_n \sinh \gamma_n z$ in order that it should vanish at $z = 0$. By an analysis similar to the above, we obtain the following variational expression for $(Z_{11} - Z_{12})^{-1}$:

$$(Z_{11} - Z_{12})^{-1} = \frac{-j \displaystyle\iint_t^{t+d} G_1(x|x')\mathcal{E}(x)\mathcal{E}(x')\,dx\,dx'}{\left(\displaystyle\int_t^{t+d} \mathcal{E}\phi_1\,dx\right)^2} \tag{77}$$

where

$$-jG_1(x|x') = \sum_{3,5,\ldots}^{\infty} Y_n\phi_n(x)\phi_n(x') + \frac{a}{d} \sum_{1,3,\ldots}^{\infty} Y_{0n}\psi_n(x)\psi_n(x') \coth \gamma_n l$$

The aperture electric field in (77), although written as $\mathcal{E}(x)$, is not the same as in (76).

We can obtain considerable information about the thick inductive iris from (76) and (77) without actually solving the equations. We note that $\tanh \gamma_n l$ and $\coth \gamma_n l$ approach unity rapidly as $n$ increases. If $l$ is not too small so that $\gamma_n l > 2$ for $n > 1$, we may replace $\tanh \gamma_n l$ by unity for $n > 1$. We may then obtain another simple equivalent circuit for the thick diaphragm. Consider the equivalent circuit illustrated in Fig. 8.13. With an open circuit at the symmetry plane, the

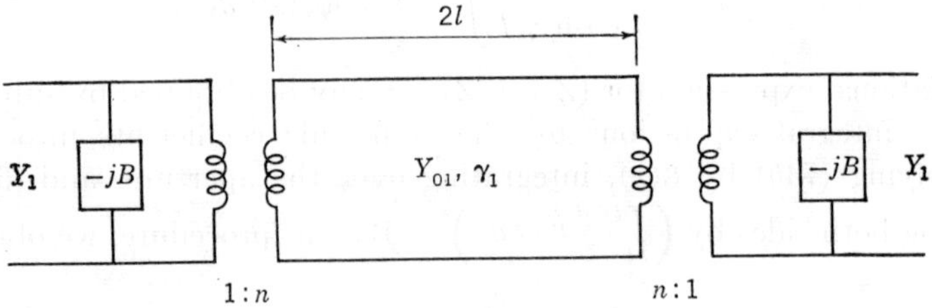


FIG. 8.13. Alternative equivalent circuit for a thick inductive window.

input admittance is $Y_{\text{in}} = -jB + n^2 Y_{01} \tanh \gamma_1 l$. Since $Y_{\text{in}}$ is also equal to $(Z_{11} + Z_{12})^{-1}$, we find upon comparison with Eq. (76) that

$$B = \frac{\displaystyle\iint_t^{t+d} \mathcal{E}(x)\mathcal{E}(x') \sum_{n=3,5,\ldots}^{\infty} \left[ |Y_n| \phi_n(x)\phi_n(x') + \frac{a}{d} |Y_{0n}| \psi_n(x)\psi_n(x') \right] dx\, dx'}{\left[ \displaystyle\int_t^{t+d} \mathcal{E}(x)\phi_1(x)\, dx \right]^2} \tag{78a}$$

$$n^2 = \frac{a}{d} \left[ \frac{\displaystyle\int_t^{t+d} \mathcal{E}(x)\psi_1(x)\, dx}{\displaystyle\int_t^{t+d} \mathcal{E}(x)\phi_1(x)\, dx} \right]^2 \tag{78b}$$

If the thickness $2l$ becomes infinite, we obtain an $H$-plane symmetrical step discontinuity. The inductive susceptance at the aperture plane is given by (78$a$) for the step, except when $\gamma_1$ is real. For $\gamma_1$ real, $Y_{01}$ is imaginary, and we have a pure susceptive termination $-jB - jn^2|Y_{01}|$ for $l$ infinite. For the thick diaphragm under open-circuit or short-circuit mid-plane terminations, the input admittance is always a pure susceptance whether $\gamma_1$ is imaginary or real. The two transformers in Fig. 8.13 may be removed if the characteristic admittance of the center transmission line is taken as $n^2 Y_{01}$.

When a short circuit is placed in the symmetry plane,

$$Y_{\text{in}} = -jB + n^2 Y_{01} \coth \gamma_1 l$$

Equation (77) now gives the solutions for $B$ and $n^2$ and results in the same set of equations as (78). If $\tanh \gamma_n l$ and $\coth \gamma_n l$ for $n > 1$ cannot be replaced by unity, the shunt susceptance $B$ becomes a function of $l$.

The reason is because of the interaction between the evanescent modes excited at each interface.

By means of wave matrices, we may readily find the over-all transmission coefficient through the thick diaphragm. With reference to

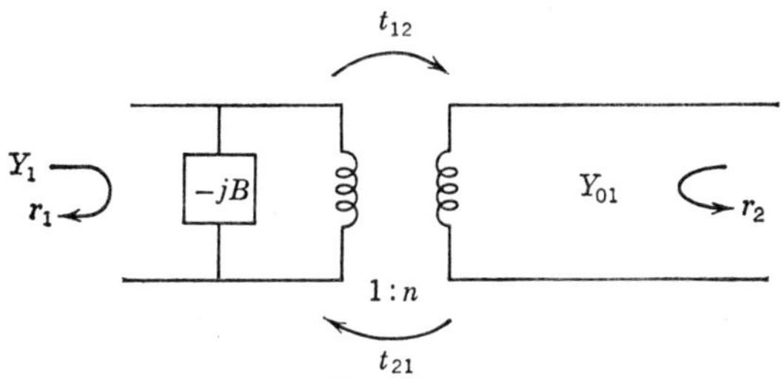


FIG. 8.14

Fig. 8.14, the following reflection and transmission coefficients, applicable to one junction in Fig. 8.13, are introduced:

$$r_1 = \frac{Y_1 - n^2Y_{01} + jB}{Y_1 + n^2Y_{01} - jB} \tag{79a}$$

$$r_2 = \frac{n^2Y_{01} - Y_1 + jB}{n^2Y_{01} + Y_1 - jB} \tag{79b}$$

$$t_{12} = (1 + r_1)n \tag{79c}$$

$$t_{21} = \frac{1 + r_2}{n} \tag{79d}$$

The over-all wave-transmission matrix which relates the amplitudes of the incident and reflected waves on the input and output sides may be found according to the methods of Sec. 3.4 and is

$$\begin{bmatrix} A_{11} & A_{12} \\ A_{21} & A_{22} \end{bmatrix} = \frac{1}{t_{12}t_{21}} \begin{bmatrix} 1 & -r_2 \\ r_1 & 1 + r_1 + r_2 \end{bmatrix} \begin{bmatrix} e^{2\gamma_1 l} & 0 \\ 0 & e^{-2\gamma_1 l} \end{bmatrix} \begin{bmatrix} 1 & -r_1 \\ r_2 & 1 + r_1 + r_2 \end{bmatrix}$$

The over-all voltage-transmission coefficient is

$$A_{11}^{-1} = \frac{t_{12}t_{21}e^{-2\gamma_1 l}}{1 - r_2{}^2e^{-4\gamma_1 l}} = T \tag{80}$$

When $\gamma_1$ is real, and $r_2{}^2e^{-4\gamma_1 l}$ is negligible compared with unity, (80) shows that $T$ is proportional to $e^{-2\gamma_1 l}$. The attenuation produced by the thick diaphragm is then directly proportional to $l$ when expressed in nepers or decibels. This is the principle of the cutoff attenuator. It should be noted that the attenuation is produced by reflection at the input and not by absorption of power. A single evanescent mode cannot propagate real power, but the combination of nonpropagating modes which decay in opposite directions does lead to a transfer of power. Such interacting evanescent modes will exist between the two end faces of the thick inductive diaphragm.

Having discussed some of the general properties of the thick inductive diaphragm, we must now return to the details of evaluating the equivalent-circuit parameters. A transformation of variables, such as was used for the thin diaphragms, will not work in the present case, since two different series occur in (76) and (77). However, one of these series consists of functions which are orthogonal over the aperture domain, and it is therefore convenient to choose

$$\mathcal{E}(x) = \sum_{m=1,3,\ldots}^{M} b_m \psi_m(x)$$

The parameter $M$ must be chosen according to the accuracy required. If we let

$$P_{mn} = \int_t^{t+d} \psi_m \phi_n \, dx = \frac{2m\pi}{d} \frac{\sin n\pi(t/a)}{m^2\pi^2/d^2 - n^2\pi^2/a^2}$$

for $n$ and $m$ both odd, then (76) becomes

$$\begin{aligned}(Z_{11} + Z_{12})^{-1} \sum_{s=1,3,\ldots}^{M} \sum_{r=1,3,\ldots}^{M} P_{s1}P_{r1}b_sb_r &= \sum_{s=1,3,\ldots}^{M} \sum_{r=1,3,\ldots}^{M} b_sb_r\left( \sum_{n=3,5,\ldots}^{\infty} Y_nP_{sn}P_{rn}\right. \\ &\quad \left. + \frac{ad}{4} \sum_{n=1,3,\ldots}^{\infty} Y_{0m} \tanh \gamma_m l \, \delta_{sm}\delta_{rm}\right) \qquad (81)\end{aligned}$$

Equating the partial derivatives with respect to $b_r$ equal to zero gives

$$(Z_{11} + Z_{12})^{-1}P_{r1} \sum_{s=1,3,\ldots}^{M} P_{s1}b_s = -j \sum_{s=1,3,\ldots}^{M} b_sg_{rs} \qquad r = 1,3, \ldots, M$$

where

$$-jg_{rs} = \sum_{n=3,5,\ldots}^{\infty} Y_nP_{sn}P_{rn} + \frac{ad}{4} \sum_{m=1,3,\ldots}^{\infty} Y_{0m} \tanh \gamma_m l \, \delta_{sm}\delta_{rm}$$

For a solution, the determinant of this set of equations must vanish. Thus denoting $(Z_{11} + Z_{12})^{-1}$ by $Y_{oc}$, we get

$$\begin{vmatrix} Y_{oc}P_{11}^2 + jg_{11} & Y_{oc}P_{11}P_{31} + jg_{13} & \cdots & Y_{oc}P_{11}P_{M1} + jg_{1M} \\ Y_{oc}P_{31}P_{11} + jg_{31} & \cdots & \cdots & Y_{oc}P_{31}P_{M1} + jg_{3M} \\ \cdots & \cdots & \cdots & \cdots \\ Y_{oc}P_{M1}P_{11} + jg_{M1} & \cdots & \cdots & Y_{oc}P_{M1}^2 + jg_{MM} \end{vmatrix} = 0$$

Performing the following operations on this determinant:

1. Dividing the $r$th row by $P_{r1}$, $r = 1, 3, \ldots, M$
2. Dividing the $s$th column by $P_{s1}$, $s = 1, 3, \ldots, M$
3. Subtracting the first row from all the other rows
4. Factoring resultant determinant into sum of two determinants

we obtain the following final result for $Y_{oc}$:

$$Y_{oc}\begin{vmatrix} 1 & 1 & \cdots & 1 \\ \dfrac{g_{31}}{P_{11}P_{31}} - \dfrac{g_{11}}{P_{11}^2} & \dfrac{g_{33}}{P_{31}^2} - \dfrac{g_{13}}{P_{11}P_{31}} & \cdots & \dfrac{g_{3M}}{P_{31}P_{M1}} - \dfrac{g_{1M}}{P_{11}P_{M1}} \\ \cdots & \cdots & \cdots & \cdots \\ \dfrac{g_{M1}}{P_{11}P_{M1}} - \dfrac{g_{11}}{P_{11}^2} & \cdots & \cdots & \dfrac{g_{MM}}{P_{M1}^2} - \dfrac{g_{1M}}{P_{11}P_{M1}} \end{vmatrix} = -j\begin{vmatrix} \dfrac{g_{11}}{P_{11}^2} & \cdots & \dfrac{g_{1M}}{P_{11}P_{M1}} \\ \dfrac{g_{31}}{P_{11}P_{31}} - \dfrac{g_{11}}{P_{11}^2} & \cdots & \dfrac{g_{3M}}{P_{31}P_{M1}} - \dfrac{g_{1M}}{P_{11}P_{M1}} \\ \cdots & \cdots & \cdots \\ \dfrac{g_{M1}}{P_{11}P_{M1}} - \dfrac{g_{11}}{P_{11}^2} & \cdots & \dfrac{g_{MM}}{P_{M1}^2} - \dfrac{g_{1M}}{P_{11}P_{M1}} \end{vmatrix} \tag{82}$$

When a two-term approximation is used for $\mathcal{E}(x)$, the evaluation of $Y_{oc} = (Z_{11} + Z_{12})^{-1}$ reduces to the evaluation of two $2 \times 2$ determinants. A similar analysis may be used for determining $(Z_{11} - Z_{12})^{-1}$ and results in an equation similar to (82), with $Y_{oc}$ replaced by $Y_{sc} = (Z_{11} - Z_{12})^{-1}$, and $g_{rs}$ replaced by $h_{rs}$, where

$$-jh_{rs} = \sum_{n=3,5,\ldots}^{\infty} Y_n P_{sn} P_{rn} + \frac{ad}{4} \sum_{m=1,3,\ldots}^{\infty} Y_{0m} \coth \gamma_m l \, \delta_{sm}\delta_{rm}$$

If we let $S_{rs} = \sum\limits_{n=3,5,\ldots}^{\infty} |Y_n| P_{rn} P_{sn}$, then $g_{rs} = h_{rs}$ for $s \neq r$, and

$$g_{rr} = S_{rr} + \frac{ad}{4} jY_{0r} \tanh \gamma_r l$$

$h_{rr} = S_{rr} + (ad/4)jY_{0r} \coth \gamma_r l$. In view of these relations, the evaluation of $Z_{11}$ and $Z_{12}$ is not unduly laborious. For results accurate to within a few per cent, a two-term approximation to $\mathcal{E}(x)$ is sufficient. The series $S_{rs}$ is suitable for direct evaluation since the terms decrease as $n^3$ and the summation is over odd values of $n$ only. The series is not directly summable since it is not an even rational function of the summation variable $n$.

## 8.5. A Narrow Inductive Strip

Figure 8.15 illustrates an inductive strip of width $2t$ and centered at $x = x_1$. The strip is assumed to be perfectly conducting and of negligible thickness in the $z$ direction. The solution to this problem will be formulated in terms of a variational expression involving the current on the obstacle.

Let an $H_{10}$ mode be incident from the region $z < 0$. The incident electric field has only a $y$ component given by $E_i = \sin(\pi x/a)e^{-\Gamma_1 z}$. Since the strip is uniform along the $y$ direction, the only higher-order modes excited are the $H_{n0}$ modes. The incident field excites a current distribution $J(x)$ on the strip. This current is directed along the $y$ axis and does

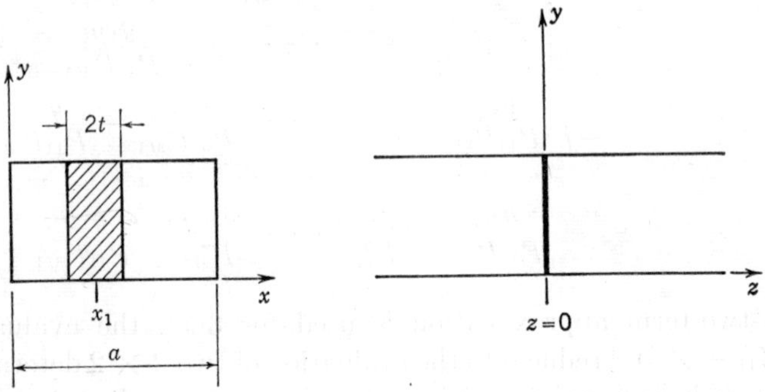


FIG. 8.15. A narrow inductive strip in a rectangular guide.

not vary with $y$ since the incident field has no variation with $y$. The scattered field in the guide may be evaluated in terms of the currents on the strip by using the Green's function for $H_{n0}$ modes given in Sec. 5.6. The scattered field is thus given by

$$E_s(x,z) = \int_S G(x,z|x')J(x')\,dx' \tag{83}$$

where $S$ denotes the surface of the strip and

$$G(x,z|x') = -\frac{j\omega\mu_0}{a}\sum_{n=1}^{\infty}\frac{1}{\Gamma_n}\sin\frac{n\pi x}{a}\sin\frac{n\pi x'}{a}e^{-\Gamma_n|z|}$$

and $\Gamma_n^2 = (n\pi/a)^2 - k_0^2$. The total field $E_t$ in the guide is the sum of the incident and scattered fields and must vanish on the perfectly conducting strip. The resulting integral equation for the current distribution $J$ is

$$\sin\frac{\pi x}{a}e^{-\Gamma_1 z} + \int_S G(x,z|x')J(x')\,dx' = 0 \qquad \text{on } S \tag{84}$$

From (83) the reflected dominant mode is seen to be given by

$$-\frac{j\omega\mu_0}{a\Gamma_1}\sin\frac{\pi x}{a}e^{\Gamma_1 z}\int_S \sin\frac{\pi x'}{a}J(x')\,dx' = R\sin\frac{\pi x}{a}e^{\Gamma_1 z} \qquad z<0 \quad (85)$$

where the reflection coefficient $R$ is defined by this equation. For $z > 0$, the total dominant mode field transmitted past the obstacle is

$$\sin\frac{\pi x}{a}e^{-\Gamma_1 z} - \frac{j\omega\mu_0}{a\Gamma_1}\sin\frac{\pi x}{a}e^{-\Gamma_1 z}\int_S \sin\frac{\pi x'}{a}J(x')\,dx' = T\sin\frac{\pi x}{a}e^{-\Gamma_1 z} \qquad z>0 \quad (86)$$

where the transmission coefficient $T$ is defined by this equation. From (85) and (86) we find that $1 + R = T$, and hence the obstacle will appear as an equivalent shunt element across a transmission line.

The integral equation (84) may be rewritten as

$$\sin\frac{\pi x}{a} - \frac{j\omega\mu_0}{a\Gamma_1}\sin\frac{\pi x}{a}\int_S \sin\frac{\pi x'}{a}J(x')\,dx' = \frac{j\omega\mu_0}{a}\sum_{n=2}^{\infty}\frac{1}{\Gamma_n}\sin\frac{n\pi x}{a}\int_S \sin\frac{n\pi x'}{a}J(x')\,dx'$$

since $z = 0$ on $S$. Using Eq. (85) now gives

$$(1+R)\sin\frac{\pi x}{a} = \frac{j\omega\mu_0}{a}\sum_{n=2}^{\infty}\frac{1}{\Gamma_n}\sin\frac{n\pi x}{a}\int_S \sin\frac{n\pi x'}{a}J(x')\,dx' \qquad (87)$$

A shunt reactance $jX$ across a transmission line with unit characteristic impedance produces a reflection coefficient given by $-jX = (1 + R)/2R$. Comparing with (87), it is seen that an expression for $jX$ may be obtained by multiplying both sides of (87) by $J(x)$, integrating over the obstacle, and then dividing both sides by $\left[\int_S J(x)\sin\frac{\pi x}{a}\,dx\right]^2$ and using (85) to obtain a factor $R$ in the denominator of the left-hand side. The indicated operations give

$$jX = -\frac{1+R}{2R} = \frac{\Gamma_1}{2}\,\frac{\sum\limits_{n=2}^{\infty}\frac{1}{\Gamma_n}\iint\limits_S J(x)J(x')\sin\frac{n\pi x}{a}\sin\frac{n\pi x'}{a}\,dx\,dx'}{\left[\int_S J(x)\sin\frac{\pi x}{a}\,dx\right]^2} \qquad (88)$$

This expression is readily shown to be a variational expression for the normalized shunt reactance $jX$. Furthermore, it is a positive-definite quadratic form, and so an approximate solution for $J$ will give an upper bound on $jX$. The variational expression (88) is also applicable to the thin-inductive-diaphragm problems considered in Sec. 8.3.

At the edge of an infinitely thin perfectly conducting strip, the normal magnetic field, and hence the tangential current density, becomes infinite as $r^{-1/2}$, where $r$ is the radial distance from the edge. However, for a sufficiently narrow strip, the integrated effect of the current should be much the same as though the current were constant over the strip. For a first approximation to $jX$ we will, therefore, assume a constant current distribution. The integrations in (88) are readily performed and give

$$\begin{aligned} jX &= \frac{\Gamma_1}{2} \frac{\displaystyle\sum_{n=2}^{\infty} \frac{1}{n^2\Gamma_n} \left[\cos \frac{n\pi}{a}(x_1 + t) - \cos \frac{n\pi}{a}(x_1 - t)\right]^2}{\left[\cos \frac{\pi}{a}(x_1 + t) - \cos \frac{\pi}{2}(x_1 - t)\right]^2} \\ &= \frac{\Gamma_1}{2} \frac{\displaystyle\sum_{n=2}^{\infty} \frac{1}{n^2\Gamma_n} \left(\sin \frac{n\pi x_1}{a} \sin \frac{n\pi t}{a}\right)^2}{\left(\sin \frac{\pi x_1}{a} \sin \frac{\pi t}{a}\right)^2} \end{aligned} \tag{89}$$

For a centered strip $x_1 = a/2$, and (89) reduces to

$$jX = \frac{\Gamma_1}{2} \csc^2 \frac{\pi t}{a} \sum_{n=3,5,\ldots}^{\infty} \frac{1}{n^2\Gamma_n} \sin^2 \frac{n\pi t}{a} \tag{90}$$

If desired, the series could be written as a dominant series,

$$\sum_{n=3,5,\ldots}^{\infty} \frac{a}{\pi n^3} \sin^2 \frac{n\pi t}{a}$$

plus a correction series $\displaystyle\sum_{n=3,5,\ldots}^{\infty} \frac{1}{n^2}\left(\frac{1}{\Gamma_n} - \frac{a}{n\pi}\right) \sin^2 \frac{n\pi t}{a}$. The first series may be readily summed. However, there is little point in treating (90) in this manner since the original series converges very fast, and only four or five terms are required. The alternative procedure suggested above leads to just as many, if not more, terms to be evaluated.

If $t$ is very small, the inductive strip is essentially equivalent to a thin inductive post of radius $t$. For this case it is worthwhile summing the series in (90). We have

$$\sum_{n=3,5,\ldots}^{\infty} \frac{1}{n^3}\sin^2\frac{n\pi t}{a} = \frac{1}{2}\sum_{n=1,3,\ldots}^{\infty} \frac{1}{n^3}\left(1-\cos\frac{2n\pi t}{a}\right) - \sin^2\frac{\pi t}{a}$$
$$= \frac{1}{2}\left(\frac{\pi t}{a}\right)^2\left(\ln\frac{a}{\pi t}+\frac{3}{2}\right) - \left(\frac{\pi t}{a}\right)^2$$

which is correct to order $(\pi t/a)^2$. Replacing $\csc^2(\pi t/a)$ by $(a/\pi t)^2$ also, (90) gives

$$jX = \frac{a\Gamma_1}{4\pi}\left[\ln\frac{a}{\pi t} - 1 + 2\left(\frac{a}{\pi t}\right)^2 \sum_{n=3,5,\ldots}^{\infty}\left(\frac{\pi}{a\Gamma_n}-\frac{1}{n}\right)\sin^2\frac{n\pi t}{a}\right] \tag{91}$$

for the inductive reactance of a thin centered post.

## REFERENCES

8.1. L. Lewin: "Advanced Theory of Waveguides," Iliffe and Sons, Ltd., London, 1951. (This reference contains many examples of waveguide discontinuities, as well as many instructive mathematical techniques for summing a variety of infinite series which arise in practice.)

8.2. H. Motz: "Electromagnetic Problems of Microwave Theory," Methuen Monograph, Methuen & Co., Ltd., London, 1951.

8.3. N. Marcuvitz and J. Schwinger: On the Representation of the Electric and Magnetic Fields Produced by Currents and Discontinuities in Waveguides, *J. Appl. Phys.*, vol. 22, pp. 806–819, June, 1951.

8.4. J. W. Miles: The Equivalent Circuit for a Plane Discontinuity in a Cylindrical Guide, *Proc. IRE*, vol. 34, pp. 728–742, October, 1946.

8.5. J. W. Miles: The Equivalent Circuit of a Corner Bend in a Rectangular Waveguide, *Proc. IRE*, vol. 35, pp. 1313–1317, November, 1947.

8.6. R. E. Collin and J. Brown: Calculation of the Equivalent Circuit of an Axially Unsymmetrical Waveguide Junction, *Jour. IEE* (*London*), vol. 103, part C, pp. 121–128, March, 1956.

8.7. N. Marcuvitz: "Waveguide Handbook," vol. 10 of MIT Rad. Lab. Series, McGraw-Hill Book Company, Inc., New York, 1951. (This is an extensive collection of theoretical and numerical results for a large variety of waveguide discontinuities.)

8.8. F. Borgnis and C. Papas: "Randwertprobleme der Mikrowellenphysik," Springer-Verlag, Berlin, 1955.

*Solution of waveguide and transmission-line discontinuity problems by conformal-mapping techniques and quasi-stationary field methods*

8.9. G. G. MacFarlane: Quasi-stationary Field Theory and Its Application to Diaphragms and Junctions in Transmission Lines and Waveguides, *Jour. IEE* (*London*), vol. 93, part IIIA, pp. 703–719, 1946.

8.10. J. R. Whinnery and H. W. Jamieson: Equivalent Circuits for Discontinuities in Transmission Lines, *Proc. IRE*, vol. 32, pp. 98–114, February, 1944.

## PROBLEMS

**8.1.** By using a suitable conformal transformation, obtain a solution for the fringing capacitance of an asymmetrical diaphragm in a parallel-plate transmission line. Assume that the lower plate is at zero potential and the upper plate and diaphragm are at a potential $V$. When used as a transmission line, find the normalized shunt susceptance across the line due to the fringing capacitance. Note that the characteristic impedance of the line per unit width is $bZ_0$.

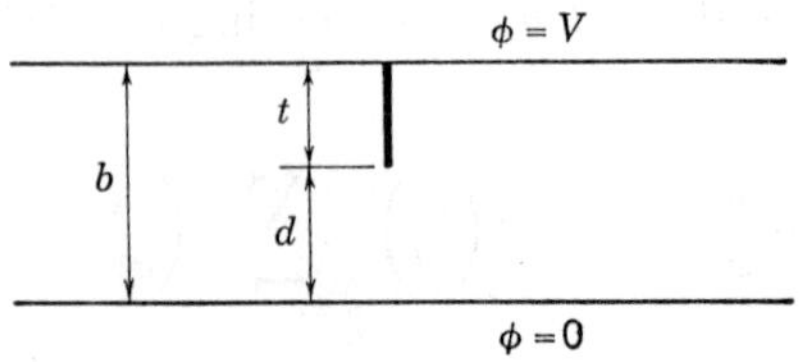


FIG. P 8.1

*Answer.* Fringing capacitance $C = (4\epsilon_0/\pi) \ln \csc (\pi d/2b)$. See additional comments on method of solution in Prob. 8.2.

**8.2.** Consider the $E$ plane step in a parallel-plate transmission line. For an incident TEM mode, set up the continuity equations for the transverse fields, and show that the equivalent circuit consists of the junction of two transmission lines with characteristic impedances unity and $d/b$, together with a shunt capacitive susceptance at the junction. By means of a suitable conformal transformation, obtain the low-frequency

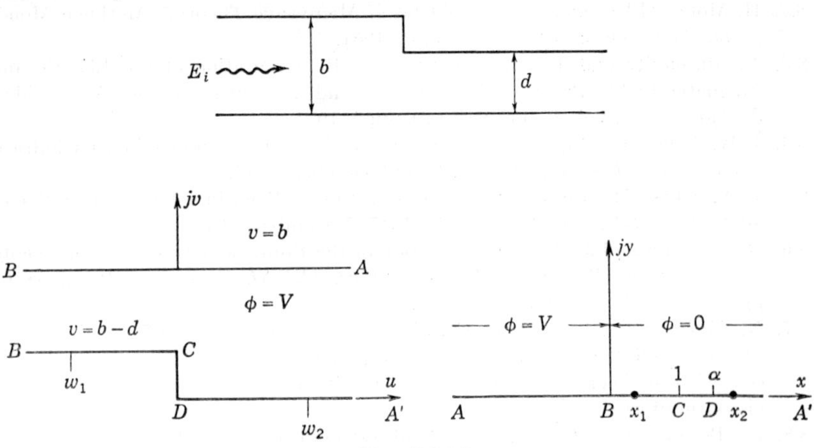


FIG. P 8.2

value of the fringing capacitance introduced across the line by the step. From the value of this capacitance, obtain the approximate value of the shunt susceptance $jB$.

*Answer*

$$B = \frac{k_0 b}{\pi} \left[ 2 \ln \frac{b^2 - d^2}{4bd} + \left( \frac{b}{d} + \frac{d}{b} \right) \ln \frac{d + b}{b - d} \right]$$

A suitable mapping function is

$$w = \frac{b}{\pi} \cosh^{-1} \frac{2Z - \alpha - 1}{\alpha - 1} - \frac{d}{\pi} \cosh^{-1} \frac{(\alpha + 1)Z - 2\alpha}{(\alpha - 1)Z} \qquad \alpha = \frac{b^2}{d^2}$$

In the $xy$ plane, $\phi = V\theta/\pi$. Total capacitance from $w_1$ to $w_2$ is

$$C(w_2 - w_1) = C(x_2 - x_1) = \frac{\epsilon_0}{\pi} \ln \frac{x_2}{x_1}$$

Fringing capacitance between $w_1$ and $w_2$ is $C(x_2 - x_1)$ minus parallel-plate capacitance

$$C_f(x_2 - x_1) = C(x_2 - x_1) - \frac{\epsilon_0 w_2}{b} + \frac{\epsilon_0}{d}[w_1 - j(b - d)]$$

Total fringing capacitance is

$$\lim_{\substack{x_2 \to \infty \\ x_1 \to 0}} C_f(x_2 - x_1) = \frac{\epsilon_0}{\pi}\left[\left(\frac{d}{b} + \frac{b}{d}\right) \ln \frac{b + d}{b - d} + 2 \ln \frac{b^2 - d^2}{4bd}\right]$$

Note that, for $x_2$ large,

$$w_2 \to \frac{b}{\pi} \cosh^{-1} \frac{2x_2}{\alpha - 1} - \frac{d}{\pi} \cosh^{-1} \frac{\alpha + 1}{\alpha - 1} \to \frac{b}{\pi} \ln \frac{4x_2}{\alpha - 1} - \frac{d}{\pi} \ln \frac{\alpha + 1 + 2\alpha^{1/2}}{\alpha - 1}$$

Also, for $x_1$ small,

$$w_1 \to \frac{b}{\pi} \cosh^{-1} \frac{\alpha + 1}{\alpha - 1} - \frac{d}{\pi} \cosh^{-1} \frac{-2\alpha}{(\alpha - 1)x_1} \to j(b - d) + \frac{b}{\pi} \ln \frac{\alpha + 1 + 2\alpha}{\alpha - 1} - \frac{d}{\pi} \ln \frac{4\alpha}{(\alpha - 1)x_1}$$

Also note that $\dfrac{\alpha + 1 + 2\alpha^{1/2}}{\alpha - 1} = \dfrac{\sqrt{\alpha} + 1}{\sqrt{\alpha} - 1}$ and that $\cosh^{-1} \theta = \ln [\theta + (\theta^2 - 1)^{1/2}]$.

**8.3.** For Prob. 8.2, set up a variational expression for the junction susceptance. Write the series occurring in the integrand as a dominant series plus a correction series which depends on the higher-mode propagation constants. Assume a constant-aperture field, and thus obtain a value for the susceptance contributed by the correction series. The total approximate susceptance at the junction is the sum of this correction term plus the quasi-static susceptance from Prob. 8.2.

*Answer.* The correction term gives a contribution, to the total susceptance, of the amount

$$\frac{2k_0 b^3}{d^2 \pi^3} \sum_{n=1}^{\infty} \sin^2 n\pi \frac{d}{b} \left[\left(1 - \frac{k_0^2 b^2}{n^2 \pi^2}\right)^{-1/2} - 1\right] \frac{1}{n^3}$$

**8.4.** Modify the theory of Probs. 8.2 and 8.3 to obtain the equivalent-circuit parameters for a symmetrical $E$ plane step in a parallel-plate transmission line as well as the equivalent-circuit parameters for symmetrical and asymmetrical $E$ plane steps in a rectangular guide.

**8.5.** Show that the higher-order $E$ modes alone cannot satisfy the boundary conditions for a capacitive diaphragm in a rectangular guide when an $H_{10}$ mode is incident. In particular, note that the normal magnetic field on the diaphragm will not vanish. Set up the boundary conditions for all of the field components for the LSE modes, and thus verify that these modes alone are sufficient for solving the capacitive-diaphragm and $E$ plane-step problems in a rectangular guide.

**8.6.** Modify the theory for the capacitive diaphragm so as to obtain the solution for the problem of a capacitive diaphragm in a rectangular guide which is filled with a lossless dielectric medium for $z \geq 0$. The diaphragm is located at $z = 0$, and the dielectric constant is such that only the $H_{10}$ mode propagates in the filled guide. Show that the equivalent circuit consists of a junction of two transmission lines of characteristic impedance unity and $\Gamma_1/\gamma_1$, where $\Gamma_1$, $\gamma_1$ are the propagation constants for the $H_{10}$ modes, together with a shunt capacitive susceptance at the junction. Note that $jB = (1 - R)/(1 + R) - \gamma_1/\Gamma_1$, where $R$ is the reflection coefficient. The variational expression obtained will give a solution for $jB$.

**8.7.** Repeat Prob. 8.6 for a symmetrical inductive diaphragm in a rectangular guide filled with a lossless dielectric medium for $z > 0$.

**8.8.** Apply the Poisson summation formula to obtain an alternative representation for the Green's function for $H_{n0}$ modes in a rectangular guide. Show that this alternative representation is the field radiated from the infinite number of images, of this line source, in the waveguide walls. In cylindrical coordinates, the electric field $E_y$ radiated by an infinite unit line current located at the origin is $(-\omega\mu_0/4)H_0^2(k_0r)$.

*Answer*

$$\frac{-\omega\mu_0}{4}\sum_{-\infty}^{\infty}\left\{H_0^2\left[k_0\sqrt{(z-z')^2+(x-x'-2na)^2}\right]\right.$$
$$\left.-H_0^2\left[k_0\sqrt{(z-z')^2+(x+x'-2na)^2}\right]\right\}$$

**8.9.** Show that the solution for two inductive strips in a waveguide of width $2a$ gives the solution to the problem of the inductive grating in free space with a TEM wave incident at an angle $\theta_i$ given by $\tan\theta_i = \pi/2a\beta_1$. Note that the $H_{10}$ mode may be decomposed into two TEM waves propagating at angles $\pm\theta_i$. From symmetry considerations the reflection and transmission coefficients for these two composite waves will be the same.

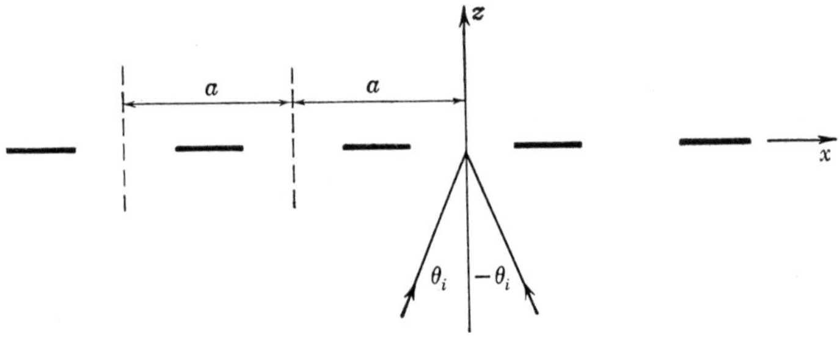


FIG. P 8.9

**8.10.** Apply Babinet's principle to obtain the solution for a capacitive strip grating in free space from the corresponding solution of the inductive-grating problem. Show that, if $B_L$ is the shunt susceptance of the inductive grating, and $B_C$ is the shunt susceptance of the capacitive grating, $B_LB_C = 4Y_0^2$, where $Y_0 = (\epsilon_0/\mu_0)^{1/2}$. If $R_1$, $T_1$ and $R_2$, $T_2$ are the reflection and transmission coefficients for the two cases, show also that $T_1 + T_2 = -(R_1 + R_2) = 1$, and $T_1T_2 = R_1R_2$. Note that in the complementary grating the aperture and screen (strips) areas are interchanged.

**8.11.** For an asymmetrical $E$ plane step in a parallel-plate transmission line, obtain a variational expression, in terms of the aperture electric field, for the position of an electric field null in the input line as a function of the short-circuit position in the output line. Assume a constant-aperture field, and find the parameters of an equivalent circuit of the form illustrated. Repeat the analysis in terms of the aperture magnetic field (assume a constant magnetic field in the aperture for a first approximation).

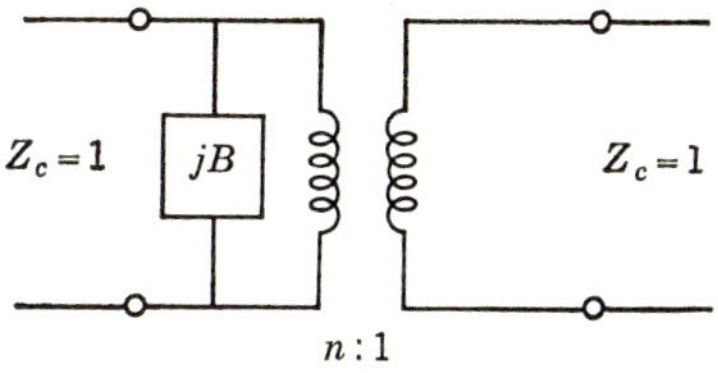


FIG. P 8.11

Compare the two solutions for the equivalent-circuit parameters for $k_0 = 2$ radians per centimeter, $2d = b = 1$ centimeter. Show that the variational expression obtained for the electric field null is equivalent to minimizing the net time-average magnetic and electric energy difference $W_m - W_e$ in the volume bounded by the short circuit and field null plane.

CHAPTER 9

# PERIODIC STRUCTURES

Periodic structures may be classified into two basic types:

1. Structures with continuous but periodically varying electrical properties, e.g., a cylindrical guide filled with a dielectric material whose dielectric constant varies in a periodic manner along the axial direction according to some periodic function such as $\kappa = \kappa_0 + \kappa_1 \cos hz$
2. Structures with periodic boundary conditions, e.g., a waveguide loaded at regular intervals by identical diaphragms

By far the most commonly occurring type is that falling in the class of structures with periodic boundary conditions. Periodic structures find application in a variety of devices such as linear particle accelerators, traveling-wave tubes, and microwave filter networks. Artificial dielectric media and diffraction gratings are further examples of periodic structures. Periodic structures such as corrugated planes have also been used as surface wave-guiding devices for antenna applications. Periodic structures have one important characteristic in common. This is the existence of discrete passbands separated by stopbands, i.e., frequency bands for which a wave propagates freely along the structure separated by frequency bands for which the wave is highly attenuated and does not propagate along the structure.

## 9.1. Floquet's Theorem

A basic theorem underlying the theory of wave propagation in periodic structures is that known as Floquet's theorem. We will illustrate this theorem by considering the problem of the propagation of an $H_{10}$ mode in an infinitely long rectangular guide, filled with a dielectric with a relative dielectric constant given by $\kappa_0 + \kappa_1 \cos (2\pi z/p)$, as illustrated in Fig. 9.1. For an $H_{10}$ mode, the field components $H_x$ and $H_z$ may be derived from the one scalar component of electric field $E_y$ by the equations

$$j\omega\mu_0 H_x = \frac{\partial E_y}{\partial z} \qquad j\omega\mu_0 H_z = -\frac{\partial E_y}{\partial x}$$

and $E_y$ satisfies the wave equation

$$\left[\frac{\partial^2}{\partial x^2} + \frac{\partial^2}{\partial z^2} + \kappa(z)k_0^2\right] E_y = 0$$

Let $E_y = \psi(z) \sin (\pi x/a)$, and hence $\psi(z)$ satisfies the equation

$$\left[\frac{d^2}{dz^2} + \left(\kappa_0 + \kappa_1 \cos \frac{2\pi z}{p}\right) k_0^2 - \frac{\pi^2}{a^2}\right] \psi(z) = 0 \tag{1}$$

If $\psi(z)$ is a solution to this equation, then $\psi(z + p)$ is also a solution, for, if we change $z$ to $z + p$ in (1), we get

$$\left[\frac{d^2}{dz^2} + \left(\kappa_0 + \kappa_1 \cos \frac{2\pi z}{p}\right) k_0^2 - \frac{\pi^2}{a^2}\right] \psi(z + p) = 0$$

since

$$\cos 2\pi \frac{z + p}{p} = \cos \frac{2\pi z}{p}$$

and hence $\psi(z + p)$ satisfies the same equation as $\psi(z)$ and is also a solution. Equation (1) is of the second order and will have two independent

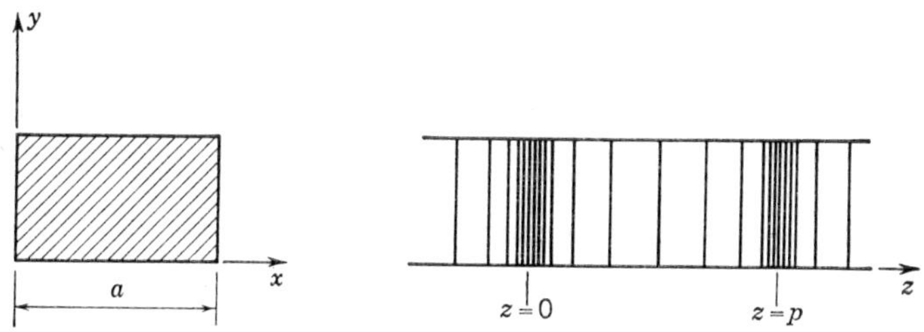


FIG. 9.1. Rectangular guide filled with a medium having a periodically varying dielectric constant.

solutions, which we will denote by $\psi_1(z)$ and $\psi_2(z)$. We have the result that $\psi_1(z + p)$ and $\psi_2(z + p)$ are also solutions, but, in general, $\psi_i(z + p)$ does not equal $\psi_i(z)$, since $\psi_i(z)$ is not necessarily periodic in $z$ with a period $p$.

Since a second-order equation has only two linearly independent solutions, any other solution must be a linear combination of the two fundamental solutions, and, hence, we may write

$$\psi_1(z + p) = \alpha_{11}\psi_1(z) + \alpha_{12}\psi_2(z) \tag{2a}$$

$$\psi_2(z + p) = \alpha_{21}\psi_1(z) + \alpha_{22}\psi_2(z) \tag{2b}$$

where $\alpha_{ij}$ are suitably determined coefficients. For the particular case when $\psi_i$ is periodic so that $\psi_i(z + p) = \psi_i(z)$, we have $\alpha_{11} = \alpha_{22} = 1$ and $\alpha_{12} = \alpha_{21} = 0$. In a similar fashion the general solution to (1) may be written as

$$F(z) = A\psi_1(z) + B\psi_2(z) \tag{3}$$

where $A$ and $B$ are suitably chosen coefficients. From (2) and (3), we get

$$\begin{aligned} F(z+p) &= A\psi_1(z+p) + B\psi_2(z+p) \\ &= (A\alpha_{11} + B\alpha_{21})\psi_1(z) + (A\alpha_{12} + B\alpha_{22})\psi_2(z) \end{aligned} \tag{4}$$

If $F(z)$ is to represent a wave propagating down the structure, it is necessary that

$$F(z+p) = e^{-\gamma p}F(z) \tag{5}$$

where $\gamma$ is the propagation constant. Thus (4) above must equal (3) multiplied by $e^{-\gamma p}$, and, when we equate the coefficients of $A$ and $B$, we get

$$Ae^{-\gamma p} = A\alpha_{11} + B\alpha_{21}$$
$$Be^{-\gamma p} = A\alpha_{12} + B\alpha_{22}$$

or

$$A(\alpha_{11} - e^{-\gamma p}) + B\alpha_{21} = 0 \tag{6a}$$
$$A\alpha_{12} + B(\alpha_{22} - e^{-\gamma p}) = 0 \tag{6b}$$

A nontrivial solution for $A$ and $B$ exists only if the determinant of the above equations vanishes; thus,

$$e^{-2\gamma p} - e^{-\gamma p}(\alpha_{11} + \alpha_{22}) + \alpha_{11}\alpha_{22} - \alpha_{12}\alpha_{21} = 0 \tag{7}$$

Equation (7) is a quadratic equation whose solution will determine the allowed values of the propagation constant $\gamma$. The solution for $e^{-\gamma p}$ is

$$e^{-\gamma p} = \frac{\alpha_{11} + \alpha_{22}}{2} \pm \left[\left(\frac{\alpha_{11} + \alpha_{22}}{2}\right)^2 - (\alpha_{11}\alpha_{22} - \alpha_{12}\alpha_{21})\right]^{1/2} \tag{8}$$

Let $\alpha_{11}\alpha_{22} - \alpha_{12}\alpha_{21} = \Delta$; then, since

$$\frac{(\alpha_{11} + \alpha_{22})^2}{4} - \left[\left(\frac{\alpha_{11} + \alpha_{22}}{2}\right)^2 - \Delta\right] = \Delta$$

we may write

$$\cosh\theta = \frac{\alpha_{11} + \alpha_{22}}{2\Delta^{1/2}} \qquad \sinh\theta = \left[\left(\frac{\alpha_{11} + \alpha_{22}}{2}\right)^2 \frac{1}{\Delta} - 1\right]^{1/2}$$

since $\cosh^2\theta - \sinh^2\theta = 1$. Hence the solution $e^{-\gamma p}$ is also given by

$$e^{-\gamma p} = \Delta^{1/2}e^{\pm\theta} = e^{\pm\theta + \frac{1}{2}\ln\Delta} \tag{9}$$

where $\cosh\theta = (\alpha_{11} + \alpha_{22})/2\Delta^{1/2}$. In general, the solution for $\gamma$ may be real, imaginary, or complex. We may show that, if $\gamma$ is a solution, then $-\gamma$ is also a solution, as is $\gamma \pm j2n\pi/p$, where $n$ is any integer. This result follows since $\Delta = 1$ (see Prob. 9.7).

Let $\phi(z) = e^{\gamma z}F(z)$, and we get

$$\phi(z+p) = e^{\gamma p}e^{\gamma z}F(z+p) = e^{\gamma z}F(z) = \phi(z)$$

by using relation (5). Hence $\phi$ is a periodic function of $z$ with a period $p$. Therefore the general solution of (1) is of the form $F(z) = e^{\pm\gamma z}\phi(z)$, where $\phi$ is a periodic function of $z$, and $\gamma$ plays the role of a propagation constant. This particular result is called Floquet's theorem.

We will not consider this problem any further here, since it would take us too far afield into the properties of Mathieu functions, which are the appropriate solutions to (1). The theory for periodic boundary conditions will be our main interest. The theory for problems of this type is most readily developed by matrix methods. We shall find, however, that solutions of the above form will be obtained. The more important aspects of the theory will be developed in the succeeding sections, and will be illustrated by application to a periodically loaded waveguide and the helix structure. Further examples on periodic structures will be covered in later chapters as well.

## 9.2. Some Properties of Lossless Microwave Quadrupoles

The basic element in the type of periodic structure we shall be concerned with will be a lossless quadrupole. Such a quadrupole may be

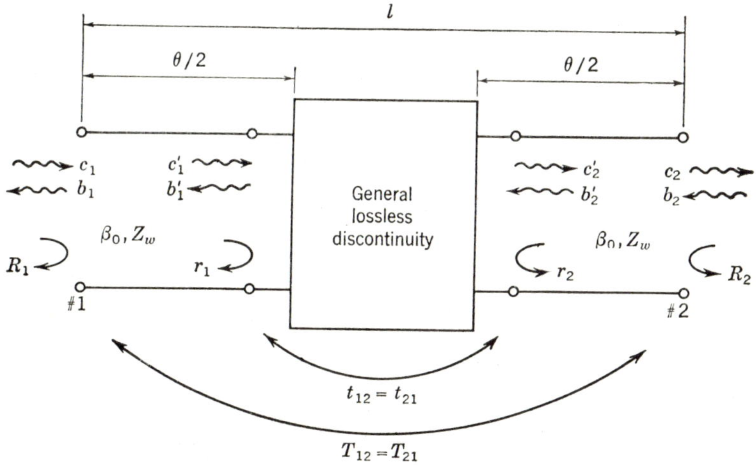


FIG. 9.2. A lossless microwave quadrupole.

described by the reflection and transmission coefficients, by a T or $\pi$ network, by a wave matrix or any other matrix which is used in low-frequency circuit theory. We shall derive a few important relations between these various methods of describing a lossless quadrupole. Consider the general discontinuity illustrated in Fig. 9.2. The electrical length of the transmission line or waveguide section on either end is $\theta/2$. It is assumed that these sections are sufficiently long so all higher-order modes which are excited by the discontinuity have decayed to a negligible value at the terminal planes. The transmission lines or waveguide sections on either side of the discontinuity are also assumed to be uniform and lossless, and to have identical electrical properties, in particular, a propagation phase constant $\beta_0$ and wave impedance $Z_w$.

Let the wave matrix describing the discontinuity be

$$[A'] = \begin{bmatrix} A'_{11} & A'_{12} \\ A'_{21} & A'_{22} \end{bmatrix} \tag{10}$$

where

$$A'_{11} = \frac{1}{t_{12}} \qquad A'_{12} = -\frac{r_2}{t_{12}} \qquad A'_{21} = \frac{r_1}{t_{12}} \qquad A'_{22} = t_{21} - \frac{r_1 r_2}{t_{12}}$$

The amplitudes $c'_1$ $b'_1$ of the forward- and backward-propagating waves on the input side are related to the amplitudes $c'_2$, $b'_2$ of the waves on the output side as follows:

$$\begin{bmatrix} c'_1 \\ b'_1 \end{bmatrix} = \begin{bmatrix} A'_{11} & A'_{12} \\ A'_{21} & A'_{22} \end{bmatrix} \begin{bmatrix} c'_2 \\ b'_2 \end{bmatrix} \tag{11}$$

The wave matrix describing the structure between the terminal planes 1 and 2 is

$$[A] = \begin{bmatrix} e^{j\theta/2} & 0 \\ 0 & e^{-j\theta/2} \end{bmatrix} \begin{bmatrix} A'_{11} & A'_{12} \\ A'_{21} & A'_{22} \end{bmatrix} \begin{bmatrix} e^{j\theta/2} & 0 \\ 0 & e^{-j\theta/2} \end{bmatrix} = \begin{bmatrix} A_{11} & A_{12} \\ A_{21} & A_{22} \end{bmatrix} \tag{12}$$

where

$$A_{11} = \frac{1}{T_{12}} = \frac{e^{j\theta}}{t_{12}} \qquad A_{12} = \frac{-R_2}{T_{12}} = \frac{-r_2}{t_{12}}$$

$$A_{21} = \frac{R_1}{T_{12}} = \frac{r_1}{t_{12}} \qquad A_{22} = \frac{T_{12}T_{21} - R_1 R_2}{T_{12}} = \frac{t_{12}t_{21} - r_1 r_2}{t_{12}} e^{-j\theta}$$

From these latter relations we may readily deduce that

$$T_{12} = t_{12}e^{-j\theta} \qquad T_{21} = t_{21}e^{-j\theta} \qquad R_1 = r_1 e^{-j\theta} \qquad R_2 = r_2 e^{-j\theta}$$

Also we have the following relations for the reflection and transmission coefficients:

$$R_1 = \frac{A_{21}}{A_{11}} \qquad R_2 = -\frac{A_{12}}{A_{11}} \qquad T_{12} = \frac{1}{A_{11}} = T_{21}$$

The equality of $t_{12}$ and $t_{21}$ and also of $T_{12}$ and $T_{21}$ follows from the condition of reciprocity. The quadrupole is reciprocal when the determinant of the $[A]$ matrix is equal to unity; i.e.,

$$A_{11}A_{22} - A_{12}A_{21} = A'_{11}A'_{22} - A'_{12}A'_{21} = 1$$

For a lossless network,

$$|R_1| = |R_2| = (1 - |T_{12}|^2)^{1/2}$$

Let $\qquad R_1 = \rho e^{j\phi_1} \qquad R_2 = \rho e^{j\phi_2} \qquad T_{12} = (1 - \rho^2)^{1/2} e^{j\alpha}$

where $\phi_1 + \phi_2 = 2\alpha \pm \pi$ from Sec. 5.7. Hence, we have

$$A_{22} = \frac{T_{12}^2 - R_1R_2}{T_{12}} = \frac{(1-\rho^2)e^{2j\alpha} + \rho^2 e^{2j\alpha}}{T_{12}} = |T_{12}|^{-1}e^{j\alpha} = A_{11}^* \tag{13a}$$

$$A_{12}A_{21} = A_{11}A_{11}^* - 1 = (1-\rho^2)^{-1} - 1 = \left|\frac{R_1}{T_{12}}\right|^2 \tag{13b}$$

With the output side matched, the input impedances at the terminal planes 1 and 2 are, respectively,

$$Z_{\text{in},1} = \frac{1+R_1}{1-R_1}Z_w = \frac{A_{11}+A_{21}}{A_{11}-A_{21}}Z_w \tag{14a}$$

$$Z_{\text{in},2} = \frac{1+R_2}{1-R_2}Z_w = \frac{A_{11}-A_{12}}{A_{11}+A_{12}}Z_w \tag{14b}$$

The structure between the two terminal planes may also be described by an equivalent T network as in Fig. 9.3. The equivalent voltages and

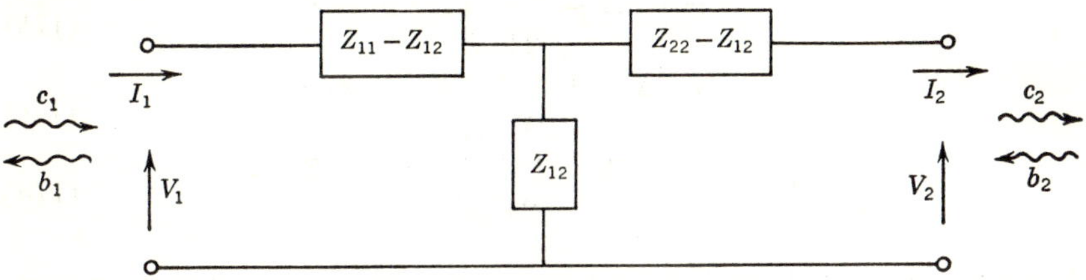


FIG. 9.3

currents $V_i$ and $I_j$ will be chosen as follows:

$$V_1 = c_1 + b_1 \qquad V_2 = c_2 + b_2 \qquad I_1 = (c_1 - b_1)Y_w \qquad I_2 = (c_2 - b_2)Y_w$$

where $c_i$ and $b_i$ are the amplitudes of the transverse electric field of the forward- and backward-traveling waves. At the terminal planes we have

$$V_1 = I_1Z_{11} - I_2Z_{12} = c_1 + b_1 = (c_1 - b_1)Y_wZ_{11} - (c_1 - b_2)Y_wZ_{12}$$
$$V_2 = I_1Z_{12} - I_2Z_{22} = c_2 + b_2 = (c_1 - b_1)Y_wZ_{12} - (c_2 - b_2)Y_wZ_{22}$$

We may rearrange these latter equations and write them in matrix form as follows:

$$\begin{bmatrix} 1 - Y_wZ_{11} & 1 + Y_wZ_{11} \\ Y_wZ_{12} & -Y_wZ_{12} \end{bmatrix} \begin{bmatrix} c_1 \\ b_1 \end{bmatrix} = \begin{bmatrix} -Y_wZ_{12} & Y_wZ_{12} \\ 1 + Y_wZ_{22} & 1 - Y_wZ_{22} \end{bmatrix} \begin{bmatrix} c_2 \\ b_2 \end{bmatrix} \tag{15}$$

We have

$$(1 - Y_wZ_{11})(-Y_wZ_{12}) - Y_wZ_{12}(1 + Y_wZ_{11}) = -2Y_wZ_{12}$$

and, hence, the inverse of the matrix on the left-hand side is

$$\frac{1}{2Y_wZ_{12}} \begin{bmatrix} Y_wZ_{12} & 1 + Y_wZ_{11} \\ Y_wZ_{12} & Y_wZ_{11}^{-1} \end{bmatrix}$$

so that multiplying both sides of (15) by this matrix gives

$$\begin{bmatrix} c_1 \\ b_1 \end{bmatrix} = \frac{1}{2Y_wZ_{12}} \begin{bmatrix} Y_wZ_{12} & Y_wZ_{11}+1 \\ Y_wZ_{12} & Y_wZ_{11}-1 \end{bmatrix} \begin{bmatrix} -Y_wZ_{12} & Y_wZ_{12} \\ 1+Y_wZ_{22} & 1-Y_wZ_{22} \end{bmatrix} \begin{bmatrix} c_2 \\ b_2 \end{bmatrix}$$

$$= \frac{1}{2Y_wZ_{12}}$$

$$\times \begin{bmatrix} Y_w{}^2\Delta + 1 + Y_w(Z_{11}+Z_{22}) & -Y_w{}^2\Delta + 1 + Y_w(Z_{11}-Z_{22}) \\ Y_w{}^2\Delta - 1 + Y_w(Z_{11}-Z_{22}) & -Y_w{}^2\Delta - 1 + Y_w(Z_{11}+Z_{22}) \end{bmatrix} \begin{bmatrix} c_2 \\ b_2 \end{bmatrix}$$

$$= \begin{bmatrix} A_{11} & A_{12} \\ A_{21} & A_{22} \end{bmatrix} \begin{bmatrix} c_2 \\ b_2 \end{bmatrix} \tag{16}$$

where $\Delta = Z_{11}Z_{22} - Z_{12}^2$. From (16) explicit expressions for the matrix elements $A_{ij}$ may be found and are

$$A_{11} = \frac{Y_w{}^2\Delta + 1 + Y_w(Z_{11}+Z_{22})}{2Y_wZ_{12}} \tag{17a}$$

$$A_{12} = \frac{-Y_w{}^2\Delta + 1 + Y_w(Z_{11}-Z_{22})}{2Y_wZ_{12}} \tag{17b}$$

$$A_{21} = \frac{Y_w{}^2\Delta - 1 + Y_w(Z_{11}-Z_{22})}{2Y_wZ_{12}} \tag{17c}$$

$$A_{22} = \frac{-Y_w{}^2\Delta - 1 + Y_w(Z_{11}+Z_{22})}{2Y_wZ_{12}} \tag{17d}$$

For a lossless structure, the impedance elements $Z_{ij}$ are pure-imaginary, and so $Z_{ij}^* = -Z_{ij}$. If all the impedance elements $Z_{ij}$ are normalized by dividing them by $Z_w$, the expressions for $A_{ij}$ are given by (17), provided $Y_w$ is put equal to unity and it is understood that the $Z_{ij}$ in these equations are normalized. The impedance of the input and output lines in Fig. 9.3 must now be taken as unity, and the currents $I_i$ as $c_i - b_i$.

For a load $Z_L$ connected at the terminal plane 2, the input impedance at the terminal plane 1 is

$$Z_{\text{in},1} = Z_{11} - \frac{Z_{12}^2}{Z_L + Z_{22}} = \frac{Z_{11}Z_L + \Delta}{Z_L + Z_{22}} \tag{18}$$

This is a bilinear transformation and will map circles into circles with straight lines as limiting cases. Such a transformation has two invariant points; i.e., there are two values of $Z_L$ in the complex $Z_L$ plane which map into two identical points in the complex $Z_{\text{in},1}$ plane. Let these invariant values be denoted by $Z_c$. Thus we must have

$$Z_c = \frac{Z_{11}Z_c + \Delta}{Z_c + Z_{22}}$$

or

$$Z_c = \frac{Z_{11} - Z_{12}}{2} \pm \left[\left(\frac{Z_{11} + Z_{22}}{2}\right)^2 - Z_{12}^2\right]^{1/2} \tag{19}$$

For a load $Z_L$ connected at terminal plane 1, the input impedance at terminal plane 2 is

$$Z_{\text{in},2} = -\frac{V_2}{I_2} = Z_{22} - \frac{Z_{12}^2}{Z_L + Z_{11}} = \frac{Z_{22}Z_L + \Delta}{Z_{11} + Z_L} \tag{20}$$

The reversal in sign in (20) is due to the assumed positive direction of current flow for $I_2$. Equation (20) leads to the following expression for the characteristic values:

$$Z_c = -\left\{\frac{Z_{11} - Z_{22}}{2} \pm \left[\left(\frac{Z_{11} + Z_{22}}{2}\right)^2 - Z_{12}^2\right]^{1/2}\right\} \tag{21}$$

These characteristic values $Z_c$ are of considerable importance since they may be taken as the characteristic impedance of the infinite periodic structure composed of an infinite number of basic sections or quadrupoles in a cascade connection.

## 9.3. Propagation in an Infinite Periodic Structure

Consider the periodic structure illustrated in Fig. 9.4 consisting of a cascade connection of an infinite number of lossless quadrupoles separated by lengths of waveguide or transmission lines of sufficient length to ensure

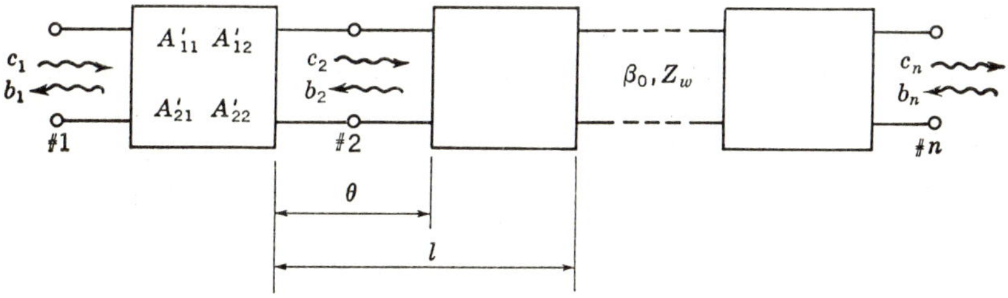


FIG. 9.4. Periodic structure consisting of a cascade connection of identical quadrupoles.

that there is no interaction between sections due to higher-order evanescent modes. This latter restriction is not necessary, and the modifications in the theory when there are higher-order mode interactions will be considered later. The amplitudes at terminal plane 2 are related to those at terminal plane 1 by the following matrix equation:

$$\begin{bmatrix} c_1 \\ b_1 \end{bmatrix} = \begin{bmatrix} A_{11} & A_{12} \\ A_{21} & A_{22} \end{bmatrix} \begin{bmatrix} c_2 \\ b_2 \end{bmatrix} \tag{22}$$

If a wave is to propagate down the structure, we must have

$$c_2 = e^{-\gamma l} c_1 \qquad b_2 = e^{-\gamma l} b_1$$

where $\gamma$ is the propagation constant, and $l$ is the physical length of each individual section. When we substitute into (22), we get two homogeneous equations for $c_1$ and $b_1$ which have a solution only if the following determinant vanishes:

$$\begin{vmatrix} A_{11} - e^{-\gamma l} & A_{12} \\ A_{21} & A_{22} - e^{-\gamma l} \end{vmatrix} = 0$$

or
$$e^{2\gamma l} - (A_{11} + A_{22})e^{\gamma l} + (A_{11}A_{22} - A_{12}A_{21}) = 0$$

Since $A_{11}A_{22} - A_{12}A_{21} = 1$, we obtain, by multiplying through by $e^{-\gamma l}$, the following solution:

$$\cosh \gamma l = \frac{A_{11} + A_{22}}{2} = \frac{1}{2}\left(\frac{1}{T_{12}} + \frac{1}{T_{12}^*}\right) = \frac{\cos \alpha}{|T_{12}|} \tag{23a}$$

where $\alpha$ is the phase angle of $T_{12}$. When $\cos \alpha < |T_{12}|$, $\gamma$ is imaginary, and the wave propagates freely, undergoing only a change in phase. When $\cos \alpha > |T_{12}|$, $\gamma$ is real, and the wave is attenuated. We also see that $-\gamma$ is a solution, and so we can have a wave propagating along the structure in the negative $z$ direction also. When $\cos \alpha = |T_{12}|$, $\gamma = 0$ or $j2n\pi/l$, and this corresponds to a pure standing wave between sections. If we substitute for $A_{11}$ and $A_{22}$ from (17), then the solution for $\gamma$ is given by

$$\cosh \gamma l = \frac{Z_{11} + Z_{22}}{2Z_{12}} \tag{23b}$$

We will now examine the wave solution in each section in more detail. We note that between each two discontinuities the wave solution is not a pure traveling wave but a combination of a traveling wave and a standing wave. The phase constant of the connecting lines is $\beta_0$, and hence the electric field of the forward- and backward-traveling waves in section 1 is proportional to $c_1 e^{-j\beta_0 z}$ and $b_1 e^{j\beta_0 z}$, respectively. The ratio of the amplitudes $b_1/c_1$ will be denoted by $R_c$, where $R_c$ is a reflection coefficient at the terminal plane 1 and is characteristic of the particular periodic structure considered. The ratio of $b_n$ to $c_n$ at the $n$th terminal plane is also $R_c$. Hence the transverse electric field in section 1 is proportional to

$$c_1(e^{-j\beta_0 z} + R_c e^{j\beta_0 z}) = c_1[(1 + R_c)\cos \beta_0 z - j(1 - R_c)\sin \beta_0 z] \tag{24a}$$

and the transverse magnetic field is proportional to

$$c_1 Y_w(e^{-j\beta_0 z} - R_c e^{j\beta_0 z}) = c_1 Y_w[(1 - R_c)\cos \beta_0 z - j(1 + R_c)\sin \beta_0 z] \tag{24b}$$

The impedance of the periodic wave at the $n$th terminal plane is

$$\frac{V_n}{I_n} = \frac{c_n + b_n}{c_n - b_n} Z_w = \frac{c_1 + b_1}{c_1 - b_1} Z_w = \frac{1 + R_c}{1 - R_c} Z_w = Z_c \tag{25}$$

and will be called the characteristic impedance of the periodic structure. It is important to note that this impedance is not unique, since it depends on the choice of location of the terminal plane. For terminals located at $z$, we find from (24) that

$$Z_c(z) = Z_w \frac{Z_c(0) - jZ_w \tan \beta_0 z}{Z_w - jZ_c(0) \tan \beta_0 z} \tag{26}$$

which is just the transmission-line equation for the transformation of impedance along a line. If we make the restriction that the only points at which we may sample the fields are at a terminal plane, then $Z_c$ is truly a unique characteristic impedance for the periodic structure. One other important fact to note is that $Z_c$ is not necessarily pure-real, but, in general, a complex quantity. The amplitudes of the waves at terminal plane 2 are $c_1e^{-\gamma l} = c_2$ and $b_1e^{-\gamma l} = b_2$, and hence the transverse electric field is proportional to $c_1e^{-\gamma l}(e^{-j\beta_0(z-l)} + R_c e^{j\beta_0(z-l)})$ in section 2, and to $c_1e^{-n\gamma l}(e^{-j\beta_0(z-nl)} + R_c e^{j\beta_0(z-nl)})$ in section $n$, since, at the $n$th terminal plane, where $z = nl$, the electric field is proportional to

$$c_n + b_n = e^{-\gamma nl}(c_1 + b_1) = e^{-\gamma nl}c_1(1 + R_c)$$

Expressions for $Z_c$ and $R_c$ are readily derived from the circuit equations for the equivalent T network in Fig. 9.3. From the circuit equations we obtain, when we substitute for $V_i$ and $I_i$, the following relations:

$$\begin{aligned} V_1 = c_1 + b_1 &= Z_{11}(c_1 - b_1)Y_w - Z_{12}(c_2 - b_2)Y_w \\ &= (c_1 - b_1)Y_w(Z_{11} - Z_{12}e^{-\gamma l}) \\ V_2 = c_2 + b_2 &= e^{-\gamma l}(c_1 + b_1) \\ &= (c_1 - b_1)Y_w(Z_{12} - Z_{22}e^{-\gamma l}) \end{aligned}$$

and, hence, $Z_c = \dfrac{c_1 + b_1}{c_1 - b_1} Z_w = Z_{11} - Z_{12}e^{-\gamma l} = Z_{12}e^{\gamma l} - Z_{22}$

Adding these two expressions for $Z_c$ and dividing by 2 gives

$$Z_c = \frac{Z_{11} - Z_{22}}{2} + Z_{12} \sinh \gamma l \tag{27}$$

From (23) we see that

$$\sinh \gamma l = (\cosh^2 \gamma l - 1)^{1/2} = \frac{1}{Z_{12}} \left[ \left( \frac{Z_{11} + Z_{22}}{2} \right)^2 - Z_{12}^2 \right]^{1/2}$$

so that the characteristic impedance given by Eq. (27) is the same as that given by (19). This is to be expected since, if the impedance is to be the same at each terminal plane, it must be invariant when transformed by the T network. For a lossless network, $Z_{ij}$ is pure-imaginary, and thus, when $\gamma$ is real, $Z_c$ is pure-imaginary, and, when $\gamma$ is imaginary, $Z_c$ is complex.

It is very important to adopt a definite convention for positive current flow if ambiguity in sign for expressions for impedances and so forth is to be avoided. We will call the complete solution for the wave that may propagate along the periodic structure a periodic wave as distinguished from the forward- and backward-traveling waves which exist in each section and make up the periodic wave. For the two partial waves which make up the periodic wave, the coefficients $c_i$ will always be used to denote the amplitude of the wave which propagates in the positive $z$ direction between sections, and $b_i$ will be used to denote the amplitude of the partial wave which propagates in the negative $z$ direction between sections. The currents associated with $c_i$ and $b_i$ will be taken as $Y_w c_i$ and $-Y_w b_i$. When $\gamma$ is imaginary, we will denote it by $j\beta_c$, where $\beta_c$ is the characteristic phase constant for the periodic structure. Using this

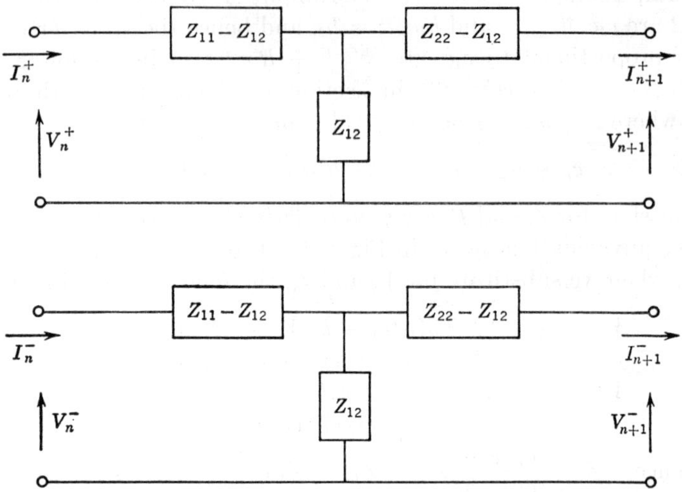


FIG. 9.5. Illustration for assumed positive directions of equivalent currents and voltages.

convention, the transverse electric field for the forward-propagating periodic wave is proportional to

$$(1 + R_c^+)c_1^+ e^{-j\beta_c nl}$$

at the $n$th terminal plane, while the transverse magnetic field is proportional to

$$(1 - R_c^+)Y_w c_1^+ e^{-j\beta_c nl} = Y_c^+(1 + R_c^+)c_1^+ e^{-j\beta_c nl}$$

at the $n$th terminal plane, where

$$R_c^+ = \frac{Z_c^+ - Z_w}{Z_c^+ + Z_w} \qquad Z_c^+ = (Y_c^+)^{-1} = \frac{Z_{11} - Z_{22}}{2} + jZ_{12}\sin\beta_c l$$

and $c_1^+$ is an arbitrary amplitude coefficient. For the periodic wave propagating in the negative $z$ direction, the transverse electric field at the

$n$th terminal plane is proportional to

$$c_1^-(1 + R_c^-)e^{j\beta_c nl}$$

while the transverse magnetic field is proportional to

$$(1 - R_c^-)c_1^- Y_w e^{j\beta_c nl} = Y_c^-(1 + R_c^-)c_1^- e^{j\beta_c nl}$$

where $R_c^- = \dfrac{Z_c^- - Z_w}{Z_c^- + Z_w} \qquad Z_c^- = (Y_c^-)^{-1} = \dfrac{Z_{11} - Z_{22}}{2} - jZ_{12}\sin\beta_c l$

With our adopted convention we note that we must distinguish between the two cases, as $Z_c$ is not the same for the periodic waves propagating in the positive $z$ direction and in the negative $z$ direction. Figure 9.5 illustrates the adopted convention as regards the positive directions of current flow.

## 9.4. Terminated Periodic Structure

Let a semi-infinite periodic structure be terminated at the $n$th terminal plane by a load impedance $Z_L$. Let the incident periodic wave at the $s$th terminal plane be

$$E_{inc} = c_1^+(1 + R_c^+)e^{-j\beta_c sl}$$

with a transverse magnetic field proportional to $Y_c^+E_{inc}$. The presence of the load gives rise to a reflected periodic wave of amplitude

$$E_{ref} = c_n^-(1 + R_c^-)e^{-j\beta_c(n-s)l}$$

at the $s$th terminal plane and with a transverse magnetic field of amplitude proportional to $Y_c^-E_{ref}$. To determine the amplitude of the reflected periodic wave, we consider the effect of the load on the partial waves of which the periodic waves are composed. At the position of the load where $s = n$, the total forward-traveling partial wave is

$$c_1^+e^{-j\beta_c nl} + c_n^-$$

while the total backward-traveling partial wave is

$$R_c^+c_1^+e^{-j\beta_c nl} + R_c^-c_n^-$$

with transverse magnetic fields proportional to

$$Y_w c_1^+ e^{-j\beta_c nl} + Y_w c_n^-$$

and

$$-Y_w R_c^+ c_1^+ e^{-j\beta_c nl} - Y_w R_c^- c_n^-$$

respectively. The load $Z_L$ gives rise to a reflection coefficient

$$R_L = \frac{Z_L - Z_w}{Z_L + Z_w}$$

for these partial waves, and, hence, we have

$$R_L(c_1^+ e^{-j\beta_c nl} + c_n^-) = R_c^+ c_1^+ e^{-j\beta_c nl} + R_c^- c_n^-$$

Solving this equation for $c_n^-$ gives

$$c_n^- = -\frac{R_L - R_c^+}{R_L - R_c^-} c_1^+ e^{-j\beta_c nl} \tag{28}$$

The incident periodic wave at the load is $c_1^+(1 + R_c^+)e^{-j\beta_c nl}$, and the reflected periodic wave at the load is, therefore,

$$\begin{aligned} c_n^-(1 + R_c^-) &= (1 + R_c^-)\frac{R_c^+ - R_L}{R_L - R_c^-} c_1^+ e^{-j\beta_c nl} \\ &= \frac{1 + R_c^-}{1 + R_c^+}\frac{R_c^+ - R_L}{R_L - R_c^-} c_1^+(1 + R_c^+)e^{-j\beta_c nl} \end{aligned} \tag{29}$$

The equivalent reflection coefficient $\Gamma_n$ for the periodic wave is thus

$$\Gamma_n = \frac{1 + R_c^-}{1 + R_c^+}\frac{R_c^+ - R_L}{R_L - R_c^-} \tag{30}$$

If we replace $Z_L$, $Z_c^+$, and $Z_c^-$ by their values given in terms of $R_L$, $R_c^+$, and $R_c^-$, we may readily show that

$$\frac{Z_L - Z_c^+}{Z_L - Z_c^-} = \frac{1 - R_c^-}{1 - R_c^+}\frac{R_L - R_c^+}{R_L - R_c^-}$$

and, hence, we get

$$\Gamma_n = -\frac{Z_c^-}{Z_c^+}\frac{Z_L - Z_c^+}{Z_L - Z_c^-} \tag{31}$$

Thus, at the $s$th terminal plane, the sum of the incident plus reflected periodic waves has a transverse electric field proportional to

$$c_n^+(1 + R_c^+)(e^{-j\beta_c l(s-n)} + \Gamma_n e^{j\beta_c l(s-n)})$$

and a transverse magnetic field proportional to

$$c_n^+(1 + R_c^+)(Y_c^+ e^{-j\beta_c l(s-n)} + \Gamma_n Y_c^- e^{j\beta_c l(s-n)})$$

where

$$c_n^+ = c_1^+ e^{-j\beta_c ln}$$

The ratio of the transverse electric field to the transverse magnetic field at the $n$th terminal plane is

$$\frac{1 + \Gamma_n}{Y_c^+ + Y_c^- \Gamma_n} = Z_L \tag{32}$$

a result readily proved from Eq. (31). If we let

$$\frac{Z_{11} - Z_{22}}{2} = \eta \tag{33}$$

$$Z_{12} \sinh \gamma l = jZ_{12} \sin \beta_c l = Z \tag{34}$$

we get

$$Z_c^+ = \eta + Z \tag{35a}$$

$$Z_c^- = \eta - Z \tag{35b}$$

$$\Gamma_n = -\frac{\eta - Z}{\eta + Z}\frac{(Z_L - \eta) - Z}{(Z_L - \eta) + Z} \tag{35c}$$

For a periodic structure composed of symmetrical sections, $Z_{11} = Z_{22}$ and $\eta = 0$; so in this case

$$Z_c^+ = -Z_c^- = Z \tag{36a}$$

$$\Gamma_n = \frac{Z_L - Z}{Z_L + Z} \tag{36b}$$

The reflection coefficient at the $s$th terminal plane is

$$\Gamma_s = \Gamma_n e^{2j\beta_c l(s-n)} \tag{37}$$

The ratio of the transverse electric field to the transverse magnetic field at the $s$th terminal plane is

$$\frac{1 + \Gamma_s}{Y_c^+ + Y_c^- \Gamma_s} = \frac{(1 + \Gamma_s)(\eta - Z)(\eta + Z)}{(\eta - Z) + \Gamma_s(\eta + Z)} = Z_s \tag{38}$$

With some algebraic manipulation, we can obtain the following results from (38):

$$Z_s - \eta = Z\frac{Z_c^- - Z_c^+\Gamma_s}{Z_c^- + Z_c^+\Gamma_s} \tag{39a}$$

and

$$\Gamma_s = -\frac{Z_c^-}{Z_c^+}\frac{(Z_s - \eta) - Z}{(Z_s - \eta) + Z} \tag{39b}$$

Substituting for $\Gamma_s$ from (37) into (39$a$) gives

$$\begin{aligned} Z_s - \eta &= Z\frac{Z_c^- - Z_c^+\Gamma_n e^{2j\beta_c l(s-n)}}{Z_c^- + Z_c^+\Gamma_n e^{2j\beta_c l(s-n)}} \\ &= Z\frac{(Z_c^- - Z_c^+\Gamma_n)/(Z_c^- + Z_c^+\Gamma_n) + jZ\tan\beta_c l(n-s)}{Z + j(Z_c^- - Z_c^+\Gamma_n)/(Z_c^- + Z_c^+\Gamma_n)\tan\beta_c l(n-s)} \\ &= Z\frac{Z_L - \eta + jZ\tan\beta_c l(n-s)}{Z + j(Z_L - \eta)\tan\beta_c l(n-s)} \end{aligned} \tag{40}$$

Equation (40) gives the input impedance $Z_s$ at the $s$th terminal plane with a load $Z_L$ connected across the $n$th terminal plane. We note that the expressions for the reflection coefficient and the transformation of impedance become formally the same as for a conventional line whenever the basic sections are symmetrical, so that $Z_{11} = Z_{22}$ and $\eta = 0$.

The use of equivalent transmission-line voltages and currents simplifies the derivation of a lot of the properties of periodic structures. The equivalent transmission-line voltages and currents are taken proportional to the total transverse electric and magnetic fields, respectively. Thus, for a periodic wave propagating in the positive $z$ direction, we have

$$V_n^+ = V_1^+ e^{-j\beta_c nl} = c_1^+(1 + R_c^+)e^{-j\beta_c nl} \tag{41a}$$
$$I_n^+ = I_1^+ e^{-j\beta_c nl} = Y_c^+ V_n^+ \tag{41b}$$

at the $n$th terminal plane, while, for a negatively propagating periodic wave, we have

$$V_n^- = V_1^- e^{j\beta_c nl} = c_1^-(1 + R_c^-)e^{j\beta_c nl} \tag{42a}$$
$$I_n^- = I_1^- e^{j\beta_c nl} = Y_c^- V_n^- \tag{42b}$$

In (42*a*) and (42*b*) the positive direction of current flow is in the positive $z$ direction. If we connect a load $Z_L$ across the $n$th terminal plane, the total voltage across the load is the sum of the voltages in the incident and reflected periodic waves:

$$V_L = V_n^+ + V_n^- = V_n^+(1 + \Gamma_n) = V_1^+ e^{-j\beta_c nl}(1 + \Gamma_n)$$

and the current flowing into the load is the sum of the currents due to the incident and reflected periodic waves:

$$\begin{aligned} I_L = I_n^+ + I_n^- &= V_n^+ Y_c^+ + V_n^- Y_c^- \\ &= V_n^+ Y_c^+ \left(1 + \frac{Y_c^-}{Y_c^+}\Gamma_n\right) \end{aligned}$$

The voltage across the load divided by the current flowing through it is equal to the load impedance, and, hence,

$$\frac{V_L}{I_L} = Z_L = \frac{1 + \Gamma_n}{Y_c^+ + Y_c^- \Gamma_n}$$

which is identical with the result obtained in (32).

From (35*c*) or (39*b*) it is seen that $Z_L = Z + \eta$ is the proper load impedance in which to terminate a periodic structure at one of the terminal planes, in order to prevent a reflected periodic wave when a periodic wave is incident from the left. For a periodic wave incident from the right along the periodic structure, the reflectionless load termination must have an impedance $Z - \eta$. These values of load impedances are equal to the characteristic impedances $Z_c^+$ and $-Z_c^-$.

Although the above load terminations result in a matched termination for the incident periodic waves, this does not mean that there is not a reflected component wave in the waveguide section in front of the terminating load impedance. The incident periodic wave is composed of forward- and backward-propagating component waves, and the load impedance $Z_L$ must differ from the impedance $Z_w$ of the waveguide or transmission line by an amount just sufficient to produce the backward-propagating component wave when the forward-propagating component wave is incident.

Often, in practice, it is desirable to excite the propagating periodic wave in a periodic structure by means of a wave incident along a uniform waveguide to which the periodic structure, together with its terminating load impedance, is connected. To prevent a reflected wave in the uniform input guide, a transition or matching section must be introduced such that the impedance seen at the input terminals to the periodic structure is transformed into the impedance $Z_w$ of the input guide. The usual techniques employed for matching an arbitrary load impedance to a guide or transmission line may be employed. An alternative procedure is to taper the periodic structure so as to provide a gradual change from the uniform waveguide to the final periodically loaded waveguide section. This is analogous to matching by means of a tapered transmission-line section.

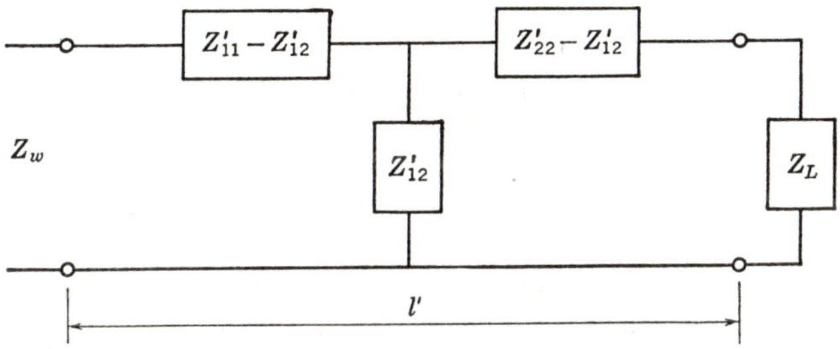


FIG. 9.6. Matching section for periodic structures.

The required parameters of a matching T section are readily obtained by a transmission-line analysis. Consider a periodic structure which is terminated in a given load impedance, such that $Z_L$ is the effective load impedance presented at the output terminals of the matching section. The parameters of the matching section will be distinguished from those of the basic sections of the periodic structure by primes, as in Fig. 9.6. If the matching T section formed a basic section in a periodic structure, it would be characterized by a propagation constant $\gamma'$ and characteristic impedances $Z_c^{+\prime}$, $Z_c^{-\prime}$, where

$$\begin{aligned}
\sinh \gamma' l' &= \frac{1}{Z'_{12}}\left[\left(\frac{Z'_{11} + Z'_{22}}{2}\right)^2 - Z'^2_{12}\right]^{1/2} \\
Z_c^{+\prime} &= \eta' + Z' \\
Z_c^{-\prime} &= \eta' - Z' \\
\eta' &= \tfrac{1}{2}(Z'_{11} - Z'_{22}) \\
Z' &= Z'_{12} \sinh \gamma' l'
\end{aligned}$$

Conventional analysis shows that the impedance seen at the input to the matching section is

$$Z_{\text{in}} = Z'_{11} - \frac{Z'^2_{12}}{Z'_{22} + Z_L} \tag{43}$$

By introducing the characteristic parameters of the matching section, (43) may be rewritten as follows [compare with (40)]:

$$Z_{\text{in}} = \eta' + Z' \frac{Z_L - \eta' + Z' \tanh \gamma' l'}{Z' + (Z_L - \eta') \tanh \gamma' l'} \tag{44}$$

To obtain a match, we must choose the parameters $Z'$, $\eta'$, and $\gamma' l'$ so that $Z_{\text{in}}$ becomes equal to $Z_w$. The particular case when $Z_L$ is real and the matching section is symmetrical is readily handled. From (44) with $\eta' = 0$ and $\gamma' l'$ replaced by $j\theta$, we get

$$Z_{\text{in}} = Z' \frac{Z_L + jZ' \tan \theta}{Z' + jZ_L \tan \theta}$$

If we make $\theta = \pi/2$, we obtain the equivalent of a quarter-wave transformer, and the required value of $Z'$ for a match is given by

$$Z' = (Z_w Z_L)^{1/2} \tag{45}$$

The required values of $Z'_{11}$ and $Z'_{12}$ are

$$Z'_{11} = Z'_{22} = 0 \tag{46a}$$
$$jZ'_{12} = (Z_w Z_L)^{1/2} \tag{46b}$$

In all the previous analysis we have considered each section to have a physical length $l$. It is not necessary to introduce a length for each section, since in all the derived formulas we always find the product $\gamma l$ occurring as a parameter. It is the phase shift per section which is of prime importance and not the phase shift per unit length. In some cases it is not possible to define a meaningful length for each section.

## 9.5. Capacitively Loaded Rectangular Waveguide

As an example, to illustrate some of the previously derived relations, we will consider a rectangular waveguide loaded by asymmetrical capaci-

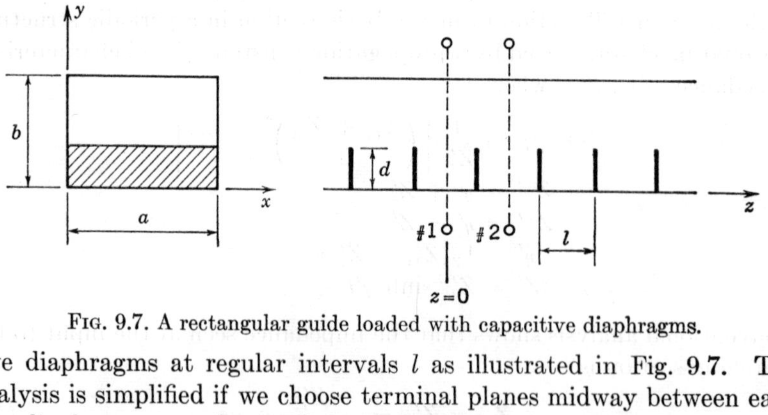


FIG. 9.7. A rectangular guide loaded with capacitive diaphragms.

tive diaphragms at regular intervals $l$ as illustrated in Fig. 9.7. The analysis is simplified if we choose terminal planes midway between each two diaphragms, so that each section may be characterized by a symmetrical T network. The frequency will be chosen so that only the dominant $H_{10}$ mode propagates. Initially we will also assume that the spacing $l$ is chosen so large that all the evanescent modes excited at each diaphragm have decayed to a negligible value at the positions of the adjacent diaphragms. The first higher-order mode has a propagation constant given by

$$\Gamma_{11}^2 = \left(\frac{\pi}{a}\right)^2 + \left(\frac{\pi}{b}\right)^2 - k_0^2$$

and, provided $e^{-\Gamma_{11}l} < 0.1$, interaction with this mode may usually be neglected. For our purpose here we will also assume that the normalized shunt susceptance of the diaphragm is given to sufficient accuracy by the low-frequency formula

$$B = \frac{4b\beta_0}{\pi} \ln \sec \frac{\pi d}{2b} \tag{47}$$

where $\beta_0$ is the phase constant for the $H_{10}$ mode, and is given by

$$\beta_0^2 = k_0^2 - \left(\frac{\pi}{a}\right)^2$$

The wave impedance $Z_w$ of the guide for the $H_{10}$ mode is $Z_0k_0/\beta_0$, where $Z_0 = (\mu_0/\epsilon_0)^{1/2}$. The susceptance $B$ is normalized with respect to $Z_w$.

Using conventional transmission-line analysis, we find that the normalized input impedance at terminal plane 1 with normalized load impedance $Z_L$ connected at terminal plane 2 as in Fig. 9.7 is

$$Z_{\text{in}} = \frac{(2t - Bt^2)/(B + 2t) + jZ_L(Bt + t^2 - 1)/(B + 2t)}{j(Bt + t^2 - 1)/(B + 2t) + Z_L}$$

where $t = \tan(\beta_0 l/2)$. Comparing with the similar equation for a symmetrical T network

$$Z_{\text{in}} = \frac{Z_{11}^2 - Z_{12}^2 + Z_{11}Z_L}{Z_{11} + Z_L}$$

shows that

$$Z_{11} = \frac{j(Bt + t^2 - 1)}{B + 2t} \tag{48a}$$

$$Z_{11}^2 - Z_{12}^2 = \frac{2t - Bt^2}{B + 2t} \tag{48b}$$

From (48), the solutions for the characteristic impedance $Z_c$ and propagation phase constant $\beta_c$ for the infinite periodic structure are readily found. After suitable algebraic reductions, we get

$$Z_c = (Z_{11}^2 - Z_{12}^2)^{1/2} = \left(1 - \frac{2B}{2\sin\beta_0 l + B\cos\beta_0 l + B}\right)^{1/2} \tag{49a}$$

$$\cos\beta_c l = \frac{Z_{11}}{Z_{12}} = \cos\beta_0 l - \frac{B}{2}\sin\beta_0 l \tag{49b}$$

If $\beta_c$ is a solution of (49b), then clearly $\beta_c \pm 2n\pi/l$ is also a solution for all integer values of $n$. Before considering this eigenvalue equation in detail, we will examine the matching problem.

The periodic loaded guide may be matched to the uniform guide by means of a shunt capacitive susceptance $B'$ placed at a distance $l'/2$ in front of the first terminal plane of the periodic structure, as shown in Fig. 9.8. We will assume that the periodic loaded guide is terminated in its characteristic impedance at the output end. The input impedance at the first terminal plane to the loaded guide will then be $Z_c$. From (46), the required parameters of the matching section are $Z'_{11} = 0$, $jZ'_{12} = Z_c^{1/2}$, since all impedances are normalized relative to the uniform guide impedance $Z_w$. Also we require that $\beta'_c l' = \pi/2$ for the matching

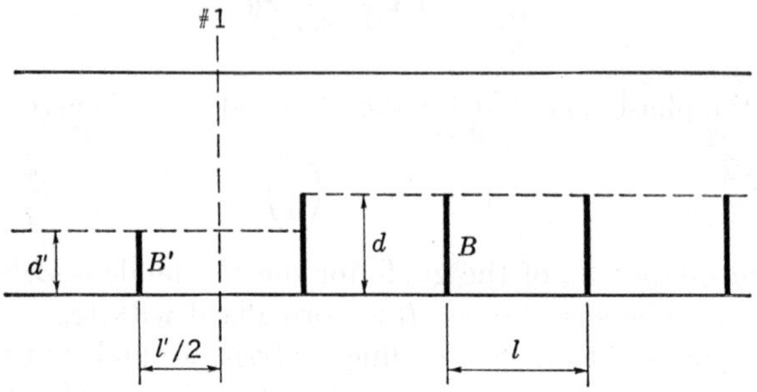


FIG. 9.8. Matching of uniform guide to periodic-loaded guide.

section. Referring to (49b), this is seen to require that $B' = 2 \cot \beta_0 l'$. From (49a), we obtain

$$(jZ'_{12})^2 = Z_c = 1 - \frac{2B'}{2 \sin \beta_0 l' + B' + B' \cos \beta_0 l}$$

These two equations may be solved for $B'$ and $l'$ when $Z_c$ has been specified. At the position of the matching diaphragm, the input admittance is $1 - jB'$. The addition of a shunt susceptance $jB'$ at this point makes the input admittance real and equal to unity. The solutions for $l'$ and $B'$ are

$$\cos \beta_0 l' = \frac{1 - Z_c}{1 + Z_c} \tag{50a}$$

$$B' = \frac{1 - Z_c}{Z_c^{1/2}} \tag{50b}$$

*The $k_0$-$\beta_c$ Diagram*

The usual way of presenting the information contained in the eigenvalue equation (49b) is to plot $\beta_c$ as a function of $k_0$. The resultant plot is called the $k_0$-$\beta_c$ or $\omega$-$\beta_c$ diagram [note that $k_0 = \omega(\mu_0\epsilon_0)^{1/2} = \omega/v_c$].

Such a plot is given in Fig. 9.9 for the following parameters: $a = 0.9$ inch, $b = 0.2$ inch, $d = 0.15$ inch, $l = 1.0$ centimeter.

The passbands and stopbands are clearly evident in Fig. 9.9. The first passband occurs in the range $1.37 < k_0 < 2.44$, the first attenuation band occurs for $3.43 < k_0 < 4.62$, the second passband occurs for $6.43 < k_0 < 7.25$, and so forth. For $k_0 > 6.19$, the first higher-order mode begins to propagate, and, consequently, the curves for $\beta_c l$ have little

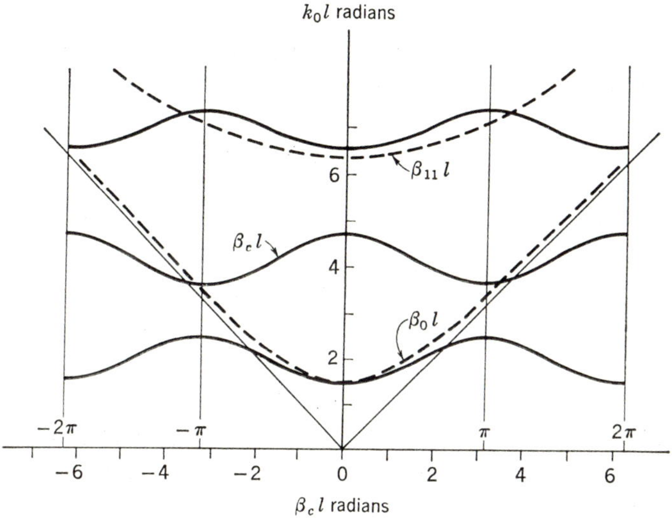


FIG. 9.9. The $k_0$-$\beta_c$ diagram for a periodic-loaded rectangular guide.

significance for $k_0$ greater than about 5. From the eigenvalue equation (49*b*), we find that the values of $\beta_0$ for $\beta_c l = 0$ are given by

$$\tan \frac{\beta_0 l}{2} = -\frac{B}{2} \qquad \text{or} \qquad \sin \frac{\beta_0 l}{2} = 0 \tag{51a}$$

while, for $\beta_c l = \pi$, the corresponding values of $\beta_0$ are given by

$$\cot \frac{\beta_0 l}{2} = \frac{B}{2} \qquad \text{or} \qquad \cos \frac{\beta_0 l}{2} = 0 \tag{51b}$$

From (51*a*), it is seen that, whenever the diaphragms are spaced by a multiple of a guide wavelength, $\beta_c l$ will equal zero. The field distribution in this case is a pure standing wave between sections, with the electric field being zero at the positions of the diaphragms. It is for this reason that the diaphragms do not produce any perturbation of the standing-wave field distribution and hence permit a solution for $\beta_c l = 2n\pi = \beta_0 l$. The other solution for $\beta_c l = 0$ or $2n\pi$, as given by (51*a*), corresponds to a field distribution for which the transverse electric field is symmetrical about but not equal to zero, at each diaphragm. A similar conclusion may be reached from (51*b*), but with the diaphragms spaced by an odd multiple of a half guide wavelength. When the standing-wave field distribution does not have a zero transverse electric field at the diaphragms, the spacing differs from a multiple of a half guide wavelength under the conditions $\beta_c l = n\pi$ because of the perturbing effect of the diaphragms. This corresponds to the other solutions given by (51).

### 9.6. Energy and Power Flow

According to Floquet's theorem, the field in the loaded guide can be represented in the following form:

$$\mathbf{E}(x,y,z) = \sum_{n=-\infty}^{\infty} a_n \mathbf{E}_n(x,y) e^{-j(\beta_c + 2n\pi/l)z} \tag{52}$$

where $a_n$ is a suitable amplitude coefficient and $\mathbf{E}_n(x,y)e^{-j(\beta_c+2n\pi/l)z}$ is the field of the *n*th mode or space harmonic. Since the field consists, in general, of an infinite series, there is no unique phase velocity. Each space harmonic has its own phase velocity $v_{pn}$ given by

$$v_{pn} = \frac{k_0 v_c}{\beta_c + 2n\pi/l} \tag{53}$$

where $v_c$ is the velocity of light in free space. Some of the space harmonics will have negative phase velocities. Replacing $\beta_c$ by $\beta_c + 2n\pi/l$ is equivalent to focusing attention on the nth- rather than the zeroth-order space-harmonic wave. As $\beta_c$ increases from zero to $2\pi/l$, the corresponding mode changes from the $n = 0$ to the $n = 1$ mode. At the same time the $n = -1$ mode goes over into the $n = 0$ mode, etc. A single space harmonic cannot exist by itself except under special situations such as when $\beta_c l = 2m\pi$ and the diaphragms are spaced by an integral multiple of a guide wavelength. In general, all the space harmonics are coupled together through the boundary conditions which the field must satisfy. Together they make up the periodic wave.

In a passband we may readily show that, with $\beta_c$ real, the time-average electric and magnetic energy stored in each section are equal. The fields at two adjacent terminal planes differ only by a factor $e^{-j\beta_c l}$, and, hence, the complex Poynting vector has the same value at each terminal plane. Thus we have

$$\iint_{1-1} \mathbf{E} \times \mathbf{H}^* \cdot \mathbf{a}_z \, da - \iint_{2-2} \mathbf{E} \times \mathbf{H}^* \cdot a_z \, da = 0$$
$$= 4j\omega(W_m - W_e) \qquad \text{and hence} \qquad W_m = W_e$$

Since the frequency dependence of the propagation phase constant $\beta_n = \beta_c + 2n\pi/l$ for the *n*th space harmonic is the same as that for $\beta_c$, each space harmonic has the same group velocity. According to Chap. 7, Eq. (91), the group velocity is given by

$$v_g = \left(\frac{d\beta_c}{d\omega}\right)^{-1} = v_c\left(\frac{d\beta_c}{dk_0}\right)^{-1} \tag{54}$$

Referring to Fig. 9.9, it is seen that, for $0 < \beta_c l < \pi$, the group velocity is positive, while, for $\pi < \beta_c l < 2\pi$, it is negative. We may also show that the power flow along the periodic structure is equal to the group velocity, divided by the period $l$, times the time-average electric plus magnetic energy stored per section.

The proof of the above property is readily established by an application of Green's theorem for two vector functions.† For two vector functions **A** and **B** which have zero divergence this theorem is

$$\iiint\limits_V (\mathbf{A}\cdot\nabla^2\mathbf{B} - \mathbf{B}\cdot\nabla^2\mathbf{A})\,dV = \oint\!\!\!\oint\limits_S [\mathbf{A}\times(\nabla\times\mathbf{B}) - \mathbf{B}\times(\nabla\times\mathbf{A})]\cdot\mathbf{n}\,da$$

where $\mathbf{n}$ is the unit outward-directed normal. Let $\mathbf{A} = \partial\mathbf{E}/\partial\omega$, $\mathbf{B} = \mathbf{E}^*$, where $\mathbf{E}$ is the total electric field of the periodic wave and is a function of $\omega$. Since $\nabla^2\mathbf{E} + k_0^2\mathbf{E} = 0$, we have

$$\nabla^2\frac{\partial\mathbf{E}}{\partial\omega} = -2\omega\mu_0\epsilon_0\mathbf{E} - k_0^2\frac{\partial\mathbf{E}}{\partial\omega}$$
$$\nabla^2\mathbf{E}^* = -k_0^2\mathbf{E}^*$$

Substituting in Green's theorem gives

$$\iiint\limits_V \left(-k_0^2\mathbf{E}^*\cdot\frac{\partial\mathbf{E}}{\partial\omega} + k_0^2\mathbf{E}^*\cdot\frac{\partial\mathbf{E}}{\partial\omega} + 2\omega\mu_0\epsilon_0\mathbf{E}\cdot\mathbf{E}^*\right)dV$$
$$= 2\omega\mu_0\epsilon_0\iiint\limits_V \mathbf{E}\cdot\mathbf{E}^*\,dV = \oint\!\!\!\oint\limits_S \left[\frac{\partial\mathbf{E}}{\partial\omega}\times(\nabla\times\mathbf{E}^*) - \mathbf{E}^*\times\left(\nabla\times\frac{\partial\mathbf{E}}{\partial\omega}\right)\right]\cdot\mathbf{n}\,da \tag{55}$$

Since $\mathbf{n}\times\mathbf{E}$ vanishes on the guide walls, the surface integral reduces to an integral over the terminal planes 1 and 2 which bound one section of the periodic structure. All losses are assumed absent, and hence

$$\oint\!\!\!\oint\limits_S \mathbf{E}\times\mathbf{H}^*\cdot d\mathbf{a} = 0 \qquad \text{or equivalently} \qquad \oint\!\!\!\oint\limits_S \mathbf{E}^*\times(\nabla\times\mathbf{E})\cdot\mathbf{n}\,da = 0$$

† A similar proof is given by S. Sensiper, Electromagnetic Wave Propagation on Helical Conductors, *MIT Research Lab. Electronics, Tech. Rept.* 194, May 16, 1951.

Differentiating with respect to $\omega$ gives

$$\oiint_S \frac{\partial \mathbf{E}^*}{\partial \omega} \times (\nabla \times \mathbf{E}) \cdot \mathbf{n}\, da = - \oiint_S \mathbf{E}^* \times \left(\nabla \times \frac{\partial \mathbf{E}}{\partial \omega}\right) \cdot \mathbf{n}\, da$$

Using this result, the right-hand side of Eq. (55) becomes

$$2 \operatorname{Re} \oiint_S \frac{\partial \mathbf{E}}{\partial \omega} \times (\nabla \times \mathbf{E}^*) \cdot \mathbf{n}\, da$$

On the terminal plane 2, the field $\mathbf{E} = \mathbf{E}_2$ is equal to $e^{-j\beta_c l}\, \mathbf{E}_1$, where $\mathbf{E}_1$ is the field at terminal plane 1. Thus we have

$$\frac{\partial \mathbf{E}_2}{\partial \omega} = e^{-j\beta_c l} \frac{\partial \mathbf{E}_1}{\partial \omega} - jl \frac{d\beta_c}{d\omega} \mathbf{E}_1 e^{-j\beta_c l}$$

Substituting into the right-hand side of Eq. (55) gives

$$\begin{aligned} 2 \operatorname{Re} \Bigg[ - \iint_{1-1} \frac{\partial \mathbf{E}_1}{\partial \omega} \times (\nabla \times \mathbf{E}_1^*) \cdot \mathbf{a}_z\, da + \iint_{2-2} \left( \frac{\partial \mathbf{E}_1}{\partial \omega} - jl \frac{d\beta_c}{d\omega} \mathbf{E}_1 \right) \\ \times (\nabla \times \mathbf{E}_1^*) \cdot \mathbf{a}_z\, da \Bigg] &= 2 \operatorname{Re} \iint_{2-2} -jl \frac{d\beta_c}{d\omega} \mathbf{E}_1 \times \nabla \times \mathbf{E}_1^* \cdot \mathbf{a}_z\, da \\ &= 2 \operatorname{Re} \omega\mu_0 l \frac{d\beta_c}{d\omega} \iint_{2-2} \mathbf{E}_1 \times \mathbf{H}_1^* \cdot \mathbf{a}_z\, da \\ &= 2\omega\mu_0\epsilon_0 \iiint_V \mathbf{E} \cdot \mathbf{E}^*\, dV = 8\omega\mu_0 W_e \\ &= 4\omega\mu_0 (W_e + W_m) \end{aligned}$$

since $W_e = W_m$. From this result the real power flow is seen to be given by

$$\begin{aligned} \operatorname{Re} \frac{1}{2} \iint_{2-2} \mathbf{E}_1 \times \mathbf{H}_1^* \cdot \mathbf{a}_z\, da &= \frac{W_e + W_m}{l\, d\beta_c/d\omega} \\ &= (W_e + W_m) \frac{v_g}{l} \end{aligned} \tag{56}$$

and is equal to the group velocity times the spatial-average electric plus magnetic energy density per section. The method of derivation is clearly applicable to any lossless periodic structure, and hence (56) is a general result for lossless structures.

## 9.7. Higher-order Mode Interaction

For close spacing of the diaphragms the first few least attenuated higher-order modes excited by one diaphragm will not have attenuated to a negligible value at the positions of the two adjacent diaphragms. The incident field on each diaphragm is therefore a combination of a dominant mode and one or more higher-order modes. The result of higher-order mode interaction is to modify the characteristic propagation phase constant $\beta_c$ from that computed on the basis of only dominant-mode interaction. The analysis in the general case will, however, be similar to that for single-mode interaction. For each interacting mode, a separate equivalent transmission line is introduced. The diaphragm

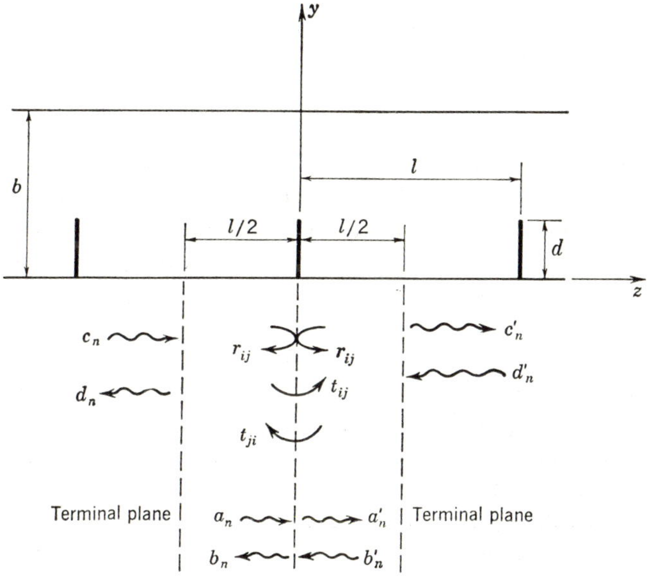


FIG. 9.10. Capacitive-loaded rectangular guide.

must now be represented by a more general network which couples the various transmission lines together. The over-all structure is still a periodic one, and solutions for periodic waves propagating along the structure may be found by the usual matrix methods. The analysis of systems of this typo has been presented by Brown.† A similar theory will be presented here but specialized to thc particular problem of a waveguide loaded with capacitive diaphragms.

Consider a rectangular guide loaded with identical capacitive diaphragms at regular intervals $l$ as in Fig. 9.10. The mode amplitudes at the input and output terminal planes of one basic section are denoted by $c_n$, $d_n$, $c'_n$, and $d'_n$, where the coefficients $c_n$, $c'_n$ pertain to forward-prop-

† J. Brown, A Theoretical Study of Some Artificial Dielectrics, Ph.D. thesis, University of London, 1954.

J. Brown, Propagation on Coupled Transmission Line Systems, *Quart. J. Mech. and Appl. Math.*, vol. 11, pp. 235–243, May, 1958.

agating modes, and $d_n$, $d'_n$ to backward-propagating modes (or attenuated modes). The transverse electric and magnetic fields to the left and right of one diaphragm may be expanded into a series of the normal waveguide modes as follows:

$$\mathbf{E}_t = \sum_n a_n \mathbf{e}_n e^{-\Gamma_n z} + \sum_n b_n \mathbf{e}_n e^{\Gamma_n z} \qquad -l \le z \le 0 \tag{57a}$$

$$\mathbf{H}_t = \sum_n a_n \mathbf{h}_n e^{-\Gamma_n z} - \sum_n b_n \mathbf{h}_n e^{\Gamma_n z} \qquad -l \le z \le 0 \tag{57b}$$

$$\mathbf{E}_t = \sum_n a'_n \mathbf{e}_n e^{-\Gamma_n z} + \sum_n b'_n \mathbf{e}_n e^{\Gamma_n z} \qquad 0 \le z \le l \tag{57c}$$

$$\mathbf{H}_t = \sum_n a'_n \mathbf{h}_n e^{-\Gamma_n z} - \sum_n b'_n \mathbf{h}_n e^{\Gamma_n z} \qquad 0 \le z \le l \tag{57d}$$

where $a_n$, $b_n$, $a'_n$, and $b'_n$ are the mode amplitudes at $z = 0$. It will be assumed that only the dominant mode ($H_{10}$ mode) propagates. The higher-order modes excited by each diaphragm are the longitudinal-section electric modes with all field components present except $E_x$. Since the mode functions $\mathbf{e}_n$ are orthogonal, and the electric field must be continuous at the diaphragm, we must have $a_n + b_n = a'_n + b'_n$. Only a finite number, say $N$, of the modes excited at one discontinuity will have an appreciable amplitude at the position of the next discontinuity. For any practical case a sufficiently accurate analysis of the loaded guide will be obtained by taking into account only the first $N$ least attenuated modes.

Each discontinuity may be characterized by a set of reflection and transmission coefficients. The following nomenclature will be adopted:

$r_{ij}$ = reflection coefficient for $i$th mode with $j$th mode incident
$t_{ij}$ = transmission coefficient through diaphragm for $i$th mode with $j$th mode incident

The indices $i, j$ take on all values from 1 to $N$. Since each discontinuity is symmetrical, the above reflection and transmission coefficients hold for modes incident from either the left or the right side of the obstacle. Also, since the transverse electric field of each mode is continuous across the diaphragm, $t_{ij} = 1 + r_{ij}$ for $i, j = 1, 2, \ldots, N$.

At $z = 0$, (57$a$) and (57$c$) give

$$\mathbf{E}_t = \sum_{n=1}^{N} a_n \mathbf{e}_n + \sum_{n=1}^{N} \mathbf{e}_n \sum_{i=1}^{N} (r_{ni} a_i + t_{ni} b'_i) \tag{58a}$$

$$\mathbf{E}_t = \sum_{n=1}^{N} b'_n \mathbf{e}_n + \sum_{n=1}^{N} \mathbf{e}_n \sum_{i=1}^{N} (r_{ni} b'_i + t_{ni} a_i) \tag{58b}$$

Let $A$ and $B$ denote column matrices with elements $a_n$, $b_n$; $n = 1, 2, \ldots, N$, respectively, while $R$ and $T$ denote the following square $N \times N$

reflection and transmission coefficient matrices:

$$R = \begin{bmatrix} r_{11} & r_{12} & \cdots & r_{1N} \\ r_{21} & \cdots & \cdots & \cdots \\ \cdots & \cdots & \cdots & \cdots \\ r_{N1} & \cdots & \cdots & r_{NN} \end{bmatrix}$$

$$T = \begin{bmatrix} t_{11} & t_{12} & \cdots & t_{1N} \\ t_{21} & \cdots & \cdots & \cdots \\ \cdots & \cdots & \cdots & \cdots \\ t_{N1} & \cdots & \cdots & t_{NN} \end{bmatrix}$$

A similar definition applies to the column matrices $A'$ and $B'$ with elements $a'_n$, $b'_n$. The relation between the mode amplitudes on adjacent sides of the diaphragm may be written in matrix form as

$$B = RA + TB' \tag{59a}$$
$$A' = RB' + TA \tag{59b}$$

since the amplitude of a reflected mode on the left is given by

$$b_n = \sum_{i=1}^{N} (r_{ni}a_i + t_{ni}b'_i)$$

and similarly for a reflected mode on the right. Solving for $A$ and $B$ in terms of $A'$ and $B'$ gives

$$A = T^{-1}A' - T^{-1}RB' \tag{60a}$$
$$B = RT^{-1}A' + (T - RT^{-1}R)B' \tag{60b}$$

or, in partitioned matrix form,

$$\begin{bmatrix} A \\ B \end{bmatrix} = \begin{bmatrix} T^{-1} & -T^{-1}R \\ RT^{-1} & T - RT^{-1}R \end{bmatrix} \begin{bmatrix} A' \\ B' \end{bmatrix} \tag{61}$$

where $T^{-1}$ is the inverse of the matrix $T$. The square matrix in (61) is the wave-amplitude transfer matrix for a single diaphragm.

A convenient set of equivalent transmission-line voltages and currents may be defined as follows:

$$V = A + B \qquad V' = A' + B'$$
$$I = A - B \qquad I' = A' - B'$$

where $V$, $V'$, $I$, and $I'$ are column matrices with elements $v_n$, $v'_n$, $i_n$, and $i'_n$; $n = 1, 2, \ldots, N$. From (60), the transfer matrix for the voltage and current matrices is readily found and is

$$\begin{bmatrix} V \\ I \end{bmatrix} = \begin{bmatrix} U & 0 \\ 2T^{-1} - 2U & U \end{bmatrix} \begin{bmatrix} V' \\ I' \end{bmatrix} \tag{62}$$

where $U$ is the unit matrix and the relation $U + R = T$ has been used to simplify the end result. The upper right-hand-side element in (62) written as $\mathbf{0}$ is an $N \times N$ matrix whose every element is zero. It will be convenient to deal with the voltage and current transfer matrix, instead of the wave-amplitude transfer matrix, because of its simpler form.

The mode amplitudes $c_n$, $d_n$, $c'_n$, $d'_n$ at the terminal planes are related to the mode amplitudes at the diaphragm as follows: $c_n = a_n e^{\Gamma_n l/2}$, $d_n = b_n e^{-\Gamma_n l/2}$, $c'_n = a'_n e^{-\Gamma_n l/2}$, $d'_n = b'_n e^{\Gamma_n l/2}$. The terminal voltages and currents will be denoted by $V_0$, $I_0$, $V'_0$, $I'_0$ and are related to $V$, $I$, $V'$, and $I'$ as follows:

$$V_0 = \begin{bmatrix} c_1 + d_1 \\ c_2 + d_2 \\ \cdots\cdots \\ c_N + d_N \end{bmatrix} = \begin{bmatrix} \cosh\frac{\Gamma_1 l}{2} & 0 & \cdots & 0 \\ 0 & \cosh\frac{\Gamma_2 l}{2} & \cdots & 0 \\ \cdots & \cdots & \cdots & \cdots \\ 0 & 0 & \cdots & \cosh\frac{\Gamma_N l}{2} \end{bmatrix} V + \begin{bmatrix} \sinh\frac{\Gamma_1 l}{2} & 0 & \cdots & 0 \\ 0 & \sinh\frac{\Gamma_2 l}{2} & \cdots & 0 \\ \cdots & \cdots & \cdots & \cdots \\ 0 & 0 & \cdots & \sinh\frac{\Gamma_N l}{2} \end{bmatrix} I \qquad (63a)$$

$$I_0 = U\sinh\frac{\Gamma l}{2}V + U\cosh\frac{\Gamma l}{2}I \qquad (63b)$$

$$V' = U\cosh\frac{\Gamma l}{2}V'_0 + U\sinh\frac{\Gamma l}{2}I'_0 \qquad (63c)$$

$$I' = U\sinh\frac{\Gamma l}{2}V'_0 + U\cosh\frac{\Gamma l}{2}I'_0 \qquad (63d)$$

where $U \cosh (\Gamma l/2)$ and $U \sinh (\Gamma l/2)$ are used to represent diagonal matrices with elements $\cosh (\Gamma_n l/2)$, $\sinh (\Gamma_n l/2)$, $n = 1, 2, \ldots, N$, as in (63$a$).

In place of the system of equations (62), we now have

$$\begin{bmatrix} V_0 \\ I_0 \end{bmatrix} = \begin{bmatrix} U\cosh\frac{\Gamma l}{2} & U\sinh\frac{\Gamma l}{2} \\ U\sinh\frac{\Gamma l}{2} & U\cosh\frac{\Gamma l}{2} \end{bmatrix} \begin{bmatrix} U & 0 \\ 2T^{-1} - 2U & U \end{bmatrix} \times \begin{bmatrix} U\cosh\frac{\Gamma l}{2} & U\sinh\frac{\Gamma l}{2} \\ U\sinh\frac{\Gamma l}{2} & U\cosh\frac{\Gamma l}{2} \end{bmatrix} \begin{bmatrix} V'_0 \\ I'_0 \end{bmatrix} \qquad (64)$$

For a propagating periodic wave we must have

$$V_0' = Ue^{-\gamma l}V_0 \qquad I_0' = Ue^{-\gamma l}I_0$$

where $\gamma$ is the characteristic propagation constant. Substituting into (64) yields the following matrix eigenvalue system:

$$\left| \begin{bmatrix} U\cosh\frac{\Gamma l}{2} & U\sinh\frac{\Gamma l}{2} \\ U\sinh\frac{\Gamma l}{2} & U\cosh\frac{\Gamma l}{2} \end{bmatrix} \begin{bmatrix} U & 0 \\ 2T^{-1} - 2U & U \end{bmatrix} \times \begin{bmatrix} U\cosh\frac{\Gamma l}{2} & U\sinh\frac{\Gamma l}{2} \\ U\sinh\frac{\Gamma l}{2} & U\cosh\frac{\Gamma l}{2} \end{bmatrix} - \begin{bmatrix} Ue^{\gamma l} & 0 \\ 0 & Ue^{\gamma l} \end{bmatrix} \right| = 0 \qquad (65)$$

where the bars signify the determinant of the matrix. The solution for $V$ and $I$ in terms of $V_0$ and $I_0$ in (63$a$) and (63$b$) is obtained by changing $l$ to $-l$ and interchanging $V$, $V_0$ and $I$, $I_0$. The end result shows at once that

$$\begin{bmatrix} U\cosh\frac{\Gamma l}{2} & U\sinh\frac{\Gamma l}{2} \\ U\sinh\frac{\Gamma l}{2} & U\cosh\frac{\Gamma l}{2} \end{bmatrix}^{-1} = \begin{bmatrix} U\cosh\frac{\Gamma l}{2} & -U\sinh\frac{\Gamma l}{2} \\ -U\sinh\frac{\Gamma l}{2} & U\cosh\frac{\Gamma l}{2} \end{bmatrix}$$

The voltage-current transfer matrix for a system of transmission lines of length $l$ is of the same form as those occurring in (63), with $l/2$ replaced by $l$. This transfer matrix is also given by the square of the transfer matrix for lines of length $l/2$, and, hence,

$$\begin{bmatrix} U\cosh\frac{\Gamma l}{2} & U\sinh\frac{\Gamma l}{2} \\ U\sinh\frac{\Gamma l}{2} & U\cosh\frac{\Gamma l}{2} \end{bmatrix}^{2} = \begin{bmatrix} U\cosh\Gamma l & U\sinh\Gamma l \\ U\sinh\Gamma l & U\cosh\Gamma l \end{bmatrix}$$

By premultiplying and postmultiplying (65) by the inverse of the hyperbolic sine and cosine matrix and using the above results, (65) may be converted to

$$\left| \begin{bmatrix} Ue^{-\gamma l} & 0 \\ (2T^{-1} - 2U)e^{-\gamma l} & Ue^{-\gamma l} \end{bmatrix} - \begin{bmatrix} U\cosh\Gamma l & -U\sinh\Gamma l \\ -U\sinh\Gamma l & U\cosh\Gamma l \end{bmatrix} \right| = 0 \qquad (66)$$

This result could have been derived directly by evaluating the transfer matrix from the position of one diaphragm to the same position of the next diaphragm.

If $\gamma$ is an eigenvalue of (66), then $-\gamma$ is also an eigenvalue. This result may be formally proved as follows. Let $V$, $I$ be an eigenvector of the matrix occurring in Eq. (66). Thus $V$, $I$ satisfy the equations

$$(Ue^{-\gamma l} - U \cosh \Gamma l)V + U \sinh \Gamma l \, I = 0 \quad (67a)$$
$$[(2T^{-1} - 2U)e^{-\gamma l} + U \sinh \Gamma l]V + (Ue^{-\gamma l} - U \cosh \Gamma l)I = 0 \quad (67b)$$

Premultiply (67*b*) by $e^{\gamma l}U \sinh \Gamma l$ to obtain

$$(U \sinh \Gamma l)(2T^{-1} - 2U + e^{\gamma l}U \sinh \Gamma l)V + (U - e^{\gamma l}U \cosh \Gamma l)U \sinh \Gamma l \, I = 0$$

since diagonal matrices commute. Substituting for $I$ from (67*a*) now gives

$$[2U \sinh \Gamma l \, T^{-1} - 2U \sinh \Gamma l + e^{\gamma l}(U \sinh \Gamma l)^2 + 2U \cosh \Gamma l - e^{-\gamma l}U - e^{\gamma l}(U \cosh \Gamma l)^2]V = 0$$

or since $(U \cosh \Gamma l)^2 - (U \sinh \Gamma l)^2 = U$, we get

$$(U \sinh \Gamma l \, T^{-1} - U \sinh \Gamma l + U \cosh \Gamma l - U \cosh \gamma l)V = 0 \quad (68)$$

Hence $\cosh \gamma l$ is an eigenvalue of the $N \times N$ square matrix

$$[U \sinh \Gamma l \, T^{-1} - U \sinh \Gamma l + U \cosh \Gamma l]$$

Since $\cosh \gamma l$ is an even function, both $\pm\gamma$ are eigenvalues.

The form of solution establishes the following two important properties of the periodic-loaded guide:

1. There exist $2N$ characteristic propagation constants $\pm\gamma_m$, $m = 1, 2, \ldots, N$, in the periodic-loaded guide with $N$ interacting modes.

2. For each eigenvalue $\gamma_m$, a separate periodic wave composed of $N$ normal waveguide modes exists.

In general, only a few of the propagation constants $\gamma_m$ are imaginary, and these are associated with propagating modes. The modes with real propagation constants are evanescent modes. These modes are excited at the input region between a uniform guide and a periodic-loaded guide and are analogous to the evanescent modes which occur in a normal uniform guide at a discontinuity. For each root $\gamma_m$ a voltage-current eigenvector $V_m$, $I_m$ may be found to within an arbitrary amplitude constant. The elements or components $v_{nm}$ and $i_{nm}$, $n = 1, 2, \ldots, N$, of $V_m$, $I_m$ completely specify the field of the $m$th periodic wave in the loaded guide.

Let $c_{nm}$, $d_{nm}$ be the mode amplitudes corresponding to the eigenvector $V_m$, $I_m$ with eigenvalue $\gamma_m$. For a propagating normal waveguide mode

$\mathbf{e}_n \times \mathbf{h}_n^* \cdot \mathbf{a}_z$ is real, while for an evanescent mode it is imaginary. The normal modes may be normalized so that

$$\frac{1}{2} \iint_S \mathbf{e}_n \times \mathbf{h}_n^* \cdot \mathbf{a}_z \, da = \begin{cases} 1 & \text{propagating mode} \\ j & \text{evanescent mode} \end{cases}$$

where the integration is over the guide cross section. Since the normal modes are orthogonal, the power flow associated with the $n$th periodic wave becomes

$$\operatorname{Re} \frac{1}{2} \sum_{n=1}^{N} (c_{nm} + d_{nm})(c_{nm}^* + d_{nm}^*) \iint_S \mathbf{e}_n \times \mathbf{h}_n^* \cdot \mathbf{a}_z \, da$$
$$= \sum_i (c_{im}c_{im}^* - d_{im}d_{im}^*) + 2 \operatorname{Im} \sum_j c_{jm}d_{jm}^* \quad (69)$$

where the first summation extends over the propagating waveguide modes, and the second summation extends over the nonpropagating modes. Real power flow occurs for evanescent waveguide modes through the interaction of the fields of the forward- and backward-attenuated modes. For a nonpropagating periodic wave the real power flow is zero.

The above theory will be applied to the capacitive-loaded rectangular guide of width $a$ and height $b$ as in Fig. 9.10. The solution for the reflection and transmission coefficient matrices may be obtained by the methods presented in Chap. 8. For simplicity only two interacting modes will be taken into account.

With a dominant $H_{10}$ incident mode, the higher-order modes excited are longitudinal-section electric modes. These modes may be derived from an $x$-directed magnetic-type Hertzian potential by means of the following equations:

$$\mathbf{E} = -j\omega\mu_0 \nabla \times \mathbf{\Pi}_h$$
$$\mathbf{H} = k_0^2 \mathbf{\Pi}_h + \nabla\nabla \cdot \mathbf{\Pi}_h$$

For the $n$th mode, a suitable form for $\Pi_h$ is $\psi_n(y)\phi(x)e^{\pm\Gamma_n z}$, where

$$\psi_n = \cos \frac{n\pi y}{b} \qquad \phi = \sin \frac{\pi x}{a} \qquad \Gamma_n^2 = \left(\frac{n\pi}{b}\right)^2 - k^2$$

and $k^2 = k_0^2 - (\pi/a)^2$. A suitable expansion for the transverse fields $E_y$ and $H_x$ is readily found to be

$$E_y = \begin{cases} a_0\phi e^{-\Gamma_0 z} + a_1\psi_1\phi e^{-\Gamma_1 z} + \sum_{n=0}^{\infty} b_n\psi_n\phi e^{\Gamma_n z} & z \le 0 \\ \sum_{n=0}^{\infty} c_n\psi_n\phi e^{-\Gamma_n z} & z \ge 0 \end{cases}$$

$$H_x = \begin{cases} -jY_0\dfrac{k^2}{k_0}\left(\dfrac{a_0\phi}{\Gamma_0}e^{-\Gamma_z} + \dfrac{a_1\psi_1\phi}{\Gamma_1}e^{-\Gamma_1 z} - \sum_{n=0}^{\infty}\dfrac{b_n\psi_n\phi}{\Gamma_n}e^{\Gamma_n z}\right) & z \le 0 \\ -jY_0\dfrac{k^2}{k_0}\sum_{n=0}^{\infty}\dfrac{c_n\psi_n\phi}{\Gamma_n}e^{-\Gamma_n z} & z \ge 0 \end{cases}$$

where $a_0$ and $a_1$ are the amplitudes of the incident modes. Let the aperture electric field be $\mathcal{E}(y)\phi(x)$ for $d < y \le b$, and zero for $0 \le y < d$. By Fourier analysis, we obtain

$$a_0 + b_0 = c_0 = \frac{1}{b}\int_d^b \mathcal{E}(y')\,dy' \tag{70a}$$

$$a_1 + b_1 = c_1 = \frac{2}{b}\int_d^b \mathcal{E}(y')\psi_1(y')\,dy' \tag{70b}$$

$$b_n = c_n = \frac{2}{b}\int_d^b \mathcal{E}(y')\psi_n(y')\,dy' \tag{70c}$$

For $b$ sufficiently small, $\Gamma_n$ may be approximated by $n\pi/b$ for $n > 0$. This approximation will be made throughout in order to simplify the analysis. The continuity condition on $H_x$ in the aperture leads to

$$\frac{a_0}{\Gamma_0} + \frac{a_1 b\psi_1}{\pi} = \frac{1}{\Gamma_0 b}\int_d^b \mathcal{E}(y')\,dy' + \frac{2}{b}\int_d^b \mathcal{E}(y')\sum_{n=1}^{\infty}\frac{1}{n}\psi_n(y)\psi_n(y')\,dy' \qquad d \le y \le b \tag{71}$$

The series is summable and yields

$$\sum_{n=1}^{\infty}\frac{1}{n}\cos\frac{n\pi y}{b}\cos\frac{n\pi y'}{b} = -\tfrac{1}{2}\ln\left|\cos\frac{\pi y}{b} - \cos\frac{\pi y'}{b}\right|$$

We now introduce a new variable $\theta$ by means of the relation

$$\alpha_1 + \alpha_2\cos\theta = \cos\frac{\pi y}{b}$$

with $\alpha_1$ and $\alpha_2$ chosen so that $y = d, b$ corresponds to $\theta = 0, \pi$. The solution for $\alpha_1$ and $\alpha_2$ is

$$\alpha_1 = -\sin^2\frac{\pi d}{2b} = \alpha_2 - 1$$

$$\alpha_2 = \cos^2\frac{\pi d}{2b}$$

A similar change in the variable $\theta'$ is also made, and thus $\mathcal{E}(y')\,dy'$ is to be replaced by $\mathfrak{F}(\theta')\,d\theta'$, where $\mathfrak{F}(\theta')$ is the aperture field as a function of $\theta'$. The integral equation (71) becomes

$$\frac{a_0}{\Gamma_0} + \frac{a_1 b}{1}(\alpha_1 + \alpha_2\cos\theta) = \left(\frac{1}{\Gamma_0 b} - \frac{\ln\alpha_2}{\pi}\right)\int_0^\pi \mathfrak{F}\,d\theta' + \frac{2}{\pi}\sum_{n=1}^{\infty}\frac{\cos n\theta}{n}\int_0^\pi \mathfrak{F}\cos n\theta'\,d\theta' \qquad (72)$$

after substituting for $\cos(\pi y/b)$ and $\cos(\pi y'/b)$ in the sum of the original series and expanding the result into a series in the new variables $\theta$, $\theta'$.

A solution to the integral equation is obtained if we choose

$$\mathfrak{F}(\theta') = A_0 + A_1\cos\theta'$$

where $A_0$ and $A_1$ are constants to be determined. Substituting into (72),

$$\frac{a_0}{\Gamma_0} + \frac{a_1 b}{\pi}(\alpha_1 + \alpha_2\cos\theta) = \frac{\pi A_0}{\Gamma_0 b}\left(1 - \frac{\Gamma_0 b}{\pi}\ln\alpha_2\right) + A_1\cos\theta$$

Equating the coefficients of $\cos\theta$ and the constant terms gives

$$a_0 + a_1\frac{b\alpha_1\Gamma_0}{\pi} = \frac{\pi A_0}{b}\left(1 - \frac{\Gamma_0 b}{\pi}\ln\alpha_2\right) \qquad (73a)$$

$$a_1\alpha_2 = \frac{\pi}{b}A_1 \qquad (73b)$$

Replacing $y$, $y'$ by the new variables $\theta$, $\theta'$ in (70) shows that

$$a_0 + b_0 = c_0 = t_{00}a_0 + t_{01}a_1 = \frac{\pi}{b}A_0$$

$$a_1 + b_1 = c_1 = t_{10}a_0 + t_{11}a_1 = \frac{\pi}{b}A_1\alpha_2 + \frac{2\pi}{b}A_0\alpha_1$$

In matrix form these relations may be written as

$$\begin{bmatrix} t_{00} & t_{01} \\ t_{10} & t_{11} \end{bmatrix}\begin{bmatrix} a_0 \\ a_1 \end{bmatrix} = \begin{bmatrix} \dfrac{\pi}{b}A_0 \\ \dfrac{\pi}{b}A_1\alpha_2 + \dfrac{2\pi}{b}A_0\alpha_1 \end{bmatrix}$$

$$= \frac{1}{1 - \dfrac{\Gamma_0 b}{\pi}\ln\alpha_2}\begin{bmatrix} 1 & b\alpha_1\dfrac{\Gamma_0}{\pi} \\ 2\alpha_1 & \alpha_2^2\left(1 - \dfrac{\Gamma_0 b}{\pi}\ln\alpha_2\right) + 2\alpha_1^2\dfrac{b\Gamma_0}{\pi} \end{bmatrix}\begin{bmatrix} a_0 \\ a_1 \end{bmatrix} \qquad (74)$$

after substituting for $A_0$ and $A_1$ from (73). From (74) the transmission coefficients $t_{ij}$ are readily identified. We get

$$t_{00} = \left(1 - \frac{\Gamma_0 b}{\pi} \ln \alpha_2\right)^{-1} = \left(1 + \frac{jB}{2}\right)^{-1}$$

where $B$ is the normalized shunt susceptance of the diaphragm as given by (47).

The inverse of the matrix $T$ is readily found and is

$$T^{-1} = (t_{00}t_{11} - t_{01}t_{10})^{-1} \begin{bmatrix} t_{11} & -t_{01} \\ -t_{10} & t_{00} \end{bmatrix}$$
$$= \begin{bmatrix} 1 - \frac{\Gamma_0 b}{\pi} \ln \alpha_2 + \frac{2\alpha_1^2}{\pi \alpha_2^2} \Gamma_0 b & -b\alpha_1 \frac{\Gamma_0}{\pi \alpha_2^2} \\ -2\frac{\alpha_1}{\alpha_2^2} & \alpha_2^{-2} \end{bmatrix} \tag{75}$$

The eigenvalue equation for the characteristic propagation constants may now be found by substituting into (68). After simplification, we obtain†

$$\left[\cos \beta_0 l + \sin \beta_0 l \left(\frac{\beta_0 b}{\pi} \ln \alpha_2 - \frac{2\alpha_1^2 \beta_0 b}{\pi \alpha_2^2}\right) - \cosh \gamma l\right]\left(\alpha_2^{-2} \sinh \frac{\pi l}{b} + e^{-\pi l/b} - \cosh \gamma l\right) + \frac{2\alpha_1^2 \beta_0 b}{\pi \alpha_2^4} \sin \beta_0 l \sinh \frac{\pi l}{b} = 0 \tag{76}$$

where $\beta_0 = |\Gamma_0|$. For determining the eigenvalue equation, it was not necessary to normalize the waveguide modes. For (69) to hold for the power-flow normalization of the waveguide modes, in the manner indicated in the discussion preceding (69), is necessary. As an example, the roots of Eq. (76) have been computed for $a = 0.9$ inch, $b = 0.2$ inch, $d = 0.15$ inch, $k_0 = 2$ radians per centimeter, and $l = 0.2$ centimeter. The results are $\gamma_0 = j\beta_c = j2.25$, $\gamma_1 = 24.4$. From Eq. (49*b*), which is based on single-mode interaction, we obtain $\gamma_0 = j2.95$. With such close spacing, it is quite apparent that the result based on single-mode interaction is considerably in error. Also it should be noted that the first evanescent periodic wave is very highly attenuated.

For small values of $l$, an alternative method of solution is available. This solution is based on a transverse resonance approach and is discussed in the next chapter. The eigenvalue equation obtained is

$$(\beta_c^2 - k^2)^{1/2} b \tanh (\beta_c^2 - k^2)^{1/2} \left(b - d + \frac{l}{\pi} \ln 2\right) = k_0 b \tan k_0 \left(d - \frac{l}{\pi} \ln 2\right) \tag{77}$$

† A similar result has been derived by Brown for a capacitive-loaded parallel-plate transmission line. See Ph.D. thesis, University of London, *op. cit.*

where $k^2 = k_0^2 - (\pi/a)^2 = \beta_0^2$. This equation gives a value of 2.18 for $\beta_c$ for the above example. This result compares favorably with that obtained from (76). For really close spacing, (77) is more accurate than (76). In practice, the combination of (49*b*) for large spacing and (77) for small spacing provides a sufficiently accurate solution for all values of $l$, and, hence, the more complicated equation (76) is not required in order to obtain good results for $\gamma_0$. However, it is required in order to obtain a solution for $\gamma_1$ and the fields for the first evanescent mode.

### 9.8. The Sheath Helix

The helix is an important structure with periodic properties and is widely used in such diverse applications as traveling-wave tubes, antennas, and delay lines. The basic property of the helix which makes it an

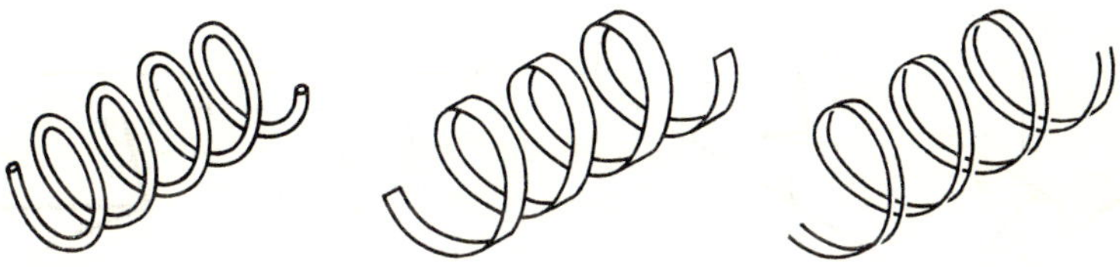

(*a*) (*b*) (*c*)

FIG. 9.11. Three typical types of helices. (*a*) Round wire helix; (*b*) tape helix; (*c*) bifilar (two-wire) helix.

essential component in the above devices is the relatively large reduction in the phase velocity of an electromagnetic wave propagating along it. To a first approximation, the reduction in phase velocity as compared with the velocity of light is by a factor equal to the pitch of the winding divided by the circumference of one turn. Figure 9.11 illustrates several typical forms of helices.

The Helmholtz equation is not separable in helical coordinates, and, consequently, a rigorous solution for electromagnetic-wave propagation along a helix has so far not been attained. Two approximate models which are amenable to solution have received considerable study. The simplest model replaces the actual helix by a sheath helix as in Fig. 9.12. The sheath helix is a cylindrical tube which is assumed to have infinite conductivity in the direction of the original winding and zero conductivity in a direction normal to the turns of the winding. It is essentially a cylindrical tube with anisotropic conductivity properties. If the actual helix consists of many turns in a length equal to a wavelength, the sheath helix will represent a good approximation. The analysis of the sheath helix has been given by Pierce.†

† J. R. Pierce, "Travelling Wave Tubes," D. Van Nostrand Company, Inc., Princeton, N.J., 1950.

The second model of a helix is an infinitely thin tape helix. The analysis of this structure has been presented by Sensiper.† The essential steps are to write a general expansion of the field in cylindrical coordinates in the regions inside and outside the helix, and then to determine the amplitude coefficients in terms of an assumed electric field in the gap, or in terms of an assumed current on the tape. The analysis becomes rather involved, and so we will consider only the sheath-helix model in detail. This does not, of course, give a complete picture of the properties of a helix in general. In particular, no information on the passband-stopband characteristics are obtained.

The developed surface of the sheath helix is illustrated in Fig. 9.12. The pitch angle is $\alpha$, and the windings of the actual helix are parallel to

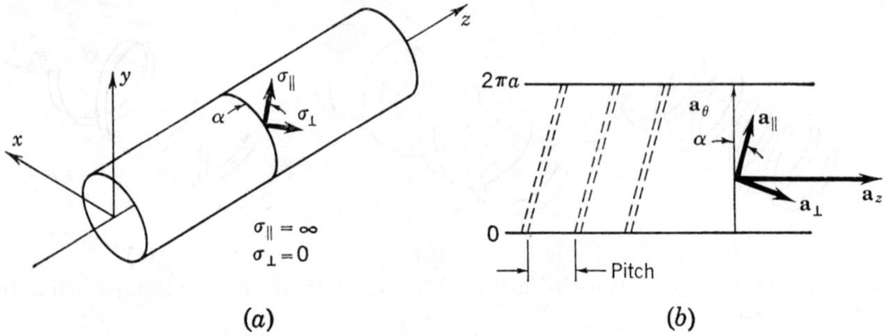


FIG. 9.12. Coordinates of the surface of a sheath helix.

the unit vector $\mathbf{a}_\parallel$. The radius of the sheath-helix tube is $a$. Let $\mathbf{E}_\parallel{}^i$, $\mathbf{E}_\parallel{}^e$ and $\mathbf{E}_\perp{}^i$, $\mathbf{E}_\perp{}^e$ be the tangential components of the electric field which, at the surface of the cylinder, are parallel to the unit vectors $\mathbf{a}_\parallel$ and $\mathbf{a}_\perp$, respectively. The superscripts $i$ and $e$ are used to distinguish between the field interior to the boundary $r = a$ and that exterior to this boundary. The boundary conditions require continuity of the tangential electric field at $r = a$ and, in particular, the vanishing of the components $\mathbf{E}_\parallel{}^i$, $\mathbf{E}_\parallel{}^e$, which are in the direction of infinite conductivity. Thus we have, at $r = a$,

$$\mathbf{E}_\parallel{}^i = \mathbf{E}_\parallel{}^e = 0 \qquad \mathbf{E}_\perp{}^i = \mathbf{E}_\perp{}^e$$

or, equivalently,

$$E_z{}^i = E_z{}^e \tag{78a}$$

$$E_\theta{}^i = E_\theta{}^e \tag{78b}$$

$$E_z{}^{i,e} = -E_\theta{}^{i,e} \cot \alpha \tag{78c}$$

The latter equation is merely the relation imposed on $E_z$ and $E_\theta$ in order that $E_\parallel = 0$. The boundary conditions on the tangential magnetic field components are

$$H_\parallel{}^i = H_\parallel{}^e$$

† S. Sensiper, Electromagnetic Wave Propagation on Helical Conductors, *MIT Research Lab. Electronics Tech. Rept.* 194, May 16, 1951.

or, equivalently,

$$H_z^i + H_\theta^i \cot\alpha = H_z^e + H_\theta^e \cot\alpha \tag{78d}$$

The component $H_\perp$ is discontinuous across the sheath by an amount equal to the current flowing on the sheath in the direction $\mathbf{a}_\parallel$. The nature of the above boundary conditions indicates that the solution for any particular mode will not be a TM or TE mode, but a combination of both. Hence, all six field components $E_r$, $E_\theta$, $E_z$, $H_r$, $H_\theta$, and $H_z$ will be present.

The general expansions for the TE and TM modes are similar to those for the circular guide. The $\theta$ and $z$ dependence may be assumed to be according to $e^{-jn\theta-\gamma z}$, where $n$ is an integer, and $\gamma$ is the propagation constant. If slow waves exist, they will have a propagation constant $\gamma = j\beta$, whose modulus $\beta$ is greater than $k_0$. In anticipation of this type of solution, the appropriate Bessel functions to use in the region $r < a$ are the modified functions of the first kind $I_n(hr)$ rather than the functions $J_n(hr)$, which occur in the solution for the circular guide. Similarly, for the external region $r > a$, the solutions which decay exponentially with $r$ are the modified Bessel functions of the second kind $K_n(hr)$. The radial wave number $h$ is given by $h = (\beta^2 - k_0^2)^{1/2}$, where $j\beta = \gamma$. If the TM field is expressed in terms of $E_z$, and the TE field in terms of $H_z$, the appropriate expansion of the field in the two regions is found to be as follows:

For $r \le a$

$$\begin{aligned}
E_z^i &= A_n I_n(hr) e^{-jn\theta-\gamma z} \\
E_r^i &= \left(\frac{\gamma}{h} A_n I_n' + \frac{n\omega\mu_0}{h^2 r} B_n I_n\right) e^{-jn\theta-\gamma z} \\
E_\theta^i &= \left(\frac{-jn\gamma}{h^2 r} A_n I_n - \frac{j\omega\mu_0}{h} B_n I_n'\right) e^{-jn\theta-\gamma z} \\
H_z^i &= B_n I_n(hr) e^{-jn\theta-\gamma z} \\
H_r^i &= \left(\frac{-n\omega\epsilon_0}{h^2 r} A_n I_n + \frac{\gamma}{h} B_n I_n'\right) e^{-jn\theta-\gamma z} \\
H_\theta^i &= \left(\frac{j\omega\epsilon_0}{h} A_n I_n' - \frac{jn\gamma}{h^2 r} B_n I_n\right) e^{-jn\theta-\gamma z}
\end{aligned}$$

For $r \ge a$

$$\begin{aligned}
E_z^e &= C_n K_n(hr) e^{-jn\theta-\gamma z} \\
E_r^e &= \left(\frac{\gamma}{h} C_n K_n' + \frac{n\omega\mu_0}{h^2 r} D_n K_n\right) e^{-jn\theta-\gamma z} \\
E_\theta^e &= \left(\frac{-jn\gamma}{h^2 r} C_n K_n - \frac{j\omega\mu_0}{h} D_n K_n'\right) e^{-jn\theta-\gamma z} \\
H_z^e &= D_n K_n(hr) e^{-jn\theta-\gamma z} \\
H_r^e &= \left(\frac{-n\omega\epsilon_0}{h^2 r} C_n K_n + \frac{\gamma}{h} D_n K_n'\right) e^{-jn\theta-\gamma z} \\
H_\theta^e &= \left(\frac{j\omega\epsilon_0}{h} C_n K_n' - \frac{jn\gamma}{h^2 r} D_n K_n\right) e^{-jn\theta-\gamma z}
\end{aligned}$$

where $A_n$, $B_n$, $C_n$, and $D_n$ are unknown amplitude constants, and the prime means differentiation of the Bessel functions with respect to the argument $hr$.

From the condition $E_z = -E_\theta \cot \alpha$, the solutions for $B_n$ and $D_n$ in terms of $A_n$ and $C_n$ are found to be

$$B_n(j\omega\mu_0 ha \cot \alpha)I'_n = A_n(h^2a - jn\gamma \cot \alpha)I_n$$
$$D_n(j\omega\mu_0 ha \cot \alpha)K'_n = C_n(h^2a - jn\gamma \cot \alpha)K_n$$

where the Bessel functions are evaluated at $r = a$. To facilitate matching the boundary conditions at $r = a$, the following two impedance functions are formed:

$$Z_i = \frac{E_\perp{}^i}{H_\parallel{}^i} = \frac{E_z{}^i \cos \alpha - E_\theta{}^i \sin \alpha}{H_z{}^i \sin \alpha + H_\theta{}^i \cos \alpha}$$
$$= \frac{E_z{}^i \sec \alpha}{H_z{}^i \sin \alpha + H_\theta{}^i \cos \alpha}$$
$$Z_e = \frac{E_\perp{}^e}{H_\parallel{}^e} = \frac{E_z{}^e \sec \alpha}{H_z{}^e \sin \alpha + H_\theta{}^e \cos \alpha}$$

The proper continuity of the fields is ensured if $Z_i = Z_e$ at $r = a$. Substituting for the field components, and simplifying the resultant expression yields the following eigenvalue equation for $\beta$ and $h$:

$$\frac{K'_n(ha)I'_n(ha)}{K_n(ha)I_n(ha)} = -\frac{(h^2a^2 + n\beta a \cot \alpha)^2}{k_0{}^2a^2h^2a^2 \cot^2 \alpha} \tag{79}$$

where $\beta^2 = h^2 + k_0{}^2$.

For the dominant mode, $n = 0$, and the eigenvalue equation reduces to

$$\frac{K'_0(ha)I'_0(ha)}{K_0(ha)I_0(ha)} = -\frac{h^2a^2}{k_0{}^2a^2 \cot^2 \alpha} \tag{80}$$

A plot of the ratio $k_0a/\beta a$ as a function of $k_0a$ is given in Fig. 9.13 for $n = 0$ and the following values of $\alpha$: 10, 15, and 20°. For $k_0a$ greater than unity, the propagation phase constant is seen to be approximately equal to $k_0 \sin \alpha$, and, hence, the reduction in phase velocity is by a factor $\sin \alpha$. The group velocity is given by $v_c(d\beta/dk_0)^{-1}$. In the region $k_0a > 1$, where $\beta = k_0 \sin \alpha$, the group velocity $v_g$ is given by $v_c \sin \alpha$ and is equal to the phase velocity $v_p$.

For $n \neq 0$, the solution for $h$ and $\beta$ is somewhat more difficult to obtain. In general, imaginary values of $h$ are not allowed, since this would lead to solutions which do not have exponential decay in the radial direction for $r > a$. Consequently $\beta$ is restricted to be less than $k_0$ for all modes. For any given value of $k_0a$, a solution for $\beta a$ can be found only for certain values of $n$. Hence, only a finite number of modes of propagation exist at any one frequency.† This property is true, in general, for

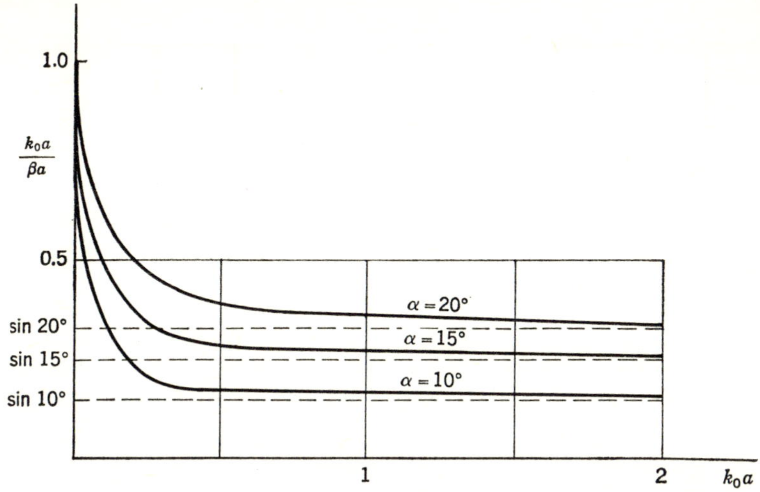


FIG. 9.13. Plot of $k_0a/\beta a$ as a function of $k_0a$ for a sheath helix.

all open-boundary waveguides, and will be examined in greater detail in Chap. 11.

## REFERENCES

9.1. L. Brillouin: "Wave Propagation in Periodic Structures," 2d ed., Dover Publications, New York, 1953.

9.2. J. C. Slater: "Microwave Electronics," D. Van Nostrand Company, Inc., Princeton, N.J., 1950.

9.3. E. L. Chu and W. W. Hansen: Disk Loaded Waveguides, *J. Appl. Phys.*, vol. 18, pp. 996–1008, 1947; vol. 20, pp. 280–285, 1949.

9.4. J. C. Slater: The Design of Linear Accelerators, *Revs. Modern Phys.*, vol. 20, pp. 473–518, July, 1948.

9.5. L. Brillouin: Wave Guides for Slow Waves, *J. Appl. Phys.*, vol. 19, pp. 1023–1041, November, 1948.

9.6. A. W. Lines, G. R. Nicoll, and A. M. Woodward: Some Properties of Waveguides with Periodic Structure, *Proc. IEE (London)*, vol. 97, part III, pp. 263–276, July, 1950.

9.7. A. H. W. Beck: "Space Charge Waves," Pergamon Press, Inc., New York, 1958. (This book has a good discussion of slow-wave structures for use in traveling-wave tubes.)

## PROBLEMS

**9.1.** A rectangular guide is loaded at regular intervals by symmetrical inductive diaphragms. Derive the eigenvalue equation for the characteristic propagation

† D. A. Watkins, "Topics in Electromagnetic Theory," John Wiley & Sons, Inc., New York, 1958.

constant for a periodic wave propagating along the loaded guide. Plot the $k_0$-$\beta_c$ diagram for $a = 0.9$ inch, $d = 0.2$ inch, $l = 1.5$ centimeters. Use the approximate formula for $B$ given in Chap. 8.

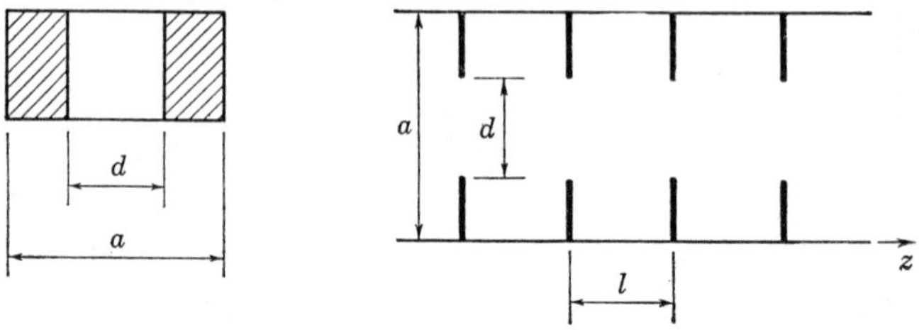


FIG. P 9.1

**9.2.** A rectangular guide is loaded by dielectric blocks of thickness $t$ and spacing $l$ and having a relative dielectric constant $\kappa$. Derive the eigenvalue equation for the propagation constant of a periodic wave. Also find the required parameters of a

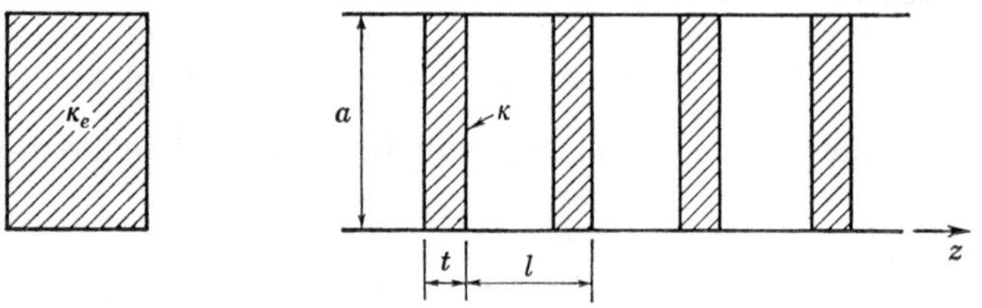


FIG. P 9.2

matching section which will match the empty guide to the periodic-loaded guide. Show that, for small spacings, the loaded guide behaves the same as a guide filled with a homogeneous dielectric material having a relative dielectric constant $\kappa_e = 1 + (\kappa - 1)t/l$. Assume $H_{10}$ mode propagation.

**9.3.** A rectangular guide is loaded by diaphragms at regular intervals $l$, each diaphragm containing a small centered circular aperture of radius $r_0$. Derive the

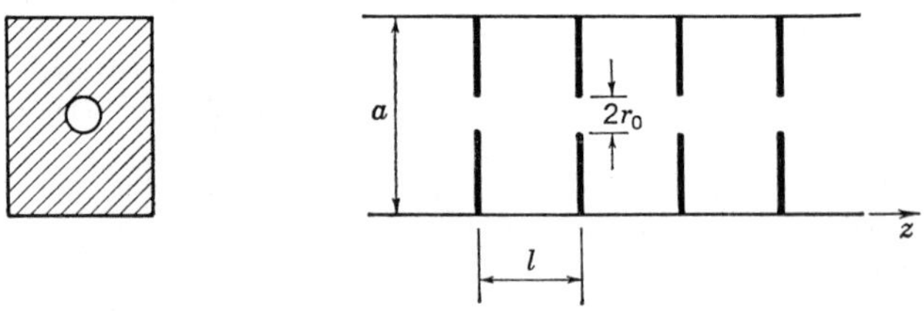


FIG. P 9.3

eigenvalue equation for the propagation constant of a periodic wave. Use the small-aperture theory of Chap. 7 to obtain an expression for the shunt susceptance of each diaphragm. For $a = 0.9$ inch, $\lambda_0 = 3.14$ centimeters, $r_0 = 0.2$ centimeter, find the spacing $l$ corresponding to the center of the passband. Plot the value of the propagation constant as a function of $k_0$ for values of $k_0$ extending through several stopbands and passbands. Note that the passbands are very narrow and separated by stopbands with relatively high attenuation. The periodic-loaded guide has the properties of a narrow band filter.

*Answer.* The center of the passband is given by $\beta_0 l = \pi/2 + \tan^{-1}(B/2)$, where $\beta_0$ is the propagation constant of the $H_{10}$ mode.

**9.4.** A structure with properties similar to those of the tape helix is the parallel-plate transmission line, which has infinite conductivity in the direction of the unit vector $\mathbf{a}_{\|}$ and zero conductivity in the direction perpendicular to this, as illustrated.

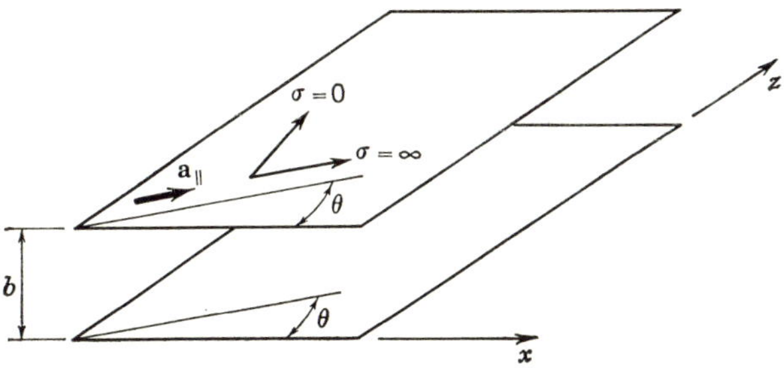


FIG. P 9.4

Obtain solutions to Maxwell's equations for a structure of this type. Note that each mode is a combination of both an $E$ and an $H$ mode. Determine the eigenvalue equation for the propagation constants. Carry through a similar analysis when $\mathbf{a}_{\|}$ makes an angle $\theta$ with the $x$ axis for the upper plate and an angle $-\theta$ for the lower plate.

**9.5.** Consider two rectangular guides propagating $H_{10}$ modes and coupled together periodically by small circular apertures of radius $r_0$, spacing $l$, and located in the center of the common side wall. Obtain the eigenvalue equation for the dominant modes of propagation which may exist in the composite guide. Show that the two normal

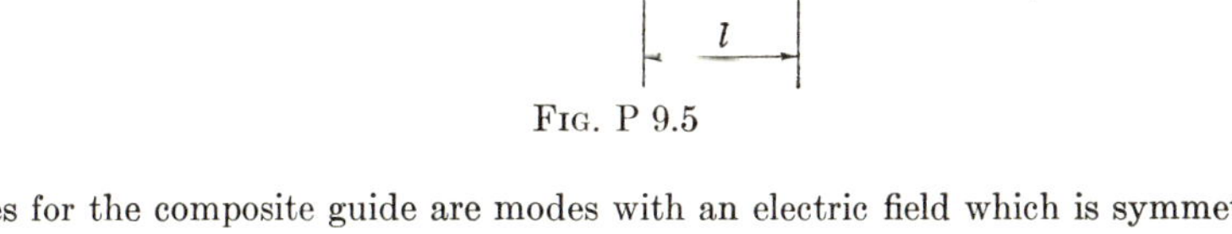


FIG. P 9.5

modes for the composite guide are modes with an electric field which is symmetrical and antisymmetrical about the common side wall. Superimpose the two normal modes such that at one transverse plane in one guide the field is zero, while in the other guide it is maximum. Note that, as this field propagates along the composite guide, the power oscillates back and forth between the two separate guides. The phenomenon is similar to that of the mechanical one of two coupled pendulums.

**9.6.** A circular guide is located by dielectric washers of thickness $t$, inner radius $a$, and spacing $l$. The dielectric constant of the material is $\kappa$. Assume that the spacing $l$

is much smaller than a wavelength, so that the loaded portion of the guide $a \le r \le b$ may be assumed to be filled with an anisotropic dielectric material with a dielectric constant $\kappa_1 = 1 + (\kappa - 1)\frac{t}{l}$ along a transverse direction, and $\kappa_2 = \left(1 - \frac{\kappa - 1}{\kappa}\frac{t}{l}\right)^{-1}$ along the $z$ direction. Obtain the solutions for circularly symmetric modes (no variation with $\theta$) in the composite structure.

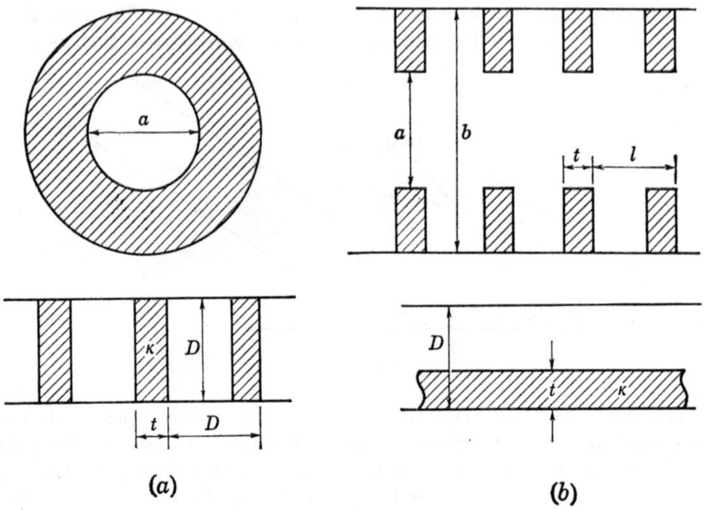


FIG. P 9.6

The equivalent dielectric constants are obtained from the static solution for the parallel-plate capacitor loaded as illustrated. Show that the capacitance per unit area is $C = \kappa_e\epsilon_0/D$, where the equivalent dielectric constant $\kappa_e$ is equal to $1 + (\kappa - 1)t/D$ for arrangement $a$ and to $\left(1 - \frac{\kappa - 1}{\kappa}\frac{t}{D}\right)^{-1}$ for arrangement $b$.

**9.7.** In Eq. (9) show that $\Delta = 1$. Begin by proving that the Wronskian determinant $W = \psi_1\psi_2' - \psi_2\psi_1'$ is equal to a constant. From Eqs. (2) the result

$$\psi_1(z + p)\psi_2'(z + p) - \psi_1'(z + p)\psi_2(z + p) = (\alpha_{11}\alpha_{22} - \alpha_{12}\alpha_{21})[\psi_1(z)\psi_2'(z) - \psi_1'(z)\psi_2(z)]$$

may be obtained. Since the Wronskian is constant, it follows that $\Delta = 1$. To prove that $W$ is a constant, use the differential equation (1) to obtain

$$\psi_2\psi_1'' - \psi_1\psi_2'' = 0$$

where the prime means differentiation with respect to $z$. By integrating once by parts the desired result is obtained.

CHAPTER 10

# INTEGRAL TRANSFORM AND FUNCTION-THEORETIC TECHNIQUES

There exists a class of two-part boundary-value problems which may be rigorously solved by means of integral transforms such as the Fourier transform or its equivalent, the two-sided Laplace transform. Among the problems which fall into this class are:

1. Radiation of acoustic or electromagnetic waves from an infinitely thin circular guide which occupies the half space $z \leq 0$
2. Radiation from two parallel plates of infinite width and extending over $-\infty \leq z \leq 0$
3. Diffraction by a perfectly conducting half plane
4. Reflection and transmission of waves at the interface between free space and an infinite array of uniformly spaced, infinitely thin parallel plates
5. Reflection at the junction of a uniform guide and a similar guide having a resistive wall for $z \geq 0$
6. Bifurcated waveguides, e.g., a rectangular guide bifurcated or divided into two uniform regions for $z \geq 0$ by an infinitely thin conducting wall
7. A semidiaphragm of the capacitive or inductive type in a rectangular guide

All the above problems are characterized by the existence of two uniform regions, one for $z \leq 0$, and the other for $z \geq 0$. At the boundary plane $z = 0$, the only discontinuity in the medium is along an infinitely thin contour such as the circular edge of a circular guide or the straight-line edge of a half plane. The solution to all the above problems may be formulated as an integral equation of the Wiener-Hopf type.† The solution to this type of integral equation may be obtained by means of an application of Fourier or bilateral Laplace transforms, together with certain function-theoretic considerations. For those problems which have a solution consisting of an infinite series of discrete eigenfunctions

† R. E. A. C. Paley and N. Wiener, Fourier Transforms in the Complex Domain, *Am. Math. Soc. Colloq. Publs.*, vol. 19, 1934.

with discrete eigenvalues, two other closely related methods of solution are also available. For one the fields on either side of the interface $z = 0$ are expanded into an infinite series of the proper eigenfunctions for each region. Imposition of the continuity conditions for the field components at the interface leads to an infinite set of linear equations for the unknown amplitude coefficients. A solution to the system of equations which arise may be obtained by a contour-integration technique. The other method of solution is more closely related to the formal method of Wiener and Hopf, in that the solution is formulated in terms of a suitable transform. From known properties of the solution in each region it is possible to construct directly the transform function which will solve the problem.†

The concepts and techniques involved will be developed by considering a number of specific examples. The first example to be considered is a relatively simple electrostatic problem, which, however, will serve to illustrate most of the important basic ideas involved, without the obscuring effects of too many complicated mathematical manipulations. This will be followed by a solution of a series of more elaborate problems. In addition to the problems solved here, references are given at the end of the chapter, where solutions to several other typical problems may be found as well as general discussions on the theory.

## 10.1. An Electrostatic Problem

Consider the two-dimensional electrostatic problem illustrated in Fig. 10.1. The two infinite outer plates are kept at zero potential. The

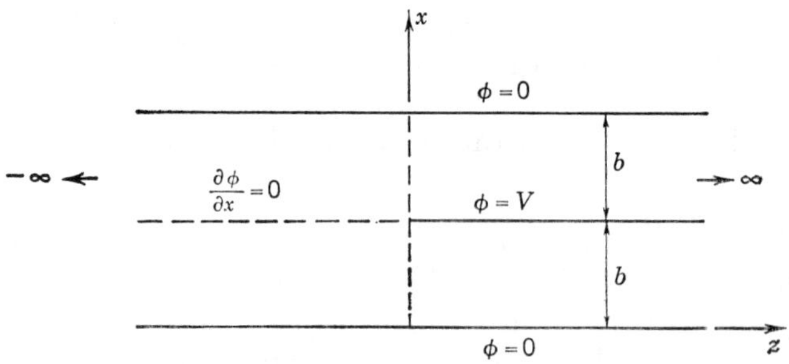


FIG. 10.1. Bifurcation of a parallel-plate region.

region $z \geq 0$ is divided or bifurcated into two identical regions by an infinitely thin conducting plate held at a potential $V$. We require a solution for the electrostatic potential function $\phi(x,z)$ which vanishes on the outer plates, reduces to $V$ on the center plate for $z \geq 0$, and is regular

† S. N. Karp, An Application of Sturm-Liouville Theory to a Class of Two Part Boundary Value Problems, *Proc. Cambridge Phil. Soc.*, vol. 53, part 2, pp. 368–381, April, 1957.

at infinity. The solution for $\phi$ will be obtained by the following three methods:

1. Solution of an infinite set of linear algebraic equations
2. Direct construction of a transform function which will provide the solution for $\phi$ upon inversion
3. Solution of a "Wiener-Hopf" integral equation

A rigorous solution may be readily found by means of a conformal transformation also, but this method is not under consideration here.

The potential $\phi$ is a solution of Laplace's equation

$$\frac{\partial^2\phi}{\partial x^2} + \frac{\partial^2\phi}{\partial z^2} = 0$$

In view of the symmetry involved, $\partial\phi/\partial n = \partial\phi/\partial x = 0$ for $x = b$, $-\infty \leq z \leq 0$. It is only necessary to solve the reduced problem in the region $x \leq b$, since, from symmetry considerations, $\phi(2b - x)$ is the solution for $b \leq x \leq 2b$, if $\phi(x)$ is the solution for $0 \leq x \leq b$. The potential $\phi$ satisfies the boundary conditions

$$\begin{array}{lll} \phi = 0 & x = 0 & \text{all } z \\ \phi = V & x = b & z \geq 0 \\ \dfrac{\partial\phi}{\partial x} = 0 & x = b & z \leq 0 \\ \phi \to 0 & & z \to -\infty \\ \phi \to \dfrac{Vx}{b} & & z \to +\infty \end{array}$$

In addition, at the edge of the bifurcation, $\phi$ is finite, but $\partial\phi/\partial r$ is asymptotic to $r^{-1/2}$, where $r$ is the radial distance from the edge; that is, $r = [z^2 + (x - b)^2]^{1/2}$.

*Method* 1

A suitable eigenfunction expansion of $\phi$ in the region $z \leq 0$ is

$$\phi(x,z) = \sum_{n=1}^{\infty} a_n \sin \frac{(2n-1)\pi x}{2b} e^{(2n-1)\pi z/2b} \tag{1a}$$

while, for $z \geq 0$, a suitable expansion of $\phi$ is

$$\phi(x,z) = \frac{Vx}{b} + \sum_{m=1}^{\infty} b_m \sin \frac{m\pi x}{b} e^{-m\pi z/b} \tag{1b}$$

In the above equations, $a_n$ and $b_m$ are unknown amplitude coefficients. At $z = 0$, both $\phi$ and $\partial\phi/\partial z$ must be continuous; hence,

$$\sum_{n=1}^{\infty} a_n \sin \frac{2n-1}{2b} \pi x = \frac{Vx}{b} + \sum_{m=1}^{\infty} b_m \sin \frac{m\pi x}{b} \tag{2a}$$

$$\sum_{n=1}^{\infty} \frac{(2n-1)\pi}{2b} a_n \sin \frac{2n-1}{2b} \pi x = -\sum_{m=1}^{\infty} \frac{m\pi}{b} b_m \sin \frac{m\pi x}{b} \tag{2b}$$

The linear term $Vx/b$ may be developed into a Fourier series in terms of either the $\sin [(2n - 1)\pi x/2b]$ functions or the $\sin (m\pi x/b)$ functions. An expansion in terms of the first set is preferable since the resultant series will be uniformly convergent at $x = b$. By conventional Fourier analysis, it is readily found that

$$\frac{Vx}{b} = \sum_{n=1}^{\infty} \frac{2V}{b^2} \frac{\sin [(2n-1)\pi/2]}{[(2n-1)\pi/2b]^2} \sin \frac{2n-1}{2b} \pi x \tag{3}$$

A system of equations to determine the unknown coefficients $b_m$ may be obtained as follows. Multiplying (2a) by $\sin [(2n - 1)\pi x/2b]$ and integrating over $0 \le x \le b$ gives

$$a_n \frac{b}{2} - \frac{V \sin [(2n-1)\pi/2]}{b[(2n-1)\pi/2b]^2} = \sum_{m=1}^{\infty} b_m \frac{m\pi}{b} \frac{\cos m\pi \sin [(2n-1)\pi/2]}{[(2n-1)\pi/2b]^2 - (m\pi/b)^2} \tag{4a}$$

and similarly, from (2b),

$$\frac{(2n-1)\pi}{2b} a_n \frac{b}{2} = -\sum_{m=1}^{\infty} b_m \left(\frac{m\pi}{b}\right)^2 \frac{\cos m\pi \sin [(2n-1)\pi/2]}{[(2n-1)\pi/2b]^2 - (m\pi/b)^2} \tag{4b}$$

For convenience, new coefficients $B_m$ given by $B_m = b_m(m\pi/b) \cos m\pi$ will be introduced. Multiplying (4a) by $-(2n - 1)\pi/2b$ and adding the resultant equation to (4b), the coefficients $a_n$ are eliminated, and we get

$$\sum_{m=1}^{\infty} \frac{B_m}{m\pi/b - (2n-1)\pi/2b} - \frac{V}{b} \frac{2b}{(2n-1)\pi} = 0 \qquad n = 1, 2, 3, \ldots \tag{5}$$

A system of equations of the above form may be solved by constructing a function $f(w)$ of the complex variable $w$ which will generate a set of equations formally identical with (5) when integrated around a suitable contour in the complex $w$ plane.

In the Mathematical Appendix, series of the type $\sum_{-\infty}^{\infty} \frac{e^{-jn\theta}}{n^2 + a^2}$ are summed. The method used is to replace the sum by a contour integral $\oint_C \frac{e^{-jw\theta}\, dw}{(w^2 + a^2)(e^{j2w\pi} - 1)}$, where $C$ is a rectangular contour, as in Fig. 10.2, which recedes to infinity. The factor $e^{j2w\pi} - 1$ is introduced to produce poles at $w = \pm n; n = 0, 1, 2, \ldots$. Since the contour integral vanishes, the sum of the residues of the integrand equals zero. The residues at the simple poles $\pm n$ give rise to the series which is to be summed. The sum of the series is then given in terms of the residues at $w = \pm ja$.

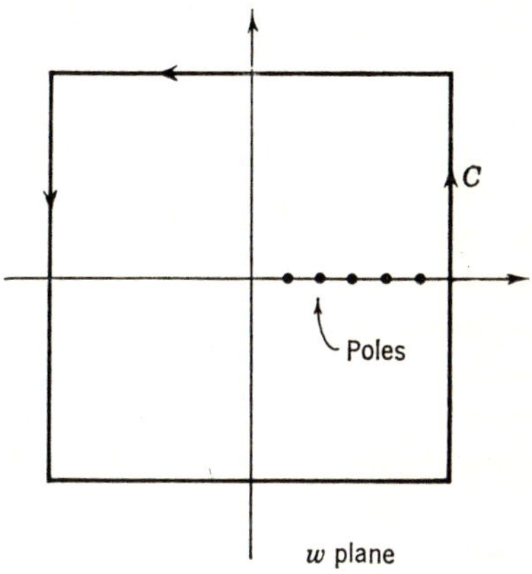

FIG. 10.2

In the set of equations (5) we know the sum of the series, but we do not know the form of each term, i.e., the coefficients $B_m$. We may, however, construct a suitable contour integral which will generate the required system of equations. Consider a contour integral of the form

$$\oint_C \frac{f(w)\, dw}{w - (2n - 1)\pi/2b} \qquad n = 1, 2, \ldots \tag{6}$$

To generate a series of the form (5), $f(w)$ must have simple poles at $w = m\pi/b$; $m = 1, 2, \ldots$. In addition, $f(w)$ must have zeros at $w = (2n - 1)\pi/2b$ in order to cancel the corresponding simple poles in the integrand. To produce a term of the form $(V/b)2b/(2n - 1)\pi$, we must also have a simple pole at $w = 0$. The contour $C$ is chosen as a rectangular contour which passes between the poles of the integrand. If, as $C$ recedes to infinity, the contour integral vanishes, then the sum of the residues must also be equal to zero. With the assumed simple poles the residue expansion of the integral gives

$$\sum_{m=1}^{\infty} \frac{r(m\pi/b)}{m\pi/b - (2n - 1)\pi/2b} - \frac{r(0)2b}{(2n - 1)\pi} = 0 \qquad n = 1, 2, \ldots \tag{7}$$

where $r(m\pi/b)$ is the residue of $f(w)$ at the simple pole $w = m\pi/b$, that is,

$$r\left(\frac{m\pi}{b}\right) = \lim_{w \to m\pi/b} \left(w - \frac{m\pi}{b}\right) f(w)$$

The system of equations (7) is formally the same as that in (5) if we identify $B_m$ with $r(m\pi/b)$ and normalize $f(w)$ such that $r(0) = V/b$. Provided the function $f(w)$ can be constructed, the solution to (5) may be found. In view of the required poles and zeros, it is seen that a suitable form for $f(w)$ is

$$f(w) = \frac{p(w) \prod_{n=1}^{\infty} \left[1 - \frac{w2b}{(2n-1)\pi}\right] e^{2wb/(2n-1)\pi}}{w \prod_{m=1}^{\infty} \left(1 - \frac{wb}{m\pi}\right) e^{wb/m\pi}} \tag{8}$$

In (8), $p(w)$ is an integral function, i.e., a function with no poles in the finite $w$ plane, which is still to be determined. The exponential factors occurring in the infinite products are required in order to ensure the uniform convergence of these products. The necessity of these factors may be seen from a consideration of the logarithmic derivative of the infinite product. Let

$$F(w) = \prod_{m=1}^{\infty} \left(1 - \frac{w}{m}\right) e^{w/m}$$

and form the logarithmic derivative $F'(w)/F(w)$. We have

$$\frac{F'}{F} = \frac{d(\ln F)}{dw} = \sum_{m=1}^{\infty} \left(\frac{1}{w-m} + \frac{1}{m}\right)$$

If the exponential factors $e^{w/m}$ were not included in the infinite product, the terms $1/m$ would not arise, and the infinite series for $F'/F$ would not be uniformly convergent. By including these factors, the series becomes uniformly convergent, and may be integrated term by term to yield the infinite-product function $F(w)$.

To determine $p(w)$, we must consider the asymptotic form of $f(w)$ as $w$ approaches infinity. The function $p(w)$ must be chosen so that $f(w)$ is of algebraic growth at infinity, and, in particular, so that $f(w) \sim w^{-\alpha}$ with $\alpha > 1$, in order for the contour integral to vanish. In order to examine the behavior of $f(w)$ at infinity, it will be convenient to rewrite $f(w)$ in terms of gamma functions. To obtain the results we want the following infinite-product expansions of the gamma function $\Gamma(u)$, and the $\sin u\pi$ functions are required,

$$\Gamma(u) = \frac{e^{-\gamma u}}{u} \left[\prod_{n=1}^{\infty} \left(1 + \frac{u}{n}\right) e^{-u/n}\right]^{-1} \tag{9a}$$

where $\gamma$ is Euler's constant and is equal to 0.5772, and

$$\sin u\pi = u\pi \prod_{n=1}^{\infty} \left(1 + \frac{u}{n}\right) e^{-u/n} \left(1 - \frac{u}{n}\right) e^{u/n} = u\pi \prod_{n=1}^{\infty} \left(1 - \frac{u^2}{n^2}\right) \tag{9b}$$

From the above results we may obtain the following relation:

$$\Gamma(u)\Gamma(-u) = -\frac{\pi}{u \sin \pi u} \tag{10}$$

The infinite product in the numerator of $f(w)$ may be rewritten as follows:

$$\prod_{n=1}^{\infty} \left[1 - \frac{2wb}{(2n-1)\pi}\right] e^{2wb/(2n-1)\pi} = \prod_{n=1,3,\ldots}^{\infty} \left(1 - \frac{2wb}{n\pi}\right) e^{2wb/n\pi}$$

$$= \frac{\displaystyle\prod_{n=1}^{\infty} \left(1 - \frac{2wb}{n\pi}\right) e^{2wb/n\pi}}{\displaystyle\prod_{n=1}^{\infty} \left(1 - \frac{wb}{n\pi}\right) e^{wb/n\pi}}$$

Using the result $\Gamma(u)\Gamma(-u) = -\pi/(u \sin \pi u)$, together with the infinite-product expansion of the gamma function, the above infinite product may be easily shown to be equal to

$$e^{\gamma wb/\pi} \frac{\sin 2wb}{\sin wb} \frac{\Gamma(2wb/\pi)}{\Gamma(wb/\pi)}$$

Treating the infinite product in the denominator of $f(w)$ in a similar way, we finally arrive at the following equivalent expression for $f(w)$:

$$f(w) = \frac{2\pi p(w)}{w} \cot wb \frac{\Gamma(2wb/\pi)}{[\Gamma(wb/\pi)]^2} \tag{11}$$

As $w$ tends to infinity, the gamma function has the following asymptotic value:

$$\Gamma(w) \sim \sqrt{2\pi}\, e^{w \ln w + 1 - w} w^{-1/2}$$

which is valid for all $w$ except in the vicinity of the negative-real axis, where the gamma function has poles at the negative integer values of $w$. Equation (10) may be used to obtain the asymptotic value of $\Gamma(w)$ on the negative-real axis. Utilizing the above result shows that the asymptotic form of $f(w)$ is

$$f(w) \sim \left(\frac{b}{w}\right)^{1/2} p(w) \cot wb\, e^{(2wb/\pi) \ln 2 - 1} \tag{12}$$

To ensure algebraic growth at infinity, we must choose $p(w)$ equal to $e^{-(2wb/\pi)\ln 2}$ times a polynomial in $w$. However, in order for the contour integral to vanish, the polynomial must be a constant chosen so that $r(0) = V/b$. For the present we will let $V$ be determined by $r(0)$ and choose $f(w)$ as follows:

$$f(w) = \frac{2\pi}{w} e^{-(2wb/\pi)\ln 2} \cot wb \frac{\Gamma(2wb/\pi)}{[\Gamma(wb/\pi)]^2} \tag{13}$$

The only poles that $f(w)$ has are the poles of the cotangent function at $w = m\pi/b$; $m = 0, 1, 2, \ldots$. All the other poles are canceled by zeros arising from the gamma functions. For $w$ approaching $m\pi/b$, we have

$$\lim_{w\to m\pi/b} \left(w - \frac{m\pi}{b}\right) \cot wb = \frac{1}{b}$$

The residue expansion (7) yields the following solution for the coefficients $B_m$ and $b_m$:

$$B_m = b_m \frac{m\pi}{b} \cos m\pi = \frac{2e^{-2m\ln 2}}{m} \frac{\Gamma(2m)}{[\Gamma(m)]^2} \qquad m = 1, 2, \ldots \tag{14}$$

For $m$ very large, the coefficients $B_m$ are approximately given by

$$B_m \sim e^{-1}(m\pi)^{-\frac{1}{2}} \tag{15}$$

This result follows directly from the asymptotic form of $f(w)$ given by (12).

The potential $V$ of the center plate is given by

$$V = br(0) = b \tag{16}$$

since $r(0)$ is equal to unity. For any other value of potential, it is necessary only to multiply the whole solution by a constant.

The solution for the coefficients $a_n$ may be obtained from the residue expansion of the following contour integral:

$$\oint_C \frac{f(w)\,dw}{w + (2n-1)\pi/2b} \qquad n = 1, 2, \ldots \tag{17}$$

We first multiply (4$a$) by $(2n - 1)\pi/2b$, and add the resulting equations to (4$b$) to obtain

$$\frac{(2n-1)\pi}{2\sin[(2n-1)\pi/2]} a_n + \frac{V}{b} \frac{2b}{(2n-1)\pi} + \sum_{m=1}^{\infty} \frac{B_m}{m\pi/b + (2n-1)\pi/2b} = 0 \qquad n = 1, 2, \ldots \tag{18}$$

The contour integral (17) vanishes as $C$ recedes to infinity, and, hence, the residue expansion of the integral, in terms of the residues at the poles

$w = m\pi/b$, $m = 0, 1, 2, \ldots$, and $w = -(2n-1)\pi/2b$, gives

$$f\left(-\frac{2n-1}{2b}\pi\right) + \frac{r(0)2b}{(2n-1)\pi} + \sum_{m=1}^{\infty} \frac{r(m\pi/b)}{m\pi/b + (2n-1)\pi/2b} = 0 \qquad n = 1, 2, \ldots \tag{19}$$

Comparing this result with (18) shows that the solution for $a_n$ is

$$a_n = -\frac{2\sin[(2n-1)\pi/2]}{(2n-1)\pi} f\left(-\frac{2n-1}{2b}\pi\right) \tag{20}$$

From (8), $f\left(-\frac{2n-1}{2b}\pi\right)$ is found to be given by

$$f\left(-\frac{2n-1}{2b}\pi\right) = -\frac{b}{2\pi} e^{(2n-1)\ln 2} \frac{[\Gamma(n-1/2)]^2}{\Gamma(2n-1)} \tag{21}$$

upon using the infinite-product expansion for the gamma functions. For $n$ large, we obtain the approximate result

$$f\left(-\frac{2n-1}{2b}\pi\right) \sim -eb\left(\frac{2\pi}{2n-1}\right)^{1/2}$$

by using the asymptotic expansion of the gamma function. Hence, the coefficients $a_n$ have the asymptotic value

$$a_n \sim 2eb\sqrt{\frac{2}{\pi}} \frac{\sin[(2n-1)\pi/2]}{(2n-1)^{3/2}} \tag{22}$$

From the asymptotic value of the coefficients $a_n$ and $b_n$ as $n$ tends to infinity, we may readily verify that $\partial\phi/\partial z$ and $\partial\phi/\partial x$ have the proper singularity at the edge of the bifurcation. For example, the series expansion for $\partial\phi/\partial z$ for $z \geq 0$, $x = b$, has the same asymptotic behavior for $z$ approaching zero as the following series:

$$\sum_{n=1}^{\infty} \frac{e^{-n\pi|z|/b}}{n^{1/2}}$$

Using the following result,

$$\int_1^{\infty} e^{-\pi z u/b} u^{-1/2}\, du = \left(\frac{b}{z}\right)^{1/2} - 2 + \frac{2\pi z}{3b} + \cdots < \sum_{n=1}^{\infty} e^{-n\pi z/b} n^{-1/2} < \int_0^{\infty} e^{-\pi z u/b} u^{-1/2}\, du = \left(\frac{b}{z}\right)^{1/2}$$

shows that the series, and hence $\partial\phi/\partial z$, is asymptotic to $|z|^{-1/2}$ as $z$ tends to zero, i.e., as the edge of the bifurcation is approached. This edge condition would not hold if $f(w)$, and hence $a_n$ and $b_n$, were not asymptotic to $w^{-\frac{1}{2}}$ as $w$ approached infinity. The behavior of $f(w)$ at infinity governs the behavior of $\phi$ at $z = 0$. When the solution to our problem has been obtained by means of an integral transform, the correlation between the behavior of $\phi$ at the origin and $f(w)$ at infinity will be seen to be a natural consequence arising from the basic properties of Fourier and Laplace transforms.

The above method of solution appears to have been applied only to equations of the form given in (4). For a system of equations of a more elaborate form, a similar technique could be used, provided a function $f(w)$ could be constructed so as to yield a residue expansion which would be formally identical with the equations to be solved. If each term in the series, apart from a term such as $m\pi/b \pm (2n - 1)\pi/2b$ in the denominator, is not representable as a product of two factors depending on $n$ alone and $m$ alone, respectively, it is generally not feasible to construct the function $f(w)$ which will provide the required solution.

*Method 2*

The second method of solution to be illustrated is an integral transform solution. As a first step, the variable $z$ is suppressed by taking the bilateral Laplace transform of $\phi$. A solution for the transform function which satisfies the required boundary conditions and has poles which will yield the known eigenfunction expansion of $\phi$ will then be constructed. Finally, the solution for $\phi$ is obtained by inverting the transform and evaluating the inversion integral in terms of its residues. The residue expansion of the integral is the eigenfunction expansion of the potential $\phi$.

Let $F(x,w)$ be the bilateral Laplace transform of $\phi(x,z)$, that is,

$$F(x,w) = \int_{-\infty}^{\infty} \phi(x,z)e^{-wz}\,dz \tag{23}$$

The transform $F(x,w)$ may be split into two separate parts as follows:

$$F_+(x,w) = \int_0^{\infty} \phi(x,z)e^{-wz}\,dz \tag{24a}$$

$$F_-(x,w) = \int_{-\infty}^{0} \phi(x,z)e^{-wz}\,dz \tag{24b}$$

For $z \geq 0$, the potential $\phi(x,z)$ is asymptotic to $Vx/b$ for $z$ large, and, hence, (24a) converges uniformly for all $w$ in the half plane $\text{Re}\,w > 0$. Thus (24a) determines $F_+$ as an analytic function of $w$ in the half plane $\text{Re}\,w > 0$. All the poles of $F_+$ must, therefore, lie in the left half plane $\text{Re}\,w < 0$. We may invert (24a) to get

$$\phi(x,z) = \frac{1}{2\pi j}\int_{C_+} e^{wz}F_+(x,w)\,dw \qquad z \geq 0 \tag{25}$$

For $z$ less than zero, (25) is equal to zero, as may be readily seen since, for $z < 0$, the contour may be closed in the right half plane; and, since no poles are enclosed, the contour integral vanishes. In (25), $C_+$ is a contour running parallel to the imaginary $w$ axis in the right half plane $\operatorname{Re} w > 0$. For $z > 0$, the contour may be closed in the left half plane, and the integral evaluated in terms of its residues at the poles. This residue expansion will give the eigenfunction expansion of $\phi$ for $z > 0$.

For $z < 0$, the potential $\phi$ is asymptotic to $a_1 \sin(\pi x/2b)e^{-\pi|z|/2b}$ for $z$ large and negative. Hence (24$b$) converges uniformly for all $w$ in the

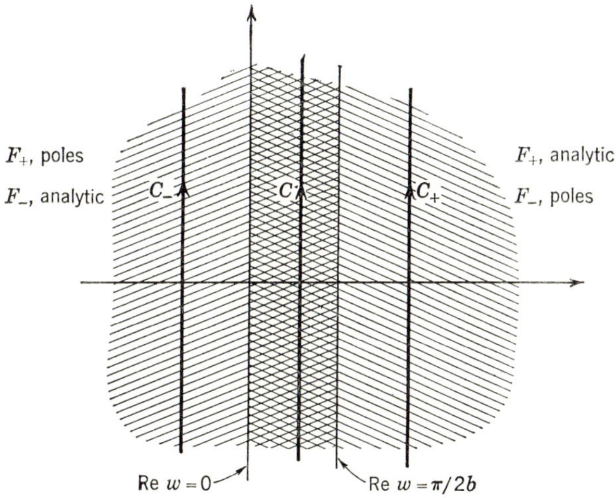


FIG. 10.3. Regions of analyticity for $F_+$ and $F_-$ and the common inversion contour $C$.

half plane $\operatorname{Re} w < \pi/2b$, and determines $F_-$ as an analytic function of $w$ in that half plane. Inverting (24$b$) gives

$$\phi(x,z) = \frac{1}{2\pi j}\int_{C_-} e^{wz}F_-(x,w)\,dw \qquad z \le 0 \tag{26}$$

where $C_-$ is a contour parallel to the imaginary $w$ axis and in the half plane $\operatorname{Re} w < \pi/2b$. Since $z < 0$, we may close the contour in the right half plane, and evaluate the integral in terms of its residues at the poles of $F_-(x,w)$. This series gives the eigenfunction expansion of $\phi$ in the region $z \le 0$. If $z > 0$ in (26), the integral is equal to zero because of the analytic properties of $F_-(x,w)$ in the left half plane.

The two functions $F_+$ and $F_-$ have a common region of analyticity $0 < \operatorname{Re} w < \pi/2b$, as illustrated in Fig. 10.3. Therefore, we may choose $C_+ = C_- = C$, where $C$ is a common contour within the common strip of analyticity and running parallel to the imaginary $w$ axis. We may therefore invert (23) directly without separating $F$ into two parts, and, hence,

$$\phi(x,z) = \frac{1}{2\pi j} \int_C e^{wz} F(x,w)\, dw \qquad \text{all } z \tag{27}$$

The bilateral Laplace transform of the equation

$$\frac{\partial^2 \phi}{\partial z^2} + \frac{\partial^2 \phi}{\partial x^2} = 0$$

with respect to the variable $z$ gives

$$\frac{d^2 F(x,w)}{dx^2} + w^2 F(x,w) = 0 \tag{28}$$

valid for $0 < \operatorname{Re} w < \pi/2b$. This result follows by integrating the second partial derivative with respect to $z$ twice by parts. The integrated terms $e^{-wz}\, \partial\phi/\partial z$, $w\phi e^{-wz}$ vanish at $z = \pm\infty$ for $w$ in the specified region because of the asymptotic behavior of $\phi$ for $z$ large.

A general solution to (28) which vanishes for $x = 0$ is

$$F = g(w) \sin wx \tag{29}$$

where $g(w)$ is a function of $w$ to be determined. Substituting into the inversion integral (27), we get

$$\phi(x,z) = \frac{1}{2\pi j} \int_C e^{wz} g(w) \sin wx\, dw \tag{30}$$

The function $g$ may be determined uniquely from the known properties of $\phi$.

When $z$ is negative, we may close the contour $C$ in the right-half $w$ plane. In order to obtain the eigenfunction expansion (1$a$) for $\phi$, we see that $g$ must have simple poles at $w = (2n - 1)\pi/2b$; $n = 1, 2, \ldots$. For $z \geq 0$, we may close the contour in the left-half $w$ plane. In order for the residue expansion of the integral to yield a solution for $\phi$ of the form given by (1$b$), the function $g$ must have simple poles at $w = -m\pi/b$, $m = 1, 2, \ldots$, in the left-half $w$ plane. In addition, we require $g$ to have a double pole at the origin, so that $g \sin wx$ will have a simple pole at $w = 0$ and give rise to a term of the form $Vx/b$. Finally, the asymptotic form of $g$ for $w$ approaching infinity must be such that the function $\phi$ will be asymptotic to $z^{1/2}$ as $z$ tends to zero. It is only because our knowledge of $\phi$ is as complete as it is that we are able to construct directly the required transform function $g(w)$. When this is not the case, we have to proceed by the more formal method of Wiener and Hopf, as given in the next section as method 3.

For $x = b$, we have

$$g(w) \sin wb = \int_{-\infty}^{\infty} e^{-wz}\phi(b,z)\, dz$$

If we replace $wz$ by $u$, and $dz$ by $du/w$, and take the limit as $w$ tends to infinity, we get

$$\lim_{w\to\infty} \frac{1}{w} \int_{-\infty}^{\infty} e^{-u}\phi\left(b, \frac{u}{w}\right) du$$

Interchanging the order of integration and the limiting operation, we get

$$\begin{aligned} \lim_{w\to\infty} g(w) \sin wb &= \int_{-\infty}^{\infty} e^{-u} \left[ \lim_{w\to\infty} \frac{1}{w}\, \phi\left(b, \frac{u}{w}\right)\right] du \\ &= w^{-(\alpha+1)} \int_{-\infty}^{\infty} a_0 u^{\alpha} e^{-u}\, du \end{aligned}$$

provided the limiting form of $\phi(b,z)$ as $z$ tends to zero is $a_0 z^{\alpha}$. Thus, if $\phi(b,z)$ is asymptotic to $z^{1/2}$ as $z$ tends to zero, the transform of $\phi$ will be asymptotic to $w^{-3/2}$ as $w$ tends to infinity.

In view of the required properties of the function $g$, it is seen that a suitable transform function is given by

$$g(w) = -\frac{f(-w)}{w \cos wb} \tag{31}$$

where $f(w)$ is the function given by (8) or its equivalent (11). The function $f(-w)$ has poles at $w = -m\pi/b$, $m = 0, 1, 2, \ldots$, and zeros at $w = -(2n - 1)\pi/2b$, $n = 1, 2, \ldots$. Hence, $g(w)$ will have simple poles at $w = -m\pi/b$, $m = 1, 2, \ldots$, and $w = (2n - 1)\pi/2b$, $n = 1, 2, \ldots$, and a double pole at $w = 0$. Furthermore, $f(-w)$ is asymptotic to $w^{-1/2}$, and, hence, the function $g(w)$ given by (31) satisfies all the required conditions. It is interesting to note the close relationship between the function $f(w)$ used to solve the system of algebraic equations (4) and the transform function $g(w)$.

The solution for $\phi$ may now be completed by substituting for $g$ from (31) into (30), and evaluating the inversion integral in terms of its residues in the left-half $w$ plane for $z \geq 0$, and in the right-half $w$ plane for $z \leq 0$. The residue at the origin gives simply $x$, and this makes the potential of the center plate equal to $b$. To obtain a potential $V$, we must multiply $g$ by the constant $V/b$. The rest of the residue expansion is the same as the series (1$a$) and (1$b$), with $b_n$ and $a_n$ given by (14) and (20), respectively. In view of the specified poles for the function $g(w)$, the boundary conditions on $\phi$ for $x = b$ are satisfied for all values of $z$.

*Method* 3

The third method of solution to be illustrated is the method of Wiener and Hopf. In this method we find a suitable Green's function, which represents the potential due to a unit line charge at $x = b$, $z = z'$ on the center plate first. The next step is to use the Green's theorem to obtain an integral equation for the charge distribution. This integral equation is of the Wiener-Hopf type and may be solved by means of a bilateral Laplace transform (or a Fourier transform). Since the method is somewhat long and involves several basic concepts, it will be advantageous to present the essential steps in the solution first. Having done this, we will proceed to work out the remaining mathematical details.

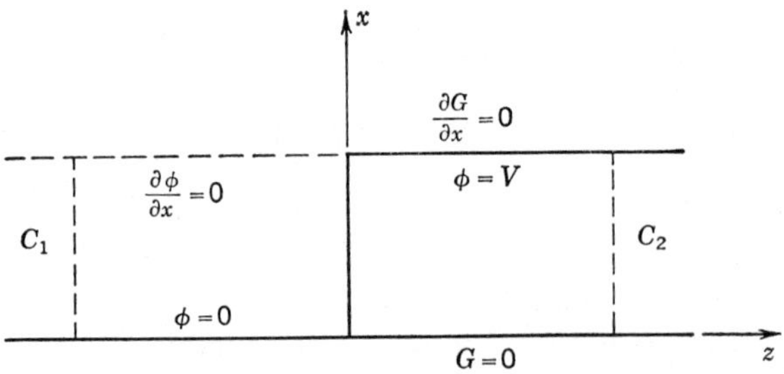


FIG. 10.4. Boundary-value problem for Green's function.

Referring to Fig. 10.4, the first step is to construct a Green's function $G(x, x', z - z')$ which is a solution of the equation

$$\frac{\partial^2 G}{\partial x^2} + \frac{\partial^2 G}{\partial z^2} = \delta(x - x')\delta(z - z') \tag{32}$$

and satisfies the boundary conditions

$$\begin{aligned} G &= 0 \qquad x = 0 \qquad \text{all } z \\ \frac{\partial G}{\partial x} &= 0 \qquad x = b \qquad \text{all } z \\ G &\rightarrow 0 \qquad\qquad\qquad \text{as } |z| \rightarrow \infty \end{aligned}$$

This Green's function is a function of the variable $z - z'$. Having determined the Green's function, we next use Green's second identity to obtain

$$\iint_S (\phi \nabla^2 G - G \nabla^2 \phi)\, dx\, dz = \oint_C \left( \phi \frac{\partial G}{\partial n} - G \frac{\partial \phi}{\partial n} \right) dl$$

where $C$ is a contour consisting of the $z$ axis, the line $x = b$, and the two segments $C_1$ and $C_2$ at $z = -\infty, +\infty$, as in Fig. 10.4. The potential $\phi$ is a solution of Laplace's equation, while $G$ is a solution of Eq. (32), and, hence, the surface integral becomes simply $\phi(x',z')$. The contour integral reduces to an integral over $z$ from zero to infinity for $x = b$, only by virtue of the boundary conditions satisfied by $\phi$ and $G$. The integral over the segments $C_1$ and $C_2$ vanishes, since both $\phi$ and $G$ tend to zero for $z \to -\infty$, while, for $z \to +\infty$, the potential $\phi$ approaches $Vx/b$, and $G$ vanishes. Thus, we get

$$\phi(x',z') = -\int_0^\infty G(b, x', z - z') \frac{\partial\phi(b,z)}{\partial x} dz$$

or, interchanging primed and unprimed coordinates,

$$\phi(x,z) = -\int_0^\infty G(b, x, z - z')\rho(z') dz' \tag{33}$$

where $\rho(z') = \partial\phi(b,z')/\partial x'$ and is proportional to the charge density on the plate at $x = b$, $z \geq 0$. Since $\rho$ is identically zero for $z' \leq 0$, the integration in (33) may be extended over the whole range $z' = -\infty$ to $z' = +\infty$. Putting $x = b$, we obtain the fundamental integral equation for $\rho$:

$$\int_{-\infty}^\infty G(z - z')\rho(z') dz' = -\phi(b,z) = \begin{cases} -\phi_-(z) & z \leq 0 \\ -\phi_+(z) = -V & z \geq 0 \end{cases} \tag{34}$$

where $\rho(z') = 0$, $z' < 0$.

By being able to extend the integration from minus to plus infinity, the faltung theorem may be used to obtain a simple expression for the bilateral Laplace transform of (34). The following transform functions are introduced:

$$\mathcal{G}(w) = \int_{-\infty}^\infty e^{-wz}G(z) dz \tag{35a}$$

$$Q(w) = Q_+(w) = \int_0^\infty e^{-wz}\rho(z) dz \tag{35b}$$

$$\psi_+(w) = \int_0^\infty e^{-wz}\phi_+(z) dz = \int_0^\infty Ve^{-wz} dz = \frac{V}{w} \tag{35c}$$

$$\psi_-(w) = \int_{-\infty}^0 \phi_-(z)e^{-wz} dz \tag{35d}$$

$$\psi = \psi_+ + \psi_- \tag{35e}$$

For $|z|$ large, $G(z)$ is asymptotic to $e^{-\pi|z|/2b}$, and, hence, the transform $\mathcal{G}(w)$ is an analytic function for all $w$ in the strip $-(\pi/2b) < \operatorname{Re} w < \pi/2b$, since in this strip the integral (35a) is uniformly convergent. Since $\phi_+$ is asymptotic to $V$ for $z \to +\infty$, and $\partial\phi/\partial x'$ is asymptotic to $\partial(Vx'/b)/\partial x' = V/b$, the transforms $Q_+$ and $\psi_+$ are analytic for all $w$ in the half plane $\operatorname{Re} w > 0$. As $z \to -\infty$, the potential $\phi_-$ is asymptotic to $e^{-\pi|z|/2b}$, and hence $\psi_-$ is analytic for all $w$ in the half plane $\operatorname{Re} w < \pi/2b$. It is seen that all the transform functions in (35) are analytic in the common strip $0 < \operatorname{Re} w < \pi/2b$. We may therefore take the bilateral Laplace transform of (34) to get

$$\mathcal{G}(w)Q_+(w) = -\psi_+(w) - \psi_-(w) \tag{36}$$

A common strip of analyticity is necessary in order for the transform equation (36) to be valid. The faltung theorem states that the transform of an integral of the form

$$\int_{-\infty}^{\infty} G(z - z')\rho(z')\,dz'$$

is the product of the transforms of $G(z)$ and $\rho(z)$, and this result was used to obtain the left-hand side of (36).

The next essential step in the method is to factor and rewrite the transform equation (36) in the following form:

$$H_+(w) = H_-(w)$$

where $H_+$ is an analytic function with no poles in the right-half $w$ plane and $H_-$ has no poles in the left-half $w$ plane. In addition $H_+$ and $H_-$ have a common strip of analyticity in which they are both equal. An analytic function $h(w)$ may now be defined as equal to $H_+$ in the right half plane and equal to $H_-$ in the left half plane. Thus, $H_-$ provides the analytic continuation of $H_+$ into the left half plane. The function $h(w)$, as defined, is free of poles in the whole finite $w$ plane and must therefore be equal to an integral function of $w$. If also $H_+$ and $H_-$ are bounded at infinity, then Liouville's theorem states that $H_+$, $H_-$, and $h$ are just equal to a constant. At any rate, sufficient knowledge of the properties of the solution is generally available to determine the integral function $h(w)$, which may be a finite polynomial in $w$ in general.

Returning to (36), we factor $\mathcal{G}$ into the form $\mathcal{G}_+/\mathcal{G}_-$, where $\mathcal{G}_+$ has no poles in the half plane $\operatorname{Re} w > \pi/2b$, and $\mathcal{G}_-$ has no poles in the half plane $\operatorname{Re} w < \pi/2b$, and also such that both $\mathcal{G}_+$ and $\mathcal{G}_-$ have algebraic growth rather than exponential growth at infinity. We may now rewrite (36) as

$$\begin{aligned} \mathcal{G}_+Q_+ &= -\mathcal{G}_-\psi_- - \mathcal{G}_-\psi_+ \\ &= -\mathcal{G}_-\psi_- - \frac{\mathcal{G}_-V}{w} \end{aligned} \tag{37}$$

upon substituting for $\psi_+$ from (35*c*). The left-hand side of (37) is analytic (free of poles) for $w$ in the half plane $\operatorname{Re} w > 0$. The function $\mathcal{G}_-\psi_-$ is free of poles in the left half plane $\operatorname{Re} w < \pi/2b$. However, $\mathcal{G}_-V/w$ has a pole at the origin, so that the two sides of (37) do not have a common strip of analyticity. To remove this deficiency, we add $\mathcal{G}_-(0)V/w$ to both sides to get

$$\mathcal{G}_+Q_+ + \mathcal{G}_-(0)\frac{V}{w} = -\mathcal{G}_-\psi_- - [\mathcal{G}_-(w) - \mathcal{G}_-(0)]\frac{V}{w} \tag{38}$$

In (38), the left-hand side is free of poles for all $w$ in the half plane $\operatorname{Re} w > 0$, while the right-hand side is now free of poles for all $w$ in the half plane Re $w < \pi/2b$. Both sides are analytic in the common strip $0 < \text{Re } w < \pi/2b$. Together they define an analytic function $h(w)$ with no poles in the finite $w$ plane. In particular, both $\mathcal{G}_+$ and $Q_+$ are found to be bounded at infinity, and, hence, both sides of (38) must be equal to a constant $A$. In fact, we find that $\mathcal{G}_+$ is asymptotic to $w^{-1/2}$, and, since $\rho$ is asymptotic to $z^{-1/2}$ as $z$ tends to zero, $Q_+$ is asymptotic to $w^{-1/2}$ also as $w$ tends to infinity. From (38) we now get

$$Q_+ = \frac{-\mathcal{G}_-(0)V/w + A}{\mathcal{G}_+(w)}$$

and it is apparent that $A$ must equal zero if $Q_+$ is to be asymptotic to $w^{-1/2}$. The solution for $Q_+$ is then

$$Q_+ = -\frac{V}{w}\frac{\mathcal{G}_-(0)}{\mathcal{G}_+(w)} \tag{39}$$

Having determined the transform $Q_+$, we return to (36) to obtain the solution for $\psi$, which is

$$\psi = -\mathcal{G}(w)Q_+(w) = \frac{\mathcal{G}_+}{\mathcal{G}_-}\frac{V}{w}\frac{\mathcal{G}_-(0)}{\mathcal{G}_+} = \frac{V}{w}\frac{\mathcal{G}_-(0)}{\mathcal{G}_-(w)} \tag{40}$$

The solution for $\phi(b,z)$ may be obtained by inverting the transform $\psi$. Again a residue expansion of the inversion integral leads to the series expansion for $\phi$.

Having outlined the basic concepts and essential steps in the solution, we must now go back and carry through the detailed analysis to obtain the Green's function, its transform, and the factors $\mathcal{G}_+$ and $\mathcal{G}_-$. The Green's function is a solution of (32). Since we require the transform of $G$, we will take the bilateral Laplace transform of (32) with respect to the variable $z - z'$, since $G$ is a function of $z - z'$, and solve for $\mathcal{G}(w)$ directly. We obtain from (32)

$$\frac{d^2\mathcal{G}(w)}{dx^2} + w^2\mathcal{G}(w) = \delta(x - x') \tag{41}$$

where $\mathcal{G}$ is given by (35*a*). Two independent solutions to the homogeneous equation $(d^2/dx^2 + w^2)\mathcal{G} = 0$ are

$$\mathcal{G}_1 = \sin wx \qquad \mathcal{G}_2 = \cos w(b - x)$$

These solutions satisfy the required boundary conditions at $x = 0$, $b$, respectively. The Wronskian determinant is

$$\mathcal{G}_1\frac{d\mathcal{G}_2}{dx} - \mathcal{G}_2\frac{d\mathcal{G}_1}{dx} = -w\cos wb$$

Using the method of Sec. 2.3, the solution for $\mathcal{G}$ is found to be given by

$$\begin{aligned}\mathcal{G} = (-w \cos wb)^{-1} \left[ \int_{x_1}^{x} \mathcal{G}_2(x)\mathcal{G}_1(x_0)\delta(x' - x_0)\, dx_0 \right. \\ \left. + \int_{x}^{x_2} \mathcal{G}_1(x)\mathcal{G}_2(x_0)\delta(x' - x_0)\, dx_0 \right] \\ = (-w \cos wb)^{-1} \begin{cases} \sin wx' \cos w(b - x) & x \geq x' \\ \sin wx \cos w(b - x') & x \leq x' \end{cases}\end{aligned} \tag{42}$$

provided we choose $x_1$ and $x_2$ so that $x_1 < x' < x$, $x < x' < x_2$. The above solution satisfies the required boundary conditions at $x = 0$, $b$. At $x' = b$, we get

$$\mathcal{G}(x,b,w) = -(w \cos wb)^{-1} \sin wx \tag{43}$$

If the transform of (32) was taken with respect to $z$ alone, we would obtain in place of (42) the result $e^{-wz'}\mathcal{G}(x,x',w)$. The inverse transform yields the solution for $G$ in the form

$$G(x, x', z - z') = \frac{1}{2\pi j} \int_C e^{w(z-z')}\mathcal{G}(x,x',w)\, dw \tag{44}$$

which is clearly a function of $z - z'$. Replacing $z - z'$ by a new variable $\lambda$ and taking the transform of (44) results in the function $\mathcal{G}$ as given by (42).

The transform equation (36) was obtained from the integral equation (34) by an application of the faltung theorem, which states that the Fourier or bilateral Laplace transform of an integral of the faltung (convolution) type is equal to the product of the transforms of the two functions occurring in the integrand. This theorem is proved as follows. Consider the following integral:

$$I = \int_{-\infty}^{\infty} G(z - z')\rho(z')\, dz'$$

Replacing $G(z - z')$ by (44) and $\rho(z')$ by

$$\frac{1}{2\pi j} \int_C e^{w_0 z'} Q(w_0)\, dw_0$$

and interchanging the order of integration gives

$$I = \frac{1}{(2\pi j)^2} \int_C \int_C \mathcal{G}(w)Q(w_0) \int_{-\infty}^{\infty} e^{wz}e^{-z'(w-w_0)}\, dz'\, dw\, dw_0$$

Since

$$\int_{-\infty}^{\infty} e^{-z'(w-w_0)}\, dz' = 2\pi j\delta(w - w_0)$$

we get

$$\begin{aligned}I &= \frac{1}{2\pi j} \int_C \int_C \mathcal{G}(w)Q(w_0)e^{wz}\delta(w - w_0)\, dw\, dw_0 \\ &= \frac{1}{2\pi j} \int_C \mathcal{G}(w)Q(w)e^{wz}\, dw\end{aligned}$$

The bilateral Laplace transform of $I$ is obtained by multiplying by $e^{-\lambda z}$ and integrating over $z$ to get

$$\frac{1}{2\pi j}\int_C\int_{-\infty}^{\infty}\mathcal{G}(w)Q(w)e^{-z(\lambda-w)}\,dz\,dw = \mathcal{G}(\lambda)Q(\lambda)$$

This is the final result which was to be proved.

To obtain (37) and (38) we have to factor

$$\mathcal{G}(w) = \mathcal{G}(b,b,w) = -w^{-1}\tan wb$$

into the form $\mathcal{G}_+/\mathcal{G}_-$, with $\mathcal{G}_+$ analytic for all $w$ in the half plane $\mathrm{Re}\, w > -\pi/2b$, and $\mathcal{G}_-$ analytic for all $w$ in the half plane $\mathrm{Re}\, w < \pi/2b$. The required factorization may be carried out by inspection when $\sin wb$ and $\cos wb$ are expressed as infinite products. We have

$$-w^{-1}\tan wb = -b(wb)^{-1}\sin wb(\cos wb)^{-1}$$

$$= -b\,\frac{\prod_{n=1}^{\infty}\left(1-\frac{wb}{n\pi}\right)e^{wb/n\pi}\prod_{n=1}^{\infty}\left(1+\frac{wb}{n\pi}\right)e^{-wb/n\pi}}{\prod_{n=1}^{\infty}\left[1-\frac{2wb}{(2n-1)\pi}\right]e^{2wb/(2n-1)\pi}\prod_{n=1}^{\infty}\left[1+\frac{2wb}{(2n-1)\pi}\right]e^{-2wb/(2n-1)\pi}} \tag{45}$$

From this result it is readily seen that a suitable choice for $\mathcal{G}_+$ and $\mathcal{G}_-$ is

$$\mathcal{G}_+(w) = -bp(w)\,\frac{\prod_{n=1}^{\infty}\left(1+\frac{wb}{n\pi}\right)e^{-wb/n\pi}}{\prod_{n=1}^{\infty}\left[1+\frac{2wb}{(2n-1)\pi}\right]e^{-2wb/(2n-1)\pi}} \tag{46a}$$

$$\mathcal{G}_-(w) = p(w)\,\frac{\prod_{n=1}^{\infty}\left[1-\frac{2wb}{(2n-1)\pi}\right]e^{2wb/(2n-1)\pi}}{\prod_{n=1}^{\infty}\left(1-\frac{wb}{n\pi}\right)e^{wb/n\pi}} \tag{46b}$$

where $p(w)$ is chosen so that $\mathcal{G}_+$ and $\mathcal{G}_-$ have algebraic growth at infinity. By expressing $\mathcal{G}_+$ in terms of gamma functions, we readily find that $\mathcal{G}_+$ is asymptotic to $w^{-1/2}p(w)\exp[(2wb/\pi)\ln 2]$, and, hence, $p(w)$ must be chosen as follows:

$$p(w) = \exp\left(-\frac{2wb}{\pi}\ln 2\right)$$

From (46) we find that $\mathcal{G}_-(0)$ is equal to unity.

From (40) the solution for the transform $\psi(w)$ of $\phi(b,z)$ may be found. The transform of $\phi(x,z)$ is given by

$$\psi(w)\,\frac{\sin wx}{\sin wb}$$

as may be seen from (33), (34), and (43). Comparing (46*a*) with (8), we see that $\mathcal{G}_+(w) = b/wf(-w)$. Hence the transform of $\phi(x,z)$ becomes

$$\begin{aligned}\frac{\psi(w)\sin wx}{\sin wb} = \frac{V}{w}\frac{\mathcal{G}_+}{\mathcal{G}_-}\frac{1}{\mathcal{G}_+}\frac{\sin wx}{\sin wb} &= -V\,\frac{\tan wb}{wb}\,\frac{\sin wx}{\sin wb}\,f(-w)\\ &= -\frac{V}{b}\,\frac{\sin wx}{w\cos wb}\,f(-w) \qquad (47)\end{aligned}$$

This is the same result as given by (29) and (31), provided $V$ is chosen equal to $b$ in (47), as was done in (31). The eigenfunction expansion of $\phi$ may be readily obtained from the residue expansion of the inversion integral for $\phi(x,z)$, that is, from

$$\phi(x,z) = \frac{1}{2\pi j}\int_C e^{wz}\,\frac{\psi(w)\sin wx}{\sin wb}\,dw \qquad (48)$$

It is apparent that, although the three methods of solution which have been presented are distinct, there is a rather close relationship among them. For problems of the above type, i.e., those whose solution consists of an expansion in terms of a discrete set of eigenfunctions, the second method based on a direct construction of the required transform function is the simplest and most direct. However, this method does demand a knowledge of the infinite set of eigenfunctions and their eigenvalues required for the complete representation of the solution. This knowledge is usually available, and, hence, the second method of solution turns out to be a very useful one for many problems of practical interest in connection with waveguides.

*Further Comments on the Wiener-Hopf Method*

The inhomogeneous Wiener-Hopf equation is an equation of the form

$$\phi(z) = v(z) + \int_0^\infty G(z - z')\phi(z')\,dz' \qquad (49)$$

where $v(z)$ is a known function and the kernel or Green's function $G(z - z')$ is a function of $z - z'$ and is also known. By introducing the functions

$$\phi_+(z) = \begin{cases}\phi(z) & z > 0\\ 0 & z < 0\end{cases}$$

$$\phi_-(z) = \begin{cases}0 & z > 0\\ \phi(z) & z < 0\end{cases}$$

we may extend the integration over all $z'$ to get

$$\phi = \phi_+ + \phi_- = v + \int_{-\infty}^{\infty} G(z - z')\phi_+(z')\,dz' \tag{50}$$

Provided we can demonstrate that the following transforms have a common strip of analyticity, we may take the bilateral Laplace transform of (50),

$$\begin{aligned}
\psi_+ &= \int_0^\infty e^{-wz}\phi_+(z)\,dz \\
\psi_- &= \int_{-\infty}^0 e^{-wz}\phi_-(z)\,dz \\
\mathcal{G} &= \int_{-\infty}^{\infty} e^{-wz}G(z)\,dz \\
V_+ &= \int_0^\infty v_+(z)e^{-wz}\,dz \\
V_- &= \int_{-\infty}^0 v(z)e^{-wz}\,dz
\end{aligned}$$

Using the faltung theorem gives

$$\psi_+ + \psi_- = V_+ + V_- + \mathcal{G}\psi_+$$

or

$$(1 - \mathcal{G})\psi_+ - V_+ = V_- - \psi_- \tag{51}$$

The next step requires a factorization of $1 - \mathcal{G}$ to give

$$1 - \mathcal{G} = \frac{H_+(W)}{H_-(W)}$$

where $H_+$ is analytic in a right half plane and $H_-$ is analytic in a left half plane. We now obtain from (51)

$$H_+\psi_+ - V_+H_- = H_-(V_- - \psi_-) \tag{52}$$

As a next step, we must factor the known function $V_+H_-$ into the sum of two factors $S_+$ and $S_-$, where $S_+$ is analytic in a right half plane and $S_-$ is analytic in a left half plane. In place of (52), we now have

$$H_+\psi_+ - S_+ = H_-(V_- - \psi_-) + S_- \tag{53}$$

The left-hand side is analytic in a right half plane, while the right side is analytic in a left half plane. In addition, if the two sides have a common strip of analyticity, then $H_+\psi_+ - S_+$ provides the analytic continuation of $H_-(V_- - \psi_-) + S_-$ into the right half plane. Together they define an integral function $h(w)$, whose form can generally be determined from the known asymptotic properties of the transforms. Having determined $h(w)$, we may solve for $\psi_+$ to obtain

$$\psi_+ = \frac{h + S_+}{H_+} \tag{54}$$

The solution for $\psi$ may now be found from (51) and is

$$\psi = V + \mathcal{G}\frac{h + S_+}{H_+} \tag{55}$$

The solution for $\phi$ is obtained from the inversion of the transform $\psi$.

The solution for the homogeneous Wiener-Hopf equation ($v = 0$) may be obtained in a similar way. Other equations of somewhat different form may be solved in an analogous manner also. In fact, the integral equation (34) differs from the type (49), but could be solved since the function $\phi_+(z)$ was known.

The most difficult task in the solution is carrying out the factorization of the term $1 - \mathcal{G}$ and $V_+H_-$. This can only be done by inspection for certain cases. Fortunately, however, a general method of carrying out the required factorization does exist. This method is based on the Cauchy integral representation leading to the Laurent expansion of a function which is analytic within a circular annulus or strip. The technique is fully described elsewhere, and, since we are not going to make use of the method, we omit the derivation.†

## 10.2. An Infinite Array of Parallel Metallic Plates

A knowledge of the reflection and transmission coefficients at the interface between free space and an infinite array of parallel metallic plates as illustrated in Fig. 10.5 is of importance in at least three applications. These are:

1. Parallel-plate microwave lens medium
2. Corrugated-plane surface waveguides for antenna applications
3. Strip artificial dielectric medium

In the first application, knowledge of the reflection and transmission coefficients for an incident TEM wave is all that is required. This solution has been given by Carlson and Heins for both polarizations, and is based on the Wiener-Hopf method. The solution to be presented here is that based on a direct construction of a suitable transform function.

If a short-circuiting plane is located at $z = l$, the resulting structure is that of a corrugated plane, along which surface waves may be guided in the $x$ direction. The surface wave is obtained by letting the angle of incidence $\theta_i$ become complex. The propagation constant in the $x$ direction may be determined by a simple transmission-line analysis, once the equivalent-circuit parameters of the interface are known. If the spacing $s$ is small compared with a wavelength, and the short-circuit position $l$ is large compared with $s$, then only the parameters of the dominant mode

† P. M. Morse and H. Feshbach, "Methods of Theoretical Physics," sec. 8.5, McGraw-Hill Book Company, Inc., New York, 1953.

are required in order to analyze the properties of the structure. When these conditions are not fulfilled, one or more higher-order modes must be taken into account as well, since the least attenuated of these will have a field which interacts appreciably with the short circuit.

The two-dimensional strip-type artificial dielectric medium is treated in Chap. 12, and so we will postpone all comments on this structure until later.

For the parallel-plate medium illustrated in Fig. 10.5, it is assumed that each plate is infinitely thin and perfectly conducting. Only the case of a parallel-polarized incident TEM wave, with the plane of incidence restricted to the $xz$ plane, will be considered. The spacing $s$ is limited to

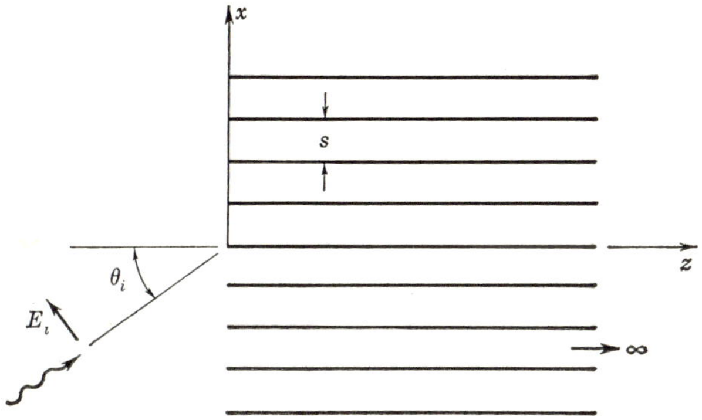


FIG. 10.5. Infinite array of equispaced parallel metallic plates.

less than $\lambda_0/2$, and so only the dominant modes on the two sides of the interface propagate. As a consequence of these assumptions, the field may be derived from a single scalar function $\phi(x,z)$, which is proportional to the $y$ component of the magnetic field. The only field components present are $H_y$, $E_x$, $E_z$, and these are given by

$$H_y = \phi \qquad E_x = j\frac{Z_0}{k_0}\frac{\partial\phi}{\partial z} \qquad E_z = -j\frac{Z_0}{k_0}\frac{\partial\phi}{\partial x}$$

Let the incident magnetic field for $z < 0$ be

$$\phi_i = A_0 e^{-jhx-\Gamma_0 z} \tag{56a}$$

where $h = k_0 \sin\theta_i$, $\Gamma_0 = jk_0\cos\theta_i$. The dominant plus higher-order reflected modes for $z < 0$ are given by

$$\phi_r = -R_1 A_0 e^{-jhx+\Gamma_0 z} + \sum_{n=-\infty}^{\infty}{}' B_n e^{-j(h+2n\pi/s)+\Gamma_n z} \tag{56b}$$

where $\Gamma_n^2 = (h + 2n\pi/s)^2 - k_0^2$, and $R_1$ is the reflection coefficient. A negative sign is introduced in front of $R_1$ in (56b) so that $R_1$ will be the usual reflection coefficient for the transverse electric field $E_x$. The form of solution given by (56) follows from a consideration of the periodic nature of the structure and the condition that $\phi$ be a solution of the two-dimensional scalar Helmholtz equation. The incident field has a progressive phase delay along the $x$ axis according to the factor $e^{-jhx}$. The scattered field must have this same progressive phase delay, and hence can be represented as the product of the factor $e^{-jhx}$ and a periodic function of $x$ with period $s$, in accordance with Floquet's theorem. The prime on the summation in (56b) means omission of the term $n = 0$, which has been separated out.

In the region $z > 0$, the solutions for $\phi$ are the usual waveguide modes ($E$ modes), and hence

$$\phi_t = \sum_{n=0}^{\infty} C_n \cos \frac{n\pi x}{s} e^{-\gamma_n z} \qquad 0 \le x \le s \tag{57}$$

For any other range of $x$, the solution is given by

$$\phi_t(x + ms, z) = e^{-jhmx}\phi_t(x,z)$$

where $m$ is any integer. In (57), $\gamma_n$ is given by $\gamma_n^2 = (n\pi/s)^2 - k_0^2$. The coefficient $C_0$ is equal to $T_{12}e^{-jhs/2}A_0\Gamma_0/\gamma_0$, where $T_{12}$ is the transverse electric field transmission coefficient. Since the phase of the incident wave varies with $x$, a phase reference point $x = s/2$ has been chosen in the definition of $T_{12}$.

The total field is a solution of the equation

$$\frac{\partial^2\phi}{\partial z^2} + \frac{\partial^2\phi}{\partial x^2} + k_0^2\phi = 0 \tag{58}$$

and satisfies the periodicity condition

$$e^{-jhms}\phi(x,z) = \phi(x + ms, z) \tag{59}$$

for all $z$ and $m$ an arbitrary integer. In addition, $\partial\phi/\partial x$ satisfies this periodicity condition as well as the condition $\partial\phi/\partial x = 0$ on the perfectly conducting plates in the half space $z > 0$. By virtue of these boundary conditions, the boundary-value problem reduces to that illustrated in Fig. 10.6. The range of $x$ involved is $0_+ \le x \le s_-$, and, since $\phi$ is discontinuous across each conducting plate by an amount equal to the current on each plate, the periodicity condition (59) cannot be applied to $\phi$ for $z \ge 0$, since this condition really means

$$e^{-jhs}\phi(0_+,z) = \phi(s_+,z)$$

The region $x = s_+$ is outside the domain of $x$ in the reduced problem of Fig. 10.6. However, $\partial\phi/\partial x = 0$ for $x = 0$, $s$; $z > 0$, and so the periodicity condition does hold for $\partial\phi/\partial x$ for all values of $z$. At the edges of the plates, $\phi$ is finite, but $\partial\phi/\partial z$ is asymptotic to $z^{-1/2}$. This edge condition governs the growth of the transform of $\phi$ at infinity.

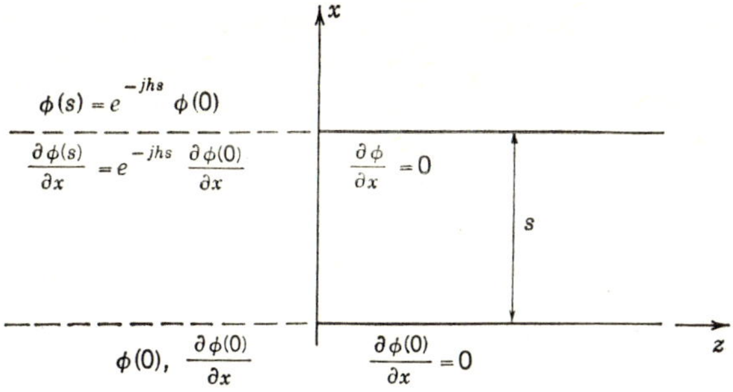


FIG. 10.6. Reduced boundary-value problem.

The solution for $\phi$ which satisfies the preceding conditions and which has the eigenfunction expansions (56) and (57) will be determined by using the known properties of $\phi$ to construct a bilateral Laplace transform, from which $\phi$ may be found by suitable contour integration.

*Transform Solution*

Let the bilateral Laplace transform of the total field $\phi$ be

$$\psi(x,w) = \int_{-\infty}^{\infty} e^{-wz}\phi(x,z)\,dz \tag{60}$$

For convenience, a transform function $g(w)$, defined as follows, will also be introduced:

$$g(w) = e^{jhs}\int_{-\infty}^{\infty} \phi(s_-,z)e^{-wz}\,dz \tag{61}$$

For $z < 0$, $\phi$ is asymptotic to $A_0e^{-\Gamma_0 z} - R_1A_0e^{\Gamma_0 z}$. Hence, the transform $\psi_-$ given below is analytic for all $w$ in the half plane $\mathrm{Re}\,w < 0$,

$$\psi_-(x,w) = \int_{-\infty}^{0} e^{-wz}\phi(z)\,dz \tag{62}$$

In the region $z > 0$, it will be assumed that small losses are present, so that $\gamma_0 = j|\gamma_0'| + \gamma_0''$, where $\gamma_0''$ is a small real and positive attenuation constant. The effect of small losses on the higher-order modes is neglected, since these decay according to $e^{-\gamma_n z}$, with $\gamma_n$ real anyway. At the end of the analysis, $\gamma_0''$ will be made zero. Thus, as $z$ approaches plus infinity, $\phi$ is asymptotic to $C_0e^{-\gamma_0'' z}$, and, hence,

$$\psi_+(x,w) = \int_{0}^{\infty} e^{-wz}\phi(x,z)\,dz \tag{63}$$

is analytic for all $w$ in the half plane $\mathrm{Re}\,w > -\gamma_0''$.

The assumption of small losses for $z > 0$ was made in order to obtain a finite-width strip in the $w$ plane in which both $\psi_-$ and $\psi_+$ are analytic. We may therefore invert (60) to get

$$\phi(x,z) = \frac{1}{2\pi j} \int_C e^{wz} \psi(x,w)\, dw \tag{64}$$

where $C$ is a contour running parallel to the imaginary $w$ axis and in the common strip of analyticity, i.e., in the strip $-\gamma_0'' < \text{Re}\, w < 0$. The transform of (58) gives

$$\frac{d^2\psi}{dx^2} + u^2\psi = 0 \tag{65}$$

where $u^2 = w^2 + k_0^2$. The transform $\psi$ must satisfy the boundary conditions

$$\left.\frac{d\psi}{dx}\right|_{x=s_-} = e^{-jhs} \left.\frac{d\psi}{dx}\right|_{x=0-} \tag{66a}$$

$$\frac{d\psi}{dx} = 0 \qquad x = 0,\, s \qquad z > 0 \tag{66b}$$

$$\psi(s_-) = e^{-jhs}\psi(0) \qquad z < 0 \tag{66c}$$

A general solution of (65) is $\psi(x,w) = A(w) \sin ux + B(w) \cos ux$. Condition (66$a$) gives

$$Au \cos us - Bu \sin us = e^{-jhs}uA$$

and, hence,
$$B = -\frac{A(e^{-jhs} - \cos us)}{\sin us}$$

From (61), we get

$$g(w) = e^{jhs}(A \sin us + B \cos us) = \frac{Ae^{jhs}}{\sin us} (\sin^2 us - e^{-jhs} \cos us + \cos^2 us)$$

Solving for $A$ gives

$$A = \frac{g \sin us}{e^{jhs} - \cos us}$$

The transform $\psi$ is given in terms of $g$ as follows:

$$\psi(x,w) = \frac{g(w)}{e^{jhs} - \cos us} [\cos u(s - x) - e^{-jhs} \cos ux]$$

after substituting for $A$ and $B$. Inverting this transform, we get

$$\phi(x,z) = \frac{1}{2\pi j} \int_C \frac{e^{wz} g(w)}{e^{jhs} - \cos us} [\cos u(s - x) - e^{-jhs} \cos ux]\, dw \tag{67}$$

From the condition $\phi(s_-,z) = e^{-jhs}\phi(0_+,z)$ for $z < 0$, we obtain the following condition on $g(w)$ [equivalent to (66$c$)]:

$$\int_C \frac{e^{wz} g(w)}{e^{jhs} - \cos us} (\cos hs - \cos us)\, dw = 0 \qquad z < 0 \tag{68}$$

The condition $\partial\psi/\partial x = 0$, $x = 0, s$; $z > 0$, imposes the further condition

$$\int_C \frac{e^{wz}g(w)}{e^{jhs} - \cos us} (u \sin us)\, dw = 0 \qquad z > 0 \tag{69}$$

Besides satisfying the conditions (68) and (69), the function $g(w)$ must be such that it will yield the solution for $\phi$ in the form of the eigenfunction expansions (56) and (57).

In (67), the contour $C$ may be closed in the right-half $w$ plane for $z < 0$. Hence, $g$ must have simple poles at $w = \Gamma_n$, $n = \pm 1, \pm 2, \ldots$, in the right half plane in order to give rise to the reflected evanescent modes for $z < 0$. Simple poles at $w = \pm\Gamma_0$ on the imaginary axis are required in order to obtain the incident and reflected dominant modes for $z < 0$. For $z > 0$, we may close the contour in the left half plane, and $g$ must have simple poles at $w = -\gamma_n$, $n = 0, 1, 2, \ldots$, in order to produce the required eigenfunctions for $z > 0$.

For (68) to hold, $g$ must cancel the poles of the factor $(e^{jhs} - \cos us)^{-1}$ in the right half plane, in order that the integrand shall be analytic in the right half plane, and cause the integral to vanish identically for $z < 0$. The poles of $g$ in the right half plane do not cause any difficulty, since the numerator of (68) contains the factor $\cos hs - \cos us$, which vanishes for $w = \pm\Gamma_n$; $n = \pm 1, \pm 2, \ldots$. Similarly, (69) will hold, provided $g$ cancels the poles of $(e^{jhs} - \cos us)^{-1}$ in the left half plane as well. Again the poles of $g$ cause no difficulty, since $\sin us = 0$ for $w = \gamma_n$ in the left-half $w$ plane.

Taking all the above considerations into account, it is readily seen that a suitable form for the transform function $g$ is

$$g(w) = \frac{p(w)(e^{jhs} - \cos us)}{(w^2 - \Gamma_0{}^2)(w + \gamma_0) \prod_{n=1}^{\infty} \dfrac{\gamma_n + w}{n\pi/s} e^{-ws/n\pi} \prod_{n=1}^{\infty} \dfrac{(\Gamma_n - w)(\Gamma_{-n} - w)}{(2n\pi/s)^2} e^{ws/n\pi}} \tag{70}$$

where $p(w)$ is an integral function yet to be determined. An examination of the logarithmic derivative of the infinite products shows that the exponential factors and the terms $n\pi/s$ and $(2n\pi/s)^2$ in the denominators ensure the uniform convergence of these products, since $\gamma_n \to n\pi/s$, and $\Gamma_n, \Gamma_{-n} \to 2n\pi/s$, for $n$ large.

The integral function $p(w)$ is to be chosen so that $g$ has algebraic growth at infinity and is asymptotic to $w^{-3/2}$. This will ensure that $\phi$ has the proper behavior at the edge of the plates. For $n$ large, we have $\Gamma_n \to 2n\pi/s + h$, $\Gamma_{-n} \to 2n\pi/s - h$, $\gamma_n \to n\pi/s$. Hence, the term $e^{-ws/n\pi}(\gamma_n + w)/(n\pi/s)$ approaches $e^{-ws/n\pi}(1 + ws/n\pi)$. The other term

$e^{ws/n\pi}(\Gamma_n - w)(\Gamma_{-n} - w)/(2n\pi/s)^2$ approaches

$$e^{(w+h)s/2n\pi}\left(1 - \frac{w+h}{2n\pi}s\right)e^{(w-h)s/2n\pi}\left(1 - \frac{w-h}{2n\pi}s\right)$$

Thus, the transform function $g$ differs only by a bounded function of $w$ from $g_1(w)$, where $g_1(w)$ is given by

$$g_1(w) = p(w)(e^{jhs} - \cos us)\left[w^3 \prod_{n=1}^{\infty}\left(1 + \frac{ws}{n\pi}\right)e^{-ws/n\pi}\right.$$
$$\left. \times \prod_{n=1}^{\infty}\left(1 - \frac{w+h}{2n\pi}s\right)e^{(w+h)s/2n\pi}\prod_{n=1}^{\infty}\left(1 - \frac{w-h}{2n\pi}s\right)e^{(w-h)s/2n\pi}\right]^{-1}$$
$$= p(w)(e^{jhs} - \cos us)\pi s\Gamma\left(\frac{ws}{\pi}\right)\left[w^2 \sin\frac{(w+h)s}{2}\sin\frac{(w-h)s}{2}\right.$$
$$\left.\Gamma\left(\frac{w+h}{2\pi}s\right)\Gamma\left(\frac{w-h}{2\pi}s\right)\right]^{-1}$$

upon using the infinite-product expansion for the gamma function. The asymptotic form of $g_1$ is readily found from the known asymptotic form of the gamma function, and is

$$g_1 \sim \frac{Kp(w)(e^{jhs} - \cos us)e^{-(ws/\pi)\ln 2}}{w^{3/2}\sin[(w+h)s/2]\sin[(w-h)s/2]}$$

where $K$ is a suitable constant. In order for $g$ to have algebraic growth at infinity and to be asymptotic to $w^{-3/2}$, we must choose $p(w)$ equal to a constant times $\exp[(ws/\pi)\ln 2]$. The reflection and transmission coefficients depend only on the ratio of amplitudes, and, hence, the exact value of the constant $K$ is not required.

We may now return to (67) and evaluate $\phi$ after substituting for $g$ from (70) and replacing $p(w)$ by $\exp[(ws/\pi)\ln 2]$. For $z < 0$, the contour $C$ may be closed in the right half plane. The incident dominant mode is given by the residue at $w = -\Gamma_0$, and is

$$\phi_i = A_0 e^{-jhx-\Gamma_0 z} = \frac{j \sin hs\, e^{-(\Gamma_0 s/\pi)\ln 2}e^{-jhx-\Gamma_0 z}}{\left[\prod_{n=1}^{\infty}\frac{(\gamma_n - \Gamma_0)(\Gamma_n + \Gamma_0)(\Gamma_{-n} + \Gamma_0)}{4(n\pi/s)^3}\right]2\Gamma_0(\gamma_0 - \Gamma_0)} \tag{71}$$

An additional minus sign arises because the integration is carried out in a clockwise sense when the contour $C$ is closed in the right half plane. The reflected dominant mode is given by the residue at $w = \Gamma_0$, and is

$$\phi_{r0} = -R_1 A_0 e^{-jhx+\Gamma_0 z}$$
$$= \frac{-j \sin hs\, e^{(\Gamma_0 s/\pi) \ln 2} e^{-jhx+\Gamma_0 z}}{\left[\prod_{n=1}^{\infty} \frac{(\gamma_n + \Gamma_0)(\Gamma_n - \Gamma_0)(\Gamma_{-n} - \Gamma_0)}{4(n\pi/s)^3}\right] 2\Gamma_0(\gamma_0 + \Gamma_0)} \tag{72}$$

Comparing (71) and (72), we obtain the solution for the reflection coefficient $R_1$:

$$R_1 = \rho e^{j\theta_1} = e^{-2\Gamma_0(s/\pi)\ln 2} \frac{\gamma_0 - \Gamma_0}{\gamma_0 + \Gamma_0} \prod_{n=1}^{\infty} \frac{(\gamma_n - \Gamma_0)(\Gamma_n + \Gamma_0)(\Gamma_{-n} + \Gamma_0)}{(\gamma_n + \Gamma_0)(\Gamma_n - \Gamma_0)(\Gamma_{-n} - \Gamma_0)} \tag{73}$$

If we now assume that $\gamma_0''$ is zero, so that $\gamma_0$ as well as $\Gamma_0$ are pure-imaginary, we see that

$$\rho = |R_1| = \frac{\gamma_0 - \Gamma_0}{\gamma_0 + \Gamma_0} \tag{74a}$$

$$\theta_1 = -2\left[|\Gamma_0| \frac{s}{\pi} \ln 2 + \sum_{n=1}^{\infty} \left(\tan^{-1} \frac{|\Gamma_0|}{\gamma_n} - \tan^{-1} \frac{|\Gamma_0|}{\Gamma_n} - \tan^{-1} \frac{|\Gamma_0|}{\Gamma_{-n}}\right)\right] \tag{74b}$$

For $z > 0$, the contour is closed in the left half plane. The dominant-mode transmitted field is given by the residue at $w = -\gamma_0$. From the relation $\gamma_0 C_0 = T_{12} A_0 \Gamma_0 e^{-jhs/2}$, the transmission coefficient $T_{12}$ is found, where $C_0$ is the amplitude of the transverse magnetic field of the transmitted wave. Thus,

$$T_{12} = e^{(\gamma_0 - \Gamma_0)(s/\pi)\ln 2} \sec \frac{hs}{2} \frac{2\gamma_0}{\gamma_0 + \Gamma_0} \prod_{n=1}^{\infty} \frac{(\gamma_n - \Gamma_0)(\Gamma_n + \Gamma_0)(\Gamma_{-n} + \Gamma_0)}{(\gamma_n - \gamma_0)(\Gamma_n + \gamma_0)(\Gamma_{-n} + \gamma_0)} \tag{75}$$

If, instead of having a dominant mode incident from $z < 0$, we have a dominant mode incident from the parallel-plate region, a similar analysis may be carried out. The only modification in the transform function $g$ which is required is replacing the factor $w + \Gamma_0$ by $w - \gamma_0$. This eliminates the incident wave for $z < 0$, and introduces an incident wave in the region $z > 0$. It is found that the reflection coefficient $R_2$ and the transmission coefficient $T_{21}$ are given by

$$R_2 = -\rho e^{j\theta_2} \tag{76a}$$

where $\rho$ is given by Eq. (74$a$) and

$$\theta_2 = 2\left[|\gamma_0| \frac{s}{\pi} \ln 2 + \sum_{n=1}^{\infty} \left(\tan^{-1} \frac{|\gamma_0|}{\gamma_n} - \tan^{-1} \frac{|\gamma_0|}{\Gamma_n} - \tan^{-1} \frac{|\gamma_0|}{\Gamma_{-n}}\right)\right] \tag{76b}$$

$$T_{21} = \frac{\Gamma_0}{\gamma_0} T_{12} \tag{76c}$$

Comparing (75) with (74*b*) and (76*b*) shows that

$$\angle T_{12} = \angle T_{21} = \frac{\theta_1 + \theta_2}{2} \tag{77}$$

a result that may be deduced by more elementary means from the properties of a lossless two-terminal-pair network. The transmitted power must be equal to the difference between the incident and reflected power, and this relation leads directly to a simple expression for the modulus of $T_{12}$ and $T_{21}$, that is,

$$|T_{12}T_{21}| = 1 - \rho^2 \tag{78}$$

When the spacing $s$ is less than about $0.2\lambda_0$, then the series in (74*b*) and (76*b*) contributes only a small amount to the phase angles $\theta_1$ and $\theta_2$.

### 10.3. Application to Capacitive-loaded Parallel-plate Transmission Line

If conducting planes are placed at $z = d$ and $z = -(b - d)$ as in Fig. 10.7, the resultant structure becomes that of a parallel-plate transmission line of height $b$ and loaded at regular intervals $s$ by asymmetrical

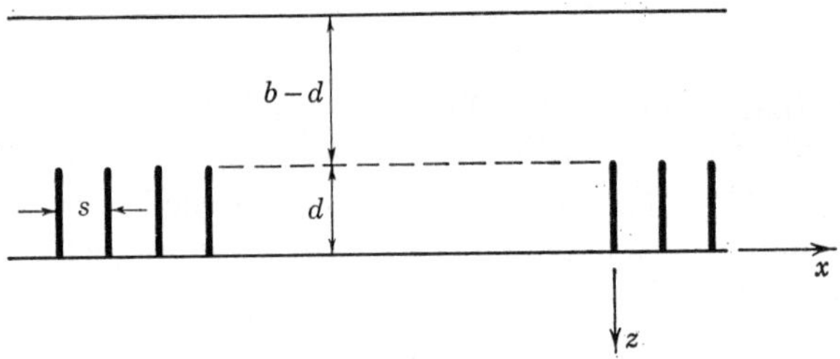


FIG. 10.7. Capacitive-loaded parallel-plate transmission line.

capacitive diaphragms of height $d$. Provided $s$ is small compared with a wavelength, and $b - d$ and $d$ are large compared with the spacing $s$, the solution for the propagation constant of the periodic wave in the loaded transmission line may be readily found from the solution of the parallel-plate medium given in Sec. 10.2.

Referring to Fig. 10.5, let the incident parallel-polarized TEM wave be $\phi_i = A_0 e^{-jhx - \Gamma_0 z}$. A dominant mode is transmitted into the parallel-plate region. There is also a reflected TEM mode in the free-space region. For a small spacing $s$, all the higher-order modes are evanescent in the $z$ direction and, hence, localized to the region near the interface. A conducting plane placed at $z = d$ interacts with the dominant mode, only provided $d$ is sufficiently large. With the conducting plane in place, a perfect standing wave along the $z$ direction exists in the free-space region. For certain values of $z$, the $x$ component of electric field will therefore be zero. At these positions, a second conducting plane may be placed to obtain the parallel-plate transmission line of Fig. 10.7.

The interface reflection and transmission coefficients are those given by (73) to (78). For $s \ll \lambda_0$, approximate values of these are given by

$$R_1 = \frac{\gamma_0 - \Gamma_0}{\gamma_0 + \Gamma_0} e^{-2\Gamma_0(s/\pi)\ln 2} \tag{79a}$$

$$R_2 = -\frac{\gamma_0 - \Gamma_0}{\gamma_0 + \Gamma_0} e^{-2\gamma_0(s/\pi)\ln 2} \tag{79b}$$

$$T_{12} = \frac{2\gamma_0}{\gamma_0 + \Gamma_0} e^{(\gamma_0 - \Gamma_0)(s/\pi)\ln 2} \tag{79c}$$

$$T_{21} = \frac{\Gamma_0}{\gamma_0} T_{12} \tag{79d}$$

The form of these reflection and transmission coefficients shows that the equivalent transmission-line circuit of a transverse section consists of two sections of short-circuited transmission lines of lengths $b - d + (s/\pi)\ln 2$ and $d - (s/\pi)\ln 2$, with characteristic impedances proportional to $\Gamma_0/\gamma_0$ and unity. This equivalent circuit is illustrated in Fig. 10.8. For specified values of $b$, $d$, and $s$, the value of $\Gamma_0$ is determined by the condition that the above circuit of a transverse section should be of resonant length. By conventional transmission-line theory, it is found that the resonance condition is determined by

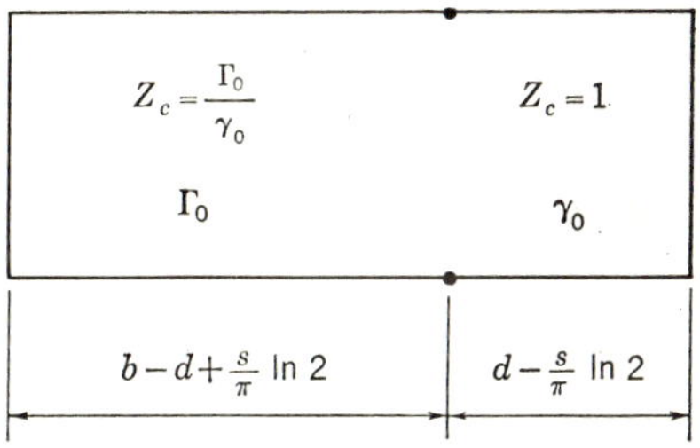

FIG. 10.8. Equivalent circuit of transverse section.

$$\Gamma_0 \tanh \Gamma_0 \left(b - d + \frac{s}{\pi}\ln 2\right) = k_0 \tan k_0 \left(d - \frac{s}{\pi}\ln 2\right) \tag{80}$$

since $\gamma_0 = jk_0$. The propagation phase constant along the $x$ direction is $h = k_0 \sin \theta_i$, and $h$ is related to $\Gamma_0$ as follows:

$$h^2 = \Gamma_0^2 + k_0^2 \tag{81}$$

The loading of the transmission line by capacitive diaphragms has an effect similar to increasing the equivalent dielectric constant of the medium. This results in $h^2$ being greater than $k_0^2$, and, hence, $\Gamma_0$ is real and the angle $\theta_i$ complex. The parameter $h$ may be considered as the characteristic propagation phase constant for the periodically loaded line, and (80) as the eigenvalue equation for $h$.

For larger spacings of the diaphragms, the above analysis is no longer valid. We may now, however, apply the analysis of Sec. 9.5. The resultant eigenvalue equation for $h$ is

$$\cos hs = \cos k_0 s - \frac{B}{2} \sin k_0 s \tag{82}$$

where $B$ is the normalized shunt susceptance of a single diaphragm. An approximate value of $B$ is given by the low-frequency formula

$$B = \frac{8bk_0}{2\pi} \ln \sec \frac{\pi d}{2b} \tag{83}$$

The combination of (80) and (82) leads to a reasonably good approximation to $h$ for all values of spacing $s$. All the above formulas become applicable to the capacitively loaded rectangular guide if $k_0^2$ is replaced by $k_0^2 - (\pi/a)^2$ throughout. This result follows from the same considerations which were used to obtain the solution for a capacitive diaphragm in a rectangular guide in Chap. 8. Image theory may be used

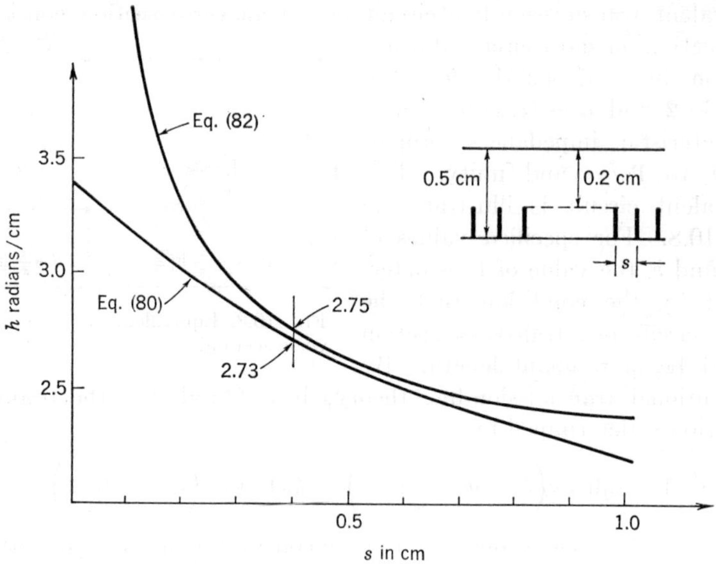


FIG. 10.9. A plot of characteristic propagation constant $h$ versus spacing $s$.

to obtain the solution when the diaphragms are symmetrically placed along both plates or centered midway between the two plates of a transmission line.

As an example, $h$ has been computed by the two methods for $0 < s < 1$ centimeter, $k_0 = 2$, $b = 0.5$ centimeter, $d = 0.3$ centimeter, and the results are plotted in Fig. 10.9. It is seen that, in the range $0.4 < s < 0.6$, the two methods give essentially the same results (within 1 per cent). For $s > 0.5$ we use (82), while for $s < 0.5$ we use (80), and thus obtain accurate results for the complete range of $s$ considered. Since $h$ is larger than $k_0$, the phase velocity $v_p = v_c k_0/h$ has been reduced to a value less than that of light in free space. The maximum reduction is in the ratio 2:3.39 for a value of $s$ approaching zero. An increase in the diaphragm height from 0.3 to 0.45 centimeter results in a maximum reduction of phase velocity by a factor 3.76 for $s$ approaching zero.

## 10.4. Inductive Semidiaphragm in a Rectangular Guide

The special case of an inductive diaphragm of the asymmetrical type and extending halfway across the guide as in Fig. 10.10 may be solved by any one of the three methods presented earlier. The similar problem of the capacitive semidiaphragm, as well as the grating problems which reduce to the above waveguide problems, may be solved in the same way. To date, the extension to the general diaphragm or strip-grating problems has not been achieved.

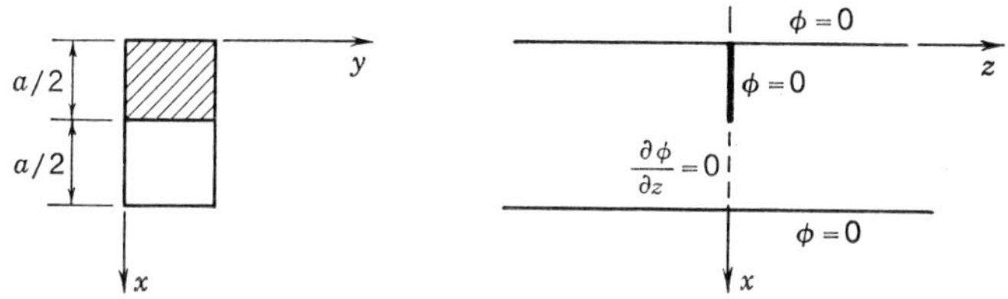


FIG. 10.10. Inductive semidiaphragm.

The problem under consideration may be simplified by considering the two special cases of even and odd excitation. For the odd case, let the electric field of the incident $H_{10}$ modes be

$$E_y = -A_1 \sin \frac{\pi x}{a} e^{-\Gamma_1 z} \qquad z < 0 \tag{84a}$$

$$E_y = A_1 \sin \frac{\pi x}{a} e^{\Gamma_1 z} \qquad z > 0 \tag{84b}$$

In view of the symmetry involved, the scattered field will be of the following form:

$$E_y = -A_1 R_0 \sin \frac{\pi x}{a} e^{\Gamma_1 z} - \sum_{n=2}^{\infty} A_n \sin \frac{n\pi x}{a} e^{\Gamma_n z} \qquad z < 0 \tag{85a}$$

$$E_y = A_1 R_0 \sin \frac{\pi x}{a} e^{-\Gamma_1 z} + \sum_{n=2}^{\infty} A_n \sin \frac{n\pi x}{a} e^{-\Gamma_n z} \qquad z > 0 \tag{85b}$$

where $\Gamma_n^2 = (n\pi/a)^2 - k_0^2$. At the position of the diaphragm $z = 0$, the electric field must be continuous and must vanish on the diaphragm $0 \le x \le a/2$. Since the sin $(n\pi x/a)$ functions are orthogonal, this is possible only if $R_0 = -1$ and $A_n = 0$, $n > 1$. Hence, for odd excitation, we may close the aperture by an electric wall since the total electric field vanishes at $z = 0$.

For the case of even excitation, let the incident electric field be

$$E_y = A_1 \sin \frac{\pi x}{a} e^{-\Gamma_1 z} \qquad z < 0 \tag{86a}$$

$$E_y = A_1 \sin \frac{\pi x}{a} e^{\Gamma_1 z} \qquad z > 0 \tag{86b}$$

Symmetry considerations show that the scattered field will be of the following form:

$$E_y = R_e A_1 \sin \frac{\pi x}{a} e^{\Gamma_1 z} + \sum_{n=2}^{\infty} B_n \sin \frac{n\pi x}{a} e^{\Gamma_n z} \qquad z < 0 \tag{87a}$$

$$E_y = R_e A_1 \sin \frac{\pi x}{a} e^{-\Gamma_1 z} + \sum_{n=2}^{\infty} B_n \sin \frac{n\pi x}{a} e^{-\Gamma_n z} \qquad z > 0 \tag{87b}$$

The transverse electric field is an even function of $z$, and, hence, the transverse magnetic field $H_x$, which is proportional to $\partial E_y/\partial z$, will be an odd function of $z$. At the aperture $z = 0$, $a/2 \le x \le a$, the transverse magnetic field vanishes, and, hence, the aperture may be closed by a magnetic wall. The reflection coefficient $R_e$ will not be equal to $-1$ in the present case, since the boundary at $z = 0$ is not a homogeneous one. The magnitude of $R_e$ will, however, equal unity, since no power can be transmitted through either the diaphragm or the magnetic wall. For the purpose of evaluating $R_e$, we need to solve the reduced problem of a semi-infinite guide closed by a combination of an electric wall for $0 \le x \le a/2$ and a magnetic wall for $a/2 \le x \le a$. The general solution is obtained by superimposing the solutions for the odd and even cases, i.e., by addition of (84) and (85) to (86) and (87). The incident field for $z < 0$ will have an amplitude of $2A_1$, while for $z > 0$ there is no incident field. The reflected field has an amplitude of $(R_e - 1)A_1$, and, hence, the equivalent reflection coefficient for an $H_{10}$ mode incident from one side only is

$$R = \frac{R_e - 1}{2} \tag{88}$$

The normalized shunt susceptance $B$ of the diaphragm is given by $-2jR/(1 + R)$. If we let $R_e = e^{j\theta}$, we find that an equivalent expression for $B$ is

$$B = 2 \tan \frac{\theta}{2} \tag{89}$$

The electric field $E_y$ for $z \ge 0$ may be represented by contour integrals in terms of suitable transform functions. For $0 \le x \le a/2$, let

$$E_y = \int_C (e^{-wz} - e^{wz}) A(w) \sin ux \, dw \tag{90a}$$

while, for $a/2 \le x \le a$,

$$E_y = \int_C (e^{-wz} + e^{wz}) B(w) \sin u(a - x) \, dw \tag{90b}$$

where $u^2 = w^2 + k_0^2$. This form of solution is seen to satisfy the required boundary conditions at $z = 0$,

$$E_y = 0 \qquad 0 \le x \le \frac{a}{2}$$

$$\frac{\partial E_y}{\partial z} = 0 \qquad \frac{a}{2} \le x \le a$$

as well as the boundary conditions at $x = 0, a; z > 0$. It is also a solution of the two-dimensional scalar Helmholtz equation. Continuity of $E_y$ at $x = a/2$ for all $z$ gives

$$\int_C A(e^{-wz} - e^{wz}) \sin \frac{ua}{2} \, dw = \int_C B(e^{-wz} + e^{wz}) \sin \frac{ua}{2} \, dw \tag{91}$$

We may also represent $E_y$ at $x = a/2$ by means of the following contour integral:

$$E_y\left(\frac{a}{2}, z\right) = \int_C e^{-wz} F(w) \, dw \qquad z > 0 \tag{92}$$

where $F(w)$ is a suitable transform function. To be consistent with (91), we require that $F(w)$ be such that

$$\int_C e^{wz} F(w) \, dw = 0 \qquad z > 0$$

which is equivalent to

$$\int_C e^{-wz} F(w) \, dw = 0 \qquad z < 0$$

Thus, $F(w)$ must be free of poles in the half plane $\operatorname{Re} w < 0$, since the contour may be closed in this half plane, and there must be no residue of the integrand inside the contour. A form for $F(w)$ which satisfies this criterion may be found, since $E_y$ is equal to zero for $z < 0$. Comparing (92) with (91) shows that

$$A(w) \sin \frac{ua}{2} = F(w) \tag{93a}$$

$$B(w) \sin \frac{ua}{2} = F(w) \tag{93b}$$

The solution for $E_y$ may be written as follows:

$$E_y = \int_C F(w) \frac{\sin ux}{\sin (ua/2)} (e^{-wz} - e^{wz}) \, dw \qquad 0 \le x \le \frac{a}{2} \tag{94a}$$

$$E_y = \int_C F(w) \frac{\sin u(a - x)}{\sin (ua/2)} (e^{-wz} + e^{wz}) \, dw \qquad \frac{a}{2} \le x \le a \tag{94b}$$

Continuity of the derivative $\partial E_y/\partial x$ at $x = a/2$ imposes the following further condition on the function $F(w)$:

$$\int_C F(w)u \cot \frac{ua}{2} e^{-wz}\, dw = 0 \qquad z > 0 \tag{95}$$

If the diaphragm were not equal to $a/2$ in width, the continuity of the derivative would lead to an equation involving both $e^{-wz}$ and $e^{wz}$. This appears to be the reason why the method of solution being carried out works only for the special case of the semidiaphragm.

The contour in (95) may be closed in the right-half $w$ plane, and the integral will vanish, provided $F(w)$ has zeros at $u = 2n\pi/a$ or

$$w = \left[\left(\frac{2n\pi}{a}\right)^2 - k_0^2\right]^{1/2}$$

$n = 1, 2, 3, \ldots$, in the right half plane, so as to cancel the poles of the cotangent function. Also, in addition, $F(w)$ must be free of poles in the right half plane, except possibly for poles at the zeros of the cotangent function, i.e., at $w = [(n\pi/a)^2 - k_0^2]^{1/2}$, $n = 1, 2, \ldots$. For convenience, let

$$\Gamma_n^2 = \left(\frac{n\pi}{a}\right)^2 - k_0^2$$

$$\Gamma_{2n}^2 = \left(\frac{2n\pi}{a}\right)^2 - k_0^2$$

At $x = a/2$, the eigenfunction expansion of $E_y$ is of the form

$$(A_1 e^{\Gamma_1 z} + R_e A_1 e^{-\Gamma_1 z}) \sin \frac{\pi}{2} + \sum_{n=3,5,\ldots}^{\infty} B_n \sin \frac{n\pi}{2} e^{-\Gamma_n z}$$

Therefore, $F(w)$ must have poles at $w = \Gamma_n$, $n = 1, 3, 5, \ldots$, in order that (92) shall give a solution for $E_y$ of the above form. A pole at $w = -\Gamma_1$ is also required, in order to give rise to the incident wave. Since we are assuming that only the dominant mode propagates, all $\Gamma_n$ except $\Gamma_1$ are real. The poles at $\pm\Gamma_1$ lie on the imaginary $w$ axis, and the contour $C$ passes around these poles on the left-hand side. The function $F(w)$ must have an asymptotic form for large $w$ that will result in a finite value for $E_y$ at the edge of the diaphragm. The derivative $\partial E_y/\partial z$ is asymptotic to $z^{-1/2}$ at the edge of the diaphragm, and this condition governs the growth of $F(w)$ at infinity.

Taking into account all the stipulated conditions on the function $F(w)$ leads to the following suitable form for $F(w)$:

$$F(w) = \frac{p(w)}{w + \Gamma_1} \frac{\prod_{n=1}^{\infty} (w - \Gamma_{2n}) \frac{a}{2n\pi} e^{wa/2n\pi}}{\prod_{n=1,3,\ldots}^{\infty} (w - \Gamma_n) \frac{a}{n\pi} e^{wa/n\pi}} \tag{96}$$

where $p(w)$ is an integral function to be determined so that $F(w)$ will have algebraic growth at infinity, in particular so that $F(w) \to w^{-3/2}$ as $w$ approaches infinity. The asymptotic form of $F(w)$ may be found in the usual manner and is

$$F(w) \sim \frac{p(w)}{w^{3/2}} e^{-(wa/\pi) \ln 2}$$

and hence $p(w)$ is chosen as $e^{(wa/\pi) \ln 2}$.

The solution for $E_y$ may now be written down, and is

$$E_y = \int_C (e^{-wz} - e^{wz}) \frac{\sin ux}{\sin (ua/2)} e^{(wa/\pi) \ln 2} \times \frac{\left[\prod_{n=1}^{\infty} (w - \Gamma_{2n}) \frac{a}{2n\pi} e^{wa/2n\pi}\right]^2}{(w + \Gamma_1 \prod_{n=1}^{\infty} (w - \Gamma_n) \frac{a}{n\pi} e^{wa/n\pi}} dw \qquad 0 \le x \le \frac{a}{2} \tag{97}$$

after multiplying the numerator and denominator of (96) by

$$\prod_{n=2,4,\ldots}^{\infty} (w - \Gamma_n) \frac{a}{n\pi} e^{wa/n\pi}$$

For $a/2 \le x \le a$, the solution is given by (97) with $e^{wz}$ and $\sin ux$ replaced by $-e^{wz}$ and $\sin u(a - x)$, respectively. For $z > 0$, the contour $C$ for the part of the solution involving the exponential function $e^{wz}$ may be closed in the left half plane. In this half plane, the integrand has poles at the zeros of $\sin (ua/2)$, that is, at $w = -\Gamma_{2n}$, and this leads to the remainder of the eigenfunction expansion of $E_y$, the first part of the expansion having been obtained from the residues at the poles occurring at $w = \Gamma_n$ in the right half plane. The solution for $E_y$ is seen to consist of two parts, one part which is an even function of $x$ about the plane $x = a/2$ and arising from the poles of $F(w)$ in the right half plane, the other part which is an odd function of $x$ about the plane $x = a/2$ and arising from the poles of the integrand in the left half plane. The even part consists of the terms $n = 1, 3, 5, \ldots$, while the odd part consists of the remaining terms $n = 2, 4, 6, \ldots$, in the expansion (86) and (87).

The residue of the integrand at $w = -\Gamma_1$ gives rise to the incident wave, while the residue at $w = \Gamma_1$ corresponds to the reflected wave. These residues are readily evaluated from (97). It is important to note, however, that the exponential convergence factors occurring in the infinite products in the numerator and denominator in (97) cannot be canceled since the resultant infinite product is not a uniformly convergent product; that is, $\prod_{n=1}^{\infty} \frac{(w - \Gamma_{2n})^2}{w - \Gamma_n} \frac{a}{4n\pi}$ is not uniformly convergent. It is preferable to rewrite (97), using the alternative form for $F(w)$ given in (96). The integrand in (97) may then be written as

$$(e^{-wz} - e^{wz}) \frac{\sin ux}{\sin (ua/2)} \frac{e^{(wa/\pi)\ln 2} \prod_{n=1}^{\infty} (w - \Gamma_{2n}) \frac{a}{2n\pi} e^{wa/2n\pi}}{(w + \Gamma_1) \prod_{n=1,3,\ldots}^{\infty} (w - \Gamma_n) \frac{a}{n\pi} e^{wa/n\pi}}$$

Using this form for the integrand, the ratio of the residues at $w = \Gamma_1$ and $w = -\Gamma_1$ is found to be

$$\begin{aligned} R_e &= -e^{(2\Gamma_1 a/\pi)(\ln 2 - 1)} \prod_{n=1}^{\infty} \frac{\Gamma_{2n} - \Gamma_1}{\Gamma_{2n} + \Gamma_1} e^{\Gamma_1 a/n\pi} \prod_{n=3,5,\ldots}^{\infty} \left( \frac{\Gamma_n + \Gamma_1}{\Gamma_n - \Gamma_1} e^{-\Gamma_1 a/n\pi} \right) \\ &= -e^{(2\Gamma_1 a/\pi)\ln 2} \\ &\quad \times \prod_{n=1}^{\infty} \left[ \frac{(\Gamma_{2n} - \Gamma_1)(\Gamma_{2n+1} + \Gamma_1)}{(\Gamma_{2n} + \Gamma_1)(\Gamma_{2n+1} - \Gamma_1)} e^{(\Gamma_1 a/\pi)[1/n - 1/(2n+1)]} \right] \end{aligned} \tag{98}$$

after replacing $n$ by $2n + 1$ in the second infinite product. Since all $\Gamma_n$ except $\Gamma_1$ are real, it is seen that the modulus of $R_e$ is unity. The phase angle of $R_e$ is given by

$$\begin{aligned} \angle R_e = \pi + \frac{2\beta_1 a}{\pi} (\ln 2 - 1) + \sum_{n=1}^{\infty} \frac{\beta_1 a}{\pi} \left( \frac{1}{n} - \frac{1}{2n+1} \right) \\ - 2 \sum_{n=1}^{\infty} \left( \tan^{-1} \frac{\beta_1}{\Gamma_{2n}} - \tan^{-1} \frac{\beta_1}{\Gamma_{2n+1}} \right) \end{aligned} \tag{99}$$

where $\beta_1 = |\Gamma_1|$. Using the result†

$$\ln 2 = 1 - \sum_{n=1}^{\infty} \frac{1}{2n(2n+1)}$$

and also the relation $\tan^{-1} a = \sin^{-1} [a/(1 + a^2)^{1/2}]$, (99) may be rewritten as follows:

$$\angle R_e = \theta = \pi - 2 \sum_{n=1}^{\infty} \left[ \sin^{-1} \frac{\beta_1 a/\pi}{4n^2 - 1} - \sin^{-1} \frac{\beta_1 a/\pi}{(4n^2 + 4n)^{1/2}} \right] \qquad (100)$$

From this result and (88), the solution for the shunt susceptance $B$ of the semidiaphragm may be obtained.

## 10.5. Application to $H$-plane Bifurcation

Figure 10.11 illustrates a rectangular guide bifurcated by an infinitely thin perfectly conducting plane at $x = a/2$, $z \leq 0$. Since the bifurcating plane is located in a plane perpendicular to the $xz$ plane which contains the $H$ vector, it is referred to as an $H$-plane bifurcation. The

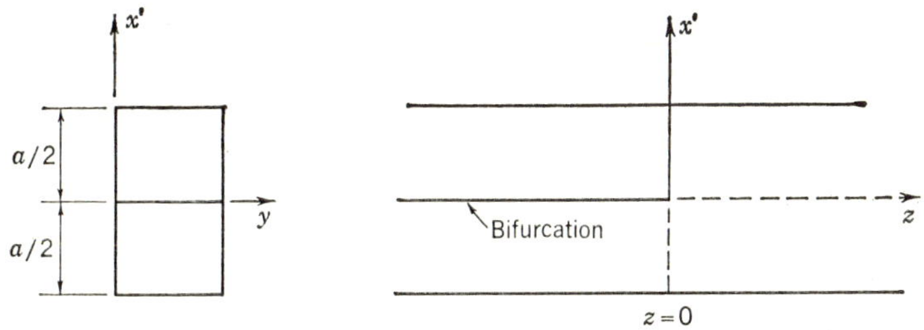


FIG. 10.11. $H$-plane bifurcation of a rectangular guide.

solution to this problem may be found in terms of the solution for the inductive semidiaphragm by an application of the Schwarz reflection principle. It will be convenient to choose a new set of coordinates $x'$, $y$, $z$, where $x' = x - a/2$.

As noted in Sec. 10.4, the solution to the problem of a rectangular guide closed by an electric wall for $0 \leq x \leq a/2$ and a magnetic wall for $a/2 \leq x \leq a$ is the sum of two parts. One part is an even function about the symmetry plane $x = a/2$ or $x' = 0$, while the other part is an odd function about $x' = 0$. Let the even part be $\phi_e(x',z)$ and the odd part be $\phi_0(x',z)$ in the region $-a/2 \leq x' \leq 0$, $z \geq 0$. The sum $\phi_e + \phi_0$ is the solution given by (97). Along the symmetry plane $x' = 0$, $z \geq 0$, the

† E. T. Copson, "Theory of Functions of a Complex Variable," p. 228, probs. 10 and 11, Oxford University Press, New York, 1935.

component functions satisfy the boundary conditions

$$\frac{\partial \phi_e}{\partial x'} = \phi_0 = 0 \qquad x' = 0$$

The Schwarz reflection principle states that the analytical continuation of a function across a plane is given by an odd reflection if the function

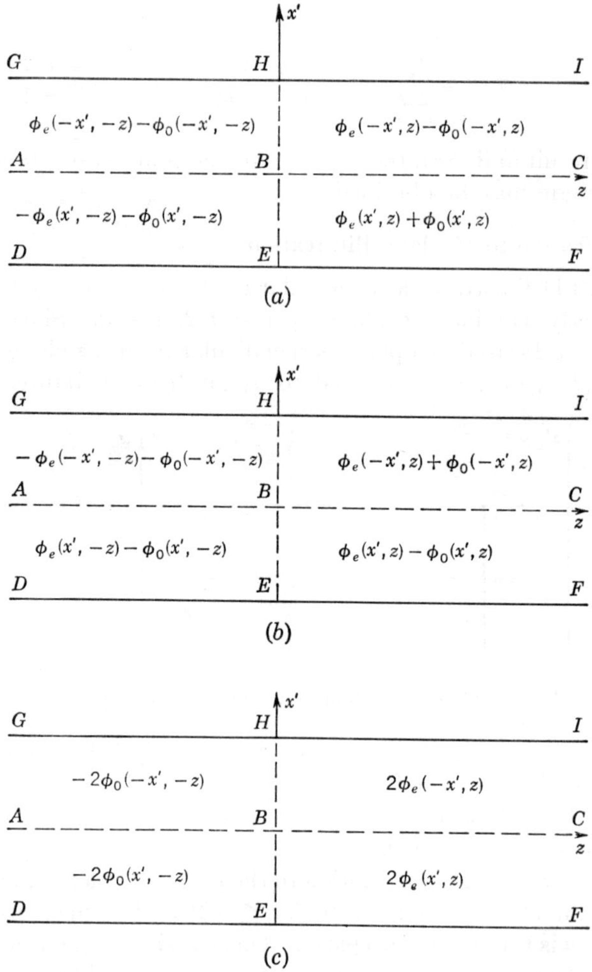


FIG. 10.12. Illustration of application of the Schwarz reflection principle.

vanishes on the plane or by an even reflection if the normal derivative of the function vanishes on the plane. With reference to Fig. 10.12*a*, let $\phi_e(x',z) + \phi_0(x',z)$ be the solution for the lower right-hand rectangular region *CBEF*. The derivative $\partial\phi_e/\partial x'$ is zero at $x' = 0$, and so the analytic continuation of $\phi_e$ into the upper region *IHBC* is given by an even reflection. The analytic continuation of $\phi_0$ is given by an odd reflection since $\phi_0 = 0$ at $x' = 0$. This solution for $z \geq 0$ is the solution given by (97). Therefore, along the segment $BE$, $\phi_e + \phi_0$ is zero. The analytic continuation of this solution into the lower left-hand region $ABED$ is therefore given by an odd reflection, and is $-\phi_e(x',-z) - \phi_0(x',-z)$. Along the segment $BH$, the solution $\phi_e(-x',z) - \phi_0(-x',z)$ has a vanishing normal derivative, and its analytic continuation into the region $ABHG$ is given by an even reflection about the line $BH$. The function as constructed above is continuous with continuous derivatives at all points except along the line $AB$. Along this line both the function and its normal derivative are discontinuous. The solution illustrated in Fig. 10.12$a$ solves the rather uninteresting problem of mixed boundary values along the plane $AB$.

In order to solve the bifurcation problem, we need to construct another function which is analytic everywhere except along $AB$, where it must cancel the discontinuity in the function of Fig. 10.12$a$. We begin with $\phi_e - \phi_0$ in the region $CBEF$, as illustrated in Fig. 10.12$b$. This function is continued into the upper region $IHBC$ as $\phi_e + \phi_0$. Taking into account that $\phi_e + \phi_0$ vanishes along $BH$, the analytic continuation into the region $ABHG$ is seen to be $-\phi_e - \phi_0$. Similarly, the analytic continuation of $\phi_e - \phi_0$ across the segment $BE$ is seen to be given by an even reflection, and is $\phi_e - \phi_0$. The final function is analytic at all points except along $AB$. If we now superimpose the two solutions $a$ and $b$, we obtain the solution illustrated in Fig. 10.12$c$. This solution is analytic everywhere except along $AB$. However, along $AB$ the solution vanishes, and we may therefore place a conducting plane along $AB$. This provides the solution of the bifurcation problem. As would be expected from symmetry considerations, the solution for $z \geq 0$ consists only of the functions $\sin(n\pi x/a)$ for $n$ odd. The reflection coefficient for an $H_{10}$ mode incident on the bifurcation from the region $z > 0$ is $R_e$ and is given by (98). The modulus of $R_e$ is unity unless the guide width $a$ is such that the $H_{20}$ mode can propagate, i.e., unless $a > \lambda$.

## REFERENCES

*Problems involving function-theoretic solution of an infinite set of algebraic equations*

10.1. L. Brillouin: Waveguides for Slow Waves, *J. Appl. Phys.*, vol. 19, pp. 1023–1041, November, 1948.

10.2. F. Berz: Reflection and Refraction of Microwaves at a Set of Parallel Metallic Plates, *Proc. IEE (London)*, vol. 98, part III, pp. 47–55, January, 1951.

10.3. E. A. N. Whitehead: The Theory of Parallel Plate Media for Microwave Lenses, *Proc. IEE (London)*, vol. 98, part III, pp. 133–140, January, 1951.

10.4. R. A. Hurd and H. Gruenberg: $H$-plane Bifurcation of Rectangular Waveguides, *Can. J. Phys.*, vol. 32, pp. 694–701, November, 1954.

10.5. R. A. Hurd: The Propagation of an Electromagnetic Wave along an Infinite Corrugated Surface, *Can. J. Phys.*, vol. 32, pp. 727–734, December, 1954.

*Problems solved by the Wiener-Hopf method*

10.6. A. E. Heins: The Radiation and Transmission Properties of a Pair of Semi-infinite Parallel Plates, *Quart. Appl. Math.*, vol. 6, part I, pp. 157–166, July, 1948; part II, pp. 215–220, October, 1948.
10.7. J. F. Carlson and A. E. Heins: The Reflection of an Electromagnetic Wave by an Infinite Set of Plates, *Quart. Appl. Math.*, vol. 4, pp. 313–329, January, 1947; vol. 5, pp. 82–88, April, 1947.
10.8. L. A. Vajnshtejn: Propagation in Semi-infinite Waveguides, six papers translated by J. Shmoys, *N.Y. Univ. Inst. Math. Sci. Rept. EM*-63.
10.9. J. D. Pearson: Diffraction of Electromagnetic Waves by a Semi-infinite Circular Waveguide, *Proc. Cambridge Phil. Soc.*, vol. 49, pp. 659–667, 1953.
10.10. V. M. Papadopoulos: Scattering by a Semi-infinite Resistive Strip of Dominant Mode Propagation in an Infinite Rectangular Waveguide, *Proc. Cambridge Phil. Soc.*, vol. 52, pp. 553–563, July, 1956.
10.11. H. Levine and J. Schwinger: On the Radiation of Sound from an Unflanged Circular Pipe, *Phys. Rev.*, vol. 73, pp. 383–406, 1948.

*Diaphragms in waveguides*

10.12. L. A. Vajnshtejn: Irises in Waveguides, *J. Tech. Phys.*, vol. 25, no. 5, pp. 841–846, 1955.
10.13. S. N. Karp and W. E. Williams: Equivalence Relations in Diffraction Theory, *Proc. Cambridge Phil. Soc.*, vol. 53, pp. 683–690, 1957.
10.14. G. L. Baldwin and A. E. Heins: On the Diffraction of a Plane Wave by an Infinite Grating, *Carnegie Inst. Technol. Tech. Rept.* 10.

*On the theory of the Wiener-Hopf method*

10.15. S. N. Karp: Wiener-Hopf Techniques and Mixed Boundary Value Problems, *Communs. Pure and Appl. Math.*, vol. 3, pp. 411–426, December, 1950.
10.16. A. E. Heins: The Scope and Limitations of the Method of Wiener and Hopf, *Communs. Pure and Appl. Math.*, vol. 9, pp. 447–466, August, 1956.
10.17. P. M. Morse and H. Feshbach: "Methods of Theoretical Physics," McGraw-Hill Book Company, Inc., New York, 1953.
10.18. B. Noble: "Methods Based on the Wiener-Hopf Technique," Pergamon Press, Inc., New York, 1958. (This recent book gives an excellent treatment of the Wiener-Hopf technique, and includes a good selection of illustrative problems that can be solved.)

## PROBLEMS

**10.1.** An infinitely wide parallel-plate transmission line is bifurcated by an infinitely thin perfect magnetic conductor in the region $z \geq 0$. Find the reflection coefficient for a TEM wave incident from $z < 0$. Assume that the dimension $b$ is such that the first mode propagates in the bifurcated region $z > 0$. Use the function-theoretic method for the solution of the infinite set of algebraic equations that determine the mode amplitudes.

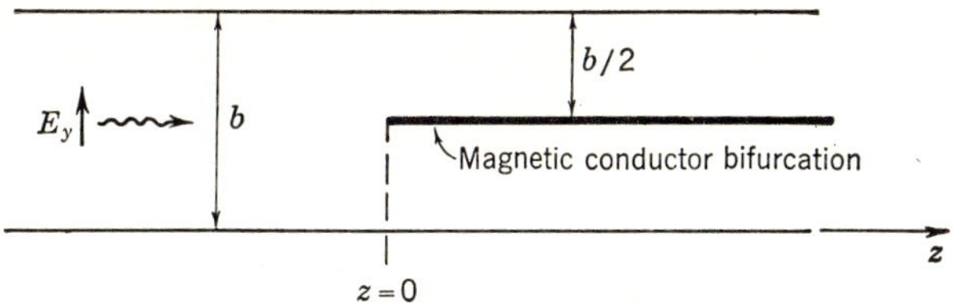


FIG. P 10.1

**10.2.** Repeat Prob. 10.1, using the method of direct construction of a suitable transform function, and also using the Wiener-Hopf method.

**10.3.** Solve the $H$-plane bifurcation problem of Sec. 10.5, using the method of direct construction of a suitable transform function.

**10.4.** Find the reflection and transmission coefficients at the interface of an infinite set of parallel perfectly conducting plates as illustrated in Fig. 10.5. Assume an incident perpendicular-polarized TEM wave. Also assume that only the $H_{10}$ mode propagates in the parallel-plate region. Note that, for an angle of incidence greater than some minimum value, there will be more than one propagating reflected wave in the region $z < 0$.

**10.5.** A rectangular guide is loaded by a thin dielectric slab of thickness $t$ for $z \geq 0$. The relative dielectric constant is $\kappa$. Maxwell's curl equation for $\mathbf{H}$ may be written as $\nabla \times \mathbf{H} = j\omega\epsilon_0\mathbf{E} + j\omega\epsilon_0\chi_e\mathbf{E}$, where $\chi_e = \kappa - 1$. The term $j\omega\epsilon_0\chi_e\mathbf{E}$ may be considered as the polarization current density in the slab. For a sufficiently thin slab, the electric

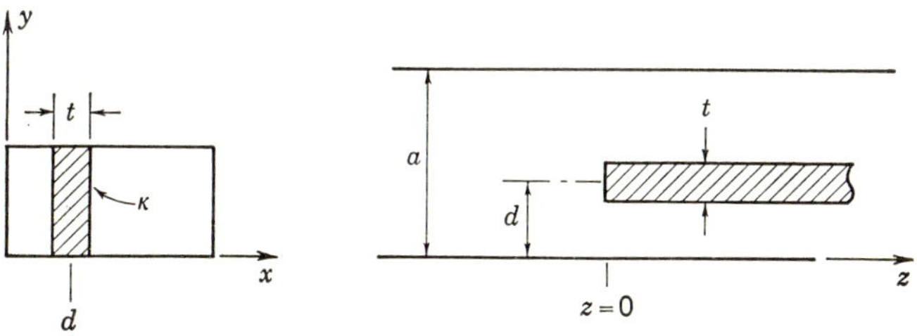


FIG. P 10.5

field of an $H_{n0}$ mode may be considered to be constant across the slab. The approximate effect of the slab is thus the same as that of an infinitely thin current sheet $j\omega\epsilon_0\chi_e t\mathbf{E}_0$, where $\mathbf{E}_0$ is the electric field of an $H_{n0}$ mode at the center of the slab, $x = d$. Using this approximation, show that the eigenvalue equation for $H_{n0}$ modes in the loaded guide is

$$\left.\frac{\partial E_y}{\partial x}\right|_{d-}^{d+} = \chi_e k_0^2 t E_y(d)$$

or

$$\cot hc + \cot hd = -\frac{k_0^2\chi_e t}{h}$$

and

$$\gamma^2 = h^2 - k_0^2$$

where $\gamma$ is the propagation constant, $h$ is the transverse wave number, and $k_0^2 = \omega^2\mu_0\epsilon_0$. Obtain a solution for the reflection and transmission coefficients for an $H_{n0}$ mode incident from $z < 0$. Use the method of direct construction of a transform function. Note that both $E_y$ and $\partial E_y/\partial z$ are finite at $z = 0$, $x = d$, and that this condition governs the growth of the transform function at infinity.

**10.6.** Solve the problem of a capacitive semidiaphragm in a parallel-plate transmission line of height $b$. Assume that only the TEM mode propagates. For an

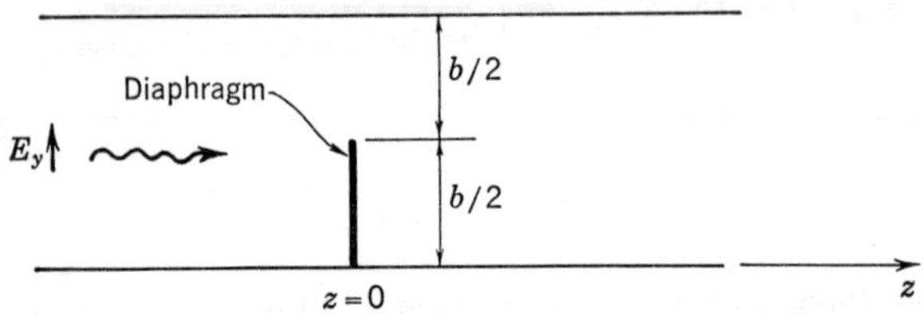


FIG. P 10.6

incident TEM mode, show that the capacitive susceptance of the diaphragm is given by $B = 2 \tan (\theta/2)$, where

$$\theta = 2 \sum_{n=1}^{\infty} \left( \sin^{-1} \frac{b/\lambda_0}{n - \frac{1}{2}} - \sin^{-1} \frac{b/\lambda_0}{n} \right)$$

**10.7.** Use the Schwarz reflection principle to obtain the solution to Prob. 10.1 from the solution of Prob. 10.6.

HINT: Construct the solution in terms of the solution for the transverse magnetic field $H_x$ of the preceding problem with the aperture closed by a magnetic wall. Assume for the present case that all modes in the bifurcated region are evanescent.

*Answer.* The reflection coefficient is $R = -e^{-j\theta}$, where $\theta$ is given in Prob. 10.6.

**10.8.** A rectangular guide has a resistive wall at $x = a$, $z > 0$, as in Prob. 5.4. An $H_{10}$ mode is incident from $z < 0$. Obtain the solution for the reflection and transmission coefficients. Compare this problem with Prob. 10.5 when the substitution $j\omega\epsilon_0(\kappa - 1) = \sigma$ is made.

CHAPTER 11

# SURFACE WAVEGUIDES

In addition to the closed cylindrical conducting tube and the conventional TEM wave-transmission line, there exists a class of open-boundary structures which are capable of guiding an electromagnetic wave. These structures are capable of supporting a mode which is intimately bound to the surface of the structure. The field is characterized by an exponential decay away from the surface and having the usual propagation function $e^{-j\beta z}$ along the axis of the structure. For this reason

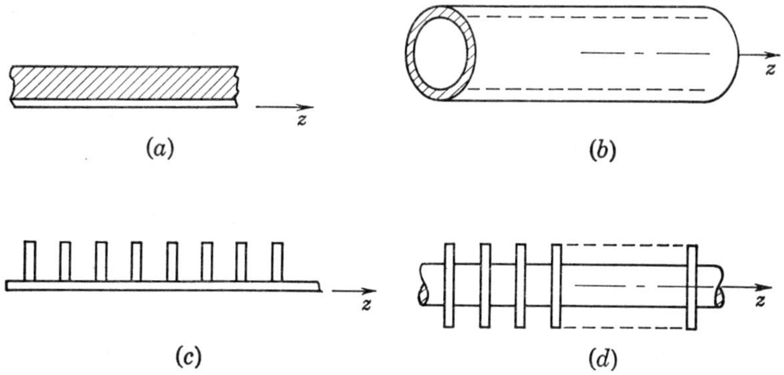


FIG. 11.1. Typical surface waveguides. (a) Dielectric-coated plane; (b) dielectric-coated wire; (c) corrugated plane; (d) corrugated cylinder.

this discrete eigenvalue solution to the wave equation is called a surface wave, and the structure which guides this wave may be appropriately referred to as a surface waveguide. Typical structures that are capable of supporting a surface wave are illustrated in Fig. 11.1. These consist of dielectric-coated planes and wires and corrugated planes and wires. In addition, a surface-wave solution exists for the interface between two different media, as well as for a variety of other structures of the periodic type with open boundaries, and also for stratified plane and cylindrical structures.

Although surface waveguides have several features which are similar to those of the cylindrical conducting-tube guide, they have many characteristics which are quite different. Some of the outstanding differences are:

1. The possibility of a surface-wave mode of propagation with no low-frequency cutoff
2. The nonexistence of an infinite number of discrete modes of propagation at a given frequency
3. The existence of a finite number of discrete modes, together with an eigenfunction solution with a continuous eigenvalue spectrum
4. The possibility of mode solutions with a phase velocity less than that of light

In this chapter some of the more important characteristics of surface waveguides will be studied by examining the solutions for typical plane and cylindrical structures. In addition, the problem of excitation of surface waves by a line source above a grounded dielectric sheet will be examined. This latter problem will provide an introduction to the role played by the continuous eigenfunction spectrum in determining the total field in the region around a surface waveguide. To evaluate the integral for the radiated field, the saddle-point method of approximate integration will be used. This is a technique that finds wide use in radiation problems.

## 11.1. Surface Waves along a Plane Interface

The guiding of plane waves by a plane interface between two different dielectric media was initially investigated by Uller† in 1903, and later by Zenneck in 1907. This surface wave is associated with the Brewster-angle phenomenon of total transmission across a dielectric interface. It is often referred to as a Zenneck wave.

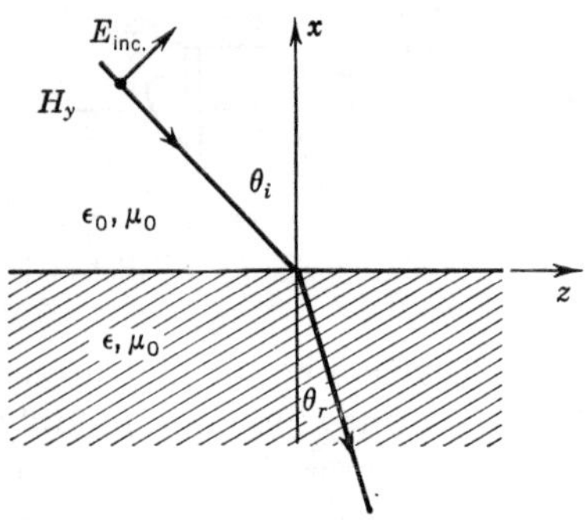

FIG. 11.2. A plane interface between free space and a dielectric-filled region.

Consider a plane interface $x = 0$, separating the air region $x > 0$ from the region $x < 0$, occupied by a lossless isotropic dielectric with permittivity $\epsilon = \kappa\epsilon_0$, where $\kappa$ is the relative dielectric constant. Let a parallel-polarized wave be incident on the interface from the air region at an angle $\theta_i$ to the interface normal, as in Fig. 11.2. The incident plane wave has the following field components, $H_y$, $E_x$, $E_z$, when the plane of incidence is the $xz$ plane.

The above problem is readily solved by introducing the wave impedances and using a transmission-line analogy.‡ The incident magnetic

† K. Uller, dissertation, Rostock, 1903.
J. Zenneck, *Ann. Phys.*, vol. 23, 1907.
‡ See Sec. 3.5.

field may be taken as

$$H_y = A \exp[-jk_0(z \sin \theta_i - x \cos \theta_i)] \qquad x > 0 \tag{1a}$$

where $k_0^2 = \omega^2\mu_0\epsilon_0$, and $A$ is an amplitude constant. In the region $x > 0$ a reflected wave with a magnetic field

$$-RA \exp[-jk_0(z \sin \theta_i + x \cos \theta_i)] \tag{1b}$$

will also exist, in general. In the dielectric region, a transmitted wave

$$TA \exp[-jk(z \sin \theta_r - x \cos \theta_r)] \tag{1c}$$

will exist, where $k^2 = \omega^2\mu_0\epsilon$, $\theta_r$ is the angle of refraction, and $R$, $T$ are the reflection and transmission coefficients, respectively. The components of the electric field are given by

$$j\omega\epsilon E_x = -\frac{\partial H_y}{\partial z} \tag{2a}$$

$$j\omega\epsilon E_z = \frac{\partial H_y}{\partial x} \tag{2b}$$

The wave impedances in the two regions are

$$Z_1 = \frac{E_z}{H_y} = Z_0 \cos \theta_i \qquad x > 0 \tag{3a}$$

$$Z_2 = Z \cos \theta_r \qquad x < 0 \tag{3b}$$

where $Z_0 = (\mu_0/\epsilon_0)^{1/2}$, and $Z = (\mu_0/\epsilon)^{1/2}$. In terms of these, the reflection coefficient $R$ is given by

$$R = \frac{Z_2 - Z_1}{Z_2 + Z_1} = \frac{\cos \theta_r - \kappa^{1/2} \cos \theta_i}{\cos \theta_r + \kappa^{1/2} \cos \theta_i} \tag{4}$$

The propagation constant along the $z$ direction must be the same in both media (Snell's law), so that $k_0 \sin \theta_i = k \sin \theta_r$. The Brewster angle for which $R$ vanishes is thus given by $\cos \theta_r = \kappa^{1/2} \cos \theta_i$, or, eliminating $\theta_r$,

$$\sin \theta_i = \left(\frac{\kappa}{\kappa + 1}\right)^{1/2} \tag{5a}$$

The corresponding angle in the dielectric medium is given by

$$\sin \theta_r = (\kappa + 1)^{-1/2} \tag{5b}$$

For this particular angle of incidence, the solution for the magnetic field reduces to

$$H_y = \begin{cases} A \exp[-jk_0(\kappa + 1)^{-1/2}(\kappa^{1/2} z - x)] & x > 0 \\ A \exp[-jk_0(\kappa + 1)^{-1/2}(\kappa^{1/2} z - \kappa x)] & x < 0 \end{cases} \tag{6}$$

This solution is the prototype solution for a surface wave guided by a plane interface.

The above solution is for a parallel-polarized plane wave, and may be classified as a transverse magnetic (TM) wave with respect to the $z$ axis. In the case when the dielectric has finite losses, no real angle of incidence can be found such that there is no reflection for the TM wave. However, we do not need to restrict the angle of incidence to real values only. As soon as complex angles of incidence are considered, it is found that a solution for an inhomogeneous plane wave of the TM type exists such that there is no reflection at the interface.

We will now consider the extension of the previous TM wave solution to the case of a dielectric with finite but small losses. Let the permittivity of the dielectric medium be $\epsilon = \epsilon' - j\epsilon''$, where $\epsilon'' \ll \epsilon'$ and represents the loss in the material. Instead of introducing the angles of incidence and refraction explicitly, the incident and transmitted waves will be represented as follows:

$$H_y = A \exp(jh_1 x - j\beta z) \qquad x > 0 \tag{7a}$$
$$H_y = A \exp(jh_2 x - j\beta z) \qquad x < 0 \tag{7b}$$

where $h_1^2 + \beta^2 = k_0^2$, and $h_2^2 + \beta^2 = k^2$, in order that the field shall satisfy the wave equation. The wave impedances are now given by

$$Z_1 = \frac{h_1}{k_0} Z_0 \tag{8a}$$

$$Z_2 = \frac{h_2}{k} Z \tag{8b}$$

For the surface wave, the reflection coefficient must vanish, and hence $Z_1 = Z_2$. The solutions for the transverse wave numbers $h_1$ and $h_2$ and the propagation constant $\beta$ are hence determined by the two following equations:

$$h_1^2 - k_0^2 = h_2^2 - k^2 \tag{9a}$$
$$h_1 k Z_0 = h_2 k_0 Z$$

or

$$h_1(\epsilon' - j\epsilon'') = \epsilon_0 h_2 \tag{9b}$$

For convenience, let $\epsilon' - j\epsilon'' = (\kappa' - j\kappa'')\epsilon_0 = \kappa\epsilon_0$. From (9$a$), we obtain $h_1 = [h_2^2 - (\kappa' - j\kappa'' - 1)k_0^2]^{1/2}$, which may be combined with (9$b$) to yield

$$h_2 = (\kappa' - j\kappa'')k_0(\kappa' - j\kappa'' + 1)^{-1/2} \tag{10}$$

Since $\kappa''$ is small compared with $\kappa'$, an approximate solution for $h_2$ is readily obtained by using the binomial expansion to get

$$h_2 \approx \frac{k_0}{(\kappa' + 1)^{1/2}} \left( \kappa' - j\tfrac{1}{2}\kappa'' \frac{\kappa' + 2}{\kappa' + 1} \right) \tag{11a}$$

The corresponding solution for $h_1$ is obtained from (9b) and (10):

$$h_1 \approx \frac{k_0}{(\kappa' + 1)^{1/2}} \left[ 1 + \frac{j\kappa''}{2(\kappa' + 1)} \right] \tag{11b}$$

For small losses, it is seen that the real parts of $h_1$ and $h_2$ are the same as in the lossless case, i.e., in the solution given by (6).

It is interesting to note from (11) that the imaginary parts of $h_1$ and $h_2$ are of opposite sign, and, hence, the surface wave has the characteristic exponential decay in the direction perpendicular and away from the interface on both sides. The solution for the magnetic field in the present case is

$$H_y = \begin{cases} A \exp\left[-j\beta z + (jh_1' - h_1'')x\right] & x > 0 \\ A \exp\left[-j\beta z + (jh_2' + h_2'')x\right] & x < 0 \end{cases} \tag{12}$$

where $h_1'$, $h_1''$ and $h_2'$, $h_2''$ are the real and imaginary parts of $h_1$ and $h_2$ as given by (11). The propagation constant $\beta$ is given by

$$\beta^2 = k_0^2 - h_1^2 = k_0^2 - (h_1' + jh_1'')^2$$

or, since $h_1'' \ll h_1'$,

$$\beta' - j\beta'' = \beta \approx (k_0^2 - h_1'^2)^{1/2} \left(1 - j\frac{h_1'h_1''}{k_0^2 - h_1'^2}\right) \tag{13}$$

For small losses, $\beta''$ is small, and the attenuation of the wave is also small. At the same time $h_1''$ is small, so that the wave is rather loosely bound to the interface.

The constant-phase planes for the wave on the air side are determined by the equation

$$\beta' z - h_1' x = \text{constant} \tag{14}$$

where $\beta'$ is the real part of $\beta$, as may be seen from (12). These planes make an angle $\theta_1$ with respect to the $x$ axis, where

$$\theta_1 = \tan^{-1} \frac{h_1'}{\beta'} = \cot \kappa' \tag{15}$$

a result which is identical with that for the lossless case. The planes of constant amplitude are determined by

$$\beta'' z + h_1'' x = \text{constant} \tag{16}$$

where $\beta''$ is the imaginary part of $\beta$. The constant-amplitude and constant-phase planes are seen to be noncoincident. Further details of this problem are given by Stratton, and hence we will not pursue it any further here.†

† J. A. Stratton, "Electromagnetic Theory," sec. 9.13, McGraw-Hill Book Company, Inc., New York, 1941.

A solution similar to the above for a perpendicular-polarized (transverse electric, TE) wave does not exist unless the two mediums have different permeabilities. When the permeabilities are the same, the wave admittances are proportional to $h_1$ and $h_2$ in the two mediums. In order for the reflection coefficient to vanish, $h_1$ must equal $h_2$. However, this is impossible since $h_1^2 - k_0^2$ must also be equal to $h_2^2 - k^2$. The solution for the case of unequal permeabilities may be constructed from the above solution by an application of the duality principle and is left as a problem at the end of the chapter.

## 11.2. Surface Waves along an Impedance Plane

By examining the possibility of surface-wave propagation along a plane which exhibits a complex surface impedance, further insight into the type of surface required to support a surface wave may be obtained. The analysis of this section will show that a TM wave requires a surface with a surface impedance having an inductive reactive term, while, in order to support a TE surface wave, the reactive part of the surface impedance must be capacitive. Since naturally occurring surfaces have an inductive reactance term in the surface impedance, the TM surface wave is of more importance than the TE wave.

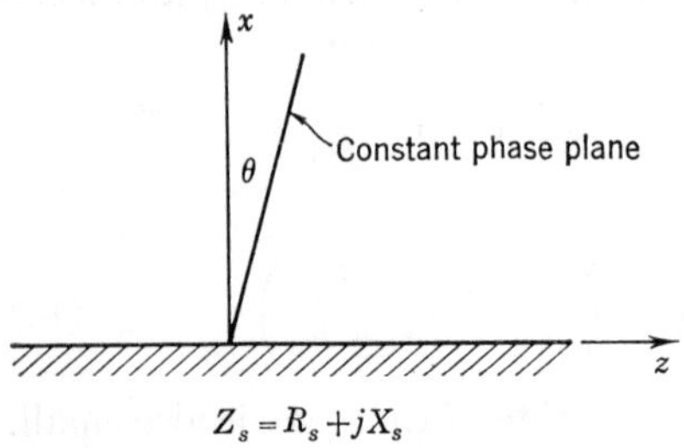

FIG. 11.3. Illustration of a plane impedance surface.

Consider a plane surface in the $x = 0$ plane which has a normalized surface impedance $Z_s = R_s + jX_s$ that is independent of the angle of incidence for incident inhomogeneous plane waves. The surface impedance is normalized with respect to the intrinsic impedance $Z_0$ of free space. With reference to Fig. 11.3, let the magnetic field $H_y$ of the assumed TM surface wave have the following form in the region $x > 0$:

$$H_y = A \exp (jhx - j\beta z) \tag{17}$$

where $h^2 + \beta^2 = k_0^2$. The electric field components are given by

$$j\omega\epsilon_0 E_z = \frac{\partial H_y}{\partial x} \qquad j\omega\epsilon_0 E_x = -\frac{\partial H_y}{\partial z}$$

The wave impedance of the wave is

$$Z_1 = \frac{E_z}{H_y} = \frac{h}{\omega\epsilon_0} = \frac{h}{k_0} Z_0$$

and must be equal to the surface impedance of the plane at $x = 0$ if the field is to satisfy the required boundary conditions on this plane without a reflected field being present. Equating the normalized wave impedance $h/k_0$ to the surface impedance gives the required solution for $h$:

$$h = h' + jh'' = k_0 Z_s = k_0 R_s + jk_0 X_s \tag{18}$$

The solution for the propagation constant $\beta$ is

$$\beta = \beta' - j\beta'' = (k_0^2 - h^2)^{1/2} = k_0(1 + X_s^2 - R_s^2 - 2jR_sX_s)^{1/2} \tag{19}$$

Equation (18) for the transverse wave number $h$ clearly shows that the field will have an exponential decay normal to the surface only if the reactive part of the surface impedance is inductive. Also, the larger the inductive reactance of the surface is, the more closely will the wave be bound to the surface. In order that the attenuation of the wave in the $z$ direction should be small, the product $R_sX_s$ should be small, so that, from (19), $\beta''$ will be small. A closely bound surface wave may be propagated along the surface with a small attenuation only if $X_s$ is made large and $R_s$ is made small.

The solution for the magnetic field is

$$H_y = A \exp\left[(jh' - h'')x - (j\beta' + \beta'')z\right] \tag{20}$$

where $h'$, $h''$, $\beta'$, and $\beta''$ are given by (18) and (19). The constant-phase planes are determined by the equation

$$h'x - \beta'z = \text{constant}$$

and are inclined toward the surface at an angle $\theta$ relative to the $x$ axis, where $\tan\theta = h'/\beta'$. The wavefront is tilted by an amount just sufficient to give a component of the Poynting vector in the direction of the impedance surface, to account for the power loss in the resistive part $R_s$ of the surface impedance. The power flow per unit area into the surface at any point $z$ is

$$\begin{aligned} P_R &= \tfrac{1}{2}\,\mathrm{Re}\int_z^{z+1} E_z H_y^*\,dz = \tfrac{1}{2}R_s Z_0 \int_z^{z+1} H_y H_y^*\,dz \\ &= \tfrac{1}{4}R_s Z_0 A A^* e^{-2\beta'' z}\,\frac{1 - e^{-2\beta''}}{\beta''} \\ &\approx \tfrac{1}{2}R_s Z_0 A A^* e^{-2\beta'' z} \end{aligned} \tag{21}$$

where $\beta''$ is small. The power flow, per unit width of the surface, in the $z$ direction is

$$P = \tfrac{1}{2}\,\mathrm{Re}\int_0^\infty E_x H_y^*\,dx = \frac{1}{4}\frac{\beta' Z_0}{h'' k_0} A A^* e^{-2\beta'' z} \tag{22}$$

The phase velocity of the wave in the $z$ direction is determined by the real part of $\beta$. From (19), we obtain

$$\beta' \approx k_0(1 + X_s^2)^{1/2}$$

when $R_s \ll X_s$, and hence the phase velocity $v_p$ is given approximately by

$$v_p = \frac{\omega}{\beta'} \approx v_c(1 + X_s^2)^{-1/2} \tag{23}$$

where $v_c$ is the velocity of plane waves in free space. The phase velocity is seen to be less than the free-space velocity for a surface wave supported by a highly inductive surface with a small resistance. When the surface resistance is equal to or greater than the reactance, the phase velocity is equal to or greater than the free-space velocity.

The above analysis is directly applicable to the problem of a surface wave guided along a conducting plane. In Sec. 3.6 it was shown that, for metals at radio frequencies, the surface impedance is independent of the angle of incidence, and has an inductive part equal to the resistive part. The normalized surface impedance of a metal interface is given by

$$Z_s = (1 + j)\left(\frac{k_0}{2\sigma Z_0}\right)^{1/2} \tag{24}$$

where $\sigma$ is the conductivity of the metal. The transverse wave number $h$ is thus given by

$$h = k_0(1 + j)\left(\frac{k_0}{2\sigma Z_0}\right)^{1/2} \tag{25a}$$

while the propagation constant $\beta$ is given by

$$\beta = k_0\left(1 - \frac{jk_0}{\sigma Z_0}\right)^{1/2} \approx k_0 - \frac{jk_0^2}{2\sigma Z_0} \tag{25b}$$

since $k_0/\sigma Z_0$ is very small for any good conductor. For metals, $\sigma$ is of the order of $10^7$ mhos per meter, $Z_0 = 377$ ohms, and, for $\lambda_0 = 3.14$ centimeters, $k_0 = 200$ radians per meter. Hence $\beta''$ is approximately $5 \times 10^{-6}$ neper per meter. This amount of attenuation is generally entirely negligible. The imaginary part $h''$ of $h$ is approximately equal to $3 \times 10^{-2}$ neper per meter. Thus the wave amplitude does not decrease to a negligible value before a distance of several thousand wavelengths from the surface has been reached. The wave is too loosely bound to the surface to be of much use in a transmission system of reasonable dimensions.

It has been seen that a conducting plane can support a surface wave of the TM type but that this wave is very loosely bound to the surface.

To obtain a practical surface waveguide, it is necessary to increase the inductive reactance of the surface without increasing the surface resistivity significantly. Two methods for achieving this end result are to coat the conductor with a thin layer of high-dielectric-constant material or to corrugate the surface. These structures will be considered in the two following sections, but first we will examine the solution for a TE surface wave guided by an impedance plane.

The TE surface wave is similar in structure to the TM wave, but with the role of electric field and magnetic field interchanged. The field is described by the following equations:

$$E_y = A \exp (jhx - j\beta z)$$

$$j\omega\mu_0 H_z = -\frac{\partial E_y}{\partial x} \qquad j\omega\mu_0 H_x = \frac{\partial E_y}{\partial z}$$

where $h^2 + \beta^2 = k_0^2$. The wave admittance is equal to

$$-\frac{H_z}{E_y} = \frac{Y_0 h}{k_0}$$

where $Y_0 = Z_0^{-1}$. The normalized wave admittance must be equal to the normalized surface admittance of the guiding plane, and, hence,

$$h = k_0 Y_s = k_0(G_s + jB_s)$$

Clearly it is necessary that the surface susceptance $B_s$ be positive if the imaginary part $h''$ of $h$ should be positive. Thus, to obtain a field that decays exponentially with distance from the surface, the surface must have a capacitive susceptance. Dielectric-coated surfaces can be designed so as to exhibit a capacitive susceptance term in the surface admittance and will, therefore, be capable of guiding a TE surface wave.

## 11.3. Conducting Plane with a Thin Dielectric Coating

The propagation of a TM surface wave along a conducting plane coated with a thin layer of dielectric has been investigated by Attwood.† The results of this analysis have shown that a thin film of dielectric leads to a large increase in the concentration of the field near the surface. The attenuation constant of such a surface waveguide can be considerably less than that of a conventional rectangular waveguide when the parameters of the dielectric film are properly chosen.

Figure 11.4 illustrates a plane conducting surface, coated with a layer of dielectric of thickness $t$ and having a relative complex dielectric constant $\kappa = \kappa' - j\kappa''$. Attwood solves this problem by neglecting the

† S. S. Attwood, Surface Wave Propagation over a Coated Plane Conductor, *J. Appl. Phys.*, vol. 22, pp. 504–509, April, 1954.

losses initially, in order to obtain expressions for the field components. The attenuation constant is then evaluated by finding the power dissipated in the dielectric and the conducting plane. Since the losses are small, this procedure is valid. However, we will follow an alternative approach and find the transverse wave numbers $h_1$ in free space, $h_2$ in the dielectric, and the propagation constant $\beta$ directly, by including the effect of the losses from the start. This approach is chosen in order to show more clearly how the surface reactance is increased relative to the surface resistance, by coating the conductor with a thin dielectric layer.

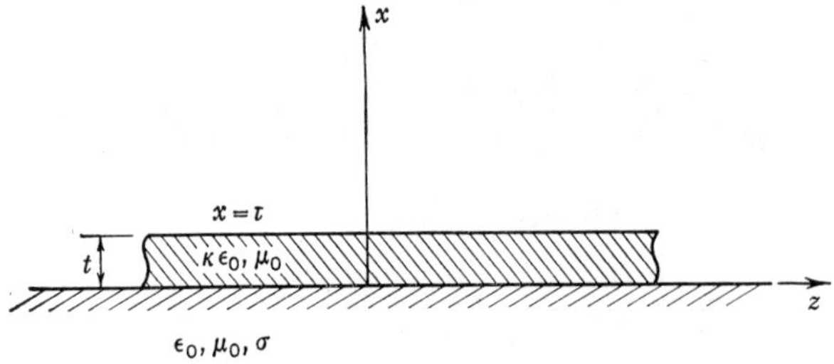


FIG. 11.4. A conducting plane coated with a thin layer of dielectric.

In the free-space region, the magnetic field $H_y$ has the form

$$H_y = A \exp (jh_1x - j\beta z) \qquad x > t$$

where $h_1^2 + \beta^2 = k_0^2$. In the dielectric, the field consists of two waves propagating in the positive and negative $x$ direction, and hence has the form

$$H_y = [B \exp jh_2x + C \exp (-jh_2x)]e^{-j\beta z} \qquad 0 \le x \le t$$

where $h_2^2 + \beta^2 = k^2 = \kappa k_0^2$. The wave numbers $h_1$, $h_2$ and the propagation constant $\beta$ may be found by matching the tangential fields at the dielectric–free-space interface, and by imposing the boundary conditions at the conducting plane. The required eigenvalue equation is more readily derived from the equivalent transmission-line circuit for propagation along the $x$ axis. The wave impedances of the wave in free space and the dielectric region are, respectively,

$$Z_1 = \frac{Z_0h_1}{k_0}$$

$$Z_2 = \frac{Z_0h_2}{\kappa k_0}$$

The normalized impedance of the conductor is

$$Z_s = R_s + jX_s = (1 + j)\left(\frac{k_0}{2\sigma Z_0}\right)^{1/2}$$

Figure 11.5 illustrates the equivalent circuit for propagation in the $x$ direction. The surface-wave solution requires that the normalized input impedance at the dielectric interface $aa$ be equal to $Z_1/Z_0$, so that there will be no reflection at this interface. Conventional transmission-line theory thus gives

$$\frac{Z_1}{Z_0} = Z_{\text{in}} = \frac{Z_2}{Z_0}\frac{Z_s + j(Z_2/Z_0)\tan h_2 t}{Z_2/Z_0 + jZ_s \tan h_2 t} \tag{26}$$

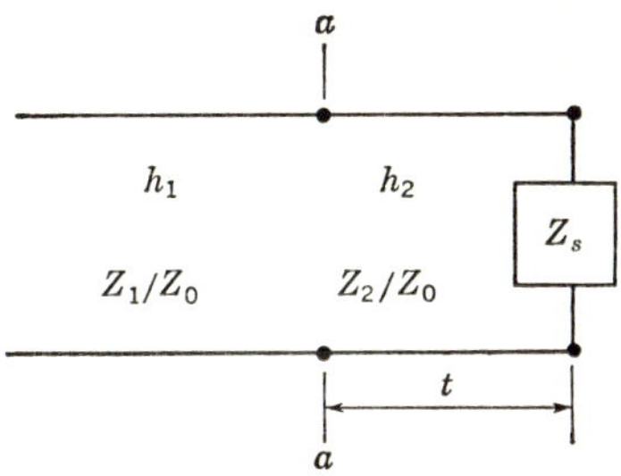

FIG. 11.5. Equivalent transmission-line circuit for coated conducting plane.

In this equation a number of approximations may be made, since $Z_s$ is very small at radio frequencies, and it is assumed that the thickness $t$ is chosen so that $h_2 t$ is also very small. Substituting for the wave impedances we obtain from (26)

$$h_1 \approx k_0\left(Z_s + j\frac{h_2{}^2 t}{\kappa k_0} - jZ_s{}^2\kappa k_0 t\right)$$

Replacing $h_2{}^2$ by $h_1{}^2 + (\kappa - 1)k_0{}^2$, and $Z_s{}^2$ by $2jR_s{}^2$, we get

$$h_1 = k_0\left(R_s + 2R_s{}^2\kappa k_0 t + j\frac{h_1{}^2 t}{\kappa k_0} + jk_0 t\frac{\kappa - 1}{\kappa} + jX_s\right)$$

The wave numbers $h_1$ and $h_2$ also satisfy the relations $h_1{}^2 + \beta^2 = k_0{}^2$, $h_2{}^2 + \beta^2 = \kappa k_0{}^2$. For small losses, $\beta$ is nearly equal to $k_0$, and hence $h_1{}^2$ is small. The wave number $h_2$, however, is of the same order of magnitude as $k_0$, and, hence, since $h_2 t$ is assumed to be small, then $k_0 t$ is also small. In view of these considerations, we may drop the terms involving $R_s{}^2$ and $h_1{}^2$ to get the relatively simple equation

$$\begin{aligned} h_1 &= k_0\left(R_s + jX_s + jk_0 t\frac{\kappa - 1}{\kappa}\right) \\ &\approx k_0\left[\left(R_s + \frac{\kappa'' k_0 t}{\kappa'^2}\right) + j\left(X_s + \frac{\kappa' - 1}{\kappa'}k_0 t\right)\right] \end{aligned} \tag{27}$$

The surface resistance is thus effectively increased by the small term $\kappa'' k_0 t/\kappa'^2$, while the effective surface reactance is increased by the relatively large additional term $(\kappa' - 1)k_0 t/\kappa'$.

As a typical example, consider a copper plane ($\sigma = 5.8 \times 10^7$ mhos per meter), coated with a film of polystyrene ($\kappa = 2.56 - j0.002$) of thickness 0.001 meter at a wavelength of 0.0314 meter. The terms occurring in (27) have the following values for this example:

$$R_s = 6.7 \times 10^{-5}$$
$$\frac{\kappa'' k_0 t}{\kappa'^2} = 6 \times 10^{-5}$$
$$X_s = 6.7 \times 10^{-5}$$
$$\frac{\kappa' - 1}{\kappa'} k_0 t = 0.122 = 1.82 \times 10^3 X_s$$

The last two terms that were dropped in arriving at (27) have the following values:

$$2R_s^2 \kappa k_0 t = (47 - j0.037) \times 10^{-10} \qquad \frac{h_1^2 t}{\kappa k_0} \approx 1.2 \times 10^{-3} + j9 \times 10^{-7}$$

These terms are seen to be negligible, and justify the approximations made. The imaginary part $h_1''$ of $h_1$ is equal to 24.4 nepers per meter, and hence the field has decayed to 10 per cent of its value at the surface in a distance of three wavelengths from the surface. The addition of a thin film of dielectric leads to a surprisingly large increase in the decay constant away from the surface.

From (27) it is seen that a conductor coated with a thin layer of dielectric may be conveniently viewed as an impedance surface with a normalized surface impedance $Z_d$ given by

$$Z_d = R_s\left(1 + \frac{\kappa'' k_0 t}{R_s \kappa'^2}\right) + j\left(X_s + \frac{\kappa' - 1}{\kappa'} k_0 t\right) \tag{28}$$

The propagation factor $\beta$ is given by

$$\begin{aligned}\beta = \beta' - j\beta'' &= (k_0^2 - h_1^2)^{1/2} = (k_0^2 - h_1'^2 + h_1''^2 - 2jh_1'h_1'')^{1/2} \\ &\approx (k_0^2 + h_1''^2)^{1/2} - j\frac{h_1'h_1''}{k_0}\end{aligned}$$

since $h_1' \ll h_1'' \ll k_0$. Substituting from (27), we get

$$\beta' \approx k_0\left[1 + \left(X_s + k_0 t \frac{\kappa' - 1}{\kappa'}\right)^2\right]^{1/2} \tag{29a}$$

$$\beta'' \approx k_0\left(R_s + \frac{\kappa'' k_0 t}{\kappa'^2}\right)\left(X_s + k_0 t \frac{\kappa' - 1}{\kappa'}\right) \tag{29b}$$

For the example considered earlier, the attenuation in the $z$ direction is $3.1 \times 10^{-3}$ neper per meter or $2.7 \times 10^{-2}$ decibel per meter. This compares favorably with the attenuation of conventional rectangular guides.

The preceding analysis is valid only for low-loss thin dielectric sheets over a low-loss conducting plane. For sheets with a thickness that is an appreciable fraction of a wavelength, the eigenvalue equation for the transverse wave number must be derived without making the approximation $h_2 t \ll 1$. The propagation of surface waves along thick dielectric slabs is considered in a latter section, and so the case of thick films will not be considered here.

## 11.4. Surface Waves along a Corrugated Plane

The propagation of surface waves along a corrugated plane has been studied by several investigators.† Corrugated planes supporting surface waves have found application in end-fire antennas and are, therefore, of some practical interest.‡ Only the simplest case, that of infinitely thin teeth, as illustrated in Fig. 11.6, will be considered here. The corrugated plane is infinite in extent, and assumed to be perfectly conducting. The corrugations are spaced by a distance $S$ and have a depth $d$.

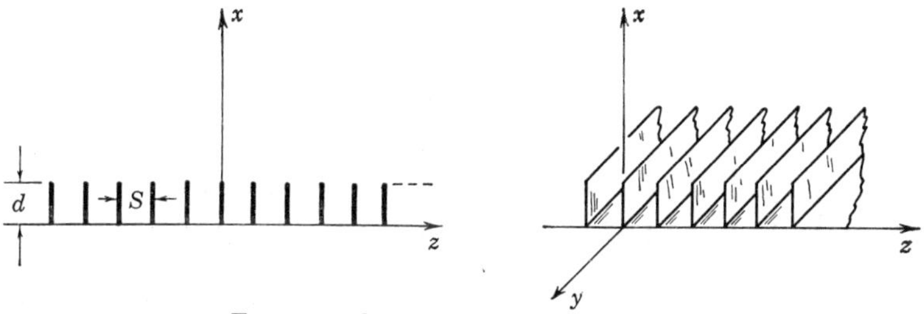


FIG. 11.6. Corrugated-plane conductor.

Consider first the solution for a TM surface wave propagating along the $z$ direction. The magnetic field has a single component $H_y$, from which the electric field components $E_x$ and $E_z$ may be found. Since the structure is a periodic one, the field has the form governed by Floquet's theorem. Above the corrugations, a suitable expansion for $H_y$ is

$$H_y = \sum_{n=-\infty}^{\infty} A_n \exp\left[-p_n x - j\left(\beta + \frac{2n\pi}{S}\right)z\right] \qquad x > d \tag{30}$$

where $(\beta + 2n\pi/S)^2 = k_0^2 + p_n^2$. In the corrugated region, $H_y$ must have a form such that the electric field derivable from $H_y$ should vanish on the edge of the corrugations, as well as on the conducting plane at

† W. Rotman, A Study of Single Surface Corrugated Guides, *Proc. IRE*, vol. 39, pp. 952–959, August, 1951.

R. S. Elliott, On the Theory of Corrugated Plane Surfaces, *IRE Trans.*, vol. AP-2, pp. 71–81, April, 1954.

R. A. Hurd, The Propagation of an Electromagnetic Wave along an Infinite Corrugated Surface, *Can. J. Phys.*, vol. 32, pp. 727–734, December, 1954.

R. W. Hougardy and R. C. Hansen, Oblique Surface Waves over a Corrugated Conductor, *IRE Trans.*, vol. AP-6, pp. 370–376, October, 1958.

‡ D. K. Reynolds and W. S. Lucke, Corrugated End Fire Antennas, *Proc. Natl. Electronics Conf.*, vol. 6, pp. 16–28, September, 1950.

$x = 0$. A suitable expansion for $H_y$ is readily found to be

$$H_y = \sum_{n=0}^{\infty} B_n \cos \frac{n\pi z}{S} \cosh h_n x \qquad 0 \le z \le S \qquad 0 \le x \le d \tag{31}$$

where $h_n^2 = (n\pi/S)^2 - k_0^2$. In the $N$th corrugation, the field is related to that in the region $0 \le z \le S$ as follows:

$$H_y(x + NS) = e^{-j\beta NS} H_y(x) \qquad \text{all } N \tag{32}$$

It will be assumed that the spacing $S$ is chosen so small that all $h_n$ except $h_0 = jk_0$ are real. Thus, only the TEM mode in the mode expansion (31) propagates in the $x$ direction between corrugations. The solution to the above boundary-value problem may be found by the methods similar to those discussed in Chap. 10. However, we may obtain the eigenvalue equation for $\beta$ by a simple transmission-line analysis. Consider a system of parallel plates of spacing $S$ and semi-infinite in length, as in Fig. 11.7.

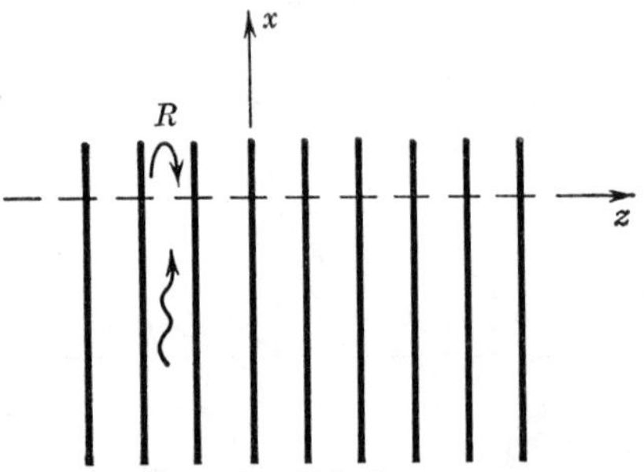

FIG. 11.7. Parallel-plate system.

Let a TEM wave with a magnetic field $H_y$ equal to

$$H_y = Ce^{-jk_0 x} e^{-j\beta NS} \qquad (N - 1)S < z < NS$$

in the $N$th corrugation be incident on the interface from the parallel-plate region. At the interface there is a reflected TEM wave,

$$H_y = -RCe^{jk_0 x} e^{-j\beta NS}$$

in the $N$th section, a dominant-mode transmitted field,

$$H_y = Ae^{-p_0(x-d)} e^{-j\beta z}$$

plus higher-order evanescent modes that decay exponentially away from the interface on both sides. For the surface wave, $p_0$ is a positive-real wave number, and, hence, there is no power flow away from the interface in the positive $x$ direction. There will, therefore, be complete reflection, and the modulus of the reflection coefficient $|R|$ must be equal to unity. In the parallel-plate region there will be a perfect standing wave, and the electric field $E_z$ will vanish at regular intervals spaced by a half wavelength. Along any one of these planes a conducting plane may be placed, and the resulting structure is a corrugated plane. When the spacing $S$ is sufficiently small, the conducting plane may be placed quite close to the interface without interacting appreciably with the evanescent modes.

If $R = e^{-j\phi}$ is the reflection coefficient for $E_z$ at the interface, the reflection coefficient $R'$ a distance $d$ from the interface is given by $e^{-j\phi-2jk_0d}$. A conducting plane may be placed at the plane where $R' = -1$, and, hence,

$$2k_0d + \phi = n\pi \qquad n = 1, 3, \ldots$$

The reflection coefficient at the interface was found in Sec. 10.2, and is given by

$$R = e^{-j\phi}$$

where

$$\phi = 2\left[\tan^{-1}\frac{k_0}{p_0} - k_0\frac{S}{\pi}\ln 2 + \sum_{n=1}^{\infty}\left(\tan^{-1}\frac{k_0}{p_n} + \tan^{-1}\frac{k_0}{p_{-n}} - \tan^{-1}\frac{k_0}{h_n}\right)\right] \tag{33}$$

in the present notation. For spacings $d$ less than about $0.2\lambda_0$, the sum over $n$ in (33) contributes a negligible amount to $\phi$. In this case,

$$\begin{aligned}\phi &= 2\left(\tan^{-1}\frac{k_0}{p_0} - k_0\frac{S}{\pi}\ln 2\right)\\ &= n\pi - 2k_0d\end{aligned}$$

and, solving for $p_0$, we get

$$p_0 = k_0 \tan k_0\left(d - \frac{S}{\pi}\ln 2\right) \tag{34}$$

A solution for positive $p_0$ exists only for certain ranges of $k_0$, and, hence, the corrugated plane exhibits the usual passband-stopband characteristics common to all periodic structures. Since the evanescent modes are TM or $E$ modes, they give rise to a net storage of electric energy at the interface, and the parallel-plate system is effectively terminated in a capacitive susceptance. Thus the position of the first minimum may occur less than $\lambda_0/4$ from the interface. For the wave on the free-space side of the corrugations, the corrugated plane appears as a short-circuited transmission line of electrical length $k_0[d - (S/\pi)\ln 2]$ for small spacing $S$. As (34) shows, the solution for $p_0$ requires a value of $d$ that will make the effective surface impedance an inductive reactance.

The propagation constant $\beta$ is given by

$$\beta = (k_0^2 + p_0^2)^{1/2}$$

and, hence, the phase velocity $v_p = \omega/\beta$ is less than the free-space velocity $v_c$. Thus the corrugated plane supports a TM surface wave of the "slow wave" type.

The corrugated plane is uniform in the $y$ direction, and the eigenvalue equation for a surface wave propagating obliquely across the corrugations may be found by a procedure similar to that used for obtaining the solution for a capacitive diaphragm in a rectangular guide from the solution for a capacitive diaphragm in a parallel-plate transmission line. As shown in Sec. 8.2, the mode of propagation will have all field components except $E_y$ present, and may be classified as a longitudinal-section electric mode with respect to the $z$ axis. If the propagation constant along the $y$ axis is $jl$ (propagation factor $e^{-jly}$), the eigenvalue equation may be obtained from (33) or (34) by replacing $k_0^2$ by $k_0^2 - l^2$ throughout, including the terms $p_n$ and $h_n$.

For small spacings (34) is valid, but for larger spacings (33) must be used. However, (33) gives a transcendental solution for $p_0$, since the terms $p_n$, $p_{-n}$ are not known before $\beta$ has been found. The series in (33) represents only a small contribution to $\phi$, and so a solution by the method of successive approximations may be applied. Equation (34) is used to find an approximate value of $\beta$, in order to find approximate values for $p_n$ and $p_{-n}$. Equation (33) may then be used to solve for a second approximation to $\beta$ and $p_0$.

If the TM surface wave is considered to be propagating (attenuated) in the positive $x$ direction, its wave impedance in this direction is capacitive, as in the usual waveguide situation. In the negative direction the wave impedance is inductive, and a surface exhibiting an inductive reactance is required in order to support this surface wave. For a TE surface wave, the wave impedance in the direction toward the surface is capacitive, and the surface reactance must be capacitive to support the wave. For TE modes, there would be a net storage of magnetic energy at the interface between a system of parallel plates and free space. This gives rise to an inductive reactance. In order to obtain a capacitive surface reactance, a short circuit must be placed at a suitable distance $d$ from the interface, and the spacing $S$ must be chosen so that the dominant mode in the parallel-plate region can propagate. The dominant mode is the $TE_{10}$ mode, and hence $S$ must be greater than $\lambda_0/2$. For this reason the interface will not support a TE surface wave. The edges of the plates or corrugations act as a diffraction grating, and, whenever the period is greater than $\lambda_0/2$, a diffraction-grating mode that propagates away from the interface always exists. If, however, the region between the plates was filled with a high-dielectric-constant material, then the spacing could be reduced to less than $\lambda_0/2$, and it should then be possible to support a TE surface wave.

*Attenuation Due to Conductor Loss*

A first approximation to the attenuation constant of the surface wave may be obtained by considering the field of the dominant modes only and evaluating the power loss in the conductor. The fields, in the case of zero loss, are approximately

$$
\begin{aligned}
H_y &= Ae^{-p_0(x-d)}e^{-j\beta z} && x > d \\
E_x &= \frac{\beta}{k_0} Z_0 H_y && x > d \\
H_y &= A \sec k_0 d \cos k_0 x && x < d \qquad 0 \le z \le S
\end{aligned}
$$

The power flow along the $z$ axis per unit width along the $y$ axis is

$$
P = \frac{\beta Z_0}{2k_0} \int_d^\infty H_y H_y^* \, dx = \frac{AA^* \beta Z_0}{4k_0 p_0} \tag{35}
$$

The power loss per unit width in one corrugation is

$$
\begin{aligned}
P_L &= \tfrac{1}{2} R_s Z_0 \left( 2 \int_0^d H_y H_y^* \, dx + \int_0^S H_y H_y^* \, dz \right) \\
&= \tfrac{1}{2} R_s Z_0 A A^* \sec^2 k_0 d \left( S + d + \frac{\sin 2k_0 d}{2k_0} \right)
\end{aligned} \tag{36}
$$

since the current density on the conductors is equal to $|H_y|$. There are $1/S$ sections per meter, and so the power loss per meter is $P_L/S$. The attenuation constant $\alpha$ is given by $\alpha = P_L/2PS$, where $P_L/S$ is the power loss per meter in the $z$ direction, and $P$ is the power flow in the $z$ direction. From (35) and (36), we obtain

$$
\alpha = \frac{R_s k_0 p_0 [S + d + (\sin 2k_0 d)/2k_0]}{\beta S \cos^2 k_0 d} \qquad \text{nepers per meter} \tag{37}
$$

where $R_s$ is the normalized surface resistivity, $p_0$ is a solution of (33) or (34), and $\beta^2 = k_0^2 + p_0^2$. To obtain a low attenuation, the surface wave should be loosely bound to the surface, in which case $p_0$ is small.

As an example, consider a corrugated copper plane with $\lambda_0 = 0.0314$ meter, $S = 0.2\lambda_0$, $d = 0.159\lambda_0$, $\sigma = 5.8 \times 10^7$ mhos per meter. From (34), the first-order solution for $p_0$ is found to be 176 nepers per meter. Using this value of $p_0$ to compute $\beta$, $p_n$, $p_{-n}$, and $h_n$, a second approximation to $p_0$ may be computed from (33) by equating $\phi$ to $\pi - 2k_0 d$. This yields a value of 182 nepers per meter for $p_0$. The first approximation, as given by (34), is seen to be quite good in this example. Using (37), the attenuation constant $\alpha$ is found to be 0.066 neper per meter. This is about 20 times greater than that for the dielectric-coated plane considered earlier. However, the wave in the present example is much more tightly bound to the surface, since $p_0$ is about seven times larger than for the dielectric-coated-plane example. A reduction of $p_0$ in the present case would reduce the attenuation proportionately.

### 11.5. Surface Waves along Dielectric Slabs

Surface waves of both the TM and TE type may be guided along a dielectric slab or sheet. Let the thickness of the sheet be $2t$, and let the relative dielectric constant be $\kappa$, as in Fig. 11.8. The dielectric will also be assumed to have negligible loss, so that $\kappa$ is real. The permeability is taken equal to $\mu_0$. The TM modes have a single component of magnetic field $H_y$ and electric field components $E_x$ and $E_z$. The TE modes have the roles of electric field and magnetic field interchanged, and hence have components $E_y$, $H_x$, $H_z$. Propagation of the modes is in the $z$ direction with a propagation factor $e^{-j\beta z}$. For both types of modes, the field decays away from the surface according to a factor $e^{-p(|x|-t)}$. Since the

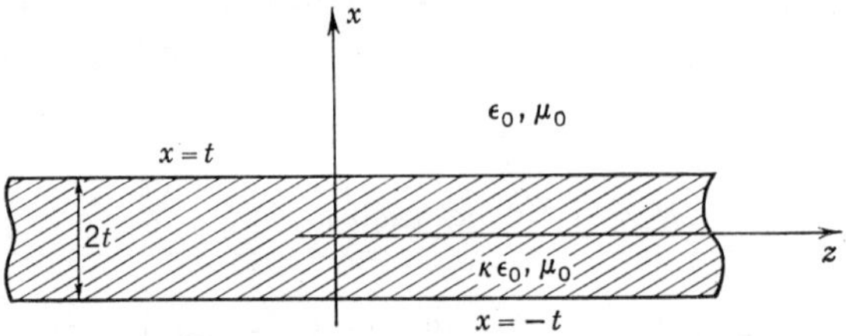


FIG. 11.8. A plane-dielectric-sheet surface waveguide.

fields satisfy the wave equation, $\beta^2 = k_0^2 + p^2$, and hence the guide wavelength and phase velocity are both less than that for plane waves in free space.

*TM Modes*

The solutions for $H_y$ in the dielectric may be of the symmetrical type (even) or antisymmetrical type (odd), where "even" and "odd" refer to the way that $H_y$ varies with $x$ about the symmetry plane $x = 0$. For the even solutions

$$\left.\frac{\partial H_y}{\partial x}\right|_{x=0} = 0$$

and, since $E_z$ is proportional to this derivative, the tangential electric field vanishes on the plane $x = 0$ for the even solutions. A conducting plane may be placed along $x = 0$, and the resultant solution is for a TM mode along a dielectric slab of thickness $t$ and placed on a conducting plane. The odd solutions correspond to a magnetic wall along the symmetry plane $x = 0$.

For the even solutions, $H_y$ will be of the following form in the dielectric region:

$$H_y = B \cos hx e^{-j\beta z} \qquad |x| \leq t$$

where $\beta^2 = k_0^2 - h^2$. Outside the dielectric, $H_y$ may be represented by

$$H_y = A \exp\left[-p(|x| - t) - j\beta z\right] \qquad |x| \geq t$$

where $\beta^2 = k_0^2 + p^2$. At $x = \pm t$, $H_y$ and $E_z$ must be continuous. Since $j\omega\epsilon E_z = \partial H_y/\partial x$, we obtain the two following equations:

$$A = B \cos ht \tag{38a}$$
$$\kappa p A = hB \sin ht \tag{38b}$$

Dividing one equation by the other yields the eigenvalue equation

$$\kappa p = h \tan ht \tag{39}$$

which, together with the relation obtained by equating the two expressions for $\beta^2$,

$$(\kappa - 1)k_0^2 = p^2 + h^2 \tag{40}$$

permits a solution for $h$ and $p$ to be found. A solution may readily be obtained by graphical means. For this purpose it is convenient to rewrite (39) and (40) as follows:

$$pt = ht \tan ht \tag{41a}$$
$$(pt)^2 + (ht)^2 = (\kappa - 1)(k_0 t)^2 \tag{41b}$$

In this latter form the equations are independent of frequency, provided the ratio of thickness $2t$ to free-space wavelength $\lambda_0$ is kept constant.

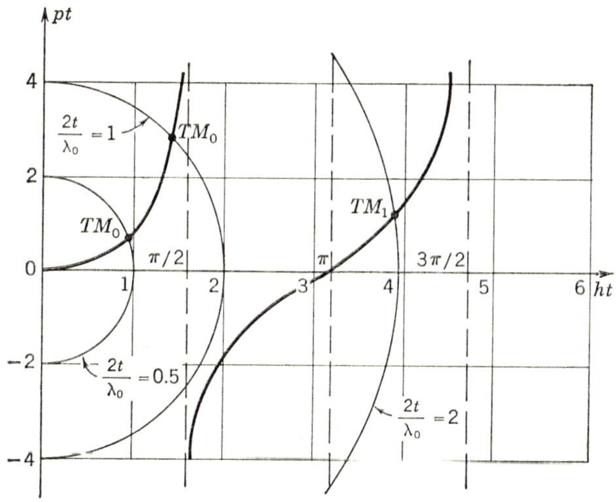


FIG. 11.9. Graphical solution for eigenvalues for even TM modes.

Equation (41$a$) determines one relation between $ht$ and $pt$, and may be plotted on an $ht$-$pt$ plane. The other equation, (41$b$), is a circle of radius $(\kappa - 1)^{1/2}k_0 t$ in the same plane. The points of intersection between the two curves determine the eigenvalues $h$ and $p$. A typical plot is given in Fig. 11.9 for a polystyrene sheet with $\kappa = 2.56$.

Several interesting properties of the modes may be deduced from the graphical solution given in Fig. 11.9. The two sets of curves will always have at least one point of intersection, even for $t/\lambda_0$ approaching zero. Hence, the first even mode, which we will label as the $TM_0$ mode, has no low-frequency cutoff. Since $p$ must be positive for a surface mode, only those points of intersection that lie in the intervals $0 \le ht \le \pi/2$, $\pi \le ht \le 3\pi/2$, or, in general, $n\pi \le ht \le n\pi + \pi/2$, where $n$ is any integer, correspond to surface-wave solutions. For sufficiently large values of $t$, there are several modes of propagation. All the higher modes will, however, have a low-frequency cutoff. Points of intersection that occur in the range intervals $n\pi + \pi/2 \le ht \le (n+1)\pi$, where $n$ is any integer, lead to solutions with negative values of $p$. These modes increase exponentially away from the surface, and hence must be excluded on physical grounds since the field amplitude becomes infinite. The higher-order modes are cut off for that value of $t/\lambda_0$ which causes the circle to intersect the tangent curves (41$a$) along the $ht$ axis. The cutoff values are readily seen to be given by $ht = n\pi$, and hence, for the $n$th mode, the value of $t/\lambda_0$ at cutoff is, from (41$b$),

$$\left(\frac{t}{\lambda_0}\right)_{co} = \frac{n}{2(\kappa - 1)^{1/2}} \qquad n = 0, 1, 2, \ldots \tag{42}$$

The surface-wave modes along a dielectric sheet are often called "trapped" modes. The reason for this terminology is that these modes may be considered as a plane wave incident from the dielectric region onto the air-dielectric interface at an angle greater than the critical angle $\theta_c$, where $\sin\theta_c = \kappa^{-1/2}$. For incident angles greater than $\theta_c$, there is complete reflection at the interface, and the field outside the dielectric is evanescent. Thus the mode consists of plane waves propagating along a zigzag path along the $z$ axis and undergoing complete reflection at each interface. In other words, the field is trapped within the dielectric sheet.

For the odd modes, the magnetic field is of the following form:

$$H_y = \begin{cases} A \exp\left[-p(x - t) - j\beta z\right] & x \ge t \\ A \csc ht \sin hx \, e^{-j\beta z} & -t \le x \le t \\ -A \exp\left[p(x + t) - j\beta z\right] & x \le -t \end{cases}$$

Matching the tangential fields at $x = \pm t$ leads to the eigenvalue equation for the odd modes:

$$\kappa pt = -ht \cot ht \tag{43a}$$

$$(pt)^2 + (ht)^2 = (\kappa - 1)(k_0 t)^2 \tag{43b}$$

The solutions for $p$ and $h$ are readily determined by a graphical procedure similar to that used for the even modes. A typical graphical solution is illustrated in Fig. 11.10 for a polystyrene sheet with $\kappa = 2.56$. From this plot it is clear that all the odd modes have a lower-frequency cutoff.

Only those points of intersection that occur in the range intervals $n\pi - \pi/2 \le ht \le n\pi$, where $n$ is any integer, give rise to surface waves with $p$ positive. The values of $t/\lambda_0$ at cutoff are given by

$$\left(\frac{t}{\lambda_0}\right)_{co} = \frac{2n+1}{2(\kappa-1)^{1/2}} \qquad n = 0, 1, 2, \ldots \tag{44}$$

The equations for the cutoff parameter for the even and odd modes may be combined to give

$$\left(\frac{2t}{\lambda_0}\right)_{co} = \frac{n}{2(\kappa-1)^{1/2}} \qquad n = 0, 1, 2, \ldots \tag{45}$$

where the even integers $n = 0, 2, 4, \ldots$ now refer to the even modes, while the odd integers $n = 1, 3, 5, \ldots$ refer to the odd modes.

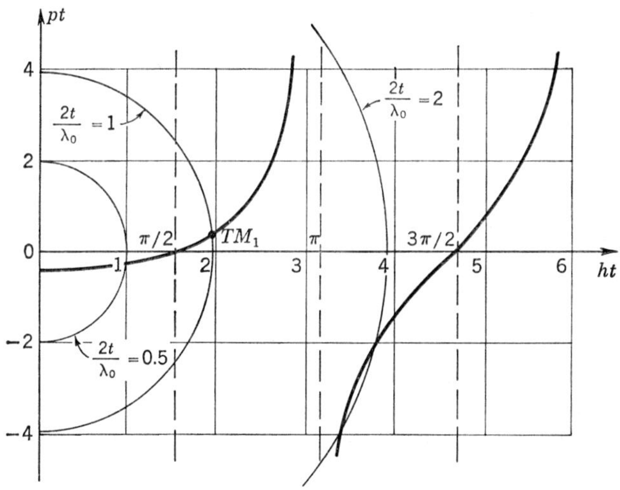


FIG. 11.10. Graphical solution for eigenvalues for odd TM modes.

*TE Modes*

The solution for TE modes is formally the same as that for the TM modes. The single electric field component $E_y$ will have the following form:

$$E_y = \begin{cases} A \exp\left[-p(|x| - t) - j\beta z\right] & |x| \ge t \\ A \sec ht \cos hx \, e^{-j\beta z} & |x| \le t \end{cases}$$

for the even modes, and

$$E_y = \begin{cases} A \exp\left[-p(x - t) - j\beta z\right] & x \ge t \\ A \csc ht \sin hx \, e^{-j\beta z} & |x| \le t \\ -A \exp\left[p(x + t) - j\beta z\right] & x \le -t \end{cases}$$

for the odd modes. The magnetic field components are given by

$$j\omega\mu_0 H_x = \frac{\partial E_y}{\partial z}$$

$$j\omega\mu_0 H_z = -\frac{\partial E_y}{\partial x}$$

By matching the tangential fields at $x = \pm t$, the eigenvalue equations for $p$ and $h$ are obtained:

$$pt = ht \tan ht \qquad \text{even modes} \tag{46a}$$

$$pt = -ht \cot ht \qquad \text{odd modes} \tag{46b}$$

$$(pt)^2 + (ht)^2 = (\kappa - 1)(k_0 t)^2 \tag{46c}$$

$$\beta^2 = k_0^2 + p^2 = \kappa k_0^2 - h^2 \tag{46d}$$

Apart from the absence of the factor $\kappa$ in (46$a$) and (46$b$), these equations are the same as for the TM modes. Consequently, a similar graphical solution may be used. Again it is only the $TE_0$, that is, the first even mode, that has no low-frequency cutoff. The value of $2t/\lambda_0$ at cutoff is given by the same equation as for the TM modes, i.e., by (45). For a dielectric slab of thickness $t$ above a conducting plane, only the odd solutions are possible, since $E_z$ does not vanish at $x = 0$ for the even modes. Thus, for a grounded dielectric slab, a TE surface mode with no low-frequency cutoff does not exist.

*The Continuous Eigenvalue Spectrum*

For any plane-dielectric sheet of given thickness, only a finite number of surface-wave modes exist. These modes obviously do not form a complete eigenfunction set in which the field from an arbitrary source can be expanded. It is, therefore, clear that the surface-wave modes do not represent the complete solution for the fields which may exist along a dielectric sheet. In order to simplify the problem, we will confine our attention to the TM waves along a grounded dielectric sheet of thickness $t$. Similar results will, however, hold for the TE surface waves along a grounded dielectric sheet, as well as for single sheets located in free space. The eigenvalue equations, (41), which apply for TM waves along a grounded dielectric sheet, has been studied in detail by Brown.† Similar results will be derived here by using a transmission-line analogy.

Consider a plane TM wave incident at an angle $\theta_i$ on a grounded dielectric sheet, as in Fig. 11.11$a$. In the dielectric, a standing wave exists because of reflection at the conducting plane. Above the dielectric, a reflected wave also exists, so that the field is a standing wave along the

† J. Brown, The Type of Wave Which May Exist near a Guiding Surface, *Proc. IEE (London)*, vol. 100, part III, pp. 363–364, November, 1953.

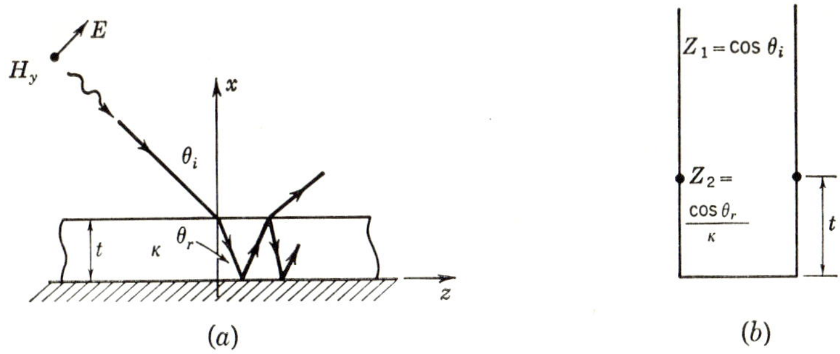


FIG. 11.11. TM wave incident on a grounded dielectric sheet.

$x$ axis in this region also. The magnetic field in the two regions may be represented as follows:

$$\begin{aligned} H_y &= A \exp[-jk_0(-x\cos\theta_i + z\sin\theta_i)] \\ &\quad - RA\exp[-jk_0(x\cos\theta_i + z\sin\theta_i)] \qquad x \geq t \qquad (47a) \\ H_y &= B\cos(kx\cos\theta_r)\exp(-jk_0 z\sin\theta_i) \qquad 0 \leq x \leq t \qquad (47b) \end{aligned}$$

where $k = \kappa^{1/2}k_0$, $k\sin\theta_r = k_0\sin\theta_i$, and $\theta_r$ is the angle of refraction. The normalized wave impedance in the negative $x$ direction in the two regions is

$$Z_1 = \frac{E_z}{Z_0 H_y} = \cos\theta_i \qquad x > t \qquad (48a)$$

$$Z_2 = \frac{\cos\theta_r}{\kappa^{1/2}} = \frac{(\kappa - \sin^2\theta_i)^{1/2}}{\kappa} \qquad 0 \leq x \leq t \qquad (48b)$$

For propagation in the $x$ direction, the equivalent transmission-line circuit is illustrated in Fig. 11.11$b$. The propagation factors in the two lines are $k_0\cos\theta_i$ and $k\cos\theta_r$. At the dielectric interface, the normalized input impedance is

$$Z_{\text{in}} = j\kappa^{-1/2}\cos\theta_r\tan(kt\cos\theta_r) \qquad (49)$$

and hence the reflection coefficient $R$ is given by

$$R = \frac{j\cos\theta_r\tan(kt\cos\theta_r) - \kappa^{1/2}\cos\theta_i}{j\cos\theta_r\tan(kt\cos\theta_r) + \kappa^{1/2}\cos\theta_i} \qquad (50)$$

and has a modulus of unity.

For convenience, let $k_0\sin\theta_i = \beta$, $jk_0\cos\theta_i = p$, and $k\cos\theta_r = h$. In place of (50), we now have

$$R = \frac{h\tan ht + \kappa p}{h\tan ht - \kappa p} \qquad (51)$$

From the wave equation, the following relations are obtained:

$$\beta^2 = k_0^2 + p^2 = \kappa k_0^2 - h^2 \qquad (52)$$

A solution for standing waves along the $x$ axis exists for all real angles of incidence $0 \le \theta_i \le \pi/2$. The corresponding ranges of $p$ and $\beta$ are

$$\begin{gathered} jk_0 \ge p \ge j0 \\ 0 \le \beta \le k_0 \end{gathered}$$

This partial solution is seen to have a continuous spectrum for $\beta$ in the range zero to $k_0$.

The standing-wave solution along the $x$ direction will also exist for imaginary values of $p$ greater than $jk_0$. This range of $p$ corresponds to imaginary angles of incidence. Hence, for $p > jk_0$, the corresponding range for $\beta$ is, from (52),

$$-j\infty < \beta < j0$$

and gives rise to a continuous spectrum of waves that is attenuated, since $\beta$ is imaginary, along the $z$ direction.

The next case to examine is that for real values of $p$. In place of a sinusoidal variation with $x$, we now obtain a field described by a hyperbolic sine function of $x$ above the dielectric. Such a solution must be discarded on physical grounds since it becomes infinite for $x$ approaching infinity. It is only for certain real values of $p$ that a solution can exist. These solutions are the surface-wave modes discussed earlier. From (51) it is seen that the reflection coefficient becomes infinite whenever $h \tan ht = \kappa p$, which is the eigenvalue equation (39). Thus the surface wave may be considered to be produced by the pole in the expression for the reflection coefficient. Since $|R|$ tends to infinity at the pole, the amplitude of the incident wave may tend to zero, and still produce a reflected wave with finite amplitude. The answer to the question of whether a surface-wave mode should be associated with a zero or a pole of the reflection coefficient depends on whether the field above the dielectric is considered to be a reflected wave or an incident wave, since the sign of the wave impedance changes accordingly. If the field is considered as an incident wave, the solution corresponds to a zero of the reflection coefficient and not to a pole, as is the case if the wave is considered as a reflected one. Zucker has resolved this difficulty by considering the excitation problem, from which it is concluded that the pole association is the correct one, since a pole of the reflection coefficient occurs on that sheet of the Riemann surface for which the integral for the radiated field converges properly.†

The complete solution for the physical field consists of one or more surface waves with $\beta > k_0$, a continuous spectrum of waves with $0 < \beta < k_0$, and a continuous spectrum of evanescent waves with $\beta = -j\alpha$

† F. J. Zucker, Two Notes on Surface Wave Nomenclature and Classification, paper presented at the URSI Symposium, Toronto, Canada, June, 1958.

and $0 < \alpha < \infty$. The field radiated by an arbitrary source will, in general, contribute to all three of the above types of fields.

In addition to the roots leading to surface waves, the eigenvalue equations (41), (43), and (46) have an infinity of solutions with complex roots. The corresponding modes are referred to as "leaky modes," since they correspond to a flow of power away from the surface. These modes, however, do not satisfy the radiation conditions at infinity, and hence do not belong to the proper eigenvalue spectrum. Nevertheless it is possible to utilize these modes to partially represent the radiated field from an elementary source.† In deforming the contour of integration for the radiated field into a path of "steepest descent," some of these complex poles are crossed, and their residues then contribute to the expansion of the field. Further discussion of these leaky modes will be given in the section on excitation of surface waves by a line source. A meaningful discussion is difficult to give without a consideration of the total field radiated by a given source and is, therefore, not given in the present section.

## 11.6. Surface Waves on Cylindrical Structures

The possibility of a surface-wave mode of propagation along a round wire of conducting material was demonstrated theoretically by Sommerfeld in 1899. The mode is a TM wave with components $H_\theta$, $E_r$, $E_z$, and is axially symmetrical, i.e., independent of the angular coordinate $\theta$. Sommerfeld's solution is given by Stratton, and, consequently, will not be repeated here.‡ The mode is found to exist only for a finite conductivity and is loosely bound to the surface. This surface-wave mode has been studied more recently by Goubau as to its suitability for a practical transmission-line system.§ The results of Goubau's studies are that a single small-diameter conducting wire will propagate the axially symmetric surface-wave mode with a low attenuation, but that the field extends a considerable distance outside the wire before decaying to a negligible value. The practical use of this surface waveguide is, therefore, limited to the high frequencies $f > 10^{10}$ cycles per second. The situation here is much the same as for the Zenneck wave above a conducting plane. In order to confine the field to the surface of the guide, the wire may be coated with a thin layer of dielectric or corrugated. These modifications have also been studied by Goubau in the same reference.

† N. Marcuvitz, On Field Representations in Terms of Leaky Modes, *IRE Trans.*, vol. AP-4, pp. 192–194, July, 1956.

‡ J. A. Stratton, "Electromagnetic Theory," McGraw-Hill Book Company, Inc., New York, 1941.

§ G. Goubau, Surface Waves and Their Applications to Transmission Lines, *J. Appl. Phys.*, vol. 21, pp. 1119–1128, November, 1950.

The axially symmetric surface wave is frequently called the Sommerfeld-Goubau wave, in honor of the principal investigators. In addition, higher-order modes, with angular dependence, may exist on a conducting wire. These modes, however, attenuate so rapidly with distance that they are of no importance except as part of the near-zone field at the region of the exciting source.

The first analysis of a dielectric-coated wire was made by Harms in 1907.† The solution for the axially symmetric mode will be given here for a lossless dielectric coating on a perfectly conducting cylinder, as illustrated in Fig. 11.12. The radius of the wire is $a$, the outside radius of the dielectric sleeve is $b$, and the thickness of the sleeve is $b - a = t$. For an axially symmetric mode or field, a TM solution is possible. For fields that vary with the angle $\theta$, each mode is a combination of a TM

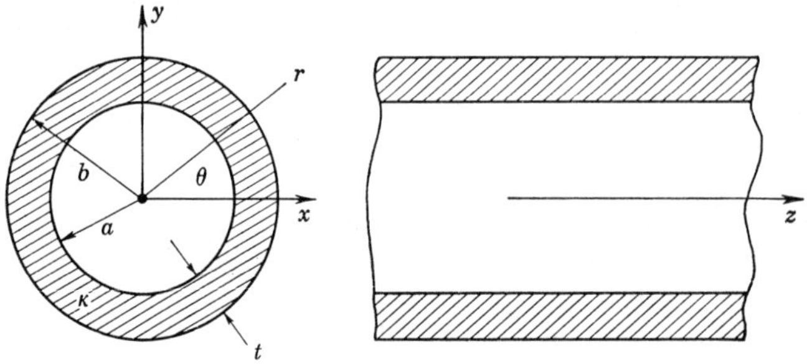


FIG. 11.12. Dielectric-coated-wire surface waveguide.

and TE mode with all six field components present. For the TM mode, with no variation with the angle coordinate, and with propagation along the $z$ axis according to $e^{-j\beta z}$, Maxwell's equations reduce to

$$E_z = \psi(r)e^{-j\beta z} \tag{53a}$$

$$E_r = -\frac{j\beta}{k_c^2}\frac{d\psi}{dr}e^{-j\beta z} \tag{53b}$$

$$H = -\frac{j\omega\epsilon}{k_c^2}\frac{d\psi}{dr}e^{-j\beta z} \tag{53c}$$

$$\frac{d^2\psi}{dr^2} + \frac{1}{r}\frac{d\psi}{dr} + k_c^2\psi = 0 \tag{53d}$$

where $k_c^2 = k_0^2 - \beta^2$ in the air region $r > b$, and is equal to $\kappa k_0^2 - \beta^2$ in the dielectric region $a < r < b$. The radial function $\psi(r)$ is a solution of Bessel's equation of order 0. In the region $a < r < b$, a suitable solution is a linear combination of the Bessel functions of the first and second kinds. In the region $r > b$, the solution is a modified Bessel function of the second kind, that is, $K_0[r(\beta^2 - k_0^2)^{1/2}]$, since this is the only solu-

† F. Harms, *Ann. Phys.*, vol. 23, pp. 44–60, 1907.

tion which decays exponentially for large $r$. Thus we have

$$\psi = \begin{cases} AK_0[r(\beta^2 - k_0^2)^{1/2}] & r > b \\ BJ_0[r(k^2 - \beta^2)^{1/2}] + CY_0[r(k^2 - \beta^2)^{1/2}] & a < r < b \end{cases}$$

where $k^2 = \kappa k_0^2$.

At $r = a$, $E_z = 0$, and hence $\psi(a) = 0$ and the constant $C$ is given by

$$C = -\frac{BJ_0[a(k^2 - \beta^2)^{1/2}]}{Y_0[a(k^2 - \beta^2)^{1/2}]}$$

The eigenvalue equation for $\beta$ is obtained by matching the fields tangential to the surface at $r = b$, and is

$$\frac{K_1(pb)}{pK_0(pb)} = \frac{\kappa}{h}\frac{J_0(ha)Y_1(hb) - J_1(hb)Y_0(ha)}{J_0(hb)Y_0(ha) - J_0(ha)Y_0(hb)} \tag{54}$$

where $p^2 = \beta^2 - k_0^2$ and $h^2 = k^2 - \beta^2$.

The eigenvalue equation is transcendental, and graphical methods would have to be used to solve it, in general. However, for surface waveguides of practical interest, the thickness $t$ is very small, the radius $a$ and, hence, $pa$ and $ha$ are small, and $p^2$ is small compared with $k_0^2$, so that $\beta$ is only moderately larger than $k_0$. In (54), the small-argument approximations

$$K_0(x) \to -\left(0.577 + \ln\frac{x}{2}\right) = -\ln 0.89x$$

$$K_1(x) \to \frac{1}{x}$$

may be used. Following Goubau, the right-hand side is simplified by expanding $J_0(ha)$, etc., in a Taylor series about $r = b$, to get

$$J_0(ha) = J_0(hb) - \frac{dJ_0(hb)}{dr}(b - a) = J_0(hb) + htJ_1(hb)$$

and similarly for $Y_0(ha)$. With these approximations, the eigenvalue equation reduces to

$$\kappa p^2 b \ln 0.89pb = -(\kappa - 1)k_0^2 t + p^2 t \approx -(\kappa - 1)k_0^2 t \tag{55}$$

since $p^2 \ll k_0^2$.

As an example, consider a copper wire with $b = 0.001$ meter, $t = 10^{-4}$ meter, $\kappa = 2.56$, and $\lambda_0 = 3.14 \times 10^{-2}$ meter. From (55), we get $p = 25.8$ nepers per meter, and, hence, $\beta = 202$ radians per meter. For large $r$, we have

$$K_0(pr) = \left(\frac{\pi}{2pr}\right)^{1/2} e^{-pr}$$

and thus, in a distance corresponding to a few wavelengths, the field has decayed to a negligible value. The approximations made in arriving at (55) are clearly valid for this example.

The attenuation may be computed by evaluating the losses in the conductor and the dielectric sleeve separately, adding the two, and dividing by twice the power flow along the guide in the $z$ direction. Numerical values are given by Goubau.†

*Surface Waves along a Dielectric Rod*

As a last example of a structure capable of supporting a surface wave, we consider a dielectric rod of radius $a$, as in Fig. 11.13. Mode propagation along the dielectric rod was studied by Hondros and Debye as

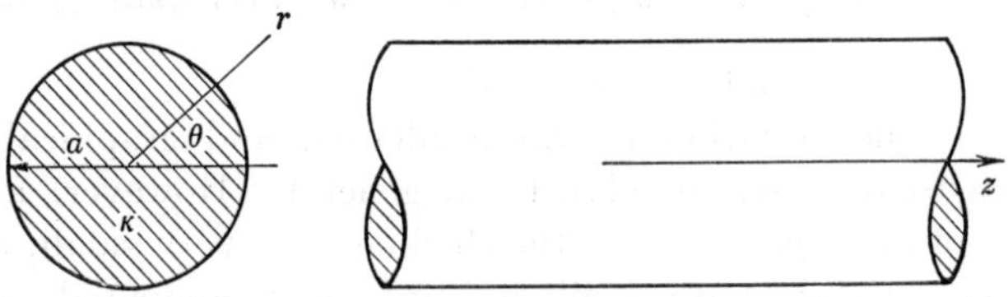


FIG. 11.13. Dielectric-rod surface waveguide.

long ago as 1910, and more recently by Elsasser and Chandler as well as others.‡ Dielectric rods are similar in behavior to dielectric sheets in that a number of surface-wave modes exist. Pure TM or TE modes are possible only if the field is independent of the angular coordinate $\theta$. As the radius of the rod increases, the number of TM and TE modes also increases. These modes do, however, have a cutoff point such that, below some minimum value of $a/\lambda_0$, the mode cannot exist any longer. When the field depends on the angular coordinate, pure TM or TE modes no longer exist. All modes with angular dependence are a combination of a TM and a TE mode, and are classified as hybrid EH or HE modes, depending on whether the TM or TE mode predominates, respectively. All these modes, with the exception of the $HE_{11}$ mode, exhibit cutoff phenomena similar to that of the axially symmetric modes. Since the $HE_{11}$ mode has no low-frequency cutoff, it is the dominant mode. This mode is widely used in the dielectric-rod

† G. Goubau, Single Conductor Surface Wave Transmission Lines, *Proc. IRE*, vol. 39, pp. 619–624, June, 1951.

‡ Hondros and Debye, *Ann. Phys.*, vol. 32, pp. 465–476, 1910.

W. M. Elsasser, Attenuation in a Dielectric Circular Rod, *J. Appl. Phys.*, vol. 20, pp. 1193–1196, December, 1949.

C. H. Chandler, Investigation of Dielectric Rod as Waveguide, *J. Appl. Phys.*, vol. 20, pp. 1188–1192, December, 1949.

antenna.† For small-diameter rods, the field extends for a considerable distance beyond the surface, and the axial propagation constant $\beta$ is only slightly larger than $k_0$. As the radius increases, the field is confined closer and closer to the rod, and $\beta$ approaches $\kappa^{1/2}k_0$ in the limit of infinite radius. Since $\beta > k_0$, the phase velocity is less than that of plane waves in free space. In all of the above respects, the dielectric rod does not differ from the plane-dielectric sheet.

The field expansion around the dielectric rod is similar to that for the sheath helix, with the exception that, in the dielectric region $r < a$, the field varies according to the Bessel functions of the first kind. By analogy with the set of equations in Sec. 9.8 for the sheath helix, we have

$$\left.\begin{aligned}
E_z &= A_n J_n(hr) e^{-jn\theta - j\beta z} \\
E_r &= \left(-\frac{j\beta}{h} A_n J_n' - \frac{n\omega\mu_0}{h^2 r} B_n J_n\right) e^{-jn\theta - j\beta z} \\
E_\theta &= \left(-\frac{n\beta}{h^2 r} A_n J_n + \frac{j\omega\mu_0}{h} B_n J_n'\right) e^{-jn\theta - j\beta z} \\
H_z &= B_n J_n(hr) e^{-jn\theta - j\beta z} \\
H_r &= \left(\frac{n\omega\epsilon}{h^2 r} A_n J_n - \frac{j\beta}{h} B_n J_n'\right) e^{-jn\theta - j\beta z} \\
H_\theta &= \left(-\frac{j\omega\epsilon}{h} A_n J_n' - \frac{n\beta}{h^2 r} B_n J_n\right) e^{-jn\theta - j\beta z}
\end{aligned}\right\} \quad r < a$$

$$\left.\begin{aligned}
E_z &= C_n K_n(pr) e^{-jn\theta - j\beta z} \\
E_r &= \left(\frac{j\beta}{p} C_n K_n' + \frac{n\omega\mu_0}{p^2 r} D_n K_n\right) e^{-jn\theta - j\beta z} \\
E_\theta &= \left(\frac{n\beta}{p^2 r} C_n K_n - \frac{j\omega\mu_0}{p} D_n K_n'\right) e^{-jn\theta - j\beta z} \\
H_z &= D_n K_n(pr) e^{-jn\theta - j\beta z} \\
H_r &= \left(-\frac{n\omega\epsilon_0}{p^2 r} C_n K_n + \frac{j\beta}{p} D_n K_n'\right) e^{-jn\theta - j\beta z} \\
H_\theta &= \left(\frac{j\omega\epsilon_0}{p} C_n K_n' + \frac{n\beta}{p^2 r} D_n K_n\right) e^{-jn\theta - j\beta z}
\end{aligned}\right\} \quad r > a$$

where $\beta^2 = k_0^2 + p^2 = k_0^2 - h^2$; $A_n$, $B_n$, $C_n$, $D_n$ are amplitude constants; and the prime indicates differentiation with respect to the arguments $hr$ and $pr$.

Imposition of the boundary conditions at $r = a$ leads to equations for determining the relative amplitudes of the coefficients, and also the

† D. G. Kiely, "Dielectric Aerials," Methuen & Co., Ltd., London, 1953.

eigenvalue equation. The eigenvalue equation is

$$\left[\frac{\kappa J_n'(u_1)}{u_1 J_n(u_1)} - \frac{K_n'(u_2)}{u_2 K_n(u_2)}\right]\left[\frac{J_n'(u_1)}{u_1 J_n(u_1)} + \frac{K_n'(u_2)}{u_2 K_n(u_2)}\right] = \left[\frac{n\beta}{k_0}\frac{(u_2^2 + u_1^2)}{u_1^2 u_2^2}\right]^2 \tag{56}$$

where $u_1 = ha$, $u_2 = pa$. When $n = 0$, the right-hand side vanishes, and each factor on the left-hand side must equal zero. These two factors give the eigenvalue equations for the axially symmetric TM and TE modes:

$$\frac{\kappa J_0'(u_1)}{u_1 J_0(u_1)} = -\frac{K_0'(u_2)}{u_2 K_0(u_2)} \qquad \text{TM modes} \tag{57a}$$

$$\frac{J_0'(u_1)}{u_1 J_0(u_1)} = -\frac{K_0'(u_2)}{u_2 K_0(u_2)} \qquad \text{TE modes} \tag{57b}$$

In addition, $u_1$ and $u_2$ are related by the equation

$$u_1^2 + u_2^2 = \kappa(\ - 1)(k_0 a)^2 \tag{58}$$

The solution to the above equations must be obtained graphically, and the computations are very laborious to carry out. The ratio of $\beta$

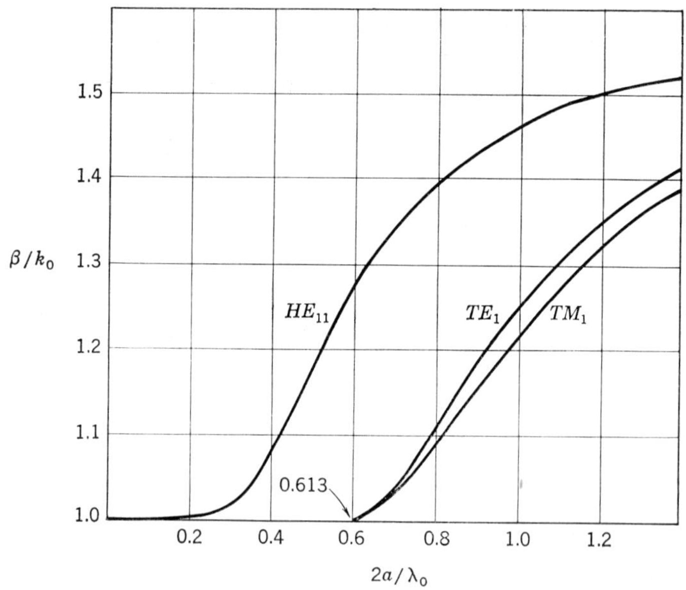


FIG. 11.14. Ratio of $\beta$ to $k_0$ for first three surface-wave modes on a polystyrene rod ($\kappa = 2.56$).

to $k_0$ as a function of $2a/\lambda_0$ for a polystyrene rod is given in Fig. 11.14 for the axially symmetric $TM_1$ and $TE_1$ modes and the $HE_{11}$ dipole mode. These curves were constructed from the numerical data given by

Elsasser.† Both the $TM_1$ and $TE_1$ modes are cut off for $2a < 0.613\lambda_0$, while the $HE_{11}$ mode has no low-frequency cutoff.

## 11.7. Field Orthogonality Properties

On any surface waveguide of the type considered in this chapter the various surface-wave modes are orthogonal to one another, as well as orthogonal to the continuous eigenvalue field, i.e., radiation field. The orthogonality properties may be proved by means of the Lorentz reciprocity theorem in a way like that for conventional waveguides.‡

Consider a surface waveguide as in Fig. 11.15. Let a current element **J** be located in the vicinity of $z = 0$. Choose a closed surface $S$, consisting of two transverse planes at $z_1$ and $z_2$ and the surface of a cylinder surrounding the guide at infinity. The current element radiates one or more surface-wave modes and a radiation field. The total field may be represented as follows:

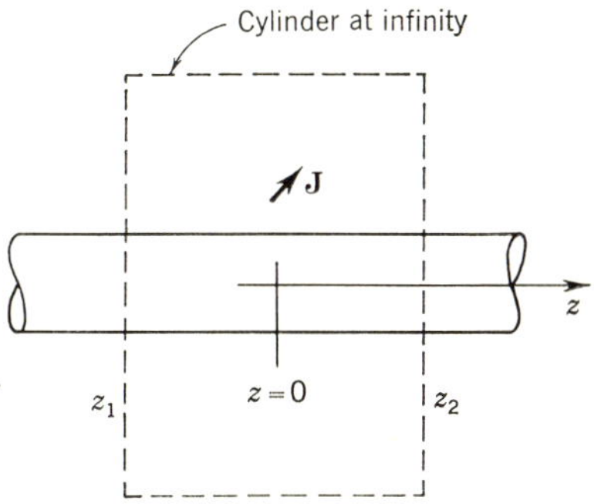


FIG. 11.15. Surface waveguide with current source.

$$\left.\begin{aligned} \mathbf{E} &= \sum_n a_n \mathbf{E}_n^+ + \mathbf{E}_R^+ \\ \mathbf{H} &= \sum_n a_n \mathbf{H}_n^+ + \mathbf{H}_R^+ \end{aligned}\right\} \quad z > 0 \tag{59a}$$

$$\left.\begin{aligned} \mathbf{E} &= \sum_n b_n \mathbf{E}_n^- + \mathbf{E}_R^- \\ \mathbf{H} &= \sum_n b_n \mathbf{H}_n^- + \mathbf{H}_R^- \end{aligned}\right\} \quad z < 0 \tag{59b}$$

where $\mathbf{E}_R$, $\mathbf{H}_R$ is the radiation field and $\mathbf{E}_n$, $\mathbf{H}_n$ is the $n$th surface-wave mode field. The transverse fields of the surface-wave modes will be represented as follows:

$$\mathbf{E}_{nt}^+ = \mathbf{e}_n e^{-\Gamma_n z} \tag{60a}$$
$$\mathbf{E}_{nt}^- = \mathbf{e}_n e^{\Gamma_n z} \tag{60b}$$
$$\mathbf{H}_{nt}^+ = \mathbf{h}_n e^{-\Gamma_n z} \tag{60c}$$
$$\mathbf{H}_{nt}^- = -\mathbf{h}_n e^{-\Gamma_n z} \tag{60d}$$

where $\mathbf{e}_n$ and $\mathbf{h}_n$ are transverse vector functions of $x$ and $y$.

† *Loc. cit.*

‡ G. Goubau, On the Excitation of Surface Waves, *Proc. IRE*, vol. 40, pp. 865–868, July, 1952.

The total field is a solution of the equations

$$\nabla \times \mathbf{E} = -j\omega\mu\mathbf{H} \qquad \nabla \times \mathbf{H} = j\omega\epsilon\mathbf{E} + \mathbf{J}$$

while each surface-wave mode is a solution of the source-free field equations. Let $\mathbf{E}_1$, $\mathbf{H}_1$ be a solution to the source-free equations. The Lorentz reciprocity theorem gives

$$\oint\!\!\!\oint_S [\mathbf{E} \times \mathbf{H}_1 - \mathbf{E}_1 \times \mathbf{H}] \cdot \mathbf{n}\, da = \iiint \mathbf{J} \cdot \mathbf{E}_1\, dV \tag{61}$$

If $\mathbf{E}$, $\mathbf{H}$ is the $n$th surface-wave mode and $\mathbf{E}_1$, $\mathbf{H}_1$ the $m$th surface-wave mode, and we take $\mathbf{J} = 0$, then (61) gives

$$\iint \mathbf{e}_n \times \mathbf{h}_m \cdot \mathbf{a}_z\, da = \iint \mathbf{e}_m \times \mathbf{h}_n \cdot \mathbf{a}_z\, da = 0 \tag{62}$$

where the integration is over a transverse plane. The derivation of (62) is the same as for conventional waveguides. Equation (62) provides an orthogonality relation for the surface-wave modes.

Next let $\mathbf{E}$, $\mathbf{H}$ be the total radiated field and let $\mathbf{E}_1$, $\mathbf{H}_1$ be the $n$th surface-wave mode field having transverse components $\mathbf{e}_n e^{-\Gamma_n z}$, $\mathbf{h}_n e^{-\Gamma_n z}$. In (61), the integral over the cylinder at infinity is zero, since the radiation field is bounded, and the surface-wave modes decay exponentially as infinity is approached. The integral over the transverse planes reduces to

$$-e^{-\Gamma_n z_1} \iint_{z_1} (\mathbf{E}_R^- \times \mathbf{h}_n - \mathbf{e}_n \times \mathbf{H}_R^-) \cdot \mathbf{a}_z\, da$$
$$+ e^{-\Gamma_n z_2} \iint_{z_2} (\mathbf{E}_R^+ \times \mathbf{h}_n - \mathbf{e}_n \times \mathbf{H}_R^+) \cdot \mathbf{a}_z\, da$$
$$- 2b_n \iint \mathbf{e}_n \times \mathbf{h}_n \cdot \mathbf{a}_z\, da = \iiint \mathbf{E}_n^+ \cdot \mathbf{J}\, dV \tag{63}$$

because of the orthogonality property (62). Since $z_1$ and $z_2$ are arbitrary, the first two terms on the left-hand side must each be equal to a constant. Furthermore, the radiation field has a continuous eigenvalue spectrum, and so its dependence on $z$ cannot annul the factors $e^{-\Gamma_n z_1}$ and $e^{-\Gamma_n z_2}$. Therefore, the constants have to be equal to zero, and (63) gives

$$-2b_n \iint \mathbf{e}_n \times \mathbf{h}_n \cdot \mathbf{a}_z\, da = \iiint_V \mathbf{E}_n^+ \cdot \mathbf{J}\, dV \tag{64}$$

$$\iint_{z_1} \mathbf{E}_R^- \times \mathbf{h}_n \cdot \mathbf{a}_z\, da = \iint_{z_2} \mathbf{E}_R^+ \times \mathbf{h}_n \cdot \mathbf{a}_z\, da$$
$$= \iint_{z_1} \mathbf{e}_n \times \mathbf{H}_R^- \cdot \mathbf{a}_z\, da = \iint_{z_2} \mathbf{e}_n \times \mathbf{H}_R^+ \cdot \mathbf{a}_z\, da = 0 \tag{65}$$

If we begin with $\mathbf{E}_1$, $\mathbf{H}_1$, corresponding to the surface-wave mode $\mathbf{E}_n^-$, $\mathbf{H}_n^-$, we obtain in a similar way

$$-2a_n \iint \mathbf{e}_n \times \mathbf{h}_n \cdot \mathbf{a}_z\, da = \iiint \mathbf{E}_n^- \cdot \mathbf{J}\, dV \tag{66}$$

Equations (64) and (66) provide the required relations for determining the surface-wave mode amplitudes $a_n$ and $b_n$.

A short-current element of length $dl$ is equivalent to an oscillating dipole $\mathbf{P}$, where $j\omega P = J\,dl$. If we have a small current loop $J\boldsymbol{\tau}$, where $\boldsymbol{\tau}$ is the unit tangent to the contour around which $J$ flows, we have

$$\oint_C I\mathbf{E}_n \cdot \boldsymbol{\tau}\,dl = I \iint_{S_0} \nabla \times \mathbf{E}_n \cdot \mathbf{n}\,da = -j\omega\mu_0 I \iint_{S_0} \mathbf{H}_n \cdot \mathbf{n}\,da$$
$$= -j\omega\mu_0 \mathbf{H}_n \cdot \mathbf{M}$$

for a small loop of area $S_0$, where the magnetic dipole moment $\mathbf{M}$ is given by

$$\mathbf{M} = I \iint_{S_0} \mathbf{n}\,da$$

and $I$ is the total loop current. In the presence of small electric and magnetic dipoles, Eqs. (64) and (66) thus become

$$2b_n \int\int \mathbf{e}_n \times \mathbf{h}_n \cdot \mathbf{a}_z\,da = -j\omega \mathbf{E}_n^+ \cdot \mathbf{P} + j\omega\mu_0 \mathbf{H}_n^+ \cdot \mathbf{M} \qquad (67a)$$
$$2a_n \int\int \mathbf{e}_n \times \mathbf{h}_n \cdot \mathbf{a}_z\,da = -j\omega \mathbf{E}_n^- \cdot \mathbf{P} + j\omega\mu_0 \mathbf{H}_n^- \cdot \mathbf{M} \qquad (67b)$$

These results are of importance in determining the fields radiated by a given system of sources in the presence of a surface-wave guide. In particular, they show at once whether a given source will or will not excite a particular surface-wave mode.

## 11.8. Excitation of Surface Waves

The problem of exciting surface waves has received considerable attention in recent years. Many papers and reports have been published on the subject, and a representative list of these is given at the end of the chapter. The problem differs somewhat from that of exciting a propagating mode in an ordinary closed-boundary guide, in that some of the power radiated by the source goes into the radiation field. The efficiency of an antenna in exciting a surface-wave mode is defined as the ratio of the power radiated as a surface wave to the total power radiated. Several investigators have shown that launching efficiencies of 80 per cent and more can be obtained.

A typical launching device is a flared horn with an aperture field chosen as nearly like the transverse field of the surface wave as possible. Other devices such as slots in conducting planes (these are equivalent to magnetic current sources), dipoles, and line sources have also been studied, and will give a good efficiency when properly oriented with respect to the surface waveguide.

In this section, the problem of a line source above a grounded dielectric sheet will be studied, in order to show the significance of the continuous eigenvalue spectrum. In addition, a discussion of the leaky modes is included. Various authors have examined this problem, and the material presented here is similar to theirs.† Our analysis parallels that given by Barone in several respects.

Figure 11.16 illustrates a lossless dielectric sheet, of thickness $t$ and with a dielectric constant $\kappa$, located on a perfectly conducting plane at

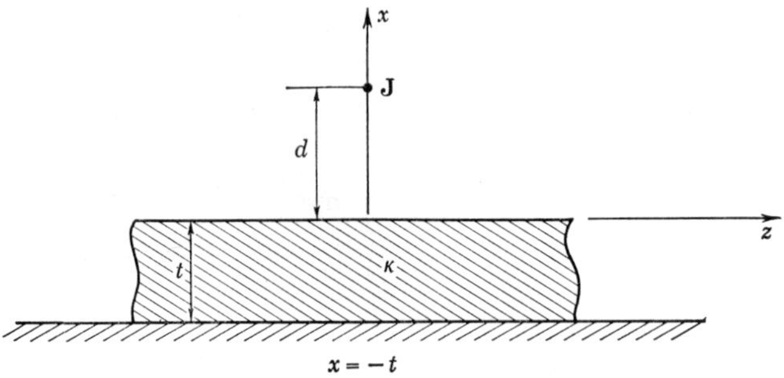


FIG. 11.16. Current line source above a grounded dielectric sheet.

$x = -t$. An electric current line source is located at $x = d$, $z = 0$ and is parallel to the $y$ axis. The current source may be represented as follows:

$$\mathbf{J} = \mathbf{a}_y \delta(z)\delta(x - d) \tag{68}$$

Since the current has no variation with $y$, the radiated field is also independent of $y$. The vector potential has only a $y$ component $\psi(x,z)$, and the electric and magnetic fields are given by

$$E_y = -j\omega\psi$$
$$\mu_0 H_x = -\frac{\partial \psi}{\partial z}$$
$$\mu_0 H_z = \frac{\partial \psi}{\partial x}$$

The function $\psi$ is a solution of

$$\frac{\partial^2 \psi}{\partial x^2} + \frac{\partial^2 \psi}{\partial z^2} + k_0^2 \psi = -\mu_0 \delta(z)\delta(x - d) \qquad x > 0 \tag{69a}$$

$$\frac{\partial^2 \psi}{\partial x^2} + \frac{\partial^2 \psi}{\partial z^2} + \kappa k_0^2 \psi = 0 \qquad -t < x < 0 \tag{69b}$$

† R. M. Whitmer, Fields in Non-metallic Waveguides, *Proc. IRE*, vol. 36, pp. 1105–1109, September, 1948.

C. T. Tai, Effect of a Grounded Slab on Radiation from a Line Source, *J. Appl. Phys.*, vol. 22, pp. 405–414, April, 1951.

S. Barone, Leaky Wave Contribution to the Field of a Line Source above a Dielectric Slab, *Polytech. Inst. Brooklyn MRI Research Rept.* R-532-56, November, 1956.

and must satisfy the boundary conditions that $\psi$ and $\partial\psi/\partial x$ be continuous at $x = 0$. In addition, $\psi$ is bounded at infinity and must represent outward-propagating waves. The solution for $\psi$ will be obtained by means of a bilateral Laplace transform.

Let

$$g(x,\gamma) = \int_{-\infty}^{\infty} \psi(x,z)e^{\gamma z}\,dz \tag{70}$$

multiply (69) by $e^{\gamma z}$, and integrate from $-\infty$ to $\infty$ to get

$$\frac{d^2g}{dx^2} + (\gamma^2 + k_0^2)g = -\mu_0\delta(x - d) \qquad x > 0 \tag{71$a$}$$

$$\frac{d^2g}{dx^2} + (\gamma^2 + \kappa k_0^2)g = 0 \qquad -t < x < 0 \tag{71$b$}$$

with $g$ and $dg/dx$ continuous at $x = 0$. In carrying out the integration with respect to $z$, the second derivative of $\psi$ with respect to $z$ is integrated by parts twice. The integrated terms vanish at infinity, provided we assume that $k_0$ has a small imaginary part, that is, $k_0 = k_0' - jk_0''$, corresponding to a small loss in the medium. Under these conditions, both $e^{\gamma z}\psi$ and $e^{\gamma z}\,\partial\psi/\partial z$ vanish at infinity for $\text{Re}\,\gamma = 0$.

A suitable form for the function $g$ is

$$\begin{aligned} g_1 &= C_1e^{-jl(x-d)} && x \ge d \\ g_2 &= C_2e^{jl(x-d)} + RC_2e^{-jl(x+d)} && 0 \le x \le d \\ g_3 &= C_3e^{-jld}(1 + R)\csc ht \sin h(x + t) && -t \le x \le 0 \end{aligned}$$

where $h^2 = \gamma^2 + \kappa k_0^2$, $l^2 = \gamma^2 + k_0^2$, and $C_1$, $C_2$, $C_3$, $R$ are constants to be determined. The above form of solution for $g$ has been chosen by a consideration of the equivalent transmission-line problem for propagation along the $x$ axis. Continuity of $g$ at $x = 0$ gives $C_2 = C_3$, while continuity of $dg/dx$ at $x = 0$ gives

$$R = \frac{jl - h\cot ht}{jl + h\cot ht} \tag{72}$$

Thus $R$ is seen to be the usual reflection coefficient at the air-dielectric interface. At $x = d$, $g$ is continuous, but $dg/dx$ is discontinuous. Integrating (71$a$) from $x = d_-$ to $x = d_+$ gives

$$\left.\frac{dg}{dx}\right|_{d_-}^{d_+} = -\mu_0$$

and this specifies the discontinuity in the derivative at the source. The boundary conditions at $x = d$ are now readily found to give

$$C_1 = \frac{\mu_0}{2jl}(1 + Re^{-2jld}) \qquad C_2 = \frac{\mu_0}{2jl}$$

Hence, the solution for $g$ becomes

$$g_1 = \frac{\mu_0}{2jl}\,(e^{-jl(x-d)} + Re^{-jl(x+d)}) \qquad x \geq d \tag{73a}$$

$$g_2 = \frac{\mu_0}{2jl}\,(e^{jl(x-d)} + Re^{-jl(x+d)}) \qquad 0 \leq x \leq d \tag{73b}$$

$$g_3 = \frac{\mu_0}{2jl}\,[e^{-jld}(1 + R)\csc ht \sin h(x + t)] \qquad -t \leq x \leq 0 \tag{73c}$$

Equations (73$a$) and (73$b$) may be combined to give

$$g = \frac{\mu_0}{2jl}\,(e^{-jl|x-d|} + Re^{-jl|x+d|}) \qquad x \geq 0 \tag{74}$$

Inverting this transform gives the solution for $\psi$ as follows:

$$\psi(x,z) = -\frac{\mu_0}{4\pi}\int_C \left(\frac{e^{-jl|x-d|}}{l} + \frac{Re^{-jl|x+d|}}{l}\right) e^{-\gamma z}\, d\gamma \tag{75}$$

where $C$ is a contour to be determined so that $\psi$ has the proper behavior at infinity, $R$ is given by (72), and $l = (\gamma^2 + k_0^2)^{1/2}$.

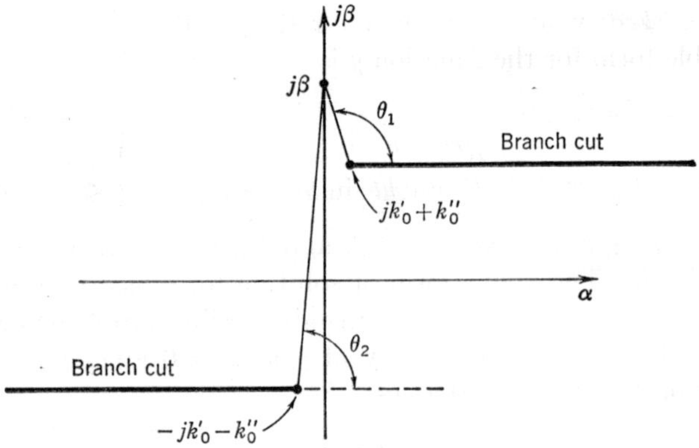


FIG. 11.17. Branch cuts in $\gamma$ plane.

For the reflection coefficient $R$, the transverse wave number is given by $h = (\gamma^2 + \kappa k_0^2)^{1/2}$. It does not matter which sign of the square root is chosen since $h \cot ht$ is an even function of $h$. For the transverse wave number $l = (\gamma^2 + k_0^2)^{1/2}$, the branch that leads to outward-propagating or attenuated waves must be chosen. This requires that

$$\operatorname{Re} l > 0$$
$$\operatorname{Im} l < 0$$

The integrand in (75) is a two-valued function of $\gamma$ because of the two branches of the function $l$. The branch points occur at

$$\pm jk_0 = \pm jk_0' \pm k_0''$$

The complex $\gamma$ plane must be cut by two branch lines running from the branch points to infinity (or joining the two branch points directly) if the integrand is to be single-valued. As far as the contour $C$ is concerned, the branch cuts may be chosen quite arbitrarily, as long as they do not intersect the contour $C$. One possible construction is shown in Fig. 11.17. For any point $j\beta$ along the imaginary axis of the complex $\gamma = j\beta + \alpha$ plane, the phase angles of the two factors $(\gamma + jk_0)^{1/2}$ and $(\gamma - jk_0)^{1/2}$ lie in the range

$$\frac{3\pi}{4} > \text{phase } (\gamma - jk_0)^{1/2} > \frac{\pi}{4}$$

$$-\frac{\pi}{4} < \text{phase } (\gamma + jk_0)^{1/2} < \frac{\pi}{4}$$

since

$$\frac{3\pi}{2} > \theta_1 > \frac{\pi}{2}$$

$$-\frac{\pi}{2} < \theta_2 < \frac{\pi}{2}$$

as $\beta$ ranges from $-\infty$ to $\infty$. The sum of the two angles $\theta_1 + \theta_2$ is always greater than $\pi$, but less than $2\pi$, for all values of $\gamma$ along the imaginary axis. Hence, the phase angle of $+(\gamma^2 + k_0^2)^{1/2}$ is always greater than $\pi/2$ but less than $\pi$. In order to satisfy the specified requirements on $l$, the negative root or branch must be chosen so that the phase angle of $l = -(\gamma - jk_0)^{1/2}(\gamma + jk_0)^{1/2}$ is in the range $3\pi/2$ to $2\pi$ for all values of $\gamma$ on the imaginary axis. With this branch chosen as the proper branch, the convergence of the integral (75) at infinity is ensured. Replacing $l$ by $-l$, we have, in place of (75), the more convenient expressions

$$\psi(x,z) = \frac{\mu_0}{4\pi} \int_C \left[ \frac{\exp j(\gamma^2 + k_0^2)^{1/2}|x - d|}{(\gamma^2 + k_0^2)^{1/2}} + \frac{R \exp j(\gamma^2 + k_0^2)^{1/2}(x + d)}{(\gamma^2 + k_0^2)^{1/2}} \right] e^{-\gamma z}\, d\gamma \tag{76}$$

where

$$R = \frac{(\gamma^2 + k^2)^{1/2} \cot (\gamma^2 + k^2)^{1/2} t + j(\gamma^2 + k_0^2)^{1/2}}{-(\gamma^2 + k^2)^{1/2} \cot (\gamma^2 + k^2)^{1/2} t + j(\gamma^2 + k_0^2)^{1/2}} \tag{77}$$

and $k^2 - \kappa k_0^2$. In (76) and (77), the proper branch is now the positive root of $(\gamma^2 + k_0^2)^{1/2}$. The integrand has poles whenever $R$ becomes infinite, i.e., when

$$(\gamma^2 + k^2)^{1/2} \cot (\gamma^2 + k^2)^{1/2} t = j(\gamma^2 + k_0^2)^{1/2} \tag{78}$$

For $\gamma = j\beta$ and $|\beta| > k_0$, we get

$$-(k^2 - \beta^2)^{1/2} \cot (k^2 - \beta^2)^{1/2} t = (\beta^2 - k_0^2)^{1/2} \tag{79}$$

The roots of this equation lead to the surface-wave modes. These roots lie on the $j\beta$ axis between $jk_0$ and $jk$ when $k_0$ and $k$ are taken real, which will be done for determining the roots of (79). For $z > 0$, the contour $C$ may be closed in the right-half $\gamma$ plane, and the roots along the positive $j\beta$ axis must be enclosed, in order to give rise to the surface-wave modes. The roots along the negative $j\beta$ axis are excluded, since these give rise to a propagation factor of the form $e^{j\beta z}$, which is not appropriate for $z > 0$. These roots, however, are included when the contour $C$ is closed in the left-half $\gamma$ plane for $z < 0$. The correct contour for evaluating (76) is thus the one illustrated in Fig. 11.18. For $z > 0$, this contour may be closed in the right-half $\gamma$ plane. Since the branch cut cannot

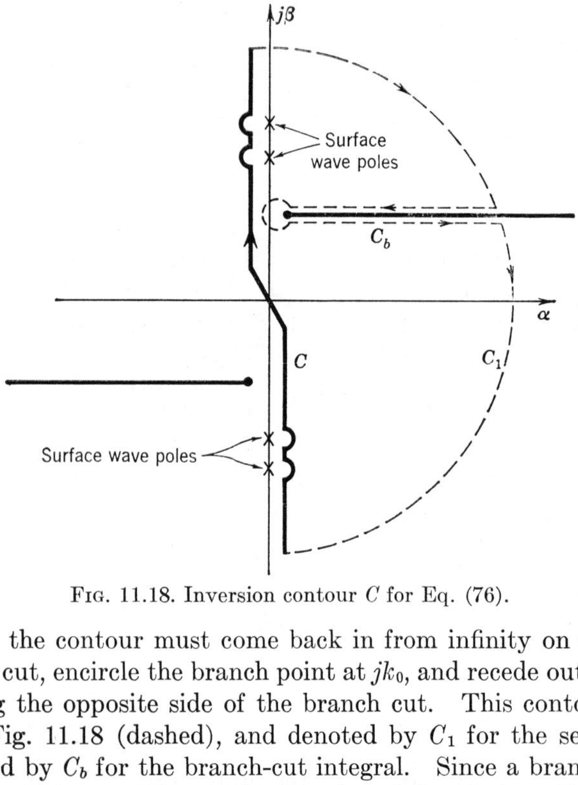


FIG. 11.18. Inversion contour $C$ for Eq. (76).

be crossed, the contour must come back in from infinity on one side of the branch cut, encircle the branch point at $jk_0$, and recede out to infinity again along the opposite side of the branch cut. This contour is illustrated in Fig. 11.18 (dashed), and denoted by $C_1$ for the semicircle at infinity, and by $C_b$ for the branch-cut integral. Since a branch cut was introduced, the integrand is single-valued, and Cauchy's residue theorem applies. The integral along $C + C_1 + C_b$ is equal to $-2\pi j$ times the enclosed residues. The negative sign arises because the contour is traversed in the clockwise sense. The original integral along $C$ is thus equal to

$$\int_C = -\int_{C_1} - \int_{C_b} - 2\pi j \sum \text{residues}$$

If the integral along the semicircle $C_1$ vanishes, then the original integral is equal to the branch-cut integral plus the residues contributed by the enclosed poles. The branch-cut integral represents the radiation field with a continuous eigenvalue spectrum, while the residues at the enclosed poles are the surface-wave modes.

Although the branch cut is arbitrary, as far as the original integral (76) along $C$ is concerned, it must be chosen more carefully if the integral along $C_1$ is to vanish. Let us assume that the branch cut has been chosen as in Fig. 11.19. On the semicircle above the branch cut, the phase angle of $(\gamma^2 + k_0^2)^{1/2}$ is more than $\pi/2$ and less than $\pi$. So $\exp j(\gamma^2 + k_0^2)^{1/2}x$ decays exponentially with increasing positive values of $x$, and (76) will vanish on this portion of the semicircle. As the point $P_1$ moves down

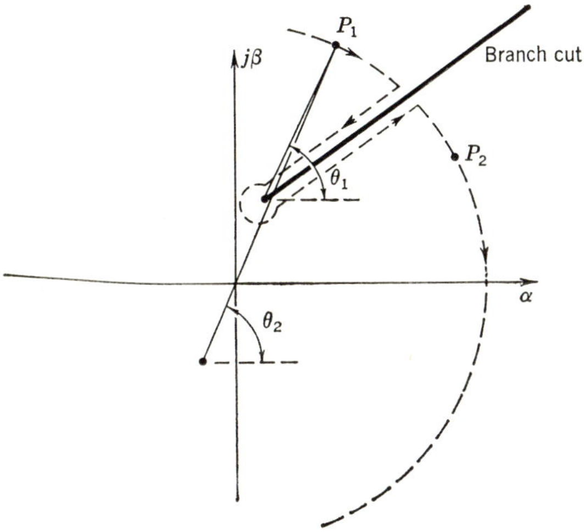


FIG. 11.19. An alternative branch cut.

$C_1$, and in along the branch cut and back out to $P_2$, the angle $\theta_1$ increases to a value greater than $2\pi$, while $\theta_2$ remains less than $\pi/2$. Over a portion of the semicircle below the branch cut, the phase angle of $(\gamma^2 + k_0^2)^{1/2}$ is greater than $\pi$, and hence $\exp j(\gamma^2 + k_0^2)^{1/2}x$ is exponentially increasing for positive values of $x$. The exponential term in (76) becomes, for large values of $\gamma$ and $x$,

$$\exp [(-j\alpha + \beta)x - (\alpha + j\beta)z]$$

and, unless $\beta x < \alpha z$, this term increases without limit as $\gamma$ tends to infinity. The difficulty with the above choice of branch cut is that this choice does not correspond to a cut which separates the proper branch, for which

$$\text{Im } (\gamma^2 + k_0^2)^{1/2} > 0 \tag{80}$$

from the improper branch, for which

$$\text{Im } (\gamma^2 + k_0^2)^{1/2} < 0 \tag{81}$$

It is only on the proper branch that the integral (76) will converge at infinity. The branch cut separating the proper and improper branches occurs along the curve Im $(\gamma^2 + k_0^2)^{1/2} = 0$. If $k_0 = k_0' - jk_0''$, this curve is given by

$$\alpha\beta = k_0'k_0'' \tag{82}$$

This curve is a portion of a hyperbola running from the branch point and down toward the real axis in the $\gamma$ plane, as illustrated in Fig. 11.20*a*. As the losses in the medium decrease, $k_0''$ decreases, and, in the limit, the hyperbola becomes the portion of the $\beta$ axis between 0 and $k_0$ and the real $\alpha$ axis in the right half plane. Along the continuation of the hyperbola from the branch points along the $j\beta$ axis, the real part of $(\gamma^2 + k_0^2)^{1/2}$

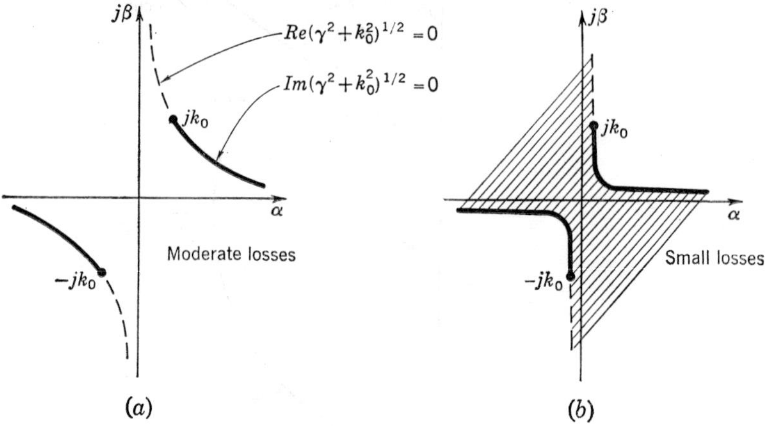


FIG. 11.20. Illustration of proper branch cut.

is zero. In the region between the two hyperbolas (crosshatched region in Fig. 11.20*b*), the real part of $(\gamma^2 + k_0^2)^{1/2}$ is negative, while the imaginary part is positive. The original contour $C$ must lie entirely in this region in order to represent outgoing waves.

With the branch cut chosen as in Fig. 11.20, the integral (76) will vanish on the semicircle in the right half plane for $z > 0$. The branch-cut integral for $k_0$ real gives an eigenvalue spectrum with $\gamma$ in the range

$$0 \le \gamma = \alpha < \infty$$
$$0 \le \gamma = j\beta \le jk_0$$

in accordance with the results of Sec. 11.5.

The integrand in (76) may be conveniently viewed as defined on a two-sheeted Riemann surface. Only on the upper sheet or surface (the proper sheet), for which condition (80) holds, does the integral (76) converge. On the improper sheet, condition (81) applies, and the integral does not converge properly at infinity. The branch cuts of Fig. 11.20

provide the only cuts by which passage from one sheet to the other can be executed.

In addition to the surface-wave poles of (78), which occur for $(\gamma^2 + k_0^2)^{1/2}$ equal to a positive-imaginary quantity, there are a number of roots for $(\gamma^2 + k_0^2)^{1/2}$ equal to a pure negative-imaginary quantity plus an infinite number of complex roots with negative-imaginary parts. All these latter roots or poles will lie on the improper sheet of the Riemann surface. It is only the surface-wave poles that lie on the proper sheet, since only on the proper sheet does condition (80) hold. The complex roots give rise to a class of modes called leaky modes. In evaluating (76) approximately by the saddle-point method of integration, the contour $C$ is deformed into a contour of steepest descent. Part of this contour may lie in the improper sheet of the Riemann surface, and the leaky-wave modes then contribute to the expansion of the field. Since the integral is too complicated to be evaluated rigorously, approximate methods must be used. The saddle-point method of integration is one approximate technique which is capable of giving good results for the far-zone radiation field. Before carrying out the saddle-point integration, the roots of (78) will be determined.

*Roots of Eq.* (78)

For $\gamma = j\beta$ and $k_0 < |\beta|$, let $k^2 - \beta^2 = h^2$, and $j(k_0^2 - \beta^2)^{1/2} = -p$. Thus we have

$$pt = -ht \cot ht \tag{83a}$$

and

$$(pt)^2 + (ht)^2 = (\kappa - 1)(k_0t)^2 = a^2 \tag{83b}$$

Only the roots corresponding to positive values of $p$ lie on the proper sheet of the Riemann surface. The solution of (83) is readily obtained graphically. For $\beta > k$, $h$ is imaginary, and we have

$$pt = |h|t \coth |h|t$$
$$(pt)^2 - (|h|t)^2 = (\kappa - 1)(k_0t)^2$$

These curves are plotted as a function of $|h|$ to the left of the origin in Fig. 11.21. There will be no root for imaginary values of $h$ unless $(\kappa - 1)^{1/2}k_0t$ is less than unity. When this condition holds, an improper pole occurs for negative $p$, but there will not be a solution for a surface wave at the same time. A surface-wave mode exists only if

$$a = (\kappa - 1)^{1/2}k_0t$$

is greater than $\pi/2$. As $k_0t$ increases, the number of surface-wave modes and improper modes continually increases.

To determine the complex roots, let $pt$ be the complex variable

$$w = u + jv$$

and let $ht$ be the complex variable $Z = x + jy$. The variables $x$, $y$, $Z$ used here should not be confused with the rectangular coordinates. Equation (83$a$) becomes $w = -Z \cot Z$, or, in component form,

$$u = -\frac{x \tan x\,(1 - \tanh^2 y) + y \tanh y\,(1 + \tan^2 x)}{\tan^2 x + \tanh^2 y} \tag{84a}$$

$$v = -\frac{y \tan x\,(1 - \tanh^2 y) + x \tanh y\,(1 + \tan^2 x)}{\tan^2 x + \tanh^2 y} \tag{84b}$$

From (84), it is found that $u$ is negative for all values of $x$ and $y$. Consequently, the real part of $p$ is always negative for the complex roots, and,

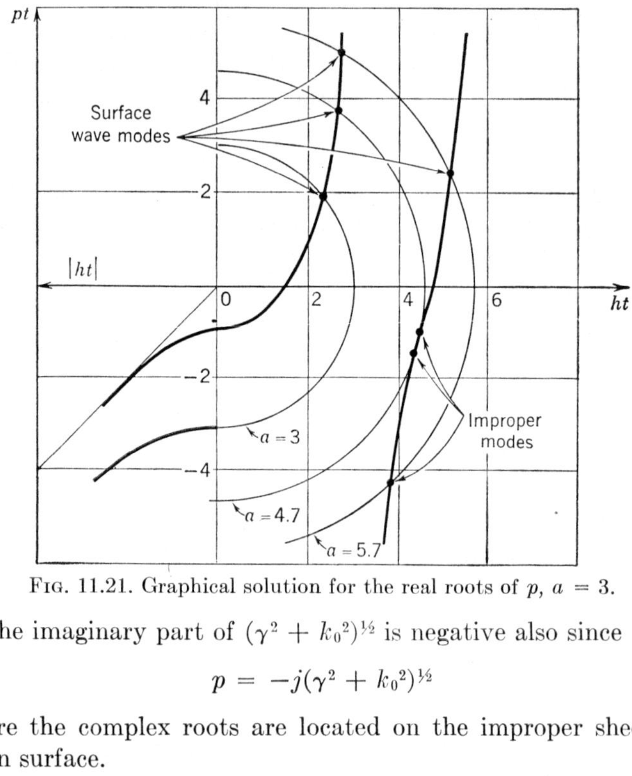


FIG. 11.21. Graphical solution for the real roots of $p$, $a = 3$.

hence, the imaginary part of $(\gamma^2 + k_0^2)^{1/2}$ is negative also since

$$p = -j(\gamma^2 + k_0^2)^{1/2}$$

Therefore the complex roots are located on the improper sheet of the Riemann surface.

By eliminating the variable $w = pt$ in (83$a$) and (83$b$), we obtain the equation $Z = \pm a \sin Z$, or, in component form,

$$\frac{x}{\sin x} = \pm a \cosh y \tag{85a}$$

$$\cos x = \pm \frac{1}{a} \frac{y}{\sinh y} \tag{85b}$$

A simultaneous solution of these two equations determines the complex roots for $Z$. From (84) the corresponding value of $w$ may be found. The values of $\gamma$ are found from the relation

$$w = -jt(\gamma^2 + k_0^2)^{1/2} \tag{86}$$

From (85), it is seen that, if $x_1 + jy_1$ is a solution, then $x_1 - jy_1$, $-x_1 + jy_1$, and $-x_1 - jy_1$ are also solutions. Hence we need consider only positive values of $x$ and $y$.

By plotting the two equations (85$a$) and (85$b$) on the $xy$ plane, the roots are readily determined. Since $a$ is greater than $\pi/2$, in order for at least one surface-wave mode to exist, a typical plot is as illustrated in Fig. 11.22. The roots are seen to be located in the vicinity of $x = n\pi - \pi/2$,

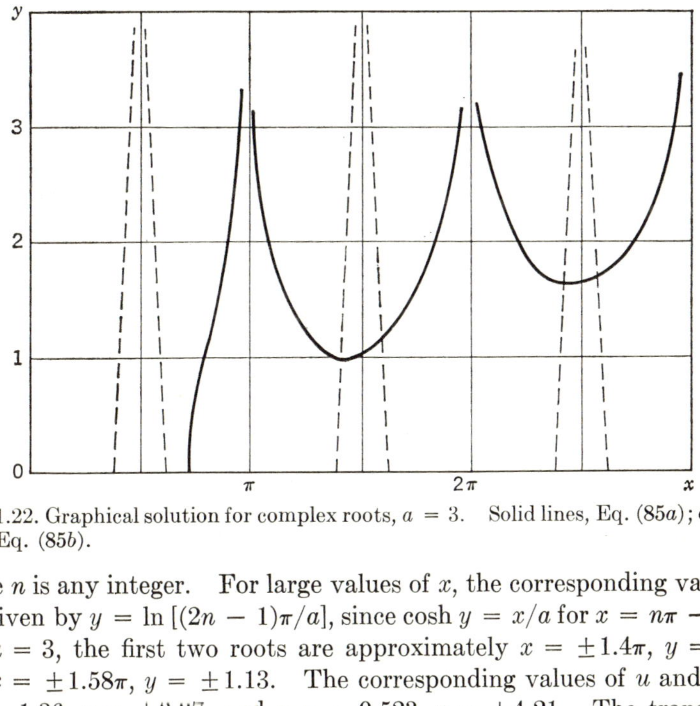


FIG. 11.22. Graphical solution for complex roots, $a = 3$. Solid lines, Eq. (85$a$); dashed lines, Eq. (85$b$).

where $n$ is any integer. For large values of $x$, the corresponding value of $y$ is given by $y = \ln [(2n - 1)\pi/a]$, since $\cosh y = x/a$ for $x = n\pi - \pi/2$. For $a = 3$, the first two roots are approximately $x = \pm 1.4\pi$, $y = \pm 1$, and $x = \pm 1.58\pi$, $y = \pm 1.13$. The corresponding values of $u$ and $v$ are $u = -1.36$, $v = \pm 3.37$, and $u = -0.523$, $v = \pm 4.21$. The transverse wave number in the $x$ direction above the dielectric sheet is

$$j(\gamma^2 + k_0^2)^{1/2} = -\frac{w}{t} = -\frac{u + jv}{t}$$

and is seen to give rise to exponentially increasing waves in the $x$ direction.

*The Saddle-point Method of Integration*

The saddle-point method (also known as the method of steepest descent) of approximate integration was introduced by Debye for obtaining asymptotic expansions of the Hankel functions. The basic principles involved will be described in connection with the first term in (76). This term is, apart from the factor $\mu_0/4\pi$,

$$\int_C \frac{\exp\,[j(\gamma^2 + k_0{}^2)^{1/2}|x - d| - \gamma z]}{(\gamma^2 + k_0{}^2)^{1/2}}\,d\gamma \tag{87}$$

and represents the direct field radiated by the line source. The integral is well known and proportional to the Hankel function of the second kind and zero order, i.e., equal to $-j\pi H_0{}^2(k_0r)$, where $r = [(x - d)^2 + z^2]^{1/2}$. In order to evaluate (87) asymptotically for large $r$ by the saddle-point method, it is convenient to change from rectangular coordinates $x$, $z$ to cylindrical coordinates $r$, $\theta$ and also to change the variable of integration $\gamma$ as follows:

$$\gamma = jk_0 \sin \phi \tag{88a}$$
$$(\gamma^2 + k_0{}^2)^{1/2} = -k_0 \cos \phi \tag{88b}$$
$$d\gamma = jk_0 \cos \phi \, d\phi \tag{88c}$$
$$\phi = \sigma + j\eta \tag{88d}$$
$$z = r \sin \theta \tag{88e}$$
$$x - d = r \cos \theta \tag{88f}$$

The integral (87) now becomes

$$-j \int_C e^{-jk_0 r \cos(\phi - \theta)}\,d\phi \tag{89}$$

The details of the saddle-point integration are greatly simplified by the above transformations.

Equation (88$a$) represents a mapping of the complex $\gamma$ plane into a strip of the complex $\phi$ plane. The two sheets of the Riemann surface map into a connected strip of width $2\pi$ along the $\sigma$ axis. From (88$a$), we have

$$\gamma = \alpha + j\beta = jk_0 \sin (\sigma + j\eta)$$

or

$$\alpha = -k_0 \cos \sigma \sinh \eta \tag{90a}$$
$$\beta = k_0 \sin \sigma \cosh \eta \tag{90b}$$

This mapping is illustrated in Fig. 11.23. The quadrants in which $\beta$ and $\alpha$ are positive or negative are readily found from (90), and are specified in the figure. On the proper sheet of the Riemann surface (top sheet), we have Im $(\gamma^2 + k_0{}^2)^{1/2} > 0$, and hence, from (88$b$), we find that the corresponding part of the $\phi$ plane is determined by the condition that $\sin \sigma \sinh \eta > 0$. The four quadrants of the top and bottom sheets of the $\gamma$ plane map into the regions designated $T_i$ and $B_i = 1, 2, 3, 4$, respectively, in Fig. 11.23. The negative sign in (88$b$) was chosen so that the quadrants $T_1$, $T_2$ and $T_3$, $T_4$ would map into adjacent strips. The hyperbolas giving the branch cuts in the $\gamma$ plane as in Fig. 11.20$b$ map into the dashed lines in Fig. 11.23. The crosshatched regions are corresponding regions in the two figures. It should be noted that there are no branch cuts in the $\phi$ plane since (88$a$) maps both sheets of the Riemann surface into a connected strip. The original contour of integration is denoted by $C$ in Fig. 11.23. This contour may be deformed into any

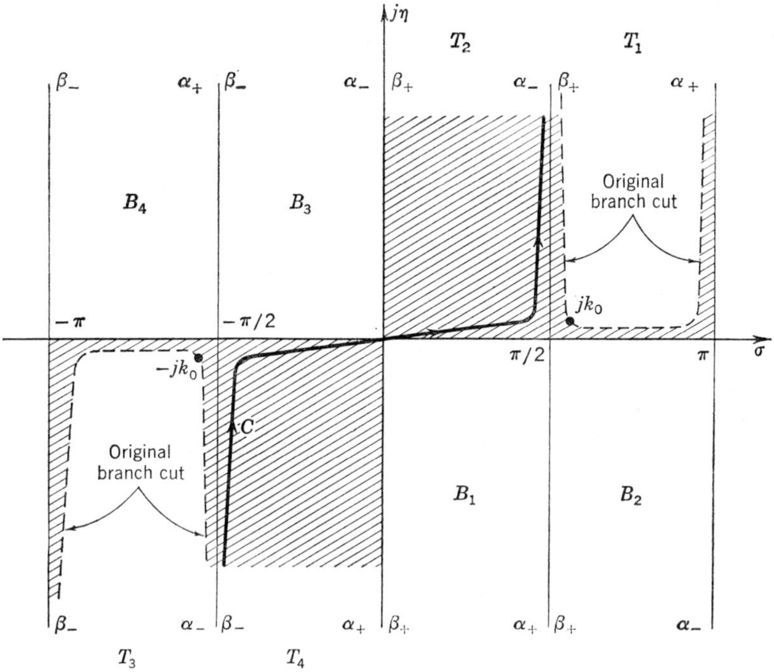


FIG. 11.23. Mapping of two sheets of $\gamma$ plane onto a strip of the $\phi$ plane.

other convenient contour in the crosshatched region without changing the value of the integral (89), since no poles will be swept across.

In the integrand for (89), let

$$f(\phi) = -jk_0 r \cos (\phi - \theta) \tag{91}$$

and the derivative with respect to $\phi$,

$$\frac{df}{d\phi} = jk_0 r \sin (\phi - \theta)$$

is equal to zero at $\phi = \theta$ or $\sigma = \theta$, $\eta = 0$. Since a complex function cannot have a maximum or a minimum, the stationary points are saddle points.† In the vicinity of the saddle point, a Taylor expansion gives

† E. A. Guillemin, "The Mathematics of Circuit Analysis," chap. VI, art. 14; chap. VII, art. 26, John Wiley & Sons, Inc., New York, 1949.

$$
\begin{aligned}
f(\phi) &\approx f(\theta) + \left.\frac{df}{d\phi}\right|_\theta (\phi - \theta) + \frac{1}{2}\left.\frac{d^2f}{d\phi^2}\right|_\theta (\phi - \theta)^2 \\
&= -jk_0r + \frac{jk_0r}{2}(\sigma - \theta + j\eta)^2
\end{aligned} \tag{92}
$$

since the first derivative vanishes and the second derivative equals $jk_0r$ at $\theta$. Let $\rho$, $\omega$ be the radial and angular coordinates with origin at $\theta$ in the $\phi$ plane. Thus $\phi - \theta = \sigma - \theta + j\eta = \rho e^{j\omega}$, and, if we let $f_1 + jf_2 = f(\phi) - f(\theta)$, we have $f_1 + jf_2 = (jk_0r/2)\rho^2 e^{2j\omega}$, and, hence,

$$f_1 = \frac{-k_0r}{2}\rho^2 \sin 2\omega \tag{93a}$$

$$f_2 = \frac{k_0r}{2}\rho^2 \cos 2\omega \tag{93b}$$

The contours $f_1 =$ constant are orthogonal to the contours $f_2 =$ constant. These contours are sketched in Fig. 11.24. The maximum rate of

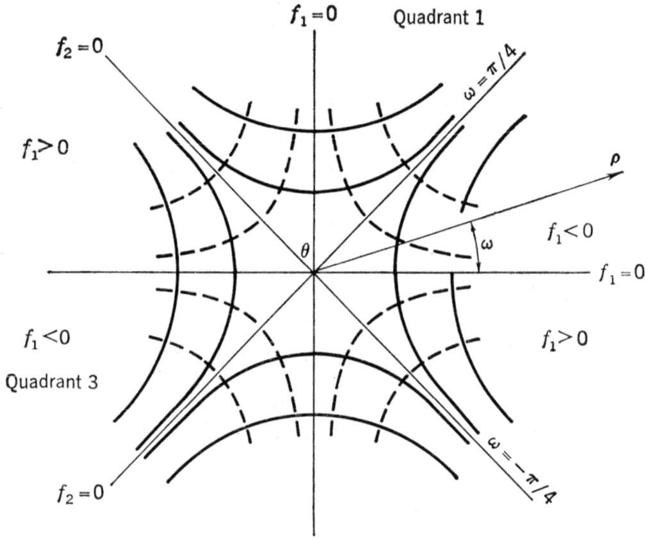


FIG. 11.24. Solid lines, $f_2 =$ constant contours; dashed lines, $f_1 =$ constant contours.

change of $f_1$ with $\rho$ occurs for $\omega = \pi/4$ and $\pi/4 + \pi$, and also for $\omega = 3\pi/4$ and $-\pi/4$. Along these paths of steepest descent, $f_2 = 0$. In the first and third quadrants, $f_1$ is negative and increases rapidly in magnitude as we move away from the saddle point along the directions $\omega = \pi/4$ and $\omega = \pi/4 + \pi$. In the second and fourth quadrants, $f_1$ increases in the positive direction at the maximum rate along the contours $\omega = 3\pi/4$ and $-\pi/4$. If $f_1$ is considered to represent the elevation of the ground with respect to a reference datum at $\phi = \theta$, it is seen that the ground has the characteristics normally associated with a saddle, since it rises rapidly in two directions and falls rapidly in the two perpendicular directions. It is from this analogy that the terminology saddle point, lines of steepest descent, etc., are obtained.

In view of the properties of the function $f$, it is clear that it is desirable to deform the contour $C$ in (89) into a steepest-descent contour passing through the saddle point at $\phi = \theta$, since, along this path, $f_1$ is negative and hence $e^{f(\phi)}$ will decrease rapidly as we move away from the saddle point. Thus, the major contribution to the integral comes from a small region or range $0 < \rho < \rho_1$ along the steepest-descent contour.

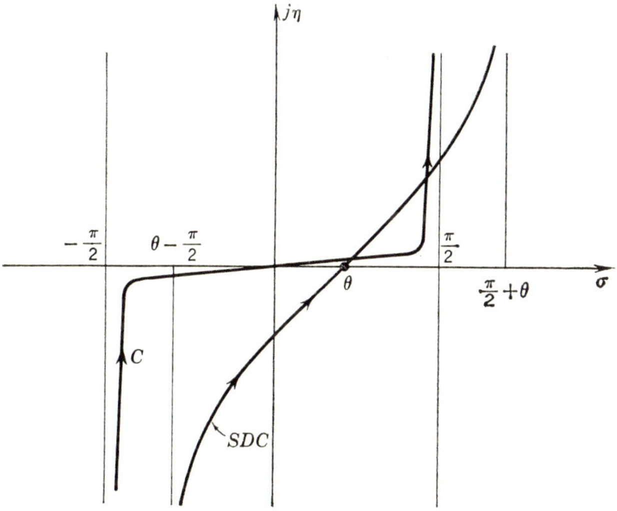


FIG. 11.25. Deformation of contour $C$ into steepest-descent contour SDC.

The steepest-descent contour is the contour along which $f_2 = 0$. For general values of $\phi$, $f_1 + jf_2 = f(\phi) - f(\theta)$, and, hence,

$$f_2 = k_0 r[1 - \cos(\sigma - \theta)\cosh\eta]$$

and the steepest-descent contour is determined by

$$\cos(\sigma - \theta)\cosh\eta = 1 \tag{94}$$

This contour is illustrated as the contour SDC in Fig. 11.25. It passes through the saddle point at an angle $\pi/4$ in accordance with (93*b*), which is valid near the saddle point.

Near the saddle point,

$$f(\phi) \approx -jk_0 r + \frac{jk_0 r}{2}\rho^2 e^{2j\omega} = -jk_0 r - \frac{k_0 r}{2}\rho^2$$

along the steepest-descent contour $\omega = \pi/4$ and $\pi/4 + \pi$. Along this path $\phi = \theta + \rho e^{j\pi/4}$ in the first quadrant, and $\phi = \theta + \rho e^{j(\pi+\pi/4)}$ in the third quadrant. Hence, we have

$$\begin{aligned} d\phi &= e^{j\pi/4}\,d\rho && \text{first quadrant} \\ d\phi &= -e^{j\pi/4}\,d\rho && \text{third quadrant} \end{aligned}$$

The integral becomes

$$-j\int_C e^f\,d\phi = -j\int_{\text{SDC}} e^f\,d\phi = -je^{-jk_0r+j\pi/4}\left(\int_{\rho_1}^{0} - e^{-k_0r\rho^2/2}\,d\rho + \int_0^{\rho_1} e^{-k_0r\rho^2/2}\,d\rho\right) \tag{95}$$

since $e^{-k_0r\rho^2/2}$ is negligible for $\rho > \rho_1$, provided $k_0r$ is sufficiently large. Combining terms in (95) gives

$$-j\int_C e^f\,d\phi = -2je^{-j(k_0r-\pi/4)}\int_0^{\rho_1} e^{-k_0r\rho^2/2}\,d\rho \tag{96}$$

Because of the rapid decay of the exponential, the integral from 0 to $\rho_1$ does not differ much from

$$\int_0^\infty e^{-k_0r\rho^2/2}\,d\rho = \left(\frac{\pi}{2k_0r}\right)^{1/2} \tag{97}$$

Therefore the original integral (87) is given by

$$-j\left(\frac{2\pi}{k_0r}\right)^{1/2} e^{-j(k_0r-\pi/4)}$$

for large $k_0r$. The asymptotic expansion of the Hankel function $H_0{}^2(k_0r)$ is $(2/\pi k_0r)^{1/2}e^{-j(k_0r-\pi/4)}$, and thus (87) is seen to be equal to $-j\pi H_0{}^2(k_0r)$. The validity of the saddle-point integration depends on $k_0r\rho^2/2$ being large for small values of $\rho$, and, hence, can only be used when $k_0r$ is large, i.e., to evaluate the field far from the source. Additional terms in the asymptotic expansion may be obtained by following the method suggested in Prob. 11.9. Multiplying the above result by $\mu_0/4\pi$, we obtain

$$\psi_{\text{direct}} = \frac{-j\mu_0}{4\pi}\left(\frac{2\pi}{k_0r}\right)^{1/2} e^{-j(k_0r-\pi/4)} \qquad k_0r \gg 1 \tag{98}$$

for the far-zone direct-radiated field. In (98) the radial coordinate $r$ is equal to $[(x-d)^2 + z^2]^{1/2}$.

*Evaluation of Reflected Field*

The second term in (76), which is the field reflected from the dielectric sheet above the ground plane, may be evaluated by the saddle-point method also. With the transformation (88), and $-d$ replaced by $d$, this term becomes

$$\frac{-j\mu_0}{4\pi}\int_C R(\phi)\exp\left[-jk_0 r\cos(\phi-\theta)\right]d\phi \tag{99}$$

where $$R(\phi)=\frac{j\cos\phi-h\cot ht}{j\cos\phi+h\cot ht}$$

and $h=(\kappa-\sin^2\phi)^{1/2}$. The radial coordinate $r$ now has its origin at $z=0$, $x=-d$. As before, the contour $C$ is deformed into a steepest-descent contour SDC passing through the saddle point $\phi=\theta$. In deforming the contour $C$, some of the poles of $R(\phi)$ may be encountered.

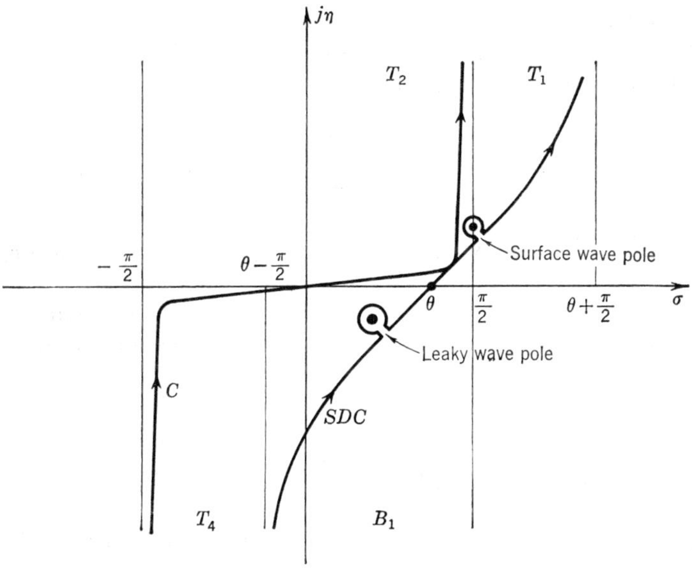


FIG. 11.26. Steepest-descent contour for reflected field.

When this happens, the contour should be warped as in Fig. 11.26, so as not to pass over the poles. In addition to the integral along the steepest-descent contour, the usual residue terms contribute to the field from the integration around the poles. The surface-wave poles lie on the lines $\sigma=\pi/2$, $\eta>0$ and $\sigma=-\pi/2$, $\eta<0$. The leaky-wave poles lie in the bottom sheet of the Riemann surface, and a finite number of these may be encountered as the contour $C$ is deformed into the section labeled $B_1$. When the leaky-wave poles contribute, they do not become infinite, since, in expression (99), the real part $f_1$ of $-jk_0 r\cos(\phi-\theta)$ is negative at the location of these poles; that is, $f_1=-k_0 r\sin(\sigma-\theta)\sinh\eta$ is negative. From Fig. 11.26, it is also seen that the surface-wave and leaky-wave residues only contribute for a range of $\theta$ greater than some minimum value $\theta_c$, since no poles are crossed for small values of $\theta$. Physically this means that the surface-wave and leaky-wave modes are not significant in determining the far-zone field except in the vicinity of the surface.

This behavior can certainly be appreciated for the surface wave since it decays exponentially in a direction perpendicular and away from the guiding surface.

Let $\phi_s$ be the value of $\phi$ at a surface-wave pole, and let $\phi_L$ denote a leaky-wave pole. Also let the residue of $R(\phi)$ at a pole be denoted by $F(\phi_s)$. Thus (99) may be written as

$$\frac{-j\mu_0}{4\pi}\int_{\text{SDC}} R(\phi)e^{f(\phi)}\,d\phi - \frac{\mu_0}{2}\sum_s F(\phi_s)e^{f(\phi_s)} - \frac{\mu_0}{2}\sum_L F(\phi_L)e^{f(\phi_L)}$$

The series arises from integration about the circles surrounding the poles. The integral along the steepest-descent contour is readily evaluated when $R(\phi)$ does not have a pole in the vicinity of the saddle point by expanding $R$ in a Taylor series about $\theta$. Thus

$$R(\phi) = R(\theta) + \sum_{n=1}^{\infty} R^n(\theta)\,\frac{(\phi-\theta)^n}{n!}$$

where $R^n(\theta) = \left.\dfrac{d^nR}{d\phi^n}\right|_\theta$. Along the steepest-descent contour near $\theta$,

$$\begin{array}{lll} \phi - \theta = \rho e^{j\pi/4} & d\phi = e^{j\pi/4}\,d\rho & \text{first quadrant} \\ \phi - \theta = -\rho e^{j\pi/4} & d\phi = -e^{j\pi/4}\,d\rho & \text{third quadrant} \end{array}$$

Hence the integral becomes

$$-\frac{j\mu_0 e^{-j(k_0r-\pi/4)}}{4\pi}\sum_{n=0}^{\infty}\frac{R^n(\theta)}{n!}\,e^{jn\pi/4}[1+(-1)^n]\int_0^{\rho_1}\rho^n e^{-k_0r\rho^2/2}\,d\rho$$

The integral over $\rho$ does not change much if $\rho_1$ is extended to infinity. The result

$$\int_0^{\infty}\rho^n e^{-k_0r\rho^2/2}\,d\rho = \frac{\Gamma[(n+1)/2]}{2(k_0r/2)^{(n+1)/2}} \qquad n > -1 \tag{100}$$

where $\Gamma$ is the gamma function, may then be used to evaluate the integral. Thus, for $k_0r$ large and for no poles of $R(\phi)$ near $\theta$, the reflected field is

$$\begin{aligned}\psi_{ref} = -\frac{\mu_0}{2}\Bigg\{&\sum_s F(\phi_s)\exp\left[-jk_0r\cos(\phi_s-\theta)\right] \\ &+ \sum_L F(\phi_L)\exp\left[-jk_0r\cos(\phi_L-\theta)\right]\Bigg\} \\ &- \frac{j\mu_0}{4\pi}e^{-j(k_0r-\pi/4)}\sum_{n=0,2,\ldots}\frac{R^n(\theta)}{n!}\,e^{jn\pi/4}\left(\frac{2}{k_0r}\right)^{(n+1)/2}\Gamma\left(\frac{n+1}{2}\right)\end{aligned} \tag{101}$$

where $r^2 = (x+d)^2 + z^2$. For $k_0r$ large, the series converges rapidly.

When $R(\phi)$ has a pole near the saddle point, the Taylor-series expansion of $R$ is no longer valid in a sufficiently large enough region around the point $\theta$. A Laurent-series expansion must be used, and this requires a further modification of the preceding method. Let $R(\phi)$ have a simple pole at $\phi = \phi_0$ with residue $F(\phi_0)$ and with $\phi_0$ located near the saddle point $\phi = \theta$. In place of the Taylor-series expansion, we have

$$R(\phi) = \frac{F(\phi_0)}{\phi - \phi_0} + R_1(\phi) \tag{102}$$

where $R_1(\phi)$ is analytic in the region around $\theta$ and may be developed in a Taylor series. For $R_1$, we have

$$R_1(\phi) = \sum_{n=0}^{\infty} a_n(\phi - \theta)^n$$

where

$$a_n = \frac{1}{n!}\frac{d^n}{d\phi^n}\left[R(\phi) - \frac{F(\phi_0)}{\phi - \phi_0}\right]\bigg|_{\phi=\theta}$$

The portion of the integral for the radiated field involving $R_1$ leads to a contribution of the form (101).

The remaining integral to be evaluated is

$$-\frac{j\mu_0 F(\phi_0)}{4\pi}\int_C \frac{\exp\,[-jk_0 r \cos\,(\phi - \theta)]}{\phi - \phi_0}\,d\phi \tag{103}$$

The asymptotic evaluation of integrals of this type has been treated by Clemmow and van der Waerden [11.16, 11.17] and more recently by Oberhettinger. The principal results obtained by Oberhettinger will be presented here.†

Consider a Laplace-type integral of the form

$$f(u) = \int_0^\infty t^\alpha g(t)e^{-ut}\,dt \tag{104}$$

where $g(t)$ is analytic at the origin and has its nearest pole at $t = t_0$, where $t_0$ is complex. Thus $g(t)$ is analytic in some sector $\theta_1 <$ phase $t$ $< \theta_2$. In (104), $\alpha$ may be a positive or negative constant with $\operatorname{Re}\alpha > -1$. If $t_0$ is a pole of order $m$, a Laurent-series development gives

$$g(t) = b_{-m}(t - t_0)^{-m} + \cdots + b_{-1}(t - t_0)^{-1} + g_1(t)$$

where $g_1(t)$ is analytic for $|t| < t_1$, and $t_1$ is the next singular point with $|t_1| > |t_0|$. The function $g_1$ may be expanded about the origin in a

† F. Oberhettinger, On a Modification of Watson's Lemma, *J. Research Natl. Bur. Standards*, vol. 63B, pp. 15–17, July–September, 1959.

Taylor series:

$$g_1(t) = \sum_{n=0}^{\infty} c_n t^n$$

where

$$c_n = \frac{1}{n!}\frac{d^n}{dt^n}\left[g(t) - \sum_{s=1}^{m} b_{-s}(t-t_0)^{-s}\right]\bigg|_{t=0}$$

The asymptotic behavior of $f(u)$ as $u \to \infty$ is governed by the behavior of $t^\alpha g(t)$ at the origin.

When the series expansion for $g(t)$ is substituted into (104) and integrated, term by term, we obtain

$$\begin{aligned} f(u) = {} & \sum_{n=0}^{\infty} c_n \Gamma(\alpha+n+1)u^{-\alpha-n-1} \\ & + \sum_{s=1}^{m} b_{-s}\Gamma(1+\alpha)(-t_0)^{(\alpha-s)/2}u^{(s-\alpha-2)/2}e^{-ut_0/2}W_{-s/2-\alpha/2,\,\alpha/2-s/2+\frac{1}{2}}(-ut_0) \end{aligned} \tag{105}$$

where $\Gamma$ is the gamma function, and $W_{a,b}(x)$ is Whittaker's confluent hypergeometric function.† This asymptotic series is valid for $u \to \infty$ in the sector $-\pi/2 - \theta_2 < \text{phase } u < \pi/2 - \theta_1$.

Some special cases of interest in diffraction work are

(1) $$W_{-\frac{1}{4},-\frac{1}{4}}(x) = \sqrt{\pi}\, x^{\frac{1}{4}} e^{x/2} \operatorname{erfc} \sqrt{x} \tag{106a}$$

where erfc $y$ is the complement of the error function erf $y$ and is given by

$$\operatorname{erfc} y = \frac{2}{\sqrt{\pi}} \int_y^\infty e^{-t^2}\, dt$$

$$\operatorname{erf} y = 1 - \operatorname{erfc} y = \frac{2}{\sqrt{\pi}} \int_0^y e^{-t^2}\, dt$$

(2) $$W_{-\frac{1}{2},0}(x) = -x^{-\frac{1}{2}} e^{x/2} \operatorname{Ei}(-x) \tag{106b}$$

where Ei $(-x)$ is the exponential integral

$$\operatorname{Ei}(-x) = -\int_x^\infty t^{-1} e^{-t}\, dt$$

(3) $$W_{0,\nu}(x) = \left(\frac{x}{\pi}\right)^{\frac{1}{2}} K_\nu\left(\frac{x}{2}\right) \tag{106c}$$

where $K_\nu$ is the modified Bessel function of the second kind.

In order to use the above results, we must convert our integral (103) into the form (104). As before, the major contribution comes from the

† A. Erdelyi, "Higher Transcendental Functions," vol. I, p. 274, McGraw-Hill Book Company, Inc., New York, 1953.

portion of the contour in the vicinity of the saddle point $\theta$. We may approximate $\cos(\phi - \theta)$ by $1 - (\phi - \theta)^2/2$. Replacing $d\phi$ by $e^{j\pi/4}\,d\rho$ in the first quadrant and by $-e^{j\pi/4}\,d\rho$ in the third quadrant, as in (95), and noting that

$$\phi - \phi_0 = \phi - \theta - (\phi_0 - \theta) = \rho e^{j\pi/4} - (\phi_0 - \theta)$$

in the first quadrant and equals

$$-\rho e^{j\pi/4} - (\phi_0 - \theta)$$

in the third quadrant, we get, in place of expression (103),

$$-\frac{j\mu_0 F(\phi_0)}{4\pi} e^{-jk_0 r}\left[\int_0^{\rho_1} \frac{e^{-k_0 r\rho^2/2}}{\rho - e^{-j\pi/4}(\phi_0 - \theta)}\,d\rho + \int_{\rho_1}^{0} \frac{e^{-k_0 r\rho^2/2}}{\rho + e^{-j\pi/4}(\phi_0 - \theta)}\,d\rho\right]$$
$$= -\frac{j\mu_0 F(\phi_0) e^{-jk_0 r - j\pi/4}(\phi_0 - \theta)}{2\pi}\int_0^{\rho_1} \frac{e^{-k_0 r\rho^2/2}}{\rho^2 + j(\phi_0 - \theta)^2}\,d\rho \qquad (107)$$

after combining terms. Since the exponential decays rapidly, we may extend the integral over $\rho$ to infinity, i.e., extend $\rho_1$ to infinity. We now make one further change in variables and put $\rho^2 = t$. Hence, we obtain for the integral $I$

$$I = \int_0^{\infty} \frac{e^{-k_0 r\rho^2/2}}{\rho^2 + j(\phi_0 - \theta)^2}\,d\rho = \frac{1}{2}\int_0^{\infty} \frac{e^{-k_0 r t/2}}{t^{1/2}[t + j(\phi_0 - \theta)^2]}\,dt \qquad (108)$$

This latter integral is of the form (104) with $\alpha = -\frac{1}{2}$, $t_0 = -j(\phi_0 - \theta)^2$, and the pole is of order 1, that is, $m = 1$. In this case, the Whittaker function reduces to the result given by (106$a$). Hence, we have

$$I = \frac{\pi}{2} e^{jk_0 r(\phi_0 - \theta)^2/2}[j(\phi_0 - \theta)^2]^{1/2}\,\mathrm{erfc}\left[\frac{jk_0 r}{2}(\phi_0 - \theta)^2\right]^{1/2} \qquad (109)$$

The error function may be expanded asymptotically to give†

$$\mathrm{erfc}\, x = \frac{e^{-x^2}}{x\sqrt{\pi}} \sum_{n=0}^{\infty} (-1)^n \frac{\Gamma(n + \frac{1}{2})}{\Gamma(\frac{1}{2})} x^{-2n} \qquad (110)$$

valid for $x \to \infty$ in the sector $-3\pi/4 < \text{phase } x < 3\pi/4$. If the phase angle of $x$ is outside this range, it is noted that both erf $x$ and erfc $x$ are odd functions of $x$, and that

$$\mathrm{erfc}\, x = 1 - \mathrm{erf}\, x = 1 + \mathrm{erf}\,(-x) = 2 - \mathrm{erfc}\,(-x) \qquad (111)$$

and this result may be used to obtain an asymptotic expansion for $x$ in the vicinity of the negative-real axis.

† *Ibid.*

In our case, $\phi_0$ lies either on the $\sigma = \pi/2$, $\eta > 0$ axis or else in the range $\eta < 0$. In both cases, the phase angle of $(jk_0r/2)^{1/2}(\phi_0 - \theta)$ is in the range $-3\pi/4$ to $3\pi/4$, and, hence, by using Eq. (110), we obtain

$$I = -\frac{j\sqrt{\pi}}{(2k_0r)^{1/2}(\phi_0 - \theta)^2} \sum_{n=0}^{\infty} (-1)^n \frac{\Gamma(n + \frac{1}{2})}{\Gamma(\frac{1}{2})} \left[\frac{jk_0r(\phi_0 - \theta)^2}{2}\right]^{-n} \tag{112}$$

Substituting (112) into (107), we finally obtain our end result for the original integral (103):

$$\frac{-\mu_0 F(\phi_0)}{2\pi} e^{-jk_0r - j\pi/4} \frac{\sqrt{\pi}}{(2k_0r)^{1/2}(\phi_0 - \theta)} \left[1 + \frac{j}{k_0r(\phi_0 - \theta)^2} \cdots\right] \tag{113}$$

When $\phi_0$ is located close to the saddle point $\theta$, this expression represents a significant modification of the radiation field, and will produce a radiation lobe in the direction for which $|\phi_0 - \theta|$ is a minimum.

## REFERENCES

*Surface waves in general*

11.1. H. M. Barlow and A. E. Karbowiak: An Investigation of the Characteristics of Cylindrical Surface Waves, *Proc. IEE (London)*, vol. 100, part III, pp. 321–328, November, 1953.

11.2. H. M. Barlow and A. L. Cullen: Surface Waves, *Proc. IEE (London)*, vol. 100, part III, pp. 329–347, November, 1953.

11.3. H. M. Barlow and A. E. Karbowiak: An Experimental Investigation of the Properties of Corrugated Cylindrical Surface Wave Guides, *Proc. IEE (London)*, vol. 101, part III, pp. 182–188, May, 1954.

11.4. H. M. Barlow and A. E. Karbowiak: An Experimental Investigation of Axial Cylindrical Surface Waves Supported by Capacitive Surfaces, *Proc. IEE (London)*, vol. 102, part B, pp. 313–322, May, 1955.

11.5. L. A. Vajnshtejn: Electromagnetic Surface Waves on a Comblike Structure, *Zhur. Tekh. Fiz.*, vol. 26, pp. 385–397, 1956.

11.6. S. P. Schlesinger and D. D. King: Dielectric Image Lines, *IRE Trans.*, vol. MTT-6, pp. 291–299, July, 1958.

*Excitation of surface waves*

11.7. A. L. Cullen: The Excitation of Plane Surface Waves, *Proc. IEE (London)*, vol. 101, part IV, pp. 225–234, August, 1954.

11.8. G. J. Rich: The Launching of a Plane Surface Wave, *Proc. IEE (London)*, vol. 102, part B, pp. 237–246, March, 1955.

11.9. D. B. Brick: The Radiation of a Hertzian Dipole over a Coated Conductor, *Proc. IEE (London)*, vol. 102, part C, pp. 104–121, 1955.

11.10. W. M. G. Fernando and H. M. Barlow: An Investigation of the Properties of Radial Cylindrical Surface Waves Launched over Flat Reactive Surfaces, *Proc. IEE (London)*, vol. 103, part B, pp. 307–318, May, 1956.

11.11. J. R. Wait: Excitation of Surface Waves on Conducting, Stratified, Dielectric Coated and Corrugated Surfaces, *J. Research Natl. Bur. Standards*, vol. 59, pp. 365–377, December, 1957.

11.12. R. H. DuHamel and J. W. Duncan: Launching Efficiency of Wires and Slots for a Dielectric Rod Waveguide, *IRE Trans.*, vol. MTT-6, pp. 277–284, July, 1958.

11.13. J. W. Duncan: The Efficiency of Excitation of a Surface Wave on a Dielectric Cylinder, *IRE Trans.*, vol. MTT-7, pp. 257–268, April, 1959.

11.14. B. Friedman and W. E. Williams: Excitation of Surface Waves, *Proc. IEE (London)*, vol. 105, part C, pp. 252–258, March, 1958.

11.15. J. Brown: Some Theoretical Results for Surface Wave Launchers, paper presented at URSI Symposium, Toronto, Canada, June, 1959.

*Saddle-point integration*

11.16. B. L. van der Waerden: On the Method of Saddle Points, *Appl. Sci. Research*, vol. 2B, pp. 33–43, 1950.

11.17. P. C. Clemmow: Some Extensions to the Method of Integration by Steepest Descents, *Quart. J. Mech. and Appl. Math.*, vol. 3, pp. 241–260, 1950.

## PROBLEMS

**11.1.** Obtain the solution for a TE surface wave guided by a plane interface between free space and a medium with parameters $\epsilon$, $\mu$, where $\mu > \mu_0$, and $\mu/\mu_0 > \epsilon/\epsilon_0$. Assume that $\epsilon$ has a small negative-imaginary part, but that $\mu$ is real.

**11.2.** A vertical line source above an impedance plane or a grounded dielectric sheet radiates cylindrical waves having no angular variation. Obtain the solution for the Zenneck wave above an impedance plane, and also for a grounded dielectric sheet of thickness $t$. The wave is described by the following function, $H_0{}^2(\beta r)e^{-px}$, where $r^2 = y^2 + z^2$, and $H_0{}^2$ is the Hankel function of the second kind.

**11.3.** For a corrugated plane, find the minimum value of the short-circuit position $d$ such that the first evanescent mode excited at the interface has decayed to less than 0.1 of its value at the interface, at the position of the short circuit. Assume a spacing $S$ between corrugations equal to $0.2\lambda_0$ and $0.4\lambda_0$.

**11.4.** Find the solution for a TM surface wave on a lossless dielectric tube of inner radius $a$ and outer radius $b$ as illustrated.

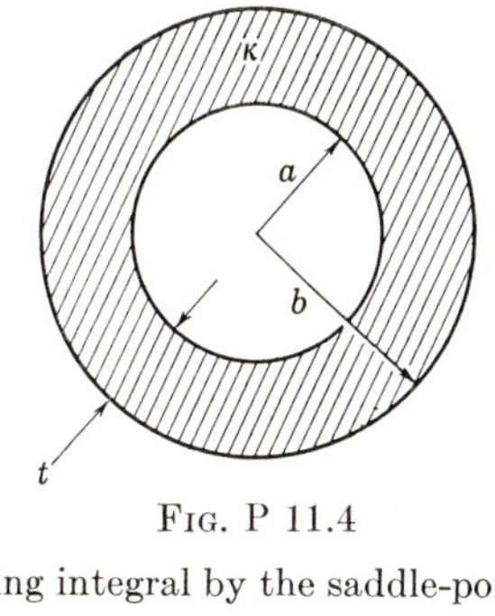


FIG. P 11.4

**11.5.** Evaluate the following integral by the saddle-point method:

$$\int_{-j\infty}^{j\infty} \frac{\exp\left[jx(\gamma^2 + k^2)^{1/2} - \gamma z\right]}{(\gamma^2 + k^2)^{1/2}(\gamma - 4jk)}\, d\gamma$$

**11.6.** Find the field radiated by a line source located a distance $d$ above a plane having a surface impedance $jX_s$. Evaluate the integral by the saddle-point method.

**11.7.** Repeat Prob. 11.6 for a surface with reactance $-jX_s$. Note that a surface-wave mode is now excited.

**11.8.** For Prob. 11.7, evaluate the total power radiated as a surface wave and also that radiated as a continuous spectrum. The power radiated as a continuous spectrum may be obtained by integrating the far-zone field, with respect to $\theta$, from zero to $\pi/2$, and multiplying by 2. Find the best position of the line source in order to obtain the maximum power in the surface wave.

**11.9.** Show that the change in variable $\cos(\phi - \theta) = 1 - jS^2$, with $-\infty < S < \infty$, corresponds to integration along the steepest-descent contour in the $\phi$ plane. Show that the integral (89) becomes

$$-je^{-j(k_0r-\pi/4)} \int_{-\infty}^{\infty} \frac{e^{-k_0rS^2}}{(1 - jS^2/2)^{1/2}} \, dS$$

Expand $(1 - jS^2/2)^{-1/2}$ in a power series valid for $|S| < \sqrt{2}$, and integrate the first two terms by using (100) to obtain the first two terms in the asymptotic evaluation of the integral.

*Answer*

$$-j\left(\frac{2\pi}{k_0r}\right)^{1/2} e^{-j(k_0r-\pi/4)} \left[1 + \frac{j}{4}\frac{\Gamma(3/2)}{(\pi k_0r)^{1/2}}\right]$$

To obtain this result, the integration is taken from $-\infty$ to $\infty$, and thus an error is made, since the power series in $S^2$ is not valid for $|S|^2 > 2$. However, the error may be shown to be negligible for large $k_0r$ (see Ref. 11.16).

CHAPTER 12

# ARTIFICIAL DIELECTRICS

At microwave frequencies, the wavelength is sufficiently short so that many of the techniques and principles used in optics may be carried over directly. One of the more obvious optical devices to find wide application is the lens. Lenses are used in a variety of antenna systems at microwave frequencies. The dielectric material for a microwave lens must have a small loss and an index of refraction preferably no smaller than about 1.5. One suitable material that is frequently used is polystyrene, which has an index of refraction equal to 1.6. A major disadvantage of solid dielectric materials is the excessive bulk and weight of the lens, since in many cases lens diameters greater than a meter are required. In 1948, Kock suggested the use of artificial dielectrics to replace the actual dielectric material, in order to overcome these disadvantages.†

An artificial dielectric is a large-scale model of an actual dielectric, obtained by arranging a large number of identical conducting obstacles in a regular three-dimensional pattern. The obstacles are supported by a lightweight binder or filler material such as expanded polystyrene (Polyfoam). Under the action of an external applied electric field, the charges on each conducting obstacle are displaced so as to set up an induced field that will cancel the applied field at the obstacle surface. The obstacle is electrically neutral so that the dominant part of the induced field is a dipole field. Each obstacle thus simulates the behavior of a molecule (or group of molecules) in an ordinary dielectric in that it exhibits a dipole moment. The combined effect of all the obstacles in the lattice is to produce a net average dipole polarization $\mathbf{P}$ per unit volume. If $\mathbf{E}$ is the average net field in the medium, the displacement $\mathbf{D}$ is given by

$$\mathbf{D} = \epsilon_0\mathbf{E} + \mathbf{P} = \epsilon\mathbf{E}$$

and it is apparent that the effective permittivity $\epsilon$ will be greater than $\epsilon_0$.

When a high-frequency magnetic field is applied to a conducting obstacle, an induced circulating current is set up, such that the induced magnetic field produced will cancel the normal component of the applied

† W. E. Kock, Metallic Delay Lenses, *Bell System Tech. J.*, vol. 27, pp. 58–82, 1948.

field at the obstacle surface. These induced currents are equivalent to a magnetic dipole, and, therefore, in general, an artificial dielectric exhibits magnetic dipole polarization as well as electric dipole polarization. The induced magnetic dipoles always oppose the inducing field, so that the artificial dielectric is always diamagnetic when obstacles having magnetic dipole moments are used. It should be noted that certain types of obstacles, properly oriented with respect to the applied field, will not have an induced magnetic dipole. If $\epsilon$ and $\mu$ are the permittivity and permeability of the artificial dielectric, the index of refraction $\eta$ is given by

$$\eta = \left(\frac{\epsilon\mu}{\epsilon_0\mu_0}\right)^{1/2}$$

Since $\mu$ is less than $\mu_0$, the presence of magnetic polarization reduces the value of $\eta$, and is for this reason usually an undesirable property.

Unless the obstacles have spherical symmetry and a cubical or random lattice pattern is used, the dipole polarization for a unit applied field will be different in different directions. The artificial dielectric will then have anisotropic properties. To date, little use of such anisotropic properties has been found, except in wave polarizers, i.e., devices for converting linear polarization to circular polarization and vice versa.

Typical structures used for artificial dielectrics are illustrated in Fig. 12.1. Figure 12.1*a* illustrates a cubical array of conducting spheres, while in Fig. 12.1*b* the spheres are replaced by thin flat conducting disks. For the disk medium, no magnetic polarization is produced if the incident field is a TEM wave with the $H$ vector in the plane of the disks. Figure 12.1*c* and *d* illustrate two-dimensional structures consisting of thin flat conducting strips and conducting cylinders or rods, respectively. With the electric field perpendicular to the strips or rods, capacitive loading of the medium is produced, and $\epsilon$ is increased. Artificial dielectrics for which the index of refraction $\eta$ is greater than unity are known as phase-delay media, since the phase velocity in the medium is less than that in free space. For the two-dimensional structures of Fig. 12.1, inductive loading is produced when the applied electric field is parallel to the strips or rods. For this polarization, $\mu$ is less than $\mu_0$, and $\eta$ is less than unity. The structure is therefore referred to as a phase-advance dielectric.

The term artificial dielectric is commonly used to incorporate structures such as the parallel-plate array and stacks of square waveguides placed side by side, in addition to structures made up of arrays of discrete obstacles. Only the latter class of structures, and, in particular, the sphere and disk media and the two-dimensional strip medium, will be examined in detail in this chapter.

There are three basic approaches used in the analysis of artificial dielectrics. The simplest is the Lorentz theory, which was originally developed for the classical theory of real dielectrics. This theory considers only dipole interaction between obstacles, and produces accurate results for obstacle spacings limited to less than about $0.1\lambda_0$ and for obstacle sizes small compared with the spacing. When the obstacle size is not small compared with the spacing, the dipole term is not a sufficiently good approximation of the induced field. The second

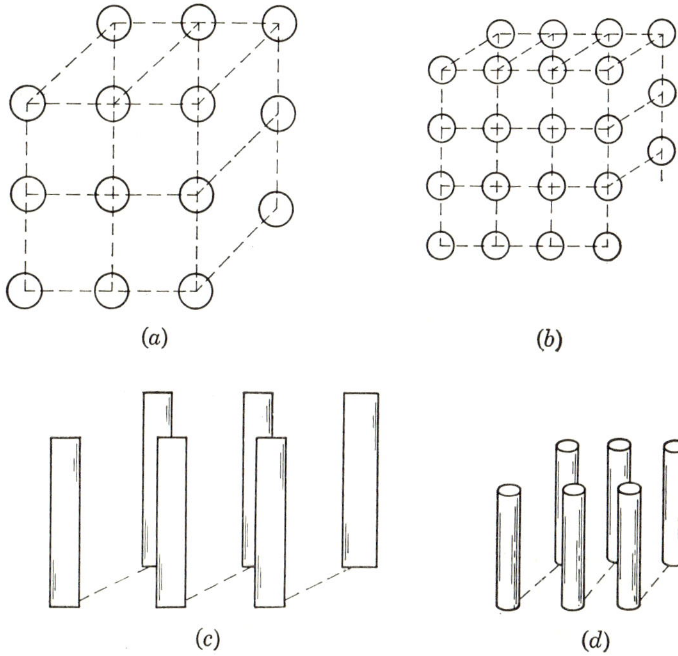


FIG. 12.1. Typical artificial dielectric structures. (*a*) Three-dimensional sphere medium; (*b*) three-dimensional disk medium; (*c*) two-dimensional strip medium; (*d*) two-dimensional rod medium.

approach is one step more accurate than the Lorentz theory, and consists of a rigorous static field solution. Such a solution will take into account all the multipoles in the expansion of the induced field and will give good results provided the obstacle spacing is small compared with the wavelength, say no greater than $0.1\lambda_0$. Finally, the third method is to attempt a solution of Maxwell's equations. In this latter method, it is often possible to reduce the problem to that of a periodically loaded transmission line, for which the methods of analysis presented in Chap. 9 are applicable.

Many people have contributed to the development of artificial dielectrics since the pioneer work of Kock. It would require too much space to review this work here on an individual basis. However, a number of the more important papers that have been published are listed at the end of the chapter.

## 12.1. Lorentz Theory

The Lorentz theory is a static field theory which provides a solution that takes into account only the dipole term in the induced field. Consider a three-dimensional regular array of identical conducting obstacles, as in Fig. 12.2. The lattice spacings are $a$, $b$, and $c$ along the $x$, $y$, and $z$ directions, respectively. Let an external electrostatic field $\mathbf{E}_0$ be applied along the $y$ direction. The obstacles will be assumed to have their principal axes of polarization coincident with the coordinate axes, so that

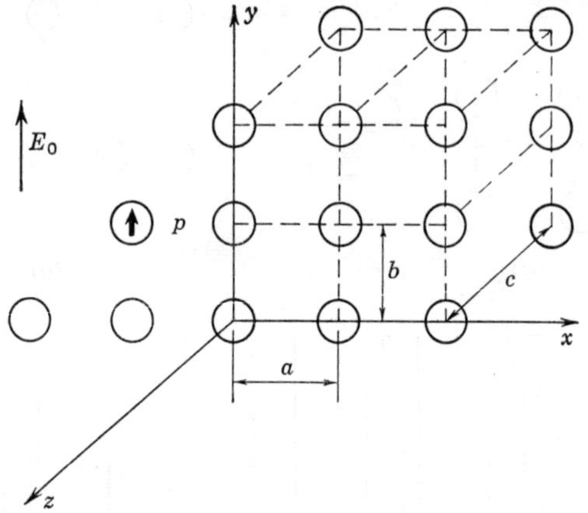


FIG. 12.2. A three-dimensional artificial dielectric.

the induced dipole moment $p$ of one obstacle is in the $y$ direction for an applied field in the $y$ direction.

In the course of the analysis, several different fields will be introduced. For convenience, these are summarized below:

$\mathbf{E}_0 = \mathbf{a}_y E_0$ = external applied field

$\mathbf{E}_1$ = dipole field produced by obstacle located at the origin

$\mathbf{E}_p$ = dipole field produced by all the obstacles in the infinite array

$\mathbf{E}_e = \mathbf{E}_0 + \mathbf{E}_p - \mathbf{E}_1$ = effective field acting to polarize the obstacle at the origin

$\mathbf{E}_i = \mathbf{E}_p - \mathbf{E}_1$ = interaction field

$\mathbf{E}_{pa}$, $\mathbf{E}_{1a}$, $\mathbf{E}_{ea}$, $\mathbf{E}_{ia}$ = fields defined as above but averaged over the volume of a unit cell with dimensions $a$, $b$, $c$

The $y$ component of the above fields is indicated by an additional subscript $y$.

Since the Lorentz theory considers only dipole interaction between obstacles, the results will be valid only for obstacles with dimensions small compared with their spacing. The field acting to polarize any one obstacle may therefore be assumed to be a uniform field equal to the field at the center of the obstacle. From considerations of symmetry, it is found that the field acting to polarize the obstacle at the origin has only a $y$ component, $E_{ey}$. The induced dipole moment is proportional to the field, and, hence,

$$p = \alpha_e \epsilon_0 E_{ey} \tag{1}$$

In this equation the proportionality constant $\alpha_e$ is called the polarizability, and is a characteristic parameter of the obstacle. For nonsymmetrical obstacles, $\alpha_e$ is different along different directions and, hence, is a tensor quantity in general. For an arbitrary obstacle, three principal axes will exist, such that the induced dipole moment is parallel to the applied field. In the present case, these principal axes are assumed to coincide with the coordinate axes as stated earlier. The number $N$ of obstacles per unit volume is equal to $(abc)^{-1}$, and, hence, the dipole polarization per unit volume is

$$P = Np = \frac{p}{abc} = N\alpha_e \epsilon_0 E_{ey} \tag{2}$$

The effective polarizing field $E_{ey}$ is equal to $E_0 + E_{iy}$, where $E_{iy}$ is the interaction field, i.e., the field due to all the neighboring obstacles. The interaction field is proportional to $p$ and may be expressed by

$$E_{iy} = \frac{Cp}{\epsilon_0} \tag{3}$$

where $C$ is called the interaction constant. In (3) it is to be understood that $E_{iy}$ is the field at the origin or center of the obstacle under consideration. Substituting into (1) and (2), we get

$$p = \alpha_e \epsilon_0 \left( E_0 + \frac{Cp}{\epsilon_0} \right)$$

or

$$p = \frac{\alpha_e \epsilon_0 E_0}{1 - \alpha_e C} \tag{4a}$$

$$P = \frac{N\alpha_e \epsilon_0 E_0}{1 - \alpha_e C} \tag{4b}$$

The $y$ component of the average displacement flux is defined by

$$D_{ay} = \epsilon_0 E_{ay} + P = \kappa \epsilon_0 E_{ay} \tag{5}$$

where $E_{ay}$ is the average value of the $y$ component of the total field in the medium, and $\kappa$ is the relative dielectric constant in the $y$ direction.

From (5), we get

$$\kappa = 1 + \frac{P}{\epsilon_0 E_{ay}} = 1 + \frac{P}{\epsilon_0(E_0 + E_{pay})} \tag{6}$$

For obstacles that are symmetrical about the coordinate planes passing through the center of the obstacle, the average field produced by all the induced dipoles is zero. With the applied field in the $y$ direction, conducting planes may be inserted into the lattice at $y = \pm b/2 + mb$, $m = 0, \pm 1, \pm 2, \ldots$, without disturbing the field distribution. If the applied field is $\mathbf{a}_y E_0$, the potential of the conducting planes at $y = \pm b/2$ may be chosen as $-E_0 b/2$ and $E_0 b/2$, as in Fig. 12.3. From symmetry considerations, it follows that, on planes at $x = \pm a/2$ and $z = \pm c/2$, the total electric field has no component normal to these planes. Also the field in every unit cell throughout the lattice is identical.

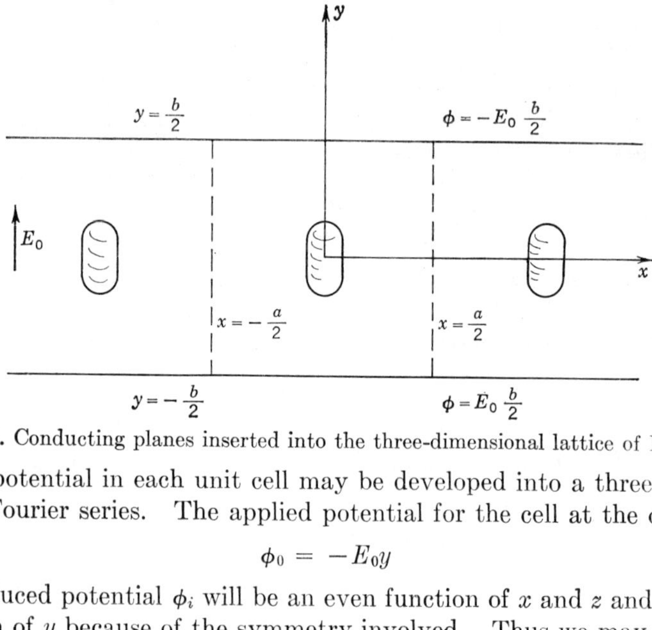


FIG. 12.3. Conducting planes inserted into the three-dimensional lattice of Fig. 12.2.

The potential in each unit cell may be developed into a three-dimensional Fourier series. The applied potential for the cell at the origin is

$$\phi_0 = -E_0 y$$

The induced potential $\phi_i$ will be an even function of $x$ and $z$ and an odd function of $y$ because of the symmetry involved. Thus we may write

$$\phi_i = \sum_{n=1}^{\infty}\sum_{m=1}^{\infty}\sum_{s=1}^{\infty} A_{nms} \cos\frac{2n\pi x}{a} \cos\frac{2s\pi z}{c} \sin\frac{2m\pi y}{b}$$

where $A_{nms}$ is an amplitude coefficient. The $y$ component of induced field is thus given by

$$E_{py} = -\frac{\partial \phi_i}{\partial y} = -\sum_{n=1}^{\infty}\sum_{m=1}^{\infty}\sum_{s=1}^{\infty} A_{nms}\frac{2m\pi}{b} \cos\frac{2n\pi x}{a} \cos\frac{2m\pi y}{b} \cos\frac{2s\pi z}{c}$$

and corresponds to the total field produced by all the obstacles in the three-dimensional array. It is clear from this expression that the average induced field in a unit cell is zero since the cosine terms have a zero average.

In view of the above discussion we may place $E_{pay}$ equal to zero in (6). Substituting for $p$ now yields the classic expression known as the Clausius-Mossotti relation:

$$\kappa = 1 + \frac{N\alpha_e}{1 - \alpha_e C} \tag{7}$$

This relation permits $\kappa$ to be evaluated when the obstacle polarizability $\alpha_e$ and the interaction constant $C$ are known. Practical artificial dielectrics consist of obstacles supported in a low-dielectric-constant binder, and (7) must then be interpreted as the dielectric constant relative to that of the binder or supporting medium. For simplicity, we will, however, in the remainder of this chapter assume that the obstacles are supported in a medium with a dielectric constant of unity.

In the above analysis we have assumed a $y$-directed applied field. A similar analysis will hold for $x$- and $z$-directed fields, but the dielectric constant along the $x$ and $z$ directions will not necessarily be the same, since the polarizability and interaction constant are different in different directions for general structures. In other words, the medium will have anisotropic properties in general.

## 12.2. Electrostatic Solution

As noted in the previous section, the problem of a regular three-dimensional lattice of conducting obstacles can be reduced to that of a single obstacle located at the center of a unit cell as in Fig. 12.4*b*. On the top and bottom faces, the potential is kept at $-E_0b/2$ and $E_0b/2$. This potential difference sets up the applied field $\mathbf{a}_y E_0$. On the side walls, $\partial\phi/\partial n$ is equal to zero, where $\phi$ is the sum of the applied potential $\phi_0$ and the induced potential $\phi_i$.

Figure 12.4*a* illustrates a unit cell with the obstacle removed. The capacitance $C_0$ of this cell is given by

$$C_0 = \epsilon_0 \frac{ac}{b} \tag{8}$$

For the unloaded cell, the charge on each plate of area $ac$ is equal to the potential difference between the plates times $C_0$, or $C_0E_0b$. For the loaded cell of Fig. 12.4*b*, the charge on each plate is increased by an amount equal to the total charge that is displaced on the conducting obstacle. Let this charge be $-q$ on the upper plate, and $q$ on the lower plate. The capacitance of the unit cell is thus increased to a value

$$C_0' = \frac{Q + q}{E_0 b} = C_0 + \frac{q}{E_0 b}$$

The effective dielectric constant of the artificial dielectric is equal to the dielectric constant of a homogeneous dielectric that would give this same increase in capacitance. Hence, we have

$$\kappa = \frac{C_0'}{C_0} = 1 + \frac{q}{C_0 E_0 b} = 1 + \frac{q}{Q} \tag{9}$$

The electrostatic approach thus aims at finding the change in the capacitance of a unit cell due to the presence of the obstacle. Although the problem is readily formulated, the solution is not always easy to

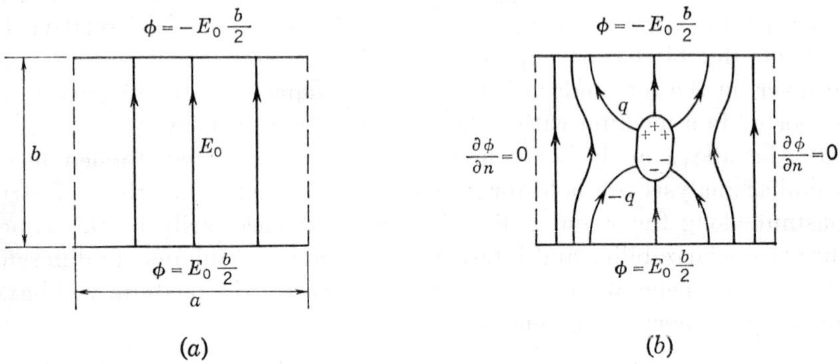


(a) (b)

FIG. 12.4. Unit cell of artificial dielectric medium. (a) Unloaded cell; (b) loaded cell.

obtain. Experimental methods may, of course, be used. One such method makes use of an electrolytic tank with insulators corresponding to the walls on which $\partial\phi/\partial n$ equals zero. In the electrolytic tank, the current density is the analog of the electric field.†

In addition to the change in capacitance of a unit cell, the change in the inductance caused by magnetic dipole polarization must also be determined. At low frequencies, there is no magnetic polarization for nonferrous obstacles. However, since artificial dielectrics are normally used at high frequencies, the field does not penetrate into the conductor, and the normal magnetic field at the surface must be zero. At low frequencies, the equivalent magnetic body is a perfectly diamagnetic body with $\mu = 0$, since, for such a body, the normal magnetic field at the surface vanishes. If in Fig. 12.4b the applied magnetic field is $H_{0x}$ in the $x$ direction, this is equivalent to passing a current of density $H_{0x}$ in the $z$ direction on the upper plate at $y = b/2$, and in the negative $z$

† S. B. Cohn, Electrolytic Tank Measurements for Microwave Metallic Delay Lens Media, *J. Appl. Phys.*, vol. 21, pp. 674–680, July, 1950.

direction on the lower plate at $y = -b/2$. When the obstacle is absent, the inductance $L_0$ of the unit cell is $L_0 = \mu_0 bc/a$. With the equivalent diamagnetic body present, the inductance is reduced to a new value $L_0'$, and the permeability of the medium may be defined by

$$\mu = \frac{aL_0'}{bc} \tag{10}$$

The induced magnetic field may be derived from a scalar magnetic potential $\phi_m$, where $\phi_m = 0$ on the sides at $x = \pm a/2$, and the normal derivative vanishes on the other four sides. From the known value of the resultant magnetic field, the inductance $L_0'$ of the unit cell may be found by well-known standard methods.

The change in inductance may also be determined experimentally by means of an electrolytic tank. Conducting plates are used for the two sides at $x = \pm a/2$, and insulating plates (dielectric sheets) for the other sides of the unit cell. The analog of the perfectly diamagnetic body is a body with zero conductivity, and, hence, a dielectric body having the same shape as the obstacle in the artificial dielectric may be used. The ratio of the resistance of the unit cell with the dielectric body absent to that with the body present is equal to the ratio $L_0'/L_0$. The index of refraction of the medium is given by

$$\eta = \left(\frac{L_0'C_0'}{L_0C_0}\right)^{1/2}$$

## 12.3. Evaluation of Interaction Constants

In this section, the interaction constant for various types of lattices will be evaluated. Some of the results in this section are not of interest in connection with artificial dielectrics, but are included because of their application to aperture coupling in rectangular waveguides.

The first case to be considered is a three-dimensional array of $y$-directed unit dipoles located at $x = na$, $y = mb$, $z = sc$, where $n$, $m$, $s$ are integers extending from plus infinity to minus infinity. The dipole at $x = y = z = 0$ is excluded. The $y$ component of the electric field at the origin is the interaction field and is equal to $C/\epsilon_0$, according to (3) for unit dipoles. The potential due to a unit dipole at $x_0$, $y_0$, $z_0$ is given by

$$\phi = \frac{\mathbf{a}_y \cdot \mathbf{r}}{4\pi\epsilon_0 r^3}$$

where $r = [(x - x_0)^2 + (y - y_0)^2 + (z - z_0)^2]^{1/2}$. For the array of dipoles,

$$\phi = \frac{1}{4\pi\epsilon_0} \sum_{n=-\infty}^{\infty} \sum_{s=-\infty}^{\infty} \sum_{m=-\infty}^{\infty}{}' \frac{y - mb}{[(x - na)^2 + (y - mb)^2 + (z - sc)^2]^{3/2}}$$

where the prime indicates omission of the term $n = m = s = 0$. The interaction constant is thus given by

$$C = \epsilon_0 E_i = -\epsilon_0 \left.\frac{\partial \phi}{\partial y}\right|_{x=y=z=0}$$
$$= \frac{1}{4\pi} \sum_{n=-\infty}^{\infty} \sum_{s=-\infty}^{\infty} \sum_{m=-\infty}^{\infty}{}' \frac{2(mb)^2 - (na)^2 - (sc)^2}{[(mb)^2 + (na)^2 + (sc)^2]^{5/2}} \tag{11}$$

The above series has been summed by Brown, using the Poisson summation formula.† However, the following alternative approach is instructive. By image theory, conducting planes may be inserted at $y = \pm b/2$. We then require the potential distribution for unit dipoles placed at $x = na$, $z = sc$, $y = 0$, for $n, s = 0, \pm 1, \pm 2, \ldots$, and $n = s \neq 0$. First consider a $y$-directed dipole at the origin, as in Fig. 12.5. This dipole may be considered as a positive charge $q\delta(x)\delta(z)\delta(y - \tau)$ at $x = z = 0$, $y = \tau$ and a negative charge $-q\delta(x)\delta(z)\delta(y + \tau)$ at $x = z = 0$, $y = -\tau$. Using cylindrical coordinates $y$, $r$, $\theta$, and noting that $\phi$ is independent of $\theta$, Poisson's equation becomes

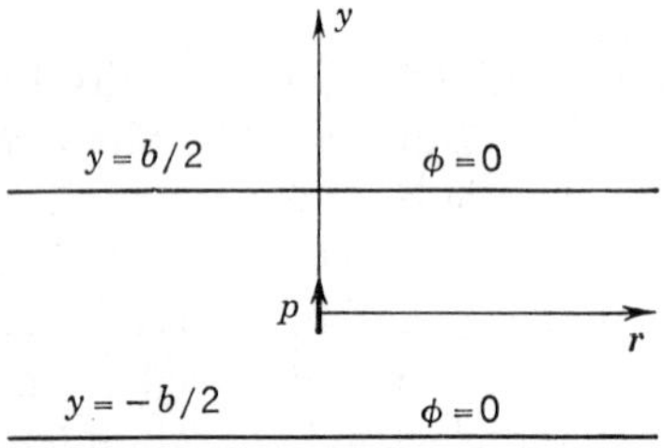

FIG. 12.5. A dipole between parallel conducting planes.

$$\frac{\partial^2 \phi}{\partial y^2} + \frac{1}{r}\frac{\partial}{\partial r} r \frac{\partial \phi}{\partial r} = -\frac{\rho}{\epsilon_0} = -\frac{q}{\epsilon_0}\delta(x)\delta(z)[\delta(y - \tau) - \delta(y + \tau)] \tag{12}$$

The potential $\phi$ is an odd function of $y$, and vanishes on the conducting planes at $y = \pm b/2$. Let

$$\phi = \sum_{m=1}^{\infty} \phi_m = \sum_{m=1}^{\infty} A_m(r) \sin \frac{2m\pi y}{b}$$

and substitute into Eq. (12) to get

$$\sum_{m=1}^{\infty} \left[\frac{1}{r}\frac{d}{dr} r \frac{d}{dr} A_m - \left(\frac{2m\pi}{b}\right)^2 A_m\right] \sin \frac{2m\pi y}{b}$$
$$= -\frac{q}{\epsilon_0}\delta(x)\delta(z)[\delta(y - \tau) - \delta(y + \tau)]$$

† M. M. Z. Kharadly and W. Jackson, The Properties of Artificial Dielectrics Comprising Arrays of Conducting Particles, *Proc. IEE* (*London*), vol. 100, part III, pp. 199–212, July, 1953.

Multiply both sides by $\sin (2m\pi y/b)$, and integrate over $y$ from $-b/2$ to $b/2$, to obtain

$$\frac{b}{2}\left[\frac{1}{r}\frac{d}{dr}r\frac{dA_m}{dr} - \left(\frac{2m\pi}{b}\right)^2 A_m\right] = -\frac{2q}{\epsilon_0}\delta(x)\delta(z)\sin\frac{2m\pi\tau}{b}$$

The dipole moment $p$ is equal to $2q\tau$, and, if we let $\tau$ tend to zero such that

$$\lim_{\tau\to 0} 2q\tau = p$$

we obtain

$$\frac{1}{r}\frac{d}{dr}r\frac{dA_m}{dr} - \left(\frac{2m\pi}{b}\right)^2 A_m = -\frac{p}{\epsilon_0}\frac{2m\pi}{b}\frac{2}{b}\delta(x)\delta(z) \tag{13}$$

For all values of $x$ and $z$ not equal to zero, the solutions to (13) are the modified Bessel functions of the first and second kind. Only the second kind of modified Bessel function vanishes as $r = (x^2 + z^2)^{1/2}$ becomes large, and, hence, this is the only solution allowed. We now have

$$A_m(r) = a_m K_0\left(\frac{2m\pi r}{b}\right) \tag{14}$$

where $a_m$ is an amplitude constant to be determined. For small $r$, we have

$$K_0\left(\frac{2m\pi r}{b}\right) \approx -\left(\gamma + \ln\frac{m\pi r}{b}\right)$$

where $\gamma$ is Euler's constant. For large $r$, $K_0$ decays exponentially according to the relation

$$K_0\left(\frac{2m\pi r}{b}\right) \approx \left(\frac{b}{4mr}\right)^{1/2} e^{-2m\pi r/b}$$

Equation (13) is the solution to the problem of a nonuniform line source of density $4m\pi p/b^2$ and having a variation $\sin (2m\pi y/b)$ with $y$. Very near this line source, i.e., for $r$ very small, the potential $\phi_m$, which is a solution of (13), must have the usual line-source logarithmic singularity. Hence,

$$\lim_{r\to 0}\phi_m = -\frac{4m\pi p}{b^2}\frac{\ln r}{2\pi\epsilon_0}$$

and since $\lim\limits_{r\to 0} a_m K_0\left(\dfrac{2m\pi r}{b}\right) = -a_m \ln r$, the solution for $a_m$ is given by

$$a_m = \frac{2mp}{\epsilon_0 b^2}$$

The solution for the potential from the dipole is thus given by

$$\phi = \sum_{m=1}^{\infty} \frac{2mp}{\epsilon_0 b^2} \sin \frac{2m\pi y}{b} K_0\left(\frac{2m\pi r}{b}\right) \tag{15}$$

For a dipole of unit strength $p = 1$, the $y$ component of electric field is given by

$$E_y = -\frac{\partial \phi}{\partial y} = -\frac{1}{\pi b \epsilon_0} \sum_{m=1}^{\infty} \left(\frac{2m\pi}{b}\right)^2 \cos \frac{2m\pi y}{b} K_0\left(\frac{2m\pi r}{b}\right) \tag{16}$$

For the two-dimensional array of unit dipoles, excluding the dipole at the origin, the $y$ component of electric field at the origin is

$$\begin{aligned}\epsilon_0 E_y &= -\frac{1}{\pi b} \sum_{n=-\infty}^{\infty} \sum_{s=-\infty}^{\infty} \sum_{m=1}^{\infty}{}' \left(\frac{2m\pi}{b}\right)^2 K_0 \left\{\frac{2m\pi}{b} [(na)^2 + (sc)^2]^{1/2}\right\} \\ &= -\frac{2}{\pi b} \sum_{s=1}^{\infty} \sum_{m=1}^{\infty} \left(\frac{2m\pi}{b}\right)^2 K_0\left(\frac{2m\pi sc}{b}\right) - \frac{2}{\pi b} \sum_{n=1}^{\infty} \sum_{m=1}^{\infty} \left(\frac{2m\pi}{b}\right)^2 \\ &\quad \times K_0\left(\frac{2m\pi na}{b}\right) - \frac{4}{\pi b} \sum_{n=1}^{\infty} \sum_{s=1}^{\infty} \sum_{m=1}^{\infty} \left(\frac{2m\pi}{b}\right)^2 \\ &\quad \times K_0 \left\{\frac{2m\pi}{b} [(na)^2 + (sc)^2]^{1/2}\right\}\end{aligned} \tag{17}$$

Since $K_0$ decays exponentially at a rapid rate, only the first terms in the first two sums are required for sufficient accuracy in most cases. Thus,

$$E_y \approx \frac{-8\pi}{\epsilon_0 b^3} \left[K_0\left(\frac{2\pi c}{b}\right) + K_0\left(\frac{2\pi a}{b}\right)\right] \tag{18}$$

when $a$ and $c$ do not differ greatly. However, this is not the interaction field at the origin, since excluding the term corresponding to the dipole at the origin is equivalent to deleting all the dipoles along the line $x = z = 0$ and located at $y = mb$, $m = \pm 1, \pm 2, \ldots$. We must add to (18) the contribution from Eq. (11) with $n = s = 0$, that is,

$$\frac{1}{\pi \epsilon_0} \sum_{m=1}^{\infty} \frac{1}{(mb)^3} = \frac{1.201}{\pi \epsilon_0 b^3}$$

to obtain the total interaction field. Hence, the interaction constant $C$ is given by

$$C = \frac{1.201}{\pi b^3} - \frac{8\pi}{b^3} \left[K_0\left(\frac{2\pi c}{b}\right) + K_0\left(\frac{2\pi a}{b}\right)\right] \tag{19}$$

For a cubical lattice $a = b = c$, Lorentz obtained the value $C = (3b^3)^{-1}$ by replacing the sums by integrals. This result does not differ much from (19) for a cubical lattice. For $x$-directed dipoles, $a$ and $b$ should be interchanged in (19), while, for $z$-directed dipoles, $b$ and $c$ are interchanged.

*Summation by Means of Poisson's Formula*

The Poisson summation formula gives

$$\sum_{m=-\infty}^{\infty} f(\alpha m) = \frac{1}{\alpha} \sum_{m=-\infty}^{\infty} F\left(\frac{2m\pi}{\alpha}\right) \tag{20}$$

where $F(w)$ is the Fourier transform of $f(z)$; that is,

$$F(w) = \int_{-\infty}^{\infty} e^{jwz} f(z)\, dz \tag{21}$$

The following results will be required:

$$\int_{-\infty}^{\infty} e^{jwz}(z^2 + r^2)^{-\nu-\frac{1}{2}}\, dz = 2\left(\frac{|w|}{2r}\right)^{\nu} \pi^{\frac{1}{2}} \frac{K_\nu(|w|r)}{\Gamma(\nu + \frac{1}{2})} \tag{22}$$

The transform of $z^2(z^2 + r^2)^{-\frac{5}{2}}$ may be obtained from (22) by putting $\nu = 2$ and differentiating both sides twice with respect to $w$. Combining this with (22) and using the Bessel-function recurrence relations permits us to obtain

$$\int_{-\infty}^{\infty} e^{jwz} \frac{2z^2 - r^2}{(z^2 + r^2)^{\frac{5}{2}}}\, dz = -2w^2 K_0(|w|r) \tag{23}$$

Equation (11) for the interaction constant $C$ may be written as

$$\begin{aligned} C = \frac{1}{4\pi} \Bigg\{ 4 \sum_{m=1}^{\infty} \frac{1}{(mb)^3} &+ 4 \sum_{n=1}^{\infty} \sum_{s=1}^{\infty} \sum_{m=-\infty}^{\infty} \frac{2(mb)^2 - (na)^2 - (sc)^2}{[(mb)^2 + (na)^2 + (sc)^2]^{\frac{5}{2}}} \\ &+ 2 \sum_{n=1}^{\infty} \sum_{m=-\infty}^{\infty} \frac{2(mb)^2 - (na)^2}{[(mb)^2 + (na)^2]^{\frac{5}{2}}} \\ &+ 2 \sum_{s=1}^{\infty} \sum_{m=-\infty}^{\infty} \frac{2(mb)^2 - (sc)^2}{[(mb)^2 + (sc)^2]^{\frac{5}{2}}} \Bigg\} \end{aligned} \tag{24}$$

The sums over $m$ may be totaled by using (20) and (23). For example,

$$\sum_{m=-\infty}^{\infty} \frac{2(mb)^2 - (na)^2}{[(mb)^2 + (na)^2]^{\frac{5}{2}}} = -\frac{2}{b} \sum_{m=-\infty}^{\infty} \left(\frac{2m\pi}{b}\right)^2 K_0\left(\frac{2|m|\pi na}{b}\right) \tag{25}$$

In (25), the $m = 0$ term on the right-hand side is zero. When the remaining sums in (24) are handled in the same way, we obtain the final result given by (19), upon making the same approximations. The results should obviously be the same, since summing over $m$ in (24) is equivalent to solving the problem of a single layer of dipoles between parallel plates by the method of images.

*Interaction Constant for Two-dimensional Lattices*

Consider a two-dimensional lattice of unit dipoles located at $x = na$, $y = mb$; $n, m = 0, \pm 1, \pm 2, \ldots$, as in Fig. 12.6. For $y$-directed

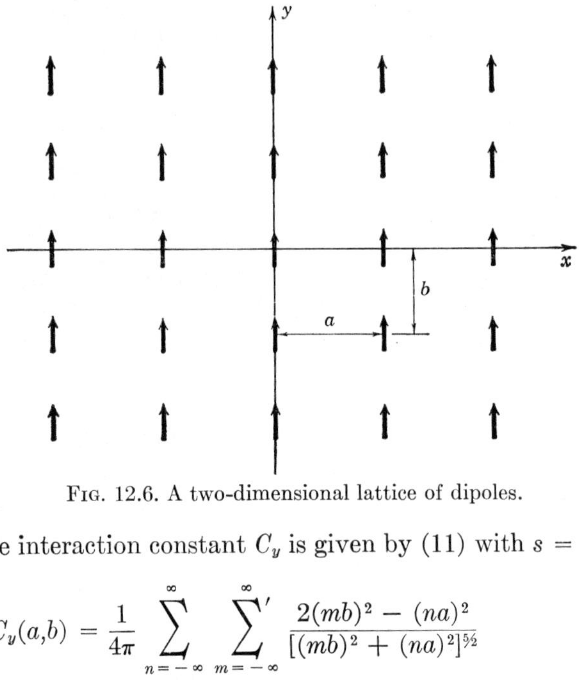


FIG. 12.6. A two-dimensional lattice of dipoles.

dipoles, the interaction constant $C_y$ is given by (11) with $s = 0$. Hence,

$$C_y(a,b) = \frac{1}{4\pi} \sum_{n=-\infty}^{\infty} {\sum_{m=-\infty}^{\infty}}' \frac{2(mb)^2 - (na)^2}{[(mb)^2 + (na)^2]^{5/2}}$$

$$= \frac{1}{\pi} \sum_{m=1}^{\infty} \frac{1}{(mb)^3} + \frac{1}{2\pi} \sum_{n=1}^{\infty} \sum_{m=-\infty}^{\infty} \frac{2(mb)^2 - (na)^2}{[(mb)^2 + (na)^2]^{5/2}} \tag{26}$$

Applying the Poisson summation formula gives

$$C_y(a,b) = \frac{1.2}{\pi b^3} - \frac{2}{\pi b} \sum_{n=1}^{\infty} \sum_{m=1}^{\infty} \left(\frac{2m\pi}{b}\right)^2 K_0\left(\frac{2m\pi na}{b}\right)$$

$$\approx \frac{1.2}{\pi b^3} - \frac{8\pi}{b^3} K_0\left(\frac{2\pi a}{b}\right) \tag{27}$$

For $x$-directed dipoles, the interaction constant $C_x(a,b)$ is given by (27) with $a$ and $b$ interchanged, i.e.,

$$C_x(a,b) = C_y(b,a) \tag{28}$$

For $z$-directed dipoles the interaction constant $C_z(a,b)$ is given by

$$C_z(a,b) = -\frac{1}{4\pi} \sum_{n=-\infty}^{\infty} \sum_{m=-\infty}^{\infty}{}' \frac{(mb)^2 + (na)^2}{[(mb)^2 + (na)^2]^{5/2}} \tag{29}$$

This series may be obtained by analogy with (11) by omitting the sum over $m$ in (11) and replacing $sc$ by $mb$. In place of (29), we may write

$$C_z(a,b) = -\frac{1}{4\pi} \sum_{n=-\infty}^{\infty} \left\{ \sum_{m=-\infty}^{\infty}{}' \frac{2(mb)^2 - (na)^2}{[(mb)^2 + (na)^2]^{5/2}} + \sum_{m=-\infty}^{\infty}{}' \frac{2(na)^2 - (mb)^2}{[(na)^2 + (mb)^2]^{5/2}} \right\}$$

Comparing with (26) gives

$$\begin{aligned} C_z(a,b) &= -C_y(b,a) - C_y(a,b) = -C_x(a,b) - C_y(a,b) \\ &= -\frac{1.2}{\pi b^3} - \frac{1.2}{\pi a^3} + \frac{8\pi}{b^3} K_0\left(\frac{2\pi a}{b}\right) + \frac{8\pi}{a^3} K_0\left(\frac{2\pi b}{a}\right) \end{aligned} \tag{30}$$

*Two-dimensional Lattice for Waveguide Aperture Coupling*

A small circular aperture in a transverse wall in a rectangular waveguide is equivalent to an $x$-directed magnetic dipole for $H_{10}$ mode excitation. The image of this dipole in the guide walls must be chosen so that

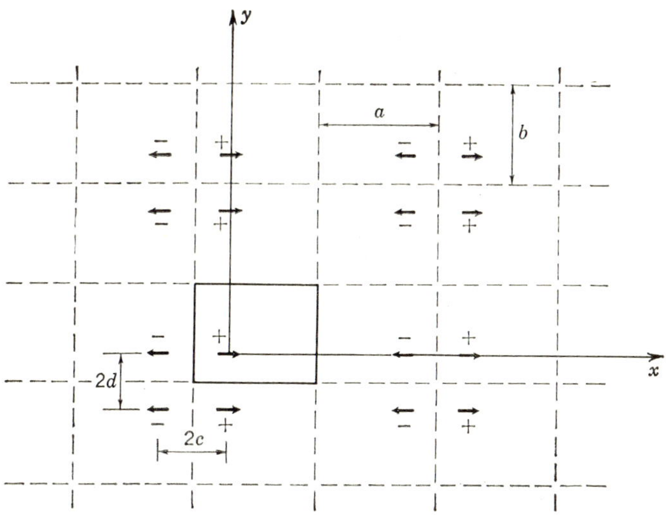


FIG. 12.7. Images of magnetic dipoles in waveguide walls.

the field produced has a vanishing normal component of magnetic field on the guide walls. The resulting lattice of dipoles is illustrated in Fig. 12.7. The positive $x$-directed dipoles are located at $x = 2na$, $y = 2mb$, and $x = 2na$, $y = 2mb - d$, where $n, m = 0, \pm 1, \pm 2, \ldots$. The negative-directed dipoles are located at $x = 2na - c$, $y = 2mb$, and $x = 2na - c$, $y = 2mb - d$. The $x$ component of magnetic field produced by a unit $x$-directed magnetic dipole at $(x_0,y_0,z_0)$ is given by

$$H_x = \frac{1}{4\pi} \frac{2(x - x_0)^2 - (y - y_0)^2 - (z - z_0)^2}{[(x - x_0)^2 + (y - y_0)^2 + (z - z_0)^2]^{5/2}}$$

The field at the origin due to all the image dipoles is the interaction field. For unit dipoles, this field is equal to the interaction constant $C_m$. Thus, we have

$$C_m = \frac{1}{4\pi} (S_1 + S_2 - S_3 - S_4) \tag{31}$$

where
$$S_1 = \sum_{n=-\infty}^{\infty} \sum_{m=-\infty}^{\infty}{}' \frac{2(2na)^2 - (2mb)^2}{[(2na)^2 + (2mb)^2]^{5/2}}$$

$$S_2 = \sum_{n=-\infty}^{\infty} \sum_{m=-\infty}^{\infty} \frac{2(2na)^2 - (2mb - d)^2}{[(2na)^2 + (2mb - d)^2]^{5/2}}$$

$$S_3 = \sum_{n=-\infty}^{\infty} \sum_{m=-\infty}^{\infty} \frac{2(2na - c)^2 - (2mb - d)^2}{[(2na - c)^2 + (2mb - d)^2]^{5/2}}$$

$$S_4 = \sum_{n=-\infty}^{\infty} \sum_{m=-\infty}^{\infty} \frac{2(2na - c)^2 - (2mb)^2}{[(2na - c)^2 + (2mb)^2]^{5/2}}$$

The series for $S_1$ and $S_2$ are of the type already summed and are equal to the following:

$$S_1 = \frac{1.201}{2a^3} - \frac{4}{a} \sum_{m=1}^{\infty} \sum_{n=1}^{\infty} \left(\frac{n\pi}{a}\right)^2 K_0\left(\frac{2n\pi mb}{a}\right) \tag{32a}$$

$$S_2 = -\frac{2}{a} \sum_{n=1}^{\infty} \left(\frac{n\pi}{a}\right)^2 K_0\left(\frac{n\pi d}{a}\right) - \frac{4}{a} \sum_{n=1}^{\infty} \sum_{m=1}^{\infty} \left(\frac{n\pi}{a}\right)^2 K_0\left(\frac{n\pi}{a} |2mb - d|\right) \tag{32b}$$

To sum $S_3$ and $S_4$, it is noted that, if $F(w)$ is the Fourier transform of $f(z)$, then the Fourier transform of $f(z + \tau)$ is $e^{-jw\tau}F(w)$. Hence, we obtain

$$S_3 = -\frac{1}{a} \sum_{m=-\infty}^{\infty} \sum_{n=-\infty}^{\infty} e^{jn\pi c/a} \left(\frac{n\pi}{a}\right)^2 K_0\left(\frac{|n|\pi}{a} |2mb - d|\right)$$

$$= -\frac{2}{a} \sum_{n=1}^{\infty} \left(\frac{n\pi}{a}\right)^2 \cos \frac{n\pi c}{a} K_0\left(\frac{n\pi d}{a}\right)$$

$$- \frac{4}{a} \sum_{m=1}^{\infty} \sum_{n=1}^{\infty} \left(\frac{n\pi}{a}\right)^2 \cos \frac{n\pi c}{a} K_0\left(\frac{n\pi}{a} |2mb - d|\right) \quad (33a)$$

The series $S_4$ is equal to $S_3$ with $d$ placed equal to zero. However, this makes the $m = 0$ term infinite, and so this term must be summed directly. The final result is

$$S_4 = 2 \sum_{n=-\infty}^{\infty} (2na - c)^{-3} - \frac{4}{a} \sum_{m=1}^{\infty} \sum_{n=1}^{\infty} \left(\frac{n\pi}{a}\right)^2 \cos \frac{n\pi c}{a} K_0\left(\frac{2n\pi b}{a}\right) \quad (33b)$$

In practice, only a few terms of the series are required for good accuracy.

When using the above theory for a circular aperture in a transverse wall of a rectangular guide, it must be kept in mind that it is essentially a static field theory and will yield accurate results only when the image apertures are spaced at a small distance compared with the wavelength. This is usually not the case except for an aperture close to the guide wall ($c$ or $d$ small). On the other hand, the effect of the image dipoles is small for small apertures not located too close to the guide wall. Thus, the theory provides a useful correction for those cases when the aperture is close to a guide wall, which is just the situation for which a correction is needed most.

## 12.4. Sphere-and Disk-type Artificial Dielectrics

In order to evaluate the index of refraction of an artificial dielectric by the Lorentz theory, the polarizability of the obstacle is required. For ellipsoids and the degenerate forms such as prolate and oblate spheroids, spheres, and circular and elliptical disks, the method presented in Sec. 7.3 may be used. A dielectric obstacle becomes equivalent to a conducting body when the permittivity $\epsilon$ is made to approach infinity. To evaluate the magnetic dipole moment, it is noted that, when $\mu$ is made to approach zero for a permeable body, it becomes a perfect diamagnetic body. The normal component of magnetic field at the surface must vanish for a perfect diamagnetic body, and this is the same boundary condition as required for the high-frequency magnetic field at the surface of a perfect conductor. The electric polarizability $\alpha_e$ and magnetic polarizability $\alpha_m$ for typical obstacles are given in Table 12.1.

TABLE 12.1. POLARIZABILITY OF ELLIPSOIDS AND THEIR DEGENERATE FORMS†

Ellipsoid

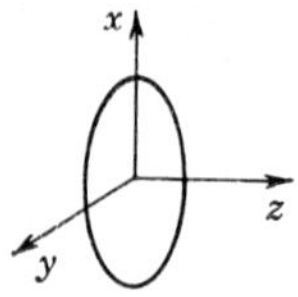


$l_1 > l_2 > l_3$

$$\alpha_{ex} = \frac{V}{L_1} \qquad \alpha_{mx} = \frac{V}{L_1 - 1}$$
$$\alpha_{ey} = \frac{V}{L_2} \qquad \alpha_{my} = \frac{V}{L_2 - 1}$$
$$\alpha_{ez} = \frac{V}{L_3} \qquad \alpha_{mz} = \frac{V}{L_3 - 1}$$

$$L_i = \frac{l_1 l_2 l_3}{2} \int_0^\infty (S + l_i^2)^{-1}[(S + l_1^2)(S + l_2^2)(S + l_3^2)]^{-1/2}\, dS$$
$$L_1 + L_2 + L_3 = 1$$

Prolate spheroid

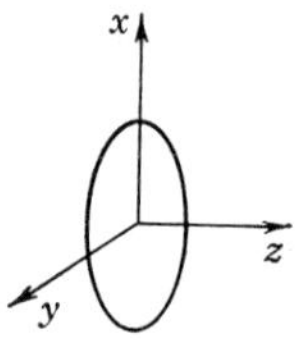


$l_1 > l_2 = l_3$

$$\alpha_{ex} = \frac{V}{L} \qquad \alpha_{mx} = \frac{-V}{1 - L}$$
$$\alpha_{ey} = \alpha_{ez} = \frac{2V}{1 - L} \qquad \alpha_{my} = \alpha_{mz} = \frac{-2V}{1 + L}$$

$$L = \frac{l_2^2}{2 l_1^2 e^3}\left(\ln \frac{1 + e}{1 - e} - 2e\right)$$

Oblate spheroid

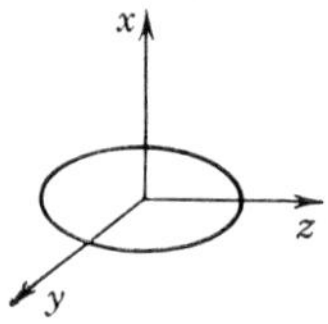


$l_1 < l_2 = l_3$

$$\alpha_{ex} = \frac{V}{L} \qquad \alpha_{mx} = \frac{-V}{1 - L}$$
$$\alpha_{ey} = \alpha_{ez} = \frac{2V}{1 - L} \qquad \alpha_{my} = \alpha_{mz} = \frac{-2V}{1 + L}$$

$$L = \frac{l_1 l_2^2}{(l_2^2 - l_1^2)^{3/2}}\left[\frac{(l_2^2 - l_1^2)^{1/2}}{l_1} - \tan^{-1}\frac{(l_2^2 - l_1^2)^{1/2}}{l_1}\right]$$

Sphere

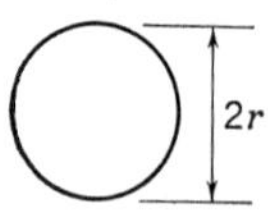


$l_1 = l_2 = l_3 = r$

$$\alpha_{ex} = \alpha_{ey} = \alpha_{ez} = 4\pi r^3 \qquad \alpha_{mx} = \alpha_{my} = \alpha_{mz} = -2\pi r^3$$

TABLE 12.1. POLARIZABILITY OF ELLIPSOIDS AND THEIR DEGENERATE FORMS† (*Continued*)

Elliptic disk

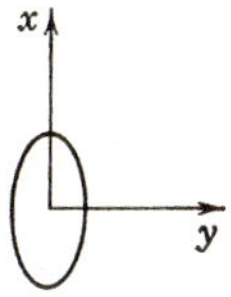


$l_1 > l_2$
$l_3 = 0$

$$\alpha_{ex} = \tfrac{4}{3}\pi l_1^3 e^2[K(e) - E(e)]^{-1}$$
$$\alpha_{ey} = \tfrac{4}{3}\pi l_1^3 e^2[(1 - e^2)^{-1}E(e) - K(e)]^{-1}$$
$$\alpha_{ez} = 0$$
$$\alpha_{mx} = \alpha_{my} = 0$$
$$\alpha_{mz} = -\frac{4\pi l_1^3(1 - e^2)}{3E(e)}$$

$K(e)$ and $E(e)$ are complete elliptic integrals of the first and second kind, respectively

Circular disk

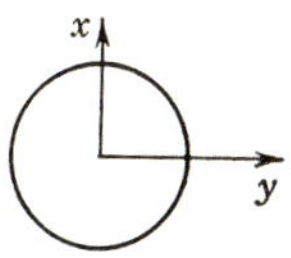


$l_1 = l_2 = r$
$l_3 = 0$

$$\alpha_{ex} = \alpha_{ey} = \tfrac{16}{3}r^3 \qquad \alpha_{mx} = \alpha_{my} = 0$$
$$\alpha_{ez} = 0 \qquad \alpha_{mz} = -\tfrac{8}{3}r^3$$

† Semiaxes are $l_1$, $l_2$, $l_3$ along the $x$, $y$, $z$ axes, respectively; eccentricity is

$$e = \left(\frac{1 - l_2^2}{l_1^2}\right)^{1/2}$$

volume $V = \tfrac{4}{3}\pi l_1 l_2 l_3$; $\alpha_{ex}$, $\alpha_{mx}$ = polarizability for fields applied along the $x$ axis; similarly for $\alpha_{ey}$, $\alpha_{my}$, $\alpha_{ez}$, and $\alpha_{mz}$.

Consider a cubic array of conducting spheres with radius $r$. Let the spacing be denoted by $a$. The number $N$ of spheres per unit volume is equal to $a^{-3}$. From the table, $\alpha_e = 4\pi r^3$, and $\alpha_m = -2\pi r^3$. The interaction constant $C$ is given by (19) as

$$C = \frac{1}{\pi a^3}[1.201 - 16\pi^2 K_0(2\pi)] = \frac{1.06}{\pi a^3}$$

From (7), the permittivity $\epsilon = \kappa\epsilon_0$ is found to be

$$\epsilon = \epsilon_0 \frac{1 + 8.4r^3/a^3}{1 - 4.24r^3/a^3}$$

For a TEM wave applied to the lattice, induced magnetic dipoles are also produced. The effective magnetic field acting to polarize the obstacle is the sum of the applied field $H_0$ and the interaction field $H_i$. By analogy with the electric case, we have

$$m = \alpha_m(H_0 + H_i)$$
$$H_i = Cm$$

and, hence,

$$m = \frac{\alpha_m H_0}{1 - \alpha_m C} \qquad (34)$$

The magnetic polarization per unit volume is $M = Nm$, and the permeability $\mu$ is given by

$$B = \mu H_0 = \mu_0(H_0 + M)$$

or

$$\mu = \mu_0 + \frac{\mu_0 N \alpha_m}{1 - \alpha_m C} \tag{35}$$

For the cubic array of spheres, we have

$$\mu = \mu_0 \frac{1 - 4.2r^3/a^3}{1 + 2.12r^3/a^3}$$

The effective index of refraction for the medium is

$$\eta = \left(\frac{\epsilon\mu}{\epsilon_0\mu_0}\right)^{1/2} = \left[\frac{(1 + 8.4r^3/a^3)(1 - 4.2r^3/a^3)}{(1 - 4.2r^3/a^3)(1 + 2.12r^3/a^3)}\right]^{1/2} \tag{36}$$

and is only moderately greater than unity because of the diamagnetic effect present. As an example, for $r = a/4$, we obtain $\eta = 1.048$.

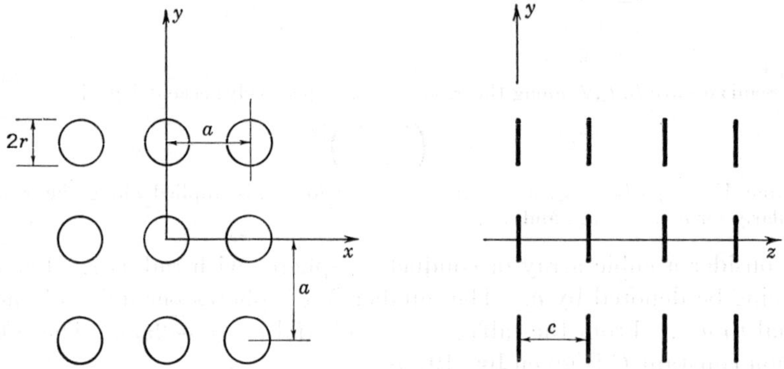


FIG. 12.8. Tetragonal array of circular disks.

The tetragonal array of circular disks illustrated in Fig. 12.8 represents a more useful medium because of the much greater index of refraction that can be obtained. The spacing of the disks in the $xy$ plane is $a$, and successive planes of disks are spaced a distance $c$ apart along the $z$ axis. A TEM wave is assumed to propagate through the lattice, along the $z$ direction. This field does not have a component of magnetic field along the $z$ axis, i.e., normal to the disk face, and, hence, there are no induced magnetic dipoles. The electric polarizability is $\alpha_e = 16r^3/3$, and, from (19), the interaction constant is found to be

$$C = \frac{1.2}{\pi a^3} - \frac{8\pi}{a^3}\left[K_0(2\pi) + K_0\left(\frac{2\pi c}{a}\right)\right]$$

The relative dielectric constant is given by (7) with $N = (a^2c)^{-1}$. The relative dielectric constant $\kappa$ is plotted as a function of $c/a$ for several values of $r/a$ in Fig. 12.9. The results of a low-frequency investigation carried out by Kharadly shows that the three-dimensional Lorentz theory gives accurate results for $c/a$ greater than about 0.8 only.† Brown has considered only the interaction between disks in a single plane. The interaction constant to be used is now that given by (27) and is equal to $0.36/a^3$. The resulting formula for $\kappa$ is

$$\kappa = 1 + \frac{8a}{c[1.5(a/r)^3 - 2.88]} \tag{37}$$

and has been found to be accurate for $c/a$ greater than 0.6.

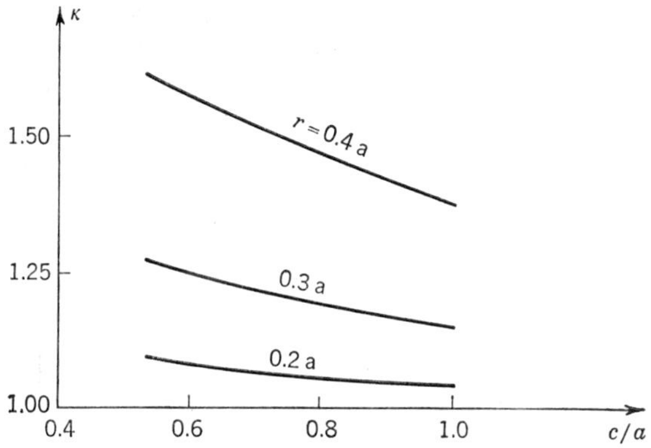


FIG. 12.9. Dielectric constant for tetragonal array of circular disks.

For small values of $c/a$, Brown has computed the dipole moment of a row of disks along the $z$ axis and used the two-dimensional form of the Clausius-Mossotti relation to obtain‡

$$\kappa = 1 + \frac{N\alpha_0}{1 - 0.5N\alpha_0} \tag{38}$$

where $N = a^{-2}$ and is the number of rows of disks per unit area, and $\alpha_0$ is the polarizability per unit length for a single row of disks and is given by

$$\alpha_0 = 2\pi r^2 \left[\left(1 - \frac{0.22c}{r}\right)^2 + \frac{0.007c^3}{r^3}\left(1 + \frac{0.22c}{r}\right)\right]$$

As $c$ approaches zero, $\alpha_0$ approaches $2\pi r^2$, which is the polarizability per unit length of a conducting cylinder. The combination of (37) for large values of $c/a$ and (38) for small values of $c/a$ provides good results for the whole range of $c/a$ values that are likely to be encountered in practice. It must be kept in mind, however, that the theory is a static field theory

† *Ibid.*

‡ J. Brown and W. Jackson, The Relative Permittivity of Tetragonal Arrays of Perfectly Conducting Thin Discs, *Proc. IEE (London)*, vol. 102, part B, pp. 37–42, January, 1955.

valid only for plane waves incident in a direction perpendicular to the plane of the disks, and with the spacing between disks less than about $0.1\lambda_0$.

## 12.5. Transmission-line Approach for Disk Medium

For a plane wave propagating through a three-dimensional array of disks, each plane of disks may be considered as a single discontinuity. For infinitely thin disks, this discontinuity is equivalent to a shunt susceptance $jB$ connected across a transmission line. The three-dimensional array of disks is, in turn, equivalent to a transmission line loaded at regular intervals by identical shunt susceptances. An evaluation of the propagation constant for this equivalent periodic-loaded transmission line leads to an expression for the equivalent index of refraction of the medium.

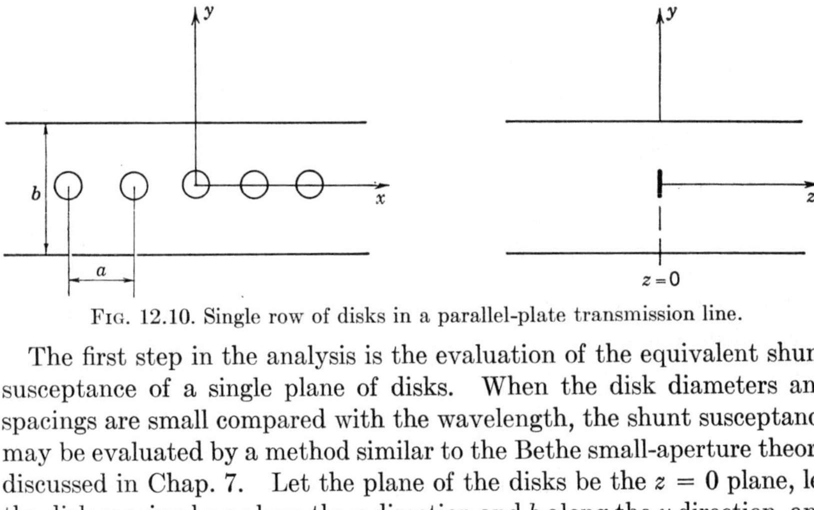


FIG. 12.10. Single row of disks in a parallel-plate transmission line.

The first step in the analysis is the evaluation of the equivalent shunt susceptance of a single plane of disks. When the disk diameters and spacings are small compared with the wavelength, the shunt susceptance may be evaluated by a method similar to the Bethe small-aperture theory discussed in Chap. 7. Let the plane of the disks be the $z = 0$ plane, let the disk spacing be $a$ along the $x$ direction and $b$ along the $y$ direction, and let the disk radius be $r$. The configuration is the same as in Fig. 12.8 with the exception of unequal spacings along the $x$ and $y$ directions. Let a plane wave with components $E_y$, $H_x$, $H_z$ be incident at an angle $\theta_i$ with respect to the $z$ axis. This incident field may be represented as follows:

$$\mathbf{E}_{inc} = a_0\mathbf{a}_y e^{-jhx-\Gamma_0 z} \tag{39a}$$

$$\mathbf{H}_{inc} = a_0(j\omega\mu_0)^{-1}\mathbf{a}_y \times \nabla e^{-jhx-\Gamma_0 z} \tag{39b}$$

where $h = k_0 \sin \theta_i$, $\Gamma_0 = (h^2 - k_0{}^2)^{1/2} = jk_0 \cos \theta_i$, and $a_0$ is an amplitude constant. A consideration of the nature of the incident field and the symmetry of the structure shows that conducting planes parallel with the $xz$ plane can be placed at $y = \pm b/2$. Hence we need consider only a single row of disks in a parallel-plate transmission line, as in Fig. 12.10.

For $z < 0$, there will be a dominant reflected mode given by

$$\mathbf{E}_{ref} = \mathbf{a}_y R a_0 e^{-jhx+\Gamma_0 z} \tag{40a}$$
$$\mathbf{H}_{ref} = R a_0 (j\omega\mu_0)^{-1} \mathbf{a}_y \times \nabla e^{-jhx+\Gamma_0 z} \tag{40b}$$

In addition, there will be a double infinite number of higher-order evanescent modes excited. These modes have the following $x$ and $z$ dependence:

$$e^{-j(h+2n\pi/a)x+\Gamma_{nm}z} \qquad n, m/2 = 0, \pm 1, \ldots$$

where $\Gamma_{nm}^2 = (h + 2n\pi/a)^2 + (m\pi/b)^2 - k_0^2$. The term $m\pi/b$ arises because of variation of the fields with $y$. All $\Gamma_{nm}$ are real for $n \neq 0$ when the spacing $a$ is less than one-half wavelength. Since we will not require expressions for these modes in order to evaluate the reflection coefficient $R$, we will simply denote them as follows:

$$\mathbf{E}_{nm}^{\mp} = \mathbf{e}_{nm}(x,y)e^{\pm\Gamma_{nm}z} \mp \mathbf{e}_{znm}(x,y)e^{\pm\Gamma_{nm}z} \tag{41a}$$
$$\mathbf{H}_{nm}^{\mp} = [\mp \mathbf{h}_{nm}(x,y) + \mathbf{h}_{znm}(x,y)]e^{\pm\Gamma_{nm}z} \tag{41b}$$

where $\mathbf{e}_{nm}$, $\mathbf{h}_{nm}$ are transverse vectors and $\mathbf{e}_{znm}$, $\mathbf{h}_{znm}$ are $z$-directed vectors. These vector functions all have the periodicity property

$$\mathbf{e}_{nm}(x + sa, y) = e^{-jhsa}\mathbf{e}_{nm}(x,y) \tag{42}$$

as required by Floquet's theorem. Because of the periodicity property, these modes form an orthogonal set over the range $sa < x \leq (s + 1)a$, where $s$ is any integer, and such that

$$\int_{-b/2}^{b/2} \int_{-a/2}^{a/2} \mathbf{e}_{nm} \times \mathbf{h}_{rs}^* \cdot \mathbf{a}_z \, dx\, dy = 0 \qquad n \neq r,\ m \neq s \tag{43}$$

where $\mathbf{h}_{rs}^*$ is the complex conjugate of $\mathbf{h}_{rs}$.

If we consider, first of all, odd excitation with incident fields for $z < 0$, as in (39), and with incident fields

$$\mathbf{E}_{inc} = -a_0 \mathbf{a}_y e^{-jhx+\Gamma_0 z}$$
$$\mathbf{H}_{inc} = -a_0 (j\omega\mu_0)^{-1} \mathbf{a}_y \times \nabla e^{-jhx+\Gamma_0 z}$$

for $z > 0$, the total transverse electric field in the aperture plane will vanish. The reflection coefficient in this case is simply $R_0 = -1$ (see discussion of Babinet's principle in Chap. 1).

For even excitation with incident fields chosen as

$$\mathbf{E}_{inc} = a_0 \mathbf{a}_y e^{-jhx+\Gamma_0 z}$$
$$\mathbf{H}_{inc} = a_0 (j\omega\mu_0)^{-1} \mathbf{a}_y \times \nabla e^{-jhx+\Gamma_0 z}$$

for $z > 0$, and as in (39) for $z < 0$, the total transverse magnetic field vanishes in the aperture plane $z = 0$ in the region between the disks. Let the reflection coefficient in this case be $R_e$. If we now superimpose the two solutions, the total incident field for $z > 0$ is zero, while that for $z < 0$ has an amplitude $2a_0$. The total reflected field for $z < 0$ has an amplitude $(R_e + R_0)a_0 = (R_e - 1)a_0 = R2a_0$, and, hence, the reflection coefficient $R$ is given by

$$R = \frac{R_e - 1}{2} \tag{44}$$

To find $R_e$, we must solve the problem of a row of conducting disks placed on a magnetic wall in the $z = 0$ plane as in Fig. 12.11. The field

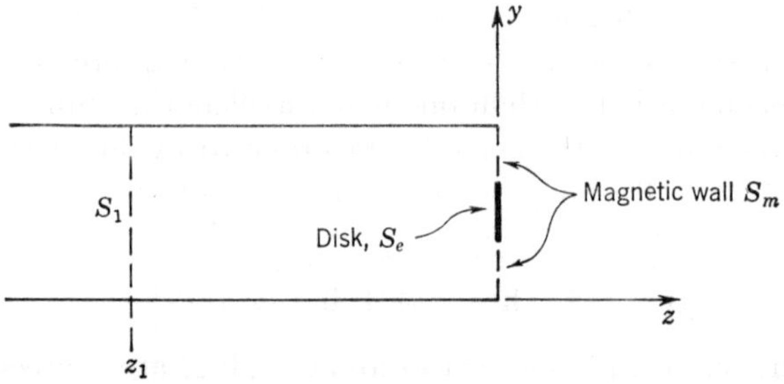


FIG. 12.11. Reduced boundary-value problem for even excitation. $S_m$ is magnetic wall; $S_e$ is electric wall (disk).

in the guide may be represented as a dominant-mode standing-wave field

$$\mathbf{E}_1 = a_0\mathbf{a}_y e^{-jhx} \cos \beta_0 z \tag{45a}$$
$$\mathbf{H}_1 = (j\omega\mu_0)^{-1} a_0 \mathbf{a}_y \times \nabla e^{-jhx} \cos \beta_0 z \tag{45b}$$

plus an additional dominant-mode reflected field

$$\mathbf{E}_2 = b_0 \mathbf{a}_y e^{-jhx+\Gamma_0 z} \tag{46a}$$
$$\mathbf{H}_2 = b_0 (j\omega\mu_0)^{-1} \mathbf{a}_y \times \nabla e^{-jhx+\Gamma_0 z} \tag{46b}$$

together with an infinite number of evanescent modes propagating in the negative $z$ direction:

$$\mathbf{E}_3 = \sum_{n,m} b_{nm} \mathbf{E}_{nm}^- \tag{47a}$$

$$\mathbf{H}_3 = \sum_{n,m} b_{nm} \mathbf{H}_{nm}^- \tag{47b}$$

In the above, $\beta_0 = |\Gamma_0|$, and the standing-wave field has a vanishing transverse magnetic field at $z = 0$ and, hence, satisfies the boundary conditions on the magnetic wall. The field $\mathbf{E}_1$, $\mathbf{H}_1$ is considered as the incident field, and the field $\mathbf{E}_2 + \mathbf{E}_3 = \mathbf{E}_s$, $\mathbf{H}_2 + \mathbf{H}_3 = \mathbf{H}_s$ as the total scattered field.

Let $\mathbf{E}'_{nm}$, $\mathbf{H}'_{nm}$ be a normal-mode standing-wave field:

$$\mathbf{E}'_{nm} = \mathbf{e}^*_{nm}(e^{-\Gamma_{nm}z} + e^{\Gamma_{nm}z}) + \mathbf{e}^*_{znm}(e^{-\Gamma_{nm}z} - e^{\Gamma_{nm}z}) \tag{48a}$$
$$\mathbf{H}'_{nm} = \mathbf{h}^*_{nm}(e^{-\Gamma_{nm}z} - e^{\Gamma_{nm}z}) + \mathbf{h}^*_{znm}(e^{-\Gamma_{nm}z} + e^{\Gamma_{nm}z}) \tag{48b}$$

The Lorentz reciprocity theorem gives

$$\oiint_S (\mathbf{E}'_{nm} \times \mathbf{H}_s - \mathbf{E}_s \times \mathbf{H}'_{nm}) \cdot \mathbf{n}\, da = 0$$

where $S$ is the closed surface consisting of cross-sectional planes at $z = 0$ and $z = z_1$, the conducting planes at $y = \pm b/2$, and planes parallel to the $yz$ plane at $x = \pm a/2$. The integral clearly vanishes on the conducting planes at $y = \pm b/2$. On the planes at $x = \pm a/2$ the integral will also vanish, since the field $\mathbf{E}_s$, $\mathbf{H}_s$ satisfies the periodicity condition (42), while $\mathbf{E}'_{nm}(x + sa, y, z) = \mathbf{E}'_{nm}(x,y,z)e^{jhsa}$, and similarly for $\mathbf{H}'_{nm}$; and the normal $\mathbf{n}$ is directed in the opposite sense on the two planes. On the cross-sectional plane at $z = z_1$, the integral reduces to

$$-2b_{nm} \int_{-b/2}^{b/2} \int_{-a/2}^{a/2} \mathbf{e}^*_{nm} \times \mathbf{h}_{nm} \cdot \mathbf{a}_z\, dx\, dy$$

provided $\mathbf{n}$ is chosen as the inward normal. On the cross-sectional plane at $z = 0$, we have

$$\mathbf{n} \times \mathbf{H}'_{nm} = 0 \qquad \mathbf{n} \times \mathbf{H}_s = 0 \qquad \text{on } S_m$$

since $\mathbf{n} \times \mathbf{H}_1 = 0$ on both $S_m$ and $S_e$. On $S_e$, we have $\mathbf{n} \times \mathbf{H}_s = \mathbf{J}$, where $\mathbf{n}$ is the normal pointing away from the disk and $\mathbf{J}$ is the current on the disk or electric wall $S_e$. Using these results, the Lorentz reciprocity principle is found to give

$$b_{nm} \int_{-b/2}^{b/2} \int_{-a/2}^{a/2} \mathbf{e}^*_{nm} \times \mathbf{h}_{nm} \cdot \mathbf{a}_z\, dx\, dy = -\iint_{S_e} \mathbf{e}^*_{nm} \cdot \mathbf{J}\, dx\, dy \tag{49}$$

an equation which permits the unknown amplitude coefficients $b_{nm}$ for the scattered field to be evaluated. In particular, if we choose for the field $\mathbf{E}'_{nm}$, $\mathbf{H}'_{nm}$ a dominant-mode standing wave

$$\begin{aligned}\mathbf{E}'_{00} &= \mathbf{a}_y e^{jhx}(e^{-\Gamma_0 z} + e^{\Gamma_0 z})\\ \mathbf{H}'_{00} &= (j\omega\mu_0)^{-1}\mathbf{a}_y \times \nabla e^{jhx}(e^{-\Gamma_0 z} + e^{\Gamma_0 z})\end{aligned}$$

we obtain

$$b_0 \int_{-b/2}^{b/2} \int_{-a/2}^{a/2} (j\omega\mu_0)^{-1}\Gamma_0\, dx\, dy = -\iint_{S_e} \mathbf{a}_y e^{jhx} \cdot \mathbf{J}\, dx\, dy \tag{50}$$

The integral on the right-hand side of (49) and (50) may be converted into a form representing coupling with an electric dipole $\mathbf{P}$ and a magnetic dipole $\mathbf{M}$, in a manner similar to that used for the aperture problem. We obtain

$$\iint_{S_e} \mathbf{E}'_{nm} \cdot \mathbf{J}\, dx\, dy = j\omega \mathbf{E}'_{nm}(0) \cdot \mathbf{P} - j\omega\mu_0 \mathbf{H}'_{nm}(0) \cdot \mathbf{M} \tag{51}$$

where the fields are evaluated at the center of the disk, and the dipole moments are given by

$$j\omega \mathbf{P} = \iint_{S_e} \mathbf{J}\, dx\, dy \tag{52a}$$

$$\mathbf{M} = \frac{1}{2} \iint_{S_e} \mathbf{r} \times \mathbf{J}\, dx\, dy \tag{52b}$$

The magnetic dipole is in the $z$ direction, while the electric dipole lies in the $xy$ plane. Using these results in (50), we obtain

$$b_0 = -\frac{\omega\mu_0}{\beta_0 ab}(j\omega \mathbf{a}_y \cdot \mathbf{P} + jh\mathbf{a}_z \cdot \mathbf{M}) \tag{53}$$

The dipole moments **P** and **M** in (52) and (53) are those of one-half of the disk, since the currents flowing on one side only are involved. For small disks with large spacings, the interaction field is negligible, and **P** and **M** may be evaluated in terms of the polarizabilities of the disk and the strength of the incident field at the center of the disk. The electric and magnetic polarizabilities of one-half of a circular disk of radius $r$ are respectively

$$\alpha_e' = \tfrac{8}{3} r^3$$
$$\alpha_m' = -\tfrac{4}{3} r^3$$

The field given by (45) may be considered as the field acting to polarize the disk. For this field, we obtain

$$\mathbf{P} = \tfrac{8}{3} r^3 \epsilon_0 a_0 \mathbf{a}_y$$

$$\mathbf{M} = -\tfrac{4}{3} r^3 \frac{h}{\omega\mu_0} a_0 \mathbf{a}_z$$

and, hence,

$$b_0 = \frac{jh^2}{ab\beta_0}\frac{4}{3} r^3 a_0 - \frac{jk_0^2}{ab\beta_0}\frac{8}{3} r^3 a_0 \tag{54}$$

The total reflected field has an amplitude $b_0 + a_0/2$, while the incident field has an amplitude $a_0/2$, as may be seen by decomposing the standing-wave field (45) into an incident and a reflected field. The reflection coefficient $R_e$ is thus given by

$$R_e = \frac{a_0 + 2b_0}{a_0} = 1 + \frac{2b_0}{a_0}$$

and, hence, the reflection coefficient $R$ is, from (44),

$$R = \frac{b_0}{a_0}$$

A shunt susceptance $jB$ on a transmission line produces a reflection coefficient

$$R = \frac{1 - jB - 1}{2 + jB} = \frac{-jB}{2 + jB} \approx \frac{-jB}{2}$$

when $B$ is very small. From (54) we now get

$$B = \frac{8r^3}{3ab\beta_0} (2k_0^2 - h^2) = B_c - B_L \tag{55}$$

where the capacitive part of $B$ is given by

$$B_c = \frac{16k_0^2r^3}{3ab\beta_0} \tag{56a}$$

while the inductive part is given by

$$B_L = \frac{8h^2r^3}{3ab\beta_0} \tag{56b}$$

The capacitive part arises from the electric polarization of the disk, while the inductive part arises from the magnetic dipole polarization. When the angle of incidence $\theta_i = 0$, then $h = 0$, and the inductive part $B_L$ vanishes. For this special case,

$$B_c = \frac{16Z_0}{3ab} \omega\epsilon_0 r^3$$

where $Z_0 = (\mu_0/\epsilon_0)^{1/2}$, and is seen to have the usual frequency dependence associated with a capacitive susceptance.

More accurate results are obtained if the interaction field is also taken into account as part of the field acting to polarize the disk. This interaction field arises mainly from the nearest neighbors. If the disk spacings $a$ and $b$ are small compared with a wavelength, the phase of the incident wave is essentially constant over the region of several disks. Under these conditions, the interaction field may be approximated by a static interaction field. The electric and magnetic dipole moments are thus increased by factors $(1 - \alpha_e C_e)^{-1}$ and $(1 - \alpha_m C_m)^{-1}$, respectively, where $C_e$ and $C_m$ are the interaction constants for the electric and magnetic cases. From (27) and (30), we have

$$C_e = \frac{1.2}{\pi b^3} - \frac{8\pi}{b^3} K_0 \left(\frac{2\pi a}{b}\right) \tag{57a}$$

$$C_m = -\frac{1.2}{\pi b^3} - \frac{1.2}{\pi a^3} + \frac{8\pi}{b^3} K_0 \left(\frac{2\pi a}{b}\right) + \frac{8\pi}{a^3} K_0 \left(\frac{2\pi b}{a}\right) \tag{57b}$$

The capacitive and inductive parts of the shunt susceptance $B$ must now be replaced by the following:

$$B_c = \frac{16k_0^2r^3}{3ab\beta_0(1 - \alpha_e C_e)} \tag{58a}$$

$$B_L = \frac{8h^2r^3}{3ab\beta_0(1 - \alpha_m C_m)} \tag{58b}$$

where $\alpha_e = 16r^3/3$, $\alpha_m = -8r^3/3$.

Having determined the shunt susceptance of a single plane of disks, we may now replace the artificial dielectric medium by a transmission line having a propagation phase constant $\beta_0$ and loaded at regular intervals by shunt susceptances $B$, as in Fig. 12.12. Periodic structures of

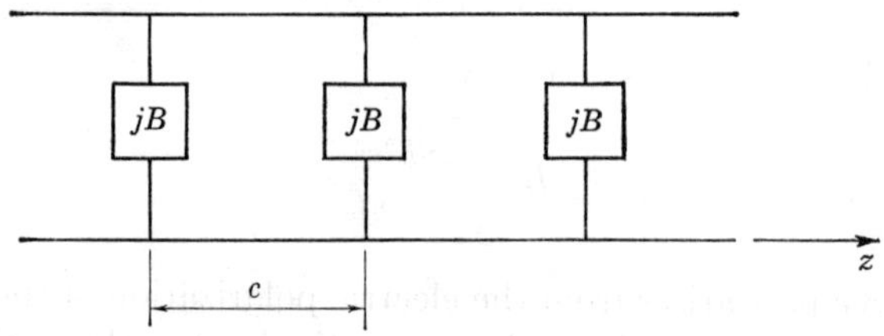


FIG. 12.12. Equivalent transmission-line circuit for three-dimensional disk medium.

this type were analyzed in Chap. 9, where it was found that the propagation phase constant $\beta_c$ of the periodic wave is given by

$$\cos \beta_c c = \cos \beta_0 c - \frac{B}{2} \sin \beta_0 c \tag{59}$$

Equation (59) neglects higher-order mode interaction between successive planes of disks. The least attenuated evanescent mode has a propagation constant

$$\Gamma_{-10} = \left[\left(\frac{\pi}{a} - h\right)^2 - k_0^2\right]^{1/2}$$

Provided $e^{-\Gamma_{-10}c} < 0.1$, interaction with higher-order modes may be neglected. For $\beta_0 c$ small, Eq. (59) gives

$$\beta_c^2 = \beta_0^2 + \frac{\beta_0 B}{c} \tag{60}$$

upon using the small-argument approximations for the trigonometric functions.

According to the Clausius-Mossotti theory, the disk-type artificial dielectric medium has a dielectric constant different from unity in the $x$ and $y$ directions, and a permeability $\mu$ different from $\mu_0$ in the $z$ direction only. Let the permittivity in the $y$ direction be $\kappa\epsilon_0$. In an anisotropic medium of this type, the solution for a plane wave with components $E_y$, $H_x$, $H_z$ is readily found to be

$$\begin{aligned} E_y &= e^{-jhx-j\beta z} \\ j\omega\mu_0 H_x &= -j\beta E_y \\ j\omega\mu H_z &= jhE_y \end{aligned}$$

and $\beta^2 = \kappa k_0^2 - (\mu_0/\mu)h^2$, since $E_y$ is a solution of

$$\frac{\partial^2 E_y}{\partial z^2} + \frac{\mu_0}{\mu}\frac{\partial^2 E_y}{\partial x^2} + \kappa k_0^2 E_y = 0$$

If we now identify $\beta$ with $\beta_c$, we obtain from (60)

$$\beta_c^2 - \beta_0^2 = \beta^2 - \beta_0^2 = (\kappa - 1)k_0^2 - \left(\frac{\mu_0}{\mu} - 1\right)h^2 = \frac{\beta_0 B}{c}$$

where $\beta_0^2 = k_0^2 - h^2$. Substituting for $B$ and equating the capacitive and inductive terms separately gives the following solutions for $\kappa$ and $\mu$:

$$\kappa = 1 + \frac{16r^3}{3abc(1 - \alpha_e C_e)} \tag{61a}$$

$$\frac{\mu}{\mu_0} = \left[1 + \frac{8r^3}{3abc(1 - \alpha_m C_m)}\right]^{-1} \approx 1 - \frac{8r^3}{3abc(1 - \alpha_m C_m)} \tag{61b}$$

These results are precisely those given by the Clausius-Mossotti theory when interaction between disks in a single plane only is taken into account [see (37)]. The validity of this theory has thus been demonstrated under the conditions that the spacing between disks is small. However, (59) is more general since it is valid for large values of $c$ as well. As noted by Brown, the theory breaks down for $c/a$ less than about 0.6 for plane waves incident normally. The reason is clear from the above transmission-line analysis, since this corresponds to a situation where higher-order mode interaction between successive rows of disks can no longer be neglected.

At the shorter wavelengths, it is desirable to use disks with diameters and spacings that are an appreciable fraction of a wavelength because of the easier construction afforded and greater economy in the number of elements required. The preceding theory is not applicable in this case for several reasons, such as:

1. The incident field varies over the region of a disk sufficiently so that the polarizing field cannot be considered constant.
2. The interaction field is not well approximated by the static field. In actual fact, the radiation field from each disk will be more important than the local induction field varying as the inverse third power of the distance.

A more satisfactory theory for this general case has recently been developed. It consists of a higher-order approximation to the dipole moments of a circular disk, and hence a more accurate value for $B$, together with the use of dynamic-field-interaction constants.† The determination of the latter is outlined in Probs. 12.7 and 12.8. The more accurate expressions for the polarizabilities of a circular disk (due to Eggimann) are

$$\alpha_e = \frac{16}{3} r^3 \left[ 1 + \frac{(k_0 r)^2}{15} (8 - 1.5 \sin^2 \theta_i) \right]$$

$$\alpha_m = -\frac{8}{3} r^3 \left[ 1 - \frac{(k_0 r)^2}{10} (2 + \sin^2 \theta_i) \right]$$

## 12.6. Two-dimensional Strip Medium

Figure 12.13 illustrates a two-dimensional conducting-strip artificial dielectric medium. The strips are infinitely long in the $y$ direction, and

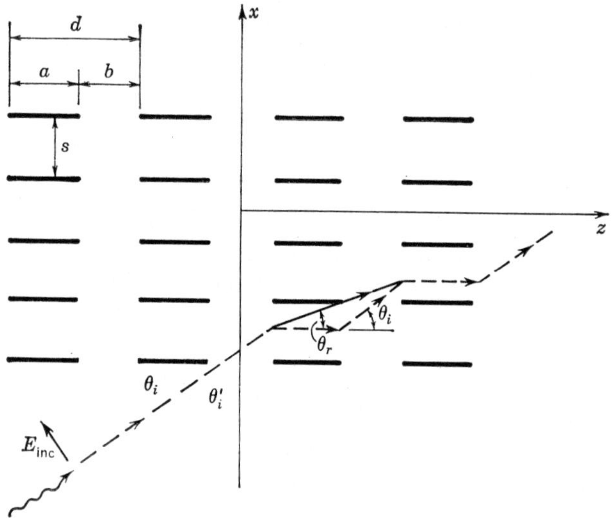


FIG. 12.13. Two-dimensional strip-type artificial dielectric medium.

have negligible thickness. The width of the strips is $a$, and the spacings along the $x$ and $z$ directions are $s$ and $d$, respectively.

For a parallel-polarized wave propagating obliquely through the medium, with the plane of propagation in the $xz$ plane, the medium exhibits anisotropic dielectric properties. This is readily seen to be so, since, for an electric field applied entirely along the $x$ direction, the strips produce no perturbation, and, hence, the dielectric constant in the $x$

† R. E. Collin and W. Eggimann, Evaluation of Dynamic Interaction Fields in a Two Dimensional Lattice, *Case Inst. Technol. Sci. Rept.* 12, March, 1960. See also *Sci. Repts.* 13–15 issued on contract AF-19(604) 3887.

direction is unity. For a field applied along the $z$ direction, a maximum perturbation is produced, and the effective dielectric constant in this direction will be quite large. No magnetic polarization is produced, since there is no component of magnetic field normal to the face of the strips. The medium is a phase-delay medium. For the opposite polarization, the medium exhibits magnetic anisotropy. The index of refraction will be less than unity because of the inductive loading, and, hence, the medium functions as a phase-advance dielectric for this other polarization.

A low-frequency solution may be obtained by using the Clausius-Mossotti theory and the dipole moment of the strips per unit length.

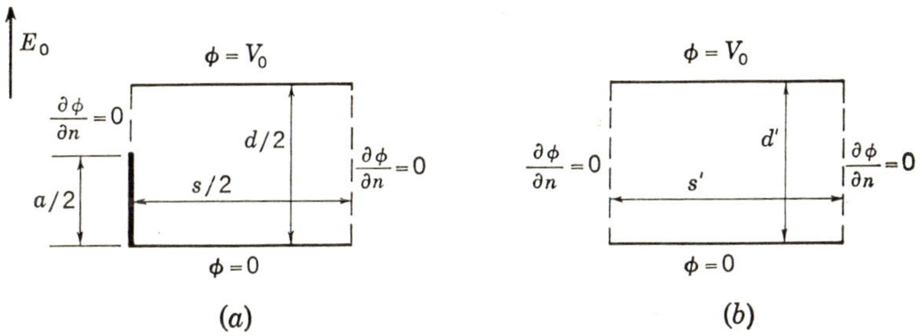


FIG. 12.14. (a) Reduced electrostatic problem for strip medium; (b) conformal mapping of boundary in (a).

A more accurate result is, however, obtained by solving the electrostatic problem with a static field applied along the $z$ direction. The problem may be reduced to that illustrated in Fig. 12.14a by a consideration of the symmetry involved. By a suitable conformal mapping, the configuration in Fig. 12.14b may be obtained. When the strip width $a$ is zero, the capacitance of a unit quarter cell is

$$C_0 = \frac{\epsilon_0 s}{d}$$

When $a \neq 0$, the capacitance of a unit quarter cell is the same as that of the cell illustrated in Fig. 12.14b, and is

$$C_0' = \frac{\epsilon_0 s'}{d'}$$

The effective dielectric constant $\kappa$ in the $z$ direction is, thus,

$$\kappa = \frac{C_0'}{C_0} = \frac{s'd}{sd'}$$

The conformal-mapping solution has been given by Howes and Whitehead and has also been derived independently by Kolettis.† The solution is transcendental in nature, and is given by

$$\kappa = \frac{K(k_1)K(k')}{K(k_1')K(k)} = \frac{2d}{s}\frac{K(k_1)}{K(k_1')} \tag{62}$$

where $K$ is the complete elliptic integral of the first kind, and $k'$, $k_1'$ are the complementary moduli $(1 - k^2)^{1/2}$, $(1 - k_1{}^2)^{1/2}$. The modulus $k_1$ is given by

$$k_1 = \frac{1 + k^2x}{k(1 + x)} - \left\{\left[\frac{1 + k^2x}{k(1 + x)}\right]^2 - 1\right\}^{1/2}$$

and $x$ is a solution of

$$\frac{a}{d}K(k') = \int_0^{(x^2-1)^{1/2}/k'x} \frac{d\lambda}{[(1 - \lambda^2)(1 - k'^2\lambda^2)]^{1/2}}$$

with $k$ and $k'$ determined so that

$$\frac{K(k')}{K(k)} = \frac{2d}{s}$$

When $s$ is small compared with $d$, the solution simplifies to

$$\kappa = \frac{d}{b + (s/\pi)\ln 4} \tag{63}$$

When the spacings $s$ and $d$ are not small compared with a wavelength, a solution based on Maxwell's equations must be used. The starting point for a solution is the determination of the reflection and transmission coefficients of a single interface. When these are known, the artificial dielectric medium may be described in terms of a periodic cascade connection of equivalent circuits. An analysis of this periodic network yields the characteristic propagation phase constant from which the equivalent dielectric constant of the medium may be deduced. Two approaches are possible. For one approach, the $x$ axis is considered as the direction of propagation. The reflection and transmission coefficients required are those for a plane strip grating in free space. When the spacing $s$ is large compared with $d$, this approach is useful, since higher-order mode interaction between adjacent gratings is negligible, and only the reflection and transmission coefficients for the dominant TEM mode are required. These may be found from the equivalent capacitive shunt susceptance of the grating.

† J. Howes and E. A. N. Whitehead, The Refractive Index of a Dielectric Loaded with Thin Metal Strips, *Elliot Bros. Research Rept.* 119, July, 1949.

N. Kolettis, Conformal Mapping Solution for Equivalent Relative Permittivity of a Strip Artificial Dielectric Medium, *Case Inst. Technol. Sci. Rept.* 2, January, 1959; work performed under contract AF-19(604) 3887.

By considering propagation to take place along the $z$ direction, the reflection and transmission coefficients required are those for the interface between free space and an infinite array of equispaced parallel plates. This problem has been rigorously solved in Chap. 10. This second approach is well suited to the case when $s$ is small compared with $d$, since, under these conditions, the evanescent modes decay rapidly, and there is negligible higher-order mode interaction between adjacent interfaces. A combination of the two approaches yields accurate results for the equivalent dielectric constant for all values of $s$ and $d$ of interest in practice.

The case of propagation through the medium in the $x$ direction, $\theta_i' = 0$, has been examined by Cohn and Brown.† For this special case, no anisotropic effects exist. The more general case of oblique propagation through the medium has been analyzed by the present author.‡ When $\theta_i' = 0$, the shunt susceptance of the strip grating is given by the formulas presented in Chap. 8 for a capacitive diaphragm in a parallel-plate transmission line. For oblique angles of incidence, the normalized shunt susceptance is given with reasonable accuracy by§

$$B = \frac{4d\cos\theta_i'}{\lambda_0}\left\{\ln\csc\frac{\pi b}{2d} + \frac{1}{2}\frac{(1-\alpha^2)^2[(1-\alpha^2/4)(A_+ + A_-) + 4\alpha^2 A_+A_-]}{(1-\alpha^2/4)+\alpha^2(1+\alpha^2/2-\alpha^4/8)(A_+ + A_-) + 2\alpha^6 A_+A_-}\right\} \tag{64}$$

where $\alpha = \sin \pi b/2d$, and

$$A_\pm = \left[1 \pm \frac{2d}{\lambda_0}\sin\theta_i' - \left(\frac{d\cos\theta_i'}{\lambda_0}\right)^2\right]^{-1} - 1$$

The equivalent circuit for propagation in the $x$ direction is illustrated in Fig. 12.15. The propagation phase constant for the transmission line is $k_0 \cos\theta_i'$, and the characteristic phase constant $\beta_c'$ for the periodic wave propagating in the $x$ direction is determined by a solution of

$$\cos\beta_c's = \cos(k_0 s\cos\theta_i') - \frac{B}{2}\sin(k_0 s\cos\theta_i') \tag{65}$$

Numerical data computed from (65) are presented in the section follow-

† S. B. Cohn, Analysis of the Metal Strip Delay Structure for Microwave Lenses, *J. Appl. Phys.*, vol. 20, pp. 257–262, March, 1949.

J. Brown, The Design of Metallic Delay Dielectrics, *Proc. IEE (London)*, vol. 97, part III, pp. 45–48, January, 1950.

‡ *Case Inst. Technol. Sci. Repts.*, no. 1, November, 1958; no. 4, June, 1959. Work supported by Air Force Cambridge Research Center Contract AF 19(604) 3887.

§ N. Marcuvitz, "Waveguide Handbook," vol. 10 of MIT Rad. Lab. Series, sec. 5.18, eq. (1*a*), McGraw-Hill Book Company, Inc., New York, 1951.

ing the solution based on an expansion in terms of modes propagating in the $z$ direction.

The reflection and transmission coefficients at the interface between free space and a set of parallel plates are given in Chap. 10, Eqs. (74), (75), and (76). These results show that the interface may be represented

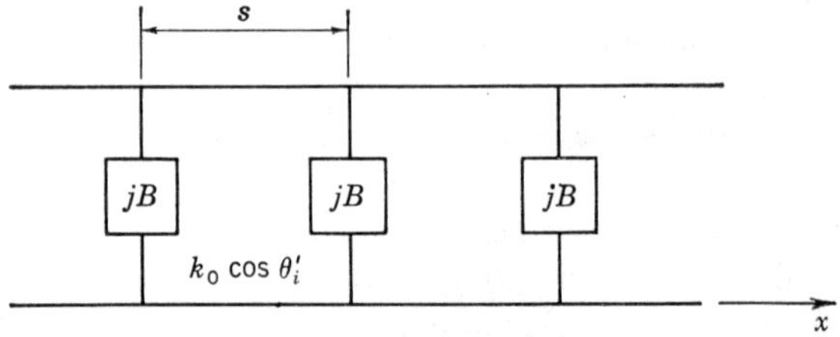


FIG. 12.15. Equivalent circuit for strip-type artificial dielectric for modes propagating in the $x$ direction.

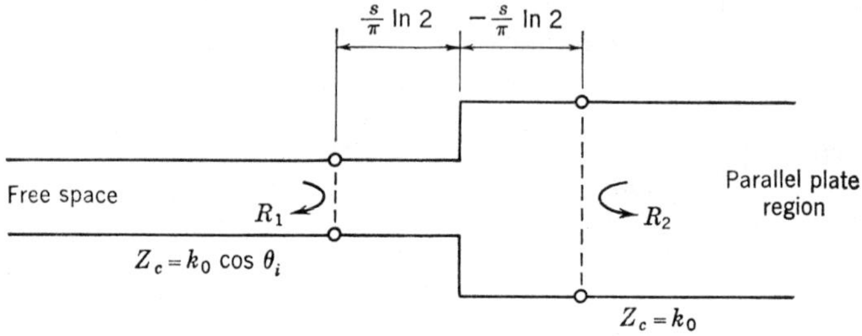


FIG. 12.16. Equivalent circuit of parallel-plate interface.

by the equivalent circuit illustrated in Fig. 12.16 when $s$ is small. From Fig. 12.16, it is seen that

$$R_1 = \frac{1 - \cos \theta_i}{1 + \cos \theta_i} \exp \left[ -jk_0 \left( \frac{s}{\pi} \ln 4 \right) \cos \theta_i \right]$$

$$R_2 = \frac{\cos \theta_i - 1}{\cos \theta_i + 1} \exp \left( jk_0 \frac{s}{\pi} \ln 4 \right)$$

which are the same results as given in Chap. 10 for $s$ small. The equivalent circuit for the artificial dielectric is readily constructed, and is illustrated in Fig. 12.17. An analysis of this periodic structure shows that the characteristic phase constant $\beta_c$ for a periodic wave propagating through the medium in the $z$ direction is a solution of the following equation:

$$(1 - \rho^2) \cos \beta_c d = \cos \left[ k_0(a + b \cos \theta_i) - k_0 \frac{s}{\pi} (\ln 4)(1 - \cos \theta_i) \right]$$
$$- \rho^2 \cos \left[ k_0(a - b \cos \theta_i) - k_0 \frac{s}{\pi} (\ln 4)(1 + \cos \theta_i) \right] \quad (66)$$

where $\rho = (1 - \cos\theta_i)/(1 + \cos\theta_i)$, and $\theta_i$ is the angle of propagation for the TEM wave in the region between strips, as in Fig. 12.13.

In order to obtain an expression for the equivalent dielectric constant of the strip medium, propagation through the artificial dielectric medium must be correlated with propagation through a uniform medium having a dielectric constant $\kappa$ in the $z$ direction and unity in the $x$ and $y$ directions. Such a medium is known as a uniaxial medium, and has a single optic axis in the $z$ direction. Propagation of a TEM wave in such a medium is

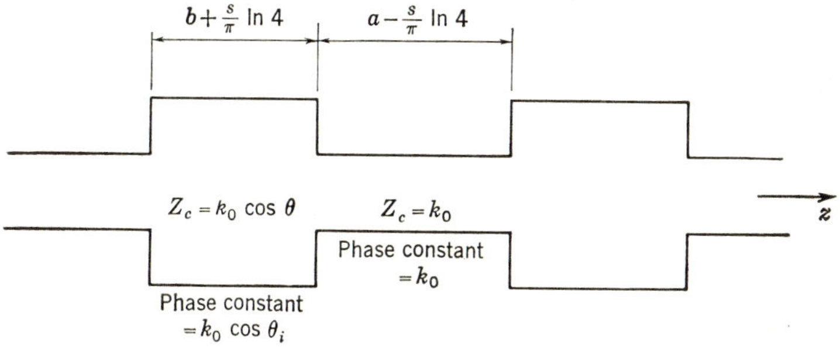


FIG. 12.17. Equivalent circuit for strip-type artificial dielectric for modes propagating in the $z$ direction.

discussed in Sec. 3.7. For the extraordinary wave with a propagation factor $e^{-j\beta\mathbf{n}\cdot\mathbf{r}}$, it was found that

$$\beta^2 = \frac{\kappa k_0^2}{n_x^2 + \kappa n_z^2} \tag{67}$$

where $n_x$ and $n_z$ are the components of the wave normal. The angle of propagation through the medium is given by

$$\tan\theta_r = \frac{n_x}{n_z} \tag{68}$$

where $\theta_r$ is measured relative to the $z$ axis. For the solution given by (66), we have

$$\beta_c = \beta n_z \qquad \text{and} \qquad k_0 \sin\theta_i = \beta n_x$$

Substituting into (67) and (68), we get

$$\kappa = \frac{k_0^2 \sin^2\theta_i}{k_0^2 - \beta_c^2} \tag{69}$$

$$\tan\theta_r = \frac{k_0 \sin\theta_i}{\beta_c} \tag{70}$$

Equation (70) gives the effective angle of propagation through the medium.

Certain limiting solutions for $\kappa$ may be obtained from (66). When $d < 0.1\lambda_0$, the cosine terms may be expanded according to the relation $\cos u = 1 - u^2/2$, and (66) together with (69) gives

$$\kappa = \frac{d}{b + (s/\pi) \ln 4}$$

which is the same as the static solution given by (63) for $s \ll d$. For this case, $\kappa$ is independent of the effective angle of propagation $\theta_r$ through the medium. For larger values of $d$, it is found that $\kappa$ varies with the angle of propagation through the medium. This effect may be considered as caused by the graininess of the lattice structure. As $\theta_i$ approaches zero, $\rho$ tends to zero, and the limiting value of $\kappa$ is again the above static value, independent of the condition $d < 0.1\lambda_0$. This limiting solution is obtained by placing $\rho$ equal to zero in (66) and solving for $\beta_c d$ to get

$$(\beta_c d)^2 = (k_0 d)^2 \left[1 - (1 - \cos \theta_i) \left(\frac{b}{d} + \frac{s}{\pi d} \ln 4\right)\right]^2$$

Expanding by the binomial expansion and substituting into (69) now gives the end result. When $\theta_i = \pi/2$, the eigenvalue equation (66) reduces to

$$\cos \beta_c d = \cos k_0 \left(a - \frac{s}{\pi} \ln 4\right) - \frac{k_0}{2} \left(b + \frac{s}{\pi} \ln 4\right) \sin k_0 \left(a - \frac{s}{\pi} \ln 4\right) \tag{71}$$

Although $\theta_i = \pi/2$, the effective angle of propagation $\theta_r$ through the medium is less than $\pi/2$. When $\theta_r = \pi/2$, $\beta_c d = 0$, and $\theta_i$ is complex with $\sin \theta_i$ greater than unity.

Returning to the first solution based on modes propagating in the $x$ direction, we have

$$\beta n_x = \beta'_c \qquad \beta n_z = k_0 \sin \theta'_i = k_0 \cos \theta_i$$

Corresponding to (69) and (70), we now get

$$\kappa = \frac{\beta'^2_c}{k_0^2 \cos^2 \theta'_i} = \frac{\beta'^2_c}{k_0^2 \sin^2 \theta_i} \tag{72}$$

$$\tan \theta_r = \frac{\beta'_c}{k_0 \cos \theta_i} = \kappa^{1/2} \tan \theta_i \tag{73}$$

As noted earlier, (72) gives accurate results for $s$ large, while (69) gives accurate results for small values of $s$. A combination of the two methods of solution yields good results for all values of $s$.

*Some Numerical Results*

In Fig. 12.18, the effective dielectric constant $\kappa$ is plotted as a function of the effective angle of propagation $\theta_r$ through the medium. These values of $\kappa$ are based on the solution given by Eqs. (66) and (69). In the same figure, the static solution (63) is also plotted. For $a/d$ greater

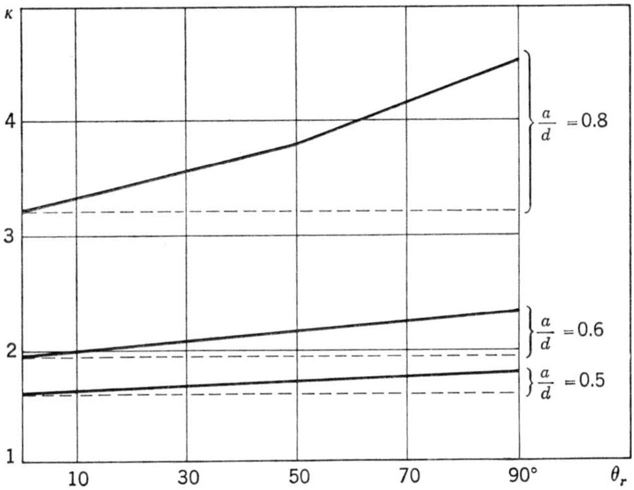


FIG. 12.18. Effective dielectric constant $\kappa$ as a function of effective angle of propagation $\theta_r$. Broken curves are static values obtained from Eq. (63), $s = 0.1\lambda_0$, $d = 0.4\lambda_0$.

than 0.5, it is seen that $\kappa$ varies considerably with the angle of propagation through the medium.

In order to compare the results of the two methods of solution, (69) and (72), $\kappa$ has been evaluated for $d = 0.4\lambda_0$ as a function of the spacing $s$ in wavelengths, for fixed values of $\theta_r$, and with $a = 0.75d$. The derived formulas do not give the end results directly, and so $\kappa$ was evaluated first as a function of $\theta_i$ for fixed values of $s/\lambda_0$, then these curves were replotted to give $\kappa$ as a function of $\theta_r$ for fixed values of $s/\lambda_0$, and finally these curves were converted into the desired family of curves. The results are plotted for $\theta_r = 30$, 50, 70, and 90° in Fig. 12.19*a*, *b*, *c*, and *d*. For $s > 0.2\lambda_0$ the first method gives good results, while for $s < 0.2\lambda_0$ the second method gives good results. By joining the two curves smoothly in the transition region around $s = 0.2\lambda_0$, accurate results for all values of $s$ are obtained.

Since the strip medium is a periodic one, it will exhibit the usual passband-stopband behavior. For $\theta_r$ equal to 70 and 90°, it is noted that $\kappa$ increases rapidly for $s/\lambda_0$ greater than some minimum value. This occurs because the edge of the first passband is being approached. The behavior may be understood qualitatively from the $\beta'_c$–$k_0 \cos \theta'_i$

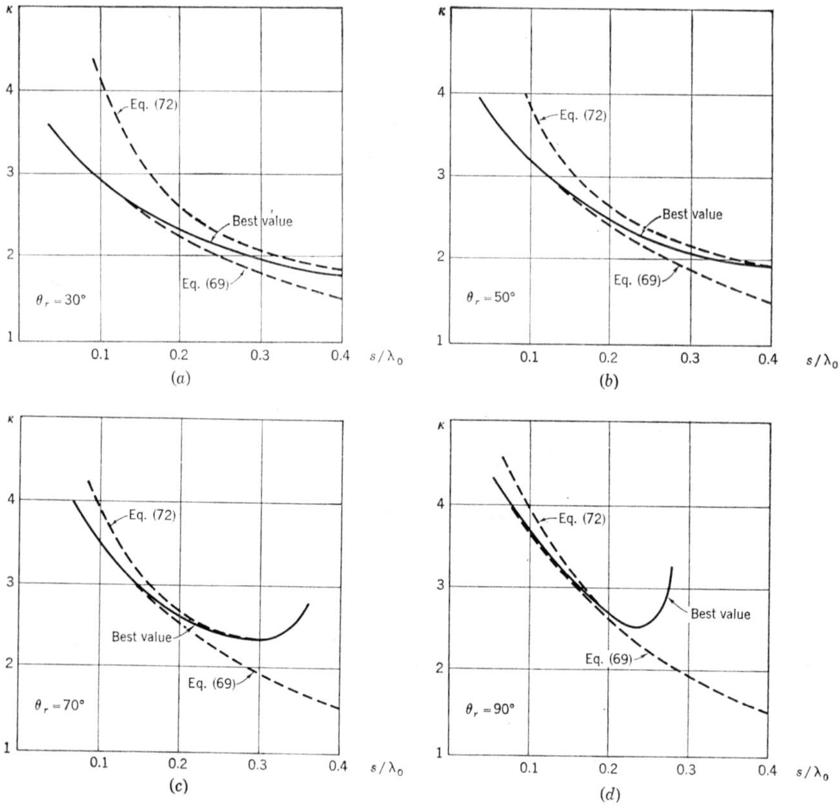


FIG. 12.19. Equivalent dielectric constant from Eqs. (69) and (72), together with the best value, $d = 0.4\lambda_0$, $a = 0.75d$.

diagram for the capacitive-loaded transmission line of Fig. 12.15. This diagram is sketched in Fig. 12.20. The effective dielectric constant $\kappa$ is given by

$$\kappa = \frac{(\beta'_c s)^2}{k_0{}^2 s^2 \cos^2 \theta'_i}$$

and is seen to be equal to the square of the cotangent of the angle of the line joining the origin $O$ to the point $P$ on the curve in Fig. 12.20. As $P$ moves up to $P_1$, the angle increases and $\kappa$ decreases. Beyond $P_1$ and up to the edge of the passband the angle decreases, and hence $\kappa$

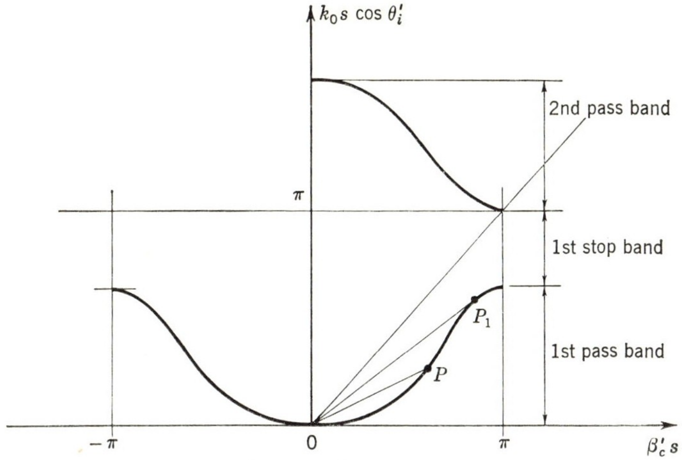


FIG. 12.20. $\beta_c'$-$k_0 \cos \theta_i'$ diagram for Eq. (65).

has a rapid increase in value just before the edge of the passband is reached.

## REFERENCES


12.1. O. M. Stuetzer: Development of Artificial Microwave Optics in Germany, *Proc. IRE,* vol. 38, pp. 1053–1056, September, 1950.

12.2. S. B. Cohn: The Electric and Magnetic Constants of Metallic Delay Media Containing Obstacles of Arbitrary Shape and Thickness, *J. Appl. Phys.,* vol. 22, pp. 628–734, May, 1951.

12.3. S. B. Cohn: Microwave Measurements on Metallic Delay Media, *Proc. IRE,* vol. 41, pp. 1177–1183, September, 1953.

12.4. S. B. Cohn: Analysis of the Metal Strip Delay Structure for Microwave Lenses, *J. Appl. Phys.,* vol. 20, pp. 257–262, March, 1949.

12.5. S. B. Cohn: Experimental Verification of the Metal Strip Delay Lens Theory, *J. Appl. Phys.,* vol. 24, pp. 839–841, July, 1953.

12.6. S. B. Cohn: Artificial Dielectrics for Microwaves, in Modern Advances in Microwave Techniques, *Polytech. Inst. Brooklyn Symposium Proc.,* vol. 4, pp. 465–480, November, 1954. (This reference contains a good source of additional references.)

12.7. J. Brown: Artificial Dielectrics Having Refractive Indices Less than Unity, *Proc. IEE (London),* vol. 100, part 4, pp. 51–62, 1953.

12.8. J. Brown and W. Jackson: The Relative Permittivity of Tetragonal Arrays of Perfectly Conducting Thin Discs, *Proc. IEE (London),* vol. 102, part B, pp. 37–42, January, 1955.

12.9. J. Brown and W. Jackson: The Properties of Artificial Dielectrics at Cm. Wavelengths, *Proc. IEE (London),* vol. 102, part B, pp. 11–16, January, 1955.

12.10. M. M. Z. Kharadly and W. Jackson: The Properties of Artificial Dielectrics Comprising Arrays of Conducting Elements, *Proc. IEE (London),* vol. 100, part III, pp. 199–212, July, 1953.

12.11. M. M. Z. Kharadly: Some Experiments on Artificial Dielectrics at Centimeter Wavelengths, *Proc. IEE (London),* vol. 102, part B, pp. 17–25, January, 1955

12.12. R. W. Corkum: Isotropic Artificial Dielectrics, *Proc. IRE*, vol. 40, pp. 574–587, May, 1952.
12.13. E. R. Wicher: The Influence of Magnetic Fields upon the Propagation of Electromagnetic Waves in Artificial Dielectrics, *J. Appl. Phys.*, vol. 22, pp. 1327–1329, November, 1951.
12.14. G. Estrin: The Effective Permeability of an Array of Thin Conducting Discs, *J. Appl. Phys.*, vol. 21, pp. 667–670, July 1950.
12.15. G. Estrin: The Effects of Anisotropy in a Three Dimensional Array of Conducting Discs, *Proc. IRE*, vol. 39, pp. 821–826, July, 1951.
12.16. A. L. Mikaelyan: Methods of Calculating the Permittivity and Permeability of Artificial Media, *Radiotekhnika*, vol. 10, pp. 23–36, 1955.
12.17. W. B. Swift and T. J. Higgins: Determination of the Design Constants of Artificial Dielectric UHF Lenses by Use of Physical Analogy, *Proc. Natl. Electronic Conf.*, vol. 9, pp. 825–832, 1953. (This article contains a good source of references on artificial dielectrics.)
12.18. H. S. Bennett: The Electromagnetic Transmission Characteristics of the Two Dimensional Lattice Medium, *J. Appl. Phys.*, vol. 24, pp. 785–810, June, 1953.
12.19. L. Lewin: Electrical Constants of Spherical Conducting Particles in a Dielectric, *Jour. IEE (London)*, vol. 94, part III, pp. 65–68, January, 1947.
12.20. Z. A. Kaprielian: Anisotropic Effects in Geometrically Isotropic Lattices, *J. Appl. Phys.*, vol. 29, pp. 1052–1063, July, 1958.
12.21. R. E. Collin: A Simple Artificial Anisotropic Dielectric Medium, *IRE Trans.*, vol. MTT-6, pp. 206–209, April, 1958.
12.22. H. E. J. Neugebauer: Clausius-Mossotti Equation for Certain Types of Anisotropic Crystals, *Can. J. Phys.*, vol. 32, pp. 1–8, January, 1954.

## PROBLEMS

**12.1.** Replace the sums in (11) by integrals, and integrate from $b/2 \to \infty$, $a/2 \to \infty$, $c/2 \to \infty$; multiply by 8, and show that the interaction constant $C$ is equal to $(3b^3)^{-1}$ for $a = b = c$.

HINT: Put $mb = y$, $dm = (1/b)\,dy$, etc., and integrate with respect to $y$ first since the integrand is equal to $\partial\phi/\partial y$.

**12.2.** Find the interaction constant for a single row of dipoles located midway between two infinite parallel conducting plates as illustrated.

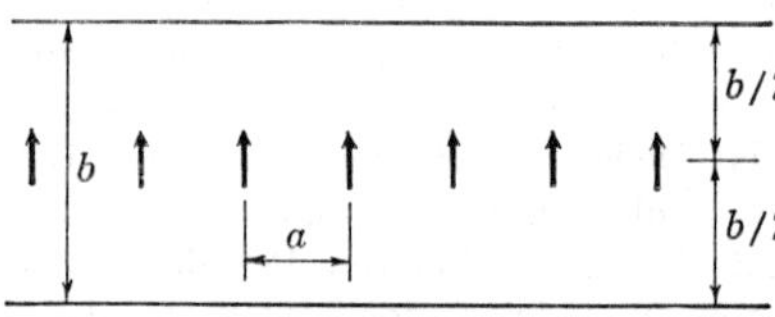


FIG. P 12.2

**12.3.** Repeat the analysis of Prob. 12.2, but with the dipole axis parallel with the conducting planes.

**12.4.** An artificial dielectric medium is made up of parallel conducting plates spaced a distance $c$ apart along the $z$ axis. In each plate a rectangular array of circular holes of radius $r$ is cut. Analyze the properties of this medium, and compare them with the properties of the circular-disk medium. Consider both normal and oblique propagation through the medium.

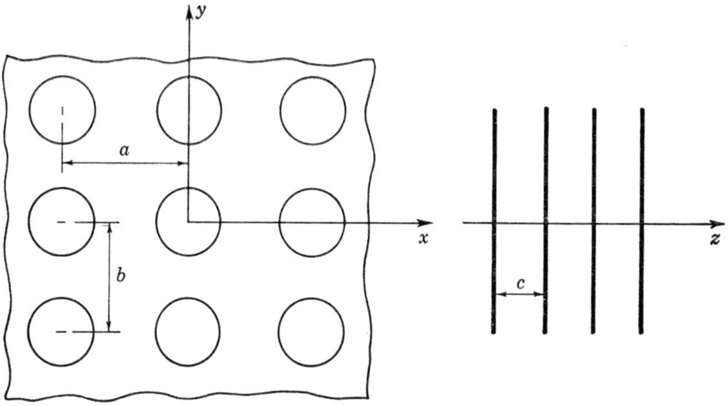


FIG. P 12.4

HINT: The shunt susceptance of one perforated plate may be obtained by duality from the susceptance of a plane of disks.

**12.5.** Consider a two-dimensional conducting-strip artificial dielectric medium with a TEM wave propagating obliquely through the lattice and having the electric field parallel to the strips. Analyze the anisotropic properties of the medium. Each inductive grating has a normalized susceptance $B_L$ given by $B_L = 4/B_c$, where $B_c$ is given by Eq. (64). The medium may be represented by an equivalent transmission line loaded at regular intervals by shunt inductive susceptances. Note that there is no propagation through the medium below some lower cutoff frequency $f_c$.

**12.6.** Find the relative dielectric constant of an artificial dielectric medium comprised of a cubical array of high-dielectric-constant spheres embedded in a low-dielectric-constant binder. What value of dielectric constant do the spheres have to have in order to give a medium with an effective dielectric constant 2.25 for sphere diameters equal to 0.6 of the spacing between spheres? Assume that the dielectric constant of the supporting medium is 1.02.

**12.7.** Consider a two-dimensional array of $y$-directed electric dipoles $Pe^{j\omega t}$, located at $x = na$, $n = 0, \pm 1, \pm 2, \ldots$ ; $y = mb$, $m = 0, \pm 1, \pm 2, \ldots$ ; $z = 0$. Find the $y$ component of the dynamic interaction field, acting on the dipole at the origin, due to all the neighboring dipoles. Assume that the dipoles are excited by a plane wave with $E_y = Ae^{-jhx-\Gamma_0 z}$, $\Gamma_0 = (h^2 - k_0^2)^{1/2}$, and, hence, the row of dipoles located at $x = na$ will have a phase $e^{-jhna}$ relative to the row located at $x = 0$.

HINT: Find a vector potential with a single $y$ component $A_y$ first, where $A_y$ is a solution of

$$\nabla^2 A_y + k_0^2 A_y = -j\omega\mu_0 P\delta(x)\delta(y)\delta(z)$$

Solve this equation in cylindrical coordinates. Note that conducting planes can be placed at $y = \pm b/2$. The steps to be followed in obtaining the complete solution are similar to those used in Sec. 7.2 to evaluate the mutual flux linkage in a loop antenna.

*Answer.* The required vector potential is

$$A_y = \frac{j\omega\mu_0 P}{2\pi b}\left[\frac{\pi}{a}\sum_{m=-\infty}^{\infty}\frac{e^{-\Gamma_m|z|}}{\Gamma_m}e^{-j(h+2m\pi/a)x} - K_0(jk_0\rho)\right.$$

$$\left.+ 2\sum_{m=-\infty}^{\infty}{}'\sum_{n=1}^{\infty}e^{-jhma}\cos\frac{2n\pi y}{b}K_0\{\gamma_n[z^2+(ma-x)^2]^{1/2}\}\right] + \frac{j\omega\mu_0 P}{4\pi}\sum_{m=-\infty}^{\infty}{}'\frac{e^{-jk_0r_m}}{r_m}$$

$$r_m{}^2 = \rho^2 + (mb-y)^2 \qquad \Gamma_m{}^2 = \left(h+\frac{2m\pi}{a}\right)^2 - k_0{}^2 \qquad \gamma_n{}^2 = \left(\frac{2n\pi}{b}\right)^2 - k_0{}^2$$

$$\rho^2 = x^2 + z^2$$

The interaction field is

$$E_{yi} = \frac{k_0{}^2P}{2ab\epsilon_0}\left[\frac{a}{\pi}\left(\gamma - \ln\frac{4\pi}{k_0a}\right) + j\left(\frac{a}{2} - \frac{1}{|\Gamma_0|}\right) + \sum_{m=1}^{\infty}\left(\frac{1}{\Gamma_m} + \frac{1}{\Gamma_{-m}} - \frac{a}{m\pi}\right)\right]$$

$$- \frac{2P}{\pi b\epsilon_0}\sum_{n=1}^{\infty}\sum_{m=1}^{\infty}\gamma_n{}^2K_0(\gamma_n ma) + \frac{P}{\pi\epsilon_0 b^3}\left[\left(1.2 - \frac{k_0{}^2b^2}{2}\ln k_0b + \frac{k_0{}^2b^2}{4} + \frac{3k_0{}^4b^4}{288}\right)\right.$$

$$\left. - j\left(\frac{\pi k_0{}^2b^2}{4} - \frac{k_0{}^3b^3}{6}\right)\right]$$

**12.8.** For the two-dimensional array of electric dipoles considered in Prob. 12.7, find the $z$ component of the magnetic interaction field $H_{zi}$ acting on the dipole at the origin.

*Answer*

$$H_{zi} = \frac{\omega P}{2\pi b}\left[-j\frac{\pi h}{|\Gamma_0|a} + \frac{\pi}{a}\sum_{m=1}^{\infty}\left(\frac{h+2m\pi/a}{\Gamma_m} + \frac{h-2m\pi/a}{\Gamma_{-m}}\right)\right.$$

$$\left. + 4\sum_{n=1}^{\infty}\sum_{m=1}^{\infty}\gamma_nK_1(\gamma_n ma)\sin hma\right]$$

This result shows that, in general, there will be coupling between the electric and magnetic dipoles in an artificial dielectric medium. The evaluation of the dynamic magnetic interaction field from $z$-directed magnetic dipoles is similar to the evaluation of mutual flux linkage in a loop antenna carried out in Sec. 7.2.

**12.9.** Consider a row of thin circular disks between two conducting planes at $y = \pm b/2$, as in Fig. 12.10. Let the incident field be

$$E_y = A_1e^{-jhx-\Gamma_0z} \qquad h = k_0\sin\theta_i \qquad \Gamma_0 = jk_0\cos\theta_i$$
$$H_z = A_1Y_0\sin\theta_i\, e^{-jhx-\Gamma_0z}$$

plus an $x$ component of magnetic field. This incident field induces an electric dipole

$$P_n = {}^{16}\!/\!_3 r^3\epsilon_0A_1e^{-jhna} = Pe^{-jhna}$$

and a magnetic dipole

$$M_n = -{}^{8}\!/\!_3 r^3A_1Y_0\sin\theta_i\, e^{-jhna} = Me^{-jhna}$$

in the disk located at $x = na$, where $r$ is the disk radius. The field scattered by the $y$-directed electric dipoles may be found from a $y$-directed vector potential $A_y$, where

$$\nabla^2 A_y + k_0^2 A_y = -j\omega\mu_0 P \sum_{n=-\infty}^{\infty} e^{-jhna}\delta(x - na)\delta(y)\delta(z)$$

Show that a suitable solution for $A_y$ is

$$A_y = \sum_{n=-\infty}^{\infty} \sum_{m=0,2,\ldots}^{\infty} a_{nm} \cos\frac{m\pi y}{b} e^{-j(h+2n\pi/a)x}e^{-\Gamma_{nm}z}$$

where $\Gamma_{nm}^2 = (h + 2n\pi/a)^2 + (m\pi/b)^2 - k_0^2$.

Multiply both sides of the differential equation for $A_y$ by $e^{j(h+2n\pi/a)x}$ and integrate over $-a/2 \le x \le a/2$, $-b/2 \le y \le b/2$ to get

$$a_{00}ab\left(\frac{\partial^2}{\partial z^2} - \Gamma_{00}^2\right)e^{-\Gamma_{00}|z|} = -j\omega\mu_0 P\delta(z)$$

Next integrate over $-\tau < z < \tau$ and let $\tau \to 0$ to obtain

$$a_{00} = \frac{j\omega\mu_0 P}{2ab\Gamma_{00}}$$

The field scattered by the $z$-directed magnetic dipoles may be derived from a $z$-directed magnetic Hertzian potential $\Pi_{mz}$, where

$$\nabla^2\Pi_{mz} + k_0^2\,\Pi_{mz} = -M \sum_{n=-\infty}^{\infty} e^{-jhna}\delta(x - na)\delta(y)\delta(z)$$

Show that a suitable solution for $\Pi_{mz}$ is

$$\sum_{n=-\infty}^{\infty} \sum_{m=0,2,\ldots}^{\infty} b_{nm} \cos\frac{m\pi y}{b} e^{-j(h+2n\pi/a)x}e^{-\Gamma_{nm}z}$$

By analogy with the solution for $a_{00}$ show that $b_{00} = M/2ab\Gamma_{00}$. Hence show that the reflection coefficient for the incident wave is

$$R = -j\omega\frac{a_{00}}{A_1} + \omega\mu_0 h\frac{b_{00}}{A_1}$$

by evaluating the dominant-mode electric field associated with $A_y$ and $\Pi_{mz}$. Show that $R$ reduces to

$$R = \frac{-j4r^3}{3ab\beta_0}(2k_0^2 - h^2) \qquad \beta_0 = |\Gamma_{00}| = |\Gamma_0|$$

Compare this method with that used in Sec. 12.5.

# MATHEMATICAL APPENDIX

## A.1. Vector Analysis

In the first three parts of this section, our frame of reference will be a rectangular $xyz$ coordinate frame. In subsection $d$, we will consider orthogonal curvilinear coordinate systems.

A vector quantity is an entity which requires both a magnitude and a direction to be specified at each point in space. The usual procedure is to give the components of the vector along three mutually orthogonal axes. A coordinate frame $xyz$ is said to be right-handed or dextral if rotation of the $x$ axis into the $y$ axis defines a sense of rotation that would advance a right-hand screw in the positive $z$ direction. If the sense of rotation given above would advance a right-hand screw along the negative $z$ direction the coordinate system is said to be left-handed or sinistral. For our rectangular coordinate frame, the unit vectors along the coordinate axis will be represented by $\mathbf{a}_x$, $\mathbf{a}_y$, $\mathbf{a}_z$.

### *a. Vector Algebra*

**Addition and Subtraction.** The addition or subtraction of two or more vectors is defined as the vector whose components are the sum or difference of the corresponding components of each individual vector; i.e.,

$$\mathbf{A} \pm \mathbf{B} = (A_x \pm B_x)\mathbf{a}_x + (A_y \pm B_y)\mathbf{a}_y + (A_z \pm B_z)\mathbf{a}_z$$

**Scalar and Vector Products.** The scalar product of a vector $\mathbf{A}$ with a vector $\mathbf{B}$ is written as $\mathbf{A} \cdot \mathbf{B}$ and is defined by

$$\mathbf{A} \cdot \mathbf{B} = |A|\,|B| \cos \theta$$

where $\theta$ is the angle between the two vectors. In rectangular coordinates, we have

$$\mathbf{A} \cdot \mathbf{B} = A_xB_x + A_yB_y + A_zB_z$$

and this is the projection of $\mathbf{A}$ onto $\mathbf{B}$ multiplied by the modulus of $\mathbf{B}$.

The vector, or cross, product of a vector $\mathbf{A}$ with a vector $\mathbf{B}$ is written as $\mathbf{A} \times \mathbf{B}$ and is a vector of magnitude $|A|\,|B| \sin \theta$, and perpendicular to both $\mathbf{A}$ and $\mathbf{B}$ with a positive direction, defined such that rotation of $\mathbf{A}$ into $\mathbf{B}$ would cause a right-hand screw to advance along it. In rectangular coordinates, we have

$$\mathbf{A} \times \mathbf{B} = \begin{vmatrix} \mathbf{a}_x & \mathbf{a}_y & \mathbf{a}_z \\ A_x & A_y & A_z \\ B_x & B_y & B_z \end{vmatrix} = -\mathbf{B} \times \mathbf{A}$$

**Scalar and Vector Higher-order Products.** The scalar triple or box product is given by

$$\mathbf{A} \cdot \mathbf{B} \times \mathbf{C} = \mathbf{A} \times \mathbf{B} \cdot \mathbf{C} = \begin{vmatrix} A_x & A_y & A_z \\ B_x & B_y & B_z \\ C_x & C_y & C_z \end{vmatrix}$$

The vector triple product is written as $\mathbf{A} \times (\mathbf{B} \times \mathbf{C})$ and is equal to $(\mathbf{A} \cdot \mathbf{C})\mathbf{B} - (\mathbf{A} \cdot \mathbf{B})\mathbf{C}$. It is important to place the parentheses correctly since, in general, $\mathbf{A} \times (\mathbf{B} \times \mathbf{C}) \neq (\mathbf{A} \times \mathbf{B}) \times \mathbf{C}$.

By an application of the above fundamental rules, we may readily obtain the following results:

$$(\mathbf{A} \times \mathbf{B}) \cdot (\mathbf{C} \times \mathbf{D}) = \begin{vmatrix} \mathbf{A} \cdot \mathbf{C} & \mathbf{A} \cdot \mathbf{D} \\ \mathbf{B} \cdot \mathbf{C} & \mathbf{B} \cdot \mathbf{D} \end{vmatrix}$$

$$(\mathbf{A} \times \mathbf{B}) \times (\mathbf{C} \times \mathbf{D}) = (\mathbf{A} \cdot \mathbf{B} \times \mathbf{D})\mathbf{C} - (\mathbf{A} \cdot \mathbf{B} \times \mathbf{C})\mathbf{D}$$

### *b. Differential Invariants*

An invariant quantity or simply an invariant is a quantity which has a unique value at a given point in space, independent of the coordinate system to which it is referred. The fundamental vector differential operator is "del," or "nabla," and is written as

$$\nabla = \mathbf{a}_x \frac{\partial}{\partial x} + \mathbf{a}_y \frac{\partial}{\partial y} + \mathbf{a}_z \frac{\partial}{\partial z}$$

This operator behaves like a vector in the sense that it combines with vectors according to the rules of vector algebra. In addition, it also performs certain differentiations with respect to the coordinates. In common with all differential operators it is a noncommutative operator; i.e., it must always occur on the left-hand side of the quantity it operates on. The manipulation of various expressions involving this vector differential operator is greatly facilitated if the differentiation operation is suppressed until the expression has been expanded into a simpler form by treating $\nabla$ as a vector.

**Gradient and Directional Derivative.** The derivative of a scalar function $\phi(x,y,z)$ with respect to distance $l$ along a continuous curve $x(l)$, $y(l)$, $z(l)$ is

$$\frac{d\phi}{dl} = \frac{\partial \phi}{\partial x}\frac{dx}{dl} + \frac{\partial \phi}{\partial y}\frac{dy}{dl} + \frac{\partial \phi}{\partial z}\frac{dz}{dl}$$

and is the directional derivative of $\phi$ along the curve. The unit tangent to the curve is $\boldsymbol{\tau}$, where

$$\boldsymbol{\tau} = \mathbf{a}_x \frac{dx}{dl} + \mathbf{a}_y \frac{dy}{dl} + \mathbf{a}_z \frac{dz}{dl}$$

The gradient of $\phi$ is given by

$$\nabla\phi = \mathbf{a}_x \frac{\partial \phi}{\partial x} + \mathbf{a}_y \frac{\partial \phi}{\partial y} + \mathbf{a}_z \frac{\partial \phi}{\partial z}$$

and, hence, the directional derivative is the projection of $\nabla\phi$ on $\boldsymbol{\tau}$; that is,

$$\frac{d\phi}{dl} = \nabla\phi \cdot \boldsymbol{\tau}$$

The directional derivative has its maximum value when $\boldsymbol{\tau}$ is parallel to $\nabla\phi$. Hence $\nabla\phi$ is in the direction of the maximum rate of change of $\phi$.

**Divergence and Curl.** The divergence of a vector may be defined as

$$\nabla \cdot \mathbf{A} = \lim_{\Delta V \to 0} \frac{\oiint_S \mathbf{A} \cdot d\mathbf{a}}{\Delta V}$$

The divergence of a vector field is a measure of the net outflow of flux through a closed surface $S$ surrounding a volume $\Delta V$, per unit volume. In rectangular coordinates,

$$\nabla \cdot \mathbf{A} = \frac{\partial A_x}{\partial x} + \frac{\partial A_y}{\partial y} + \frac{\partial A_z}{\partial z}$$

The curl of a vector is a measure of the net circulation of the vector field around the boundary of an infinitesimal element of area. Mathematically,

$$(\text{curl } \mathbf{A})_n = (\nabla \times \mathbf{A})_n = \lim_{\Delta S \to 0} \frac{\oint_C \mathbf{A} \cdot d\mathbf{l}}{\Delta S}$$

where $(\nabla \times \mathbf{A})_n$ is the component of the curl of $\mathbf{A}$ in the direction of the normal to the surface $\Delta S$, and $C$ is the boundary of $\Delta S$. In rectangular coordinates, the curl of $\mathbf{A}$ is given by the expansion of the following determinant:

$$\nabla \times \mathbf{A} = \begin{vmatrix} \mathbf{a}_x & \mathbf{a}_y & \mathbf{a}_z \\ \dfrac{\partial}{\partial x} & \dfrac{\partial}{\partial y} & \dfrac{\partial}{\partial z} \\ A_x & A_y & A_z \end{vmatrix}$$

**Some Differential Identities.** The following identity is readily established:

$$\nabla \times \nabla \times \mathbf{A} = \nabla\nabla \cdot \mathbf{A} - \nabla^2\mathbf{A}$$

Treating $\nabla$ as a vector and expanding the triple vector product, but always keeping $\nabla$ in front of $\mathbf{A}$, gives the result at once. Other useful identities are

$$\begin{aligned} \nabla \times \nabla\phi &= 0 \\ \nabla \cdot \nabla \times \mathbf{A} &= 0 \\ \nabla \cdot (\mathbf{A} \times \mathbf{B}) &= \mathbf{B} \cdot \nabla \times \mathbf{A} - \mathbf{A} \cdot \nabla \times \mathbf{B} \\ \nabla \times (\mathbf{A} \times \mathbf{B}) &= \mathbf{A}\nabla \cdot \mathbf{B} - \mathbf{B}\nabla \cdot \mathbf{A} + (\mathbf{B} \cdot \nabla)\mathbf{A} - (\mathbf{A} \cdot \nabla)\mathbf{B} \end{aligned}$$

The above expressions are readily obtained, as, for example,

$$\begin{aligned} \nabla \cdot (\mathbf{A} \times \mathbf{B}) &= \nabla_A \cdot (\mathbf{A} \times \mathbf{B}) + \nabla_B \cdot (\mathbf{A} \times \mathbf{B}) \\ &\quad - (\nabla_A \times \mathbf{A}) \cdot \mathbf{B} - \nabla_B \cdot (\mathbf{B} \times \mathbf{A}) = \mathbf{B} \cdot (\nabla \times \mathbf{A}) - \mathbf{A} \cdot \nabla \times \mathbf{B} \end{aligned}$$

where $\nabla_A$ means operation on $\mathbf{A}$ only, and similarly for $\nabla_B$. Each term may be treated as a scalar triple product, and the dot and cross may be interchanged.

### *c. Integral Invariants and Transformations*

**Gauss' Law or Divergence Theorem.** This theorem states that the volume integral of the divergence of a vector is equal to the surface integral of the normal outward component of the vector over a closed surface $S$ surrounding the volume $V$; that is,

$$\iiint_V \nabla \cdot \mathbf{A}\, dV = \iint_S \mathbf{A} \cdot d\mathbf{a}$$

**Green's Identities.** Green's first identity is

$$\iiint_V (\nabla v \cdot \nabla u + u \nabla^2 v)\, dV = \iint_S u \nabla v \cdot d\mathbf{a}$$

where $u$ and $v$ are two scalar functions. The proof follows by applying the divergence theorem to the vector function $u \nabla v$. Interchanging $u$ and $v$ and subtracting gives Green's second identity or, as sometimes called, Green's theorem,

$$\iiint_V (u \nabla^2 v - v \nabla^2 u)\, dV = \iint_S (u \nabla v - v \nabla u) \cdot d\mathbf{a}$$

This result is of considerable use in proving the orthogonality of the eigenfunctions for the scalar wave equations. It is also of fundamental importance in the solution of certain boundary-value problems by the use of Green's functions.

**Stokes' Law.** This law gives the value of the integral of the curl of a vector over a surface $S$ in terms of a line integral around a closed contour $C$, which bounds the surface $S$:

$$\iint_S \nabla \times \mathbf{A} \cdot d\mathbf{a} = \oint_C \mathbf{A} \cdot d\mathbf{l}$$

The proof follows readily from the definition of the curl of a vector.

**Vector Form of Green's Identities.** The vector forms of Green's first and second identities are

$$\iiint_V (\nabla \times \mathbf{A} \cdot \nabla \times \mathbf{B} - \mathbf{A} \cdot \nabla \times \nabla \times \mathbf{B})\, dV = \iiint_V \nabla \cdot (\mathbf{A} \times \nabla \times \mathbf{B})\, dV$$
$$= \iint_S \mathbf{A} \times \nabla \times \mathbf{B} \cdot d\mathbf{a}$$

$$\iiint_V (\mathbf{B} \cdot \nabla \times \nabla \times \mathbf{A} - \mathbf{A} \cdot \nabla \times \nabla \times \mathbf{B})\, dV = \iint_S (\mathbf{A} \times \nabla \times \mathbf{B} - \mathbf{B} \times \nabla \times \mathbf{A}) \cdot d\mathbf{a}$$

The vector form of the second identity is used to prove the orthogonality of the eigenfunctions of the vector wave equation as well as in the solution of certain vector boundary-value problems by means of Green's functions.

The following modification of the vector form of Green's first identity may also be established:

$$\iiint_V (\nabla \times \mathbf{A} \cdot \nabla \times \mathbf{B} + \nabla \cdot \mathbf{A}\, \nabla \cdot \mathbf{B} + \mathbf{A} \cdot \nabla^2 \mathbf{B})\, dV$$
$$= \oiint_S (\mathbf{A} \times \nabla \times \mathbf{B} + \mathbf{A}\, \nabla \cdot \mathbf{B}) \cdot d\mathbf{a}$$

### *d. Orthogonal Curvilinear Coordinates*

In the present subsection we will derive the general relationships between the components of a vector in rectangular coordinates and those of a vector referred to an orthogonal curvilinear coordinate frame, and also the transformation laws for various vector differential operations. For this purpose we will relabel our $x$, $y$, $z$ axes as $x_1$, $x_2$, $x_3$ with unit vectors $\mathbf{a}_1$, $\mathbf{a}_2$, $\mathbf{a}_3$ directed along them.

**Transformation of Coordinates.** Consider three continuously differentiable functions of the three variables $x_i$,

$$u_i = u_i(x_1,x_2,x_3) \qquad i = 1, 2, 3$$

which defines a transformation from the coordinates $x_i$ to the coordinates $u_i$. The differentials $dx_i$ transform as follows:

$$dx_i = \sum_{s=1}^{3} \frac{\partial u_i}{\partial x_s} dx_s$$

which we may write in matrix form as follows:

$$\begin{bmatrix} du_1 \\ du_2 \\ du_3 \end{bmatrix} = \begin{bmatrix} \frac{\partial u_1}{\partial x_1} & \frac{\partial u_1}{\partial x_2} & \frac{\partial u_1}{\partial x_3} \\ \frac{\partial u_2}{\partial x_1} & \frac{\partial u_2}{\partial x_2} & \frac{\partial u_2}{\partial x_3} \\ \frac{\partial u_3}{\partial x_1} & \frac{\partial u_3}{\partial x_2} & \frac{\partial u_3}{dx_3} \end{bmatrix} \begin{bmatrix} dx_1 \\ dx_2 \\ dx_3 \end{bmatrix} = [T] \begin{bmatrix} dx_1 \\ dx_2 \\ dx_3 \end{bmatrix}$$

where $[T]$ is the transformation matrix. The determinant of the matrix $[T]$ is called the Jacobian and written as

$$J = |T| = \frac{\partial(u_1,u_2,u_3)}{\partial(x_1,x_2,x_3)}$$

An expansion of the triple scalar product $\nabla u_1 \cdot \nabla u_2 \times \nabla u_3$ shows that it is equal to $J$. When $J \neq 0$, we can invert the above equations and solve for the $dx_i$ as linear functions of the $du_i$, and also solve for the $x_i$ as functions of the $u_i$; thus

$$\begin{bmatrix} dx_1 \\ dx_2 \\ dx_3 \end{bmatrix} = \begin{bmatrix} \frac{\partial x_1}{\partial u_1} & \frac{\partial x_1}{\partial u_2} & \frac{\partial x_1}{\partial u_3} \\ \frac{\partial x_2}{\partial u_1} & \frac{\partial x_2}{\partial u_2} & \frac{\partial x_2}{\partial u_3} \\ \frac{\partial x_3}{\partial u_1} & \frac{\partial x_3}{\partial u_2} & \frac{\partial x_3}{\partial u_3} \end{bmatrix} \begin{bmatrix} du_1 \\ du_2 \\ du_3 \end{bmatrix} = [T_i] \begin{bmatrix} du_1 \\ du_2 \\ du_3 \end{bmatrix}$$

where $[T_i]$ is the inverse transformation matrix. If we premultiply this equation by $[T]$, we get

$$[T][dx_i] = [du_i] = [T][T_i][du_i]$$

from which we conclude that $[T][T_i] = [I]$, or the unit matrix. From this latter result we get

$$\sum_{s=1}^{3} \frac{\partial u_i}{\partial x_s}\frac{\partial x_s}{\partial u_r} = \delta_{ir}$$

and

$$\sum_{s=1}^{3} \frac{\partial x_i}{\partial u_s}\frac{\partial u_s}{\partial x_r} = \delta_{ir}$$

where $\delta_{ir} = 1$ for $i = r$, and zero otherwise.

Instead of specifying a point by the coordinates $x_{i0}$, we may equally well specify its location by the trio of numbers $u_{i0} = u_i(x_{10},x_{20},x_{30})$. When the $u_{i0}$ are given, the point in question is located at the intersection of the coordinate surfaces $u_i = u_{i0}$.

These surfaces intersect along three curves called the coordinate curves. Along any one curve, say $u_2 = u_{20}$, $u_3 = u_{30}$, only one of the $u_i$ will vary, in this case $u_1$.

At the point of intersection of the three coordinate curves, we construct a miniature rectangular coordinate frame by means of the unit vectors $\mathbf{e}_i$, directed tangent to the three coordinate curves $u_i$ increasing. We may now specify a vector by giving its components in this new reference frame. For this reason, the $u_i$ are called curvilinear coordinates. When the three coordinate curves intersect at right angles everywhere, we have an orthogonal curvilinear coordinate frame. From this point on, we will assume that our curvilinear coordinates are orthogonal, and also that the $u_i$ and $\mathbf{e}_i$ have been labeled so that $\mathbf{e}_1$, $\mathbf{e}_2$, $\mathbf{e}_3$ form a right-handed system. Since $\nabla u_i$ is a vector normal to the surface $u_i = u_{i0}$, the condition that the curvilinear coordinates should form an orthogonal system is that $\nabla u_s \cdot \nabla u_r = 0$ for $r \neq s$.

**Line Element, or Fundamental Metric.** The differentials $du_i$ are not, in general, measures of arc length $dl_i$ along the coordinate curves. We convert them to arc length by multiplying by suitable scale factors $h_i$; thus $dl_i = h_i \, du_i$. The line element, or fundamental metric, as it is called, is

$$d\mathbf{l} = \sum_{i=1}^{3} \mathbf{a}_i \, dx_i = \sum_{i=1}^{3} \mathbf{e}_i h_i \, du_i$$

and

$$dl^2 = \sum_{i=1}^{3} dx_i^2 = \sum_{i=1}^{3} h_i^2 \, du_i^2$$

For an orthogonal coordinate system, all coordinate surfaces intersect normally, and thus $\nabla u_i$, which is normal to the coordinate surface $u_i = u_{i0}$, must be tangent to the $u_i$ coordinate curve, i.e., parallel to $\mathbf{e}_i$. The element of arc length along the $u_i$ coordinate curve is $dl_i$. The directional derivative of $u_i$ along $u_i$ is, therefore,

$$\frac{du_i}{dl_i} = \nabla u_i \cdot \mathbf{e}_i$$

and hence

$$dl_i = \frac{du_i}{|\nabla u_i|} = h_i \, du_i$$

The scale factors $h_i$ are thus given by

$$h_i = |\nabla u_i|^{-1} = \left[ \sum_{s=1}^{3} \left( \frac{\partial u_i}{\partial x_s} \right)^2 \right]^{-\frac{1}{2}}$$

The element of volume is

$$\begin{aligned} dV &= h_1 h_2 h_3 \, du_1 \, du_2 \, du_3 \\ &= \frac{du_1 \, du_2 \, du_3}{|\nabla u_1| \, |\nabla u_2| \, |\nabla u_3|} = \frac{du_1 \, du_2 \, du_3}{J} \end{aligned}$$

because of the mutual orthogonality of the gradients.

When the $u_i$ are given as functions of the $x_i$, we can readily evaluate the scale factors $h_i$. It is convenient to have an alternative expression for $h_i$, which can be easily evaluated when the $x_i$ are given as functions of the $u_i$, since this will eliminate the necessity of solving for the $u_i$ as functions of the $x_i$ first. The differentials $dx_i$ are given by

$$dx_i = \sum_{s=1}^{3} \frac{\partial x_i}{\partial u_s} du_s$$

and
$$dx_i^2 = \sum_{s=1}^{3} \sum_{r=1}^{3} \frac{\partial x_i}{\partial u_s} \frac{\partial x_i}{\partial u_r} du_s\, du_r$$

and, hence,
$$dl^2 = \sum_{i=1}^{3} dx_i^2 = \sum_{i=1}^{3} \sum_{s=1}^{3} \sum_{r=1}^{3} \frac{\partial x_i}{\partial u_s} \frac{\partial x_i}{\partial u_r} du_s\, du_r$$

$$= \sum_{i=1}^{3} \sum_{s=1}^{3} \left(\frac{\partial x_i}{\partial u_s}\right)^2 du_s^2 = \sum_{s=1}^{3} du_s^2 \sum_{i=1}^{3} \left(\frac{\partial x_i}{\partial u_s}\right)^2$$

since
$$\sum_{i=1}^{3} \sum_{s=1}^{3} \sum_{r=1}^{3} \frac{\partial x_i}{\partial u_s} \frac{\partial x_i}{\partial u_r} du_s\, du_r = 0 \qquad \text{for } r \neq s$$

Therefore
$$h_s^2 = \sum_{i=1}^{3} \left(\frac{\partial x_i}{\partial u_s}\right)^2$$

an expression readily evaluated when the $x_i$ are known functions of the $u_i$. The proof that the triple summation above equals zero unless $r = s$ is as follows. We know that

$$h_s^{-2} = \sum_{i=1}^{3} \left(\frac{\partial u_s}{\partial x_i}\right)^2$$

and that
$$\sum_{p=1}^{3} \frac{\partial x_k}{\partial u_p} \frac{\partial u_p}{\partial x_j} = \delta_{kj}$$

from the properties of the transformation matrices $[T]$ and $[T_i]$. Multiply this latter result by $\partial u_s/\partial x_j$, and sum over $j$ to get

$$\sum_{p=1}^{3} \frac{\partial x_k}{\partial u_p} \sum_{j=1}^{3} \frac{\partial u_p}{\partial x_j} \frac{\partial u_s}{\partial x_j} = \sum_{p=1}^{3} \frac{\partial x_k}{\partial u_p} \nabla u_p \cdot \nabla u_s$$

$$= \frac{1}{h_s^2} \frac{\partial x_k}{\partial u_s} = \sum_{j=1}^{3} \delta_{kj} \frac{\partial u_s}{\partial x_j} = \frac{\partial u_s}{\partial x_k}$$

We now use the result $h_s^2\, \partial u_s/\partial x_k = \partial x_k/\partial u_s$ in the triple summation to get

$$\sum_{s=1}^{3} \sum_{r=1}^{3} du_s\, du_r \sum_{i=1}^{3} h_s^2 h_r^2 \frac{\partial u_s}{\partial x_i} \frac{\partial u_r}{\partial x_i} = \sum_{s=1}^{3} h_s^4\, du_s^2\, |\nabla u_s|^2 \qquad \text{since } \nabla u_s \cdot \nabla u_r = 0$$

unless $s = r$. When $s \neq r$, the triple summation vanishes, and this completes the proof.

In a cylindrical coordinate frame $r\theta z$, the line elements are given by

$$h_1 = 1 \qquad h_2 = r \qquad h_3 = 1$$

In a spherical coordinate system $r\theta\phi$, where $\theta$ is the polar angle, the line elements are

$$h_1 = 1 \qquad h_2 = r \qquad h_3 = r \sin \theta$$

**Gradient in Curvilinear Coordinates.** The total differential of a scalar function $\phi(u_1,u_2,u_3)$ is

$$d\phi = \sum_{i=1}^{3} \frac{\partial \phi}{\partial u_i} du_i$$

Since $du_i$ is a function of the $x_s$, we have

$$du_i = \sum_{s=1}^{3} \frac{\partial u_i}{\partial x_s} dx_s = \nabla u_i \cdot d\mathbf{l}$$

where $d\mathbf{l} = \mathbf{a}_1\, dx_1 + \mathbf{a}_2\, dx_2 + \mathbf{a}_3\, dx_3$. Hence, we get

$$d\phi = \sum_{i=1}^{3} \frac{\partial \phi}{\partial u_i} \nabla u_i \cdot d\mathbf{l}$$

and

$$\frac{d\phi}{dl} = \left( \sum_{i=1}^{3} \frac{\partial \phi}{\partial u_i} \nabla u_i \right) \cdot \frac{d\mathbf{l}}{dl}$$

The unit tangent along the curve is $d\mathbf{l}/dl = \boldsymbol{\tau}$, and, since the directional derivative of $\phi$ is $d\phi/dl = \nabla\phi \cdot \boldsymbol{\tau}$, we have the result that

$$\nabla \phi = \sum_{i=1}^{3} \frac{\partial \phi}{\partial u_i} \nabla u_i$$

where $\nabla u_i$ may be referred to either the $x_i$ or the $u_i$ coordinate frame. Now $\nabla u_i$ has the direction of $\mathbf{e}_i$ and a magnitude of $h_i^{-1}$, and so

$$\nabla \phi = \sum_{i=1}^{3} \frac{1}{h_i} \frac{\partial \phi}{\partial u_i} \mathbf{e}_i$$

**Components of a Vector in $u_i$ Coordinates.** The differential $du_i$ is equal to

$$\nabla u_i \cdot \sum_{s=1}^{3} \mathbf{a}_s\, dx_s$$

When only one of the $dx_s$ is different from zero, we have a partial differential

$$\partial u_i = \nabla u_i \cdot \mathbf{a}_s\, \partial x_s$$

and this is the change in $u_i$ when $x_s$ changes by $\partial x_s$. The change in $u_i$ is along the $u_i$ coordinate curve; hence,

$$\partial u_i = |\nabla u_i| \partial x_s \cos \theta$$

or

$$\cos \theta = |\nabla u_i|^{-1} \frac{\partial u_i}{\partial x_s} = h_i \frac{\partial u_i}{\partial x_s}$$

where $\cos\theta$ is the direction cosine between the $x_s$ axis and the $u_i$ coordinate curve. In proving that the triple summation, in the previous section, vanished for $r \neq s$, we obtained as an intermediate step the result

$$\frac{1}{h_s^2}\frac{\partial x_k}{\partial u_s} = \frac{\partial u_s}{\partial x_k}$$

Therefore, we have the two following equivalent expressions for the direction cosine between the unit vectors $\mathbf{a}_s$ and $\mathbf{e}_i$:

$$\cos\theta = h_i\frac{\partial u_i}{\partial x_s} = \frac{1}{h_i}\frac{\partial x_s}{\partial u_i}$$

From the above results we may readily construct the following table of direction cosines between the unit vectors $\mathbf{a}_i$ and $\mathbf{e}_s$:

| | $\mathbf{e}_1$ | $\mathbf{e}_2$ | $\mathbf{e}_3$ |
|---|---|---|---|
| $\mathbf{a}_1$ | $h_1\frac{\partial u_1}{\partial x_1} = \frac{1}{h_1}\frac{\partial x_1}{\partial u_1}$ | $h_2\frac{\partial u_2}{\partial x_1} = \frac{1}{h_2}\frac{\partial x_1}{\partial u_2}$ | $h_3\frac{\partial u_3}{\partial x_1} = \frac{1}{h_3}\frac{\partial x_1}{\partial u_3}$ |
| $\mathbf{a}_2$ | $h_1\frac{\partial u_1}{\partial x_2} = \frac{1}{h_1}\frac{\partial x_2}{\partial u_1}$ | $h_2\frac{\partial u_2}{\partial x_2} = \frac{1}{h_2}\frac{\partial x_2}{\partial u_2}$ | $h_3\frac{\partial u_3}{\partial x_2} = \frac{1}{h_3}\frac{\partial x_2}{\partial u_3}$ |
| $\mathbf{a}_3$ | $h_1\frac{\partial u_1}{\partial x_3} = \frac{1}{h_1}\frac{\partial x_3}{\partial u_1}$ | $h_2\frac{\partial u_2}{\partial x_3} = \frac{1}{h_2}\frac{\partial x_3}{\partial u_2}$ | $h_3\frac{\partial u_3}{\partial x_3} = \frac{1}{h_3}\frac{\partial x_3}{\partial u_3}$ |

In our curvilinear coordinate frame we can now express a vector $\mathbf{A}$ with rectangular components $A_i$ as follows:

$$\begin{aligned}\mathbf{A} &= \sum_{s=1}^{3}\left(A_1\frac{1}{h_s}\frac{\partial x_1}{\partial u_s} + A_2\frac{1}{h_s}\frac{\partial x_2}{\partial u_s} + A_3\frac{1}{h_s}\frac{\partial x_3}{\partial u_s}\right)\mathbf{e}_s \\ &= \sum_{s=1}^{3}\left(A_1 h_s\frac{\partial u_s}{\partial x_1} + A_2 h_s\frac{\partial u_s}{\partial x_2} + A_3 h_s\frac{\partial u_s}{\partial x_3}\right)\mathbf{e}_s \\ &= A_{u_1}\mathbf{e}_1 + A_{u_2}\mathbf{e}_2 + A_{u_3}\mathbf{e}_3\end{aligned}$$

These equations may be written in matrix form as follows:

$$\begin{aligned}\begin{bmatrix} A_{u1} \\ A_{u2} \\ A_{u3}\end{bmatrix} &= \begin{bmatrix} \frac{1}{h_1}\frac{\partial x_1}{\partial u_1} & \frac{1}{h_1}\frac{\partial x_2}{\partial u_1} & \cdots \\ \frac{1}{h_2}\frac{\partial x_1}{\partial u_2} & \cdots & \cdots \\ \cdots & \cdots & \cdots \end{bmatrix}\begin{bmatrix} A_1 \\ A_2 \\ A_3\end{bmatrix} \\ &= \begin{bmatrix} \frac{1}{h_1} & 0 & 0 \\ 0 & \frac{1}{h_2} & 0 \\ 0 & 0 & \frac{1}{h_3}\end{bmatrix}[T^*]\begin{bmatrix} A_1 \\ A_2 \\ A_3\end{bmatrix}\end{aligned}$$

and
$$\begin{bmatrix} A_{u1} \\ \\ A_{u_2} \\ A_{u_3} \end{bmatrix} = \begin{bmatrix} h_1 \dfrac{\partial u_1}{\partial x_1} & h_1 \dfrac{\partial u_1}{\partial x_2} & \cdots \\ h_2 \dfrac{\partial u_2}{\partial x_1} & \cdots & \cdots \\ \cdots & \cdots & \cdots \end{bmatrix} \begin{bmatrix} A_1 \\ \\ A_2 \\ A_3 \end{bmatrix}$$
$$= \begin{bmatrix} h_1 & 0 & 0 \\ 0 & h_2 & 0 \\ 0 & 0 & h_3 \end{bmatrix} [T] \begin{bmatrix} A_1 \\ A_2 \\ A_3 \end{bmatrix}$$

There are two expressions for the direction cosines and, consequently, two distinct but equivalent transformation laws for the vector components $A_i$. These transformation laws can be simplified so as to eliminate the diagonal matrices by redefining our vector components. There are two possibilities, and for one we choose

$$\mathfrak{A}_{ui}^* = h_i A_{u_i}$$

We call the vector $\mathfrak{A}^*$ the covariant vector corresponding to our real vector **A**. The components $h_i A_{u_i}$ are its covariant components. The components of $\mathfrak{A}^*$ transform according to the law

$$[\mathfrak{A}_{u_i}^*] = [T^*][A_i]$$

We regard the matrix $[T^*]$ as our standard matrix, and the transformation effected by it as our standard transformation. For this reason $\mathfrak{A}^*$ is called a covariant vector signifying having the same variation, i.e., having the same transformation law, as our standard.

The contravariant vector corresponding to the real vector **A** is defined as the vector $\mathfrak{A}$ with contravariant components $h_i^{-1} A_{u_i}$. The contravariant components transform as follows upon change of coordinates:

$$[\mathfrak{A}_{u_i}] = [T][A_i]$$

The name contravariant signifies variation or transformation unlike or different from our standard transformation.

The advantages of the above definitions are:

1. The transformation matrices are simpler.
2. The transformation laws are symmetrical in the $x_i$ and $u_i$ coordinates.
3. If either the $x_i$ are given as functions of $u_i$, or vice versa, we can transform the covariant or contravariant components, respectively, in a straightforward manner, since we can readily evaluate the elements in the transformation matrix. After transforming, we convert back to our real vector by multiplying the covariant and contravariant components by $h_i^{-1}$ and $h_i$, respectively.

If the $x_i$ form an orthogonal curvilinear coordinate system as well, the above transformations are still valid for the covariant and contravariant components. The real vector in the $x_i$ system is then obtained by multiplying by suitable scale factors. As an example, we have the following transformation for the differentials $dx_i$, $[du_i] = [T][dx_i]$, and, hence, the $dx_i$ transform as the components of a contravariant vector. To convert to our real vector $d\mathbf{l} = \sum_{i=1}^{3} \mathbf{a}_i \, dx_i$, we multiply the components by $h_i$ to get $d\mathbf{l} = \sum_{i=1}^{3} h_i \, du_i \mathbf{e}_i$, which is the line element in the $u_i$ coordinate frame. As a second example, we consider the transformation of the gradient of a scalar function $\phi$. The gradient has components $\partial\phi/\partial x_i$, and, if we regard these as the covariant components, then the covariant components in the $u_i$ coordinate frame are given by

$$\left[\frac{\partial \phi}{\partial u_i}\right] = [T^*]\left[\frac{\partial \phi}{\partial x_i}\right]$$

The transformation law for the $\partial\phi/\partial x_i$ is just the familiar rule for partial differentiation, and is one reason why we regard $[T^*]$ as our standard transformation matrix. The components of the real vector corresponding to the components of its covariant counterpart are given by $h_i^{-1}\,\partial\phi/\partial u_i$, and these are the components of the gradient of $\phi$ in the $u_i$ coordinate frame.

**Divergence and Curl in Orthogonal Curvilinear Coordinates.** The divergence of a vector **A** in $u_i$ coordinates is readily evaluated from the definition given in subsection *b*. Consider a small-volume element $\Delta u_1\,\Delta u_2\,\Delta u_3/J$, as illustrated in Fig. A.1. The normal component of **A** at surface 1 is $A_{u_1}\mathbf{e}_1$, and the flux passing out through this

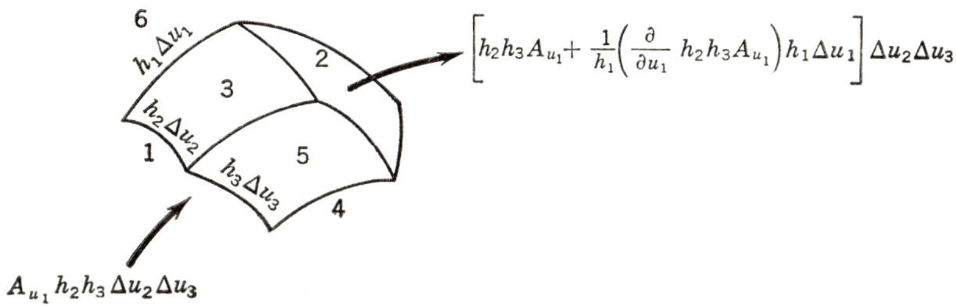


FIG. A.1. Illustration of divergence of a vector.

surface is $-A_{u_1}h_2h_3\,\Delta u_2\,\Delta u_3$. The flux through surface 2 is $A_{u_1}h_2h_3\,\Delta u_2\,\Delta u_3 + \left[\frac{1}{h_1}\frac{\partial}{\partial u_1}(h_3h_2A_{u_1})\right] h_1\,\Delta u_1\,\Delta u_2\,\Delta u_3$, and so the net outward flux from surfaces 1 and 2 is $\frac{\partial}{\partial u_1}(h_2h_3A_{u_1})\,\Delta u_1\,\Delta u_2\,\Delta u_3$. A cyclic interchange of subscripts gives the expressions for the net flux through surfaces 3 and 4 and surfaces 5 and 6. Combining all the terms gives

$$\nabla\cdot\mathbf{A} = \lim_{\Delta V\to 0}\frac{\oint\!\!\oint_S \mathbf{A}\cdot d\mathbf{a}}{\Delta V}$$
$$= \frac{1}{h_1h_2h_3}\left[\frac{\partial}{\partial u_1}(h_2h_3A_{u_1}) + \frac{\partial}{\partial u_2}(h_1h_3A_{u_2}) + \frac{\partial}{\partial u_3}(h_1h_2A_{u_3})\right]$$

The components of the curl of **A** may be evaluated in a similar fashion and, when expressed in determinant form, are given by

$$\nabla\times\mathbf{A} = \frac{1}{h_1h_2h_3}\begin{vmatrix} h_1\mathbf{e}_1 & h_2\mathbf{e}_2 & h_3\mathbf{e}_3 \\ \frac{\partial}{\partial u_1} & \frac{\partial}{\partial u_2} & \frac{\partial}{\partial u_3} \\ h_1A_{u_1} & h_2A_{u_2} & h_3A_{u_3} \end{vmatrix}$$

The Laplacian of a function is the divergence of the gradient and is written as $\nabla\cdot\nabla\phi = \nabla^2\phi$. In rectangular coordinates it is equal to

$$\nabla^2\phi = \sum_{i=1}^{3}\frac{\partial^2\phi}{\partial x_i^2}$$

The form for the Laplacian in curvilinear coordinates is obtained by replacing the components $A_{u_i}$ in the expression for $\nabla \cdot \mathbf{A}$ by those of $\nabla\phi$, that is, by $\dfrac{1}{h_i}\dfrac{\partial \phi}{\partial u_i}$.

*e. Properties of Scalar and Vector Fields*

In the present subsection we shall derive some of the more basic properties common to all scalar and vector fields. Our frame of reference will be an $xyz$ rectangular coordinate frame, but the results will hold in any coordinate frame.

**Scalar Fields.** A scalar function $\phi(x,y,z)$ with continuous first-order partial derivatives defined at almost all points in a given space represents a scalar field. The points where $\phi$ or its derivatives are discontinuous are the sources or sinks of the field. The gradient of $\phi$ is a vector which measures the maximum rate of change of the field in space. The line integral of $\nabla\phi$ between two points $p_1$ and $p_2$ connected by a curve $C$ is $\int_{p_1}^{p_2} \nabla\phi \cdot d\mathbf{l}$. The component of $\nabla\phi$ along $d\mathbf{l}$ is just the directional derivative of $\phi$ along the curve $C$, and, hence,

$$\int_{p_1}^{p_2} \nabla\phi \cdot d\mathbf{l} = \int_{p_1}^{p_2} \frac{d\phi}{dl}\, dl = \phi(p_2) - \phi(p_1)$$

independently of the path $C$ followed. For $\phi$ to represent a physical field, it must be single-valued, and, hence, the line integral around a closed contour vanishes; i.e.,

$$\oint_C \nabla\phi \cdot d\mathbf{l} = 0$$

Because of the above line-integral property, the scalar field $\phi$ is called a conservative potential field. The constant potential surfaces are the surfaces $\phi$ = constant. The vector $\nabla\phi$ is tangent to the system of lines which intersect the constant potential surfaces orthogonally. When $\nabla\phi$ has zero divergence, $\phi$ is a solution of Laplace's equation $\nabla^2\phi = 0$. If we let $u = v = \phi$ in Green's first identity, we get

$$\iiint_V (\nabla\phi)^2\, dV = \oiint_S \phi \frac{\partial \phi}{\partial n}\, da$$

when $\nabla^2\phi = 0$ in $V$. If $\phi$ is constant on $S$, the surface integral vanishes, and, hence, $\nabla\phi = 0$ also. Thus $\phi$ is constant throughout $V$ also. When $\phi$ vanishes on $S$, it vanishes everywhere within $V$ also, unless it has a singularity or a source inside $S$.

**Vector Fields.** A vector function $\mathbf{P}(x,y,z)$ represents a vector field. The points at which $\mathbf{P}$ or its derivatives are discontinuous correspond to the sources or sinks of the field or to idealized physical boundaries. When $\nabla \cdot \mathbf{P} = 0$, we say that $\mathbf{P}$ is a solenoidal or rotational vector field. Since $\nabla \cdot \nabla \times \mathbf{A} = 0$, we may derive $\mathbf{P}$ from the curl of a suitable vector potential function $\mathbf{A}$. All the lines of force in a divergenceless field close upon themselves; i.e., nowhere do they terminate in a source or sink. The relation $\mathbf{P} = \nabla \times \mathbf{A}$ determines $\mathbf{A}$ only to within an additive arbitrary vector $\nabla\phi$, which is the gradient of a scalar function, since $\nabla \times \nabla\phi = 0$. Thus we may equally well derive $\mathbf{P}$ from the curl of a vector $\mathbf{A}'$, where $\mathbf{A}' = \mathbf{A} + \nabla\phi$. The transformation to the new vector potential $\mathbf{A}'$ is called a gauge transformation.

When $\nabla \times \mathbf{P} = 0$, we call $\mathbf{P}$ an irrotational or lamellar field. From Stokes' law we have

$$\oint_C \mathbf{P} \cdot d\mathbf{l} = \iint_S \nabla \times \mathbf{P} \cdot d\mathbf{a} = 0$$

and, hence, we may derive **P** from the gradient of a scalar function as follows:

$$\mathbf{P} = -\nabla\psi$$

where $\psi$ is called a scalar potential. It may readily be shown that, given a vector function **P** which is continuous with continuous derivatives in a region $V$ and on its boundary $S$, any one of the following statements implies the rest:

1. The line integral $\int_1^2 \mathbf{P} \cdot d\mathbf{l}$ is independent of the path followed in $V$.
2. $\oint_C \mathbf{P} \cdot d\mathbf{l} = 0$ for every closed contour in $V$.
3. $\psi$ exists such that $\mathbf{P} = -\nabla\psi$.
4. $\nabla \times \mathbf{P} = 0$ throughout $V$.

Helmholtz's theorem states that a general vector field will have both a solenoidal part and a lamellar part and may be derived from a vector and a scalar potential. Let the solenoidal part be $\mathbf{P}_s$ and the lamellar part be $\mathbf{P}_l$. The divergence of **P** does not vanish, and, hence, we put

$$\nabla \cdot \mathbf{P} = \nabla \cdot \mathbf{P}_l = \rho$$

where $\rho$ is a scalar function constituting the source for $\mathbf{P}_l$. The source for $\mathbf{P}_s$ is **J** and gives the curl of $\mathbf{P}_s$, that is,

$$\nabla \times \mathbf{P} = \nabla \times \mathbf{P}_s = \mathbf{J}$$

where **J** is a vector source function. Since $\nabla \cdot \mathbf{P}_s = 0$, we may put $\mathbf{P}_s$ equal to the curl of a vector potential as follows:

$$\mathbf{P}_s = \nabla \times \mathbf{A}$$

The lamellar part $\mathbf{P}_l$ may be put equal to $-\nabla\psi$, where $\psi$ is a scalar potential. Substituting for $\mathbf{P}_s$ and $\mathbf{P}_l$ into the source equations gives

$$\nabla \times \nabla \times \mathbf{A} = \nabla\nabla \cdot \mathbf{A} - \nabla^2\mathbf{A} = \mathbf{J}$$
$$\nabla \cdot \nabla\psi = \nabla^2\psi = -\rho$$

By a suitable gauge transformation, the divergence of **A** can be made zero, and so we may take, in general,

$$\nabla^2\mathbf{A} = -\mathbf{J}$$

When the sources **J** and $\rho$ are interrelated and are functions of time as well, then the equations satisfied by the potentials are not so simple as those given above. A familiar example is the electromagnetic field.

The function $(4\pi r)^{-1}$, where $r^2 = (x - x_0)^2 + (y - y_0)^2 + (z - z_0)^2$, is a solution of Poisson's equation for a unit source located at $(x_0,y_0,z_0)$, that is,

$$\nabla^2 \frac{1}{r} = -4\pi\delta(x - x_0)\delta(y - y_0)\delta(z - z_0)$$

where the delta functions are used to represent a unit source. The delta function $\delta(x - x_0)$ is defined to be equal to zero at all points except $x = x_0$, where it becomes infinite in such a manner that

$$\int_{x-\alpha}^{x+\alpha} \delta(x - x_0)\, dx_0 = 1$$

where $\alpha$ is a small positive constant. If $\mathbf{B}(x_0)$ is an arbitrary vector function which is continuous at $x_0 = x$, we get the result

$$\int_{x_1}^{x_2} \mathbf{B}(x_0)\delta(x - x_0)\, dx_0 = \begin{cases} 0 & x \text{ not in the interval } x_1 - x_2 \\ \mathbf{B}(x) & x \text{ in the interval } x_1 - x_2 \end{cases}$$

A similar result holds for the product of the three delta functions $\delta(x - x_0)\delta(y - y_0)\delta(z - z_0)$. If the interval of integration over a given volume contains the point $(x,y,z)$, then the integral of the product of an arbitrary continuous function times the three-dimensional delta function gives the value of the function at the point $(x,y,z)$, that is,

$$\iiint_V \mathbf{B}(x_0,y_0,z_0)\delta(x - x_0)\delta(y - y_0)\delta(z - z_0)\, dV_0 = \begin{cases} 0 & (x,y,z) \text{ not in } V \\ \mathbf{B}(x,y,z) & (x,y,z) \text{ in } V \end{cases}$$

The exact analytical form of the delta function does not concern us since we will use only the above properties. It may be noted that the function $-\nabla^2 (1/4\pi r)$ has the above properties. This function is symmetrical in the variables $x$, $y$, $z$ and $x_0$, $y_0$, $z_0$, so that

$$\nabla_0{}^2 \frac{1}{r} = \nabla^2 \frac{1}{r}$$

where $\nabla_0$ signifies differentiation with respect to the variables $x_0$, $y_0$, $z_0$.

From this property of the function $(4\pi r)^{-1}$ we may obtain the solutions for the potentials $\mathbf{A}$ and $\psi$ by superposition. The results are

$$\mathbf{A} = \frac{1}{4\pi} \iiint_V \frac{\mathbf{J}(x_0,y_0,z_0)}{r}\, dV_0$$

$$\psi = \frac{1}{4\pi} \iiint_V \frac{\rho(x_0, \ldots)}{r}\, dV_0$$

which is readily demonstrated by operating on $\mathbf{A}$ and $\psi$ by $\nabla^2$ and using the delta function property of $-\nabla^2 (1/4\pi r)$.

We will now examine the properties of our general vector field $\mathbf{P} = \mathbf{P}_s + \mathbf{P}_l$ in the interior of a volume $V$ and on the surface $S$ which completely surrounds $V$. Using the delta-function property of $\frac{-1}{4\pi} \nabla^2 \frac{1}{r}$ permits us to write

$$\mathbf{P}(x,y,z) = -\frac{1}{4\pi} \nabla^2 \iiint_V \frac{P(x_0,y_0,z_0)}{r}\, dV_0$$

The $\nabla^2$ operator is equal to $\nabla\nabla\cdot - \nabla \times \nabla \times$, and so we obtain the result

$$\mathbf{P}(x,y,z) = -\frac{1}{4\pi} \nabla\nabla \cdot \iiint_V \frac{\mathbf{P}(x_0,y_0,z_0)}{r}\, dV_0 + \nabla \times \nabla \times \frac{1}{4\pi} \iiint_V \frac{\mathbf{P}(x_0, \ldots)}{r}\, dV_0$$

which will enable us to express $\mathbf{P}$ as the negative gradient of a scalar function plus the curl of a vector function.

Consider the following integral:

$$\nabla \times \iiint\limits_V \frac{P(x_0, \ldots)}{r}\, dV_0 = \iiint\limits_V \nabla \times \frac{\mathbf{P}(x_0, \ldots)}{r}\, dV_0$$

Now $\nabla \times \dfrac{\mathbf{P}(x_0, \ldots)}{r} = \mathbf{P}(x_0, \ldots) \times \nabla_0 \dfrac{1}{r} = -\nabla_0 \times \dfrac{\mathbf{P}(x_0, \ldots)}{r} + \dfrac{1}{r} \nabla_0 \times \mathbf{P}(x_0, \ldots)$

The above integral becomes

$$\iiint\limits_V \frac{\nabla_0 \times \mathbf{P}(x_0, \ldots)}{r}\, dV_0 - \iiint\limits_V \nabla_0 \times \frac{\mathbf{P}}{r}\, dV_0 = \iiint\limits_V \frac{\nabla_0 \times \mathbf{P}}{r}\, dV_0 + \oiint\limits_S \frac{\mathbf{P} \times \mathbf{n}}{r}\, dV_0$$

The reduction of the second volume integral to a surface integral is readily accomplished by applying the divergence theorem to the following vector function, $\mathbf{a} \times \mathbf{P}/r$, where $\mathbf{a}$ is an arbitrary constant vector. We get

$$\iiint\limits_V \left(\nabla_0 \cdot \mathbf{a} \times \frac{\mathbf{P}}{r}\right) dV_0 = -\mathbf{a} \cdot \iiint\limits_V \nabla_0 \times \frac{\mathbf{P}}{r}\, dV_0 = \oiint\limits_S \left(\mathbf{a} \times \frac{\mathbf{P}}{r}\right) \cdot \mathbf{n}\, da_0 = \mathbf{a} \cdot \oiint\limits_S \frac{\mathbf{P} \times \mathbf{n}}{r}\, da_0$$

and, since $\mathbf{a}$ is arbitrary, the required result follows at once by equating the two integrals which are dot-multiplied by $\mathbf{a}$.

Next we consider the following integral:

$$\begin{aligned} \nabla \cdot \iiint\limits_V \frac{\mathbf{P}(x_0, \ldots)}{r}\, dV_0 &= \iiint\limits_V \nabla \cdot \frac{\mathbf{P}(x_0, \ldots)}{r}\, dV_0 \\ &= -\iiint\limits_V \nabla_0 \cdot \frac{\mathbf{P}}{r}\, dV_0 + \iiint\limits_V \frac{\nabla_0 \cdot \mathbf{P}}{r}\, dV_0 \\ &= \iiint\limits_V \frac{\nabla_0 \cdot \mathbf{P}}{r}\, dV_0 - \oiint\limits_S \frac{\mathbf{P} \cdot \mathbf{n}}{r}\, da_0 \end{aligned}$$

since $\qquad \nabla \cdot \dfrac{\mathbf{P}}{r} = \mathbf{P} \cdot \nabla \dfrac{1}{r} = -\mathbf{P} \cdot \nabla_0 \dfrac{1}{r} = \dfrac{1}{r} \nabla_0 \cdot \mathbf{P} - \nabla_0 \cdot \dfrac{\mathbf{P}}{r}$

Collecting our results together shows that the vector field $\mathbf{P}$ is given by the following expression:

$$\mathbf{P} = \nabla \times \left(\iiint\limits_V \frac{\nabla_0 \times \mathbf{P}}{4\pi r}\, dV_0 + \oiint\limits_S \frac{\mathbf{P} \times \mathbf{n}}{4\pi r}\, da_0\right) - \nabla \left(\iiint\limits_V \frac{\nabla_0 \cdot \mathbf{P}}{4\pi r}\, dV_0 - \oiint\limits_S \frac{\mathbf{P} \cdot \mathbf{n}}{4\pi r}\, da_0\right)$$

which is the mathematical statement of Helmholtz's theorem. We note that only if both $\nabla \times \mathbf{P}$ and $\mathbf{P} \times \mathbf{n}$ are equal to zero will the field in $V$ be a pure lamellar field. Similarly, only if both $\nabla \cdot \mathbf{P}$ and $\mathbf{P} \cdot \mathbf{n}$ vanish will the field be a pure solenoidal field.

This is not a surprising situation since, if the field outside $S$ is to be zero, then either $\mathbf{P} \times \mathbf{n}$ and $\mathbf{P} \cdot \mathbf{n}$ vanishes on $S$ or else a surface distribution of sources must be placed on $S$ to account for the discontinuous change in $\mathbf{P}$ as the surface $S$ is traversed from the interior to the exterior of $V$. Thus, even though the volume sources $\mathbf{J} = \nabla \times \mathbf{P}$ and $\rho = \nabla \cdot \mathbf{P}$ vanish, the surface sources may not, and the field $\mathbf{P}$ in the interior of $V$ will have both a lamellar and a solenoidal part. These results are of fundamental importance in the theory of electromagnetic cavities or resonators.

## A.2. Dyadic Analysis

When each component of a vector $\mathbf{A}$ is linearly related to all three components of a vector $\mathbf{B}$ as follows:

$$A_i = \sum_{j=1}^{3} C_{ij} B_j \qquad i, j = x, y, z$$

then the relationship is conveniently represented by a matrix equation or by introducing the dyadic which corresponds to the matrix with elements $C_{ij}$. The dyadic is obtained by associating with each element $C_{ij}$ the unit vectors $\mathbf{a}_i$, $\mathbf{a}_j$. The dyadic will be represented by boldface type with a bar above the quantity. The dyadic $\bar{\mathbf{C}}$ is given by

$$\begin{aligned} & \mathbf{a}_x\mathbf{a}_x C_{xx} + \mathbf{a}_x\mathbf{a}_y C_{xy} + \mathbf{a}_x\mathbf{a}_z C_{xz} \\ \bar{\mathbf{C}} = & + \mathbf{a}_y\mathbf{a}_x C_{yx} + \mathbf{a}_y\mathbf{a}_y C_{yy} + \mathbf{a}_y\mathbf{a}_z C_{yz} \\ & + \mathbf{a}_z\mathbf{a}_x C_{zx} + \mathbf{a}_z\mathbf{a}_y C_{zy} + \mathbf{a}_z\mathbf{a}_z C_{zz} \end{aligned}$$

The algebra of dyadics has a very close correspondence with matrix algebra. The dot product of two dyadics $\bar{\mathbf{C}}$ and $\bar{\mathbf{D}}$ is written $\bar{\mathbf{C}} \cdot \bar{\mathbf{D}}$, and is defined as the dyadic obtained by dotting the unit vectors which appear on either side and adjacent to the dot. The result is seen to be the same as the matrix product of the two matrices which correspond to the dyadics $\bar{\mathbf{C}}$ and $\bar{\mathbf{D}}$. The product, in general, is not commutative, so that $\bar{\mathbf{C}} \cdot \bar{\mathbf{D}} \neq \bar{\mathbf{D}} \cdot \bar{\mathbf{C}}$ unless $\bar{\mathbf{C}}$ and $\bar{\mathbf{D}}$ are symmetric dyadics. A dyadic is symmetric if its associated matrix is symmetric. The dot product of a dyadic with a vector is defined in a similar way. Thus, we have $\mathbf{A} = \bar{\mathbf{C}} \cdot \mathbf{B}$ for the system of equations introduced earlier.

The sum or difference of two dyadics is the dyadic with components which are the sum or difference of the components of the individual dyadics. Multiplication by a constant multiplies each component by this constant. Corresponding to the unit matrix is a unit dyadic called the idemfactor. The idemfactor is given by

$$\bar{\mathbf{I}} = \mathbf{a}_x\mathbf{a}_x + \mathbf{a}_y\mathbf{a}_y + \mathbf{a}_z\mathbf{a}_z$$

and multiplication of a vector or dyadic by $\bar{\mathbf{I}}$ leaves the vector or dyadic unchanged. The dyadic corresponding to the transposed matrix is called the conjugate dyadic and will be represented as $\bar{\mathbf{C}}^*$, where the star indicates the conjugate. The following results follow from the corresponding property for matrix multiplication:

$$\begin{aligned} \bar{\mathbf{C}} \cdot \bar{\mathbf{D}} &= \bar{\mathbf{D}}^* \cdot \bar{\mathbf{C}}^* \\ \bar{\mathbf{C}} \cdot \bar{\mathbf{B}} &= \mathbf{B} \cdot \bar{\mathbf{C}}^* \\ \mathbf{A} \cdot (\bar{\mathbf{C}} \cdot \mathbf{B}) &= (\mathbf{A} \cdot \bar{\mathbf{C}}) \cdot \mathbf{B} \end{aligned}$$

The vector, or cross, product of two dyadics is formed by crossing the unit vectors which appear on adjacent sides of the cross in the expression $\bar{\mathbf{C}} \times \bar{\mathbf{D}}$. A similar definition applies to the cross product between a vector and a dyadic. The following useful relation is also readily proved:

$$(\mathbf{A} \cdot \bar{\mathbf{C}}) \times \mathbf{B} = \mathbf{A} \cdot (\bar{\mathbf{C}} \times \mathbf{B})$$

The dyadic $\bar{\mathbf{C}}$ may be written as a sum of three suitably defined vectors associated with the three unit vectors $\mathbf{a}_x$, $\mathbf{a}_y$, $\mathbf{a}_z$ as follows:

$$\bar{\mathbf{C}} = \mathbf{C}_1\mathbf{a}_x + \mathbf{C}_2\mathbf{a}_y + \mathbf{C}_3\mathbf{a}_z$$

where the associated vectors are given by

$$\begin{aligned} \mathbf{C}_1 &= C_{xx}\mathbf{a}_x + C_{yx}\mathbf{a}_y + C_{zx}\mathbf{a}_z \\ \mathbf{C}_2 &= C_{xy}\mathbf{a}_x + C_{yy}\mathbf{a}_y + C_{zy}\mathbf{a}_z \\ \mathbf{C}_3 &= C_{xz}\mathbf{a}_x + C_{yz}\mathbf{a}_y + C_{zz}\mathbf{a}_z \end{aligned}$$

The divergence of the dyadic $\bar{\mathbf{C}}$ is readily seen to be

$$\nabla \cdot \bar{\mathbf{C}} = (\nabla \cdot \mathbf{C}_1)\mathbf{a}_x + (\nabla \cdot \mathbf{C}_2)\mathbf{a}_y + (\nabla \cdot \mathbf{C}_3)\mathbf{a}_z$$

and is a vector. The curl of the dyadic $\bar{\mathbf{C}}$ is given by

$$\nabla \times \bar{\mathbf{C}} = (\nabla \times \mathbf{C}_1)\mathbf{a}_x + (\nabla \times \mathbf{C}_2)\mathbf{a}_y + (\nabla \times \mathbf{C}_3)\mathbf{a}_z$$

Using the concept of associated vectors allows us to readily derive the equivalent of the divergence theorem and Stokes' theorem for dyadics. The results are

$$\iiint_V \nabla \cdot \bar{\mathbf{C}}\, dV = \oiint_S d\mathbf{a} \cdot \bar{\mathbf{C}} = \oiint_S \mathbf{n} \cdot \bar{\mathbf{C}}\, da$$

$$\iint_S d\mathbf{a} \cdot \nabla \times \bar{\mathbf{C}} = \oint_C d\mathbf{l} \cdot \bar{\mathbf{C}}$$

A dyadic may be referred to another coordinate system by finding the projection of the unit vectors $\mathbf{a}_x$, $\mathbf{a}_y$, $\mathbf{a}_z$ on the unit vectors in the second frame of reference. Of particular importance is a transformation to a coordinate frame in which a dyadic $\bar{\mathbf{C}}$ will have the simple diagonal form

$$\bar{\mathbf{C}} = C_{11}\mathbf{e}_1\mathbf{e}_1 + C_{22}\mathbf{e}_2\mathbf{e}_2 + C_{33}\mathbf{e}_3\mathbf{e}_3$$

where $\mathbf{e}_1$, $\mathbf{e}_2$, $\mathbf{e}_3$ are unit vectors along the preferred coordinate curves. This is referred to as a transformation to principal axes or normal coordinates. Let $\mathbf{n}$ be a unit vector along one of the principal axes, i.e., along one of the directions defined by $\mathbf{e}_1$, $\mathbf{e}_2$, or $\mathbf{e}_3$. The dot product of $\mathbf{n}$ with $\bar{\mathbf{C}}$ must then yield simply a vector in the direction of $\mathbf{n}$, say $\lambda\mathbf{n}$. Hence, we have

$$\bar{\mathbf{C}} \cdot \mathbf{n} = \lambda\mathbf{n} = \lambda\bar{\mathbf{I}} \cdot \mathbf{n}$$

or

$$(\bar{\mathbf{C}} - \lambda\bar{\mathbf{I}}) \cdot \mathbf{n} = 0$$

This is a system of three homogeneous equations for which a solution for $n_x$, $n_y$, $n_z$ can exist only if the determinant of the unknown variables vanishes. Therefore, we must have

$$\begin{vmatrix} C_{xx} - \lambda & C_{xy} & C_{xz} \\ C_{yx} & C_{yy} - \lambda & C_{yz} \\ C_{zx} & C_{zy} & C_{zz} - \lambda \end{vmatrix} = 0$$

This equation is often called a secular equation. It is a cubic equation in $\lambda$, the solutions for $\lambda$ being called the eigenvalues or latent roots. For each root of $\lambda$, a solution for $n_x$, $n_y$, $n_z$ can be found, and this solution is called an eigenvector or modal column. When the dyadic $\bar{\mathbf{C}}$ is real and symmetric, all the eigenvalues are real, and the eigenvectors may be chosen real. When all the roots of $\lambda$ are distinct, the eigenvectors are mutually orthogonal. If two or more roots are equal, the eigenvectors can be chosen so as to form an orthogonal set in an infinite number of ways.

### A.3. Matrices

In this section we will summarize some basic definitions as well as derive some useful properties of matrices. The basic rules of matrix algebra will not be discussed.

We will write $[A]$ for the matrix

$$\begin{bmatrix} a_{11} & a_{12} & \cdots & a_{1n} \\ a_{21} & a_{22} & \cdots & \cdots \\ \cdots & \cdots & \cdots & \cdots \\ a_{n1} & \cdots & \cdots & a_{nn} \end{bmatrix}$$

whose elements are $a_{ij}$. The first subscript refers to the row and the second to the column. The determinant of $[A]$ will be written as $|A|$.

*Definitions*

**Transpose Matrix.** The matrix obtained by interchanging the rows and columns of $[A]$ is called the transpose matrix. It will be designated by $[A]_t$ and has the elements $a_{ji}$ in the $i$th row and $j$th column.

**Adjoint Matrix.** The matrix composed of the elements $A_{ji}$, where $A_{ji}$ is the cofactor of $a_{ij}$ in the determinant $|A|$, is called the adjoint matrix and will be designated by $[A]_a$.

**Inverse Matrix.** The inverse matrix $[A]^{-1}$ is defined so that

$$[A]^{-1}[A] = [A][A]^{-1} = [I]$$

where $[I]$ is the unit matrix. It has elements $A_{ji}/|A|$ and hence is given by

$$[A]^{-1} = \frac{[A]_a}{|A|}$$

**Orthogonal Matrix.** When $[A]_t = [A]^{-1}$, the matrix $[A]$ is an orthogonal matrix.

**Reciprocal Matrix.** The transpose of the inverse matrix is called the reciprocal matrix. It has elements $A_{ji}/|A|$ in the $i$th row and $j$th column.

**Symmetrical Matrix.** When $a_{ij} = a_{ji}$, the matrix is symmetrical.

**Hermitian Matrix.** When $a_{ij} = a_{ji}^*$, the matrix is a Hermitian matrix. The (*) denotes the complex conjugate value.

**Unitary Matrix.** When the matrix $[A]$ is equal to the inverse of the complex conjugate transposed matrix $([A^*]_t)^{-1}$, it is called a unitary matrix.

*Matrix Eigenvalue Problem*

The following equation is a typical matrix eigenvalue problem:

$$\begin{bmatrix} a_{11} & a_{12} & \cdots & a_{1n} \\ a_{21} & \cdots & \cdots & \cdots \\ \cdots & \cdots & \cdots & \cdots \\ \cdots & \cdots & \cdots & a_{nn} \end{bmatrix} \begin{bmatrix} e_1 \\ e_2 \\ \cdot \\ \cdot \\ \cdot \\ e_n \end{bmatrix} = \lambda \begin{bmatrix} p_{11} & p_{12} & \cdots & p_{1n} \\ p_{21} & \cdots & \cdots & \cdots \\ \cdots & \cdots & \cdots & \cdots \\ p_{n1} & \cdots & \cdots & p_{nn} \end{bmatrix} \begin{bmatrix} e_1 \\ e_2 \\ \cdot \\ \cdot \\ \cdot \\ e_n \end{bmatrix}$$

or $[A][E] = \lambda[P][E]$. The values of $\lambda$ for which this equation has a solution are called the eigenvalues or latent roots. The above equation may be written as

$$([A] - \lambda[P])[E] = 0$$

and a solution exists only if the following determinant vanishes:

$$\begin{vmatrix} a_{11} - \lambda p_{11} & a_{12} - \lambda p_{12} & \cdots & a_{1n} - \lambda p_{1n} \\ \cdots & \cdots & \cdots & \cdots \\ a_{n1} - \lambda p_{n1} & \cdots & \cdots & a_{nn} - \lambda p_{nn} \end{vmatrix} = 0$$

The above characteristic equation is of the $n$th degree in $\lambda$ and will have $n$ roots, some of which may be equal or degenerate. For each value of $\lambda$, say $\lambda_s$, a solution for $[E]$, say $[E_s]$, may be found. Each solution is called an eigenvector or modal column. When $[A]$ and $[P]$ are real and symmetric or Hermitian, all the eigenvalues are real. For the $s$th solution, we have

$$[A][E_s] = \lambda_s[P][E_s]$$
$$[A^*][E_s^*] = \lambda_s^*[P^*][E_s^*]$$

The transpose of the second equation gives

$$[E_s^*]_t[A] = \lambda_s^*[E_s^*][P]$$

when $[A]$ and $[P]$ are Hermitian. Postmultiply this latter equation by $[E_s,]$ premultiply the first by $[E_s^*]_t$, and subtract to get

$$(\lambda_s - \lambda_s^*)[E_s^*]_t[P][E_s] = 0$$

Since $[E_s^*]_t[P][E_s]$ is not equal to zero, $\lambda_s$ must equal $\lambda_s^*$, and, hence, $\lambda_s$ is real. Thus all the eigenvalues are real since $s$ is arbitrary.

When $[A]$ and $[P]$ are real and symmetric or Hermitian, the eigenvectors form an orthogonal set with respect to the weighting factors $P_{ij}$ when all the eigenvalues are distinct. The eigenvectors corresponding to degenerate roots are not uniquely determined. However, a basic set may be chosen in such a manner that the totality of eigenvectors still forms an orthogonal set with respect to the weighting factors $P_{ij}$.

We will prove the following orthogonality property for the case when $[A]$ and $[P]$ are real and symmetric and all the eigenvalues are distinct:

$$\sum_{i=1}^{n} \sum_{j=1}^{n} e_{is} e_{jr} P_{ij} = \delta_{sr}$$

where $\delta_{sr}$ is the Kronecker delta and equals unity for $r = s$, and zero otherwise. We begin with the equations

$$[A][E_s] = \lambda_s[P][E_s]$$

and its transpose for the $r$th solution,

$$[E_r]_t[A] = \lambda_r[E_r]_t[P]$$

Premultiplying the first by $[E_r]_t$, postmultiplying the second by $[E_s]$, and subtracting gives

$$(\lambda_s - \lambda_r)[E_r]_t[P][E_s] = 0$$

Since $\lambda_s \neq \lambda_r$, we get

$$[E_r]_t[P][E_s] = 0$$

which when expanded gives the orthogonality condition stated above. Since the magnitude of each eigenvector is not determined, they may always be normalized so that

$$\sum_{i=1}^{n} \sum_{j=1}^{n} e_{is}e_{js}P_{ij} = 1$$

In many cases in practice, the matrix $[P]$ is simply the unit matrix $[I]$, and the orthogonality condition simplifies to $\sum_{i=1}^{n} e_{is}e_{ir} = \delta_{sr}$.

*Cayley-Hamilton's Theorem*

This theorem states that a square matrix $[A]$ satisfies its own characteristic equation, i.e.,

$$F([A]) = ([A] - \lambda_1[I])([A] - \lambda_2[I]) \cdots ([A] - \lambda_n[I]) = 0$$

Postmultiplying by $[E_s]$ gives

$$\begin{aligned}([A] - \lambda_1[I]) \cdots ([A] - \lambda_{n-1}[I])[E_s]\,(\lambda_s - \lambda_n) \\ = ([A] - \lambda_1[I]) \cdots [E_s](\lambda_s - \lambda_s)(\lambda_s - \lambda_{s+1}) \cdots (\lambda_s - \lambda_n) = 0\end{aligned}$$

This result holds for all values of $s$. Any nonzero $n$-dimensional vector or column matrix $[B]$ may be expressed as a linear function of the $[E_s]$, say

$$[B] = \sum_{s=1}^{n} b_s[E_s]$$

Using the above result shows that $F([A])[B] = 0$, and, hence, $F([A]) = 0$, since $[B]$ is not equal to zero.

*Sylvester's Theorem*

Consider the following equation:

$$\begin{aligned}f([A]) = C_1([A] - \lambda_2[I])([A] - \lambda_3[I]) \cdots ([A] - \lambda_n[I]) \\ + C_2([A] - \lambda_1[I])([A] - \lambda_3[I]) \cdots ([A] - \lambda_n[I]) \\ + C_n([A] - \lambda_1[I]) \cdots ([A] - \lambda_{n-1}[I])\end{aligned}$$

which defines an $n$th-degree polynomial function of $[A]$. Postmultiplying by $[E_s]$ gives

$$f([A])[E_s] = C_s(\lambda_s - \lambda_1) \cdots (\lambda_s - \lambda_{s-1})(\lambda_s - \lambda_{s+1}) \cdots (\lambda_s - \lambda_n)[E_s]$$

since all terms containing the eigenvalue $\lambda_s$ vanish. Furthermore, it is seen that

$$f([A])[E_s] = f(\lambda_s)[E_s]$$

and, hence, the constants $C_s$ are given by

$$C_s = \frac{f(\lambda_s)}{\prod\limits_{\substack{i=1 \\ i \neq s}}^{n} (\lambda_s - \lambda_i)}$$

provided all the eigenvalues are distinct. Therefore $f([A])$ is given by

$$f([A]) = \sum_{s=1}^{n} f(\lambda_s) \left. \frac{\prod_{i=1}^{n} ([A] - \lambda_i[I])}{\prod_{i=1}^{n} (\lambda_s - \lambda_i)} \right|_{i \neq s}$$

a result known as Sylvester's theorem. This theorem is useful for finding the $m$th power of a matrix. For a $2 \times 2$ matrix with eigenvalues $\lambda_1$ and $\lambda_2$, we have

$$[A]^m = \lambda_1{}^m \frac{[A] - \lambda_2[I]}{\lambda_1 - \lambda_2} + \lambda_2{}^m \frac{[A] - \lambda_1[I]}{\lambda_2 - \lambda_1}$$

### A.4. Calculus of Variations

Consider the following integral in $N$-dimensional space:

$$\int_{V_N} F[x_1,x_2, \ldots ,x_N;u(x_1, \ldots ,x_N);v(x_1, \ldots ,x_N);\dot{u}_1,\dot{u}_2, \ldots ,\dot{u}_N; \dot{v}_1,\dot{v}_2, \ldots ,\dot{v}_N]\, dx_1\, dx_2 \cdots dx_N$$

where $\dot{u}_i = \partial u/\partial x_i$, $\dot{v}_i = \partial v/\partial x_i$. In the above, $u$ and $v$ are functions of the $x_i$ only. We are given the function $F$ and are required to find the two functions $u$ and $v$ so that the above integral will have a stationary value for arbitrary first-order variations in $u$ and $v$ that do not violate the boundary conditions; i.e., the values of $u$ and $v$ are specified on the boundary, and so the variations in $u$ and $v$ vanish at the ends of the ranges of integration. The functions $F$, $u$, and $v$ are assumed continuous with at least piecewise-continuous partial derivatives. The function $F$ is called the Lagrangian function. Let the first-order variations in $u$ and $v$ be $h(x_1, \ldots ,x_N)$ and $g(x_1, \ldots ,x_N)$, respectively, where $h$ and $g$ vanish on the boundary of $V_N$. The variations in $\dot{u}_i$ and $\dot{v}_i$ are $\partial h/\partial x_i$ and $\partial g/\partial x_i$, respectively. When $u$ and $v$ vary, the function $F$ is a function of $u$, $v$, $\dot{u}_i$, and $\dot{v}_i$. The first-order change in $F$ is obtained by expanding $F$ in a Taylor series about the assumed correct functional forms of $u$ and $v$; thus

$$\begin{aligned} F - F_0 = \delta F &= \frac{\partial F}{\partial u}\, du + \frac{\partial F}{\partial v}\, dv + \sum_{i=1}^{N} \left( \frac{\partial F}{\partial \dot{u}_i}\, d\dot{u}_i + \frac{\partial F}{\partial \dot{v}_i}\, d\dot{v}_i \right) \\ &= \frac{\partial F}{\partial u} h + \frac{\partial F}{\partial v} g + \sum_{i=1}^{N} \left( \frac{\partial F}{\partial \dot{u}_i} \frac{\partial h}{\partial x_i} + \frac{\partial F}{\partial \dot{v}_i} \frac{\partial g}{\partial x_i} \right) \end{aligned}$$

when higher-order terms are dropped. Hence we have

$$\begin{aligned} \delta \int_{V_N} F\, dx_1 \cdots dx_N &= \int_{V_N} \delta F\, dx_1 \cdots dx_N \\ &= \int_{V_N} \left[ \frac{\partial F}{\partial u} h + \frac{\partial F}{\partial v} g + \sum_{i=1}^{N} \left( \frac{\partial F}{\partial \dot{u}_i} \frac{\partial h}{\partial x_i} + \frac{\partial F}{\partial \dot{v}_i} \frac{\partial g}{\partial x_i} \right) \right] dx_1 \cdots dx_N \end{aligned}$$

where $\delta$ signifies the first variation. Consider the $N$-dimensional integral of a typical term like $\dfrac{\partial F}{\partial \dot{u}_i} \dfrac{\partial h}{\partial x_i}$. We have

$$\int_{V_{N-1}} \left( \int_{x_{i1}}^{x_{i2}} \frac{\partial F}{\partial \dot{u}_i} \frac{\partial h}{\partial x_i} dx_i \right) dx_1 \cdots dx_{i-1}\, dx_{i+1} \cdots dx_N$$
$$= \int_{V_{N-1}} \left( \frac{\partial F}{\partial \dot{u}_i} h \Big|_{x_{i1}}^{x_{i2}} - \int_{x_{i1}}^{x_{i2}} h \frac{\partial}{\partial x_i} \frac{\partial F}{\partial \dot{u}_i} dx_i \right) dx_1 \cdots dx_N$$

by an integration by parts once. The points $x_{i1}$ and $x_{i2}$ lie on the boundary of $V_N$, and, since $h$ vanishes on the boundary, the integrated term vanishes, and we are left with

$$-\int_{V_N} h \frac{\partial}{\partial x_i} \frac{\partial F}{\partial \dot{u}_i} dx_1 \cdots dx_N$$

Using this result, the integral of $\delta F$ over $V_N$ becomes

$$\int_{V_N} \left[ h \left( \frac{\partial F}{\partial u} - \sum_{i=1}^{N} \frac{\partial}{\partial x_i} \frac{\partial F}{\partial \dot{u}_i} \right) + g \left( \frac{\partial F}{\partial v} - \sum_{i=1}^{N} \frac{\partial}{\partial x_i} \frac{\partial F}{\partial \dot{v}_i} \right) \right] dx_1 \cdots dx_N$$

In order for the original integral to be stationary, the first-order variation must vanish. Since $g$ and $h$ are arbitrary, we conclude that

$$\frac{\partial F}{\partial u} - \sum_{i=1}^{N} \frac{\partial}{\partial x_i} \frac{\partial F}{\partial \dot{u}_i} \qquad \text{and} \qquad \frac{\partial F}{\partial v} - \sum_{i=1}^{N} \frac{\partial}{\partial x_i} \frac{\partial F}{\partial \dot{v}_i}$$

must vanish independently. These two partial differential equations determine the functional forms of $u$ and $v$ which make the integral of $F$ over $V_N$ stationary. This partial differential equation is called the Euler-Lagrange differential equation. If $F$ was a function of any number of functions $u$, $v$, $w$, etc., we would find that each function $u$, $v$, $w$, etc., satisfied a partial differential equation of the above form.

*Constraints*

When the functions $u$ and $v$ are not independent but are related by an implicit functional form

$$G(u;v;\dot{u}_1, \ldots, \dot{u}_N;\dot{v}_1, \ldots, \dot{v}_N;x_1, \ldots, x_N) = 0$$

we call $G$ an equation of constraint, since it constrains the functions $u$ and $v$ to satisfy a certain functional relationship. In our variational integral, the variations in $u$ and $v$ are now constrained in such a manner that $G = 0$ always. The integral of $G$ over $V_N$ is also identically zero, so that any multiple of $G$ may be added to $F$ without changing the value of the integral of $F$ over $V_N$. Thus we consider the variation in the following integral:

$$\int_{V_N} (F + \lambda G)\, dx_1 \cdots dx_N$$

The multiplier $\lambda$ is in general a function of $x_1, \ldots, x_N$, and is called a Lagrange multiplier. Since the constraint function has been introduced into the integral, the variation may be conducted by treating $u$ and $v$ as independent functions. Thus, we obtain the two following partial differential equations, which, together with the original constraint function $G = 0$, allow us to find a solution for $u$, $v$, and $\lambda$,

$$\frac{\partial(F + \lambda G)}{\partial u} - \sum_{i=1}^{N} \frac{\partial}{\partial x_i} \frac{\partial(F + \lambda G)}{\partial \dot{u}_i} = 0$$

$$\frac{\partial(F + \lambda G)}{\partial v} - \sum_{i=1}^{N} \frac{\partial}{\partial x_i} \frac{\partial(F + \lambda G)}{\partial \dot{v}_i} = 0$$

### A.5. Infinite Products and the Gamma Function

Let $f(z)$ be an integral function, i.e., a function with no singularities in the finite complex $z$ plane. The logarithmic derivative of this function is then a meromorphic function, i.e., a function whose only singularities are poles. Let $f'(z)/f(z)$ have simple poles at $z = z_n$ with unit residues. The partial-fraction expansion is

$$\frac{f'(z)}{f(z)} = \frac{f'(z)}{f(z)}\bigg|_{z=0} + \sum_{n} \left(\frac{1}{z - z_n} + \frac{1}{z_n}\right)$$

Integrating with respect to $z$ from 0 to $z$ gives

$$\ln f(z)\bigg|_0^z = \frac{f'(0)}{f(0)} z + \sum_{n} \left(\ln \frac{z - z_n}{-z_n} + \frac{z}{z_n}\right)$$

or

$$f(z) = f(0)e^{[f'(0)/f(0)]z} \prod_{n} \left(1 - \frac{z}{z_n}\right) e^{z/z_n}$$

which is the infinite-product expansion of the integral function $f(z)$. When $f(z)$ is an even function $f_e(z)$ of $z$, $f_e'(0)$ equals zero, and we get

$$f_e(z) = f_e(0) \prod_{n} \left(1 - \frac{z^2}{z_n{}^2}\right)$$

Consider the function cos $z$, which has zeros at $z = n\pi + \pi/2 = (n + \tfrac{1}{2})\pi$. Using the general formula gives the following infinite-product expansion:

$$\cos z = \prod_{n=0}^{\infty} \left[1 - \frac{z^2}{(n + \frac{1}{2})^2\pi^2}\right]$$

Similarly the infinite-product expansion of $(\sin z)/z$ is found to be

$$\sin z = z \prod_{n=1}^{\infty} \left(1 - \frac{z^2}{n^2\pi^2}\right)$$

*Gamma Function*

The gamma function is defined by the following integral:

$$\Gamma(z) = \int_0^\infty e^{-u}u^{z-1}\, du$$

The integral defines an analytic function of $z$ for all $z$ for which the real part is positive. An equivalent definition of the $\Gamma$ function is the following infinite-product representation:

$$\Gamma(z) = \frac{e^{-\gamma z}}{z} \left[\prod_{n=1}^{\infty} \left(1 + \frac{z}{n}\right) e^{-z/n}\right]^{-1}$$

where $\gamma = 0.57722$ and is Euler's constant. The infinite-product representation defines $\Gamma(z)$ as an analytic function throughout the complex $z$ plane, except in the vicinity of the negative-real axis, where $\Gamma(z)$ has simple poles. The following useful properties of the gamma function have been established:

1. $\Gamma(z + 1) = z\Gamma(z)$.
2. $\Gamma(n + 1) = n!$, $n$ an integer.
3. $\Gamma(\tfrac{1}{2}) = \pi^{1/2}$.
4. $2^{2z-1}\Gamma(z)\Gamma(z + \tfrac{1}{2}) = \pi^{1/2}\Gamma(2z)$.
5. $\Gamma(z)\Gamma(1 - z) = \left[z \prod_{n=1}^{\infty} \left(1 - \dfrac{z^2}{n^2}\right)\right]^{-1} = \dfrac{\pi}{\sin \pi z}$.
6. As $z \to \infty$, the asymptotic value of $\Gamma(z)$ is $(2\pi)^{1/2}e^{z \ln z}z^{-1/2}e^{-z+1}$, valid for all $z$ except in the region of the negative-real axis, where $\Gamma(z)$ has poles at $z = 0, -1, -2, \ldots$, etc. From (5) an asymptotic expansion valid for $z$ on the negative-real axis may be found.
7. For $x$ finite and $|y|$ very large, $|\Gamma(x + jy)| \sim (2\pi)^{1/2}|y|^{x-1/2}e^{-\pi|y|/2}e^{1-x}$.

### A.6. Summation of Fourier Series

The solution of many waveguide problems involves Fourier series for which it is desirable to have summation formulas. There are several methods by which a wide variety of Fourier series may be summed in closed form or converted to more rapidly converging series. Some of the more useful methods will be considered here. The type of series to be considered is of the form

$$\sum_n \frac{e^{j\alpha n}}{P(n)}$$

where $P(n)$ is a rational function of the summation variable $n$. A number of the summation formulas which will be derived will be for the exponential function $e^{j\alpha n}$. The required summation formulas for the corresponding trigonometric series may be obtained by taking the real and imaginary parts separately.

*Geometric-series Method*

The following geometric series is a useful one from which several summation formulas may be developed:

$$\sum_{n=1}^{\infty} r^n = \frac{r}{1 - r} \qquad |r| < 1$$

Let $r = \lambda e^{j\alpha}$, where $\lambda$ is a positive number less than unity. The sum of the following series follows at once:

$$\sum_{n=1}^{\infty} \lambda^n e^{j\alpha n} = \frac{\lambda e^{j\alpha}}{1 - \lambda e^{j\alpha}}$$

Taking the limit as $\lambda$ approaches unity, we get

$$\sum_{n=1}^{\infty} e^{j\alpha n} = \frac{e^{j\alpha}}{1 - e^{j\alpha}}$$

Equating the real and imaginary parts gives

$$\sum_{n=1}^{\infty} \cos n\alpha = -\tfrac{1}{2} \qquad \sum_{n=1}^{\infty} \sin n\alpha = \tfrac{1}{2} \cot \frac{\alpha}{2}$$

The above series may be differentiated with respect to $\alpha$ before taking the limit and gives

$$\sum_{n=1}^{\infty} jn\lambda^n e^{j\alpha n} = \frac{j\lambda e^{j\alpha}}{(1 - \lambda e^{j\alpha})^2}$$

Equating the real and imaginary parts and taking the limit as $\lambda$ approaches unity gives

$$\sum_{n=1}^{\infty} n \cos n\alpha = \lim_{\lambda \to 1} \operatorname{Re} \frac{\lambda e^{j\alpha}}{(1 - \lambda e^{j\alpha})^2} = -\tfrac{1}{4} \csc^2 \frac{\alpha}{2}$$

$$\sum_{n=1}^{\infty} n \sin n\alpha = 0$$

In the above series it should be noted that, for $\lambda = 1$, the points $\alpha = m\pi$ are singular points. From the integral of the series $\sum_{n=1}^{\infty} e^{j\alpha n}$, further summation formulas may be obtained. For example, we have

$$j \int_{j\infty}^{\alpha} \sum_{n=1}^{\infty} e^{j\alpha n}\, d\alpha = \sum_{n=1}^{\infty} \frac{e^{j\alpha n}}{n} = j \int_{j\infty}^{\alpha} \frac{e^{j\alpha}}{1 - e^{j\alpha}}\, d\alpha = -\ln (1 - e^{j\alpha})$$

To justify interchanging the order of integration and summation, we should include the factor $\lambda$, and take the limit as $\lambda$ approaches 1; however, the result is the same as given above.

Sometimes it is desirable to sum a series over only odd or even values of $n$. Such summations may be performed by replacing $\alpha$ by $\alpha + \pi$ and adding or subtracting series. To sum the series

$$\sum_{n=1,3,\ldots}^{\infty} \frac{e^{jn\alpha}}{n}$$

we consider

$$\sum_{n=1}^{\infty} \frac{e^{jn(\alpha+\pi)}}{n} = -\sum_{n=1,3,\ldots}^{\infty} \frac{e^{j\alpha n}}{n} + \sum_{n=2,4,\ldots}^{\infty} \frac{e^{j\alpha n}}{n} = -\ln (1 + e^{j\alpha})$$

Subtracting this series from the series with $\alpha + \pi$ replaced by $\alpha$ gives

$$\sum_{n=1,3,\ldots}^{\infty} \frac{e^{j\alpha n}}{n} = \tfrac{1}{2} \ln \frac{1 + e^{j\alpha}}{1 - e^{j\alpha}} = \tfrac{1}{2} \ln \cot \frac{\alpha}{2} + j\frac{\pi}{4}$$

Equating the real and imaginary parts gives

$$\sum_{n=1,3,\ldots}^{\infty} \frac{\cos n\alpha}{n} = -\tfrac{1}{2} \ln \tan \frac{\alpha}{2} \qquad \sum_{n=1,3,\ldots}^{\infty} \frac{\sin n\alpha}{n} = \frac{\pi}{4}$$

Series involving products of trigonometrical terms may be summed by reducing them to the exponential type first and adding or subtracting series if necessary. Thus

$$\sum_{n=1,3,\ldots}^{\infty} \frac{\sin n\alpha \sin n\beta}{n}$$

may be written as

$$\frac{1}{2} \sum_{n=1,3\ldots}^{\infty} \left[\frac{\sin n(\alpha+\beta)}{n} + \frac{\sin n(\alpha-\beta)}{n}\right] = \tfrac{1}{2} \operatorname{Im} \sum_{n=1,3,\ldots}^{\infty} \frac{e^{jn(\alpha+\beta)} + e^{jn(\alpha-\beta)}}{n} = \frac{\pi}{4}$$

if $\alpha$ and $\beta$ are positive and real and $\beta < \alpha$ and $0 < \alpha + \beta < \pi$. These conditions follow from the properties of the logarithmic function since the sum is given by

$$\tfrac{1}{4} \operatorname{Im} \left(\ln j \cot \frac{\alpha+\beta}{2} + \ln j \cot \frac{\alpha-\beta}{2}\right)$$

The summation formulas given in part 2 of Table A.1 have been obtained by the above methods. To avoid an unduly large number of entries, only summation formulas for the exponential type of series are listed.

TABLE A.1
Part 1: Maclaurin Series Expansions

| | | |
|---|---|---|
| $\ln \sin x$ | $\ln x - \frac{x^2}{6} - \frac{x^4}{180} - \frac{x^6}{2{,}835} - \cdots$ | $-\pi < x < \pi$ |
| $\ln \cos x$ | $-\frac{x^2}{2} - \frac{x^4}{12} - \frac{x^6}{45} - \frac{17x^8}{2{,}520} - \cdots$ | $-\frac{\pi}{2} < x < \frac{\pi}{2}$ |
| $\ln \tan x$ | $\ln x + \frac{x^2}{3} + \frac{7x^4}{90} + \frac{62x^6}{2{,}835} + \cdots$ | $-\frac{\pi}{2} < x < \frac{\pi}{2}$ |
| $\sin x$ | $x - \frac{x^3}{3!} + \frac{x^5}{5!} - \frac{x^7}{7!} + \cdots$ | all $x$ |
| $\cos x$ | $1 - \frac{x^2}{2!} + \frac{x^4}{4!} - \frac{x^6}{6!} + \cdots$ | all $x$ |
| $\tan x$ | $x + \frac{x^3}{3} + \frac{2x^5}{15} + \cdots$ | $\lvert x\rvert < \frac{\pi}{2}$ |
| $\cot x$ | $\frac{1}{x} - \frac{x}{3} - \frac{x^3}{45} - \cdots$ | $\lvert x\rvert < \pi$ |

TABLE A.1 (*Continued*)

| | | |
|---|---|---|
| $\sec x$ | $1 + \frac{x^2}{2!} + \frac{5x^4}{4!} + \cdots$ | $\|x\| < \frac{\pi}{2}$ |
| $\csc x$ | $\frac{1}{x} + \frac{x}{3!} + \frac{7x^3}{3 \times 5!} + \cdots$ | $\|x\| < \pi$ |
| $\ln x$ | $2\left[\frac{x-1}{x+1} + \frac{1}{3}\left(\frac{x-1}{x+1}\right)^3 + \frac{1}{5}\left(\frac{x-1}{x+1}\right)^5 + \cdots\right]$ | $x > 0$ |

Part 2: Summations by Geometric Series

| | | |
|---|---|---|
| $\sum_1^\infty e^{jnx}$ | $\frac{e^{jx}}{1 - e^{jx}} = -\frac{1}{2} + \frac{j}{2}\cot\frac{x}{2}$ | all $x$ |
| $\sum_1^\infty ne^{jnx}$ | $\frac{e^{jx}}{(1 - e^{jx})^2} = -\frac{1}{4}\csc^2\frac{x}{2}$ | all $x$ |
| $\sum_1^\infty \frac{e^{jnx}}{n}$ | $-\ln(1 - e^{jx}) = -\ln 2\sin\frac{x}{2} + j\frac{\pi - x}{2}$ | $0 < x < 2\pi$ |
| $\sum_1^\infty \frac{e^{jnx}}{n^2}$ | $\frac{\pi^2}{6} - \frac{x}{4}(2\pi - x) - j\left(x\ln x - x - \frac{x^3}{72} - \frac{x^5}{14{,}400} - \cdots\right)$ | $0 < x < 2\pi$ |
| $\sum_1^\infty \frac{1}{n^2}$ | $\frac{\pi^2}{6}$ | |
| $\sum_1^\infty \frac{e^{jnx}}{n^3}$ | $j\left(\frac{\pi^2 x}{6} - \frac{\pi x^2}{4} + \frac{x^3}{12}\right) + \left(\frac{x^2}{2}\ln x - \frac{3x^2}{4} - \frac{x^4}{288} - \cdots\right) + \sum_1^\infty \frac{1}{n^3}$ | $0 < x < 2\pi$ |
| $\sum_{1,3,\ldots}^\infty e^{jnx}$ | $\frac{e^{jx}}{1 - e^{2jx}} = \frac{j}{2}\csc x$ | all $x$ |
| $\sum_{1,3,\ldots}^\infty ne^{jnx}$ | $-\frac{1}{2}\cot x\csc x$ | all $x$ |

TABLE A.1 (*Continued*)

| | | |
|---|---|---|
| $\sum_{1,3,\ldots}^{\infty} \frac{e^{jnx}}{n}$ | $\frac{1}{2} \ln \frac{1 + e^{jx}}{1 - e^{jx}} = -\frac{1}{2} \ln \tan \frac{x}{2} + j\frac{\pi}{4}$ | $0 < x < \pi$ |
| $\sum_{1,3,\ldots}^{\infty} \frac{e^{jnx}}{n^2}$ | $\frac{\pi^2}{8} - \frac{\pi x}{4} - \frac{j}{2}\left(x \ln \frac{x}{2} - x + \frac{x^3}{36} + \frac{7x^5}{7{,}200} + \cdots\right)$ | $0 < x < \pi$ |
| $\sum_{1,3,\ldots}^{\infty} \frac{1}{n^2}$ | $\frac{\pi^2}{8}$ | |
| $\sum_{1,3,\ldots}^{\infty} \frac{e^{jnx}}{n^3}$ | $\sum_{1,3,\ldots}^{\infty} \frac{1}{n^3} + \frac{1}{2}\left(\frac{x^2}{2} \ln \frac{x}{2} - \frac{3x^2}{4} + \frac{x^4}{144} + \cdots\right) + j\left(\frac{\pi^2 x}{8} - \frac{\pi x^2}{8}\right)$ | $0 < x < \pi$ |
| $\sum_{2,4,\ldots}^{\infty} \frac{e^{jnx}}{P(n)}$ | $\sum_{1}^{\infty} \frac{e^{jnx}}{P(n)} - \sum_{1,3,\ldots}^{\infty} \frac{e^{jnx}}{P(n)}$ | |
| $\sum_{1}^{\infty} (-1)^n e^{jnx}$ | $-\frac{1}{2} - \frac{j}{2} \tan \frac{x}{2}$ | all $x$ |
| $\sum_{1}^{\infty} (-1)^n n e^{jnx}$ | $-\frac{1}{4} \sec^2 \frac{x}{2}$ | all $x$ |
| $\sum_{1}^{\infty} \frac{(-1)^n e^{jnx}}{n}$ | $-\ln 2 \cos \frac{x}{2} - j\frac{x}{2}$ | all $x$ |
| $\sum_{1}^{\infty} \frac{(-1)^n e^{jnx}}{P(n)}$ | $\sum_{1}^{\infty} \frac{e^{jn(x+\pi)}}{P(n)}$ | |
| $\sum_{1,3,\ldots}^{\infty} \frac{1}{n^3}$ | $\frac{1}{16}\left(\frac{3\pi^2}{2} + \pi^2 \ln \frac{4}{\pi} - \frac{\pi^4}{288} - \frac{7\pi^6}{345{,}600} - \cdots\right) = 0.875\zeta(x) = 1.05175$ where $\zeta(x)$ is the Riemann zeta function | |

Table A.1 (*Continued*)

Part 3: Summations by Contour Integration

$$\sum_{1}^{\infty} \frac{\cos nx}{n^2 - a^2} \qquad \frac{1}{2a^2} - \frac{\pi}{2a}\frac{\cos (x - \pi)a}{\sin \pi a} \qquad 0 \le x \le 2\pi$$

$$\sum_{1}^{\infty} \frac{\sin nx}{n(n^2 - a^2)} \qquad \frac{(x - \pi) \sin \pi a + \pi \sin (\pi - x)a}{2a^2 \sin \pi a} \qquad 0 \le x \le 2\pi$$

$$\sum_{1}^{\infty} \frac{\cos nx}{n^2(n^2 - a^2)} \qquad \frac{\pi x}{2a^2} - \frac{\pi^2}{6a^2} - \frac{x^2}{4a^2} - \frac{\pi}{2}\frac{\cos (\pi - x)a}{a^3 \sin \pi a} + \frac{1}{2a^4} \qquad 0 \le x \le 2\pi$$

$$\sum_{1}^{\infty} \frac{\cos nx}{(n^2 - a^2)(n^2 - b^2)} \qquad -\frac{1}{2a^2b^2} - \frac{\pi}{2}\left[\frac{\cos (\pi - x)a}{a(a^2 - b^2) \sin \pi a} + \frac{\cos (\pi - x)b}{b(b^2 - a^2) \sin \pi b}\right] \qquad 0 \le x \le 2\pi$$

$$\sum_{1}^{\infty} \frac{n \sin nx}{n^2 - a^2} \qquad \frac{\pi}{2}\frac{\sin (\pi - x)a}{\sin \pi a} \qquad 0 \le x \le 2\pi$$

$$\sum_{1}^{\infty} \frac{n \sin nx}{(n^2 - a^2)(n^2 - b^2)} \qquad \frac{\pi}{2}\left[\frac{\sin (\pi - x)a}{(b^2 - a^2) \sin \pi a} + \frac{\sin (\pi - x)b}{(a^2 - b^2) \sin \pi b}\right] \qquad 0 \le x \le 2\pi$$

$$\sum_{1,3,\ldots}^{\infty} \frac{\cos nx}{n^2 - a^2} \qquad \frac{\pi \sin a(\pi/2 - x)}{4a \cos (\pi a/2)} \qquad 0 \le x \le \pi$$

$$\sum_{1,3,\ldots}^{\infty} \frac{\cos nx}{n^2(n^2 - a^2)} \qquad \frac{\pi}{4a^2}\left[\frac{\sin a(\pi/2 - x)}{a \cos (\pi a/2)} - x + \frac{\pi}{2}\right] \qquad 0 \le x \le \pi$$

$$\sum_{1,3,\ldots}^{\infty} \frac{n \sin nx}{n^2 - a^2} \qquad \frac{\pi \cos a(\pi/2 - x)}{4 \cos (\pi a/2)} \qquad 0 \le x \le \pi$$

$$\sum_{1,3,\ldots}^{\infty} \frac{\sin nx}{n(n^2 - a^2)} \qquad \frac{\pi \cos a(\pi/2 - x)}{4a^2 \cos (\pi a/2)} - \frac{\pi}{4a^2} \qquad 0 \le x \le \pi$$

*Contour-integration Method*

The theory of contour integration provides a very powerful method for the summation of a variety of series, the restrictions on the series being that they be representable as a sum over both positive and negative integers, and also satisfy certain convergence criteria which will become clear after considering a few typical examples. In order for a series to be representable as a sum over both positive and negative integers, it is necessary that the series be an even function of the summation variable $n$.

Consider the following series, $\sum_{1}^{\infty} \frac{\cos n\alpha}{P(n)}$, where $P(n)$ is a rational even function of $n$. This series may be written as

$$\frac{1}{2}\sum_{-\infty}^{\infty}\frac{e^{j\alpha n}}{P(n)} - \frac{1}{2P(0)}$$

Consider the following contour integral:

$$\oint_{C_n} \frac{e^{j\alpha z}\,dz}{P(z)(e^{2\pi jz}-1)}$$

where the contour $C_n$ consists of the square with vertices at $\pm(n+\frac{1}{2})$, $\pm j(n+\frac{1}{2})$, as in Fig. A.2, and $n$ is chosen large enough so that all the zeros of $P(z)$ lie within $C_n$. The factor $e^{2\pi jz}-1$ is introduced in the denominator of the integral to bring in poles at $\pm n$. If we can show that the integral over $C_n$ vanishes as $n$ approaches infinity, then, by Cauchy's integral formula, the sum of the residues of the integrand must equal zero, and we get

$$\sum_{-\infty}^{\infty}\frac{e^{j\alpha n}}{2\pi jP(n)} + \sum\left[\text{residues of } \frac{e^{j\alpha z}}{P(z)(e^{2\pi jz}-1)} \text{ at zeros of } P(z)\right] = 0$$

since the residue at the poles of $[e^{2\pi jz}-1]^{-1}$ is $1/2\pi j$. The above formula is for the case when none of the zeros of $P(z)$ are integers. If some of the zeros are integers, it is necessary to consider the residues at the double, triple poles, etc., of the integrand. These cases are easily treated in practice when they arise. If the summation is over only the odd or the even integers, then we introduce the factors $e^{j\pi z}+1$ and $e^{j\pi z}-1$, respectively, in place of $e^{2\pi jz}-1$ in the denominator. When the terms alternate in sign, an additional exponential term such as $e^{\pm j\pi z}$ must be introduced in the numerator or denominator.

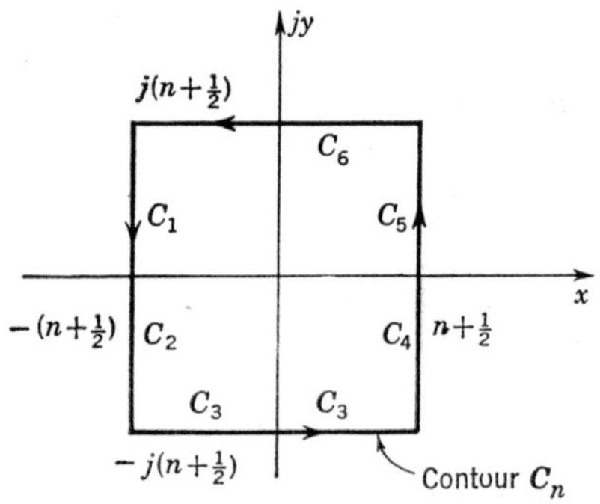


FIG. A.2

We will now show that the integral around $C_n$ vanishes, provided $P(z)$ is of the order $R^\epsilon$, $\epsilon > 1$, as $|z| = R \to \infty$, irrespective of the polar angle $\theta$ in the complex $z$ plane. Since $P(z)$ is a rational function of $z$, there exists a positive constant $M$ such that $|P(z)| \geq M$ everywhere on $C_n$ for all $n$ greater than some given finite $n$. On the segment $C_1$, we have

$$\left|\int_{C_1}\frac{e^{-j\alpha(n+\frac{1}{2})}e^{-\alpha y}\,dz}{P(z)(e^{-j2\pi(n+\frac{1}{2})}e^{-2\pi y}-1)}\right| \leq \frac{1}{M}\int_{C_1}\frac{e^{-\alpha y}|dz|}{e^{-2\pi y}+1} \leq \int_{C_1}\frac{e^{-\alpha y}}{M}|dz| \leq \frac{n+\frac{1}{2}}{M}$$

A similar result holds for the segment $C_5$. On the segment $C_2$, we have

$$\left| \int_{C_2} \frac{e^{-j\alpha(n+\frac{1}{2})}e^{-\alpha y}\,dz}{P(z)(e^{-j2\pi(n+\frac{1}{2})}e^{-2\pi y} - 1)} \right| \le \int_{C_2} \frac{e^{-\alpha y}|dz|}{M(e^{-2\pi y} + 1)}$$
$$= \frac{1}{M}\int_{C_2} \frac{e^{(2\pi-\alpha)y}|dz|}{1 + e^{2\pi y}} \le \frac{n + \frac{1}{2}}{M}$$

since $y \le 0$, provided $\alpha \le 2\pi$. The same result also holds on the segment $C_4$. In a like manner, we may show that the integrals over $C_3$ and $C_6$ are both less than $2(n + \frac{1}{2})/M$. We thus find by adding the above results that the modulus of the integral over $C_n$ is less than $(8n + 4)/M$. Hence, provided $M$ increases as fast as $n^\epsilon$, $\epsilon > 1$, the integral around $C_n$ will vanish in the limit. It is, therefore, necessary that

$$\lim_{|z|\to\infty} |P(z)| = |z|^\epsilon \qquad \epsilon > 1 \text{ for all values of } \theta$$

If the summation is over only the odd or even integers, then $\alpha$ is restricted to the range $\alpha < \pi$, in order that the integral shall converge in the lower half plane. If $\alpha$ is less than zero, then it is necessary only to change the sign of the exponential term in the denominator, and the integral around $C_n$ may again be shown to vanish in the limit as $n$ approaches infinity. The argument is the same as that used for positive $\alpha$. Or, alternatively, the result follows at once by changing $z$ to $-z$.

As an example, we will sum the following series: $\sum_1^\infty \frac{\cos n\alpha}{n^2 - a^2}$. This series may be written as $\frac{1}{2}\sum_{-\infty}^{\infty} \frac{e^{jn\alpha}}{n^2 - a^2} + \frac{1}{2a^2}$. There are poles of order 1 at $n = \pm a$, with residues $e^{j\alpha a}/2a$ and $-e^{-j\alpha a}/2a$. By the previously given formula, we get

$$\frac{1}{2}\sum_{-\infty}^{\infty} \frac{e^{jn\alpha}}{n^2 - a^2} = \frac{-\pi j}{2a}\left(\frac{e^{j\alpha a}}{e^{2\pi ja} - 1} - \frac{e^{-j\alpha a}}{e^{-2\pi ja} - 1}\right) = \frac{-\pi}{2a}\frac{\cos(\alpha - \pi)a}{\sin \pi a}$$

Thus the sum of our series is given by

$$\sum_1^\infty \frac{\cos n\alpha}{n^2 - a^2} = \frac{1}{2a^2} - \frac{\pi}{2a}\frac{\cos(\pi - \alpha)a}{\sin \pi a}$$

When $\alpha = 0$, we get

$$\sum_1^\infty \frac{1}{n^2 - a^2} = \frac{1}{2a^2} - \frac{\pi}{2a}\cot \pi a$$

which is the well-known partial-fraction expansion of the cotangent function. A further result of interest is the limit of the above sum as $a$ approaches zero. Thus, we find

$$\sum_1^\infty \frac{1}{n^2} = \lim_{a\to 0} \frac{1 - \pi a \cot \pi a}{2a^2} = \tfrac{1}{6}\pi^2$$

Further summation formulas may be obtained by integration, differentiation, adding and subtracting series, and so on. When differentiating a series, the convergence of the resulting series has to be established. Integration with respect to a parameter does not introduce any convergence difficulties. However, it is important to consider the constant term that arises upon integration. In practice, this is perhaps most readily taken into account by integrating between a fixed and a variable limit. For example,

$$j\sum_{1}^{\infty}\int_0^x \frac{e^{jnx}}{n}\,dx = \sum_{1}^{\infty}\frac{e^{jnx}}{n^2} - \sum_{1}^{\infty}\frac{1}{n^2} = -j\int_0^x \ln 2\sin\frac{x}{2}\,dx$$
$$-\int_0^x \frac{\pi - x}{2}\,dx = -\frac{x}{4}(2\pi - x) - j\left(x\ln x - x - \frac{x^3}{72}\cdots\right)$$

where $\ln 2 \sin (x/2)$ has been expanded into a Maclaurin series in order to carry out the integration. This gives us the result

$$\sum_{1}^{\infty}\frac{1}{n^2}e^{jnx} = \frac{\pi^2}{6} - \frac{x}{4}(2\pi - x) - j\left(x\ln x - x - \frac{x^3}{72}\cdots\right)$$

The summation formulas given in part 3 of Table A.1 have been obtained by the above contour-integration method. A series such as $\sum\limits_{1,3,\ldots}^{\infty}\dfrac{n\sin nx}{n^2 - a^2}$ cannot be summed by this method, since the general term decreases only as $n^{-1}$ for large $n$. However, this series is equal to $\dfrac{d}{dx}\sum\limits_{1,3,\ldots}^{\infty}\dfrac{-\cos nx}{n^2 - a^2}$. The convergence of the resulting series is the same as that of $\sum\limits_{1,3,\ldots}^{\infty}\dfrac{\sin nx}{n}$, that is, for $|x| < \pi/2$.

There does not appear to be a similar procedure which may be used when the series to be summed is an odd function of $n$. This is quite a restriction, since many of the series which occur in practice are odd functions of $n$.

*Integral Transform Methods*

Integral transforms, such as the Laplace, Fourier, and Mellin transforms, can be used in a variety of ways to sum certain types of series in closed form. They also are found useful in many cases in converting relatively complicated series into simpler ones which are more easily summed, or in converting relatively slowly converging series into much more rapidly converging ones. We will consider the application of the Laplace transform to the summation of series first. The technique will be largely developed by means of particular examples.

Consider the following general series, $\sum\limits_{0}^{\infty} f(\alpha n)$, where $f(x)$ is of such a form that its Laplace transform exists. We have by definition

$$F(p) = \int_0^\infty f(x)e^{-px}\,dx$$

This integral determines $F(p)$ as an analytic function of the complex variable

$$p = u + jv$$

whose singularities all lie to the left of some value of $u = c$ in the $p$ plane. The inversion integral gives

$$f(x) = \frac{1}{2\pi j} \int_{c-j\infty}^{c+j\infty} F(p) e^{px}\, dp$$

If we replace $x$ by $\alpha n$ and sum over $n$, we get

$$\sum_0^\infty f(\alpha n) = \frac{1}{2\pi j} \sum_0^\infty \int_{c-j\infty}^{c+j\infty} e^{p\alpha n} F(p)\, dp$$

This is permissible since the inversion integral holds identically for all values of $x$, with the exception of certain values of $x$ for which $f(x)$ may be discontinuous. In our case it is assumed that $f(x)$ is a continuous function of $x$. If, now, the inversion integral involving $F(p)$ is uniformly convergent, we may interchange the order of integration and summation to get

$$\begin{aligned} \sum_0^\infty f(\alpha n) &= \frac{1}{2\pi j} \int_{c-j\infty}^{c+j\infty} F(p) \sum_0^\infty e^{p\alpha n}\, dp \\ &= \frac{1}{2\pi j} \int_{c-j\infty}^{c+j\infty} \frac{F(p)}{1 - e^{p\alpha}}\, dp \end{aligned}$$

provided $f(x)$ is of such form that we may take $c < 0$ in order that $\sum_0^\infty e^{p\alpha n}$ should represent a convergent series. This implies that $f(x)$ is asymptotic to $e^{-\epsilon x}$, $\epsilon > 0$, as $x$ approaches infinity. In general, this condition is not satisfied, but, in practice, we can multiply $f(\alpha n)$ by $e^{-\epsilon n}$, sum the resulting series, and then take the limit as $\epsilon$ approaches zero. If we can evaluate the resulting integral, we have the sum of the series in closed form. Alternatively, we may expand the integral by the residue theorem. Let us choose for our contour the line $p = c$, $c < 0$, and the semicircle in the right half plane. Since $F(p)$ is analytic for all $u > c$, the only poles of the integrand occur at $e^{\alpha p} = 1$ or $p = \pm(2\pi nj/\alpha)$, $n = 0, 1, 2, \ldots$. The residues at these poles are $-\frac{1}{\alpha} F\left(\pm \frac{2\pi nj}{\alpha}\right)$. Hence, provided the integral around the semicircle vanishes, we get

$$\sum_0^\infty f(\alpha n) = \frac{1}{\alpha} \sum_0^\infty F\left(\frac{2\pi jn}{\alpha}\right)$$

The change in sign is due to the fact that the contour is traversed in a clockwise sense. Sometimes it is possible to close the contour in the left half plane, and the sum of the series is then given in terms of the residues at the poles of $F(p)$.

As a first example, we will sum the series $\sum_0^\infty ne^{-na}$. The Laplace transform of $xe^{-ax}$ is $(p + a)^{-2}$, and $\int_{c-j\infty}^{c+j\infty} \frac{e^{px}}{(p + a)^2}\, dp$ is uniformly convergent. Furthermore $F(p)$ is analytic for all Re $p > -a$. Therefore, we have

$$\sum_1^\infty ne^{-na} = \sum_{-\infty}^\infty (2n\pi j + a)^{-2} = \frac{-1}{4\pi^2} \sum_{-\infty}^\infty \left(n + \frac{a}{2\pi j}\right)^{-2} = -\tfrac{1}{4} \sinh^{-2} \frac{a}{2}$$

which is the same result as obtained in the first section by a different method. The above result follows when use is made of the partial-fraction expansion of $\csc^2 (a/2)$; that is

$$\sum_{-\infty}^{\infty} \frac{1}{(n+a)^2} = \pi^2 \csc^2 \pi a$$

If we close the contour in the left half plane, the only pole of the integrand is a double pole at $p = -a$. The residue at this pole is

$$\frac{d}{dp}\left[(p+a)^2 \frac{1}{(1-e^p)(p+a)^2}\right]\Bigg|_{p=-a} = \frac{e^{-a}}{(1-e^{-a})^2} = \tfrac{1}{4} \sinh^{-2} \frac{a}{2}$$

Hence, we see that closing the contour in the left half plane has the advantage of giving the sum in closed form directly.

We may extend the definition of the Laplace transform so that it is valid for negative values of $x$ as follows:

$$\mathfrak{L}f(x) = \int_{-\infty}^{0} f(x)e^{-px}\,dx \qquad x < 0$$

The inversion formula gives

$$f(x) = \frac{1}{2\pi j}\int_{c-j\infty}^{c+j\infty} e^{px}F(p)\,dp$$

where $c$ now lies to the left of all the singularities of $F(p)$. For positive values of $x$, the contour may be closed in the left half plane, and, since no singularities are enclosed, the integral vanishes along with $f(x)$. For negative values of $x$, the contour may be closed in the right half plane, and the original function $f(x)$ is recovered.

It is now possible to consider a series such as $\sum_{-\infty}^{\infty} f(\alpha n)$. We will impose on $f(x)$ the condition that it be of integrable square over $-\infty \le x \le \infty$; that is, $\int_{-\infty}^{\infty} |f(x)|^2\,dx$ exists. The existence of the integral implies that $f(x)$ has no poles on the real axis. Let us break $f(x)$ up into two parts as follows:

$$f_+(x) = \begin{cases} f(x) & x \ge 0 \\ 0 & x < 0 \end{cases} \qquad f_-(x) = \begin{cases} 0 & x > 0 \\ f(x) & x < 0 \end{cases}$$

The corresponding Laplace transforms are

$$F_+(p) = \int_0^{\infty} e^{-px}f_+(x)\,dx \qquad F_-(p) = \int_{-\infty}^{0} e^{-px}f_-(x)\,dx$$

The inverse transforms are

$$f_+ = \frac{1}{2\pi j}\int_{c_+-j\infty}^{c_++j\infty} e^{px}F_+\,dp \qquad f_- = \frac{1}{2\pi j}\int_{c_--j\infty}^{c_-+j\infty} e^{px}F_-\,dp$$

From Parseval's theorem we have that $\int_{-j\infty}^{j\infty} |F(p)|^2\,dp$ exists if $f(x)$ is of integrable square. Consequently, $F(p)$ has no poles on the imaginary axis, and we may, therefore, take $c_+ < 0$ and $c_- > 0$ in the above inversion integrals. Under these conditions, we get

$$\sum_{0}^{\infty} f(\alpha n) = \frac{1}{2\pi j} \int_{c_+ - j\infty}^{c_+ + j\infty} F_+(p)(1 - e^{\alpha p})^{-1}\, dp = \frac{1}{\alpha} \sum_{-\infty}^{\infty} F_+\left(\frac{2\pi n j}{\alpha}\right)$$

upon closing the contour in the right half plane. Similarly,

$$\sum_{-1}^{-\infty} f(\alpha n) = \frac{1}{\alpha} \sum_{-\infty}^{\infty} F_-\left(\frac{2\pi n j}{\alpha}\right)$$

upon closing the contour in the left half plane. Combining these results, we have, finally,

$$\sum_{-\infty}^{\infty} f(\alpha n) = \frac{1}{\alpha} \sum_{-\infty}^{\infty} F\left(\frac{2\pi n j}{\alpha}\right)$$

If the series on the right-hand side can be summed, or if it is more rapidly converging than the original series, some advantage will have been gained. As an example in the use of the above generalized Laplace transform, we will sum the following series, $\sum_{-\infty}^{\infty} \frac{e^{j\alpha n}}{n^2 - a^2}$. The general term, considered as a function of $x$, has poles at $x = \pm a$. Provided $a$ has a small imaginary part, this function is of integral square over $-\infty \le x \le \infty$. The Laplace transforms are

$$F_+ = \int_0^{\infty} \frac{e^{j\alpha x - px}}{x^2 - a^2}\, dx \qquad F_- = \int_{-\infty}^{0} \frac{e^{j\alpha x - px}}{x^2 - a^2}\, dx$$

$F_+$ is analytic for all Re $p \ge 0$, while $F_-$ is analytic for all Re $p \le 0$. Consequently, we may evaluate the transform for the particular value Re $p = 0$. We have

$$F(p) = \int_{-\infty}^{\infty} \frac{e^{j\alpha x - px}}{x^2 - a^2}\, dx$$

For $\alpha - v \ge 0$, we may replace $x$ by the complex variable $z = x + jy$, and evaluate the integral by contour integration, closing the contour in the upper half plane. The only pole of the integrand in the upper half plane is at $z = a$. Hence,

$$F(p) = \frac{\pi j}{a} e^{(j\alpha - p)a}$$

$p = jv$, $v \le \alpha$. For $\alpha - v \le 0$, the contour is closed in the lower half plane and gives

$$F(p) = \frac{\pi j}{a} e^{-(j\alpha - p)a} \qquad p = jv \qquad v \ge \alpha$$

These results may be combined and give $F(p) = (\pi j/a)e^{j|\alpha - v|a}$ valid for Re $p = 0$ and all $v$. It may be easily shown that $F(p)$ is of integrable square over $-j\infty \le p \le j\infty$, and thus, we may apply the previously derived summation formula. That is,

$$\sum_{-\infty}^{\infty} \frac{e^{j\alpha n}}{n^2 - a^2} = \frac{\pi j}{a} \sum_{-\infty}^{\infty} e^{j|\alpha - 2\pi n|a}$$

Restricting $\alpha$ to the range $0 \le \alpha \le 2\pi$, we have

$$\frac{\pi j}{a} \sum_{-\infty}^{\infty} e^{j|\alpha - 2n\pi|a} = \frac{\pi j}{a}\left[(e^{-j\alpha a} + e^{j\alpha a}) \sum_{1}^{\infty} e^{j2\pi na} + e^{j\alpha a}\right]$$

$$= \frac{\pi j}{a}(j \cos \alpha a \cot \pi a + j \sin \alpha a) = -\frac{\pi}{a}\frac{\cos (\pi - \alpha)a}{\sin \pi a}$$

The same result may be derived by the contour-integration method. In fact, for series of the type considered above, the contour-integration method is considerably more straightforward.

If we make the following change in notation, $p = -j\omega$, $dp = -j\,d\omega$, then our generalized Laplace transforms become Fourier transforms, and the resulting formula is one form of the Poisson summation formula. We have

$$F(\omega) = \int_{-\infty}^{\infty} e^{j\omega t} f(t)\,dt = \mathfrak{F}f(t)$$

which is the Fourier transform of $f(t)$. The inversion integral becomes

$$f(t) = \frac{1}{2\pi}\int_{-\infty}^{\infty} e^{-j\omega t} F(\omega)\,d\omega$$

By a procedure similar to that used for the Laplace transform, we obtain the following summation formula:

$$\sum_{-\infty}^{\infty} f(\alpha n) = \frac{1}{\alpha}\sum_{-\infty}^{\infty} F\left(\frac{2n\pi}{\alpha}\right)$$

where $F(\omega)$ is the Fourier transform of $f(t)$ as defined above.

The utility of the Poisson summation formula in converting a slowly convergent series into a rapidly convergent series may be appreciated from the following property of the Fourier transform. If $f(t)$ is a function which is localized near $t = 0$, then $F(\omega)$ is "spread out," or has an appreciable value for a large range of $\omega$; for example, if $f(t) = \delta(t)$, then $F(\omega) = 1$. Conversely, if $f(t)$ is spread out along the $t$ axis, then $F(\omega)$ is concentrated near $\omega = 0$. A slowly convergent series with a typical term $f(\alpha n)$ is thus converted to a rapidly converging series with a typical term $F(2n\pi/a)$, since $F$ will be small for large values of $n$ when $f(\alpha n)$ decreases slowly with increasing $n$.

Many useful summation formulas may also be derived using Mellin transforms in an analogous way. The interested reader should consult the references at the end of this section for illustrations on the use of the Mellin transforms in summing series.

*Summation of Odd Series*

In this section we shall consider a method by which many series which are odd functions of the summation variable $n$ can be reduced to much more rapidly converging series. As an example to illustrate the method, consider the following series, $\sum_{1}^{\infty} \frac{\sin n\alpha}{n^2 - a^2}$, $0 \le \alpha \le \pi$, $a \gg 1$, and $a$ not an integer. Since $a \gg 1$, the numerical evaluation of the above series must include terms up to $n = N$, where $N^2 \gg a^2$. For instance, if $a \approx 10$, at least 30 terms must be included. The oscillatory nature of the sine function further adds to the slow convergence of the series. If the above series is multiplied by tanh $np$, it becomes an even function of $n$, and the resulting series is summable by the contour-integration method. We may write the above series as

$$\sum_1^\infty \frac{\sin n\alpha}{n^2 - a^2} = \sum_1^\infty \frac{\sin n\alpha \tanh np}{n^2 - a^2} + \sum_1^\infty \frac{\sin n\alpha}{n^2 - a^2}(1 - \tanh np)$$

where the second series on the right is rapidly converging, provided $p$ is not too small, i.e., for $p = 1$; four terms are sufficient since $0.9993 < \tanh np < 1$ for $n > 4$. The first series may be written as $\frac{1}{2j}\sum_{-\infty}^{\infty} \frac{e^{jn\alpha}\tanh np}{n^2 - a^2}$. This series is readily summed by the contour-integration method. The poles of the integrand are at $w = \pm a$ and at $w = \pm j[(2n+1)/2p]\pi$, where $\tanh wp$ becomes infinite. Carrying through the algebra, we get

$$\sum_1^\infty \frac{\sin n\alpha}{n^2 - a^2} = \sum_1^\infty \frac{\sin n\alpha}{n^2 - a^2}(1 - \tanh np) + \frac{\pi}{2a}\tanh pa \frac{\sin (\pi - \alpha)a}{\sin \pi a}$$
$$- \frac{\pi}{p} \sum_{n=1,3,\ldots}^{\infty} \frac{(\pi - \alpha)\sinh (n\pi/2p)}{(n^2\pi^2/4p^2 + a^2)[\sinh (n\pi/2p)\pi]}$$

The choice of the value of $p$ is governed by the values of the parameters $\alpha$ and $a$. When $\alpha$ is small, it is advantageous to take $p$ small. The terms are asymptotic to

$$\frac{\pi}{p} e^{-n\pi\alpha/2p}\left(\frac{n^2\pi^2}{4p^2} + a^2\right)^{-1}$$

and, hence, it is desirable to have $p = \pi\alpha$, so that the exponential term is less than 0.01 for $n > 9$. However, if $\alpha$ is very small, then it is desirable to take $p > \pi\alpha$, so that the correction series involving $1 - \tanh np$ will converge rapidly enough.

When $a = 1.5$, $\alpha = 1$, and $p = \pi$, it takes 25 terms of the original series to give three-figure accuracy for the sum. The same accuracy can be obtained from one term of the correction series plus five terms of the hyperbolic-sine series. The sum is $-0.214$, and we can see that the saving in computational labor is considerable.

## REFERENCES

*Vector and dyadic analysis*

A.1. L. Brand: "Vector and Tensor Analysis," John Wiley & Sons, Inc., New York, 1947.

A.2. P. M. Morse and H. Feshbach: "Methods of Theoretical Physics," part I, McGraw-Hill Book Company, Inc., New York, 1953.

A.3. B. Spain: "Tensor Calculus," University Mathematical Texts, Interscience Publishers, Inc., New York, 1953.

*Matrices*

A.4. E. A. Guillemin: "The Mathematics of Circuit Analysis," John Wiley & Sons, Inc., New York, 1949.

A.5. A. C. Aitken: "Determinants and Matrices," University Mathematical Texts, Interscience Publishers, Inc., New York, 1948.

*Calculus of variations*

A.6. R. Weinstock: "Calculus of Variations," McGraw-Hill Book Company, Inc., New York, 1952.

A.7. R. Courant and D. Hilbert: "Methods of Mathematical Physics," English ed., Interscience Publishers, Inc., New York, 1953.

*Infinite products and gamma functions*

A.8. E. T. Copson: "Theory of Functions of a Complex Variable," Oxford University Press, New York, 1935.

A.9. E. C. Titchmarsh: "Theory of Functions," 2d ed., Oxford University Press, New York, 1939. (See also Ref. A.2.)

*Summation of Fourier series*

A.10. L. A. Pipes: The Summation of Fourier Series by Operational Methods, *J. Appl. Phys.*, vol. 21, pp. 298–301, April, 1950.

A.11. A. D. Wheelon: On the Summation of Infinite Series in Closed Form, *J. Appl. Phys.*, vol. 25, pp. 113–118, January, 1954.

A.12. G. G. Macfarlane: The Application of Mellin Transforms to the Summation of Slowly Convergent Series, *Phil. Mag.*, vol. 40, pp. 188–197, February, 1949. (See also Ref. A.7, sec. 3.3, and probs. 26–28 at end of sec. 13.96, as well as Ref. A.2.)

## PROBLEMS

**A.1.** Expand the following by treating $\nabla$ as a vector operator:

$$\nabla \times \phi \mathbf{A} \qquad \nabla \times (\mathbf{A} \times \nabla\phi) \qquad \nabla \cdot \mathbf{A} \times \nabla\phi \qquad \nabla \mathbf{B} \cdot \nabla \times \mathbf{A}$$

**A.2.** Construct a table of direction cosines between the unit vectors in a cylindrical and spherical coordinate system.

**A.3.** Find the components of the vector function

$$\mathbf{A} = x\mathbf{a}_x + xy\mathbf{a}_y + xyz\mathbf{a}_z$$

in spherical and cylindrical coordinates. Also determine the covariant and contravariant components.

**A.4.** Evaluate the divergence of the vector function $\mathbf{A}$ of Prob. A.3 in rectangular, cylindrical, and spherical coordinates, and thereby verify that the same result is obtained by all three methods.

**A.5.** Obtain the divergence and curl of the following dyadic function:

$$\bar{\mathbf{A}} = x^2\mathbf{a}_x\mathbf{a}_x + y^2\mathbf{a}_x\mathbf{a}_y + z^2\mathbf{a}_x\mathbf{a}_z + xy\mathbf{a}_y\mathbf{a}_x + xy^2\mathbf{a}_y\mathbf{a}_y + x^2y\mathbf{a}_y\mathbf{a}_z + x^2y^2z\mathbf{a}_z\mathbf{a}_x + xz^2\mathbf{a}_z\mathbf{a}_y + yz^2\mathbf{a}_z\mathbf{a}_z$$

**A.6.** Separate the following dyadic into the sum of a symmetrical and unsymmetrical dyadic:

$$\bar{\mathbf{C}} = x^2\mathbf{a}_x\mathbf{a}_x + x^2y\mathbf{a}_x\mathbf{a}_y + y^2x\mathbf{a}_y\mathbf{a}_x + y^2\mathbf{a}_y\mathbf{a}_y$$

**A.7.** Find the principal axis (coordinate transformation) that will reduce the following dyadic to diagonal form:

$$\bar{\mathbf{C}} = 4\mathbf{a}_x\mathbf{a}_x + 3\mathbf{a}_x\mathbf{a}_y + 2\mathbf{a}_y\mathbf{a}_x + \mathbf{a}_y\mathbf{a}_y$$

**A.8.** Evaluate the following matrix product by using Sylvester's theorem:

$$\begin{bmatrix} 4 & 3 \\ 2 & 2 \end{bmatrix}^4$$

**A.9.** Expand the following functions into infinite products:

$$\frac{\sin wa \cos wb}{w} \qquad \cos (w^2 + k^2)^{1/2}a \qquad (w^2 + k^2)^{1/2} \sin (w^2 + k^2)^{1/2}a$$

where $w$ is the variable.

**A.10.** Sum the following series:

$$\sum_{1,3,\ldots}^{\infty} \frac{\cos nx}{n^2} \qquad \sum_{1,2,\ldots}^{\infty} \frac{\sin nx \sin nv}{n^3} \qquad \sum_{2,4,\ldots}^{\infty} \frac{\cos nx \sin nv}{n^2}$$

**A.11.** Sum the first four series listed in part 3 of Table A.1 by the contour-integration method.

# NAME INDEX

# SUBJECT INDEX